OCCUPATIONAL SAFETY AND HEALTH

For Technologists, Engineers, and Managers

SIXTH EDITION

DAVID L. GOETSCH

President and CEO, The Institute for Workforce Transformation

PEARSON

Prentice
Hall

Upper Saddle River, New Jersey
Columbus, Ohio

Library of Congress Cataloging-in-Publication Data

Goetsch, David L.

Occupational safety and health for technologists, engineers, and managers/David L. Goetsch.—6th ed.

p. cm.

ISBN 0-13-239760-9

1. Industrial safety—United States. 2. Industrial hygiene—United States. I. Title.

T55.G586 2008

658.4'08—dc22 2007000497

Editor in Chief: Vernon R. Anthony
Acquisitions Editor: Jill Jones-Renger
Editorial Assistant: Yvette Schlarman
Production Editor: Kevin Happell
Production Coordination: GGS Book Services
Design Coordinator: Diane Ernsberger
Cover Design and Art: Aaron Dixon
Production Manager: Deidra Schwartz
Director of Marketing: David Gesell
Marketing Manager: Jimmy Stephens
Marketing Coordinator: Alicia Dysert

This book was set in Melior by GGS Book Services. It was printed and bound by Courier Westford. The cover was printed by Coral Graphics.

Pearson Education Ltd.
Pearson Education Singapore Pte. Ltd.
Pearson Education Canada, Ltd.
Pearson Education—Japan

Pearson Education Australia Pty. Limited
Pearson Education North Asia Ltd.
Pearson Educación de Mexico, S.A. de C.V.
Pearson Education Malaysia Pte. Ltd.

10 9 8 7 6 5 4 3 2 1
ISBN-13: 978-0-13-239760-5
ISBN-10: 0-13-239760-9

PREFACE

BACKGROUND

The field of occupational safety and health has undergone significant change over the past two decades. There are many reasons for this. Some of the more prominent include the following: technological changes that have introduced new hazards in the workplace; proliferation of health and safety legislation and corresponding regulations; increased pressure from regulatory agencies; realization by executives that workers in a safe and healthy workplace are typically more productive; health care and workers' compensation cost increases; increased pressure from environmental groups and the public; a growing interest in ethics and corporate responsibility; professionalization of health and safety occupations; increased pressure from labor organizations and employees in general; rapidly mounting costs associated with product safety and other types of litigation; and increasing incidents of workplace violence.

All of these factors, when taken together, have made the job of the modern safety and health professional more challenging and more important than it has ever been. These factors have also created a need for an up-to-date book on workplace safety and health that contains the latest information needed by people who will practice this profession in an age of global competition and rapid technological change.

WHY WAS THIS BOOK WRITTEN AND FOR WHOM?

This book was written to fulfill the need for an up-to-date, practical teaching resource that focuses on the needs of modern safety and health professionals practicing in the workplace. It is intended for use in universities, colleges, community colleges, and corporate training settings that offer programs, courses, workshops, and seminars in occupational safety and health. Educators in such disciplines as industrial technology, manufacturing technology, industrial engineering, engineering technology, occupational safety, management, and supervision will find this book both valuable and easy to use. The direct, straightforward presentation of material focuses on making the theories and principles of occupational safety and health practical and useful in a real-world setting. Up-to-date research has been integrated throughout in a down-to-earth manner.

ORGANIZATION OF THE BOOK

The text contains 31 chapters organized into five parts, each focusing on a major area of concern for modern safety and health professionals. The chapters are presented in an order that is compatible with the typical organization of a college-level safety and health course. A standard chapter format is used throughout the book. Each chapter begins with a list of major topics and ends with a comprehensive summary. Following the summary, most chapters include review questions, key terms and concepts, and endnotes. Within each chapter are case studies to promote classroom discussion, as well as at least one safety fact or myth. These materials are provided to encourage review, stimulate additional thought, and provide opportunities for applying what has been learned.

SUPPLEMENTS

This text is accompanied by a Companion Website at **www.prenhall.com/ goetsch.** An online Instructor's Manual is also available to instructors through this title's catalog page at **www.prenhall.com**. Instructors can search for a text by author, title, ISBN, or by selecting the appropriate discipline from the pull down menu at the top of the catalog home page. To access supplementary materials online, instructors need to request an instructor access code. Go to **www.prenhall.com**, click the **Instructor Resource Center** link, and then click Register Today for an instructor access code. Within 48 hours of registering you will receive a confirming e-mail including an instructor access code. Once you have received your code, go to the site and log on for full instructions on downloading the materials that you wish to use.

HOW THIS BOOK DIFFERS FROM OTHERS

This book was written because in the age of global competition, safety and health in the workplace have changed drastically. Many issues, concerns, and factors relating specifically to modern workplace environments have been given more attention, greater depth of coverage, and more illumination here than other textbooks. Some of the areas receiving more attention and specific occupational examples include

- The OSH Act and OSHA
- Standards and codes

- Laws and liability
- Stress-related problems
- Life safety and fire hazards
- The evolving roles of health and safety professionals
- Health and safety training
- Human factors in safety
- Environmental issues and ISO 14000 standards
- Computers, robots, and automation
- Ethics and safety
- Bloodborne pathogens in the workplace
- Product safety and liability
- Ergonomics and safety
- The relationship between safety and quality
- Workplace violence
- Workers' compensation
- Repetitive strain injuries (RSIs)
- Terrorism threats in the workplace
- Safety-friendly corporate culture

NEW TO THIS EDITION

This sixth edition of *Occupational Safety and Health* contains much new and updated material, including the following:

General Revisions

- All OSHA standards, as well as those of other regulatory agencies, were updated.
- Research material, related content, and corresponding endnotes were updated throughout.
- Older content has been retained only to the extent that it is still valid.
- The entire book was rearranged into five parts, and all chapters were grouped together in these parts and renumbered. The following content matrix shows the old and new chapter numbers as well as their organization.

New or Updated Material in This Edition

Chapter 2:   A new section on the global impact of accidents and injuries

Chapter 3:   A new section on obesity and accident causation

Chapter 8:   A new section covering the 10 accident investigation mistakes to avoid

Chapter 10: A new section on participatory ergonomics

Chapter 18: A new section on preventing are flash injuries (NFPA 70E)

Chapter 25: New sections on (a) emergency planning for workers with disabilities, and (b) safely resuming business after a disaster

Chapter 27: A new section on occupational safety and health management systems (OSHMSs)

Chapter 28: A new section on persuasion as a safety promotion tool.

Chapter 31: This is a new chapter on establishing a "safety-friendly" corporate culture in organizations

Contents Matrix (New Chapter Numbers vs. Old)

New Old

Part 1: Historical Perspective and Overview

New	Old	
1	1	Safety and Health Movement, Then and Now
2	2	Accidents and Their Effects
3	3	Theories of Accident Causation
4	27	Roles and Professional Certifications for Safety and Health Professionals
5	29	Safety, Health, and Competition in the Global Marketplace

Part 2: Laws and Regulations

New	Old	
6	4	The OSH Act, Standards, and Liability
7	5	Workers' Compensation
8	22	Accident Investigation and Reporting
9	26	Product Safety and Liability

Part 3: The Human Element

New	Old	
10	6	Ergonomic Hazards: Musculoskeletal Disorders (MSDs) and Cumulative Trauma Disorders (CTDs)
11	7	Stress and Safety
12	24	Safety and Health Training
13	30	Violence in the Workplace

Part 4: Hazard Assessment, Prevention, and Control

New	Old	
14	8	Mechanical Hazards and Machine Safeguarding
15	9	Falling, Impact, Acceleration, Lifting, and Vision Hazards
16	10	Hazards of Temperature Extremes
17	11	Pressure Hazards
18	12	Electrical Hazards
19	13	Fire Hazards and Life Safety
20	14	Industrial Hygiene and Confined Spaces
21	15	Radiation Hazards
22	16	Noise and Vibration Hazards
23	18	Computers, Automation, and Robots
24	20	Bloodborne Pathogens in the Workplace

Part 5: Management of Safety and Health

25	17	Preparing for Emergencies and Terrorism
26	19	Ethics and Safety
27	21	Hazard Analysis/Prevention and Safety Management
28	23	Promoting Safety
29	25	Environmental Safety and ISO 14000 (Environmental Management)
30	28	TSM: Total Safety Management in a Quality Management Setting
31		Establishing a Safety-Friendly Corporate Culture

ABOUT THE AUTHOR

David L. Goetsch is Provost of the joint campus of the University of West Florida and Okaloosa-Walton College in Fort Walton Beach, Florida, and professor of safety, quality, and environmental management. He also administers the state of Florida's Center for Manufacturing Competitiveness that is located on this campus. In addition, Dr. Goetsch is president and CEO of the Institute for Workforce Transformation (IWT), a private consulting firm dedicated to the continual improvement of organizational competitiveness, safety, and quality. Dr. Goetsch is cofounder of The Quality Institute, a partnership of the University of West Florida, Okaloosa-Walton College, and the Okaloosa Economic Development Council. He currently serves on the executive board of the institute.

ACKNOWLEDGMENTS

The author acknowledges the invaluable assistance of the following people in developing this book: Dr. Lissa Galbraith, Florida A&M/Florida State University, for the material she contributed on electrical and fire hazards in the first edition; Harvey Martin, health and safety manager of Metric Systems Corporation in Fort Walton Beach, Florida, for providing up-to-date research material; Faye Crawford, for word processing of the manuscript; and the following reviewers for their invaluable input: Steven A. Freeman—Iowa State University; JoDell K. Steuver—Purdue University; and Ottis E. Walizer—Minot State University. Special acknowledgment goes to Harry D. Leiman for his contributions in updating all OSHA standards.

SAFETY VERSUS HEALTH

The title of this book intentionally includes the words *safety* and *health*. Throughout the text, the titles "safety and health professional" and "safety and health manager" are used. This, too, is done by design. This approach underscores the point that the field of occupational safety has been broadened to encompass both safety and health. Consequently, managers, technical personnel, and engineers in this field must be knowledgeable about safety and health and be prepared to oversee a corporate program that encompasses both areas of responsibility.

Safety and health, although closely related, are not the same. One view is that safety is concerned with injury-causing situations, whereas health is concerned with disease-causing conditions. Another view is that safety is concerned with hazards to humans that result from sudden severe conditions; health deals with adverse reactions to prolonged exposure to dangerous, but less intense, hazards. Both of these views are generally accurate in portraying the difference between safety and health. However, the line between these two concepts is not always clearly marked.

For example, stress is a hazard that can cause both psychological and physiological problems over a prolonged period. In this case, it is a health concern. On the other hand, an overly stressed worker may be more prone to unintentionally forget safety precautions and thus may cause an accident. In this case, stress is a safety concern.

Because managers in this evolving profession are likely to be responsible for safety *and* health, it is important that they have a broad academic background covering both. This book attempts to provide that background.

This broadening of the scope of the profession does not mean that specialists in safety and health are not still needed. They are. Chapter 4 shows how today's safety and health manager is a generalist who often heads a team of specialists such as safety engineers, health physicists, industrial hygienists, occupational nurses, occupational physicians, and risk managers. In order to manage a team of specialists in these various areas, safety and health managers must have the broad and comprehensive background that this book provides.

BRIEF CONTENTS

PART 1 HISTORICAL PERSPECTIVE AND OVERVIEW 1

1 Safety and Health Movement, Then and Now 3
2 Accidents and Their Effects 22
3 Theories of Accident Causation 40
4 Roles and Professional Certifications for Safety and Health Professionals 65
5 Safety, Health, and Competition in the Global Marketplace 88

PART 2 LAWS AND REGULATIONS 109

6 The OSH Act, Standards, and Liability 111
7 Workers' Compensation 174
8 Accident Investigation and Reporting 211
9 Product Safety and Liability 232

PART 3 THE HUMAN ELEMENT 253

10 Ergonomic Hazards: Musculoskeletal Disorders (MSDs) and Cumulative Trauma Disorders (CTDs) 255
11 Stress and Safety 296
12 Safety and Health Training 310
13 Violence in the Workplace 352

PART 4 HAZARD ASSESSMENT, PREVENTION, AND CONTROL 383

14 Mechanical Hazards and Machine Safeguarding 385
15 Falling, Impact, Acceleration, Lifting, and Vision Hazards 411
16 Hazards of Temperature Extremes 452

17 Pressure Hazards   470

18 Electrical Hazards   487

19 Fire Hazards and Life Safety   511

20 Industrial Hygiene and Confined Spaces   544

21 Radiation Hazards   604

22 Noise and Vibration Hazards   626

23 Computers, Automation, and Robots   659

24 Bloodborne Pathogens in the Workplace   681

PART 5 MANAGEMENT OF SAFETY AND HEALTH   709

25 Preparing for Emergencies and Terrorism   711

26 Ethics and Safety   747

27 Hazard Analysis Prevention and Safety Management   765

28 Promoting Safety   795

29 Environmental Safety and ISO 14000 (Environmental Management)   819

30 TSM: Total Safety Management in a Quality Management Setting   861

31 Establishing a Safety-Friendly Corporate Culture   878

CONTENTS

PART 1 HISTORICAL PERSPECTIVE AND OVERVIEW   1

1   Safety and Health Movement, Then and Now   3

Developments Before the Industrial Revolution   4
Milestones in the Safety Movement   5
Tragedies That Have Changed the Safety Movement   7
Role of Organized Labor   10
Role of Specific Health Problems   11
Development of Accident Prevention Programs   13
Development of Safety Organizations   14
Safety and Health Movement Today   17
Integrated Approach to Safety and Health   17
New Materials, New Processes, and New Problems   18
Rapid Growth in the Profession   19

2   Accidents and Their Effects   22

Costs of Accidents   23
Accidental Deaths in the United States   24
Accidents versus Other Causes of Death   25
Work Accident Costs and Rates   26
Time Lost Because of Work Injuries   27
Deaths in Work Accidents   27
Work Injuries by Type of Accident   27
Death Rates by Industry   28
Parts of the Body Injured on the Job   29
Chemical Burn Injuries   29
Heat Burn Injuries   30
Repetitive Strain/Soft-Tissue Injuries   31

Estimating the Cost of Accidents   32
Global Impact of Accidents and Injuries   35

3   Theories of Accident Causation   40
Domino Theory of Accident Causation   41
Human Factors Theory of Accident Causation   43
Accident/Incident Theory of Accident Causation   46
Epidemiological Theory of Accident Causation   48
Systems Theory of Accident Causation   50
Combination Theory of Accident Causation   54
Behavioral Theory of Accident Causation   55
Drugs and Accident Causation   57
Depression and Accident Causation   58
Management Failures and Accident Causation   59
Obesity and Accident Causation   61

4   Roles and Professional Certifications for Safety and Health Professionals   65
Modern Safety and Health Teams   65
Safety and Health Manager   66
Engineers and Safety   75
Industrial Hygienist   83
Health Physicist   83
Occupational Physician   84
Occupational Health Nurse   86
Risk Manager   87
Certification of Safety and Health Professionals   88
Certified Professional Ergonomist   91
Certified Occupational Health Nurse   93

5   Safety, Health, and Competition in the Global Marketplace   88
Competitiveness Defined   99
Productivity and Competitiveness   100
Quality and Competitiveness   103
How Safety and Health Can Improve Competitiveness   104

PART 2 LAWS AND REGULATIONS

6   The OSH Act, Standards, and Liability   111
Rationale for the OSH Act   112
OSHA's Mission and Purpose   112

OSH Act Coverage   113

OSHA Standards   114

OSHA's Record Keeping and Reporting   120

Keeping Employees Informed   128

Workplace Inspections and Enforcement   128

OSHA's Enhanced Enforcement Policy   129

Citations and Penalties   131

Appeals Process   133

State-Level OSHA Programs   134

Services Available from OSHA   137

Employer Rights and Responsibilities   140

Employee Rights and Responsibilities   142

Keeping Up-to-Date on OSHA   143

Problems with OSHA   144

Other Agencies and Organizations   145

OSHA's General Industry Standards   150

OSHA's Maritime Standards   159

OSHA's Construction Standards   162

Standards and Codes   163

Laws and Liability   163

7   Workers' Compensation   174

Overview of Workers' Compensation   174

Historical Perspective   177

Workers' Compensation Legislation   179

Modern Workers' Compensation   180

Workers' Compensation Insurance   183

Resolution of Workers' Compensation Disputes   185

Injuries and Workers' Compensation   186

Disabilities and Workers' Compensation   188

Monetary Benefits of Workers' Compensation   193

Medical Treatment and Rehabilitation   193

Medical Management of Workplace Injuries   196

Administration and Case Management   197

Cost Allocation   198

Problems with Workers' Compensation   198

Spotting Workers' Compensation Fraud
    and Abuse   199

Future of Workers' Compensation   200

Cost-Reduction Strategies   202

8 Accident Investigation and Reporting 211

Types of Accident Investigations 212
When to Investigate 212
What to Investigate 213
Who Should Investigate 216
Conducting the Investigation 216
Interviewing Witnesses 219
Reporting Accidents 222
Ten Accident Investigation Mistakes to Avoid 227

9 Product Safety and Liability 232

Product Liability and the Law 233
Developing a Product Safety Program 238
Evaluating the Product Safety Program 240
Role of the Safety and Health Professional 241
Quality Management and Product Safety 242
Product Safety Program Record Keeping 244
User Feedback Collection and Analysis 246
Product Literature and Safety 246

PART 3 THE HUMAN ELEMENT

**10 Ergonomic Hazards: Musculoskeletal Disorders
(MSDs) and Cumulative Trauma Disorders (CTDs) 255**

Ergonomics Defined 256
Human Factors and Ergonomic Hazards 257
Factors Associated with Physical Stress 258
Ergonomics: A Political Football 260
OSHA's Voluntary Ergonomics Guidelines 261
Worksite Analysis Program for Ergonomics 266
Hazard Prevention and Control 270
Medical Management Program 271
Training and Education 275
Common Indicators of Problems 276
Identifying Specific Ergonomic Problems 277
Ergonomic Problem-Solving Strategies 279
Economics of Ergonomics 285
Cumulative Trauma Disorders (CTDs) 287
Participatory Ergonomics 292

11 Stress and Safety   296

Workplace Stress Defined   296
Sources of Workplace Stress   297
Human Reactions to Workplace Stress   300
Measurement of Workplace Stress   301
Shift Work, Stress, and Safety   302
Improving Safety by Reducing Workplace Stress   303
Stress in Safety Managers   305
Stress and Workers' Compensation   306

12 Safety and Health Training   000

Rationale for Safety and Health Training   311
Education and Training Requirements   315
Safety and Health Professionals as Trainers   319
Preparing Safety and Health Instruction   321
Presenting Safety and Health Instruction   324
Applying Safety and Health Instruction   331
Evaluating Safety and Health Instruction   331
Training Supervisors   333
Training New and Transferred Employees   335
Job Safety Analysis as a Training Technique   338
Training Opportunities Available   340
Illiteracy and Safety   342
English as a Second Language Training   345
OSHA Standards and Training   347

13 Violence in the Workplace   352

Occupational Safety and Workplace Violence:
    The Relationship   353
Workplace Violence: Definitions   353
Workplace Violence: Cases   354
Size of the Problem   356
Legal Considerations   358
Risk-Reduction Strategies   360
Contributing Social and Cultural Factors   362
OSHA's Voluntary Guidelines   364
Conflict Resolution and Workplace Violence   373
Do's and Don'ts for Supervisors   377
Emergency Preparedness Plan   378

PART 4 HAZARD ASSESSMENT, PREVENTION, AND CONTROL

14 **Mechanical Hazards and Machine Safeguarding   385**
    Common Mechanical Injuries   386
    Safeguarding Defined   389
    OSHA's Requirements for Machine Guarding   390
    Risk Assessment in Machine Operation   390
    Requirements for All Safeguards   392
    Point-of-Operation Guards   393
    Point-of-Operation Devices   397
    Machine Guarding Self-Assessment   399
    Feeding and Ejection Systems   399
    Robot Safeguards   400
    Control of Hazardous Energy (Lockout/Tagout Systems)   400
    General Precautions   406
    Basic Program Content   407
    Taking Corrective Action   407

15 **Falling, Impact, Acceleration, Lifting, and Vision Hazards   411**
    Causes of Falls   411
    Kinds of Falls   412
    Walking and Slipping   412
    Slip and Fall Prevention Programs   416
    OSHA Fall Protection Standards   418
    Ladder Safety   421
    What to Do After a Fall   423
    Monitor Fall Protection Equipment and Know Why It Fails   424
    Impact and Acceleration Hazards   424
    Lifting Hazards   433
    Standing Hazards   437
    Hand Protection   439
    Personal Protective Equipment   442
    Forklift Safety (Powered Industrial Trucks)   444

16 **Hazards of Temperature Extremes   452**
    Thermal Comfort   452
    Heat Stress and Strain   453
    Cold Stress   458
    Burns and Their Effects   462
    Chemical Burns   465

17 Pressure Hazards   470

Pressure Hazards Defined   470
Sources of Pressure Hazards   472
Boilers and Pressure Hazards   473
High-Temperature Water Hazards   473
Hazards of Unfired Pressure Vessels   474
Hazards of High-Pressure Systems   475
Cracking Hazards in Pressure Vessels   475
Nondestructive Testing of Pressure Vessels   477
Pressure Dangers to Humans   478
Decompression Procedures   479
Measurement of Pressure Hazards   480
Reduction of Pressure Hazards   482

18 Electrical Hazards   487

Electrical Hazards Defined   487
Sources of Electrical Hazards   491
Electrical Hazards to Humans   495
Detection of Electrical Hazards   496
Reduction of Electrical Hazards   497
OSHA's Electrical Standards   501
Electrical Safety Program   502
Electrical Hazards Self-Assessment   503
Prevention of Arc Flash Injuries (NFPA 70E)   505

19 Fire Hazards and Life Safety   511

Fire Hazards Defined   512
Sources of Fire Hazards   516
Fire Dangers to Humans   520
Detection of Fire Hazards   521
Reduction of Fire Hazards   521
Development of Fire Safety Standards   527
OSHA Fire Standards   528
Life Safety   529
Flame-Resistant Clothing   532
Fire Safety Programs   533
Explosive Hazards   535
OSHA's Firefighting Options   537
Self-Assessment in Fire Protection   539

20 Industrial Hygiene Confined Spaces   544

Overview of Industrial Hygiene   545

Industrial Hygiene Standards   545

OSH Act and Industrial Hygiene   546

Hazards in the Workplace   551

Toxic Substances Defined   553

Entry Points for Toxic Agents   553

Effects of Toxic Substances   556

Relationship of Doses and Responses   557

Airborne Contaminants   558

Effects of Airborne Toxics   559

Effects of Carcinogens   561

Asbestos Hazards   561

Indoor Air Quality and "Sick-Building" Syndrome   565

Toxic Mold and Indoor Air Quality   567

Threshold Limit Values   569

Hazard Recognition and Evaluation   572

Prevention and Control   573

NIOSH and Industrial Hygiene   577

NIOSH Guidelines for Respirators   578

Standards and Regulations   581

General Safety Precautions   584

Confined Spaces Hazards   585

OSHA Confined Space Standard   587

Confined Space Management Policy   592

OSHA Standards for Toxic and Hazardous Materials   595

OSHA's Hazard Communication Standard   596

21 Radiation Hazards   604

Ionizing Radiation: Terms and Concepts   605

Exposure of Employees to Radiation   606

Precautions and Personal Monitoring   607

Caution Signs and Labels   608

Evacuation Warning Signal   609

Instructing and Informing Personnel   609

Storage and Disposal of Radioactive Material   610

Notification of Incidents   611

Reports and Records of Overexposure   611

Notice to Employees   612

Nonionizing Radiation   615

Electromagnetic Fields in the Workplace   618

OSHA Standards for Health and Environmental Controls   621

22 Noise and Vibration Hazards   626

Hearing Loss Prevention Terms   626

Characteristics of Sound   628

Hazard Levels and Risks   630

Standards and Regulations   631

Workers' Compensation and Noise Hazards   638

Identifying and Assessing Hazardous Noise Conditions   639

Noise Control Strategies   640

Vibration Hazards   646

Other Effects of Noise Hazards   649

Corporate Policy   649

Evaluating Hearing Loss Prevention Programs   651

23 Computers, Automation, and Robots   659

Impact of Automation on the Workplace   659

VDTs in Offices and Factories   660

Human–Robot Interaction   663

Safety and Health Problems Associated with Robots   664

Safety and Health in Office Automation   667

Industrial Medicine and Robots   670

Technological Alienation in the Automated Workplace   671

Minimizing the Problems of Automation   672

Challenge for the Future   675

24 Bloodborne Pathogens in the Workplace   681

Facts about AIDS   681

Symptoms of AIDS   682

AIDS in the Workplace   683

Legal Concerns   686

AIDS Education   690

Counseling Infected Employees   692

Easing Employees' Fears about AIDS   693

Protecting Employees from AIDS   693

Hepatitis B (HBV) and Hepatitis C (HCV) in the Workplace   695

OSHA's Standard on Occupational Exposure to
    Bloodborne Pathogens   699

Preventing and Responding to Needlestick Injuries   704

PART 5 MANAGEMENT OF SAFETY AND HEALTH 709

25 Preparing for Emergencies and Terrorism   711
Rationale for Emergency Preparation   712
Emergency Planning and Community Right-to-Know Act   712
Organization and Coordination   714
OSHA Standards   715
First Aid in Emergencies   716
How to Plan for Emergencies   719
Planning for Workers with Disabilities   724
Evacuation Planning   728
Customizing Plans to Meet Local Needs   730
Emergency Response   731
Computers and Emergency Response   732
Dealing with the Psychological Trauma of Emergencies   733
Recovering from Disasters   735
Terrorism in the Workplace   737
Resuming Business after a Disaster   741

26 Ethics and Safety   747
An Ethical Dilemma   748
Ethics Defined   749
Ethical Behavior in Organizations   750
Safety and Health Professionals' Role in Ethics   751
Company's Role in Ethics   753
Handling of Ethical Dilemmas   756
Questions to Ask When Making Decisions   757
Ethics and Whistle-Blowing   758

27 Hazard Analysis/Prevention and Safety Management   765
Overview of Hazard Analysis   766
Preliminary Hazard Analysis   767
Detailed Hazard Analysis   769
Hazard Prevention and Deterrence   778
OSHA Process Safety Standard   780
Risk Assessment   783
Safety Management Concerns   784
Best Practices in Safety Management   788
Occupational Health and Safety Management Systems   790

28 Promoting Safety 795

Company Safety Policy 796
Safety Rules and Regulations 797
Employee Participation in Promoting Safety 798
Safety Training 798
Suggestion Programs 799
Visual Awareness 799
Safety Committees 802
Personal Commitment to Workplace Safety 804
Employee–Management Participation 805
Incentives 806
Competition 808
Company-Sponsored Wellness Programs 808
Teamwork Approach to Promoting Safety 809
Persuasion as a Promotional Tool 814

29 Environmental Safety and ISO 14000 (Environmental Management) 819

Safety, Health, and the Environment 819
Legislation and Regulation 820
Types of Environments 823
Role of Safety and Health Professionals 824
Hazards of the Environment 825
Hazardous Waste Reduction 830
Environmental Management System (EMS) 835
International Organization for Standardization (ISO) 841
ISO 14000 842
ISO 14000 Series of Standards 842
ISO 14001 Standard 844
ISO 14000 Success Story 856

30 TSM: Safety Management in a Quality Management Setting 861

What Is QM? 862
How Does QM Relate to Safety? 863
Safety Management in a QM Setting 865
What Is TSM? 865
Translating TSM into Action 867
Fundamental Elements of TSM 869
Rationale for TSM 874
Implementing TSM: The Model 874

31 Establishing a Safety-Friendly Corporate Culture   878
Safety-Friendly Corporate Culture Defined   878
Importance of Having a Safety-Friendly Corporate Culture   879
Globalization of Competition and Safety   880
How Corporate Cultures Are Created   882
What a Safety-Friendly Corporate Culture Looks Like   883
Ten Steps for Establishing a Safety-Friendly Corporate Culture   884

Glossary   893

Index   915

1

HISTORICAL
PERSPECTIVE AND
OVERVIEW

1   Safety and Health Movement, Then and Now   3
2   Accidents and Their Effects   22
3   Theories of Accident Causation   40
4   Roles and Professional Certification for Safety and Health Professionals   65
5   Safety, Health, and Competition in the Global Marketplace   98

SAFETY AND HEALTH MOVEMENT, THEN AND NOW

Major Topics

- Developments before the Industrial Revolution
- Milestones in the Safety Movement
- Tragedies That Have Changed the Safety Movement
- Role of Organized Labor
- Role of Specific Health Problems
- Development of Accident Prevention Programs
- Development of Safety Organizations
- Safety and Health Movement Today
- Integrated Approach to Safety and Health
- New Materials, New Processes, and New Problems
- Rapid Growth in the Profession

The **safety movement** in the United States has developed steadily since the early 1900s. In that time period, industrial accidents were commonplace in this country; for example, in 1907 over 3,200 people were killed in mining accidents. Legislation, precedent, and public opinion all favored management. There were few protections for workers' safety.

Working conditions for industrial employees today have improved significantly. The chance of a worker being killed in an industrial accident is less than half of what it was 60 years ago.[1] According to the National Safety Council (NSC), the current death rate from work-related injuries is approximately 4 per 100,000, or less than a third of the rate 50 years ago.[2]

Improvements in safety until now have been the result of pressure for legislation to promote safety and health, the steadily increasing costs associated with accidents and injuries, and the professionalization of safety as an occupation. Improvements in the future are likely to come as a result of greater awareness of the cost effectiveness and resultant competitiveness gained from a safe and healthy workforce.

3

This chapter examines the history of the safety movement in the United States and how it has developed over the years. Such a perspective will help practicing and prospective safety professionals form a better understanding of both their roots and their future.

DEVELOPMENTS BEFORE THE INDUSTRIAL REVOLUTION

It is important for students of occupational health and safety to first study the past. Understanding the past can help safety and health professionals examine the present and future with a sense of perspective and continuity. Modern developments in health and safety are neither isolated nor independent. Rather, they are part of the long continuum of developments in the safety and health movement.

The continuum begins with the days of the ancient Babylonians. During that time, circa 2000 BC, their ruler Hammurabi developed his **Code of Hammurabi**. The code encompassed all the laws of the land at that time, showed Hammurabi to be a just ruler, and set a precedent followed by other Mesopotamian kings. The significance of the code from the perspective of safety and health is that it contained clauses dealing with injuries, allowable fees for physicians, and monetary damages assessed against those who injured others.[3] This clause from the code illustrates Hammurabi's concern for the proper handling of injuries: "If a man has caused the loss of a gentleman's eye, his own eye shall be caused to be lost."[4]

This movement continued and emerged in later Egyptian civilization. As evidenced from the temples and pyramids that still remain, the Egyptians were an industrious people. Much of the labor was provided by slaves, and there is ample evidence that slaves were not treated well; that is, unless it suited the needs of the Egyptian taskmasters.

One such case occurred during the reign of Rameses II (circa 1500 BC), who undertook a major construction project, the Ramesseum. To ensure the maintenance of a workforce sufficient to build this huge temple bearing his name, Rameses created an industrial medical service to care for the workers. They were required to bathe daily in the Nile and were given regular medical examinations. Sick workers were isolated.[5]

The Romans were vitally concerned with safety and health, as can be seen from the remains of their construction projects. The Romans built aqueducts, sewerage systems, public baths, latrines, and well-ventilated houses.[6]

As civilization progressed, so did safety and health developments. In 1567, Philippus Aureolus produced a treatise on the pulmonary diseases of miners. Titled *On the Miners' Sickness and Other Miners' Diseases*, the treatise covered diseases of smelter workers and metallurgists and diseases associated with the handling of and exposure to mercury. Around the same time, Georgius Agricola published his treatise *De Re Metallica*, emphasizing the need for ventilation in mines and illustrating various devices that could be used to introduce fresh air into mines.[7]

The eighteenth century saw the contributions of Bernardino Ramazzini, who wrote *Discourse on the Diseases of Workers*. Ramazzini drew conclusive parallels between diseases suffered by workers and their occupations. He related occupational diseases to the handling of harmful materials and to irregular or unnatural movements of the body. Much of what Ramazzini wrote is still relevant today.[8]

The Industrial Revolution changed forever the methods of producing goods. According to LaDou, the changes in production brought about by the Industrial Revolution can be summarized as follows:

- Introduction of **inanimate power** (i.e., steam power) to replace people and animal power
- Substitution of machines for people
- Introduction of new methods for converting raw materials
- Organization and specialization of work, resulting in a division of labor[9]

These changes necessitated a greater focusing of attention on the safety and health of workers. Steam power increased markedly the potential for life-threatening injuries, as did machines. The new methods used for converting raw materials also introduced new risks of injuries and diseases. Specialization, by increasing the likelihood of boredom and inattentiveness, also made the workplace a more dangerous environment.

MILESTONES IN THE SAFETY MOVEMENT

Just as the United States traces its roots to Great Britain, the safety movement in this country traces its roots to England. During the Industrial Revolution, child labor in factories was common. The hours were long, the work hard, and the conditions often unhealthy and unsafe. Following an outbreak of fever among the children working in their cotton mills, the people of Manchester, England, began demanding better working conditions in the factories. Public pressure eventually forced a government response, and in 1802 the Health and Morals of Apprentices Act was passed. This was a milestone piece of legislation: it marked the beginning of governmental involvement in workplace safety.

When the industrial sector began to grow in the United States, hazardous working conditions were commonplace. Following the Civil War, the seeds of the safety movement were sown in this country. Factory inspection was introduced in Massachusetts in 1867. In 1868, the first barrier safeguard was patented. In 1869, the Pennsylvania legislature passed a mine safety law requiring two exits from all mines. The Bureau of Labor Statistics (BLS) was established in 1869 to study industrial accidents and report pertinent information about those accidents.

The following decade saw little new progress in the safety movement until 1877, when the Massachusetts legislature passed a law requiring safeguards for hazardous machinery. This year also saw passage of the Employer's Liability Law, establishing the potential for **employer liability** in workplace accidents. In 1892, the first recorded safety program was established in a Joliet, Illinois, steel plant in response to a scare caused when a flywheel exploded. Following the explosion, a committee of managers was formed to investigate and make recommendations. The committee's recommendations were used as the basis for the development of a safety program that is considered to be the first safety program in American industry.

Around 1900, Frederick Taylor began studying efficiency in manufacturing. His purpose was to identify the impact of various factors on efficiency, productivity, and profitability. Although safety was not a major focus of his work, Taylor did draw a connection between lost personnel time and management policies and procedures. This connection between safety and management represented a major step toward broad-based safety consciousness.

In 1907, the U.S. Department of the Interior created the Bureau of Mines to investigate accidents, examine health hazards, and make recommendations for improvements. Mining workers definitely welcomed this development since over 3,200 of their fellow workers were killed in mining accidents in 1907 alone.[10]

One of the most important developments in the history of the safety movement occurred in 1908 when an early form of **workers' compensation** was introduced in the United States. Workers' compensation actually had its beginnings in Germany. The practice soon spread throughout the rest of Europe. Workers' compensation as a concept made great strides in the United States when Wisconsin passed the first effective workers' compensation law in 1911. In the same year, New Jersey passed a workers' compensation law that withstood a court challenge.

The common thread among the various early approaches to workers' compensation was that they all provided some amount of compensation for on-the-job injuries regardless of who was at fault. When the workers' compensation concept was first introduced in the United States, it covered a very limited portion of the workforce and provided only minimal benefits. Today, all 50 states have some form of workers' compensation that requires the payment of a wide range of benefits to a broad base of workers. Workers' compensation is examined in more depth in Chapter 7.

The Association of Iron and Steel Electrical Engineers (AISEE), formed in the early 1900s, pressed for a national conference on safety. As a result of the AISEE's efforts, the first meeting of the **Cooperative Safety Congress (CSC)** took place in Milwaukee in 1912. What is particularly significant about this meeting is that it planted the seeds for the eventual establishment of the National Safety Council. A year after the initial meeting of the CSC, the **National Council of Industrial Safety (NCIS)** was established in Chicago. In 1915, this organization changed its name to the National Safety Council. It is now the premier safety organization in the United States.

From the end of World War I (1918) through the 1950s, safety awareness grew steadily. During this period, the federal government encouraged contractors to implement and maintain a safe work environment. Also during this period, industry in the United States arrived at two critical conclusions: (1) there is a definite connection between quality and safety, and (2) off-the-job accidents have a negative impact on productivity. The second conclusion became painfully clear to manufacturers during World War II when the call-up and deployment of troops had employers struggling to meet their labor needs. For these employers, the loss of a skilled worker due to an injury or for any other reason created an excessive hardship.[11]

The 1960s saw the passage of a flurry of legislation promoting workplace safety. The Service Contract Act of 1965, the Federal Metal and Non-metallic Mine Safety Act, the Federal Coal Mine and Safety Act, and the Contract Workers and Safety Standards Act all were passed during the 1960s. As their names indicate, these laws applied to a limited audience of workers. According to the Society of Manufacturing Engineers (SME), more significant legislation than that enacted in the 1960s was needed:

> Generally, the state legislated safety requirements only in specific industries, had inadequate safety and health standards, and had inadequate budgets for enforcement. . . . The injury and death toll due to industrial mishaps was still . . . too high. In the late 1960s, more than 14,000 employees were killed annually in connection with their jobs. . . . Work injury rates were taking an upward swing.[12]

These were the primary reasons behind passage of the **Occupational Safety and Health Act (OSH Act)** of 1970 and the Federal Mine Safety Act of 1977. These federal laws, particularly the OSH Act, represent the most significant legislation to date in the history of the safety movement. Figure 1–1 summarizes some significant milestones in the development of the safety movement in the United States.

TRAGEDIES THAT HAVE CHANGED THE SAFETY MOVEMENT

Safety and health tragedies in the workplace have greatly accelerated the pace of the safety movement in the United States. Three of the most significant events in the history of the safety and health movement were the Hawk's Nest, asbestos, and Bhopal tragedies. This section explains these three milestone events and their lasting effects on the safety and health movement in the United States.

Hawk's Nest Tragedy

In the 1930s, the public began to take notice of the health problems suffered by employees who worked in dusty environments. The Great Depression was indirectly responsible for the attention given to an occupational disease that came to be known as *silicosis*. As the economic crash spread, business after business shut down and laid off its workers. Unemployed miners and foundry workers began to experience problems finding new jobs when physical examinations

1867	Massachusetts introduces factory inspection.
1868	Patent is awarded for first barrier safeguard.
1869	Pennsylvania passes law requiring two exits from all mines, and the Bureau of Labor Statistics is formed.
1877	Massachusetts passes law requiring safeguards on hazardous machines, and the Employer's Liability Law is passed.
1892	First recorded safety program is established.
1900	Frederick Taylor conducts first systematic studies of efficiency in manufacturing.
1907	Bureau of Mines is created by U.S. Department of the Interior.
1908	Concept of workers' compensation is introduced in the United States.
1911	Wisconsin passes the first effective workers' compensation law in the United States, and New Jersey becomes the first state to uphold a workers' compensation law.
1912	First Cooperative Safety Congress meets in Milwaukee.
1913	National Council of Industrial Safety is formed.
1915	National Council of Industrial Safety changes its name to National Safety Council.
1916	Concept of negligent manufacture is established (product liability).
1936	National Silicosis Conference convened by the U.S. Secretary of Labor.
1970	Occupational Safety and Health Act passes.
1977	Federal Mine Safety Act passes.
1986	Superfund Amendments and Reauthorization Act pass.
1990	Amended Clean Air Act of 1970 passes.
1996	Total safety management (TSM) concept is introduced.
2000	U.S. firms begin to pursue ISO 14000 registration for environmental safety management.
2003	Workplace terrorism is an ongoing concern of safety and health professionals.
2007	Safety of older people reentering the workplace becomes an issue.

Figure 1–1
Milestones in the safety movement.

revealed they had lung damage from breathing silica. Cautious insurance companies recommended preemployment physicals as a way to prevent future claims based on preexisting conditions. Applicants with silica-damaged lungs were refused employment. Many of them sued. This marked the beginning of industrywide interest in what would eventually be called the "king" of occupational diseases.

Lawsuits and insurance claims generated public interest in silicosis, but it was the Hawk's Nest tragedy that solidified public opinion in favor of protecting workers from this debilitating disease.[13] A company was given a contract to drill a passageway through a mountain located in the Hawk's Nest region of West Virginia (near the city of Gauley Bridge). Workers spent as many as 10 hours per day breathing the dust created by drilling and blasting. It turned out that this particular mountain had an unusually high silica content. Silicosis is a disease that normally takes 10 to 30 years to show up in exposed workers. At Hawk's Nest, workers began dying in as little time as a year. By the time the project was completed, hundreds had died. To make matters even worse, the company often buried an employee who died from exposure to silica in a nearby field without notifying the family. Those who inquired were told that their loved one left without saying where he was going.

A fictitious account of the Gauley Bridge disaster titled *Hawk's Nest*, by Hubert Skidmore, whipped the public outcry into a frenzy, forcing Congress to respond.

This tragedy and the public outcry that resulted from it led a group of companies to form the Air Hygiene Foundation to conduct research and develop standards for working in dusty environments. Soon thereafter, the U.S. Department of Labor provided the leadership necessary to make silicosis a compensable disease under workers' compensation in most states. Today, dust-producing industries use a wide variety of administrative controls, engineering controls, and personal protective equipment to protect workers in dusty environments. However, silicosis is still a problem. Approximately one million workers in the United States are still exposed to silica every year, and 250 people die annually from silicosis.

Asbestos Menace

Asbestos was once considered a "miracle" fiber, but in 1964 Dr. Irving J. Selikoff told 400 scientists at a conference on the biological effects of asbestos that this widely used material was killing workers. This conference changed how Americans viewed not just asbestos, but workplace hazards in general. Selikoff was the first to link asbestos to lung cancer and respiratory diseases.[14]

At the time of Selikoff's findings, asbestos was one of the most widely used materials in the United States. It was found in homes, schools, offices, factories, ships, and even in the filters of cigarettes. Selikoff continued to study the effects of asbestos exposure from 1967 to 1986. During this time, he studied the mortality rate of 17,800 workers who had been exposed to asbestos. He found asbestos-related cancer in the lungs, gastrointestinal tract, larynx, pharynx, kidneys, pancreas, gall bladder, and bile ducts of workers.

Finally, in the 1970s and 1980s, asbestos became a controlled material. Regulations governing the use of asbestos were developed, and standards for exposure were established. Asbestos-related lawsuits eventually changed how industry dealt with this tragic material. In the 1960s, industry covered up or denied the truth about asbestos. Now, there is an industrywide effort to protect workers who must remove asbestos from old buildings and ships during remodeling, renovation, or demolition projects.

Bhopal Tragedy

On the morning of December 3, 1984, over 40 tons of methyl isocyanate (MIC) and other lethal gases, including hydrogen cyanide, leaked into the northern end of Bhopal, killing more than 3,000 people in its aftermath.[15] After the accident, it was discovered that the protective equipment that could have halted the impending disaster was not in full working order. The refrigeration system that should have cooled the storage tank was shut down, the scrubbing system that should have absorbed the vapor was not immediately available, and the flare system that would have burned any vapor that got past the scrubbing system was out of order.[16]

The International Medical Commission visited Bhopal to assess the situation and found that as many as 50,000 other people were exposed to the poisonous gas and may still suffer disability as a result. This disaster shocked the world. Union Carbide, the owner of the chemical plant in Bhopal, India, where the incident occurred, was accused of many things, including the following:

- *Criminal negligence.*
- *Corporate prejudice.* Choosing poverty-stricken Bhopal, India, as the location for a hazardous chemical plant on the assumption that few would care if something went wrong.
- *Avoidance.* Putting its chemical plant in Bhopal, India, to avoid the stricter safety and health standards of the United States and the Occupational Safety and Health Administration (OSHA) in particular.

In February 1989, India's Supreme Court ordered Union Carbide India, Ltd., to pay $470 million in compensatory damages. The funds were paid to the Indian government to be used to compensate the victims. This disaster provided the impetus for the passage of stricter safety legislation worldwide. In the United States, it led to passage of the Emergency Planning and Community Right-to-Know Act (EPCRA) of 1986.

ROLE OF ORGANIZED LABOR

Organized labor has played a crucial role in the development of the safety movement in the United States. From the outset of the Industrial Revolution in this country, organized labor has fought for safer working conditions and appropriate compensation for workers injured on the job. Many of the earliest developments in the safety movement were the result of long and hard-fought battles by organized labor.

Although the role of unions in promoting safety is generally acknowledged, one school of thought takes the opposite view. Proponents of this dissenting view hold that union involvement actually slowed the development of the safety movement. Their theory is that unions allowed their demands for safer working conditions to become entangled with their demands for better wages and, as a result, they met with resistance from management. Regardless of the point of view, there is no question that working conditions in the earliest years of the safety movement were often reflective of an insensitivity to safety concerns on the part of management.

Among the most important contributions of organized labor to the safety movement was their work to overturn antilabor laws relating to safety in the workplace. These laws were the fellow servant rule, the statutes defining contributory negligence, and the concept of assumption of risk.[17] The **fellow servant rule** held that employers were not liable for workplace injuries that resulted from the negligence of other employees. For example, if Worker X slipped and fell, breaking his back in the process, because Worker Y spilled oil on the floor and left it there, the employer's liability was removed. In addition, if the actions of employees contributed to their own injuries, the employer was absolved of any liability. This was the doctrine of **contributory negligence**. The concept of **assumption of risk** was based on the theory that people who accept a job assume the risks that go with it. It says employees who work voluntarily should accept the consequences of their actions on the job rather than blame the employer.

Because the overwhelming majority of industrial accidents involve negligence on the part of one or more workers, employers had little to worry about. Therefore, they had little incentive to promote a safe work environment. Organized labor played a crucial role in bringing deplorable working conditions to the attention of the general public. Public awareness and, in some cases, outrage eventually led to these **employer-biased laws** being overturned in all states except one. In New Hampshire, the fellow servant rule still applies.

ROLE OF SPECIFIC HEALTH PROBLEMS

Specific health problems that have been tied to workplace hazards have played significant roles in the development of the modern safety and health movement. These health problems contributed to public awareness of dangerous and unhealthy working conditions that, in turn, led to legislation, regulations, better work procedures, and better working conditions.

Lung disease in coal miners was a major problem in the 1800s, particularly in Great Britain, where much of the Western world's coal was mined at the time. Frequent contact with coal dust led to a widespread outbreak of anthrocosis among Great Britain's coal miners. Also known as the *black spit*, this disease persisted from the early 1800s, when it was first identified, until around 1875, when it was finally eliminated by such safety and health measures as ventilation and decreased work hours.

Discussion Case

What Is Your Opinion?

Two safety and health students are debating the issue of corporate responsibility. Tom thinks that industry has clearly demonstrated its unwillingness over the years to provide a safe and healthy work environment for employees. He offers such examples as the Gauley Bridge disaster and the Bhopal incident as evidence. Janet agrees that industry indeed has a checkered past on safety and health, but she thinks employers have learned that their workforce is a valuable asset that should be protected. Tom's response is, "Take away federal and state mandates, and industry would return to its old ways in less than a year." Join this debate. What is your opinion?

In the 1930s, Great Britain saw a resurgence of lung problems among coal miners. By the early 1940s, British scientists were using the term *coal-miner's pneumoconiosis*, or CWP, to describe a disease from which many miners suffered. Great Britain designated CWP a separate and compensable disease in 1943. However, the United States did not immediately follow suit, even though numerous outbreaks of the disease had occurred among miners in this country.

The issue was debated in the United States until Congress finally passed the Coal Mine Health and Safety Act in 1969. The events that led up to passage of this act were tragic. An explosion in a coal mine in West Virginia in 1968 killed 78 miners. This tragedy focused attention on mining health and safety, and Congress responded by passing the Coal Mine Health and Safety Act. The act was amended in 1977 and again in 1978 to broaden the scope of its coverage.

Over the years, the diseases suffered by miners were typically lung diseases caused by the inhalation of coal dust particulates. However, health problems were not limited to coal miners. Other types of miners developed a variety of diseases, the most common of which was silicosis. Once again, it took a tragic event—the Gauley Bridge disaster, discussed earlier—to focus attention on a serious workplace problem.

Congress held a series of hearings on the matter in 1936. That same year, representatives from business, industry, and government attended the National Silicosis Conference, convened by the U.S. Secretary of Labor. Among other outcomes of this conference was a finding that silica dust particulates did, in fact, cause silicosis.

Mercury poisoning is another health problem that has contributed to the evolution of the safety and health movement by focusing public attention on unsafe conditions in the workplace. The disease was first noticed among the citizens of a Japanese fishing village in the early 1930s. A disease with severe symptoms was common in Minamata, but extremely rare throughout the rest of Japan. After much investigation into the situation, it was determined that a nearby chemical plant periodically dumped methyl mercury into the bay that was the village's primary source of food. Consequently, the citizens of this small village ingested hazardous dosages of mercury every time they ate fish from the bay.

Mercury poisoning became an issue in the United States after a study was conducted in the early 1940s that focused on New York City's hat-making industry. During that time, many workers in this industry displayed the same types of symptoms as the citizens of Minamata, Japan. Because mercury nitrate was used in the production of hats, enough suspicion was aroused to warrant a study. The study linked the symptoms of workers with the use of mercury nitrate. As a result, the use of this hazardous chemical in the hat-making industry was stopped, and a suitable substitute—hydrogen peroxide—was found.

As discussed earlier, asbestos was another important substance in the evolution of the modern safety and health movement. By the time it was determined that asbestos is a hazardous material, the fibers of which can cause asbestosis or lung cancer (mesothelioma), thousands of buildings contained the substance. As these buildings began to age, the asbestos, particularly that used to insulate pipes, began to break down. As asbestos breaks down, it releases dangerous microscopic fibers into the air. These fibers are so hazardous that removing asbestos from old buildings has become a highly specialized task requiring special equipment and training.

DEVELOPMENT OF ACCIDENT PREVENTION PROGRAMS

In the modern workplace, there are many different types of **accident prevention** programs ranging from the simple to the complex. Widely used accident prevention techniques include failure minimization, fail-safe designs, isolation, lockouts, screening, personal protective equipment, redundancy, timed replacements, and many others. These techniques are individual components of broader safety programs. Such programs have evolved since the late 1800s.

In the early 1800s, employers had little concern for the safety of workers and little incentive to be concerned. Consequently, organized safety programs were nonexistent, a situation that continued for many years. However, between World War I and World War II, industry discovered the connection between quality and safety. Then, during World War II, troop call-ups and deployments created severe labor shortages. Faced with these shortages, employers could not afford to lose workers to accidents or for any other reason. This realization created a greater openness toward giving safety the serious consideration that it deserved. For example, according to the Society of Manufacturing Engineers, around this time industry began to realize the following:

- Improved engineering could prevent accidents.
- Employees were willing to learn and accept safety rules.

Figure 1–2
Three E's of safety.

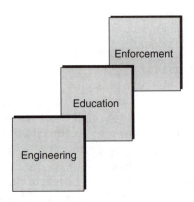

- Safety rules could be established and enforced.
- Financial savings from safety improvement could be reaped by savings in compensation and medical bills.[18]

With these realizations came the long-needed incentive for employers to begin playing an active role in creating and maintaining a safe workplace. This, in turn, led to the development of organized safety programs sponsored by management. Early safety programs were based on the **three E's of safety:** engineering, education, and enforcement (see Figure 1–2). The engineering aspects of a safety program involve making design improvements to both product and process. By altering the design of a product, the processes used to manufacture it can be simplified and, as a result, made less dangerous. In addition, the manufacturing processes for products can be engineered in ways that decrease potential hazards associated with the processes.

The education aspect of a safety program ensures that employees know how to work safely, why it is important to do so, and that safety is expected by management. Safety education typically covers the what, when, where, why, and how of safety.

The enforcement aspect of a safety program involves making sure that employees abide by safety policies, rules, regulations, practices, and procedures. Supervisors and fellow employees play a key role in the enforcement aspects of modern safety programs.

DEVELOPMENT OF SAFETY ORGANIZATIONS

Today, numerous organizations are devoted in full, or at least in part, to the promotion of safety and health in the workplace. Figure 1–3 lists organizations with workplace safety as part of their missions. Figure 1–4 lists several governmental agencies and two related organizations concerned with safety and health. These lists are extensive now, but this has not always been the case. Safety organizations in this country had humble beginnings.

Figure 1–3
Organizations concerned
with workplace safety.

Alliance for American Insurers

American Board of Industrial Hygiene

American Conference of Government Industrial
Hygienists

American Industrial Hygiene Association

American Insurance Association

American National Standards Institute

American Occupational Medical Association

American Society for Testing and Materials

American Society of Mechanical Engineers

American Society of Safety Engineers

Chemical Transportation Emergency Center

Human Factors Society

National Fire Protection Association

National Safety Council

National Safety Management Society

Society of Automotive Engineers

System Safety Society

Underwriters Laboratories Inc.

American Public Health Association*

Bureau of Labor Statistics

Bureau of National Affairs

Commerce Clearing House*

Environmental Protection Agency

National Institute for Standards and Technology (formerly National Bureau of
Standards)

National Institute for Occupational Safety and Health

Occupational Safety and Health Administration

Superintendent of Documents, U.S. Government Printing Office

U.S. Consumer Product Safety Commission

*Not a government agency.

Figure 1–4
Government agencies and other organizations concerned with workplace safety.

Safety Fact

Safety Movement and War

World War II actually had a positive effect on the modern safety and health movement. During the war, there was a shortage of able-bodied, skilled workers in factories supporting the war effort because most of these workers were in the armed services. Consequently, preserving the safety and health of the relatively few skilled workers still available was paramount. The law of supply and demand suddenly made workplace safety a significant issue, which it still is today. The military war is over, but the economic war still rages. To be competitive in this international conflict, employers today must follow the lead of their predecessors during World War II and protect their employees.

The grandfather of them all is the National Safety Council. The Society of Manufacturing Engineers traces the genesis of this organization as follows:

> The Association of Iron and Steel Electrical Engineers was organized in the first decade of the 20th century and devoted much attention to safety problems in its industry. In 1911, a request came from this association to call a national industrial safety conference. The first Cooperative Safety Congress met in Milwaukee in 1912. A year later, at a meeting in New York City, the National Council of Industrial Safety was formed. It began operation in a small office in Chicago. At its meeting in 1915, the organization's name was changed to the National Safety Council (NSC).[19]

Today, the NSC is the largest organization in the United States devoted solely to safety and health practices and procedures. Its purpose is to prevent the losses, both direct and indirect, arising out of accidents or from exposure to unhealthy environments. Although it is chartered by an act of Congress, the NSC is a nongovernmental, not-for-profit, public service organization.

The **Occupational Safety and Health Administration (OSHA)** is the government's administrative arm for the Occupational Safety and Health Act (OSH Act). Formed in 1970, OSHA sets and revokes safety and health standards, conducts inspections, investigates problems, issues citations, assesses penalties, petitions the courts to take appropriate action against unsafe employers, provides safety training, provides injury prevention consultation, and maintains a database of health and safety statistics.

Another governmental organization is the **National Institute for Occupational Safety and Health (NIOSH)**. This organization is part of the Centers for Disease Control and Prevention (CDC) of the Department of Health and Human Services. NIOSH is required to publish annually a comprehensive list of all known toxic substances. NIOSH will also provide on-site tests of potentially toxic substances so that companies know what they are handling and what precautions to take.

SAFETY AND HEALTH MOVEMENT TODAY

The safety and health movement has come a long way since the Industrial Revolution. Today, there is widespread understanding of the importance of providing a safe and healthy workplace. The tone was set during and after World War II when all the various practitioners of occupational health and safety began to see the need for cooperative efforts. These practitioners included safety engineers, safety managers, industrial hygienists, occupational health nurses, and physicians.

One of the earliest and most vocal proponents of the cooperative or integrated approach was H. G. Dyktor. He proposed the following objectives of integration:

- Learn more through sharing knowledge about health problems in the workplace, particularly those caused by toxic substances.
- Provide a greater level of expertise in evaluating health and safety problems.
- Provide a broad database that can be used to compare health and safety problems experienced by different companies in the same industry.
- Encourage accident prevention.
- Make employee health and safety a high priority.[20]

INTEGRATED APPROACH TO SAFETY AND HEALTH

This integrated approach has become the norm that typifies the safety and health movement of today. By working together and drawing on their own respective areas of expertise, safety and health professionals are better able to identify, predict, control, and correct safety and health problems.

OSHA reinforces the integrated approach by requiring companies to have a plan for doing at least the following: (1) providing appropriate medical treatment for injured or ill workers, (2) regularly examining workers who are exposed to toxic substances, and (3) having a qualified first-aid person available during all working hours.

Smaller companies may contract out the fulfillment of these requirements. Larger companies often maintain a staff of safety and health professionals. According to Hamilton and Hardy, the health and safety staff in a modern industrial company may include the following positions:

- *Industrial hygiene chemist and/or engineer.* Companies that use toxic substances may employ **industrial hygiene chemists** periodically to test the work environment and the people who work in it. In this way, unsafe conditions or hazardous levels of exposure can be identified early and corrective or preventive measures can be taken. Dust levels, ventilation, and noise levels are also monitored by individuals serving in this capacity.
- *Radiation control specialist.* Companies that use or produce radioactive materials employ **radiation control specialists** who are typically electrical

engineers or physicists. These specialists monitor the radiation levels to which workers may be exposed, test workers for levels of exposure, respond to radiation accidents, develop companywide plans for handling radiation accidents, and implement decontamination procedures when necessary.

- *Industrial safety engineer or manager.* Individuals serving as **industrial safety engineers or managers** are safety and health generalists with specialized education and training. In larger companies, they may be devoted to safety and health matters. In smaller companies, they may have other duties in addition to safety and health. In either case, they are responsible for developing and carrying out the company's overall safety and health program including accident prevention, accident investigation, and education and training.[21]

Other professionals who may be part of a company's safety and health team include occupational nurses, physicians, psychologists, counselors, educators, and dietitians.

NEW MATERIALS, NEW PROCESSES, AND NEW PROBLEMS

The job of the safety and health professional is more complex than it has ever been. The materials out of which products are made have become increasingly complex and exotic. Engineering metals now include carbon steels, alloy steels, high-strength low-alloy steels, stainless steels, managing steels, cast steels, cast irons, tungsten, molybdenum, titanium, aluminum, copper, magnesium, lead, tin, zinc, and powdered metals. Each of these metals requires its own specialized processes.

Nonmetals are more numerous and have also become more complex. Plastics, plastic alloys and blends, advanced composites, fibrous materials, elastomers, and ceramics also bring their own potential hazards to the workplace.

In addition to the more complex materials being used in modern industry and the new safety and health concerns associated with them, modern industrial processes are also becoming more complex. As these processes become automated, the potential hazards associated with them often increase. Computers; lasers; industrial robots; nontraditional processes such as explosive welding, photochemical machining, laser beam machining, ultrasonic machining, and chemical milling; automated material handling; water-jet cutting expert systems; flexible manufacturing cells; and computer-integrated manufacturing have all introduced new safety and health problems in the workplace and new challenges for the safety and health professional.

Chapter 23 is devoted to coverage of the special safety and health problems associated with computers, robots, and automation. In addition, coverage of specific aspects of these problems is provided in different chapters throughout this book.

RAPID GROWTH IN THE PROFESSION

The complexities of the modern workplace have made safety and health a growing profession. Associate and baccalaureate degree programs in industrial technology typically include industrial safety courses. Some engineering degree programs have safety and health tracks. Several colleges and universities offer full degrees in occupational safety and health.

The inevitable result of the increased attention given to safety and health is that more large companies are employing safety and health professionals and more small companies are assigning these duties to existing employees. This is a trend that is likely to continue as employers see their responsibilities for safety and health spread beyond the workplace to the environment, the community, the users of their products, and the recipients of their by-products and waste.

SUMMARY

1. Safety and health awareness has a long history. There is evidence of occupational safety and health efforts as far back as the time of the Egyptian pharaohs. The Code of Hammurabi, circa 2000 BC, contained clauses that could be interpreted as early attempts at workers' compensation. There is also evidence of concern for safety and health during the time of the Romans.

2. Milestones in the development of the safety movement in the United States include the following: first recorded safety program in 1892, creation of the Bureau of Mines in 1907, passage of the first effective workers' compensation law in the United States in 1911, and passage of OSHA in 1970.

3. Organized labor has played a crucial role in the development of the safety movement in the United States. Particularly important was the work of unions to overturn antilabor laws inhibiting safety in the workplace.

4. Specific health problems associated with the workplace have contributed to the development of the modern safety and health movement. These problems include lung diseases in miners, mercury poisoning, and lung cancer tied to asbestos.

5. Tragedies have changed the face of the safety movement at different times in the United States. The Hawk's Nest tragedy, asbestos menace, and Bhopal disaster are examples of such tragedies.

6. Widely used accident prevention techniques include failure minimization, fail-safe designs, isolation, lockouts, screening, personal protective equipment, redundancy, and timed replacements.

7. The development of the safety movement in the United States has been helped by the parallel development of safety organizations. Prominent among these are the National Safety Council, the National Safety Management Society, the American Society of Safety Engineers, and the American Industrial Hygiene Association.

8. The safety and health movement today is characterized by professionaliza-tion and integration. The safety and health team of a large company may include an industrial chemist or engineer, radiation control specialist, safety engineer or manager, occupational nurse, counselor, psychologist, and dietitian. New materials and processes are introducing new safety and health problems, making the integrated approach a practical necessity and promoting growth in the profession.

KEY TERMS AND CONCEPTS

Accident prevention

Asbestos menace

Assumption of risk

Bhopal tragedy

Code of Hammurabi

Contributory negligence

Cooperative Safety Congress (CSC)

Employer-biased laws

Employer liability

Fellow servant rule

Hawk's Nest tragedy

Inanimate power

Industrial hygiene chemist

Industrial safety engineer

Industrial safety manager

National Council of Industrial Safety (NCIS)

National Institute for Occupational Safety and Health (NIOSH)

Occupational Safety and Health Act (OSH Act)

Occupational Safety and Health Administration (OSHA)

Organized labor

Radiation control specialist

Safety movement

Three E's of safety

Workers' compensation

REVIEW QUESTIONS

1. To what cause(s) can the improvements in workplace safety made to date be attributed?
2. Explain the significance of the Code of Hammurabi in terms of the safety movement.
3. Describe the circumstances that led to the development of the first orga-nized safety program.
4. What is Frederick Taylor's connection to the safety movement?
5. Explain the development of the National Safety Council.
6. What impact did labor shortages in World War II have on the safety movement?
7. Explain how workplace tragedies have affected the safety movement. Give examples.

8. Explain the primary reasons behind the passage of the OSH Act.
9. Summarize briefly the role that organized labor has played in the advancement of the safety movement.
10. Define the following terms: fellow servant rule, contributory negligence, and assumption of risk.
11. Explain the three E's of safety.
12. Explain the term *integration* as it relates to modern safety and health.

ENDNOTES

1. S. Minter, "The Birth of OSHA," *Occupational Hazards*, July 1998, 59.
2. National Safety Council, *Accident Facts* (Chicago: National Safety Council, 1998).
3. J. LaDou, ed., *Introduction to Occupational Health and Safety* (Chicago: National Safety Council, 1997), 28.
4. Ibid., 28.
5. A. Soubiran, "Medical Services under the Pharaohs," *Abbottempo* 1: 19–23.
6. LaDou, *Occupational Health and Safety*, 31.
7. Ibid., 34.
8. Ibid., 35.
9. Ibid., 37.
10. S. Minter and V. Sutcliff, "Fighting Two Wars," *Occupational Hazards*, July 1998, 41–42.
11. Ibid., 41.
12. Ibid., 42.
13. Ibid., 41–42.
14. L. Finnegan, "Asbestos Becomes a Menace," *Occupational Hazards*, July 1998, 53–54.
15. L. Finnegan, "Reform and Reinvention," *Occupational Hazards*, July 1998, 67–68.
16. Ibid., 67.
17. Minter and Sutcliff, "Fighting Two Wars," 41.
18. Ibid., 41.
19. Ibid., 42.
20. H. G. Dyktor, "Integration of Industrial Hygiene with Industrial Medicine," *Industrial Medicine* 9, no. 4 (1940): 193.
21. A. Hamilton and H. Hardy, *Industrial Toxicology* (Boston: John Wright, 1983).

ACCIDENTS AND THEIR EFFECTS

Major Topics

- Costs of Accidents
- Accidental Deaths in the United States
- Accidents versus Other Causes of Death
- Work Accident Costs and Rates
- Time Lost Because of Work Injuries
- Deaths in Work Accidents
- Work Injuries by Type of Accident
- Death Rates by Industry
- Parts of the Body Injured on the Job
- Chemical Burn Injuries
- Heat Burn Injuries
- Repetitive Strain/Soft Tissue Injuries
- Estimating the Cost of Accidents
- Global Impact of Accidents and Injuries

There is a long history of debate in this country concerning the effect of **accidents** on industry (the workers and the companies) and the cost of preventing accidents. Historically, the prevailing view was that **accident prevention** programs were too costly. The more contemporary view is that accidents are too costly and that accident prevention makes sense economically. As a result, accident prevention, which had been advocated on a moral basis, is now justified in economic terms.

Accidents are the fourth leading cause of death in this country after heart disease, cancer, and strokes. This ranking is based on all types of accidents including motor vehicle accidents, drownings, fires, falls, natural disasters, and work-related accidents.

Although deaths from **natural disasters** tend to be more newsworthy than workplace deaths, their actual impact is substantially less. For example, natural disasters in the United States cause fewer than 100 deaths per year on average. **Workplace accidents**, on the other hand, cause more than 10,000 deaths every year in the United States.[1] The following quote from the National Safety Council (NSC) puts workplace accidents and deaths in the proper perspective, notwithstanding their apparent lack of newsworthiness.

> While you make a 10-minute speech—2 persons will be killed and about 170 will suffer a disabling injury. Costs will amount to $2,800,000. On the average, there are 11 accidental deaths and about 1,030 disabling injuries every hour during the year.[2]

This chapter provides prospective and practicing **safety and health professionals** with the information they need to have a full understanding of workplace accidents and their effect on industry in the United States. Such an understanding will help professionals play a more effective role in keeping both management and labor focused appropriately on safety and health in the workplace.

COSTS OF ACCIDENTS

To gain a proper perspective on the economics of workplace accidents, we must view them in the overall context of all accidents. The overall cost of accidents in the United States is approximately $150 billion. These costs include such factors as **lost wages**, **medical expenses**, **insurance administration**, **fire-related losses**, motor vehicle **property damage**, and **indirect costs**.

Figure 2–1 breaks down this overall amount by categories of accidents. Figure 2–2 breaks them down by cost categories. Notice in Figure 2–1 that workplace accidents rank second behind motor vehicle accidents in cost. Figure 2–2 shows that the highest cost category is wages lost by workers who are either injured or killed. The category of indirect losses from work accidents consists of costs associated with responding to accidents (i.e., giving first aid, filling out accident reports, handling production slowdowns).

Clearly, accidents on and off the job cost U.S. industry dearly. Every dollar that is spent responding to accidents is a dollar that could have been reinvested in modernization, research and development, facility upgrades, and other competitiveness-enhancing activities.

Figure 2–1
Accident costs by accident type (in billion, in a typical year).

Motor vehicle accidents	$72
Workplace accidents	48
Home accidents	18
Public accidents	12

Figure 2–2
Accident costs by categories
(in billions, in a typical year).

Wages lost	$37
Medical expenses	24
Insurance administration	29
Property damage (motor vehicle)	27
Fire losses	10
Indirect losses for work accidents	23

ACCIDENTAL DEATHS IN THE UNITED STATES

Accidental deaths in the United States result from a variety of causes, including motor vehicle accidents, falls, poisoning, drowning, fire-related injuries, suffocation (ingested object), firearms, medical complications, air transport accidents, machinery, mechanical suffocation, and the impact of falling objects. The NSC periodically computes death totals and death rates in each of these categories. The statistics for a typical year are as follows:

- *Motor vehicle accidents.* Motor vehicle accidents are the leading cause of accidental deaths in the United States each year. They include deaths resulting from accidents involving mechanically or electrically powered vehicles (excluding rail vehicles) that occur on or off the road. In a typical year, there are approximately 47,000 such deaths in the United States.
- *Falls.* This category includes all deaths from **falls** except those associated with transport vehicles. For example, a person who is killed as the result of falling while boarding a bus or train would not be included in this category. In a typical year, there are approximately 13,000 deaths in the United States from falls.
- *Poisoning.* The **poisoning** category is divided into two subcategories: (1) poisoning by solids and liquids, and (2) poisoning by gases and vapors. The first category includes deaths that result from the ingestion of drugs, medicine, widely recognized solid and liquid poisons, mushrooms, and shellfish. It does not include poisoning from spoiled food or salmonella. The second category includes deaths caused by incomplete combustion (for example, gas vapors from an oven or unlit pilot light) or from carbon monoxide (for example, exhaust fumes from an automobile). In a typical year, there are approximately 6,000 deaths in the first category and 1,000 in the second.
- *Drowning.* This category includes work-related and non-work-related **drownings** but excludes those associated with floods or other natural disasters. In a typical year, there are approximately 5,000 deaths from drowning in the United States.
- *Fire-related injuries.* This category includes deaths from burns, asphyxiation, falls, and those that result from falling objects in a fire. In a typical year, there are over 4,000 fire-related deaths in the United States.

- *Suffocation (ingested object).* This category includes deaths from the ingestion of an object that blocks the air passages. In many such deaths, the ingested object is food. In a typical year, there are approximately 4,000 suffocation deaths in the United States.
- *Firearms.* This category includes deaths that result when recreational activities involving firearms or household accidents involving firearms result in death. For example, a person killed in the home while cleaning a firearm would be included in this category. However, a person killed in combat would not be. In a typical year, there are approximately 2,000 deaths in this category.
- *Others.* This category includes deaths resulting from medical complications arising out of mistakes made by health care professionals, air transport injuries, interaction with machinery, **mechanical suffocation**, and the impact of falling objects. In a typical year, there are over 14,000 deaths in these subcategories.[3]

ACCIDENTS VERSUS OTHER CAUSES OF DEATH

Although there are more deaths every year from **heart disease**, **cancer**, and **strokes** than from accidents, these causes tend to be concentrated among people at or near retirement age. Among people 37 years of age or younger—prime working years—accidents are the number one cause of death. Figure 2–3 summarizes the causes of death for persons from 25 to 44 years of age. Notice that the leading cause is accidents.

Figure 2–3 shows that accidents represent a serious detriment to productivity, quality, and competitiveness in today's workplace. Yet accidents are the one cause of death and injury that companies can most easily control. Although it is true that companies may have some success in decreasing the incidence of heart disease and stroke among their employees through such activities as corporate wellness programs, their impact in this regard is limited. However, employers can have a significant impact on preventing accidents.

Figure 2–3
Causes of accidents (ages 25 to 44 years in a typical year).

Accidents	27,500
Cancer	20,300
Motor vehicle	16,500
Heart disease	16,000
Poison (solid, liquid)	2,700
Drowning	1,500
Falls	1,100
Fire related	900

WORK ACCIDENT COSTS AND RATES

Workplace accidents cost employers millions every year. Consider the following examples from the recent past. Arco Chemical Company was ordered to pay $3.48 million in fines as a result of failing to protect workers from an explosion at its petrochemical plant in Channelview, Texas. The steel-making division of USX paid a $3.25 million fine to settle numerous health and safety violation citations. BASF Corporation agreed to pay a fine of $1.06 million to settle OSHA citations associated with an explosion at a Cincinnati chemical plant that caused 2 deaths and 17 injuries.

These examples show the costs of fines only. In addition to fines, these employers incurred costs for safety corrections, medical treatment, survivor benefits, death and burial costs, and a variety of indirect costs. Clearly, work accidents are expensive. However, the news is not all bad. The trend in the rate of accidents is downward.

Work **accident rates** in this century are evidence of the success of the safety movement in the United States. As the amount of attention given to workplace safety and health has increased, the accident rate has decreased. According to the NSC,

> Between 1912 and 1998, accidental work deaths per 100,000 population were reduced 81 percent, from 21 to 4. In 1912, an estimated 18,000 to 21,000 workers' lives were lost. In 1998, in a workforce more than triple in size and producing 11 times the goods and services, there were approximately 10,000 work deaths.[4]

As Figure 2–1 shows, the cost of these 10,000 work deaths and work injuries was $48.5 billion. This translates into a cost of $420 per worker in the United States, computed as the value-added required per worker to offset the cost of work injuries. It translates further into $610,000 per death and $18,000 per disabling injury.[5]

Although statistics are not available to document the supposition, many safety and health professionals believe that the major cost of accidents and injuries on the job results from damage to morale. Employee morale is a less tangible factor than documentable factors such as lost time and medical costs. However, it is widely accepted among management professionals that few factors affect productivity more than employee morale. Employees with low morale do not produce up to their maximum potential. This is why so much time and money are spent every year to help supervisors and managers learn different ways to help improve employee morale.

Because few things are so detrimental to employee morale as seeing a fellow worker injured, accidents can have a devastating effect on morale. Whenever an employee is injured, his or her colleagues silently think, "That could have been me," in addition to worrying about the employee. Morale is damaged even more if the injured employee is well liked and other employees know his or her family.

TIME LOST BECAUSE OF WORK INJURIES

An important consideration when assessing the effect of accidents on industry is the amount of **lost time** due to **work injuries.**[6] According to the NSC, approximately 35,000,000 hours are lost in a typical year as a result of accidents. This is actual time lost from disabling injuries and does not include additional time lost for medical checkups after the injured employee returns to work. Accidents that occurred in previous years often continue to cause lost time in the current year.

DEATHS IN WORK ACCIDENTS

Deaths on the job have decreased markedly over the years. However, they still occur. For example, in a typical year, there are 10,400 work deaths in the United States. The causes of death in the workplace vary. They include those related to motor vehicles, falls, electric current, drowning, fires, air transport, poisoning, water transport, machinery, falling objects, rail transport, and mechanical suffocation.[7] Figure 2–4 gives a complete breakdown of the percentages for the various categories of causes.

WORK INJURIES BY TYPE OF ACCIDENT

Work injuries can be classified by the type of accident from which they resulted. The most common causes of work injuries are

- Overexertion
- Impact accidents
- Falls
- Bodily reaction (to chemicals)
- Compression

Figure 2–4
Work deaths by cause for a typical year.

Motor vehicle related	37.2%
Falls	12.5
Electric current	3.7
Drowning	3.2
Fire related	3.1
Air transport related	3.0
Poison (solid, liquid)	2.7
Water transport related	1.6
Poison (gas, vapor)	1.4
Other	31.6

- Motor vehicle accidents
- Exposure to radiation or caustics
- Rubbing or abrasions
- Exposure to extreme temperatures

Overexertion, the result of employees working beyond their physical limits, is the leading cause of work injuries. According to the NSC, almost 31 percent of all work injuries are caused by overexertion. **Impact accidents** involve a worker being struck by or against an object. The next most prominent cause of work injuries is falls.[8] The remaining accidents are distributed fairly equally among the other causes just listed.

DEATH RATES BY INDUSTRY

A variety of agencies and organizations including the Bureau of Labor Statistics, the National Center for Health Statistics, and the National Safety Council collect data on **death rates** within industrial categories.[9] Such information can be used in a variety of ways, not the least of which is in assigning workers' compensation rates. The most widely used industrial categories are agriculture, including farming, forestry, and fishing; mining/quarrying, including oil and gas drilling and extraction; construction; manufacturing; transportation/public utilities; trade, both wholesale and retail; services, including finance, insurance, and real estate; and federal, state, and local government.

When death rates are computed on the basis of the number of deaths per 100,000 workers in a given year, the industry categories rank as follows (from highest death rate to lowest):

1. Mining/quarrying
2. Agriculture
3. Construction
4. Transportation/public utilities

Discussion Case

What Is Your Opinion?

Mack Jones has been a safety engineer at Zumwalt Processing Company for almost 25 years. His son David is a recent college graduate who has been the assistant safety director at another company for just six months. Over supper last night, they had a discussion about work injuries. During the discussion, Mack said he thought back injuries were still the problem that safety professionals should worry about the most. David disagreed. He said repetitive-motion/soft tissue injuries such as carpal tunnel syndrome were a bigger problem in the modern computerized workplace. What is your opinion?

5. Government
6. Manufacturing
7. Services
8. Trade

The rankings sometimes change slightly from year to year. For example, agriculture and mining/quarrying may exchange the first and second ranking in any given year. This is also true at the low end of the rankings with services and trade. However, generally, the typical ranking is as listed.

PARTS OF THE BODY INJURED ON THE JOB

In order to develop and maintain an effective safety and health program, it is necessary to know not only the most common causes of death and injury but also the parts of the body most frequently injured. The NSC stated the following:

> Disabling work injuries in the entire nation totaled approximately 1.75 million in 1998. Of these, about 10,400 were fatal and 60,000 resulted in some permanent impairment. Injuries to the back occurred most frequently, followed by thumb and finger injuries and leg injuries.[10]

Typically, the most frequent injuries to specific parts of the body are as follows (from most frequent to least):

1. Back
2. Legs and fingers
3. Arms and multiple parts of the body
4. Trunk
5. Hands
6. Eyes, head, and feet
7. Neck, toes, and body systems

The back is the most frequently injured part of the body. Legs and fingers are injured with approximately the same frequency, as are arms and multiple parts of the body; the hands are next in frequency, followed by the eyes, the head, and feet; and neck, toes, and body systems. This ranking shows that one of the most fundamental components of a safety and health program should be instruction on how to lift without hurting the back (see Chapter 15).

CHEMICAL BURN INJURIES

Chemical burn injuries are a special category with which prospective and practicing safety professionals should be familiar. The greatest incidence of chemical burns (approximately one-third) occurs in manufacturing.[11] Other high-incidence industries are services, trade, and construction.

The chemicals that most frequently cause chemical burn injuries include acids and alkalies; soaps, detergents, and cleaning compounds; solvents and degreasers; calcium hydroxide (a chemical used in cement and plaster); potassium hydroxide (an ingredient in drain cleaners and other cleaning solutions); and sulfuric acid (battery acid). Almost 46 percent of all chemical burn injuries occur while workers are cleaning equipment, tools, and vehicles.[12]

What is particularly disturbing about chemical burn injuries is that a high percentage of them occur in spite of the use of personal protective equipment, the provision of safety instruction, and the availability of treatment facilities. In some cases, the personal protective equipment is faulty or inadequate. In others, it is not properly used in spite of instructions.

Preventing chemical burn injuries presents a special challenge to safety and health professionals. The following strategies are recommended:

- Familiarize yourself, the workers, and their supervisors with the chemicals that will be used and their inherent dangers.
- Secure the proper personal protection equipment for each type of chemical that will be used.
- Provide instruction on the proper use of personal protection equipment and then make sure that supervisors confirm that the equipment is used properly every time.
- Monitor that workers are wearing personal protection equipment and replace it when it begins to show wear.

HEAT BURN INJURIES

Heat burn injuries present a special challenge to safety and health professionals in the modern workplace. Almost 40 percent of all such injuries occur in manufacturing every year. The most frequent causes are flame (this includes smoke inhalation injuries), molten metal, petroleum asphalts, steam, and water. The most common activities associated with heat burn injuries are welding, cutting with a torch, and handling tar or asphalt.[13]

Following are several factors that contribute to heat burn injuries in the workplace. Safety and health professionals who understand these factors will be in a better position to prevent heat burn injuries.

- Employer has no health and safety policy regarding heat hazards.
- Employer fails to enforce safety procedures and practices.
- Employees are not familiar with the employer's safety policy and procedures concerning heat hazards.
- Employees fail to use or improperly use personal protection equipment.
- Employees have inadequate or worn personal protection equipment.
- Employees work in a limited space.
- Employees attempt to work too fast.
- Employees are careless.
- Employees have poorly maintained tools and equipment.[14]

These factors should be considered carefully by safety and health professionals when developing accident prevention programs. Employees should be familiar with the hazards, should know the appropriate safety precautions, and should have and use the proper personal protection equipment. Safety professionals should monitor to ensure that safety rules are being followed, that personal protection equipment is being used correctly, and that it is in good condition.

REPETITIVE STRAIN/SOFT-TISSUE INJURIES

Repetitive strain injury (RSI) is a broad and generic term that encompasses a variety of injuries resulting from cumulative trauma to the soft tissues of the body, including tendons, tendon sheaths, muscles, ligaments, joints, and nerves. Such injuries are typically associated with the soft tissues of the hands, arms, neck, and shoulders.

Carpal tunnel syndrome (CTS) is the most widely known repetitive strain injury. There are also several other repetitive strain injuries to the body's soft tissues. The carpal tunnel is the area inside the wrist through which the median nerve passes. It is formed by the wrist bones and a ligament. CTS is typically caused by repeated and cumulative stress on the median nerve. Symptoms of CTS include numbness, a tingling sensation, and pain in the fingers, hand, and/or wrist.

Stress placed on the median nerve typically results from repeated motion while the hands and fingers are bent in an unnatural position. However, sometimes the stress results from a single traumatic event such as a sharp blow to the wrist.

Evidence suggests a higher incidence of CTS among women than men. Scientists have found that the incidence rate for CTS among women was 1.96 per 1,000 full-time equivalent (FTE) workers and 1.58 per 1,000 FTE for men. The overall incidence rate for CTS is increasing at a rate of more than 15 percent per year.[15]

A common misconception about RSI is that it is synonymous with carpal tunnel syndrome. It isn't. In fact, carpal tunnel syndrome is relatively rare among RSI patients. Following is a list of the broad classifications of RSI and the types commonly associated with each classification:

Muscle and Tendon Disorders
- Tendinitis
- Muscle damage
- Tenosynovitis
- Stenosing tenosynovitis

DeQuervain's Disease
- Flexor tenosynovitis (trigger finger)
- Shoulder tendinitis

- Forearm tendinitis
- Cervical radioculopathy
- Epicondylitis
- Ganglion cysts

Tunnel Syndromes
- Carpal tunnel syndrome
- Radial tunnel syndrome

Ulnar Nerve Disorders
- Sulcus ulnaris syndrome
- Cubital tunnel syndrome
- Guyon's canal syndrome

Nerve and Circulation Disorders
- Thoracic outlet syndrome
- Raynaud's disease

Other Associated Disorders
- Reflex sympathetic dysfunction
- Focal dystonia (writer's cramp)
- Degenerative joint disorder (osteoarthritis)
- Fibromyalgia
- Dupuytren's contracture[16]

RSI and its manifestations are explained in greater detail in Chapter 10.

ESTIMATING THE COST OF ACCIDENTS

Even decision makers who support accident prevention must consider the relative costs of such efforts. Clearly, accidents are expensive. However, to be successful, safety and health professionals must be able to show that accidents are more expensive than prevention. To do this, they must be able to estimate the cost of accidents. The procedure for estimating costs set forth in this section was developed by Professor Rollin H. Simonds of Michigan State College working in conjunction with the Statistics Division of the National Safety Council.

Cost Estimation Method

Professor Simonds states that in order to have value, a cost estimate must relate directly to the specific company in question. Applying broad industry cost factors will not suffice. To arrive at company-specific figures, Simonds recommends that costs associated with an accident be divided into *insured* and *uninsured* costs.[17]

Determining the insured costs of accidents is a simple matter of examining accounting records. The next step involves calculating the uninsured costs. Simonds recommends that accidents be divided into the following four classes:

- *Class 1 accidents.* Lost workdays, permanent partial disabilities, and temporary total disabilities.
- *Class 2 accidents.* Treatment by a physician outside the company's facility.
- *Class 3 accidents.* Locally provided first aid, property damage of less than $100, or the loss of fewer than eight hours of work time.
- *Class 4 accidents.* Injuries that are so minor they do not require the attention of a physician, result in property damage of $100 or more, or cause eight or more work hours to be lost.[18]

Average uninsured costs for each class of accident can be determined by pulling the records of all accidents that occurred during a specified period and sorting the records according to class. For each accident in each class, record every cost that was not covered by insurance. Compute the total of these costs by class of accident and divide by the total number of accidents in that class to determine an average uninsured cost for each class, specific to the particular company.

Figure 2–5 is an example of how the average cost of a selected sample of Class 1 accidents can be determined. In this example, there were four Class 1 accidents in the pilot test. These four accidents cost the company a total of $554.23 in uninsured costs, or an average of $138.56 per accident. Using this information, accurate cost estimates of an accident can be figured, as can accurate predictions.

Class of Accident	Accident Number							
Class 1	1	2	3	4	5	6	7	8
Cost A	$ 16.00	$ 6.95	$ 15.17	$ 3.26				
Cost B	72.00	103.15	97.06	51.52				
Cost C	26.73	12.62	—	36.94				
Cost D	—	51.36	—	38.76				
Cost E	—	11.17	—	24.95				
Cost F	—	—	—	−13.41				
Cost G	—	—	—	—				
Total	114.73	185.25	112.23	142.02				

Grand Total: $554.23

Average Cost per Accident: $138.56 (Grand Total ÷ Number of Accidents)

Signature: _____   Date: _____

Figure 2–5
Uninsured costs worksheet.

Other Cost Estimation Methods

The costs associated with workplace accidents, injuries, and incidents fall into broad categories such as the following:

- Lost work hours
- Medical costs
- Insurance premiums and administration
- Property damage
- Fire losses
- Indirect costs

Calculating the direct costs associated with lost work hours involves compiling the total number of lost hours for the period in question and multiplying the hours times the applicable loaded labor rate. The loaded labor rate is the employee's hourly rate plus benefits. Benefits vary from company to company but typically inflate the hourly wage by 20 to 35 percent. A sample cost-of-lost-hours computation follows:

$$\text{Employee Hours Lost (4th quarter)} \times \text{Average Loaded Labor Rate} = \text{Cost}$$
$$386 \times 13.48 = \$5,203.28$$

In this example, the company lost 386 hours due to accidents on the job in the fourth quarter of its fiscal year. The employees who actually missed time at work formed a pool of people with an average loaded labor rate of $13.48 per hour ($10.78 average hourly wage plus 20 percent for benefits). The average loaded labor rate multiplied times the 386 lost hours reveals an unproductive cost of $5,203.28 to this company.

By studying records that are readily available in the company, a safety professional can also determine medical costs, insurance premiums, property damage, and fire losses for the time period in question. All these costs taken together result in a subtotal cost. This figure is then increased by a standard percentage to cover indirect costs to determine the total cost of accidents for a specific time period. The percentage used to calculate indirect costs can vary from company to company, but 20 percent is a widely used figure.

Estimating Hidden Costs

Safety professionals often use the *iceberg analogy* when talking about the real costs of accidents. Accident costs are like an iceberg in that their greatest portion is hidden from view.[19] In the case of icebergs, the larger part is hidden beneath the surface of the water. In the case of an accident, the larger part of the actual cost is also hidden beneath the surface.

According to Daniel Corcoran,

When a serious accident occurs, there is usually a great deal of activity associated with the accident. There may be a slowdown in production near the site of the accident, for instance. There also will be a need to replace the injured worker, at least temporarily, and there will be costs associated with the learning curve of the replacement worker.

	Checklist for **Estimating the Hidden Costs of Accidents**
√	Paid time to the injured employee on the day of the accident.
√	Paid time of any emergency-responder personnel involved (including ambulance driver).
√	Paid time of all employees who were interviewed as part of the accident investigation.
√	Paid time of the safety personnel who conducted the accident investigation.
√	Paid time of the human resources personnel who handled the workers' compensation and medical aspects of the accident.
√	Paid time of the supervisor involved in the accident investigation and accident response.
√	Paid time to employees near the accident working (or slowed down) temporarily as a result of the accident.
√	Paid time to employees who spent time talking about the accident as news of it spread through the company's grapevine.

Figure 2–6
Some accident costs that might be overlooked.

The supervisor and the accident investigation team probably will need to spend time conducting an investigation, and there will be a lot of time spent on the administration of paper work related to the accident.[20]

There are many different models that can be used for estimating both the direct and indirect costs of accidents. Some of these models are so complex that their usefulness is questionable. The checklist in Figure 2–6 is a simple and straight-forward tool that can be used to estimate the hidden costs of accidents.

GLOBAL IMPACT OF ACCIDENTS AND INJURIES

According to the International Labour Organization (ILO) of the United Nations, approximately 2.2 million people die every year of work-related injuries and occupational diseases.[21] Actually, due to poor record keeping and reporting in underdeveloped countries, it is estimated that this figure is low.

Rapid development and the pressure of global competition are resulting in increased workplace fatalities in China and the Pacific Rim countries.

What is missing in many of the developing countries that are becoming industrialized is a safety and health infrastructure. Such an infrastructure would include government-enforced safety and health regulations, company-sponsored safety and health training, record-keeping and reporting systems, and management practices that make occupational safety and health a fully integrated component of work processes and the competitive philosophy of organizations. Occupational safety and health must come to be seen as a strategy for sustaining economic growth and social development in emerging countries.

There is much to be done if developing countries are going to provide safe and healthy working conditions for their citizens. The ILO reports:

- Record-keeping and reporting systems in developing countries are deteriorating instead of improving. Consequently, only a fraction of the real toll of workplace accidents and injuries is being reported.
- Men in developing countries tend to die as the result of accidents, lung diseases, and work-related cancers such as those caused by asbestos. Women in developing countries suffer more from musculoskeletal disorders, communicable diseases, and psychosocial problems.
- Occupational injuries in developing countries are more prevalent in such high-risk industries as mining, construction, and agriculture.
- Younger workers in developing countries are more likely to suffer nonfatal injuries, while older workers are more likely to suffer fatal injuries.
- In developing countries, more than half of retirements are taken early to collect pensions based on work-related disabilities rather than normal retirement.

SUMMARY

1. The approximate cost of accidents in the United States is $150 billion annually. This includes the direct and indirect costs of accidents that occur on and off the job.
2. The leading causes of accidental deaths in the United States are motor vehicle accidents, falls, poisoning, drowning, fire-related injuries, suffocation, firearms, medical complications, air transport accidents, machinery-related injuries, mechanical suffocation, and the impact of falling objects.
3. The leading causes of death in the United States are heart disease, cancer, and stroke. However, these causes are concentrated among people at or near retirement age. Among people 37 years of age and younger, accidents are the number one cause of death.
4. Between 1912 and 1998, the number of accidental work deaths per 100,000 population declined by 81 percent, from 21 to 4.
5. Approximately 35,000,000 work hours are lost annually as a result of accidents. This is actual time lost from disabling injuries and does not include

additional time lost to medical checkups after the injured employee returns to work.

6. The leading causes of death in work accidents are motor vehicle related, falls, electric current, drowning, fire related, air transport related, poisoning, and water transport related.

7. The leading causes of work injuries are overexertion, impact accidents, falls, bodily reaction, compression, motor vehicle accidents, exposure to radiation or caustics, rubbing or abrasions, and exposure to extreme temperatures.

8. When death rates are computed on the basis of number of deaths per 100,000 workers, the industry categories are ranked as follows (from highest death rate to lowest): mining/quarrying, agriculture, construction, transportation/public utilities, government, manufacturing, services, and trade.

9. Typically, the ranking of injuries to specific parts of the body are as follows (from most frequently injured to least): back; legs and fingers; arms and multiple parts of body; trunk; hands; eyes, head, and feet; and neck, toes, and body systems.

10. The chemicals most frequently involved in chemical burn injuries include acids and alkalies; soaps, detergents, and cleaning compounds; solvents and degreasers; calcium hydroxide; potassium hydroxide; and sulfuric acid.

11. The most frequent causes of heat burn injuries are flame, molten metal, petroleum, asphalt, steam, and water.

12. Carpal tunnel syndrome (CTS) is an injury to the median nerve in the wrist that typically results from repeated stress placed on the nerve. Symptoms of CTS include numbness, a tingling sensation, and pain in the hand and/or wrist.

13. Repetitive strain injury (RSI) is a broad and generic term that encompasses a variety of injuries resulting from cumulative trauma to the soft tissues of the body, including tendons, tendon sheaths, muscles, ligaments, joints, and nerves. Such injuries are typically associated with the soft tissues of the hands, arms, neck, and shoulders.

14. Accident rates are especially high in developing countries because these countries are responding to the pressures of global competition without first putting a safety and health infrastructure in place (e.g., regulations, training, record keeping, etc.).

KEY TERMS AND CONCEPTS

Accident prevention	Death rates
Accident rates	Drownings
Accidents	Falls
Cancer	Fire-related losses
Carpal tunnel syndrome (CTS)	Heart disease
Chemical burn injuries	Heat burn injuries

Impact accidents   Overexertion
Indirect costs   Poisoning
Insurance administration   Property damage
Lost time   Repetitive strain injury (RSI)
Lost wages   Safety and health professionals
Mechanical suffocation   Strokes
Medical complications   Suffocation
Medical expenses   Work injuries
Natural disasters   Workplace accidents

REVIEW QUESTIONS

1. What are the leading causes of death in the United States?
2. When the overall cost of an accident is calculated, what elements make up the cost?
3. What are the five leading causes of accidental deaths in the United States?
4. What are the leading causes of death in the United States of people between the ages of 25 and 44?
5. Explain how today's rate of accidental work deaths compares with the rate in the early 1900s.
6. What are the five leading causes of work deaths?
7. What are the five leading causes of work injuries by type of accident?
8. When death rates are classified by industry type, what are the three leading industry types?
9. Rank the following body parts according to frequency of injury from highest to lowest: neck, fingers, trunk, back, and eyes.
10. Name three chemicals that frequently cause chemical burns in the workplace.
11. Identify three factors that contribute to heat burn injuries in the workplace.
12. Explain the difference between RSI and carpal tunnel syndrome.
13. Explain the reasons for high accident rates in developing countries.

ENDNOTES

1. National Safety Council, *Accident Facts* (Chicago: National Safety Council, 2005), 37.
2. Ibid., 25.
3. Ibid., 4–5.
4. Ibid., 34.
5. Ibid., 35.
6. Ibid., 35.

7. Ibid., 36.
8. Ibid., 36.
9. Ibid., 37.
10. Ibid., 38.
11. Ibid., 39.
12. Ibid., 40.
13. Ibid., 41.
14. Ibid., 41.
15. ErgoOutfitters.com, http://www.ergooutfitters.com, March 31, 2006.
16. Ibid.
17. National Safety Council, *Accident Prevention Manual for Industrial Operations: Administration and Programs* (Chicago: National Safety Council, 1997), 158.
18. Ibid., 158.
19. Daniel Corcoran, "The Hidden Value of Safety." *Occupational Health & Safety* 71, no. 6: 20–22.
20. Corcoran, "The Hidden Value of Safety," 22.
21. Occupational Health & Safety Online, "Study: 2.2 Million People Die Worldwide of Work-Related Accidents, Occupational Diseases." www.ohsonline.com, September 20, 2005, 1–3.

THEORIES OF ACCIDENT CAUSATION

Major Topics

- Domino Theory of Accident Causation
- Human Factors Theory of Accident Causation
- Accident/Incident Theory of Accident Causation
- Epidemiological Theory of Accident Causation
- Systems Theory of Accident Causation
- Combination Theory of Accident Causation
- Behavioral Theory of Accident Causation
- Drugs and Accident Causation
- Depression and Accident Causation
- Management Failures and Accident Causation
- Obesity and Accident Causation

Each year, work-related accidents cost the United States almost $50 billion.[1] This figure includes costs associated with lost wages, medical expenses, insurance costs, and indirect costs. The number of persons injured in **industrial place accidents** in a typical year is 7,128,000, or 3 per 100 persons per year.[2] In the workplace, there is one accidental death approximately every 51 minutes and one injury every 19 seconds.[3]

Why do accidents happen? This question has concerned safety and health decision makers for decades, because in order to prevent accidents we must know why they happen. Over the years, several theories of accident causation have evolved that attempt to explain why accidents occur. Models based on these theories are used to predict and prevent accidents.

The most widely known theories of accident causation are the domino theory, the human factors theory, the accident/incident theory, the epidemiological theory, the systems theory, the combination theory, and the behavioral theory. This chapter provides practicing and prospective safety professionals with the information they need to understand fully and apply these theories.

DOMINO THEORY OF ACCIDENT CAUSATION

An early pioneer of accident prevention and industrial safety was Herbert W. Heinrich, an official with the Travelers Insurance Company. In the late 1920s, after studying the reports of 75,000 industrial accidents, Heinrich concluded that

- 88 percent of industrial accidents are caused by unsafe acts committed by fellow workers.
- 10 percent of industrial accidents are caused by unsafe conditions.
- 2 percent of industrial accidents are unavoidable.[4]

Heinrich's study laid the foundation for his *Axioms of Industrial Safety* and his theory of accident causation, which came to be known as the **domino theory**. So much of Heinrich's theory has been discounted by more contemporary research that it is now considered outdated. However, because some of today's more widely accepted theories can be traced back to Heinrich's theory, students of industrial safety should be familiar with his work.

Heinrich's Axioms of Industrial Safety

Heinrich summarized what he thought health and safety decision makers should know about industrial accidents in 10 statements he called **Axioms of Industrial Safety**. These axioms can be paraphrased as follows:

1. Injuries result from a completed series of factors, one of which is the accident itself.
2. An accident can occur only as the result of an unsafe act by a person and/or a physical or mechanical hazard.
3. Most accidents are the result of unsafe behavior by people.
4. An unsafe act by a person or an unsafe condition does not always immediately result in an accident/injury.
5. The reasons why people commit unsafe acts can serve as helpful guides in selecting corrective actions.
6. The severity of an accident is largely fortuitous, and the accident that caused it is largely preventable.
7. The best accident prevention techniques are analogous with the best quality and productivity techniques.
8. Management should assume responsibility for safety because it is in the best position to get results.
9. The supervisor is the key person in the prevention of industrial accidents.
10. In addition to the direct costs of an accident (e.g., compensation, liability claims, medical costs, and hospital expenses), there are also hidden or indirect costs.[5]

According to Heinrich, these axioms encompass the fundamental body of knowledge that must be understood by decision makers interested in preventing accidents. Any accident prevention program that takes all 10 axioms into account is more likely to be effective than a program that leaves out one or more axioms.

Heinrich's Domino Theory

Perhaps you have stood up a row of dominoes, tipped the first one over, and watched as each successive domino topples the one next to it. This is how Heinrich's theory of accident causation works. According to Heinrich, there are five factors in the sequence of events leading up to an accident. These factors can be summarized as follows:

1. *Ancestry and social environment.* Negative character traits that may lead people to behave in an unsafe manner can be inherited **(ancestry)** or acquired as a result of the **social environment**.
2. *Fault of person.* Negative character traits, whether inherited or acquired, are why people behave in an unsafe manner and why hazardous conditions exist.
3. *Unsafe act/mechanical or physical hazard.* **Unsafe acts** committed by people and **mechanical** or **physical hazards** are the direct causes of accidents.
4. *Accident.* Typically, accidents that result in injury are caused by falling or being hit by moving objects.
5. *Injury.* Typical injuries resulting from accidents include lacerations and fractures.[6]

Heinrich's theory has two central points: (1) Injuries are caused by the action of **preceding factors**; and (2) removal of the **central factor** (unsafe act/**hazardous condition**) negates the action of the preceding factors and, in so doing, prevents accidents and injuries.

Domino Theory in Practice

Construction Products Company (CPC) is a distributor of lumber, pipe, and concrete products. Its customers are typically small building contractors. CPC's facility consists of an office in which orders are placed and several large warehouses. Contractors place their orders in the office. They then drive their trucks through the appropriate warehouses to be loaded by CPC personnel.

Because the contractors are small operations, most of their orders are also relatively small and can be loaded by hand. Warehouse personnel go to the appropriate bins, pull out the material needed to fill their orders, and load the materials on customers' trucks. Even though most orders are small enough to be loaded by hand, many of the materials purchased are bulky and cumbersome to handle. Because of this, CPC's loaders are required to wear such personal protection gear as hard hats, padded gloves, steel-toed boots, and lower-back-support belts.

For years CPC's management team had noticed an increase in minor injuries to warehouse personnel during the summer months. Typically, these injuries consisted of nothing worse than minor cuts, scrapes, and bruises. However, this past summer had been different. Two warehouse workers had sustained serious back injuries. These injuries have been costly to CPC both financially and in terms of employee morale.

An investigation of these accidents quickly identified a series of events and a central causal behavior that set up a *domino effect* that, in turn, resulted in the injuries. The investigation revealed that CPC's warehouses became so hot during the summer months that personal protection gear was uncomfortable. As a result, warehouse personnel simply discarded it. Failure to use appropriate personal protection gear in the summer months had always led to an increase in injuries. However, because the injuries were minor in nature, management had never paid much attention to the situation. It was probably inevitable that more serious injuries would occur eventually.

To prevent a recurrence of the summer-injury epidemic, CPC's management team decided to remove the causal factor—failure of warehouse personnel to use their personal protection gear during the summer months. To facilitate the removal of this factor, CPC's management team formed a committee consisting of one executive manager, one warehouse supervisor, and three warehouse employees.

The committee made the following recommendations: (1) provide all warehouse personnel with training on the importance and proper use of personal protection gear; (2) require warehouse supervisors to monitor the use of personal protection gear more closely; (3) establish a company policy that contains specific and progressive disciplinary measures for failure to use required personal protection gear; and (4) implement several heat reduction measures to make warehouses cooler during the summer months.

CPC's management team adopted all the committee's recommendations. In doing so, it removed the central causal factor that had historically led to an increase in injuries during the summer months.

HUMAN FACTORS THEORY OF ACCIDENT CAUSATION

The **human factors theory** of accident causation attributes accidents to a chain of events ultimately caused by **human error**. It consists of the following three broad factors that lead to human error: overload, inappropriate response, and inappropriate activities (see Figure 3–1). These factors are explained in the following paragraphs.

Overload

Overload amounts to an imbalance between a person's capacity at any given time and the load that person is carrying in a given state. A person's capacity is the product of such factors as his or her natural ability, training, state of mind,

Figure 3–1
Factors that cause human errors.

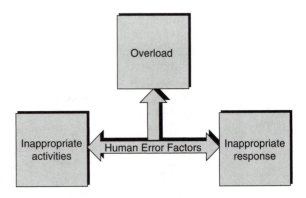

fatigue, stress, and physical condition. The load that a person is carrying consists of tasks for which he or she is responsible and added burdens resulting from **environmental factors** (noise, distractions, and so on), **internal factors** (personal problems, emotional stress, and worry), and **situational factors** (level of risk, unclear instructions, and so on). The state in which a person is acting is the product of his or her motivational and arousal levels.

Inappropriate Response and Incompatibility

How a person responds in a given situation can cause or prevent an accident. If a person detects a hazardous condition but does nothing to correct it, he or she has responded inappropriately. If a person removes a safeguard from a machine in an effort to increase output, he or she has responded inappropriately. If a person disregards an established safety procedure, he or she has responded inappropriately. Such responses can lead to accidents. In addition to **inappropriate responses**, this component includes workstation incompatibility. The incompatibility of a person's workstation with regard to size, force, reach, feel, and similar factors can lead to accidents and injuries.

Safety Fact

Pregnancy and Work

Strenuous physical work and pregnancy can be a dangerous combination. Too much strenuous labor can result in a miscarriage. The types of work to be avoided by pregnant employees include the following:

- Standing for more than three hours per day
- Operating machinery that vibrates
- Lifting heavy loads
- Working in extremes of hot or cold

Shift work and workstations that require awkward postures can also put pregnant employees at risk. The third trimester is the most risk-intensive time during pregnancy.

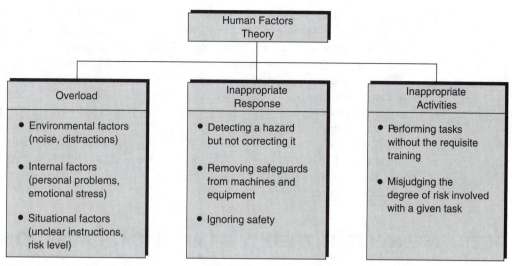

Figure 3–2
Human factors theory.

Inappropriate Activities

Human error can be the result of **inappropriate activities**. An example of an inappropriate activity is a person who undertakes a task that he or she doesn't know how to do. Another example is a person who misjudges the degree of risk involved in a given task and proceeds based on that misjudgment. Such inappropriate activities can lead to accidents and injuries. Figure 3–2 summarizes the various components of the human factors theory.[7]

Human Factors Theory in Practice

Kitchenware Manufacturing Incorporated (KMI) produces aluminum kitchenware for commercial settings. After 10 years of steady, respectable growth in the U.S. market, KMI suddenly saw its sales triple in less than six months. This rapid growth was the result of KMI's successful entry into European and Asian markets.

The growth in sales, although welcomed by both management and employees, quickly overloaded and, before long, overwhelmed the company's production facility. KMI responded by adding a second shift of production personnel and approving unlimited overtime for highly skilled personnel. Shortly after the upturn in production, KMI began to experience a disturbing increase in accidents and injuries. During his accident investigations, KMI's safety manager noticed that human error figured prominently in the accidents. He grouped all the human errors identified into three categories: (1) overload, (2) inappropriate response, and (3) inappropriate activities.

In the category of *overload*, he found that the rush to fill orders was pushing production personnel beyond their personal limits in some cases, and beyond their capabilities in others. Stress, insufficient training of new employees, and fatigue all contributed to the overload. In the category of *inappropriate response*, the safety manager determined that many of KMI's production personnel had removed safeguards from their machines in an attempt to speed up production. All the machines involved in accidents had had safeguards removed.

In the category of *inappropriate activities*, the safety manager found that new employees were being assigned duties for which they weren't yet fully trained. As a result, they often misjudged the amount of risk associated with their work tasks.

With enough accident investigations completed to identify a pattern of human error, the safety manager prepared a presentation containing a set of recommendations for corrective measures for KMI's executive management team. His recommendations were designed to prevent human-error-oriented accidents without slowing production.

ACCIDENT/INCIDENT THEORY OF ACCIDENT CAUSATION

The **accident/incident theory** is an extension of the human factors theory. It was developed by Dan Petersen and is sometimes referred to as the Petersen accident/incident theory.[8] Petersen introduced such new elements as **ergonomic traps**, the decision to err, and systems failures, while retaining much of the human factors theory. A model based on his theory is shown in Figure 3–3.

In this model, overload, ergonomic traps, or a decision to err lead to human error. The decision to err may be conscious and based on logic, or it may be unconscious. A variety of pressures such as deadlines, peer pressure, and budget factors can lead to **unsafe behavior**. Another factor that can influence such a decision is the "It won't happen to me" syndrome.

The systems failure component is an important contribution of Petersen's theory. First, it shows the potential for a **causal relationship** between management decisions or management behavior and safety. Second, it establishes management's role in accident prevention as well as the broader concepts of safety and health in the workplace.

Following are just some of the different ways that systems can fail, according to Petersen's theory:

- Management does not establish a comprehensive safety policy.
- Responsibility and authority with regard to safety are not clearly defined.
- Safety procedures such as measurement, inspection, correction, and investigation are ignored or given insufficient attention.
- Employees do not receive proper orientation.
- Employees are not given sufficient safety training.

Accident/Incident Theory in Practice

Poultry Processing Corporation (PPC) processes chickens and turkeys for grocery chains. Poultry processing is a labor-intensive enterprise involving a great deal of handwork. A variety of different knives, shears, and cleavers are used.

Figure 3–3
Accident/incident theory.

Much of the work is monotonous and repetitive. Selected parts of the overall process must be done in cold conditions.

PPC has gone to great lengths to ensure that workstations are ergonomically sound, that personal protection gear is used as appropriate, and that adequate precautions are taken to prevent illness and injuries. As a result, PPC is an award-winning company in the area of workplace safety and health.

Consequently, the poultry-processing industry was shocked when a class action lawsuit was filed against PPC on behalf of over 50 employees, all of whom claimed to be suffering from carpal tunnel syndrome. Because of PPC's excellent safety and health record, most observers felt sure that the company would be vindicated in the end.

The company's policies and procedures relating to safety and health were investigated thoroughly by consultants brought in by both PPC and the attorney for the plaintiffs. Over 100 witnesses gave depositions, and several preliminary hearings were held. By the time the trial finally rolled around, both sides had accumulated mountains of paper and filing cabinets full of evidence. Then, suddenly and without advance notice, PPC offered a substantial financial settlement, which the plaintiffs accepted.

It was one of PPC's outside consultants who discovered what had caused the increased incidence of carpal tunnel syndrome. The company had always used a centralized approach to managing safety and health. Responsibility for such tasks as measurement, inspection, correction, and investigation was assigned to the safety manager, Joe Don Huttle. Huttle had an excellent record during his 20 years in the poultry-processing industry, with the last 5 spent at PPC. In fact, he was so well respected in the industry that his peers had elected him president of a statewide safety organization. This, as it turned out, is where PPC's troubles began.

When Huttle took it over, the safety organization had experienced a three-year decline in membership and was struggling to stay afloat financially. He had been elected as "the man who could save the organization." Intending to do just that, Huttle went right to work. For months at a time he worked seven days a week, often spending as much as two weeks at a time on the road. When he was in his office at PPC, Huttle was either on the telephone or doing paperwork for the safety organization.

Within six months, he had reversed the organization's downhill slide, but not without paying a price at home. During the same six-month period, his duties at PPC were badly neglected. Measurement of individual and group safety performance had come to a standstill. The same was true of inspection, correction, investigation, and reporting.

It was during this time of neglect that the increased incidence of carpal tunnel syndrome occurred. Safety precautions that Huttle had instituted to guard against this particular problem were no longer observed properly once the workers realized that he had stopped observing and correcting them. Measurement and inspection may also have prevented the injuries had Huttle maintained his normal schedule of these activities.

PPC's consultant, in a confidential report to executive managers, cited the *accident/incident theory* in explaining his view of why the injuries occurred. In this report, the consultant said that Huttle was guilty of applying "it won't happen here" logic when he made a conscious decision to neglect his duties at PPC in favor of his duties with the professional organization. Of course, the employees themselves were guilty of not following clearly established procedures. However, because Huttle's neglect was also a major contributing factor, PPC decided to settle out of court.

EPIDEMIOLOGICAL THEORY OF ACCIDENT CAUSATION

Traditionally, safety theories and programs have focused on accidents and the resulting injuries. However, the current trend is toward a broader perspective that also encompasses the issue of industrial hygiene. **Industrial hygiene** concerns environmental factors that can lead to sickness, disease, or other forms of impaired health.

This trend has, in turn, led to the development of an epidemiological theory of accident causation. Epidemiology is the study of causal relationships between environmental factors and disease. The **epidemiological theory** holds that the models used for studying and determining these relationships can also be used to study causal relationships between environmental factors and accidents or diseases.[9]

Figure 3–4 illustrates the epidemiological theory of accident causation. The key components are **predispositional characteristics** and **situational characteristics**. These characteristics, taken together, can either result in or prevent conditions that may result in an accident. For example, if an employee who is particularly susceptible to peer pressure (predispositional characteristic) is pressured by his coworkers (situational characteristic) to speed up his operation, the result will be an increased probability of an accident.

Epidemiological Theory in Practice

Jane Andrews was the newest member of the loading unit for Parcel Delivery Service (PDS). She and the other members of her unit were responsible for loading 50 trucks every morning. It was physically demanding work, and she was the first woman ever selected by PDS to work in the loading unit. She had

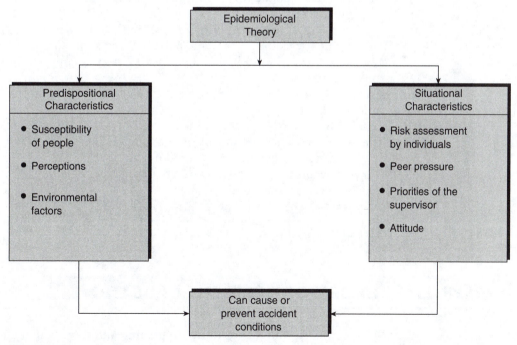

Figure 3–4
Epidemiological theory.

gotten the job as part of the company's upward mobility program. She was excited about her new position because within PDS, the loading unit was considered a springboard to advancement. Consequently, she was eager to do well. The responsibility she felt toward other female employees at PDS only intensified her anxiety. Andrews felt that if she failed, other women might not get a chance to try in the future.

Before beginning work in the loading unit, employees must complete two days of training on proper lifting techniques. The use of back-support belts is mandatory for all loading dock personnel. Consequently, Andrews became concerned when the supervisor called her aside on her first day in the unit and told her to forget what she had learned in training. He said, "Jane, nobody wants a back injury, so be careful. But the key to success in this unit is speed. The lifting techniques they teach in that workshop will just slow you down. You've got the job, and I'm glad you're here. But you won't last long if you can't keep up."

Andrews was torn between following safety procedures and making a good impression on her new supervisor. At first, she made an effort to use proper lifting techniques. However, when several of her coworkers complained that she wasn't keeping up, the supervisor told Andrews to "keep up or get out of the way." Feeling the pressure, she started taking the same shortcuts she had seen her coworkers use. Positive results were immediate, and Andrews received several nods of approval from fellow workers and a "good job" from the supervisor. Before long, Andrews had won the approval and respect of her colleagues.

However, after two months of working in the loading unit, she began to experience persistent lower-back pain. Andrews felt sure that her hurried lifting techniques were to blame, but she valued the approval of her supervisor and fellow workers too much to do anything that might slow her down. Finally, one day while loading a truck, Andrews fell to the pavement in pain and could not get up. Her back throbbed with intense pain, and her legs were numb. She had to be rushed to the emergency room of the local hospital. By the time Andrews checked out of the hospital a week later, she had undergone major surgery to repair two ruptured discs.

Jane Andrews's situation can be explained by the *epidemiological theory* of accident causation. The predispositional factor was her susceptibility to peer pressure from her coworkers and supervisor. The applicable situational factors were peer pressure and the priorities of the supervisor. These factors, taken together, caused the accident.

SYSTEMS THEORY OF ACCIDENT CAUSATION

A *system* is a group of regularly interacting and interrelated components that together form a unified whole. This definition is the basis for the **systems theory** of accident causation. This theory views a situation in which an accident may occur as a system comprised of the following components: person (host), machine (agency), and **environment**.[10] The likelihood of an accident occurring is determined by how these components interact. Changes in the patterns of interaction can increase or reduce the probability of an accident.

For example, an experienced employee who operates a numerically controlled five-axis machining center in a shop environment may take a two-week vacation. Her temporary replacement may be less experienced. This change in one component of the system (person/host) increases the probability of an accident. Such a simple example is easily understood. However, not all changes in patterns of interaction are this simple. Some are so subtle that their analysis may require a team of people, each with a different type of expertise.

The primary components of the systems model are the person/machine/environment, information, decisions, risks, and the task to be performed.[11] Each of the components has a bearing on the probability that an accident will occur. The systems model is illustrated in Figure 3–5.

As this model shows, even as a person interacts with a machine within an environment, three activities take place between the system and the task to be performed. Every time a task must be performed, there is the risk that an accident may occur. Sometimes the risks are great; at other times, they are small. This is where information collection and decision making come in.

Based on the information that has been collected by observing and mentally noting the current circumstances, the person weighs the risks and decides whether to perform the task under existing circumstances. For example, say a machine operator is working on a rush order that is behind schedule. An important safety device has malfunctioned on his machine. Simply taking it off will interrupt work for only five minutes, but it will also increase the probability of an accident. However, replacing it could take up to an hour. Should the operator remove the safety guard and proceed with the task or take the time to

Figure 3–5
Systems theory model.

Discussion Case

What Is Your Opinion?

"All accidents, one way or another, are the result of human error." "No, accidents are the result of a combination of things. I like the combination theory." "You're both wrong. Accidents are best explained by the domino theory." So the debate went in Dr. Jameson's class at Burton State University. What is your opinion concerning the various theories of accident causation?

replace it? The operator and his supervisor may assess the situation (collect information), weigh the risks, and make a decision to proceed. If their information was right and their assessment of the risks accurate, the task will probably be accomplished without an accident.

However, the environment in which the machine operator is working is unusually hectic, and the pressure to complete an order that is already behind schedule is intense. These factors are **stressors** that can cloud the judgment of those collecting information, weighing risks, and making the decision. When stressors are introduced between points 1 and 3 in Figure 3–5, the likelihood of an accident increases.

For this reason, Firenzie recommends that five factors be considered before beginning the process of collecting information, weighing risks, and making a decision:

- Job requirements
- The workers' abilities and limitations
- The gain if the task is successfully accomplished
- The loss if the task is attempted but fails
- The loss if the task is not attempted[12]

These factors can help a person achieve the proper perspective before collecting information, weighing risks, and making a decision. It is particularly important to consider these factors when stressors such as noise, time constraints, or pressure from a supervisor may tend to cloud one's judgment.

Systems Theory in Practice

Precision Tooling Company (PTC) specializes in difficult orders that are produced in small lots, and in making corrections to parts that otherwise would wind up as expensive rejects in the scrap bin. In short, PTC specializes in doing the types of work that other companies cannot, or will not, do. Most of PTC's work comes in the form of subcontracts from larger manufacturing companies. Consequently, living up to its reputation as a high-performance, on-time company is important to PTC.

Because much of its work consists of small batches of parts to be reworked, PTC still uses several manually operated machines. The least experienced machinists operate these machines. This causes two problems. The first problem is that it is difficult for even a master machinist to hold to modern tolerance levels on these old machines. Consequently, apprentice machinists find holding to precise tolerances quite a challenge. The second problem is that the machines are so old that they frequently break down.

Complaints from apprentice machinists about the old machines are frequent. However, their supervisors consider time on the old "ulcer makers" to be one of the rites of passage that upstart machinists must endure. Their attitude is, "We had to do it, so why shouldn't you?" This was where things stood at PTC when the company won the Johnson contract.

PTC had been trying for years to become a preferred supplier for H. R. Johnson Company. PTC's big chance finally came when Johnson's manufacturing division incorrectly produced 10,000 copies of a critical part before noticing the problem. Simply scrapping the part and starting over was an expensive solution. Johnson's vice president for manufacturing decided to give PTC a chance.

PTC's management team was ecstatic! Finally, they had an opportunity to partner with H. R. Johnson Company. If PTC could perform well on this one, even more lucrative contracts were sure to follow. The top managers called a companywide meeting of all employees. Attendance was mandatory. The CEO explained the situation as follows:

> Ladies and gentlemen, we are faced with a great opportunity. I've just signed a contract with H. R. Johnson Company to rework 10,000 parts that their manufacturing folks produced improperly. The rework tasks are not that complicated, but every part has got to go through several manual operations at the front end of the rework process. This means our manual machining unit is going to have to supply the heroes on this job. I've promised the manufacturing VP at Johnson that we would have his parts ready in 90 days. I know that's a lot to do in so short a period of time, but Johnson is in a real bind here. If we can produce on this one, they won't forget us in the future.

This put PTC's apprentice machinists on the spot. If PTC didn't perform on this contract, it would be their fault. They cursed their old machines and got to work. The CEO had said the rework tasks would not be "that complicated," but, as it turned out, the processes weren't that simple either. The problem was tolerances. Holding to the tolerances specified in the Johnson contract took extra time and a special effort on every single part. Before long, the manual machining unit was behind schedule, and management was getting nervous. The situation was made even worse by the continual breakdowns and equipment failures experienced. The harder the unit supervisor pushed, the more stressed the employees and machines became.

Predictably, it wasn't long before safety procedures were forgotten, and unreasonable risks were being taken. The pressure from management, the inexperience of the apprentice machinists, and the constant equipment failures finally took their toll. In a hurry to get back on schedule, and fearing that his machine would break down again, one machinist got careless and ran his hand into the cutter on his milling machine. By the time the machine had been shut down, his hand was badly mutilated. In the aftershock of this accident, PTC was unable to meet the agreed-upon completion schedule. Unfortunately, PTC did not make the kind of impression on H. R. Johnson's management team that it had hoped.

This accident can be explained by the systems theory. The *person-machine-environment* chain has a direct application in this case. The person involved was relatively inexperienced. The machine involved was old and prone to breakdowns. The environment was especially stressful and pressure packed. These three factors, taken together, resulted in this serious and tragic accident.

COMBINATION THEORY OF ACCIDENT CAUSATION

There is often a degree of difference between any theory of accident causation and reality. The various models presented with their corresponding theories in this chapter attempt to explain why accidents occur. For some accidents, a given model may be very accurate. For others, it may be less so. Often the cause of an accident cannot be adequately explained by just one model or theory. Thus, according to the **combination theory**, the actual cause may combine parts of several different models. Safety personnel should use these theories as appropriate both for accident prevention and accident investigation. However, they should avoid the tendency to try to apply one model to all accidents.

Combination Theory in Practice

Crestview Grain Corporation (CGC) maintains 10 large silos for storing corn, rice, wheat, barley, and various other grains. Because stored grain generates fine dust and gases, ventilation of the silos is important. Consequently, all of CGC's silos have several large vents. Each of these vents uses a filter similar to the type used in home air conditioners that must be changed periodically.

There is an element of risk involved in changing the vent filters because of two potential hazards. The first hazard comes from unvented dust and gases that can make breathing difficult, or even dangerous. The second hazard is the grain itself. Each silo has a catwalk that runs around its inside circumference near the top. These catwalks give employees access to the vents that are also near the top of each silo. The catwalks are almost 100 feet above ground level, they are narrow, and the guardrails on them are only knee high. A fall from a catwalk into the grain below would probably be fatal.

Consequently, CGC has well-defined rules that employees are to follow when changing filters. Because these rules are strictly enforced, there had never been an accident in one of CGC's silos—that is, not until the Juan Perez tragedy occurred. Perez was not new to the company. At the time of his accident, he had worked at CGC for over five years. However, he was new to the job of silo maintenance. His inexperience, as it turned out, would prove fatal.

It was time to change the vent filters in silo number 4. Perez had never changed vent filters himself. He hadn't been in the job long enough. However, he had served as the required "second man" when his supervisor, Bao Chu Lai, had changed the filters in silos 1, 2, and 3. Because Chu Lai was at home recuperating from heart surgery and would be out for another four weeks, Perez decided to change the filters himself. Changing the filters was a simple enough task, and Perez had always thought the second-man concept was overdoing it a little. He believed in taking reasonable precautions as much as the next person, but in his opinion, CGC was paranoid about safety.

Perez collected his safety harness, respirator, and four new vent filters. Then he climbed the external ladder to the entrance–exit platform near the top of silo number 4. Before going in, Perez donned his respirator and strapped on his safety harness. Opening the hatch cover, he stepped inside the silo onto the catwalk. Following procedure, Perez attached a lifeline to his safety harness, picked up the new vent filters, and headed for the first vent. He changed the first two filters without incident. It was while he was changing the third filter that tragedy struck.

The filter in the third vent was wedged in tightly. After several attempts to pull it out, Perez became frustrated and gave the filter a good jerk. When the filter suddenly broke loose, the momentum propelled Perez backward and he toppled off the catwalk. At first it appeared that his lifeline would hold, but without a second person to pull him up or call for help, Perez was suspended by only the lifeline for over 20 minutes. He finally panicked, and in his struggle to pull himself up, knocked open the buckle of his safety harness. The buckle gave way, and Perez fell over 50 feet into the grain below. The impact knocked off his respirator, the grain quickly enveloped him, and Perez was asphyxiated.

The accident investigation that followed revealed that several factors combined to cause the fatal accident—the combination theory. The most critical of these factors were as follows:

- Absence of the supervisor
- Inexperience of Perez
- A conscious decision by Perez to disregard CGC's safety procedures
- A faulty buckling mechanism on the safety harness
- An unsafe design (only a knee-high guardrail on the catwalk)

BEHAVIORAL THEORY OF ACCIDENT CAUSATION

The behavioral theory of accident causation and prevention is often referred to as **behavior-based safety (BBS)**. BBS has both proponents and critics. One of the most prominent proponents of BBS is E. Scott Geller, a senior partner of Safety Performance Solutions, Inc. and a professor of psychology. It is appropriate that Geller is a professional psychologist because BBS is the application of behavioral theories from the field of psychology to the field of occupational safety.

According to Geller, there are seven basic principles of BBS: (1) intervention that is focused on employee behavior; (2) identification of external factors that will help understand and improve employee behavior (from the perspective of safety in the workplace); (3) direct behavior with activators or events antecedent to the desired behavior, and motivation of the employee to behave as desired with incentives and rewards that will follow the desired behavior; (4) focus on the positive consequences that will result from the desired behavior as a way to motivate employees; (5) application of the scientific method to improve attempts at behavioral interventions; (6) use of theory to integrate information rather than to limit possibilities; and (7) planned interventions with the feelings and attitudes of the individual employee in mind.[13]

Those who have studied psychology will recognize BBS as an innovative and practical application of standard behavioral theory to the field of occupational safety. These theories are relevant in any situation in which certain types of human behaviors are desired while others are to be avoided. Positive reinforcement in the form of incentives and rewards is used to promote the desired (safe) behaviors and to discourage undesirable (unsafe) behaviors.

Proponents of BBS use the "ABC" model to summarize the concept of understanding human behavior and developing appropriate interventions when the behavior is undesirable (unsafe). Geller explains the model as follows:

> Behavior-based safety trainers and consultants teach the ABC model (or three-term contingency) as a framework to understand and analyze behavior or to develop interventions for improving behavior. As given in BBS principle 3 . . . the "A" stands for activators or antecedent events that precede behavior ("B") and "C" refers to the consequences following behavior or produced by it. Activators direct behavior, whereas consequences motivate behavior.[14]

Two other advocates of BBS, Bruce Fern and Lori Alzamora, propose the expansion of the ABC model to ABCO.[15] The "O" stands for outcomes. They explain the addition as follows:

> "*Outcome*" refers to the longer-term results of engaging in safe or unsafe behavior. For example, an antecedent of a sign requiring employees to wear safety goggles could produce the behavior of putting on the goggles, the consequence of avoiding an eye injury, and the outcome of being able to continue working and enjoying time with the family. On the other hand, the consequence of not wearing goggles could be an eye injury with a potential outcome of blindness, time off the job, and a reduced quality of life. Failure to address the issue of outcomes represents a lost opportunity to give employees a good reason for engaging in safe behaviors.[16]

Behavioral Theory in Action

Mark Potter is the safety manager for Excello Corporation. Several months ago, he became concerned because employees seemed to have developed a lax attitude toward wearing hard hats. What really troubled Potter was that there is more than the usual potential for head injuries because of the type of work done in Excello's plant, and he had personally witnessed two near misses in less than a week. An advocate of behavior-based safety (BBS), he decided to apply the ABC model in turning around this unsafe behavior pattern.

His first step was to remove all the old "Hard Hat Area" signs from the plant and replace them with newer, more noticeable signs. Then he scheduled a brief seminar on head injuries and cycled all employees through it over a two-week period. The seminar took an unusual approach. It told a story of two employees. One was in a hospital bed surrounded by family members he did not even recognize. The other was shown enjoying a family outing with happy family members. The clear message of the video was "the difference between these two employees is a hard hat." These two activities were the antecedents to the behavior he hoped to produce (all employees wearing hard hats when in a hard hat area).

The video contained a powerful message and it had the desired effect. Within days, employees were once again disciplining themselves to wear their hard hats (the desired behavior). The consequence was that near misses stopped and no head injuries have occurred at Excello in months. The outcome of this is that Excello's employees have been able to continue enjoying the fruits of their labor and the company of loved ones.

DRUGS AND ACCIDENT CAUSATION

One of the most pernicious causes of accidents on the job is chemicals—but not the kind industrial hygienists generally concern themselves with.[17] The chemicals alluded to here are the illicit drugs and alcohol used by employees. Drugs and alcohol are the root cause or contributing cause of many accidents on the job every year. Consequently, safety professionals need to be on guard for employees who are drug and alcohol abusers.

According to Stephen Minter,

> The workplace cannot be separated from the society around it, and substance abuse continues to be a serious and costly health and safety issue for employers. According to surveys by the Department of Health and Human Services, some 77 percent of drug users are employed—more than 9 million workers. An estimated 6.5 percent of full-time and 8.6 of part-time workers use illicit drugs. More than a third of all workers between the ages of 18 and 25 are binge drinkers. . . . Alcoholism alone causes 500 million lost days annually (125 million days are lost each year due to work-related injuries). . . . Some 20 percent of workers report that they have been put in danger or injured, or had to work harder, redo work or cover for a co-worker, as a result of a co-worker's drinking.[18]

These discouraging statistics are why so many companies implement drug-free workplace programs. In fact, since 1989 federal contractors have been required to do so. Such programs typically include the following components: drug-free workplace policy, supervisory training, employee education, employee assistance programs, and alcohol and drug testing.

Establishing drug-free workplace programs is typically the responsibility of the human resources department. However, safety and health professionals should be aware of the workplace problems that can be caused by alcohol and drug abuse. Further, if a cross-functional team of representatives from various departments is convened by the human resources department for the purpose of developing a drug-free workplace program, the chief safety and health professional for the organization should be a member of that team.

DEPRESSION AND ACCIDENT CAUSATION

An invisible problem in today's workplace is **clinical depression**. People who suffer from clinical depression are seriously impaired and, as a result, they pose a clear and present safety risk to themselves, fellow workers, and their employer.[19] Mental health professionals estimate that up to 10 percent of the adult population in the United States suffers from clinical depression. This translates to 1 in every 20 people on the job.

The causes of clinical depression are many and varied, but the most common causes are biological (too few or too many of the brain chemicals known as **neurotransmitters**), cognitive (negative thought processes), genetic (family history of depression), and concurring illnesses (strokes, cancer, heart disease, Alzheimer's, and other diseases can increase the incidence of depression).

According to Todd Nighswonger,

Depression results in more than 200 million lost workdays and costs the U.S. economy $43.7 billion annually. Much of that cost is hidden, including $23.8 billion lost to U.S. businesses in absenteeism and lost productivity. Beyond productivity issues, studies suggest that depressed workers may be more prone to accidents. Stephen Heidel, M.D., MBA, an occupational psychiatrist in San Diego, notes a lack of concentration, fatigue, failing memory and slow reaction time as reasons that workers who are depressed may not work safely.[20]

Warning Signs

Safety and health professionals are not mental health professionals and should not attempt to play that role. However, they should be alert to the warning signs of clinical depression in employees. These signs are as follows:

- Persistent dreary moods (sadness, anxiety, nervousness)
- Signs of too little sleep
- Sleeping on the job or persistent drowsiness
- Sudden weight loss or gain
- General loss of interest, especially in areas of previous interest
- Restlessness, inability to concentrate, or irritability
- Chronic physical problems (headaches, digestive disorders, etc.)
- Forgetfulness or an inability to make simple decisions

- Persistent feelings of guilt
- Feelings of low self-worth
- Focus on death or talk of suicide

Safety and health professionals who recognize any or all of these symptoms in an employee should avoid the natural human tendency to *help the employee deal with the problems*. Rather, the appropriate action is to get the employee into the hands of competent mental health professionals right away. The best way to do this is to approach the employee's supervisor and recommend that he or she refer the employee to the organization's employee assistance program (EAP) or to the human resources department. If the supervisor is uncomfortable approaching the employee in question or does not know how to go about it, recommend that he or she use the following statement suggested by the Society for Human Resource Management:

> I'm concerned that recently you've been late to work often and are not meeting your performance objectives. I'd like to see you get back on track. I don't know whether this is the case for you, but if personal issues are affecting your work, you can speak confidentially to one of our employee assistance counselors. The service was set up to help employees. Our conversation today and appointments with the counselor will be kept confidential. Whether or not you contact this service, you will still be expected to meet your performance goals.[21]

Sources of Help

Because clinical depression in employees has become such an all-pervasive problem that increases the risk of accidents and injuries on the job, safety and health professionals need to learn all they can about this problem and keep up-to-date with the latest information concerning it. The following sources may help:

Employee Assistance Professionals Association, 703-522-6272,
    http://www.eapa.org.
National Institute of Mental Health, 800-421-4211,
    http://www.nimh.nih.gov.
National Mental Health Association, 800-969-NMHA,
    http://www.nmha.org.

MANAGEMENT FAILURES AND ACCIDENT CAUSATION

One of the leading causes of accidents in the workplace is the failure of management to do its part to ensure a safe and healthy work environment.[22] Different levels of management have different levels of responsibility. The level of management with the most direct, hands-on, day-to-day responsibility for workplace safety and health is the supervisory level. Supervisors play a critical role in making sure that employees work in a safe and healthy environment.

Role of the Supervisor in Workplace Safety and Health

Safety and health professionals cannot do their jobs effectively without the full cooperation and day-to-day assistance of first-line supervisors. Supervisors and safety professionals must be partners when it comes to providing a safe and healthy workplace for employees. Supervisors should be assigned responsibility for the work environment and for the safety of employees in their units. Safety and health professionals should be readily available to help supervisors fulfill this responsibility.

Key responsibilities of supervisors relating to workplace safety and health include the following:

- Orienting new employees to the safe way to do their jobs
- Ensuring that new and experienced employees receive the safety and health training they need on a continual basis
- Monitoring employee performance and enforcing safety rules and regulations
- Assisting safety and health professionals in conducting accident investigations
- Assisting safety and health professionals in developing accident reports
- Keeping up-to-date on safety issues
- Setting a positive example for employees that says "the safe way is the right way"

Typical Management Failures That Cause Accidents

Management failures represent a major cause of accidents on the job. If management is serious about providing a safe and healthy work environment for employees it must (1) show employees that safe and healthy work practices are expected by including such practices in job descriptions, monitoring employee work practices, and setting an example of safe and healthy work practices; (2) provide training in how to work safely, including orientation training for new employees as well as ongoing updated training for experienced employees; (3) include safe and healthy work practices as criteria in the periodic performance appraisals of employees; and (4) reinforce safe and healthy work practices by rewarding and recognizing employees who use them. Common examples of management failures include the following:

> *Poor housekeeping or improper use of tools, equipment, or facilities.* Management either has not developed the necessary requirements, or has but does not enforce them. The management failure in this case could be lack of safety procedures (failure to let employees know the expectations), lack of training (failure to give employees the knowledge and skills they need to work safely), or failure to properly supervise (failure to monitor employee actions).
>
> *Pressure to meet deadlines.* Sometimes management has developed a good safety and health policy, established good safety and health procedures, built safety and health expectations into job descriptions

and performance appraisals, and provided the necessary training only to put all this aside when a rush order comes in. This may be the most problematic of the many different types of management failures that can occur because it can undermine all of the organization's safety and health efforts. When management allows safety and health procedures to be ignored or, worse yet, encourages them to be ignored to speed up production in the short run, employees soon get the message that safety and health are important only when there is no rush. This is an example of management failing to set the proper example.

OBESITY AND ACCIDENT CAUSATION

Researchers at Ohio State University found that extremely obese people are more likely than normal-weight people to injure themselves.[23] This is bad news for what an Australian study conducted at Queensland University calls sedentary workplaces—those that involve a lot of sitting at desks. It is bad news because the Australian study concluded that the more people sit at desks during the workday, the more likely they are to be overweight.

Obesity has long been associated with such chronic diseases as high blood pressure, coronary heart disease, diabetes, and certain types of cancer, but these studies now tie it to workplace injuries too. The ramifications for safety and health professionals are profound. The World Health Organization estimates that there are more than 300 million obese people worldwide. In the industrialized nations of the world—nations such as the United States—the number of people considered obese is growing rapidly. These studies used a standard body mass index (BMI) score of 30 or above to define obesity.

In the study conducted by Ohio State University, researchers collected data on more than 2,500 adults. The data show that 26 percent of extremely obese male subjects reported personal injuries. The percentage for extremely obese women was only slightly lower at approximately 22 percent. Researchers compared these percentages with those for normal-weight people and found a noticeable difference in the percentages of males and females who reported injuries (17 percent for males and 12 percent for women). Researchers used a BMI of 18.5 to 24.9 to define "normal weight."

The most common causes of injuries to obese people were the result of overexertion (35.2 percent) and falls (29.9) percent. Underweight people—BMI of 18.5 or less—reported the fewest number of injuries. According to the study's author, Huiyun Xiang, "There is undeniably a link between obesity and injury risk in adults. Efforts to promote optimal body weight may reduce not only the risk of chronic diseases, but also the risk of unintentional injuries."[24]

SUMMARY

1. The domino theory of accident causation was one of the earliest developed. The theory posits that injuries result from a series of factors, one of which is an accident. The theory is operationalized in 10 statements called the Axioms of Industrial Safety. According to this theory, there are five factors in the sequence of events leading to an accident: ancestry and social environment, fault of person, unsafe act/mechanical or physical hazard, accident, and injury.

2. The human factors theory of accident causation attributes accidents to a chain of events ultimately caused by human error. It consists of three broad factors that lead to human error: overload, inappropriate response, and inappropriate activities.

3. The accident/incident theory of accident causation is an extension of the human factors theory. It introduces such new elements as ergonomic traps, the decision to err, and systems failures.

4. The epidemiological theory of accident causation holds that the models used for studying and determining the relationships between environmental factors and disease can be used to study causal relationships between environmental factors and accidents.

5. The systems theory of accident causation views any situation in which an accident may occur as a system with three components: person (host), machine (agency), and environment.

6. The combination theory of accident causation posits that no one model or theory can explain all accidents. Factors from two or more models may be part of the cause.

7. There are seven principles of behavior-based safety: intervention; identification of internal factors; motivation to behave in the desired manner; focus on the positive consequences of appropriate behavior; application of the scientific method integration of information; and planned interventions.

8. Drugs and alcohol are the root or a contributing cause of many workplace accidents every year. Approximately 77 percent of drug users are employed, and more than a third of all workers between 18 and 25 are binge drinkers. Alcoholism alone causes 500 million lost days annually.

9. Clinical depression is an invisible problem in the workplace. However, it can be a major cause of accidents. One in 20 people suffer from clinical depression, which is the root cause of more than 200 million lost workdays annually.

10. Management failures are another leading cause of accidents on the job. If management is serious about workplace safety and health, it must establish expectations, provide training, evaluate employee performance with safety in mind, and reinforce safe and healthy behavior.

11. There is a strong correlation between obesity and injuries, suggesting a need to promote optimal body weight as an injury prevention strategy.

KEY TERMS AND CONCEPTS

Accident/incident theory	Inappropriate responses
Ancestry	Industrial hygiene
Axioms of Industrial Safety	Industrial place accidents
Behavior-based safety (BBS)	Internal factors
Causal relationship	Mechanical hazards
Central factor	Neurotransmitters
Clinical depression	Obesity
Combination theory	Overload
Domino theory	Physical hazards
Environment	Preceding factors
Environmental factors	Predispositional characteristics
Epidemiological theory	Situational characteristics
Ergonomic traps	Situational factors
Fault of person	Social environment
Hazardous condition	Stressors
Human error	Systems theory
Human factors theory	Unsafe acts
Inappropriate activities	Unsafe behavior

REVIEW QUESTIONS

1. Explain the domino theory of accident causation, including its origin and its impact on more modern theories.
2. What were the findings of Herbert W. Heinrich's 1920s study of the causes of industrial accidents?
3. List five of Heinrich's Axioms of Industrial Safety.
4. Explain the following concepts in the domino theory: preceding factor; central factor.
5. What are the three broad factors that lead to human error in the human factors theory? Briefly explain each.
6. Explain the systems failure component of the accident/incident theory.
7. What are the key components of the epidemiological theory? How does their interaction affect accident causation?
8. Explain the systems theory of accident causation.
9. What impact do stressors have in the systems theory?
10. List five factors to consider before making workplace decisions that involve risk.

11. Explain the principles of behavior-based safety.
12. What is the role of the safety and health professional with regard to handling employees who might be drug or alcohol abusers?
13. List the warning signs of clinical depression.
14. What must management do if it is serious about providing a safe and healthy work environment for employees?
15. Explain the connection between obesity and injuries.

ENDNOTES

1. National Safety Council, *Accident Facts* (Chicago: National Safety Council, 2005), 23.
2. Ibid., 24.
3. Ibid., 26.
4. Industrial Foundation for Accident Prevention, www.ifap.asn.au, March 31, 2006.
5. Ibid.
6. Ibid.
7. Ibid.
8. Ibid.
9. M. A. Topf, "Chicken/Egg/Chegg!", *Occupational Health Safety* 68, no. 6: 60–66.
10. D. L. Goetsch, *Implementing Total Safety Management* (Upper Saddle River, NJ: Prentice Hall, 1998), 227.
11. Ibid.
12. Ibid.
13. E. S. Geller, "Behavior-Based Safety: Confusion, Controversy, and Clarification," *Occupational Health & Safety* 68, no. 1: 40–49.
14. Ibid., 44.
15. B. Fern and L. P. Alzamora, "How and Why Behavioral Safety Needs to Change," *Occupational Health & Safety* 68, no. 9: 69.
16. Ibid.
17. Stepher G. Minter, "The Safety Threat from Within," *Occupational Hazards* 64, no. 4: 8.
18. Ibid.
19. Todd Nighswonger, "Depression: The Unseen Safety Risk," *Occupational Hazards* 64, no. 4: 38–42.
20. Ibid., 40.
21. Ibid., 42.
22. Retrieved from http://online.misu.kodak.edu/19577/AccCautrac.htm.
23. Occupational Health & Safety Online, "Obesity Studies Focus on Injuries, Sedentary Workplaces." Retrieved from http://www.ohsonline.com/stevens/ohspub.nsf/d3d5b4f938b22b6e8625670c006dbc58/d9, July 25, 2005, 1–3.
24. Ibid., 3.

ROLES AND PROFESSIONAL CERTIFICATIONS FOR SAFETY AND HEALTH PROFESSIONALS

Major Topics

- Modern Safety and Health Teams
- Safety and Health Manager
- Engineers and Safety
- Industrial Hygienist
- Health Physicist
- Occupational Physician
- Occupational Health Nurse
- Risk Manager
- Certification of Safety

This book was designed for use by prospective and practicing safety and health managers. People with such titles are typically responsible to higher management for the safety and health of a company's workforce. Modern safety and health managers seldom work alone. Rather, they usually head a team of specialists that may include engineers, physicists, industrial hygienists, occupational physicians, and occupational health nurses.

It is important for safety and health managers today to understand not only their roles but also the roles of all members of the safety and health team. This chapter provides the information that prospective and practicing safety and health managers need to know about the roles of safety personnel in the age of high technology and the certifications that they need.

MODERN SAFETY AND HEALTH TEAMS

The issues that concern modern safety and health managers are multifaceted and complex. They include such diverse issues as stress; explosives; laws, standards, and codes; radiation; AIDS; product safety and liability; ergonomics;

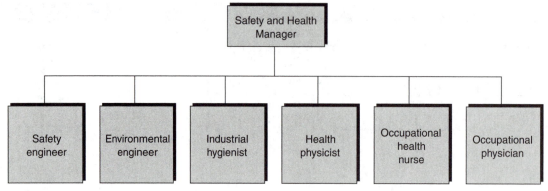

Figure 4–1
A modern safety and health team.

ethics; automation; workers' compensation; and an ever-changing multitude of others.

It would be unreasonable to expect one person to be an expert in all the many complex and diverse issues faced in the modern workplace. For this reason, the practice of safety and health management in the age of high technology has become a team sport. Figure 4–1 illustrates the types of positions that may comprise a safety and health team. In the remaining sections of this chapter, the roles, duties, responsibilities, and relationships of members of the safety and health team are described. In small companies, one person may have to wear several of these hats and contract for nursing services.

SAFETY AND HEALTH MANAGER

The most important member of the safety and health team is its manager. Companies that are committed to providing a safe and healthy workplace employ a **safety and health manager** at an appropriate level in the corporate hierarchy. The manager's position in the hierarchy is an indication of the company's commitment and priorities. This, more than anything else, sets the tone for a company's safety and health program.

In times past, companies with a highly placed safety and health manager were rare. However, passage of the OSH Act in 1970 (see Chapter 6) began to change this. The OSH Act, more than any other single factor, put teeth in the job descriptions of safety and health professionals. OSHA standards, on-site inspections, and penalties have encouraged a greater commitment to safety and health. Environmental, liability, and workers' compensation issues have also had an impact, as has the growing awareness that providing a safe and healthy workplace is the right thing to do from both an ethical and a business perspective.

Job of the Safety and Health Manager

The job of the safety and health manager is complex and diverse. Figure 4–2 is an example of a job description for such a position. The description attests to the diverse nature of the job. Duties range from hazard analysis, accident reporting, standards and compliance, record keeping, and training to emergency planning and so on.

The minimum educational requirement set by Poultry Processing, Inc. (PPI) is an occupational certificate from a community college with the full

---

POULTRY PROCESSING, INC.

Highway 90 West
Crestview, FL 32536

Vacancy Announcement

Position Title: Safety and Health Manager

Position Description: The Safety and Health Manager for PPI is responsible for establishing, implementing, and managing the company's overall safety and health program. The position reports to the local plant manager. Specific duties include the following:

- Establish and maintain a comprehensive companywide safety and health program.
- Assess and analyze all departments, processes, and materials for potential hazards.
- Work with appropriate personnel to develop, implement, monitor, and evaluate accident prevention and hazard control strategies.
- Ensure companywide compliance with all applicable laws, standards, and codes.
- Coordinate the activities of all members of the company's safety and health program.
- Plan, implement, and broker, as appropriate, safety and health–related training.
- Maintain all required safety and health–related records and reports.
- Conduct accident investigations as necessary.
- Develop and maintain a companywide emergency action plan (EAP).
- Establish and maintain an ongoing safety promotion effort.
- Analyze the company's products from the perspectives of safety, health, and liability.

Qualifications Required: The following qualifications have been established by the PPI management team with input from all levels and all departments.

- *Minimum Education.* Applicants must have at least a one-year community college certificate or an associate of science or applied science degree in industrial safety or a closely related degree (AS holders will begin work at a salary 15 percent higher than a certificate graduate).
- *Preferred Education.* Applicants with a bachelor's degree in any of the following major fields of study will be given first priority: industrial safety and health, industrial technology, industrial management, manufacturing technology, engineering technology, and related. Degree programs in these fields must include at least one three-semester or five-quarter-hour course in industrial or occupational safety and health.

Figure 4–2
Safety and health manager job description.

associate of science or applied science degree preferred. Preference is given to applicants with a bachelor's degree in specifically identified fields.

Role in the Company Hierarchy

The safety and health manager described in Figure 4–2 reports to PPI's local plant manager and has line authority over all other members of the safety and health team. This and the duties set forth in the job description are evidence of the company's commitment to safety and health and that PPI is large enough to have a dedicated safety and health manager.

In some companies, the safety and health manager may also have other duties such as production manager or personnel manager. In these cases, the other members of the safety and health team, like those shown in Figure 4–1, are not normally company employees. Rather, they are available to the company on a part-time or consultative basis as needed. The safety and health manager's role in a company depends in part on whether his or her safety and health duties are full time or are in addition to other duties.

Another role determinant is the issue of authority. Does the safety and health manager have line or staff authority? **Line authority** means that the safety and health manager has authority over and supervises certain employees (i.e., other safety and health personnel). **Staff authority** means that the safety and health manager is the staff person responsible for a certain function, but he or she has no line authority over others involved with that function.

Those occupying staff positions operate like internal consultants—that is, they may recommend, suggest, and promote, but they do not have the authority to order or mandate. This is typically the case with safety and health managers. Even managers with line authority over other safety and health personnel typically have a staff relationship with other functional managers (e.g., personnel, production, or purchasing). For example, consider the following safety and health–related situations:

- A machine operator continually creates unsafe conditions by refusing to practice good housekeeping.
- A certain process is associated with an inordinately high number of accidents.
- A new machine is being purchased that has been proven to be unsafe at other companies.

In the first example, the safety and health manager could recommend that the employee be disciplined but could not normally undertake or administer disciplinary measures. In the second example, the safety and health manager could recommend that the process be shut down until a thorough analysis could be conducted, hazards identified, and corrective measures taken. However, the manager would rarely have the authority to order the process to be shut down. In the final example, the safety and health manager could recommend that an alternative machine be purchased, but he or she would not normally have the authority to stop the purchase.

Maintaining a safe and healthy workplace while playing the role of internal consultant is often the greatest challenge of safety and health managers. It requires managers to be resourceful, clever, astute with regard to corporate politics, good at building relationships, persuasive, adept at trading for favors, credible, and talented in the development and use of influence.

Problems Safety and Health Managers Face

As if the diversity and complexity of the job were not enough, there are a number of predictable problems that safety and health managers are likely to face. These problems are discussed in the following paragraphs.

Lack of Commitment

Top management may go along with having a companywide safety and health program because they see it as a necessary evil. The less enthusiastic may even see safety and health as a collection of government regulations that interfere with profits. Although this is less often true now than it has been in the past, safety and health professionals should be prepared to confront a less than wholehearted commitment in some companies.

Production versus Safety

Industrial firms are in business to make a profit. They do this by producing or processing products. Therefore, anything that interferes with production or processing is likely to be looked on unfavorably. At times, a health or safety measure will be viewed by some as interfering with productivity. A common example is removal of safety devices from machines as a way to speed production. Another is running machines until the last possible moment before a shift change rather than shutting down with enough time left to perform routine maintenance and housekeeping tasks.

The modern marketplace has expanded globally and, therefore, become intensely competitive. To survive and succeed, today's industrial firm must continually improve its productivity, quality, cost, image, response time, and service. This sometimes puts professionals who are responsible for safety and health at odds with others who are responsible for productivity, quality, cost, and response time.

Sometimes, this cannot be avoided. At other times, it is the fault of a management team that is less than fully committed to safety and health. However, sometimes the fault rests squarely on the shoulders of the safety and health manager. This is because one of the most important responsibilities of this person is to convince higher management, middle management, supervisors, and employees that, in the long run, the safe and healthy way of doing business is also the competitive, profitable way of doing business. The next section explains several strategies for making this point.

Companywide Commitment to Safety and Health

In many cases, safety and health managers have been their own worst enemy when it comes to gaining a companywide commitment. The most successful are those who understand the goals of improved productivity, quality, cost, image, service, and response time and are able to convey the message that a safe and healthy workplace is the best way to accomplish these goals. The least successful are those who earn a reputation for being grumpy in-house bureaucrats who quote government regulations chapter and verse but know little and care even less about profits. Unfortunately, in the past, there have been too many safety and health managers who fall into the latter category. This is not the way to gain a companywide commitment to safety and health, but it is a sure way to engender resentment.

Lack of Resources

Safety and health managers are like other managers in an organization in that they must compete for the resources needed to do the job. Often they find that their departments rank lower in priority than the production and operations departments (at least until a disaster occurs). Safety and health managers need to become proficient in showing the financial benefits of a safe workplace.

Today's safety and health manager must understand the bottom-line concerns of management, supervisors, and employees and be able to use these concerns to gain a commitment to safety and health. Figure 4–3 illustrates the essential message that **competitiveness** comes from continually improving a company's productivity, quality, cost, image, service, and response time. These continual improvements can be achieved and maintained best in a safe and healthy work environment.

Safety and health managers should use this message to gain a commitment from management and employees. Following are some strategies that can be used to get the point across.

Productivity, Quality, Cost, and Response Time

These four factors, taken together, are the key to productivity in the age of high technology and global competitiveness. The most productive company is the one that generates the most output with the least input. Output is the company's product. Input is any resource—time, talent, money, technology, and so on—needed to produce the product. **Quality** is a measure of reliability and customer satisfaction. **Cost** is the amount of money required to purchase the item. If all other factors are equal, customers will select the product that costs less. **Response time** is the amount of time that elapses between an order being placed and the product being delivered.

To compete in the global marketplace, industrial companies must continually improve these four factors. At the most fundamental level, successfully competing in the global marketplace means having the best people and the best

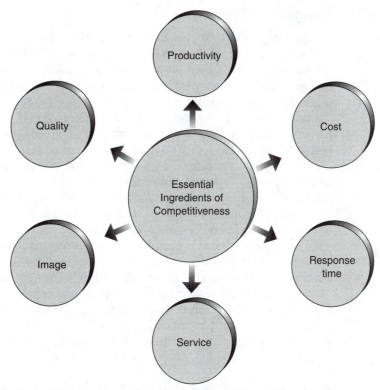

Figure 4–3
Factors that produce competitiveness.

technology and getting the most out of both by applying the best management strategies.

Safety and health managers who understand this can use their knowledge to gain a commitment to their programs. In attempting to do so, the following five points are helpful:

- If it is important to attract and keep the best people, a safe and healthy workplace will help.
- If it is important to get the most out of talented people, it must be important to keep them safe and healthy so that they are functioning at peak performance levels (for example, the best technician in the world can't help when he or she is out of action as the result of an injury or illness).
- Employees cannot concentrate fully on quality when they are concerned for their safety and health.
- Keeping industrial technologies up-to-date requires the continual investment of funds. Profits that are siphoned off to pay the costs associated with accidents, emergencies, and health problems cannot be reinvested in the latest technologies needed to stay competitive.
- With the skyrocketing costs of medical care, workers' compensation, and litigation, it costs less to prevent accidents than to pay for them.

Image and Service

Image and **service** are also important factors in the competitiveness equation. Of the two, image relates more directly to safety and health.

In today's intensely competitive marketplace, a company's image, internal and external, can be a deciding factor in its ability to succeed. Companies that establish a solid internal image in terms of safety and health will find it easier to attract and keep the best employees. Companies that establish a solid external image with regard to environmental and product safety issues will find it easier to attract and retain customers.

Gaining a full and real commitment to safety and health in the workplace is one of the most important roles of the safety and health manager. Traditionally, safety and health managers have argued their cases from the perspectives of ethics or government mandates. The ethical argument is as valid now as it has always been and should continue to be used.

However, in today's competition-driven workplace, managers responsible for the bottom line may resent arguments that are based on government mandates and regulations. On the other hand, these same managers may respond positively if they can be shown that resources invested in safety and health can actually improve a company's competitiveness. Using the points made earlier about productivity, quality, cost, image, and response time can go a long way in helping to gain management commitment to safety and health.

Education and Training for Safety and Health Managers

Advances in technology, new federal legislation, the potential for costly litigation, and a proliferation of standards have combined to make the job of safety and health professionals more complex than ever before. These factors have correspondingly increased the importance of education and training for safety and health managers. The ideal formula for safety and health professionals is formal education prior to entering the profession supplemented by in-service training on a lifelong basis afterward.

Universities, colleges, and community colleges across the country have responded to the need for formal education for safety and health managers as well as other safety and health personnel. Many community colleges offer occupational certificates and **associate degrees** in applied science or science degrees with such program titles as industrial safety, occupational safety, environmental technology, safety and health management, and industrial hygiene.

Universities have responded to the need for formal education by making safety and health–related courses either optional or required parts of such **bachelor's** (or baccalaureate) **degree** programs as industrial technology, manufacturing technology, engineering technology, industrial engineering technology, industrial management, and industrial engineering. Some universities offer bachelor's degrees in industrial safety and health, occupational safety management, and industrial hygiene.

Formal education provides the foundation of knowledge needed to enter the profession. Having begun a career as a safety and health manager, the next challenge is keeping up as the laws, regulations, standards, and overall body of knowledge relating to safety and health grow, change, and evolve.

In-service training, ongoing interaction with professional colleagues, and continued reading of professional literature are effective ways to stay current. New safety and health managers should move immediately to get themselves "plugged into" the profession. This means joining the appropriate professional organizations, becoming familiar with related government agencies, and establishing links with relevant standards organizations. The next section covers agencies and organizations that can be particularly helpful to safety and health managers.

Helpful Agencies and Organizations

Numerous agencies and organizations are available to help the safety and health manager keep up-to-date. These agencies and organizations provide databases, training, and professional literature. There are professional societies, trade associations, scientific organizations, certification boards, service organizations, and emergency service organizations. Those listed here represent only a portion of those available.

- *Certification boards.* Professional certification is an excellent way to establish one's status in the field of safety and health. To qualify to take a certification examination, safety and health managers must have the required education and experience and submit letters of recommendation as specified by the **certification board**. Figure 4–4 contains the names and addresses of certification boards of interest to safety and health managers. Certification is covered in greater detail later in this chapter.
- *Professional societies.* **Professional societies** are typically formed to promote professionalism, adding to the body of knowledge, and forming networks among colleagues in a given field. Numerous professional societies focus on various safety and health issues. Figure 4–5 summarizes some of these.
- *Scientific standards and testing organizations.* Scientific **standards and testing organizations** conduct research, run tests, and establish standards that identify the acceptable levels for materials, substances, conditions, and mechanisms to which people might be exposed in the modern workplace. Figure 4–6 summarizes those of critical importance to safety and health managers.
- *Government agencies.* Many **government agencies** are concerned with various aspects of workplace-related safety and health. Some of the most helpful agencies for the safety and health manager are listed in Figure 4–7.
- *Trade associations.* The purpose of a **trade association** is to promote the trade that it represents. Consequently, material produced by trade associations can be somewhat self-serving. Even so, trade associations can be

Figure 4–4
Professional certification boards.

- **Board of Certified Safety Professionals of America**
  208 Burwash Ave.
  Savoy, IL 61874
  217-359-9265
  http://www.bcsp.org/

- **American Board of Industrial Hygiene**
  6015 W. St. Joseph, Suite 102
  Lansing, MI 48917-3980
  517-321-2638
  http://www.abih.org/

- **Board of Certification in Professional Engineering**
  PO Box 2811
  Bellingham, WA 98227-2811
  360-671-7601
  http://www.bcpe.org/

- **The Institute of Industrial Engineers**
  25 Technology Park
  Norcross, GA 30092
  770-449-0460
  http://www.iienet.org

- **American Board for Occupational Health Nurses**
  201 East Ogden Ave., Suite 114
  Hinsdale, IL 60521
  630-789-5799
  http://www.abohn.org

Figure 4–5
Professional societies.

American Academy of Industrial Hygiene
302 S. Waverly Rd.
Lansing, MI 48917

American Industrial Hygiene Association
475 Wolf Ledges Pkwy.
Akron, OH 44311

American Occupational Medical Association
2340 S. Arlington Heights Rd.
Arlington Heights, IL 60005

American Society of Safety Engineers
1900 E. Oakton St.
Des Plaines, IL 60016

National Safety Council
1121 Spring Lake Dr.
Itasca, IL 60143

Society of Toxicology
1133 I St. NW, Suite 800
Washington, DC 20005

Figure 4–6
Scientific standards and testing organizations.

American National Standards Institute (ANSI)
1430 Broadway
New York, NY 10018

International Atomic Energy Agency
Wagramstrasse 5
A01400 Vienna, Austria

National Fire Protection Association (NFPA)
Batterymarch Park
Quincy, MA 02269

Underwriters Laboratories Inc. (UL)
333 Pfingsten Rd.
New York, NY 10017

American Society of Mechanical Engineers
(ASME)
345 E. 47th St.
New York, NY 10017

American Society for Testing and Materials
(ASTM)
1916 Race St.
Philadelphia, PA 19103

valuable sources of information and training for safety and health managers. Figure 4–8 lists some of the trade associations that can be particularly helpful.

ENGINEERS AND SAFETY

Engineers can make a significant contribution to safety. Correspondingly, they can cause, inadvertently or through incompetence, accidents that result in serious injury and property damage. The engineer has more potential to affect safety in the workplace than any other person does. The following example illustrates this:

A car-pooler transports himself and three fellow employees to work each day. He is not a particularly safe driver and does not insist that his passengers use seatbelts. After running a red light, he crashes into the side of a building while swerving to avoid another vehicle. The two passengers wearing seatbelts are not hurt, but the driver and one other passenger, neither of whom were wearing seatbelts, are critically injured.

This brief story illustrates an accident that has two things in common with many workplace accidents. The first is a careless worker—in this case, the driver. The second is other workers who do not follow prescribed safety rules—in

Bureau of Mines
Department of the Interior
2401 E St. NW
Washington, DC 20241

National Institute for Occupational Safety and
    Health
Department of Health and Human Services
Parklawn Bldg.
5600 Fishers La. Rd. 1401
Rockville, MD 20857

Center for Devices and Radiological Health
Food and Drug Administration
8757 Georgia Ave.
Silver Springs, MD 20910

Consumer Product Safety Commission
1111 18th St. NW
Washington, DC 20207

Department of Labor
200 Constitution Ave. NW
Washington, DC 20210

Environmental Protection Agency
401 M St. SW
Washington, DC 20460

Mine Safety and Health Administration
Department of Labor
4015 Wilson Blvd.
Arlington, VA 22203

National Bureau of Standards Headquarters
Route I-270 and Quince Orchard Rd.
Gaithersburg, MD 20899

National Center for Toxicological Research
Food and Drug Administration
5600 Fishers La.
Rockville, MD 20857

National Technical Information Service
Department of Commerce
5285 Port Royal Rd.
Springfield, VA 22161

Nuclear Regulatory Commission
1717 H St. NW
Washington, DC 20555

Occupational Safety and Health Commission
Department of Labor
200 Constitution Ave. NW
Washington, DC 20210

Occupational Safety and Health Review
    Committee
1825 K St. NW
Washington, DC 20006

Office of Energy Research
Department of Energy
1000 Independence Ave. SW
Washington, DC 20585

Office of Hazardous Materials Transportation
400 7th St. SW
Washington, DC 20590

U.S. Fire Administration
Federal Emergency Management Agency
16825 S. Seton Ave.
Emmitsburg, MD 21727

Figure 4–7
Government agencies.

this case, failing to use seatbelts. Employees such as these can and do cause many workplace accidents, but even the most careless employee cannot cause a fraction of the problems caused by a careless engineer. The following example illustrates this point:

An engineer is charged with the responsibility for designing a new seatbelt that is comfortable, functional, inexpensive, and easy for factory workers to install. He designs a belt that meets all of these requirements, and it is installed in 10,000 new cars. As the cars are bought and accidents begin to occur, it becomes apparent that the new seatbelt fails in crashes involving speeds over 36 miles per hour. The engineer who designed the belt took all factors into consideration except one: safety.

Figure 4–8
Trade associations.
Note: Complete addresses are available in the reference section of most college and university libraries. Most have Web sites that can be found on the Internet.

Alliance of American Insurers

American Foundrymen's Association

American Insurance Association

American Iron and Steel Institute

American Metal Stamping Association

American Petroleum Institute

American Welding Society

Associated General Contractors of America

Compressed Gas Association

Industrial Safety Equipment Association

Institute of Makers of Explosives

Lead Industrial Association

National Electrical Manufacturers Association

National LP-Gas Association

National Machine Tool Builders Association

Scaffolding, Shoring, and Forming Institute

Soap and Detergent Association

Technical Association of the Pulp and Paper Industry

This brief story illustrates how far-reaching an engineer's impact can be. With a poorly designed seatbelt installed in 10,000 automobiles, the engineer has inadvertently endangered the lives of as many as 40,000 people (estimating a maximum of four passengers per automobile).

The engineer's opportunity for both good and bad comes during the design process. The process is basically the same regardless of whether the product being designed is a small toy, an industrial machine, an automobile, a nuclear power plant, a ship, a jumbo jetliner, or a space vehicle. Safety and health professionals should be familiar with the design process so that they can more fully understand the role of engineers concerning workplace safety.

Not all engineers are design engineers. However, engineers involved in design are usually in the aerospace, electrical, mechanical, and nuclear fields. The following paragraphs give an overview of these design-oriented engineering fields as seen in the course descriptions of a college catalog.

Aerospace Engineering

The Bachelor of Science in Engineering (Aerospace Engineering) program incorporates a solid foundation of physical and mathematical fundamentals which provides the basis for the development of the engineering principles essential to the understanding of both atmospheric and extra-atmospheric flight. Aerodynamics, light-weight structures, flight propulsion, and related subjects typical of aeronautical engineering are included. Other courses introduce problems associated with space flight and its requirements. Integration of fundamental principles with useful applications is made in design work in the junior and senior years. Thus, the program prepares the student to contribute to future technological growth, which promises exciting and demanding careers in aerospace engineering. . . . Examples of concentration areas are: aerodynamics; design; flight propulsion; flight structures; space technology; stability, control, and guidance.[1]

Electrical Engineering

Electrical Engineering is a science-oriented branch of engineering primarily concerned with all phases and development of the transmission and utilization of electric power and intelligence. The study of electrical engineering can be conveniently divided into the academic areas of circuits, electronics, electromagnetics, electric energy systems, communications, control, and computer engineering.[2]

Mechanical Engineering

Mechanical Engineering is the professional field that is concerned with motion and the processes whereby other energy forms are converted into motion. Mechanical engineers are the people who are responsible for conceiving, designing, manufacturing, testing, and marketing devices and systems that alter, transfer, transform, and utilize the energy forms that ultimately cause motion. Thus mechanical engineers . . . are the people who make the engines that power ships, trains, automobiles, and spacecraft; they design the power plants which convert the energy in fuels, atoms, waterfalls, and sunlight into useful mechanical forms; and they construct intelligent machines and robots as well as the gears, cams, bearings, and couplings that facilitate and control all kinds of mechanical motion.[3]

Industrial Engineering

Industrial Engineering is concerned with the design, improvement, and installation of integrated systems which include people, material, equipment, and energy. This field of engineering draws upon specialized knowledge and skills in the mathematical,

Safety Fact

The Calculus Controversy in Safety and Health Programs

Many of the programs in colleges and universities that prepare safety and health professionals require calculus. Some practicing professionals in the field think that college programs require too much math and too little management, business, and international safety. These professionals argue that calculus has no value to them on the job, but that more business and management courses would. The opposing faction in the field continues to defend calculus as necessary and important. A third faction supports calculus and business management courses, claiming that the safety and health professional needs both.

physical, and social sciences in concert with the principles and methods of engineering analysis and design to specify, predict, and evaluate the results to be obtained from such systems.

Nuclear Engineering

Nuclear Engineering Sciences comprises those fields of engineering and science directly concerned with the release, control, and safe utilization of nuclear energy. Applications range over such broad topics as the design, development, and operation of nuclear reactor power systems to the applications of radiation in medicine, space, industry, and other related areas. The nuclear engineer, by virtue of his/her engineering and science-based training, is in a unique position to contribute to the many diverse aspects of this major component of the energy radiation field.[4]

Design Process

Professor William S. Chalk describes the **design process**:

The design process is a plan of action for reaching a goal. The plan, sometimes labeled problem-solving strategy, is used by engineers, designers, drafters, scientists, technologists, and a multitude of professionals.[5]

The design process proceeds in five sequential steps:

1. *Problem identification.* Engineers draft a description of the problem. This involves gathering information, considering constraints, reviewing specifications, and combining all of these into a clear and concise description of the problem.

2. *Synthesis.* Engineers combine or synthesize systematic, scientific procedures with creative techniques to develop initial solutions to the problem identified in Step 1. At this point, several possible solutions may be considered.

3. *Analysis and evaluation.* All potential solutions developed in the previous step are subjected to scientific analysis and careful evaluation. Such questions as the following are asked: Will the proposed solution satisfy the functional requirements? Will it meet all specifications? Can it be produced quickly and economically?

4. *Document and communicate.* Engineering drawings, detailed calculations, and written specifications are prepared. These document the design and communicate its various components to interested parties. It is common to revise the design at this point based on feedback from different reviewers.

5. *Produce and deliver.* Shop or detail drawings are developed, and the design is produced, usually as a prototype. The prototype is analyzed and tested. Design changes are made if necessary. The product is then produced and delivered.

The design process gives engineers unparalleled opportunities to contribute significantly to safety in the workplace and in the marketplace by producing products with safety built into them. However, in too many cases, the design process does not serve this purpose. There are two primary reasons for this:

- In analyzing and evaluating designs, engineers consider such factors as function, cost, life span, and manufacturability. All too often safety is not even considered or is only a secondary consideration.
- Even when engineers do consider safety in analyzing and evaluating designs, many are insufficiently prepared to do so effectively.

Engineers who design products may complete their entire college curriculum without taking even one safety course. Safety courses, when available to engineering students in design-oriented disciplines, often tend to be electives. This limits the contributions that design engineers can make to both product and workplace safety. Figure 4–9 is a typical core-course listing for a mechanical engineering student.

Safety Engineer

The title **safety engineer** is often a misnomer in the modern workplace. It implies that the person filling the position is a degreed engineer with formal education and/or special training in workplace safety. Although this is sometimes the case, typically the title is given to the person who has overall responsibility for the company's safety program (the safety manager) or to a member of the company's safety team. This person is responsible for the traditional aspects of the safety program, such as preventing mechanical injuries; falls, impact, and acceleration injuries; heat and temperature injuries; electrical accidents; fire-related accidents; and so on.

In the former case, the person should be given a title that includes the term *manager*. In the latter case, the title "*safety engineer*" is appropriate. However, persons with academic credentials in areas other than engineering should be encouraged to seek such positions because they are likely to be at least as well

Figure 4–9
Typical required core courses for mechanical and industrial engineering degrees.

Engineering mechanics: Statics
Engineering mechanics: Dynamics
Mechanics of materials
Kinematics and dynamics of machinery
Manufacturing processes
Control of mechanical engineering systems
Mechanical vibrations
Machine analysis and design
Thermodynamics
Heat transfer
Fluid dynamics

prepared and possibly even better prepared than persons with engineering degrees. These other educational disciplines include industrial technology, industrial engineering technology, manufacturing technology, engineering technology, industrial management, and industrial safety technology (bachelor's or associate degree).

There are signs that engineering schools are becoming more sensitive to safety and health issues. Graduate degrees in such areas as nuclear physics and nuclear engineering now often require safety courses. The federal government sponsors postgraduate studies in safety. However, the following quote summarizes clearly and succinctly the current status of safety engineering:

> Four states now have registration of professional engineers in a Safety Engineer discipline. Registration gives the registrant the right to use the title "Safety Engineer," but the enabling law has no other requirement that the services of such an engineer be used. The contracting office of one military service responsible for development of advanced high-tech systems does require certain hazard analyses and documents to be signed off and approved at specific points in the designs. Such approval will be valid only under the signature of a registered safety engineer or other engineer shown to have had extensive experience in safety programs. The principal problem is that in a new advanced design project, there may be 400 engineers with no training in accident avoidance who may make critical errors and only one or two safety engineers to find them.[6]

Industrial Engineers and Safety

Industrial engineers are the most likely candidates from among the various engineering disciplines to work as safety engineers. Their knowledge of industrial systems, both manual and automated, can make them valuable members of a design team, particularly one that designs industrial systems and technologies. They can also contribute after the fact as a member of a company's safety team by helping design job and plant layouts for both efficiency and safety.

The industrial engineering discipline can be described as follows:

> Industrial growth has created unusual opportunities for the industrial and systems engineer. Automation and the emphasis on increased productivity coupled with higher levels of systems sophistication are providing impetus to the demand for engineering graduates with a broad interdisciplinary background. The industrial engineering option prepares the student for industrial practice in such areas as product design, process design, plant operation, production control, quality control, facilities planning, work system analysis and evaluation, and economic analysis of operational systems.[7]

Although industrial engineers are more likely to work as safety engineers than are engineers from other disciplines, they are not much more likely to have safety courses as a required part of their program of study. However, their focus on industrial systems and the integration of people and technology does give industrial engineers a solid foundation for additional learning through either in-service training or graduate work.

Environmental Engineers and Safety

A relatively new discipline (when compared with more traditional disciplines such as mechanical engineering) is environmental engineering. This discipline may be described as follows:

> Environmental Engineering Sciences is a field in which the application of engineering and scientific principles is used to protect and preserve human health and the well-being of the environment. It embraces the broad field of the general environment including air and water quality, solid and hazardous wastes, water resources and management, radiological health, environmental biology and chemistry, systems ecology, and water and waste-water treatment.[8]

With the addition of health concerns to the more traditional safety concerns, **environmental engineers** will be sought as members of corporate safety and health teams. The course work they take is particularly relevant since all of it relates either directly or indirectly to health. Figure 4–10 shows the types of courses typically required of environmental engineering students.

A person with the type of formal education shown in Figure 4–10 would be a valuable addition to the safety and health team of any modern industrial firm. Environmental engineers typically report to the overall safety and health manager and are responsible for those elements of the program relating to hazardous waste management, atmospheric pollution, indoor air pollution, water pollution, and wastewater management.

Chemical Engineers and Safety

Increasingly, industrial companies are seeking **chemical engineers** to fill the industrial hygiene role on the safety and health team. Their formal education makes people in this discipline well equipped to serve in this capacity. Chemical engineering may be described as follows:

> Although chemical engineering has existed as a field of engineering for only about 80 years, its name is no longer completely descriptive of this dynamic, growing

Figure 4–10
Typical required core courses for environmental engineering students.

Environmental biology
Environmental chemistry
Water chemistry
Atmospheric pollution
Solid waste management
Water and wastewater
Hazardous waste control
Environmental resources management
Air pollution control design
Hydraulic systems design

profession. The work of the chemical engineer is neither restricted to the chemical industry nor limited to chemical changes or chemistry. Instead, modern chemical engineers, who are also called process engineers, are concerned with all the physical and chemical changes of matter to produce economically a product or result that is useful to mankind. Such a broad background has made the chemical engineer extremely versatile and capable of working in a wide variety of industries: chemical, petroleum, aerospace, nuclear, materials, microelectronics, sanitation, food processing, and computer technology. The chemical industry alone provides an opportunity for the chemical engineer to participate in the research, development, design, or operation of plants for the production of new synthetic fibers, plastics, chemical fertilizers, vitamins, antibiotics, rocket fuels, nuclear fuels, paper pulp, photographic products, paints, fuel cells, transistors, and the thousands of chemicals that are used as intermediates in the manufacture of the above products.[9]

INDUSTRIAL HYGIENIST

Industrial hygiene is defined by the American Industrial Hygiene Association as the "science and art devoted to the recognition, evaluation, and control of those environmental factors or stresses, arising in and from the workplace, which may cause sickness, impaired health and well-being, or significant discomfort and inefficiency among workers or among citizens of the community."[10] The National Safety Council (NSC) describes the job of the **industrial hygienist** as follows:

> An industrial hygienist is a person having a college or university degree or degrees in engineering, chemistry, physics, medicine, or related physical and biological sciences who, by virtue of special studies and training, has acquired competence in industrial hygiene. Such special studies and training must have been sufficient in all of the above cognate sciences to provide the abilities: (a) to recognize environmental factors and to understand their effect on humans and their well-being; (b) to evaluate, on the basis of experience and with the aid of quantitative measurement techniques, the magnitude of these stresses in terms of ability to impair human health and well-being; and (c) to prescribe methods to eliminate, control, or reduce such stresses when necessary to alleviate their effects.[11]

Industrial hygienists are primarily concerned about the following types of hazards: solvents, particulates, noise, dermatoses, radiation, temperature, ergonomics, toxic substances, biological substances, ventilation, gas, and vapors. In a safety and health team, the industrial hygienist typically reports to the safety and health manager.

HEALTH PHYSICIST

Health physicists are concerned primarily with radiation in the workplace. Consequently, they are employed by companies that generate or use nuclear power. Their primary duties include the following: monitoring radiation inside and outside the facility, measuring the radioactivity levels of biological samples, developing the radiation components of the company's emergency action plan, and supervising the decontamination of workers and the workplace when necessary.

Nuclear engineering and nuclear physics are the two most widely pursued fields of study for health physicists. A study conducted by Moeller and Eliassen gave the following breakdown of the academic preparation of practicing health physicists:

Associate degree	5.41 percent
Baccalaureate degree	28.38 percent
Master's degree	42.43 percent
Doctorate degree	19.46 percent[12]

The remaining practitioners are nondegreed personnel who have completed various types of noncollege credit training. This breakdown shows that graduate study is particularly important for health physicists. Professionals in this field may be certified by the American Board of Health Physics (ABHP).

OCCUPATIONAL PHYSICIAN

Occupational medicine as a specialized field dates back to World War II, when the United States experienced unprecedented industrial expansion. Production of manufactured goods skyrocketed, and workplace-related medical needs followed suit. Occupational medicine was not classified as a medical specialty by the American Board of Preventive Medicine until 1955, however. The NSC describes the main concerns of the **occupational physician** as follows:

- Appraisal, maintenance, restoration, and improvement of the workers' health through application of the principles of preventive medicine, emergency medical care, rehabilitation, and environmental medicine.
- Promotion of a productive and fulfilling interaction of the worker and the job, via application of principles of human behavior.
- Active appreciation of the social, economic, and administrative needs and responsibilities of both the worker and work community.
- Team approach to safety and health, involving cooperation of the physician with occupational or industrial hygienists, occupational health nurses, safety personnel, and other specialties.[13]

Occupational physicians are fully degreed and licensed medical doctors. In addition they must have completed postgraduate work in the following areas of safety:

biostatistics and epidemiology, industrial toxicology, work physiology, radiation (ionizing and nonionizing), noise and hearing conservation, effects of certain environmental conditions such as high altitude and high pressures (hyperbaric and hypobaric factors), principles of occupational safety, fundamentals of industrial hygiene, occupational aspects of dermatology, psychiatric and psychological factors, occupational respiratory diseases, biological monitoring, ergonomics, basic personnel management functions, record and data collection, governmental regulations, general environmental health (air, water, ground pollution, and waste management control).[14]

Bernardino Ramazzini is widely thought of as being the first occupational physician. This is primarily as a result of his study of the work-related problems of workers in Modena, Italy, and a subsequent book he authored titled *The Diseases of Workers* (1700).[15] The first leading occupational physician in the United States was Alice Hamilton, MD. According to the NSC,

> In 1910 Dr. Hamilton became managing director of the Illinois Occupational Disease Commission in the United States. Members of this commission were given one year to study and report on health hazards in Illinois industries. Dr. Hamilton investigated lead poisoning in workers who manufactured white lead carbonate, ceramics, railway carriages, automobiles, batteries, printing and mining, and smelting of the metal itself. She fought with company officials for changes in their plants; she lectured to the public; she worked with the Hull House community; she wrote books and magazine articles on occupational disease; she was appointed to Harvard Medical School in 1921 as assistant professor of medicine (the first woman to hold a teaching position there); and she was the only woman member of the League of Nations Health Committee in the 1920s.[16]

Pioneers such as Ramazzini and Hamilton paved the way for the approximately 3,000 occupational physicians practicing today. Whereas in the past the primary role of the occupational physician was treatment, today the primary role is prevention. This means more analytical, diagnostic, and intervention-related work than in the past and a much more proactive rather than passive approach.

One major difference in the role of modern occupational physicians compared with those in the past is in their relationship with employers. In the past, the occupational physician tended to be an in-house physician. Over 80 percent of the Fortune 100 companies still have at least one in-house occupational physician. However, the trend is away from this approach to the contracted approach in all but the largest companies. Companies contract with a private physician, clinic, or hospital to provide specific medical services. The health care provider is typically placed on a retainer in much the same manner as an attorney.

This contracted approach presents a new challenge for the safety and health manager. Including the occupational physician in planning, analyzing, assessing, monitoring, and other prevention-related duties can be more difficult when the physician is not in-house and readily available. For this reason, the safety and health manager must work with higher management to develop a contract that builds in time for prevention activities and compensates the health care provider appropriately.

The NSC makes the following recommendations concerning occupational physicians (OPs):

> Whether your company uses an in-house OP, an OP that works as a consultant or a clinic that provides all health care, remember:

- There should be a written medical program available to all management and employees.
- The OP should understand the workplace and the chemicals used and produced.
- Periodic tours of all facilities are necessary for an understanding of possible work-related injuries, and also to aid in job accommodation.
- The OP should be familiar with OSHA and NIOSH health mandates.
- The OP should be the leader of other medical personnel.
- And, the OP should understand what the company expects and what the OP expects from the company.[17]

OCCUPATIONAL HEALTH NURSE

Occupational health nurses have long been important members of corporate safety and health programs. According to the American Association of Occupational Health Nurses (AAOHN),

> Occupational health nursing is the application of nursing principles in conserving the health of workers in all occupations. It involves prevention, recognition, and treatment of illness and injury, and requires special skills and knowledge in the areas of health education and counselling, environmental health, rehabilitation, and human relations.[18]

Like occupational physicians, today's occupational health nurses have seen their profession evolve over the years. The shift in emphasis is away from after-the-fact treatment to prevention-related activities such as analysis, monitoring, counseling, and education.

The AAOHN defines the objectives of occupational nurses as follows:

- To adapt the nursing program to meet the specific needs of the individual company
- To give competent nursing care for all employees
- To ensure that adequate resources are available to support the nursing program
- To seek out competent medical direction if it is not available on-site
- To establish and maintain an adequate system of records relating to workplace health care
- To plan, prepare, promote, present, and broker educational activities for employees
- To establish and maintain positive working relationships with all departments within the company

Discussion Case

What Is Your Opinion?

"What do you mean, we will need a team of safety and health professionals? Can't we just hire a safety manager or a safety engineer?" asked the frustrated CEO. The consultant smiled understandingly, but stood firm. "Look," said the consultant, "you have just acquired two new companies. One is a chemical processing plant. The other is a manufacturer that produces hazardous by-products. Both have enormous profit potential, but right now those profits are being drained off by legal and medical expenses. The little you invest in the safety and health personnel that I'm recommending will pay off many times over in the future." "But I don't like to hire support personnel," countered the CEO. "If they don't make a direct contribution, why hire them?" Who is right in this discussion? What is your opinion?

- To maintain positive working relationships with all components of the local health care community
- To monitor and evaluate the nursing program on a continual basis and adjust accordingly[19]

Like all members of the safety and health team, occupational nurses are concerned about cost containment. Cost containment is the driving force behind the emphasis on activities promoting health and accident prevention.

Occupational nurses typically report to an occupational physician when he or she is part of the on-site safety and health team. In companies that contract for off-site physician services, occupational nurses report to the overall safety and health manager.

RISK MANAGER

Organizations are at risk every time they open their doors for business. On any given day, an employee may be injured, a customer may have an accident, or a consumer may be injured using the organization's product. *Risk* is defined as a specific contingency or peril. Because the situations that put organizations at risk can be so expensive when they do occur, many organizations employ risk managers.

Risk management consists of the various activities and strategies that an organization can use to protect itself from situations, circumstances, or events that may undermine its security. You are, yourself, a risk manager. You take action every day to protect your personal and economic security. Take, for instance, the act of driving an automobile—a risky undertaking. Every time you drive, you put yourself at risk of incurring injuries, medical bills, lawsuits, and property damage.

You manage the risk associated with driving using two broad strategies: **reduction** and **transference**. The risk associated with driving an automobile can be reduced by wearing a seatbelt, driving defensively, and obeying traffic laws. The remaining risk is managed by transferring it to an insurance company by purchasing a policy that covers both collision and liability.

The same approach—managing risk by using reduction and transference strategies—applies in the workplace. Risk managers work closely with safety and health personnel to reduce the risk of accidents and injuries on the job. They also work closely with insurance companies to achieve the most effective transference possible.

CERTIFICATION OF SAFETY AND HEALTH PROFESSIONALS

Professional certification is an excellent way to establish credentials in the safety, health, and environmental profession. The most widely pursued accreditations are

- Certified Safety Professional, awarded by the Board of Certified Safety Professionals (BCSP)
- Certified Industrial Hygienist, awarded by the American Board of Industrial Hygiene
- Certified Professional Ergonomist, awarded by the Board of Certification in Professional Ergonomics
- Certified Occupational Health Nurse, awarded by the American Board for Occupational Health Nurses

Certified Safety Professional

To quality for the Certified Safety Professional title, applicants must follow these steps:

1. Apply to the Board of Certified Safety Professionals
2. Meet an academic requirement
3. Meet a professional safety experience requirement
4. Pass the Safety Fundamentals Examination
5. Pass the Comprehensive Practice Examination[20]

Academic Requirement

The model educational background for a candidate for the **Certified Safety Professional (CSP)** is a bachelor's degree in safety from a program accredited by the Accreditation Board for Engineering and Technology (ABET). Because many people enter the safety profession from other educational backgrounds, candidates for the CSP may substitute other degrees plus professional safety experience for an accredited BA in safety. A CSP candidate must meet one of the following minimum educational qualifications:

- Associate degree in safety and health, or
- Bachelor's degree in any field

Experience Requirement

In addition to the academic requirement, CSP candidates must have four years of professional safety experience in addition to any experience used to meet the academic requirement. Professional safety experience must meet all the following criteria to be considered acceptable by the Board of Certified Safety Professionals:

- The professional safety function must be the primary function of the position. Collateral duties in safety are not considered the primary function.
- The position's primary responsibility must be the prevention of harm to people, property, and the environment, rather than responsibility for responding to harmful events.
- The professional safety function must be at least 50 percent of the position duties.
- The position must be full time (defined by the BCSP as at least 35 hours per week).
- The position must be at the professional level. This is determined by evaluating the degree of responsible charge and reliance by peers, employers, or clients and on the person's ability to defend analytical approaches used in professional practice and the recommendations made for controlling hazards through engineering or administrative approaches.
- The position must have breadth of duties. This is determined by evaluating the variety of hazards about which a candidate must advise and the range of skills involved in recognizing, evaluating, and controlling hazards. Examples of skills are analysis, synthesis, design, investigation, planning, administration, and communication.

Examination Requirements

The process to achieve the Certified Safety Professional designation typically involves passing two examinations: Safety Fundamentals and Comprehensive Practice.

- *Safety Fundamentals Examination.* The Safety Fundamentals Examination covers basic knowledge appropriate to professional safety practice. Candidates who meet the academic standard (achieve 48 points through an associate or bachelor's degree plus experience) may sit for the Safety Fundamentals Examination. Upon passing this examination, candidates receive the Associate Safety Professional (ASP) title to denote their progress toward the CSP.
- *Comprehensive Practice Examination.* All CSP candidates must acquire 96 points and pass the Comprehensive Practice Examination. To take this

examination, a candidate must meet both the academic and experience requirements and have passed or waived the Safety Fundamentals Examination. The total credit for academic degrees at all levels plus the months of professional safety experience must equal or exceed 96 points. After passing this examination, a candidate receives the Certified Safety Professional title. The address, telephone numbers, and Web site for the Board of Certified Safety Professionals are shown in Figure 4–4.

Certified Industrial Hygienist

In the mid-1950s, a group of industrial hygienists from a national organization recommended that a voluntary certification program be established for industrial hygiene practitioners. In 1960, an independent corporation was established from the two national membership organizations—American Industrial Hygiene Association (AIHA) and American Council of Government Industrial Hygienists (ACGIH)—to establish a national examination process to certify a minimum level of knowledge in industrial hygiene.[21] Because the program was voluntary, it did not restrict the practice of individuals calling themselves "industrial hygienists." Indeed, today there are many competent persons practicing the profession of industrial hygiene who have not even sought certification. However, since its establishment, the program has proven to be a hallmark of achievement that provides an indicator of success in the field. It measures to a defined standard the knowledge of a practicing industrial hygienist in 16 technical areas of practice (called *rubrics*).

The intended purpose of the examination for **Certified Industrial Hygienist (CIH)** is to ensure that professionals working in this field have the skills and knowledge needed in the practice of industrial hygiene. The intent of the board is that the examinations are fair, reasonable, current, and representative of a broad range of industrial hygiene topics. Successful completion of the examination, meeting the educational requirements, and meeting the comprehensive, professional-level industrial hygiene experience requirements are the means by which the board attempts to identify those practitioners who have demonstrated the necessary skills to be an industrial hygienist.

Certification Process

What is the process for becoming certified? First, there is the obvious need for technical knowledge. There are various methods for gathering this information, but no one system has been identified as being the most effective. Review courses are available to prepare aspiring applicants. Conventional wisdom tends to support knowledge gained through experience and the watchful guidance of a competent mentor. Many individuals take the examination to find out their weak areas and study up on those until they manage to pass.

The professional reference questionnaire (PRQ) is important as well. The board's requirement for experience is based on activity on a professional or apprentice level. This causes a sticking point for many who find they are in a position with titles such as "industrial hygienist" or "project manager" when their scope of practice relies on strict adherence to a regulatory interpretation or exercise of very little independent judgment. Others with titles of "IH technician" or "specialist" may be acting totally independently and practicing with a scope of extreme variability and many unknowns. This should be portrayed in the PRQ to be fair to the applicant. Details are important to the board in making their evaluation of an applicant.

The American Board of Industrial Hygiene (ABIH) also has a process established for reviewing each application and preparing each examination. Examinations are prepared in minute detail. Each question is evaluated by a group of practicing Certified Industrial Hygienists to ensure its correctness and relevance to the practice of IH. Each item is rated on difficulty for its target audience (CORE or Comprehensive), and this is used to set the passing score for each test. Each question is also rated by professional testers to ensure its validity as a question for an examination. Questions are selected for use on an examination based on the latest survey of the practice of industrial hygiene, in both rubric areas and domains of practice, to achieve a balance indicative of the current practice and some historical knowledge. After each presentation of the examination, questions are again reviewed for validity. The examination itself is the subject of an effort to ensure that it adheres to the standardized evaluation method.

Certification Maintenance

After a candidate has managed to meet the established standard for certification, maintenance becomes the issue. The board requires that all Certified Industrial Hygienists demonstrate that they are active in the field and have continued to improve their knowledge. Seven categories of practice are noted for the accumulation of "points" toward the 40 required every five years. Some portion of these points are gathered for active practice, technical committee work, publications, education and meetings, teaching, retest, or other work. This requires that some program of continuing education must be pursued. The address, telephone number, and Web site for the American Board of Industrial Hygiene is shown in Figure 4–4.

Certified Professional Ergonomist

The examination for **Certified Professional Ergonomist (CPE)** is administered by the Board of Certification in Professional Ergonomics. Details concerning the examination are explained in the following paragraphs.[22]

To take the examination, individuals must meet the following requirements:

- *Academic requirements.* Applicants should have graduated from a regionally accredited college or university with a master's degree, or equivalent, in one of the correlative fields of ergonomics, such as biomechanics, human factors/ergonomics, industrial engineering, industrial hygiene, kinesiology, psychology, or systems engineering. Not everyone trained in these fields necessarily has the capabilities required for certification. The board uses other criteria to determine whether an applicant's ergonomics education has been sufficiently broad.

- *Work experience.* Applicants must have completed at least four years of ergonomic work experience. Appropriateness of work experience is determined from the applicant's employment history and evidence of participation in projects requiring ergonomic expertise.
- *Work product.* One work sample must be submitted with the application. This work sample must demonstrate a breadth of ergonomic knowledge and the ability to use ergonomic methods successfully. The work sample must demonstrate, at a minimum, the candidate's capabilities in the application of ergonomic principles to the design of a product, system, or work environment. Allowable work products are technical reports, design papers, analysis reports, evaluation reports, patent applications, or a thorough written description of the project.

Associate-Level Certification

The Associate Ergonomics Professional (AEP) category is a precursor to the CPE designation and is available to a candidate who

- Meets the education requirements for BCPE certification
- Has passed Part 1 of the examination (on basic knowledge of human factors/ergonomics)
- Is currently working toward fulfilling the BCPE requirement of four years' practical experience as a human factors and ergonomics professional

Bachelor's Degree Certification

A credential is available that recognizes a candidate for achieving the following levels of knowledge, skills, and experience in ergonomics practice:

- A bachelor's degree from a recognized university
- At least 200 hours of ergonomics training
- At least two full years practicing ergonomics
- A satisfactory score on the four-hour, two-part, multiple-choice examination on ergonomics foundations and ergonomics practice methods

This certification may be obtained by candidates with a BA in engineering, health care/rehabilitation, industrial hygiene, and psychology. The address, telephone number, and Web site for the Board of Certification in Professional Ergonomics is shown in Figure 4–4.

Certified Occupational Health Nurse

The American Board for Occupational Health Nurses (ABOHN) offers several certifications relating to safety and health:

- **Certified Occupational Health Nurse (COHN)**
- Certified Occupational Health Nurse–Specialist (COHN-S)

- Certified Occupational Health Nurse/Case Manager (COHN/CM)
- Certified Occupational Health Nurse–Specialist/Case Manager (COHN-S/CM)
- Certified Occupational Health Nurse–Safety Manager (COHN-SM)

All the certifications offered by the ABOHN require that individuals first pass either the COHN or COHN-S certification tests. The other certifications are sub-specialties.

Academic and Experience Requirements

In order to sit for the COHN certification tests, individuals must be registered nurses holding an associate degree or higher degree or their international equivalents. The focus of this credential is the nurse's role as a clinician, adviser, coordinator, and case manager. To sit for the COHN-S certification test, individuals must be registered nurses who hold a bachelor's degree or higher. The focus of this credential is on the nurse's role in direct care, management, education, consulting, and case management. The BA required for this credential need not be in nursing.

In addition to the academic requirements, those who wish to achieve COHN or COHN-S certification must have 4,000 or more hours of work experience in occupational health and 50 or more contact hours of continuing education completed during the past five years. Additional testing is required for the various subspecialty credentials.

The newest certification offered by the ABOHN is the COHN-SM or Safety Manager credential. The focus of this credential is on occupational health nurses who plan, organize, implement, and evaluate hazard control activities to help organizations reduce or eliminate risks in the workplace. In order to sit for this subspecialty test, an individual must hold the COHN or COHN-S credential, hold a position that requires at least 25 percent of the time be spent on safety management activities, have completed 50 or more hours of safety-related continuing education during the past five years, and have 1,000 or more hours of safety management experience earned during the past five years.

SUMMARY

1. The modern safety and health team is headed by a safety and health manager. Depending on the size of the company and the commitment of its management, the team may include people in the following additional positions: safety engineer, industrial hygienist, environmental engineer, health physicist, occupational physician, and occupational health nurse.
2. The job of the safety and health manager is complex and diverse, focusing on analysis, prevention, planning, evaluation, promotion, and compliance. Educational requirements range from technical certificates to graduate

degrees. Typical college majors held by practitioners include industrial safety and health technology, industrial technology, industrial engineering technology, manufacturing technology, industrial management, and engineering technology.

3. Engineers can have a significant impact on safety in the workplace and the marketplace by designing safety into products. The engineering disciplines geared most closely to design are aerospace, electrical, nuclear, and mechanical. Not all safety engineers are engineering majors. In fact, graduates of various technology degrees may have more formal education in safety than engineering graduates do.

4. Industrial hygienists are concerned with (a) recognizing the impact of environmental factors on people, (b) evaluating the potential hazards of environmental stressors, and (c) prescribing methods to eliminate stressors.

5. Health physicists are concerned primarily with radiation in the workplace. Their duties include monitoring radiation in the air, measuring radioactivity levels in biological samples, developing the radiation components of a company's emergency action plan, and supervising decontamination activities.

6. Occupational physicians are medical doctors who specialize in workplace-related health problems and injuries. In the past, occupational physicians treated injuries and illnesses as they occurred. Today's occupational physicians focus more attention on anticipating and preventing problems.

7. Occupational health nurses are concerned with conserving the health of workers through prevention, recognition, and treatment. Like occupational physicians, occupational health nurses focus more attention on anticipation and prevention than they did in the past.

8. Risk management involves the application of risk reduction strategies and transferring remaining risk to insurance companies.

9. Professional certification is an excellent way to establish credentials in the safety, health, and environmental management profession. The most widely pursued certifications are Certified Safety Professional, Certified Industrial Hygienist, Certified Professional Ergonomist, and Certified Occupational Health Nurse.

KEY TERMS AND CONCEPTS

Aerospace engineering

Analysis and evaluation

Associate degree

Bachelor's degree

Certification boards

Certified Industrial Hygienist (CIH)

Certified Occupational Health Nurse (COHN)

Certified Professional Ergonomist (CPE)

Certified Safety Professional (CSP)

Chemical engineer

Competitiveness
Cost
Design engineers
Design process
Document and communicate
Electrical engineering
Environmental engineering
Ergonomist
Formal education
Government agencies
Health physicist
Image
In-service training
Industrial engineering
Industrial hygienist
Lack of commitment
Line authority
Mechanical engineering
Nuclear engineering
Occupational health nurse

Occupational physician
Problem identification
Produce and deliver
Production versus safety
Productivity
Professional societies
Profits
Quality
Reduction
Response time
Risk management
Safety and health manager
Safety engineer
Service
Staff authority
Standards and testing
  organizations
Synthesis
Trade associations
Transference

REVIEW QUESTIONS

1. What types of positions may be included in a modern safety and health team?
2. Briefly explain the impact that such issues as workers' compensation and the environment have had on the commitment of corporate management to safety and health.
3. What is the difference between a staff and a line position?
4. Explain the types of problems that safety and health managers can expect to confront in attempting to implement their programs.
5. Briefly explain what a company must do to succeed in today's competitive global marketplace.
6. How can safety and health managers use the competitiveness issue to gain a commitment to their programs?
7. List five different college majors that can lead to a career as a safety and health manager.

8. Explain the importance of ongoing in-service training for modern safety and health managers and how to get it.
9. How can safety and health managers become certified in their profession?
10. Name three professional societies that a modern safety and health manager may join.
11. What is meant by the following statement: "If a physician makes an error, he may harm one person, but an engineer who errs may harm hundreds."
12. Explain how the design process can affect safety.
13. What types of engineers are most likely to work as design engineers?
14. Why is the title "safety engineer" sometimes a misnomer?
15. What specific strengths may industrial engineers bring to bear as safety engineers?
16. What specific strengths may environmental engineers bring to the safety and health team?
17. What specific strengths may chemical engineers bring to the safety and health team?
18. Describe the job of the industrial hygienist.
19. What is a health physicist?
20. Describe the job of the occupational physician.
21. Describe the job of the occupational health nurse.
22. Explain the concept of risk management.
23. Explain the role of the ergonomist.
24. Explain how to achieve each of the following certifications: Certified Safety Professional, Certified Industrial Hygienist, and Certified Professional Ergonomist.
25. Explain how to achieve the following certification: Certified Occupational Health Nurse–Safety Manager.

ENDNOTES

1. *The University Record*, Undergraduate Catalog of the University of Florida, 2006, 67.
2. Ibid., 70.
3. Ibid., 74.
4. Ibid., 75.
5. D. L. Goetsch, J. Nelson, and William S. Chalk, *Technical Drawing*, 5th ed. (Albany: Delmar, 2005), 791.
6. W. Hammer, *Occupational Safety Management and Engineering*, 5th ed. (Upper Saddle River, NJ: Prentice Hall, 2001), 791.
7. *The University Record*, 73.
8. Ibid., 72.
9. Ibid., 68.
10. American Industrial Hygiene Association (AIHA), *Engineering Field Reference Manual*. Retrieved from http://www.aiha.org, April 4, 2006.

11. C. Berry (revised by Barbara A. Plog), "The Industrial Hygienist," in *Fundamentals of Industrial Hygiene* (Chicago: National Safety Council, 1998), 571.
12. Ibid., 572.
13. C. Zenz, "The Occupational Physician," in *Fundamentals of Industrial Hygiene*, 607.
14. Ibid.
15. National Safety Council, "Occupational Physicians: What Is Their New Role?" Retrieved from http://www.nsc.org, April 4, 2006.
16. Ibid.
17. Ibid., 38.
18. American Association of Occupational Health Nurses (AAOHN), *A Guide for Establishing an Occupational Health Nursing Service*. Retrieved from http://www.aaoha.org, April 5, 2006.
19. Ibid., 79.
20. Board of Certified Safety Professionals (BCSP), "The CSP and Specialty Certifications," July 2000, 1–6, http://www.bcsp.org/.
21. American Board of Industrial Hygiene (ABIH), "Certified Industrial Hygienist (CIH)—What It Means," July 2000, 1–5, http://www.abih.org/.
22. Board of Certified Professional Ergonomists (BCPE), "BCPE—Frequently Asked Questions," July 2000, 1–5, http://www.bcpe.org/.

SAFETY, HEALTH, AND COMPETITION IN THE GLOBAL MARKETPLACE

Major Topics

- Competitiveness Defined
- Productivity and Competitiveness
- Quality and Competitiveness
- How Safety and Health Can Improve Competitiveness

One of the most frequently heard terms in the language of modern business and industry is *competitiveness*. To survive and prosper in today's global marketplace, industrial companies must be competitive. Companies that used to compete only with neighboring firms now find themselves competing against companies from Japan, Germany, Taiwan, Korea, France, Great Britain, China, and many other countries throughout the world.

The global marketplace is intensely competitive. It is like a sports contest that never ends. You may win today, but the race begins again tomorrow. Competing in the global marketplace has been described as the equivalent of running in a race that has no finish line.

The need to achieve peak performance levels day after day puts intense pressure on companies, and pressure runs downhill. This means that all employees from executive-level managers to workers on the shop floor feel it. It is not uncommon for this pressure to create a harried atmosphere that can increase the likelihood of accidents. It can also lead to shortcuts that increase the potential for health hazards (e.g., improper storage, handling, and use of hazardous materials).

This is unfortunate because the most competitive companies typically are also the safest and healthiest. It is critical that safety and health professionals understand the positive relationship between safety and health and competitiveness. It is also extremely important that they be able to articulate that connection effectively.

This chapter provides prospective and practicing safety and health professionals with the information they need to (1) understand the concept of global competitiveness; (2) understand its impact on a company's safety and health program; and (3) understand how to use the global competitiveness issue to gain a commitment to safety and health or to increase an existing commitment.

COMPETITIVENESS DEFINED

The Institute for Corporate Competitiveness defines **competitiveness** as "the ability to consistently succeed and prosper in the marketplace whether it is local, regional, national, or global."[1] The most competitive companies are those that do the following: (1) consistently outperform their competitors in the key areas of quality, productivity, response time, service, cost, and corporate image (Figure 5–1); and (2) continually improve all these areas. The two key concepts associated with competitiveness are **peak performance** and **continual improvement**.

Figure 5–1
Key areas that result
in competitiveness.

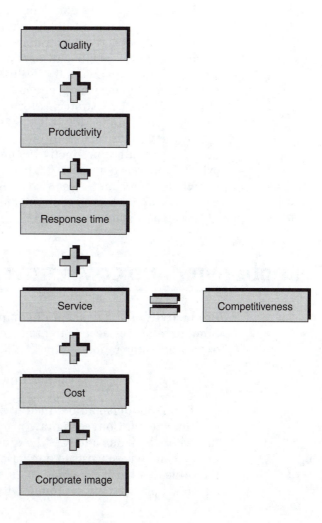

Competing in the **global marketplace** is difficult to do but easy to understand. One need only think of Olympic athletes to gain a perspective on the concept. Olympic athletes train continually to improve their performance to world-class levels. Their goal is to improve continually in practice so that they can achieve peak performance levels in competition. Those who do this consistently win. However, these athletes also know that, if they want to continue competing, there will always be another race. Consequently, the process of improvement must be continual. Winners today who rest on their laurels will be losers tomorrow. Records set today will be broken tomorrow. What is true for the Olympic athlete is also true for the modern industrial company.

The productivity record of today will be broken tomorrow. What is considered world-class quality today will be considered mediocre tomorrow. What is excellent service today will be unsatisfactory service tomorrow. The acceptable costs of today will be considered too high tomorrow. Record response time today will be too slow tomorrow. Finally, an excellent corporate image today can fade quickly and be gone tomorrow.

The marketplace is truly like a never-ending series of athletic contests. In order to compete, industrial companies must continually and consistently combine the best people and the best technologies with the best management strategies. As explained later in this chapter, a safe and healthy workplace is one of the best management strategies in this age of global competitiveness. It is also one of the best strategies for getting and keeping the best people while simultaneously protecting valuable technologies from damage.

Of the various factors influencing a company's competitiveness, the two most important are productivity and quality. The other factors are, to a large extent, functions of these two. Therefore, safety and health managers should be knowledgeable about these two concepts and be able to discuss them on an equal footing with production managers, supervisors, and representatives of higher management.

PRODUCTIVITY AND COMPETITIVENESS

Productivity is the concept of comparing **output** of goods or services to the **input** of resources needed to produce or deliver them. Productivity is typically expressed as the ratio of output to input in the following manner:

$$\text{Output/Input} = \text{Productivity}$$

To fully understand productivity, it is necessary to understand the concept of value added. Converting raw materials into usable products adds value to the materials. **Value added** is the difference between what it costs to produce a product and what it costs to purchase it (Figure 5–2). This difference represents the value that has been added to the product by the production process. Value added is increased when productivity is increased.

Figure 5–2
The value-added concept.

In Figure 5–2, Company A and Company B manufacture the same product. Each is able to sell its product for $1,300. However, Company B can manufacture the product for $400 because it is more productive than Company A, which produces the same product at a cost of $700. The productivity difference gives Company B at least two competitive advantages: (1) It can lower its price for the product below that of Company A and still make a profit; and (2) its greater profit margin gives Company B more capital to reinvest in upgrading its facility, equipment, and personnel, which in turn will improve its competitiveness even more.

This productivity difference is important because all competing companies that produce a given product probably pay approximately the same for the raw materials. Consequently, the winner most likely is the one that adds the most value to the materials, which means the one that is the most productive (produces the most output with the least input).

Following are some rules of thumb that production managers use to monitor productivity in their plants. Safety and health managers should be familiar with these rules:

• Productivity is declining when (1) output declines, and input is constant; or (2) output is constant, but input increases.
• Productivity is improving when (1) output is constant, but input decreases; or (2) output increases, and input is constant.

In considering the concept of productivity, remember that it represents only half of the equation. Nothing has been gained if productivity is improved to the detriment of quality. When productivity is improved, quality must also improve or at least remain constant.

Productivity: A Global View

In the community of industrialized nations, the United States has historically been the most productive. This country's productive capacity peaked during World War II when the United States amazed its enemies and allies alike with its enormous production capabilities.

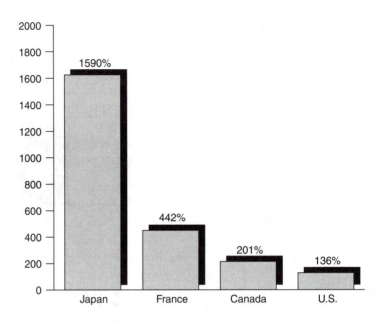

Figure 5–3
Increases in output
per hour of workers
(1950–2003).

The perception, however, is that the United States has seen productivity in the industrial sector decline steadily since World War II. Actually, this isn't true. Between 1950 and 1997, productivity in the United States increased by 129 percent.[2] The problem is that productivity in other industrialized countries, particularly Japan, increased even more during this same period (Figure 5–3). As a result, companies in the United States have found it increasingly difficult to compete in the global marketplace, particularly in the production of automobiles, computers, and consumer electronics products.

Pressure to increase productivity often results in actions that are detrimental to the safety and health of workers. In their rush to meet deadlines or quotas, workers may disassemble safeguards, stop taking the time to use appropriate personal protective gear, ignore safety rules, neglect equipment maintenance duties, improperly handle or store toxic substances, and take chances that they wouldn't take under normal conditions.

Such efforts occasionally result in short-term productivity improvements, but they invariably do more harm than good in the long run. The inevitable result is that productivity ultimately suffers, and the company finds that it won a battle only to lose the war.

It's as though a football team is losing late in a game, and the coach is getting desperate. Rather than sticking to what he knows will work best in the long run, the coach gives in to desperation and has the team's star quarterback run several trick plays that leave him vulnerable. His team wins the game, but in the process loses the star quarterback for the rest of the season to a crippling injury. The team's chances of having a championship season are forfeited for the sake of one game.

Production managers often respond to the pressures of competition in much the same way as the coach did and invariably with the same results. Safety and health professionals will probably face similar situations. Strategies for preventing this type of response are discussed later in this chapter.

QUALITY AND COMPETITIVENESS

Quality goes hand-in-hand with productivity in the competitiveness equation. Today's production company must have both. Quality without productivity results in costs that are too high to be competitive. Productivity without quality results in a shabby product that quickly tarnishes the corporate image.

Quality is a measure of the extent to which a product or service meets or exceeds customer expectations. According to Peters, it is important to define quality in terms of customer expectations.[3]

Quality: A Global View

The most important consumer product in the global marketplace is the automobile. Therefore, it is often used as the measuring stick for making quality comparisons among the leading industrialized nations. U.S. automobile makers compete with those of Japan, Europe, East Asia, Brazil, and Mexico.

The basis of comparison, used by Congress's Office of Technology Assessment, is assembly defects. Beginning in the 1970s and extending through the 1980s, Japanese automakers held center stage in the worldwide quality drama. Quality in the best U.S. plants equaled that in East Asia, Mexico, and Brazil but did not equal that in Japanese or European plants. The worst plants in the United States allowed more defects than the worst plants of all other competitors except those in Europe.

However, by adopting a variety of quality improvement strategies that, collectively, came to be known as *quality management*, automakers in the United States have been able to reemerge as quality leaders. The market share taken away from American automakers by their Japanese competitors began to return in the late 1980s and early 1990s. By working with Dr. W. Edwards Deming— the quality guru who is credited with giving birth to the so-called Japanese miracle—Ford was able to lead Chrysler and General Motors down the path to recovery and reemergence. Competition remains fierce in the global automobile market, and consumers worldwide have benefited as a result. World-class quality is an ever-changing phenomenon; continued improvement has become the norm.

Whereas the pressure to increase productivity can be detrimental to safety and health, the pressure to improve quality generally supports safety and health. In the next section, prospective and practicing safety and health professionals will see how to use productivity and quality to gain a commitment to safety and health or to increase a commitment that already exists.

HOW SAFETY AND HEALTH CAN IMPROVE COMPETITIVENESS

When the pressures of competing become intense, it is not uncommon for safety and health to be given a lower priority. Not only is this wrong from an ethical standpoint, it is also wrong from the perspective of competitiveness and profitability. In the days when a worker's most important qualifications were physical strength and stamina, there were always plenty of applicants in the labor pool. If a worker was injured, several equally qualified applicants were waiting to replace him or her.

However, with the dawning of the age of high technology and the advent of global competition, this situation changed. Mental ability became more important than physical ability; suddenly, the number of qualified applicants got smaller. In fact, whereas the workplace is becoming increasingly technical and complex, the literacy level of the labor pool is actually declining in the United States. Consider the implications for competitiveness of the following statement:

> The basic skills necessary to be productive in a modern industrial setting are increasing steadily. At the same time the national high school dropout rate continues to increase as does the number of high school graduates who are functionally illiterate in spite of their diploma. This means that while the number of high-skilled jobs in modern industry is increasing, the number of people able to fill them is on the decline. The impact this will have on industry in the United States can be summarized as follows:
>
> 1. Difficulty in filling high-skill jobs.
> 2. Lower levels of productivity and, as a result, a lower level of competitiveness.
> 3. Higher levels of waste.
> 4. Higher potential for damage to sophisticated technological systems.
> 5. Greater numbers of dissatisfied employees in the workplace.[4]

The **illiteracy problem**, coupled with rapid and continual technological change, has serious implications for the global competitiveness of U.S. companies. Modern industrial companies are like modern sports teams. To compete, they must recruit, employ, and develop the best possible personnel. Having done so, they must keep them safe and healthy to derive the benefits of their talents.

Talented people working in a safe and healthy environment will be more competitive than equally talented people who are constantly distracted by concerns for their safety and health. In addition, the most talented employees cannot help a company compete if they are slowed by injuries. The **aging workforce** also has serious implications for the global competitiveness of U.S. companies. They bring invaluable experience and a positive work ethic to the job, but they also have special needs that introduce a whole new set of safety and health concerns.

The people side of the competitiveness equation is only part of it. There is also the technology side. With technological change occurring so rapidly, reinvestment in modern equipment is essential. Today's industrial firm must invest a higher percentage of its profits in equipment upgrades and do so more frequently. Money that must be diverted to workers' compensation, medical claims, product liability litigation, and environmental cleanups is money that could have been invested in technological upgrades.

If they are able to convey these facts effectively, safety and health professionals should be able to secure a strong commitment to safety and health in the workplace. In attempting to do so, these professionals should still use the ethics and compliance arguments because these are proper and compelling. However, health and safety professionals must not stop there. They must convince companies that committing to safety and health is not just the right thing to do ethically or the smart thing to do legally, it is also the profitable thing to do in terms of competitiveness.

Nothing will help safety and health managers make this point more than the competitiveness equation: productivity, quality, response time, cost, service, and corporate image. Each of these ingredients relies directly in some way on having a safe and healthy workplace. Some of the more direct ways are summarized in the following paragraphs.

Productivity and Cost

Productivity is a function of people, technology, and management strategies. Safety and health have a direct effect on the first two. Companies with a reputation for providing a safe and healthy workplace will find it easier to attract and keep the best people. Correspondingly, these talented employees will be able to focus their skills more intently on being productive rather than worrying about accidents or health problems. On the **cost** side of the issue, companies with a record of safe and healthy practices will be better able to reinvest in equipment upgrades than those who must divert funds into such nonproductive costs as medical claims, environmental cleanups, and safety and health–related litigation. As shown in Figure 5–2, the more productive a company, the more competitive it can be in setting prices for its products.

Quality

Quality is essential to competitiveness. Fortunately, those practices that enhance quality also tend to enhance safety and health. Quality requires strict adherence to established production practices, attention to detail, and a commitment to doing things the right way. So do safety and health. For this reason, safety and health professionals should also be advocates of quality.

Response Time

Response time is like productivity in that it is a function of people, technology, and management strategies. Therefore, the same arguments that apply to productivity apply to response time. This is particularly important in the age of global competition because response time and quality, taken together, are becoming more important than the old "low-bid" approach when it comes to winning contracts. The ability to deliver a quality product on demand is a pre-requisite to participating in just-in-time contracts. These are contracts in which suppliers provide their products just in time to be used in the production process. This allows manufacturers to produce their products while carrying little or no inventory, a characteristic of today's most competitive companies.

Service

Service is an important ingredient in competitiveness. With industrial companies, **service** typically means in-field or after-delivery service. The most common example is service provided to people who purchase an automobile. Service is important because it can have a significant impact on customer satisfaction and, in turn, on corporate image. Service is not closely associated with safety and health and is, therefore, not a component to be used when trying to gain a commitment to safety and health.

Image

In a competitive world, industrial companies must be concerned about their **corporate image**. An image of being concerned about employee safety and health will help companies attract and keep the best people. An image of being concerned about product safety will help companies market their products. An image of being concerned about the environment will make a company a welcome neighbor in any community. Correspondingly, a poor image in any of these areas can undermine a company's competitiveness. A product safety problem that becomes a media story can cause a company's sales of a given product to plummet overnight. This happened to a major automobile manufacturer when it was discovered that one of its economy models had a tendency to ignite and explode in rear-end collisions. It also happened to pharmaceutical manufacturers when various headache relief products were tampered with and found to have been laced with poison. Had these manufacturers not shown

---

Safety Myth

Safety and Global Competitiveness

Maintaining a safe and healthy workplace cannot be shown to enhance a company's global competitiveness. Right? Not necessarily. Productivity is a measurement of how well various factors, when taken together, perform. These factors include employees, material, money, and motivation. All four of these factors are affected by the work environment. In fact, more than 90 percent of employees identify *quality of the work environment* as critical in terms of their motivation and performance.

Source: From "Employee Perceptions: Impact of Work Factors on Job Performance," by the Institute for Corporate Competition, Niceville, FL, Report 2003-11, June 2003.

Discussion Case
What Is Your Opinion?
"Explain this to me again. You say that making all of these modifications to our work-stations and implementing these new procedures will make us more competitive. How? These workstation modifications are expensive, and it seems to me that some of these new procedures will just slow down production." This wasn't the first time that Garner Baxter, an independent safety and health consultant, had been asked to explain the strategic advantages of a safe and healthy workplace. Can the quality of the work environment really make an organization more competitive? What is your opinion?

their concern by immediately recalling all containers that may have been affected and quickly developing tamper-proof substitutes, they could have suffered irreparable damage in the marketplace.

Image problems are not limited to product liability issues. Companies that are not careful about protecting the environment may find themselves the subject of protest demonstrations on the nightly news. Such negative publicity can harm the corporate image and translate very quickly into market losses. As corporate decision makers become more sensitive to these facts, safety and health professionals can use them to gain a commitment to their programs.

SUMMARY

1. Competitiveness is the ability to succeed and prosper in the local, regional, national, and global marketplace. The most competitive companies are those that consistently outperform their competitors in the key areas of quality, productivity, response time, service, cost, and image.
2. Productivity is a measure of output in goods and services compared to input of resources needed to produce or deliver them. Part of productivity is the concept of value added, which is measured as the difference between what it costs a company to produce a product and the competitive market price of that product.
3. Quality is a measure of the extent to which a product meets or exceeds customer expectations. It goes hand-in-hand with productivity. Quality without productivity results in costs that are too high to be competitive. Productivity without quality results in an unacceptable product.
4. Safety and health contribute to competitiveness in the following ways: (a) by helping companies attract and keep the best people; (b) by allowing employees to focus on peak performance without being distracted by concerns for their safety and health; (c) by freeing money that can be reinvested in technology updates; and (d) by protecting the corporate image.

KEY TERMS AND CONCEPTS

Aging workforce

Competitiveness

Continual improvement

Corporate image

Cost

Global marketplace

Illiteracy problem

Input

Output

Peak performance

Productivity

Quality

Response time

Service

Value added

REVIEW QUESTIONS

1. Explain why and how global competition can have a negative impact on safety and health in the workplace.
2. Define the term *competitiveness*.
3. What are the common characteristics of the most competitive companies?
4. Explain the importance of continual improvement as it relates to global competitiveness.
5. Define the terms *productivity* and *value added*. Explain their relationship.
6. State two rules of thumb that explain how to recognize declining productivity.
7. State two rules of thumb that explain how to recognize improving productivity.
8. Explain the relationship between productivity and quality as it relates to competitiveness.
9. Define the term *quality*.
10. Write a brief rebuttal to the following statement: "A safety and health program is just a bunch of bureaucratic regulations that get in the way of profits."

ENDNOTES

1. Institute for Corporate Competitiveness, *Manual of Services* (Niceville, FL: Institute for Corporate Competitiveness, 2006), 1.
2. National Council for Occupational Education and American Association of Community Colleges, *Productive America: Two-Year Colleges Unite to Improve Productivity in the Nation's Workforce* (Washington, DC: 1990), 1–12.
3. T. Peters, *Thriving on Chaos* (New York: Harper & Row, 1987), 78.
4. D. L. Goetsch, *Effective Supervision* (Upper Saddle River, NJ: Prentice Hall, 2002), 210.

2

Laws and Regulations

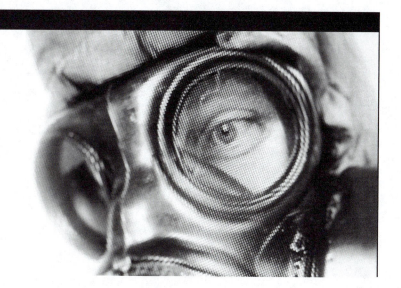

6   The OSH Act, Standards, and Liability
7   Workers' Compensation
8   Accident Investigation and Reporting
9   Product Safety and Liability

THE OSH ACT, STANDARDS, AND LIABILITY

Major Topics

- Rationale for the OSH Act
- OSHA's Mission and Purpose
- OSH Act Coverage
- OSHA Standards
- OSHA's Record Keeping and Reporting
- Keeping Employees Informed
- Workplace Inspections and Enforcement
- OSHA's Enhanced Enforcement Policy
- Citations and Penalties
- Appeals Process
- State-Level OSHA Programs
- Services Available from OSHA
- Employer Rights and Responsibilities
- Employee Rights and Responsibilities
- Keeping Up-to-Date on OSHA
- Problems with OSHA
- Other Agencies and Organizations
- OSHA's General Industry Standards
- OSHA's Maritime Standards
- OSHA's Construction Standards
- Standards and Codes
- Laws and Liability

Since the early 1970s, the amount of legislation passed—and the number of subsequent regulations—concerning workplace safety and health have increased markedly. Of all the legislation, by far the most significant has been the **Occupational Safety and Health Act (OSH Act)** of 1970. Prospective and practicing safety and health professionals must be knowledgeable about the OSH Act and the agency established by it—the **Occupational Safety and Health**

Administration (OSHA). This chapter provides students with the information they need about the OSH Act, OSHA, and other pertinent federal legislation and agencies.

RATIONALE FOR THE OSH ACT

Perhaps the most debilitating experience one can have on the job is to be involved in, or exposed to, a work-related accident or illness. Such an occurrence can be physically and psychologically incapacitating for the victim, psychologically stressful for the victim's fellow workers, and extraordinarily expensive for the victim's employer. In spite of this, until 1970, laws governing workplace safety were limited and sporadic. Finally, in 1970, Congress passed the Occupational Safety and Health Act (OSH Act) with the following stated purpose: "to assure so far as possible every working man and woman in the nation safe and healthful working conditions and to preserve our human resources."[1]

According to the U.S. Department of Labor, in developing this comprehensive and far-reaching piece of legislation, Congress considered the following statistics:

- Every year, an average of 14,000 deaths was caused by **workplace accidents**.
- Every year, 2.5 million workers were disabled in workplace accidents.
- Every year, approximately 300,000 new cases of **occupational diseases** were reported.[2]

Clearly, a comprehensive, uniform law was needed to help reduce the incidence of work-related injuries, illnesses, and deaths. The OSH Act of 1970 addressed this need. It is contained in Title 29 of the Code of Federal Regulations, Parts 1900 through 2200. The act also establishes the Occupational Safety and Health Administration (OSHA), which is part of the U.S. Department of Labor and is responsible for administering the OSH Act.

OSHA'S MISSION AND PURPOSE

According to the Department of Labor, OSHA's mission and purpose can be summarized as follows:

- Encourage employers and employees to reduce workplace hazards.
- Implement new safety and health programs.
- Improve existing safety and health programs.
- Encourage research that will lead to innovative ways of dealing with workplace safety and health problems.
- Establish the rights of employers regarding the improvement of workplace safety and health.

- Establish the rights of employees regarding the improvement of workplace safety and health.
- Monitor job-related illnesses and injuries through a system of reporting and record keeping.
- Establish training programs to increase the number of safety and health professionals and to improve their competence continually.
- Establish mandatory workplace safety and health standards and enforce those standards.
- Provide for the development and approval of state-level workplace safety and health programs.
- Monitor, analyze, and evaluate state-level safety and health programs.[3]

OSH ACT COVERAGE

The OSH Act applies to most employers. If an organization has even one employee, it is considered an employer and must comply with applicable sections of the act. This includes all types of employers from manufacturing and construction to retail and service organizations. There is no exemption for small businesses, although organizations with 10 or fewer employees are exempted from OSHA inspections and the requirement to maintain injury and illness records.

Although the OSH Act is the most comprehensive and far-reaching piece of safety and health legislation ever passed in this country, it does not cover all employers. In general, the OSH Act covers employers in all 50 states, the District of Columbia, Puerto Rico, and all other territories that fall under the jurisdiction of the U.S. government. Exempted employers are

- Persons who are self-employed
- Family farms that employ only immediate members of the family
- Federal agencies covered by other federal statutes (in cases where these other federal statutes do not cover working conditions in a specific area or areas, OSHA standards apply)
- State and local governments (except to gain OSHA's approval of a state-level safety and health plan, states must provide a program for state and local government employees that is at least equal to its private sector plan)
- Coal mines (coal mines are regulated by mining-specific laws)

Federal government agencies are required to adhere to safety and health standards that are comparable to and consistent with OSHA standards for private sector employees. OSHA evaluates the safety and health programs of federal agencies. However, OSHA cannot assess fines or monetary damages against other federal agencies as it can against private sector employers.

There are many OSHA requirements to which employers must adhere. Some apply to all employers—except those exempted—whereas others apply only to specific types of employers. These requirements cover areas of concern such as the following:

- Fire protection
- Electricity
- Sanitation
- Air quality
- Machine use, maintenance, and repair
- Posting of notices and warnings
- Reporting of accidents and illnesses
- Maintaining written compliance programs
- Employee training

In addition to these, other more important and widely applicable requirements are explained later in this chapter.

OSHA STANDARDS

The following statement by the U.S. Department of Commerce summarizes OSHA's responsibilities relating to standards:

> In carrying out its duties, OSHA is responsible for promulgating legally enforceable standards. OSHA standards may require conditions, or the adoption or use of one or more practices, means, methods, or processes reasonably necessary and appropriate to protect workers on the job. It is the responsibility of employers to become familiar with standards applicable to their establishments and to ensure that employees have and use personal protective equipment when required for safety.[4]

The general duty clause of the OSH Act requires that employers provide a workplace that is free from hazards that are likely to harm employees. This is important because the general duty clause applies when there is no specific OSHA standard for a given situation. Where OSHA standards do exist, employers are required to comply with them as written.

How Standards Are Developed

OSHA develops standards based on its perception of need and at the request of other federal agencies, state and local governments, other standards-setting agencies, labor organizations, or even individual private citizens. OSHA uses the committee approach for developing standards, both standing committees within OSHA and special ad hoc committees. Ad hoc committees are appointed to deal with issues that are beyond the scope of the standing committees.

OSHA's standing committees are the **National Advisory Committee on Occupational Safety and Health (NACOSH)** and the Advisory Committee on Construction Safety and Health. NACOSH makes recommendations on standards to the secretary of health and human services and to the secretary of labor. The Advisory Committee on Construction Safety and Health advises the secretary of labor on standards and regulations relating specifically to the construction industry.

The **National Institute for Occupational Safety and Health (NIOSH)**, like OSHA, was established by the OSH Act. Whereas OSHA is part of the Department of Labor, NIOSH is part of the Department of Health and Human Services. NIOSH has an education and research orientation. The results of this agency's research are often used to assist OSHA in developing standards.

OSHA Standards versus OSHA Regulations

OSHA issues both standards and regulations. Safety and health professionals need to know the difference between the two. OSHA standards address specific hazards such as working in confined spaces, handling hazardous waste, or working with dangerous chemicals. Regulations are more generic in some cases than standards and more specific in others. However, even when they are specific, regulations do not apply to specific hazards. Regulations do not require the rigorous review process that standards go through. This process is explained in the next section.

How Standards Are Adopted, Amended, or Revoked

OSHA can adopt, amend, or revise standards. Before any of these actions can be undertaken, OSHA must publish its intentions in the *Federal Register* in either a notice of proposed rule making or an advance notice of proposed rule making. The **notice of proposed rule making** must explain the terms of the new rule, delineate proposed changes to existing rules, or list rules that are to be revoked. The advance notice of proposed rule making may be used instead of the regular notice when it is necessary to solicit input before drafting a rule.

After publishing notice, OSHA must conduct a public hearing if one is requested. Any interested party may ask for a public hearing on a proposed rule or rule change. When this happens, OSHA must schedule the hearing and announce the time and place in the *Federal Register.*

The final step, according to the Department of Labor, is as follows:

> After the close of the comment period and public hearing, if one is held, OSHA must publish in the Federal Register the full, final text of any standard amended or adopted and the date it becomes effective, along with an explanation of the standard and the reasons for implementing it. OSHA may also publish a determination that no standard or amendment needs to be issued.[5]

How to Read an OSHA Standard

OSHA standards are typically long and complex and are written in the language of lawyers and bureaucrats, making them difficult to read. However, reading OSHA standards can be simplified somewhat if one understands the system.

OSHA standards are part of the Code of Federal Regulations (CFR) published by the Office of the Federal Register. The regulations of all federal-government agencies are published in the CFR. Title 29 contains all the standards assigned to OSHA. Title 29 is divided into several parts, each carrying a four-number designator (such as Part 1901, Part 1910). These parts are divided into sections, each carrying a numerical designation. For example, 29 CFR 1910.1 means *Title 29, Part 1910, Section 1, Code of Federal Regulations.*

The sections are divided into four different levels of subsections, each with a particular type of designator as follows:

First Level:	Alphabetically using lowercase letters in parentheses: (a) (b) (c) (d)
Second Level:	Numerically using numerals in parentheses: (1) (2) (3) (4)
Third Level:	Numerically using roman numerals in parentheses: (i) (ii) (iii) (iv)
Fourth Level:	Alphabetically using uppercase letters in parentheses: (A) (B) (C) (D)

Occasionally, the standards go beyond the fourth level of subsection. In these cases, the sequence just described is repeated with the designator shown in parentheses underlined. For example: (a), (1), (i), (A).

Understanding the system used for designating sections and subsections of OSHA standards can guide readers more quickly to the specific information needed. This helps reduce the amount of cumbersome reading needed to comply with the standards.

Temporary Emergency Standards

The procedures described in the previous section apply in all cases. However, OSHA is empowered to pass **temporary emergency standards** on an emergency basis without undergoing normal adoption procedures. Such standards remain in effect only until permanent standards can be developed.

To justify passing temporary standards on an emergency basis, OSHA must determine that workers are in imminent danger from exposure to a hazard not covered by existing standards. Once a temporary standard has been developed, it is published in the *Federal Register.* This step serves as the notification step in the permanent adoption process. At this point, the standard is subjected to all the other adoption steps outlined in the preceding section.

How to Appeal a Standard

After a standard has been passed, it becomes effective on the date prescribed. This is not necessarily the final step in the **appeals process**, however. A standard, either permanent or temporary, may be appealed by any person who is opposed to it.

An appeal must be filed with the U.S. Court of Appeals serving the geographic region in which the complainant lives or does business. Appeal paperwork must be initiated within 60 days of a standard's approval. However, the filing of one or more appeals does not delay the enforcement of a standard unless the court of appeals handling the matter mandates a delay. Typically, the new standard is enforced as passed until a ruling on the appeal is handed down.

Requesting a Variance

Occasionally, an employer may be unable to comply with a new standard by the effective date of enforcement. In such cases, the employer may petition OSHA at the state or federal level for a variance. Following are the different types of variances that can be granted.

Temporary Variance

When an employer advises that it is unable to comply with a new standard immediately but may be able to if given additional time, a **temporary variance** may be requested. OSHA may grant such a variance for up to a maximum of one year. To be granted a temporary variance, employers must demonstrate that they are making a concerted effort to comply and taking the steps necessary to protect employees while working toward compliance.

Application procedures are very specific. Prominent among the requirements are the following: (1) identification of the parts of the standard that cannot be complied with; (2) explanation of the reasons why compliance is not possible; (3) detailed explanations of the steps that have been taken so far to comply with the standard; and (4) explanation of the steps that will be taken to comply fully.

According to the U.S. Department of Labor, employers are required to keep their employees informed. They must "certify that workers have been informed of the variance application, that a copy has been given to the employees' authorized representative, and that a summary of the application has been posted wherever notices are normally posted. Employees also must be informed that they have the right to request a hearing on the application."[6]

Variances are not granted simply because an employer cannot afford to comply. For example, if a new standard requires employers to hire a particular type of specialist but there is a shortage of people with the requisite qualifications, a temporary variance might be granted. However, if the employer simply cannot afford to hire such a specialist, the variance will probably be denied. Once a temporary variance is granted, it may be renewed twice. The maximum period of each extension is six months.

Permanent Variance

Employers who feel they already provide a workplace that exceeds the requirements of a new standard may request a **permanent variance**. They present their evidence, which is inspected by OSHA. Employees must be informed of the application for a variance and notified of their right to request a hearing. Having reviewed the evidence and heard testimony (if a hearing has been held), OSHA can award or deny the variance. If a permanent variance is awarded, it comes with a detailed explanation of the employer's ongoing responsibilities regarding the variance. If, at any time, the company does not meet these responsibilities, the variance can be revoked.

Other Variances

In addition to temporary and permanent variances, an experimental variance may be awarded to companies that participate in OSHA-sponsored experiments to test the effectiveness of new health and safety procedures. Variances also may be awarded in cases where the secretary of labor determines that a variance is in the best interest of the country's national defense.

When applying for a variance, employers are required to comply with the standard until a decision has been made. If this is a problem, the employer may petition OSHA for an interim order. If granted, the employer is released from the obligation to comply until a decision is made. In such cases, employees must be informed of the order.

Typical of OSHA standards are the confined space and hazardous waste standards. Brief profiles of these standards provide an instructive look at how OSHA standards are structured and the extent of their coverage.

Confined Space Standard

This standard was developed in response to the approximately 300 work-related deaths that occur in confined spaces each year.[7] The standard applies to a broad cross section of industries that have employees working in spaces with the following characteristics: limited openings for entry or exit, poor natural ventilation, and a design not intended to accommodate continuous human occupancy. Such spaces as manholes, storage tanks, underground vaults, pipelines, silos, vats, exhaust ducts, boilers, and degreasers are typically considered confined spaces.

The key component in the standard is the *permit requirement.* Employers are required to develop an in-house program under which employees must have a permit to enter confined spaces. Through such programs, employers must do the following:

- Identify spaces that can be entered only by permit.
- Restrict access to identified spaces to ensure that only authorized personnel may enter.
- Control hazards in the identified spaces through engineering, revised work practices, and other methods.
- Continually monitor the identified spaces to ensure that any known hazards remain under control.

The standard applies to approximately 60 percent of the workers in the United States. Excluded from coverage are federal, state, and local government employees; agricultural workers; maritime and construction workers; and employees of companies with 10 or fewer workers.

Hazardous Waste Standard

This standard specifically addresses the safety of the estimated 1.75 million workers who deal with hazardous waste: hazardous waste workers in all situations including treatment, storage, handling, and disposal; firefighters; police officers; ambulance personnel; and hazardous materials response team personnel.[8]

The requirements of this standard are as follows:

- Each hazardous waste site employer must develop a safety and health program designed to identify, evaluate, and control safety and health hazards, and provide for emergency response.
- There must be preliminary evaluation of the site's characteristics prior to entry by a trained person to identify potential site hazards and to aid in the selection of appropriate employee protection methods.
- The employer must implement a site control program to prevent contamination of employees. At a minimum, the program must identify a site map, site work zones, site communications, safe work practices, and the location of the nearest medical assistance. Also required in particularly hazardous situations is the use of the buddy system so that employees can keep watch on one another and provide quick aid if needed.
- Employees must be trained before they are allowed to engage in hazardous waste operations or emergency response that could expose them to safety and health hazards.
- The employer must provide medical surveillance at least annually and at the end of employment for all employees exposed to any particular hazardous substance at or above established exposure levels or those who wear approved respirators for 30 days or more on-site.
- Engineering controls, work practices, and personal protective equipment, or a combination of these methods, must be implemented to reduce exposure below established exposure levels for the hazardous substances involved.
- There must be periodic air monitoring to identify and quantify levels of hazardous substances and to ensure that proper protective equipment is being used.
- The employer must set up an informational program with the names of key personnel and their alternates responsible for site safety and health, and the requirements of the standard.
- The employer must implement a decontamination procedure before any employee or equipment leaves an area of potential hazardous exposure; establish operating procedures to minimize exposure through contact

with exposed equipment, other employees, or used clothing; and provide showers and change rooms where needed.
- There must be an emergency response plan to handle possible on-site emergencies prior to beginning hazardous waste operations. Such plans must address personnel roles; lines of authority, training, and communications; emergency recognition and prevention; safe places of refuge; site security; evacuation routes and procedures; emergency medical treatment; and emergency alerting.
- There must be an off-site emergency response plan to better coordinate emergency action by local services and to implement appropriate control actions.[9]

OSHA'S RECORD KEEPING AND REPORTING

One of the breakthroughs of the OSH Act was the centralization and systematization of **record keeping**. This has simplified the process of collecting health and safety statistics for the purpose of monitoring problems and taking the appropriate steps to solve them.

Over the years, OSHA has made substantial changes to its record-keeping and reporting requirements. Employers had complained for years about the mandated injury and illness record-keeping system. Their complaints can be summarized as follows:

- Original system was cumbersome and complicated.
- OSHA record-keeping rule had not kept up with new and emerging issues.
- There were too many interpretations in many of the record-keeping documents.
- Record-keeping forms were too complex.
- Guidelines for record keeping were too long and difficult to understand.

In response to these complaints, OSHA initiated a dialogue among stakeholders to improve the record-keeping and reporting process. Input was solicited and received from employers, unions, trade associations, record keepers, OSHA staff, state occupational safety and health personnel, and state consultation program personnel. OSHA's goals for the new record-keeping and reporting system were as follows:

- Simplify all aspects of the process
- Improve the quality of records
- Meet the needs of a broad base of stakeholders
- Improve access for employees
- Minimize the regulatory burden
- Reduce vagueness—give clear guidance
- Promote the use of data from the new system in local safety and health programs

In recording and reporting occupational illnesses and injuries, it is important to have common definitions. The Department of Labor uses the following definitions for record-keeping and reporting purposes:

> An occupational injury is any injury such as a cut, fracture, sprain, or amputation that results from a work-related accident or from exposure involving a single incident in the work environment. An occupational illness is any abnormal condition or disorder other than one resulting from an occupational injury caused by exposure to environmental factors associated with employment. Included are acute and chronic illnesses or diseases which may be caused by inhalation, absorption, ingestion, or direct contact with toxic substances or harmful agents.[10]

Reporting Requirements

All occupational illnesses and injuries must be reported if they result in one or more of the following:

- Death of one or more workers
- One or more days away from work
- Restricted motion or restrictions to the work that an employee can do
- Loss of consciousness of one or more workers
- Transfer of an employee to another job
- Medical treatment beyond in-house first aid (if it is not on the first-aid list, it is considered medical treatment)
- Any other condition listed in Appendix B of the rule.[11]

Record-Keeping Requirements

Employers are required to keep injury and illness records for each location where they do business.[12] For example, an automobile manufacturer with plants in several states must keep records at each individual plant for that plant. Records must be maintained on an annual basis using special forms prescribed by OSHA. Computer or electronic copies can replace paper copies. Records are not sent to OSHA. Rather, they must be maintained locally for a minimum of three years. However, they must be available for inspection by OSHA at any time.

All records required by OSHA must be maintained on the following forms or forms based on them:

> *OSHA's Form 300.* Form 300 is used to record information about every work-related death and every work-related injury or illness that involves loss of consciousness, restricted work activity, job transfer, days away from work, or medical treatment beyond first aid. First aid is defined as follows:
> - Using nonprescription medication at nonprescription doses
> - Administering tetanus immunizations
> - Cleaning, flushing, or soaking wounds on the surface of the skin

- Using wound coverings such as bandages, gauze pads, and so on
- Using hot or cold therapy
- Using totally nonrigid means of support, such as elastic bandages, wraps, nonrigid backbelts, and so forth
- Using temporary immobilization devices such as splints, slings, neck collars, or back boards while transporting an accident victim
- Drilling a fingernail or toenail to relieve pressure, or draining fluids from blisters
- Using eye patches
- Using irrigation, tweezers, cotton swab, or other simple means to remove splinters or foreign material from areas other than the eye
- Using finger guards
- Using massages
- Drinking fluids to relieve heat stress

Injuries that require no more than these first-aid procedures do not have to be recorded. Form 300 is also used to record significant work-related injuries and illnesses that are diagnosed by a physician or licensed health care professional as well as those that meet any of the specific recording criteria set forth in 29 CFR Part 1904.12. Figure 6–1 is an example of a log based on OSHA's Form 300.

OSHA's Form 300A. All organizations covered by 29 CFR Part 1904 must complete **Form 300A**, even if there have been no work-related injuries or illnesses during the year in question. This form is used to summarize all injuries and illnesses that appear on OSHA's Form 300.

OSHA's Form 301. **Form 301** is used for every incidence of a recordable injury or illness. Form 301 must be completed within seven calendar days of learning that a recordable injury or illness has occurred. Figure 6–2 is an example of an incident report based on OSHA's Form 301.

Reporting and Record-Keeping Summary

Reporting and record-keeping requirements appear as part of several different OSHA standards. Not all of them apply in all cases. Following is a summary of the most widely applicable OSHA reporting and record-keeping requirements.

- *29 CFR 1903.2(a) OSHA Poster.* **OSHA Poster 2203**, which advises employees of the various provisions of the OSH Act, must be conspicuously posted in all facilities subject to OSHA regulations.
- *29 CFR 1903.16(a) Posting of OSHA Citations.* Citations issued by OSHA must be clearly posted for the information of employees in a location as close as possible to the site of the violation.
- *29 CFR 1904.2 Injury/Illness Log.* Employers are required to maintain a log and summary of all *recordable* (Form 300) injuries and illnesses of their employees.

You must record information about every work-related death and about every work-related injury or illness that involves loss of consciousness, restricted work activity or job transfer, days away from work, or medical treatment beyond first aid. You must also record significant work-related injuries and illnesses that are diagnosed by a physician or licensed health care professional. You must also record work-related injuries and illnesses that meet any of the specific recording criteria listed in 29 CFR Part 1904.8 through 1904.12. Feel free to use two lines for a single case if you need to. You must complete an Injury and Illness Incident Report (OSHA Form 301) or equivalent form for each injury or illness recorded on this form. If you're not sure whether a case is recordable, call your local OSHA office for help.

Company name:_____
City:_____ State:_____

IDENTIFY THE PERSON

(A) Case No.	(B) Employee's name	(C) Job title (e.g., Welder)

DESCRIBE THE CASE

(D) Date of injury or onset of illness	(E) Where the event occurred (e.g., Landing dock north end)	(F) Describe injury or illness, parts of body affected, and object/substance that directly injured or made person ill (e.g., Second degree burns on right forearm from acetylene torch)

CLASSIFY THE CASE

Using these four categories, check ONLY the most serious result for each case:

Death (G)	Days away from work (H)	Remained at work: Job transfer or restriction (J)	Remained at work: Other recordable cases (J)

Enter the number of days the injured or ill worker was:

On job transfer or restriction (K)	Away from work (L)

Check the "Injury" column or choose one type of illness: (M)

Injury (1)	Skin disorder (2)	Respiratory condition (3)	Poisoning (4)	Hearing Injury (3)	All other(s) (6)

Page Totals ➤

Figure 6–1
Log based on OSHA's Form 300.

Within 7 calendar days after you receive information that a recordable work-related injury or illness has occurred, you must fill out this form or an equivalent. Some state workers' compensation, insurance, or other reports may be acceptable substitutes. To be considered an equivalent form, any substitute must contain all the information asked for on this form.

According to Public Law 91-596 and 29 CFR 1904, OSHA's recordkeeping rule, you must keep this form on file for five years following the year to which it pertains.

Completed by: _____

Title _____

Phone ( ) _____ - _____

Date _____ / _____ / _____

INFORMATION ABOUT THE EMPLOYEE

1) Full Name _____
2) Street _____
   City _____ State _____ Zip _____
3) Date of birth _____ / _____ / _____
4) Date hired _____ / _____ / _____
5) ❑ Male
   ❑ Female

INFORMATION ABOUT THE PHYSICIAN OR OHER HEALTH CARE PROFESSIONAL

6) Name of physician or other health care professional
   _____
7) If treatment was given away from the worksite, where was it given?
   Facility _____
   Street _____
   City _____ State _____ Zip _____
8) Was employee treated in an emergency room?
   ❑ Yes
   ❑ No
9) Was employee hospitalized overnight as an in-patient?
   ❑ Yes
   ❑ No

INFORMATION ABOUT THE CASE

10) Case number from *Log* _____ *(Transfer the number from the Log after you record the case.)*
11) Date of injury or illness _____ / _____ / _____
12) Time employee began work _____ AM / PM
13) Time of event _____ AM / PM
    ❑ Check if time cannot be determined
14) **What was the employee doing just before the incident occurred?** Describe the activity, as well as the tools, equipment, or material the employee was using. Be specific. *Examples*: "climbing a ladder while carrying roofing materials"; "spraying chlorine from hand sprayer"; "daily computer key-entry."
15) **What happened?** Tell us how the injury occurred. *Examples*: "When ladder slipped on wet floor, worker fell 20 feet"; "Worker was sprayed with chlorine when gasket broke during replacement"; "Worker developed soreness in wrist over time."
16) **What was the injury or illness?** Tell us the part of the body that was affected and how it was affected; be more specific than "hurt," "pain," or "sore." *Examples*: "strained back"; "chemical burn, hand"; "carpal tunnel syndrome."
17) **What object or substance directly harmed the employee?** *Examples*: "concrete floor"; "chlorine"; "radial arm saw." If this question does not apply to the incident, leave it blank.
18) **If the employee died, when did death occur?**
    Date _____ / _____ / _____

Figure 6–2
Accident report based based on OSHA's Form 301.

- *29 CFR 1904.4 Supplementary Records.* Employers are required to maintain supplementary records (OSHA Form 301) that give more complete details relating to all recordable injuries and illnesses of their employees.
- *29 CFR 1904.5 Annual Summary.* Employers must complete and post an annual summary (OSHA Form 300A) of all recordable illnesses and

Safety Fact

Access to Records

"Access" is an important factor relating to record keeping. Safety and health records containing medical information should be monitored closely with access carefully controlled. However, employees must have access to their own records. In order to gain access, employees should submit a written request asking that records be made available to them or their physician. No record should be released without the written consent of the employee in question.

injuries. The summary must be posted from February 1 to April 30 every year.

- *29 CFR 1904.6 Lifetime of Records.* Employers are required to keep injury and illness records on file for three years.
- *29 CFR 1904.7 Access to Records.* Employers are required to give employees, government representatives, and former employees and their designated representatives access to their own individual injury and illness records.
- *29 CFR 1904.8 Major Incident Report.* A major incident is the death of one employee or the hospitalization of five or more employees in one incident. All such incidents must be reported to OSHA.
- *29 CFR 1904.11 Change of Ownership.* When a business changes ownership, the new owner is required to maintain the OSHA-related records of the previous owner.
- *29 CFR 1910.1020(d) Medical/Exposure Records.* Employers are required to maintain medical and/or exposure records for the duration of employment plus 30 years (unless the requirement is superseded by another OSHA standard).
- *29 CFR 1910.1020(e) Access to Medical/Exposure Records.* Employers that keep medical and/or exposure records are required to give employees access to their own individual records.
- *29 CFR 1910.1020(g)(1) Toxic Exposure.* Employees who will be exposed to toxic substances or other harmful agents in the course of their work must be notified when first hired and reminded continually on at least an annual basis thereafter of their right to access their own individual medical records.
- *29 CFR 1910.1020(g)(2) Distribution of Materials.* Employers are required to make a copy of OSHA's *Records Access Standard* (29 CFR 1910.1020) available to employees. They are also required to distribute to employees any informational materials provided by OSHA.

OSHA makes provisions for awarding record-keeping variances to companies that wish to establish their own record-keeping systems (see variance regulation 1905). Application procedures are similar to those described earlier in this chapter for standard variances. To be awarded a variance, employers must show that their record-keeping system meets or exceeds OSHA's requirements.

Incidence Rates

Two concepts can be important when completing OSHA 300 forms: *incidence rates* and *severity rates.* On occasion, it is necessary to calculate the total injury and illness incident rate of an organization in order to complete an OSHA Form 300. This calculation must include fatalities and all injuries requiring medical treatment beyond mere first aid.

The formula for determining the total injury and illness incident rate follows:

$$IR = N \times 200{,}000 \div T$$

IR = Total injury and illness incidence rate
N = Number of injuries, illnesses, and fatalities
T = Total hours worked by all employees during the period in question

The number 200,000 in the formula represents the number of hours that 100 employees work in a year (40 hours per week $\times$ 50 weeks = 2,000 hours per year per employee). Using the same basic formula with only minor substitutions, safety managers can calculate the following types of incidence rates:

- Injury rate
- Illness rate
- Fatality rate
- Lost workday cases rate
- Number of lost workdays rate
- Specific hazard rate
- Lost workday injuries rate

The *number of lost workdays rate*, which does not include holidays, weekends, or any other days that employees would not have worked anyway, takes the place of the old severity rate calculation.

Record-Keeping and Reporting Exceptions

Among the exceptions to OSHA's record-keeping and reporting requirements, the two most prominent are as follows:

- Employers with 10 or fewer employees (full- or part-time in any combination)
- Employers in one of the following categories: real estate, finance, retail trade, or insurance

There are also partial exceptions to OSHA's record-keeping and reporting requirements. Most businesses that fall into Standard Industrial Classifications (SIC) codes 52–89 are exempt from all record-keeping and reporting requirements except in the case of a fatality or an incident in which five or more employees are hospitalized. The following types of organizations are partially exempt. They are not required to maintain OSHA injury and illness records unless asked by OSHA to do so. OSHA's request must be in writing.

Advertising services

Apparel and accessory stores

Barber shops

Beauty shops

Bowling centers

Candy, nut, and confectionary stores

Child day care services

Computer and data processing services

Credit reporting and collection services

Dairy products stores

Dance studios, schools, and halls

Depository institutions

Drug and proprietary stores

Eating and drinking establishments

Educational institutions and services

Engineering, accounting, and related

Funeral service and crematories

Gasoline service stations

Hardware stores

Health services, not elsewhere classified

Holding and other investment offices

Individual and family services

Insurance agents, brokers, services

Insurance carriers

Legal services

Liquor stores

Meat and fish markets

Mailing, reproduction, and stenographic

Medical and dental laboratories

Membership organizations

Miscellaneous business services

Miscellaneous food stores

Miscellaneous personnel services

Miscellaneous shopping goods stores

Motion picture

Motorcycle dealers

Museums and art galleries

New and used car dealers

Non-depository institutions

Offices and clinics of dentists

Offices of osteopathic physicians

Offices and clinics of medical doctors

Offices of other health practitioners

Photographic studios, portrait

Producers, orchestras, entertainers

Radio, television, and computer stores

Real estate agents and managers

Retail stores, not elsewhere classified

Research, management, and related

Retail bakeries

Re-upholstery and furniture repair

Security and commodity brokers

Services, not elsewhere classified

Shoe repair and shoeshine parlors

Social services, not elsewhere classified

Title abstract offices

Used car dealers

KEEPING EMPLOYEES INFORMED

One of the most important requirements of the OSH Act is *communication.* Employers are required to keep employees informed about safety and health issues that concern them. Most of OSHA's requirements in this area concern the posting of material. Employers are required to post the following material at locations where employee information is normally displayed:

- OSHA Poster 2203, which explains employee rights and responsibilities as prescribed in the OSH Act. The state version of this poster may be used as a substitute.
- Summaries of variance requests of all types.
- Copies of all OSHA citations received for failure to meet standards. Unlike other informational material, citations must be posted near the site of the violation. They must remain until the violation is corrected or for a minimum of three days, whichever period is longer.
- OSHA Form 300A (Summary of Workplace Injuries and Illnesses). Each year the new summary must be posted by February 1 and must remain posted until April 30.

In addition to the posting requirements, employers must also provide employees who request them with copies of the OSH Act and any OSHA rules that may concern them. Employees must be given access to records of exposure to hazardous materials and medical surveillance that has been conducted.

WORKPLACE INSPECTIONS AND ENFORCEMENT

One of the methods OSHA uses for enforcing its rules is the **workplace inspection**. OSHA personnel may conduct workplace inspections unannounced, and except under special circumstances, giving an employer prior notice is a crime punishable by fine, imprisonment, or both.

When OSHA compliance officers arrive to conduct an inspection, they are required to present their credentials to the person in charge. Having done so, they are authorized to enter, at reasonable times, any site, location, or facility where work is taking place. They may inspect, at reasonable times, any condition, facility, machine, equipment, materials, and so on. Finally, they may question, in private, any employee or other person formally associated with the company.

Under special circumstances, employers may be given up to a maximum of 24 hours' notice of an inspection. These circumstances are

- When imminent danger conditions exist
- When special preparation on the part of the employer is required
- When inspection must take place at times other than during regular business hours

- When it is necessary to ensure that the employer, employee representative, and other pertinent personnel will be present
- When the local area director for OSHA advises that advance notice will result in a more effective inspection

Employers may require that OSHA have a judicially authorized warrant before conducting an inspection. The U.S. Supreme Court handed down this ruling in 1978 (*Marshall v. Barlow's Inc.*).[13] However, having obtained a legal warrant, OSHA personnel must be allowed to proceed without interference or impediment.

The OSH Act applies to approximately six million work sites in the United States. Sheer volume dictates that OSHA establish priorities for conducting inspections. These priorities are as follows: imminent danger situations, catastrophic fatal accidents, employee complaints, planned high-hazard inspections, and follow-up inspections.

After being scheduled, the inspection proceeds in the following steps:

1. The OSHA compliance officer presents his or her credentials to a company official.
2. The compliance officer conducts an **opening conference** with pertinent company officials and employee representatives. The following information is explained during the conference: why the plant was selected for inspection, the purpose of the inspection, its scope, and applicable standards.
3. After choosing the route and duration, the compliance officer makes the **inspection tour**. During the tour, the compliance officer may observe, interview pertinent personnel, examine records, take readings, and make photographs.
4. The compliance officer holds a **closing conference**, which involves open discussion between the officer and company and employee representatives. OSHA personnel advise company representatives of problems noted, actions planned as a result, and assistance available from OSHA.

OSHA'S ENHANCED ENFORCEMENT POLICY

Organizations that receive OSHA citations for high-gravity violations are subject to enhanced enforcement measures. Components of the enhanced enforcement policy include follow-up inspections, programmed inspections, public awareness, settlements, and Section 11(b) summary enforcement orders.

Follow-Up Inspections

High-gravity violations include high-gravity willful violations, multiple high-gravity serious violations, repeat violations at the originating establishment, failure-to-abate notices, and serious or willful violations related to a workplace fatality. Organizations that commit any of these types of violations will receive on-site follow-up inspections. In addition, OSHA's area directors are empowered to conduct follow-up inspections to verify compliance if there is reason to suspect that a required abatement has not occurred.

Programmed Inspections

OSHA's *site-specific targeting process* uses objective selection criteria to schedule programmed inspections. As part of this process, OSHA records the name of the overall corporate entity during inspections and prioritizes within its site-specific targeting list all branch and affiliated facilities under the entity's broad corporate umbrella that have received high-gravity violations.

Public Awareness

When an organization receives a high-gravity violation of the type listed above under "Follow-up Inspections," OSHA makes the public aware of the violation and all applicable enforcement actions taken by issuing press releases through local and national media. In addition, high-gravity violations at branch and affiliated facilities are made known to the company's corporate headquarters through official notification by OSHA of the citation and corresponding penalties.

Settlements

Provisions for high-gravity violation settlement agreements

1. Require the organization in question to hire consultants to develop a feasible process for changing the health and safety culture in the facility where the violations occurred
2. Apply the violation settlement not just to the facility in question but also corporatewide
3. Include information about other job sites of the organization in question
4. Require the organization to consent in advance to report all serious injuries and illnesses that require outside medical care and to receive OSHA inspections based on the report
5. Agree to consent to entry of a court enforcement order under Section 11(b) of the OSH Act

Section 11(b) Summary Enforcement Orders

In cases of the kinds of high-gravity violations listed earlier in this section, OSHA applies to the appropriate federal court of appeal for orders that summarily enforce the citations in question under Section 11(b) of the OSH Act. This means that once a Section 11(b) order has been entered, organizations that fail to comply may be held in contempt of court—an action OSHA pursues.

CITATIONS AND PENALTIES

Based on the findings of the compliance officer's workplace inspections, OSHA is empowered to issue citations and/or assess penalties. A citation informs the employer of OSHA violations. Penalties are typically fines assessed as the result of citations. The types of citations and their corresponding penalties are as follows:

- *Other-than-serious violation.* A violation that has a direct relationship to job safety and health, but probably would not cause death or serious physical harm. A **proposed penalty** of up to $7,000 for each violation is discretionary. A penalty for an other-than-serious violation may be adjusted downward by as much as 95 percent, depending on the employer's good faith (demonstrated efforts to comply with the act), history of previous violations, and size of business.
- *Serious violation.* A violation in which there is a high probability that death or serious physical injury may result and that the employer knew or should have known of the hazard. OSHA proposes a mandatory penalty for each serious violation. The actual penalty may be adjusted downward depending on the employer's good faith, history of prior violations, and the gravity of the alleged violation.
- *Willful violation.* A violation that the employer intentionally and knowingly commits. The employer either knows that what he or she is doing constitutes a violation or is aware that a hazardous condition exists and has made no reasonable effort to eliminate it. Penalties may be proposed for each willful violation, with a minimum penalty of $5,000 for each violation. A proposed penalty for a willful violation may be adjusted downward, depending on the size of the business and its history of previous violations. Usually, no credit is given for good faith. If an employer is convicted of a willful violation of a standard that has resulted in the death of an employee, the offense is punishable by a court-imposed fine or by imprisonment for up to six months, or both. A fine of up to $250,000 for an individual or $500,000 for a corporation may be imposed for a criminal conviction.
- *Repeat violation.* A violation of any standard, regulation, rule, or order where, upon reinspection, a substantially similar violation is found. **Repeat violations** can result in a fine for each such violation. To be the basis of a repeat citation, the original citation must be final; a citation under contest may not serve as the basis for a subsequent repeat citation.
- *Failure to abate prior violation.* Failure to correct a prior violation may bring a civil penalty for each day that the violation continues beyond the prescribed abatement date.
- *De minimis violation.* Violations of standards that have no direct bearing on safety and health. **De minimis violations** are documented like any other violation, but they are not included in citations.[14]

In addition to the citations and penalties described in the preceding paragraphs, employers may also be penalized by additional fines and/or prison if convicted of any of the following offenses: (1) falsifying records or any other

Safety Fact

OSHA's Most Frequently Given Citations

Although the order can change from year to year, the following categories of violations typically receive the most OSHA citations in a given year:

- Scaffolding
- Hazard communication
- Fall production
- Respiratory protection
- Lockout/tagout
- Powered industrial trucks (forklifts)
- Electrical wiring methods, components, and equipment
- Machine guarding
- Electrical requirements
- Ladders

information given to OSHA personnel; (2) failing to comply with posting requirements; and (3) interfering in any way with OSHA compliance officers in the performance of their duties.

Examples of OSHA Citations and Fines

Private sector companies in the United States experience more than five million workplace injuries and illness every year. In just the manufacturing and construction sectors alone, OSHA documents more than 75,000 violations annually and issues penalties amounting to more than $70 million. This section provides examples of companies that received citations for various violations and, as a result, were assessed penalties in the form of fines.

Failure to properly use machine guards. A wire manufacturing company was fined $47,700 when an employee who was operating two wire spinning machines was caught around the neck by a loop of wire and choked to death. The company was cited for failure to properly guard machine pulleys, use lockout/tagout procedures, and properly guard rotating machine parts (the spools on the wire spinning machines).[15]

Selected through OSHA's site-specific targeting program. A metal manufacturing company came under OSHA's scrutiny because it had a higher-than-average number of lost day injuries. OSHA conducted an on-site inspection and found numerous violations including the following: failure to remove dangerous electrical wiring, provide personal protection gear and ensure employees wear it, reduce hazardous noise levels, guard machines, train employers on hazardous communication, and properly store oxygen tanks. As a result of these violations, OSHA fined the company $288,000.[16]

Failure to train employees in the application of proper safety procedures.
A paper manufacturing company was fined $258,000 when it allowed
employees to apply improper and unsafe procedures when attempting
to light a boiler. As a result of these improper procedures, the boiler
exploded. Two of the three employees were killed (fire and chemical
burns). The company was also cited for failure to remove debris from
the facility, repair numerous structural hazards (loose bricks, concrete,
and glass), and provide fall protection.[17]

Failure to abate previous hazards. A woodworking company was fined
$27,800 when it failed to correct hazardous situations that had been
previously cited. The conditions that led to the fine included failure
to protect employees from paint and lacquer finish vapors, properly
store flammable materials, fit-test and train employees in the proper
use of respirators, and train employees on fire safety.[18]

APPEALS PROCESS

Employee Appeals

Employees may not contest the fact that citations were or were not awarded or
the amounts of the penalties assessed. However, they may appeal the following
aspects of OSHA's decisions regarding their workplace: (1) the amount of time
(abatement period) given an employer to correct a hazardous condition that
has been cited, and (2) an employer's request for an extension of an abatement
period. Such appeals must be filed within 10 working days of a posting.
Although opportunities for formal appeals by employees are unlimited,
employees may request an informal conference with OSHA officials to discuss
any issue relating to the findings and results of a workplace inspection.

Employer Appeals

Employers may appeal a citation, an abatement period, or the amount of a proposed penalty. Before actually filing an appeal, however, an employer may ask
for an informal meeting with OSHA's area director. The area director is
empowered to revise citations, abatement periods, and penalties in order to
settle disputed claims. If the situation is not resolved through this step,
an employer may formalize the appeal. Formal appeals are of two types:
(1) a petition for modification of abatement, or (2) a notice of contest. The
specifics of both are explained in the following paragraphs.

Petition for Modification of Abatement

The **Petition for Modification of Abatement (PMA)** is available to employers
who intend to correct the situation for which a citation was issued, but who
need more time. As a first step, the employer must make a good-faith effort to

correct the problem within the prescribed timeframe. Having done so, the employer may file a petition for modification of abatement. The petition must contain the following information:

- Descriptions of steps taken so far to comply
- Length of additional time is needed for compliance and why
- Descriptions of the steps being taken to protect employees during the interim
- Verification that the PMA has been posted for employee information and that the employee representative has been given a copy

Notice of Contest

An employer who does not wish to comply may contest a citation, an abatement period, and/or a penalty. The first step is to notify OSHA's area director in writing. This is known as filing a **notice of contest**. It must be done within 15 federal working days of receipt of a citation or penalty notice. The notice of contest must clearly describe the basis for the employer's challenge and contain all of the information about what is being challenged (i.e., amount of proposed penalty or abatement period, and so on).

Once OSHA receives a notice of contest, the area director forwards it and all pertinent materials to the **Occupational Safety and Health Review Commission (OSHRC)**. OSHRC is an independent agency that is associated with neither OSHA nor the Department of Labor. The Department of Labor describes how OSHRC handles an employer's claim:

> The commission assigns the case to an administrative law judge. The judge may disallow the contest if it is found to be legally invalid, or a hearing may be scheduled for a public place near the employer's workplace. The employer and the employees have the right to participate in the hearing; the OSHRC does not require that they be represented by attorneys. Once the administrative law judge has ruled, any party to the case may request further review by OSHRC. Any of the three OSHRC commissioners also may, at his or her own motion, bring a case before the Commission for review. Commission rulings may be appealed to the appropriate U.S. Court of Appeals.[19]

Employer appeals are common. In fact, each issue of *Occupational Hazards* carries a special section titled "Contested Cases." A review of examples of contested cases can be instructive in both process and outcomes. Two such cases follow on the next page.

STATE-LEVEL OSHA PROGRAMS

States are allowed to develop their own safety and health programs.[20] In fact, the OSH Act encourages it. As an incentive, OSHA will fund up to 50 percent of the cost of operating a state program for states with approved plans. States may develop comprehensive plans covering public and private employers or

Case Study

Lorenzo Textile Mill, Inc.

In response to an employee complaint, an OSHA compliance officer inspected Lorenzo Textile Mill, Inc., and cited the company for failure to post a copy of the OSHA cotton dust standard. Lorenzo contested the citation on the following grounds:

- Because the standard is 133 pages in length, it does not lend itself to posting.
- Because of the standard's length, a copy was placed on a table in the employees' lunchroom rather than being posted.
- Each new employee was given a copy of the standard along with an explanation of its contents.[21]

OSHA referred the case to an administrative judge. According to *Occupational Hazards*, the judge "concluded that Lorenzo had done all that could be reasonably expected to keep the standard posted; furthermore, since each employee was provided with a copy, the purpose of the posting was accomplished."[22]

Case Study

IGC Contracting, Inc.

Following an order from his supervisor, a carpenter working for IGC reluctantly removed all the guardrails from stairwell openings, floor openings, and open-side stairs at a Berkeley Heights, New Jersey, job site. Before doing so, he telephoned his union representative, who complained to OSHA. The complaint resulted in an investigation, and IGC was cited for willful violation of three standards. The proposed penalty for each violation was $5,000, for a total penalty of $15,000.[23]

IGC contested the citation, arguing that the foreman who ordered the guardrails removed had done so on his own initiative and that no management official knew of his order. IGC claimed that its superintendent would have rescinded the order if he had known about it. According to *Occupational Hazards*,

> Judge Knight found that the foreman was responsible for the day-to-day work and safety of IGC workers, and that he had exposed IGC and other subcontractors' workers to the danger of death or serious physical harm—in direct violation of OSHA safety standards—by ordering the guardrails removed.[24]

All three of the proposed penalties were upheld. However, the amount of each was reduced by $2,500 for a total penalty of $7,500. The administrative judge was following the guidelines that require consideration of the following factors in assessing penalties: gravity of the violation, company size, and the company's history of good faith. Because IGC is a small company, its penalties were reduced.

Alaska PO Box 21149 Juneau, AL 99802-1149 907-465-2700	Maryland 501 St. Paul Pl., 15th Flr. Baltimore, MD 21202 301-333-4179	Oregon 21 Labor & Ind. Bldg. Salem, OR 97310 503-378-3304
Arizona 800 W. Washington Phoenix, AZ 85007 602-542-5795	Michigan (Labor) 309 N. Washington Sq. PO Box 30015 Lansing, MI 48909 517-335-8022	Puerto Rico Prudencio Rivera Martinez Bldg. 505 Munoz Rivera Ave. Hato Rey, PR 00918 809-654-2119-22
California 395 Oyster Pt.Blvd., 3rd Flr. Wing C San Francisco, CA 94080 415-737-2960	Michigan (Public Health) 3423 N. Logan St., Box 30195 Lansing, MI 48909 517-335-8022	South Carolina PO Box 11329 Columbia, SC 29211-1329 803-734-9594
Connecticut 200 Folly Brook Blvd. Wethersfield, CT 06109 203-566-5123	Minnesota 443 Lafayette Rd. St. Paul, MN 55155 651-296-2342	Vermont 120 State St. Montpelier, VT 05602 802-828-2765
Hawaii 830 Punchbowl St. Honolulu, HI 96813 808-548-3150	Nevada 1370 S. Curry St. Carson City, NV 89710 702-885-5240	Virginia PO Box 12064 Richmond, VA 23241-0064 804-786-2376
Indiana State Office Bldg. 1013 100 N. Senate Ave. Indianapolis, IN 46204-2287 317-232-2665	New Mexico 1190 St. Francis Dr., N2200 Santa Fe, NM 87503-0968 702-885-5240	Virgin Islands Box 890, Christiansted St. Croix, VI 00820 809-773-1994
Iowa 1000 E. Grand Ave. Des Moines, IA 50319 515-281-3447	New York One Main St. Brooklyn, NY 11201 518-457-3518	Washington Gen. Admin. Bldg. Rm. 334-AX-31 Olympia, WA 98504-0631 206-753-6307
Kentucky U.S. Hwy. 127 South Frankfort, KY 40601 502-564-3070	North Carolina 4 W. Edenton St. Raleigh, NC 27601 919-733-7166	

Figure 6–3
States with approved safety and health plans.

limit their plans to coverage of public employers only. In such cases, OSHA covers employers not included in the state plan.

To develop an OSHA-approved safety and health plan, a state must have adequate legislative authority and must demonstrate the ability to develop standards-setting, enforcement, and appeals procedures within three years; public employee protection; a sufficient number of qualified enforcement personnel; and education, training, and technical assistance programs. When a state satisfies all these requirements and accomplishes all developmental steps,

> OSHA then certifies that a state has the legal, administrative, and enforcement means necessary to operate effectively. This action renders no judgement on how well or poorly a state is actually operating its program, but merely attests to the structural completeness of its program. After this certification, there is a period of at least one year to determine if a state is effectively providing safety and health protection.[25]

Figure 6–3 lists the states that currently have OSHA-approved safety and health plans. Connecticut and New York cover the public sector only. The Virgin Islands and Puerto Rico account for 2 of the 23 plans.

SERVICES AVAILABLE FROM OSHA

In addition to setting standards and inspecting for compliance, OSHA provides services to help employers meet the latest safety and health standards. The services are typically offered at no cost. Three categories of services are available from OSHA: consultation, voluntary protection programs, and training and education services.

Consultation Services

Consultation services provided by OSHA include assistance in (1) identifying hazardous conditions; (2) correcting identified hazards; and (3) developing and implementing programs to prevent injuries and illnesses. To arrange consultation services, employers contact the consultation provider in their state (see Figure 6–4).

The actual services are provided by professional safety and health consultants, who are not OSHA employees. They typically work for state agencies or universities and provide consultation services on a contract basis; OSHA provides the funding. OSHA publication 3047, titled *Consultation Services for the Employer*, may be obtained from the nearest OSHA office.

Voluntary Protection Programs

OSHA's **Voluntary Protection Programs (VPPs)** serve the following three basic purposes:

- To recognize companies that have incorporated safety and health programs into their overall management system
- To motivate companies to incorporate health and safety programs into their overall management system

State	Telephone	State	Telephone
Alabama	205-348-3033	Nebraska	401-471-4717
Alaska	907-264-2599	Nevada	701-789-0546
Arizona	602-255-5795	New Hampshire	603-271-3170
Arkansas	501-682-4522	New Jersey	609-984-3517
California	415-557-2870	New Mexico	505-827-2885
Colorado	303-491-6151	New York	518-457-5468
Connecticut	203-566-4550	North Carolina	919-733-3949
Delaware	302-571-3908	North Dakota	701-224-2348
District of Columbia	202-576-6339	Ohio	614-644-2631
Florida	850-488-3044	Oklahoma	405-528-1500
Georgia	404-894-8274	Oregon	503-378-3272
Guam	9-011 671-646-9246	Pennsylvania	800-381-1241 (Toll-free)
Hawaii	808-548-7510		412-357-2561
Idaho	208-385-3283	Puerto Rico	809-754-2134-2171
Illinois	312-917-2339	Rhode Island	401-277-2438
Indiana	317-232-2688	South Carolina	803-734-9579
Iowa	515-281-5352	South Dakota	605-688-4101
Kansas	913-296-4386	Tennessee	615-741-7036
Kentucky	502-564-6895	Texas	512-458-7254
Louisiana	504-342-9601	Utah	801-530-6868
Maine	207-289-6460	Vermont	801-828-2765
Maryland	301-333-4219	Virginia	804-367-1986
Massachusetts	616-727-3463	Virgin Islands	809-772-1315
Michigan	517-335-8250 (Health)	Washington	206-586-0961
	517-322-1814 (Safety)	West Virginia	304-348-7890
Minnesota	612-297-2393	Wisconsin	608-266-8579 (Health)
Mississippi	601-987-3961		414-512-5063 (Safety)
Missouri	314-751-3403	Wyoming	307-777-7786
Montana	406-444-6401		

Figure 6–4
State consultation project directory.

- To promote positive, cooperative relationships among employers, employees, and OSHA

OSHA currently operates three programs under the VPP umbrella. These programs are discussed in the following paragraphs.

Star Program

The **Star Program** recognizes companies that have incorporated safety and health into their regular management system so successfully that their injury rates are below the national average for their industry. This is OSHA's most strenuous program. To be part of the Star Program, a company must demonstrate

- Management commitment
- Employee participation
- An excellent work-site analysis program
- A hazard prevention and control program
- A comprehensive safety and health training program[26]

Merit Program

The **Merit Program** is less strenuous than the Star Program. It is seen as a steppingstone to recognize companies that have made a good start toward Star Program recognition. OSHA works with such companies to help them take the next step and achieve Star Program recognition.

Demonstration Program

The Department of Labor describes the **Demonstration Program** as follows: "for companies that provide Star-quality worker protection in industries where

Safety Fact

Maine OSHA's Top 200 Program

The OSHA of Maine developed one of the most innovative state-level safety programs. Maine's *Top 200 Program* departs from the traditional OSHA approach of *inspect and fine*—an approach that many safety professionals believe encourages businesses to cover up problems instead of identifying and dealing with them. Instead, Maine's OSHA identifies the 200 businesses in the state that record the most injuries each year and makes them an innovative offer. Any company on the list that agrees to identify its own safety problems and work with OSHA officials to correct them is exempt from wall-to-wall inspections. As a result, Maine's business community is identifying and solving an unprecedented number of safety problems. In the eight years prior to implementing the Top 200 Program, a total of 37,000 safety problems had been reported to Maine's OSHA. After just two years of the program, this number had increased to 174,331, and 118,051 of these had been corrected. Along the way, OSHA became a partner to the business community instead of an intrusive arm of big government.

In recognition of the innovative spirit and positive results produced by this initiative, Maine's Top 200 Program received the *Innovations in American Government Award* from the Ford Foundation and the John F. Kennedy School of Government at Harvard University.

certain Star requirements can be changed to include these companies as Star participants."[27]

Companies participating in any of the VPPs are exempt from regular programmed OSHA inspections. However, employee complaints, accidents that result in serious injury, or major chemical releases will be "handled according to routine enforcement procedures."[28]

Training and Education Services

Training and education services available from OSHA take several forms. OSHA operates a training institute in Des Plaines, Illinois, that offers a wide variety of services to safety and health personnel from the public and private sectors. The institute has a full range of facilities including classrooms and laboratories in which it offers more than 60 courses.

To promote training and education in locations other than the institute, OSHA awards grants to nonprofit organizations. Colleges, universities, and other nonprofit organizations apply for funding to cover the costs of providing workshops, seminars, or short courses on safety and health topics currently high on OSHA's list of priorities. Grant funds must be used to plan, develop, and present instruction. Grants are awarded annually and require a match of at least 20 percent of the total grant amount.

EMPLOYER RIGHTS AND RESPONSIBILITIES

OSHA is very specific in delineating the rights and responsibilities of employers regarding safety and health. These rights and responsibilities, as set forth in OSHA publication 2056, are summarized in this section.

Employer Rights

The following is a list of **employer rights** under the OSH Act. Employers have the right to

- Seek advice and consultation as needed by contacting or visiting the nearest OSHA office
- Request proper identification of the OSHA compliance officer prior to an inspection
- Be advised by the compliance officer of the reason for an inspection
- Have an opening and closing conference with the compliance officer in conjunction with an inspection
- Accompany the compliance officer on the inspection
- File a notice of contest with the OSHA area director within 15 working days of receipt of a notice of citation and proposed penalty
- Apply for a temporary variance from a standard if unable to comply because the materials, equipment, or personnel needed to make necessary changes within the required time are not available

- Apply for a permanent variance from a standard if able to furnish proof that the facilities or methods of operation provide employee protection at least as effective as that required by the standard
- Take an active role in developing safety and health standards through participation in OSHA Standards Advisory Committees, through nationally recognized standards-setting organizations, and through evidence and views presented in writing or at hearings
- Be assured of the confidentiality of any trade secrets observed by an OSHA compliance officer during an inspection
- Ask NIOSH for information concerning whether any substance in the workplace has potentially toxic effects[29]

Employer Responsibilities

In addition to the rights set forth in the previous subsection, employers have prescribed responsibilities. The following is a list of **employer responsibilities** under the OSH Act. Employers must

- Meet the general duty responsibility to provide a workplace free from hazards that are causing or are likely to cause death or serious physical harm to employees and to comply with standards, rules, and regulations issued under the OSH Act
- Be knowledgeable of mandatory standards and make copies available to employees for review upon request
- Keep employees informed about OSHA
- Continually examine workplace conditions to ensure that they conform to standards
- Minimize or reduce hazards
- Make sure employees have and use safe tools and equipment (including appropriate personal protective equipment) that is properly maintained
- Use color codes, posters, labels, or signs as appropriate to warn employees of potential hazards
- Establish or update operating procedures and communicate them so that employees follow safety and health requirements
- Provide medical examinations when required by OSHA standards
- Provide the training required by OSHA standards
- Report to the nearest OSHA office within eight hours any fatal accident or one that results in the hospitalization of three or more employees
- Keep OSHA-required records of injuries and illnesses and post a copy of OSHA Form 300 from February 1 through April 30 each year (this applies to employers with 11 or more employees)
- At a prominent location within the workplace, post OSHA Poster 2203 informing employees of their rights and responsibilities
- Provide employees, former employees, and their representatives access to the Log of Work-Related Injuries and Illnesses (OSHA Form 300) at a reasonable time and in a reasonable manner

- Give employees access to medical and exposure records
- Give the OSHA compliance officer the names of authorized employee representatives who may be asked to accompany the compliance officer during an inspection
- Not discriminate against employees who properly exercise their rights under the act
- Post OSHA citations at or near the work site involved (each citation or copy must remain posted until the violation has been abated or for three working days, whichever is longer)
- Abate cited violations within the prescribed period[30]

EMPLOYEE RIGHTS AND RESPONSIBILITIES

Employee Rights

Section 11(c) of the OSH Act delineates **employee rights**. These rights are actually protection against punishment for employees who exercise their right to pursue any of the following courses of action:

- Complain to an employer, union, OSHA, or any other government agency about job safety and health hazards
- File safety or health grievances
- Participate in a workplace safety and health committee or in union activities concerning job safety and health
- Participate in OSHA inspections, conferences, hearings, or other OSHA-related activities[31]

Employees who feel they are being treated unfairly because of actions they have taken in the interest of safety and health have 30 days in which to contact the nearest OSHA office. Upon receipt of a complaint, OSHA conducts an investigation and makes recommendations based on its findings. If an employer refuses to comply, OSHA is empowered to pursue legal remedies at no cost to the employee who filed the original complaint.

In addition to those just set forth, employees have a number of other rights. Employees may

- Expect employers to make review copies available of OSHA standards and requirements
- Ask employers for information about hazards that may be present in the workplace
- Ask employers for information on emergency procedures
- Receive safety and health training
- Be kept informed about safety and health issues
- Anonymously ask OSHA to conduct an investigation of hazardous conditions at the work site

- Be informed of actions taken by OSHA as a result of a complaint
- Observe during an OSHA inspection and respond to the questions asked by a compliance officer
- See records of hazardous materials in the workplace
- See their medical record
- Review the annual Log of Work-Related Injuries and Illnesses (OSHA Form 300)
- Have an exit briefing with the OSHA compliance officer following an OSHA inspection
- Anonymously ask NIOSH to provide information about toxicity levels of substances used in the workplace
- Challenge the abatement period given employers to correct hazards discovered in an OSHA inspection
- Participate in hearings conducted by the Occupational Safety and Health Review Commission
- Be advised when an employer requests a variance to a citation or any OSHA standard
- Testify at variance hearings
- Appeal decisions handed down at OSHA variance hearings
- Give OSHA input concerning the development, implementation, modification, and/or revocation of standards[32]

Employee Responsibilities

Employees have a number of specific responsibilities. The following list of **employee responsibilities** is adapted from OSHA 2056, 1991 (Revised). Employees must

- Read the OSHA poster at the job site and be familiar with its contents
- Comply with all applicable OSHA standards
- Follow safety and health rules and regulations prescribed by the employer and promptly use personal protective equipment while engaged in work
- Report hazardous conditions to the supervisor
- Report any job-related injury or illness to the employer and seek treatment promptly
- Cooperate with the OSHA compliance officer conducting an inspection
- Exercise their rights under the OSH Act in a responsible manner[33]

KEEPING UP-TO-DATE ON OSHA

OSHA's standards, rules, and regulations are always subject to change. The development, modification, and revocation of standards is an ongoing process. It is important for prospective and practicing safety and health professionals to stay up-to-date with the latest actions and activities of OSHA. Following is an annotated list of strategies that can be used to keep current:

- Establish contact with the nearest regional or area OSHA office and periodically request copies of new publications or contact the OSHA Publications Office at the following address:

  OSHA Publications Office
  200 Constitution Ave. NW
  Room N–3101
  Washington, DC 20210
  http://www.osha.gov/pls/publications/pubindex.list

- Review professional literature in the safety and health field. Numerous periodicals carry OSHA updates that are helpful.
- Establish and maintain relationships with other safety and health professionals for the purpose of sharing information, and do so frequently.
- Join professional organizations, review their literature, and attend their conferences.

PROBLEMS WITH OSHA

Federal agencies are seldom without their detractors. Resentment of the federal bureaucracy is intrinsic in the American mind-set. Consequently, complaints about OSHA are common. Even supporters occasionally join the ranks of the critics. Often, the criticisms leveled against OSHA are valid.

Criticisms of OSHA take many different forms. Some characterize OSHA as an overbearing bureaucracy with little or no sensitivity to the needs of employers who are struggling to survive in a competitive marketplace. At the same time, others label OSHA as timid and claim it does not do enough. At different times and in different cases, both points of view have probably been at least partially accurate.

Most criticism of OSHA comes in the aftermath of major accidents or a workplace disaster. Such events typically attract a great deal of media attention, which, in turn, draws the attention of politicians. In such cases, the criticism tends to focus on the question, "Why didn't OSHA prevent this disaster?" At congressional hearings, detractors will typically answer this question by claiming that OSHA spends too much time and too many resources dealing with matters of little consequence while ignoring real problems. Supporters of OSHA will typically answer the question by claiming that a lack of resources prevents the agency from being everywhere at once. There is a measure of validity in both answers.

On one hand, OSHA has made a significant difference in the condition of the workplace in this country. On the other hand, large, centralized bureaucratic agencies rarely achieve a high level of efficiency. This is compounded in OSHA's case by the fact that the agency is subject to the ebb and flow of congressional support, particularly in the area of funding. Consequently, OSHA is likely to continue to be an imperfect organization subject to ongoing criticism.

OTHER AGENCIES AND ORGANIZATIONS

Although OSHA is the most widely known safety and health organization in the federal government, it is not the only one. Figure 6–5 lists associations, agencies, and organizations (including OSHA) with safety and health as part of their mission.

Of those listed, the most important to modern safety and health professionals are NIOSH and OSHRC. The missions of these organizations are summarized in the following paragraphs.

Figure 6–5
Agencies, organizations, and associations dealing with safety and health.

American Public Health Association
1015 15th St. NW
Washington, DC 20005

Bureau of Labor Statistics
U.S. Department of Labor
Washington, DC 20212

Bureau of National Affairs, Inc.
Occupational Safety and Health Reporter
1231 25th St. NW
Washington, DC 20037

Commerce Clearing House
Employee Safety and Health Guide
4205 W. Peterson Ave.
Chicago, IL 60646

Environmental Protection Agency
401 M St. SW
Washington, DC 20001

Mine Safety and Health Administration
4015 Wilson Blvd.
Rm. 601
Arlington, VA 22203

National Institute for Occupational Safety and
   Health (NIOSH)
4676 Columbia Pkwy.
Cincinnati, OH 45226

Occupational Safety and Health Administration
U.S. Department of Labor
200 Constitution Ave.
Washington, DC 20210

Occupational Safety and Health Review
   Committee (OSHRC)
Washington, DC 20210

U.S. Consumer Product Safety Commission
Washington, DC 20207

NIOSH

The **National Institute for Occupational Safety and Health (NIOSH)** is part of the Department of Health and Human Services (HHS). (Recall that OSHA is part of the Department of Labor.) NIOSH has two broad functions: research and education. The main focus of the agency's research is on toxicity levels and human tolerance levels of hazardous substances. NIOSH prepares recommendations along these lines for OSHA standards dealing with hazardous substances. NIOSH studies are also published and made available to employers. Each year, NIOSH publishes updated lists of toxic materials and recommended tolerance levels. These publications represent the educational component of NIOSH's mission.

The Department of Health and Human Services describes NIOSH as follows:

> In 1973, NIOSH became a part of the Centers for Disease Control (CDC), an arm of the Public Health Service in the Department of Health and Human Services. NIOSH is unique among federal research institutions because it has the authority to conduct research in the workplace, and to respond to requests for assistance from employers and employees.

NIOSH also consults with the Department of Labor (DOL) and other federal, state, and local government agencies to promote occupational safety and health and makes recommendations to DOL about worker exposure limits.

NIOSH estimates that over 7,000 people are killed at work each year, and nearly 12 million nonfatal injuries occur in the workplace. In addition to death and injury, it is estimated that over 10 million men and women are exposed to hazardous substances in their jobs that can eventually cause fatal or debilitating diseases. To help establish priorities in developing research and control of these hazards, NIOSH developed a list of the 10 leading work-related diseases and injuries:

1. Occupational lung diseases
2. Musculoskeletal injuries
3. Occupational cancers
4. Occupational cardiovascular disease
5. Severe occupational traumatic injuries
6. Disorders of reproduction
7. Neurotoxic disorders
8. Noise-induced hearing loss
9. Dermatological conditions
10. Psychological disorders[34]

Figure 6–6 is an organizational chart showing the major divisions in NIOSH. The four principal divisions are described in the following paragraphs.

Division of Biomedical and Behavioral Science

The **Division of Biomedical and Behavioral Science (DBBS)** conducts research in the areas of toxicology, behavioral science, ergonomics, and the health consequences of various physical agents.[35] DBBS investigates problems created by

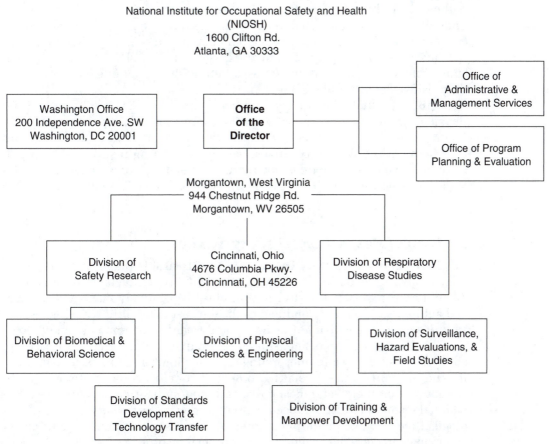

Figure 6–6
The major divisions of NIOSH.

new technologies and develops biological monitoring and diagnostic aids to ensure that the workplace is not responsible for diminished health, functional capacity, or life expectancy of workers. Consultation and data are furnished for developing the criteria for workplace exposure standards.

DBBS conducts laboratory and work-site research into the psychological, behavioral, physiological, and motivational factors relating to job stress as well as those induced by chemical and physical agents. The division assesses physical work capacity and tolerance for environmental conditions as influenced by age, gender, body type, and physical fitness. Interventions and control procedures using various approaches are developed and tested for their ability to reduce undue job stress.

The DBBS toxicology research involves dose-response and methods development studies to evaluate the effects of toxic agents and to develop techniques that utilize biomarkers as indicators of toxic exposures and early indicators of pathological change. The studies include applications of cellular biology, immunochemistry, pharmacokinetics, and neurophysiology. Through laboratory analysis of biological samples from animals exposed experimentally and humans exposed occupationally to workplace hazards, DBBS provides clinical and biochemical consultations for ascertaining the extent of exposure and for diagnosing occupational diseases.

DBBS conducts laboratory and work-site research on hazards from physical agents such as noise, vibration, and nonionizing energy sources. Studies seek to identify exposure factors that are significant to the health and well-being of the workforce. Investigations, including instrumentation and methods development, for characterizing and relating exposure factors to biological and performance changes in animal and human populations are also undertaken.

Division of Respiratory Disease Studies

The **Division of Respiratory Disease Studies (DRDS)** is the focal point for the clinical and epidemiological research that NIOSH conducts on occupational respiratory diseases.[36] The division provides legislatively mandated medical and autopsy services and conducts medical research to fulfill NIOSH's responsibilities under the Federal Mine Safety and Health Amendments Act of 1977.

The division conducts field studies of occupational respiratory diseases, in addition to designing and interpreting cross-sectional and prospective morbidity and mortality studies relating to occupational respiratory disease. Field studies are conducted at mines, mills, and other industrial plants where occupational respiratory diseases occur among workers. The division uses epidemiological techniques, including studies of morbidity and mortality, to detect common characteristics related to occupational respiratory diseases.

To formulate and implement programs that will identify factors involved in the early detection and differential rates of susceptibility to occupational respiratory diseases, DRDS conducts cell biology research to determine the role of microorganisms and environmental exposure in these diseases. The division also provides autopsy evaluations and a pathology research program. Research is conducted on immunological mechanisms and cell physiology to determine the effects of environmental exposure as it relates to occupational respiratory diseases.

DRDS provides for planning, coordinating, and processing medical examinations mandated under the Federal Mine Safety and Health Amendments Act of 1977 and operates a certification program for medical facilities and physicians who participate in the examination program. DRDS also evaluates and approves employer programs for the examination of employees in accordance with published regulations as well as arranging for the examination of employees who work at locations that lack an approved examination program. The division conducts the National Coal Workers' Autopsy Program and performs research into the postmortem identification and quantification of occupational respiratory exposures.

Division of Surveillance, Hazard Evaluations, and Field Studies

The **Division of Surveillance, Hazard Evaluations, and Field Studies (DSHEFS)** conducts surveillance of the nation's workforce and workplaces to assess the magnitude and extent of job-related illnesses, exposures, and hazardous agents.[37] DSHEFS conducts legislatively mandated health hazard evaluations and industrywide epidemiological research programs, including longitudinal studies of records and clinical and environmental field studies and surveys. DSHEFS also provides technical assistance on occupational safety and health problems to other federal, state, and local agencies; other technical groups; unions; employers; and employees.

Surveillance efforts are designed for the early detection and continuous assessment of the magnitude and extent of occupational illnesses, disabilities, deaths, and exposures to hazardous agents, using new and existing data sources from federal, state, and local agencies; labor; industry; tumor registries; physicians; and medical centers. DSHEFS also conducts evaluation and validation studies of reporting systems covering occupational illnesses, with the intent of improving methods for measuring the magnitude of occupational health problems nationwide.

Division of Training and Manpower Development

The **Division of Training and Manpower Development (DTMD)** implements Section 21 of the OSH Act, which sets forth training and education requirements.[38] DTMD develops programs to increase the numbers and competence of safety and health professionals. This continuing education program provides short-term technical training courses including seminars, independent study packages, and specialized workshops to federal, state, and local government; private industry; labor unions; and other organizations in the field of safety and health. The curriculum development component develops courseware and other training materials for NIOSH-sponsored training programs, including those represented by in-house faculty as well as those conducted by universities and other outside training organizations.

The educational resource development program continually assesses human resource needs for safety and health practitioners and researchers on a nationwide basis. To help meet the demand, DTMD administers a major training grant program to foster the development of academically based training programs for occupational physicians, occupational health nurses, industrial hygienists, toxicologists, epidemiologists, and safety professionals. DTMD also develops specific criteria for the selection of qualified organizations to conduct research training, graduate education, continuing education, and outreach programs to expand the network of knowledgeable professionals in occupational safety and health.

DTMD initiates special-emphasis projects targeted to physicians, engineers, managers, vocational education students, and science teachers to facilitate the inclusion of occupational safety and health knowledge in their formal program of study. The division establishes a collaborating relationship with the many professional societies and accrediting bodies to formalize the process of long-term commitment through professional networking.

OSHRC

The **Occupational Safety and Health Review Commission (OSHRC)** is not a government agency. Rather, it is an independent board whose members are appointed by the president and given quasi-judicial authority to handle contested OSHA citations. When a citation, proposed penalty, or abatement period issued by an OSHA area director is contested by an employer, OSHRC hears the case. OSHRC is empowered to review the evidence, approve the recommendations of the OSHA area director, reject those recommendations, or revise them by assigning substitute values. For example, if an employer contests the amount of a proposed penalty, OSHRC is empowered to accept the proposed amount, reject it completely, or change it.

Mine Safety and Health Administration (MSHA)

The mining industry is exempt from OSHA regulations. Instead, mining is regulated by the Metal and Non-Metallic Mine Safety Act. OSHA does regulate those aspects of the industry that are not directly involved in actual mining work. There is a formal memorandum of understanding between OSHA and the Mine Safety and Health Administration (MSHA), the agency that enforces the Metal and Non-Metallic Mine Safety Act.

In 1977, Congress passed the Mine Safety and Health Act, which established the Mine Safety and Health Administration (MSHA) as a functional unit within the U.S. Department of Labor. MSHA works with MESA to ensure that the two agencies do not become embroiled in jurisdictional disputes.

Federal Railroad Administration

Railroads, for the most part, fall under the jurisdiction of OSHA. The Federal Railroad Administration (FRA) exercises limited jurisdiction over railroads in situations involving working conditions. Beyond this, railroads must adhere to the standards for *General Industry* in CFR Part 1910. OSHA and FRA personnel coordinate to ensure that jurisdictional disputes do not arise.

OSHA'S GENERAL INDUSTRY STANDARDS

The most widely applicable OSHA standards are the *General Industry Standards.* These standards are found in 29 CFR 1910. Part 1910 consists of 21 subparts, each carrying an uppercase-letter designation. Subparts A and B contain no compliance requirements. The remaining subparts are described in the following subsections.

Discussion Case

What Is Your Opinion?

"The OSH Act is a nightmare! All it has accomplished is the creation of a department full of governmental bureaucrats who bully private industry." This was the opening line in a debate on government safety regulations sponsored by the Industrial Technology Department of Pomona State University. The OSHA advocate in the debate responded as follows: "The OSH Act is a model of government regulation as it should be. Had private industry been responsive to the safety and health concerns of employees, there would have been no need for government regulation." These are two widely divergent viewpoints. What is your opinion in this matter?

Subpart C: General Safety and Health Provisions

The only compliance standard in Subpart C is *Access to Employee Exposure and Medical Records.* Employers that are required to keep medical and exposure records must do the following: (1) maintain the records for the duration of employment plus 30 years, and (2) give employees access to their individual personal records.

Subpart D: Walking–Working Surfaces

Subpart D contains the standards for all surfaces on which employees walk or work. Specific sections of Subpart D are as follows:

1910.21	Definitions
1910.22	General requirements
1910.23	Guarding floor and wall openings and holes
1910.24	Fixed industrial stairs
1910.25	Portable wood ladders
1910.26	Portable metal ladders
1910.27	Fixed ladders
1910.28	Safety requirements for scaffolding
1910.29	Manually propelled mobile ladder stands and scaffolds (towers)
1910.30	Other working surfaces

Subpart E: Means of Egress

Subpart E requires employers to ensure that employees have a safe, accessible, and efficient means of escaping a building under emergency circumstances. Specific sections of Subpart E are as follows:

1910.35	Definitions
1910.36	General requirements

1910.37	Maintenance safeguards and operational features for exit routes
1910.38	Emergency action plan
1910.39	Fire protection plan

Emergency circumstances may be caused by fire, explosions, hurricanes, tornadoes, flooding, terrorist acts, earthquakes, nuclear radiation, or other acts of nature not listed here.

Subpart F: Powered Platforms

Subpart F applies to powered platforms, mechanical lifts, and vehicle-mounted work platforms. The requirements of this subpart apply only to employers who use this type of equipment in facility maintenance operations. Specific sections of Subpart F are as follows:

1910.66	Powered platforms for building maintenance
1910.67	Vehicle-mounted elevating and rotating work platforms
1910.68	Manlifts

Subpart G: Health and Environmental Controls

The most widely applicable standard in Subpart G is 1910.95 (occupational noise exposure). Other standards in this subpart pertain to situations where ionizing and/or nonionizing radiation are present. Specific sections of Subpart G are as follows:

1910.94	Ventilation
1910.95	Occupational noise exposure
1910.96	Ionizing radiation
1910.97	Nonionizing radiation
1910.98	Effective dates

Subpart H: Hazardous Materials

Four of the standards in Subpart H are widely applicable. Section 1901.106 is an extensive standard covering the use, handling, and storage of flammable and combustible liquids. Of particular concern are fire and explosions. Section 1910.107 applies to indoor spray-painting processes and processes in which paint (powder coating) is applied in powder form (e.g., electrostatic powder spray).

Section 1910.119 applies to the management of processes involving specifically named chemicals and flammable liquids and gases. Section 1910.120 contains requirements relating to emergency response operations and hazardous waste. All the standards contained in Subpart H are as follows:

| 1910.101 | Compressed gases (general requirements) |
| 1910.102 | Acetylene |

1910.103	Hydrogen
1910.104	Oxygen
1910.105	Nitrous oxide
1910.106	Flammable and combustible liquids
1910.107	Spray finishing using flammable and combustible materials
1910.108	Dip tanks containing flammable and combustible materials
1910.109	Explosive and blasting agents
1910.110	Storage and handling of liquefied petroleum gases
1910.111	Storage and handling of anhydrous ammonia
1910.119	Process safety management of highly hazardous chemicals
1910.120	Hazardous waste operations and emergency response

Subpart I: Personal Protective Equipment

Subpart I contains three of the most widely applicable standards: 1910.132 General Requirements; 1910.133 Eye and Face Protection; and 1910.134 Respiratory Protection. The most frequently cited OSHA violations relate to these and the other personal protective equipment standards. All the standards in this subpart are as follows:

1910.132	General requirements
1910.133	Eye and face protection
1910.134	Respiratory protection
1910.135	Occupational head protection
1910.136	Occupational foot protection
1910.137	Electrical protective devices
1910.138	Hand protection

Subpart J: General Environment Controls

This subpart contains standards that are widely applicable because they pertain to general housekeeping requirements. An especially important standard contained in this subpart is 1910.146: Permit-Required Confined Spaces. A confined space is one that meets any or all of the following criteria:

- Large enough and so configured that a person can enter it and perform assigned work tasks therein
- Continuous employee occupancy is not intended

The *lockout/tagout* standard is also contained in this subpart. All the standards in this subpart are as follows:

1910.141	Sanitation
1910.142	Temporary labor camps
1910.144	Safety color code for marking physical hazards
1910.145	Accident prevention signs and tags
1910.146	Permit-required confined space
1910.147	Control of hazardous energy (lockout/tagout)
1910.148	Standards organizations
1910.149	Effective dates

Subpart K: Medical and First Aid

This is a short subpart, the most important section of which pertains to eye-flushing. If employees are exposed to *injurious corrosive materials*, equipment must be provided for quickly flushing the eyes and showering the body. The standard also requires medical personnel to be readily available. "Readily available" can mean that there is a clinic or hospital nearby. If such a facility is not located nearby, employers must have a person on hand who has had first-aid training. The standard in this subpart is

1910.151	Medical seminars and first aid

Subpart L: Fire Protection

This subpart contains the bulk of OSHA's fire protection standard. These standards detail the employer's responsibilities concerning fire brigades, portable fire-suppression equipment, fixed fire-suppression equipment, and fire-alarm systems. Employers are not required to form fire brigades, but if they choose to, employers must adhere to the standard set forth in 1910.156. The standards in this subpart are as follows:

Fire Protection

1910.155	Scope, application, and definitions applicable to this subpart
1910.156	Fire brigades

Portable Fire-Suppression Equipment

1910.157	Portable fire extinguishers
1910.158	Standpipe and hose systems

Fixed Fire-Suppression Equipment

1910.159	Automatic sprinkler systems
1910.160	Fixed extinguishing systems, general

1910.161	Fixed extinguishing systems, dry chemical
1910.162	Fixed extinguishing systems, gaseous agent
1910.163	Fixed extinguishing systems, water spray and foam

Other Fire Protection Systems

| 1910.164 | Fire detection systems |
| 1910.165 | Employee alarm systems |

Subpart M: Compressed Gas/Air

This subpart contains just three sections and only one standard, 1910.169. This standard applies to compressed-air equipment that is used in drilling, cleaning, chipping, and hoisting. There are many other uses of compressed air in work settings, but 1910.169 applies only to these applications. The standard in this subpart is

| 1910.169 | Air receivers |

Subpart N: Materials Handling and Storage

This subpart contains one broad standard—1910.176—and several more specific standards relating to 1910.176. Subpart N is actually limited in scope. It applies only to the handling and storage of materials, changing rim wheels on large vehicles, and the proper use of specific equipment identified in the standards' titles. All the standards in this subpart are as follows:

1910.176	Handling materials—general
1910.177	Servicing multi-piece and single-piece rim wheels
1910.178	Powered industrial trucks
1910.179	Overhead and gantry cranes
1910.180	Crawler locomotive and truck cranes
1910.181	Derricks
1910.183	Helicopters
1910.184	Slings

Subpart O: Machinery and Machine Guarding

This subpart contains standards relating to specific types of machines. The types of machines covered are identified in the titles of the standards contained in Subpart O. These standards are as follows:

| 1910.211 | Definitions |
| 1910.212 | General requirements for all machines |

1910.213	Woodworking machinery requirements
1910.214	Cooperage machinery
1910.215	Abrasive wheel machinery
1910.216	Mills and calendars in the rubber and plastics industries
1910.217	Mechanical power presses
1910.218	Forging machines
1910.219	Mechanical power-transmission apparatus

Subpart P: Hand Tools/Portable Power Tools

This subpart contains standards relating to the use of hand tools, portable power tools, and compressed-air-powered tools. The types of tools covered in this subpart, in addition to typical hand tools, include jacks, saws, drills, sanders, grinders, planers, power lawnmowers, and other tools. The standards contained in this subpart are as follows:

1910.241	Definitions
1910.242	Hand- and portable-powered tools and equipment, general
1910.243	Guarding of portable tools and equipment
1910.244	Other portable tools and equipment

Subpart Q: Welding, Cutting, and Brazing

Welding, cutting, and brazing are widely used processes. This subpart contains the standards relating to these processes in all their various forms. The primary safety and health concerns are fire protection, employee personal protection, and ventilation. The standards contained in this subpart are as follows:

1910.251	Definitions
1910.252	General requirements
1910.253	Oxygen-fuel gas welding and cutting
1910.254	Arc welding and cutting
1910.255	Resistance welding
1910.256	Sources of standards
1910.257	Standards organizations

Subpart R: Special Industries

This subpart is different from others in Part 1910. Whereas other subparts deal with specific processes, machines, and materials, Subpart R deals with specific industries. Each separate standard relates to a different category of industry. The standards contained in this subpart are as follows:

1910.261	Pulp, paper, and paperboard mills
1910.262	Textiles
1910.263	Bakery equipment
1910.264	Laundry machinery and operations
1910.265	Sawmills
1910.266	Pulpwood logging
1910.268	Telecommunications
1910.272	Grain handling facilities

Subpart S: Electrical

This subpart contains standards divided into the following two categories: (1) design of electrical systems, and (2) safety-related work practices. These standards are excerpted directly from the National Electrical Code. Those included in Subpart S are as follows:

1910.301	Introduction
1910.302	Electric utilization systems
1910.303	General requirements
1910.304	Wiring design and protection
1910.305	Wiring methods, components, and equipment for general use
1910.306	Specific-purpose equipment and installations
1910.307	Hazardous (classified) locations
1910.308	Special systems
1910.331	Scope
1910.332	Training
1910.333	Selection and use of work practices
1910.334	Use of equipment
1910.335	Safeguards for personnel protection
1910.399	Definitions applicable to this subpart

Subpart T: Commercial Diving Operations

This subpart applies only to commercial diving enterprises. The standards contained in Subpart T are divided into six categories: (1) general, (2) personnel requirements, (3) general operations and procedures, (4) specific operations and procedures, (5) equipment procedures and requirements, and (6) record keeping. These standards are as follows:

1910.401	Scope and application
1910.402	Definitions
1910.410	Qualifications of dive teams
1910.420	Safe practices manual
1910.421	Pre-dive procedures
1910.422	Procedures during dive
1910.423	Post-dive procedures
1910.424	SCUBA diving
1910.425	Surface-supplied-air diving
1910.426	Mixed-gas diving
1910.427	Liveboating
1910.430	Equipment
1910.440	Record keeping

Subpart Z: Toxic and Hazardous Substances

This is an extensive subpart containing the standards that establish *permissible exposure limits* (PELs) for selected toxic and hazardous substances (those for which PELs have been established). All such substances have an assigned PEL which is the amount of a given airborne substance to which employees can be exposed during a specified period of time.

The standards relating to these specific toxic and hazardous substances are contained in 1910.1000 through 1910.1450 and are as follows:

1910.1000	Air contaminants
1910.1001	Asbestos
1910.1002	Coal tar pitch volatiles; interpretation of term
1910.1003	4-Nitrobiphenyl
1910.1004	Alpha-Naphthylamine
1910.1006	Methyl chloromethyl ether
1910.1007	Dichlorobenzidine (and its salts)
1910.1008	Bis-Chloromethyl ether
1910.1009	Beta-Naphthylamine
1910.1010	Benzedrine
1910.1011	4-Aminodiphenyl
1910.1012	Ethyleneimine
1910.1013	Beta-Propiolactone

1910.1014	2-Acetylaminofluorene
1910.1015	4-Dimethylaminoazobenzene
1910.1016	N-Nitrosodimethylamine
1910.1017	Vinyl chloride
1910.1018	Inorganic arsenic
1910.1020	Access to employee exposure and medical records
1910.1025	Lead
1910.1026	Chromium
1910.1027	Cadmium
1910.1028	Benzene
1910.1029	Coke oven emissions
1910.1030	Bloodborne pathogens
1910.1043	Cotton dust
1910.1044	1,2-dibromo-3-chloropropane
1910.1045	Acrylonitrile
1910.1047	Ethylene oxide
1910.1050	Methylenedianiline
1910.1051	Butadiene
1910.1052	Methylene chloride
1910.1096	Ionizing radiation
1910.1200	Hazard communication
1910.1201	Retention of DOT markings, placards, and labels
1910.1450	Occupational exposure to hazardous chemicals in laboratories

OSHA's General Industry Standards were covered in some depth in this section because they have the broadest application for students of workplace safety. The CFR also contains standards for the maritime industry. These standards are described in the sections that follow, but in less detail than was devoted to OSHA's General Standards because they are not as widely applicable.

OSHA'S MARITIME STANDARDS

OSHA's Maritime Standards apply to shipbuilding, ship-repairing, and ship-breaking operations not already covered by U.S. Coast Guard regulations. It is important to note that Coast Guard regulations do take precedence over OSHA Maritime Standards and supersede those standards in cases of overlap or conflict. Part 1915 of 29 CFR contains the standards relating to shipbuilding, ship repairing, and ship breaking. Part 1917 contains the standards for marine terminals, and Part 1918 contains longshoring standards. Part 1919 contains the gear-certification standards.

Part 1915: Shipyard Employment

Part 1915 is divided into seven subparts covering the various aspects of shipyard operations. Those subparts are as follows:

1915.1	Purpose and authority
1915.2	Scope and application
1915.3	Responsibility
1915.4	Definitions
1915.5	Reference specifications, standards, and codes
1915.6	Commercial diving operations
1915.7	Competent person

Part 1917: Marine Terminals

Part 1917 is divided into seven subparts covering the various operations associated with marine terminals. Those subparts are as follows:

Subpart A	Scope and definitions
Subpart B	Marine terminal operations
Subpart C	Cargo handling gear and equipment
Subpart D	Specialized terminals
Subpart E	Personal protection
Subpart F	Terminal facilities
Subpart G	Related terminal operations and equipment

Part 1918: Longshoring

Longshoring involves loading and unloading maritime vessels. Part 1918 contains the standards relating to longshoring and is subdivided into the following subparts:

Subpart A	General provisions
Subpart B	Gangways and gear certification
Subpart C	Means of access
Subpart D	Working surfaces

Subpart E	Opening and closing hatches
Subpart F	Ship's cargo-handling gear
Subpart G	Cargo-handling gear and equipment other than ship's gear
Subpart H	Handling cargo
Subpart I	General working conditions
Subpart J	Personal protective equipment

Gear Certification

Cargo gear- and material-handling devices must be certified as being safe for use in the handling, loading, unloading, and transport of materials. Persons who certify the gear and devices must be properly accredited. Part 1919 contains the following standards regulating the accreditation process:

Subpart C	General safety and health provisions
Subpart D	Occupational health and environment controls
Subpart E	Personal protective and life-saving equipment
Subpart F	Fire protection and prevention
Subpart G	Signs, signals, and barricades
Subpart H	Materials handling, storage, use, and disposal
Subpart I	Tools (hand and power)
Subpart J	Welding and cutting
Subpart K	Electrical
Subpart L	Scaffolding
Subpart M	Floor and wall openings
Subpart N	Cranes, derricks, hoists, elevators, and conveyors
Subpart O	Motor vehicles, mechanized equipment, and marine vessels
Subpart P	Excavations
Subpart Q	Concrete and masonry construction
Subpart S	Underground construction, caissons, and cofferdams
Subpart T	Demolition
Subpart U	Blasting and the use of explosives
Subpart V	Power transmission and distribution
Subpart W	Rollover protective structures; overhead protection
Subpart X	Stairways and ladders
Subpart Y	Commercial diving operations
Subpart Z	Toxic and hazardous substances

OSHA'S CONSTRUCTION STANDARDS

These standards apply to employers involved in construction, alteration, or repair activities. To further identify the scope of the applicability of its construction standards, OSHA took the terms *construction, alteration*, and *repair* directly from the Davis-Bacon Act. This act provides minimum wage protection for employees working on construction projects. The implication is that if the Davis-Bacon Act applies to an employer, OSHA's Construction Standards also apply.[39]

These standards are contained in Part 1926 of the CFR Subparts A–Z. OSHA does not base citations on material contained in Subparts A and B. Consequently, those subparts have no relevance here. The remaining subparts are as follows:

Subpart C	General safety and health provisions
Subpart D	Occupational health and environmental controls
Subpart E	Personal protective and life-saving equipment
Subpart F	Fire protection and prevention
Subpart G	Signs, signals, and barricades
Subpart H	Materials handling, storage, use, and disposal
Subpart I	Tools—hand and power
Subpart J	Welding and cutting
Subpart K	Electrical
Subpart L	Scaffolding
Subpart M	Fall protection
Subpart N	Cranes, derricks, hoists, elevators, and conveyors
Subpart O	Motor vehicles, mechanized equipment, and marine operations
Subpart P	Excavations
Subpart Q	Concrete and masonry construction
Subpart R	Steel erection
Subpart S	Underground construction, caissons, and cofferdams
Subpart T	Demolition
Subpart U	Blasting and the use of explosives
Subpart V	Power transmission and distribution
Subpart W	Rollover protective structures; overhead protection
Subpart X	Ladders
Subpart Y	Commercial diving operations
Subpart Z	Toxic and hazardous substances

STANDARDS AND CODES

A **standard** is an operational principle, criterion, or requirement—or a combination of these. A **code** is a set of standards, rules, or regulations relating to a specific area. Standards and codes play an important role in modern safety and health management and engineering. These written procedures detail the safe and healthy way to perform job tasks and, consequently, to make the workplace safer and healthier.

Having written standards and codes that employees carefully follow can also decrease a company's exposure to costly litigation. Courts tend to hand down harsher rulings to companies that fail to develop or adapt, implement, and enforce appropriate standards and codes. Consequently, safety and health professionals should be familiar with the standards and codes relating to their company.

Numerous organizations develop standards for different industries. These organizations can be categorized broadly as follows: the government, professional organizations, and technical/trade associations.

Organizations that fall within these broad categories develop standards and codes in a wide variety of areas including, but not limited to, the following: dust hazards, electricity, emergency electricity systems, fire protection, first aid, hazardous chemicals, instrumentation, insulation, lighting, lubrication, materials, noise/vibration, paint, power, wiring, pressure relief, product storage and handling, piping materials, piping systems, radiation exposure, safety equipment, shutdown systems, and ventilation.

Figure 6–7 is a list of organizations that develop and publish standards and codes covering a wide variety of fields and areas of concern. Figure 6–8 is a reference for applicable standards and codes providers in various areas of concern (see Figure 6–7 to decode the abbreviations).

LAWS AND LIABILITY

The body of law pertaining to workplace safety and health grows continually as a result of a steady stream of liability litigation. Often a company's safety and health professionals are key players in litigation alleging negligence on the part of the company when an accident or health problem has occurred. Because safety and health litigation has become so prevalent today, professionals in the field need to be familiar with certain fundamental legal principles relating to such litigation. These principles are explained in the following section.

Fundamental Legal Principles

The body of law that governs safety and health litigation evolves continually. However, even though cases that set new precedents and establish new principles continue to occur, a number of fundamental legal principles surface

Organization	Abbreviation
U.S. Government	
Department of Transportation	DOT
Environmental Protection Agency	EPA
Federal Aviation Administration	FAA
Hazardous Materials Regulation Board	HMRB
Mine Safety and Health Administration	MSHA
National Institute for Standards and Technology	NIST
Occupational Safety and Health Administration	OSHA
U.S. Coast Guard	USCG
Professional Associations	
American Conference of Governmental Industrial Hygienists	ACGIH
American Industrial Hygiene Association	AIHA
American Institute of Chemical Engineers	AIChE
American Society of Heating, Refrigeration and Air Conditioning Engineers	ASHRAE
American Society of Mechanical Engineers	ASME
Illumination Engineers Society	IES
Institute of Electrical and Electronic Engineers	IEEE
Instrument Society of America	ISA
Technical and Trade Associations	
Air Conditioning and Refrigeration Institute	ARI
Air Moving and Conditioning Association	AMCA
American Association of Railroads	AAR
American Gas Association	AGA

Figure 6–7
Organizations that develop standards and codes.

frequently in the court. The most important of these—as well as several related legal terms—are summarized in the following paragraphs.

Negligence

Negligence means failure to take reasonable care or failure to perform duties in ways that prevent harm to humans or damage to property. The concept of *gross negligence* means failure to exercise even slight care or intentional failure to perform duties properly, regardless of the potential consequences. *Contributory negligence* means that an injured party contributed in some way to his or her own injury. In the past, this concept was used to protect defendants against negligence charges because the courts awarded no damages to plaintiffs who had contributed in any way to their own injury. Modern court cases have rendered this approach outdated with the introduction of *comparative negligence.* This

Organization	Abbreviation
Technical and Trade Associations—cont'd	
American Petroleum Institute	API
American Water Works Association	AWWA
Chlorine Institute	CI
Compressed Gas Association	CGA
Cooling Tower Institute	CTI
Manufacturing Chemists Association	MCA
Manufacturers Standardization Society	MSS
National Electrical Manufacturers Association	NEMA
National Fluid Power Association	NFPA
Pipe Fabrication Institute	PFI
Scientific Apparatus Makers Association	SAMA
Society and Plastics Industry	SPI
Steel Structures Painting Council	SSPC
Tubular Exchanger Manufacturers Association	TEMA
Testing Organizations	
American National Standards Institute	ANSI
American Society for Testing and Materials	ASTM
National Fire Protection Association	NFPA
National Safety Council	NSC
Underwriters Laboratories Inc.	UL
Insurance Organizations	
American Insurance Association	ATA
Factory Insurance Association	FIA
Factory Mutual Systems	FMS
Industrial Risk Insurers	IRI
Oil Insurance Association	OIA

Figure 6–7 *(continued)*

concept distributes the negligence assigned to each party involved in litigation according to the findings of the court.

Liability

Liability is a duty to compensate as a result of being held responsible for an act or omission. A newer, related concept is *strict liability*. This means that a company is liable for damages caused by a product that it produces, regardless of negligence or fault.

Care

Several related concepts fall under the heading of care. *Reasonable care* is the amount that would be taken by a prudent person in exercising his or her legal obligations toward others. *Great care* means the amount of care that would be taken by an extraordinarily prudent person in exercising his or her legal

Area of Concern	Organizations That Develop and Publish Standards
Dust hazards	ANSI, NFPA, UL, NSC, ACGIH, AIHA, BM, USCG
Electricity	ANSI, NFPA, NSC, ADA, FIA, FM, OIA, API, USCG, OSHA, IEEE, NEMA
Emergency electricity	NFPA, AIA, FM, IEEE, AGA, NEMA, USCG
Fire protection	ANSI, NFPA, UL, NSC, AIA, FIA, OIA, AWWA, API, CGA, MCA, NEMA, BM, USCG, OSHA
First aid	ANSI, NFPA, NSC, AIA, ACGIH, AIHA, CI, MCA, DOT, USCG
Flammability of substances	ANSI, NFPA, UL, NSC, FIA, FM, MCA, SPI, DOT, USCG, NIST
Hazardous chemicals	ANSI, NFPA, UL, NSC, AIA, FIA, FM, ACGIH, AIHA, AIChE, CI, CGA, MCA, DOT, USCG, OSHA
Instrumentation	ANSI, ASTM, NFPA, UL, AIA, FIA, FM, OIA, IEEE, ISA, AWWA, ARI, API, CGA, SAMA, NIST
Insulation	ANSI, ASTM, UL, AIA, FM, OIA, ASHRAE, USCG
Lighting	ASTM, NFPA, UL, AIA, FM, OIA, ASHRAE, USCG
Lubrication	ANSI, NFPA, ASME, AMCA
Materials	ANSI, ASTM, NFPA, UL, NSC, AIA, FM, OIA, ISA, AWWA, CI, CGA, CTI, MCA, TEMA, USCG, HMR
Noise—vibration	ANSI, ASTM, NFPA, UL, NSC, ACGIH, AIHA, ASHRAE, ISA, ARI, AMCA, AGA, NFPA, EPA
Paint/coatings	ANSI, ASTM, UL, AIChE, AWWA, SSPC, HMRB
Power writing	ANSI, NFPA, UL, FIA, FM, OIA, IEEE, API, NEMA, USCG, OSHA
Pressure relief	NFPA, FIA, FM, OIA, API, CI, CGA, USCG, HMRB, OSHA
Product storage/handling	ANSI, NFPA, AIA, FIA, FM, OIA, AIChE, AAR, API, CI, CGA, MCA, USCG, OSHA
Piping materials	ANSI, ASTM, NFPA, UL, AIA, FIA, FM, ASME, ASHRAE, AWWA, ARI, AGA, API, CI, CGA, MSS
Piping systems	ANSI, ASTM, NFPA, UL, AIA, FIA, FM, ASME, ASHRAE, AWWA, ARI, AGA, API, CI, CGA, MSS
Radiation exposure	ANSI, NFPA, NSC, AIA, ACGIH, AIHA, ASME, ISA, DOT
Safety equipment	ANSI, UL, NSC, FM, ACGIH, AIHA, CI, CGA, MCA, BM, USCG, OSHA
Shutdown systems	NFPA, UL, AIA, FIA, OIA, API, USCG
Ventilation	ANSI, NFPA, NSC, AIA, FIA, FM, ACGIH, AIHA, ASHRAE, ISA, CI, DOT, HMRB

Figure 6–8
Reference table of standards providers.
Note: Refer to Figure 6–7 for an explanation of abbreviations.

obligations toward others. *Slight care* represents the other extreme: a measure of care less than what a prudent person would take. A final concept in this category is the *exercise of due care.* This means that all people have a legal obligation to exercise the amount of care necessary to avoid, to the extent possible, bringing harm to others or damage to their property.

Ability to Pay

The concept of **ability to pay** applies when there are a number of defendants in a case, but not all have the ability to pay financial damages. It allows the court to assess all damages against the defendant or defendants who have the ability to pay. For this reason, it is sometimes referred to as the "deep pockets" principle.

Damages

Damages are financial awards assigned to injured parties in a lawsuit. *Compensatory damages* are awarded to compensate for injuries suffered and for those that will be suffered. *Punitive damages* are awarded to ensure that a guilty party will be disinclined to engage in negligent behavior in the future.

Proximate Cause

Proximate cause is the cause of an injury or damage to property. It is that action or lack of action that ties one person's injuries to another's lack of reasonable care.

Willful/Reckless Conduct

Behavior that is even worse than gross negligence is **willful/reckless conduct**. It involves intentionally neglecting one's responsibility to exercise reasonable care.

Tort

A **tort** is an action involving a failure to exercise reasonable care that as a result may lead to civil litigation.

Foreseeability

The concept of **foreseeability** holds that a person can be held liable for actions that result in damages or injury only when risks could have been reasonably foreseen.

The types of questions around which safety and health litigation often revolve are these: Does the company keep employees informed of rules and regulations? Does the company enforce its rules and regulations? Does the company provide its employees with the necessary training? The concepts set forth in this section come into play as both sides in the litigation deal with these questions from their respective points of view.

Health and safety professionals can serve their companies best by (1) making sure that a policy and corresponding rules and regulations are in place; (2) keeping employees informed about rules and regulations; (3) encouraging proper enforcement practices; and (4) ensuring that employees get the education and training they need to perform their jobs safely.

SUMMARY

1. The impetus for passing the Occupational Safety and Health Act, or OSH Act, was that workplace accidents were causing an average of 14,000 deaths every year in the United States. Each year, 2.5 million workers were disabled in workplace accidents, and approximately 300,000 new cases of occupational diseases were reported annually.

2. The mission of OSHA is to ensure to the extent possible that every working person in the United States has a safe and healthy working environment so that valuable human resources are preserved and protected. The Department of Labor breaks down this mission statement further into the following specific purposes: (a) encourage employers and employees to reduce workplace hazards; (b) implement new safety and health programs; (c) improve existing safety and health programs; (d) encourage research that will lead to innovative ways of dealing with workplace safety and health problems; (e) establish the rights of employers and employees regarding the improvement of workplace safety and health; (f) monitor job-related illnesses and injuries through a system of reporting and record keeping; (g) establish training programs to increase the number of safety and health professionals and to continually improve their competence; (h) establish mandatory workplace safety and health standards and enforce those standards; (i) provide for the development and approval of state-level workplace safety and health programs; and (j) monitor, analyze, and evaluate state-level safety and health programs.

3. The OSH Act covers all employers and all 50 states, the District of Columbia, Puerto Rico, and all other territories that fall under the jurisdiction of the U.S. government with the following exceptions: (a) persons who are self-employed, (b) family farms that employ only immediate members of the family, (c) federal agencies covered by other federal statutes, and (d) state and local governments.

4. OSHA developed standards based on its perception of need at the request of other federal agencies, state and local governments, other standards-setting agencies, labor organizations, or even individual private citizens. OSHA uses the committee approach for developing standards. OSHA's

standing committees are the National Advisory Committee on Occupational Safety and Health and the Advisory Committee on Construction Safety and Health.

5. OSHA can take three different types of action on standards: a standard may be adopted, amended, or revoked. Before any of these actions can be undertaken, OSHA must publish its intentions in the *Federal Register*. OSHA has two options for meeting this requirement: a notice of proposed rule making and an advance notice of proposed rule making.

6. Once the standard has been passed, it becomes effective on the date prescribed. However, any person who is opposed to a standard may file an appeal in the court of appeals serving the geographic region in which the complainant lives or does business. Appeal paperwork must be initiated within 60 days of a standard's approval.

7. When an employer is unable to comply with a new standard immediately but may be able to if given time, a temporary variance may be requested. OSHA will grant such a variance up to a maximum of one year. Employers must demonstrate that they are making a concerted effort to comply and must take the steps necessary to protect employees while working toward compliance.

8. Employers who feel that their workplace already exceeds the requirements of a new standard may request a permanent variance and must present their evidence to OSHA for inspection. Employees must be informed of the application for a variance and notified of their right to request a hearing.

9. OSHA provides for the centralization and systematization of record-keeping and reporting requirements of the OSH Act to employers of 11 or more workers. Both exempt and nonexempt employers must report the following types of accidents within eight hours: (a) those that result in deaths, and (b) those that result in the hospitalization of three or more employees.

10. All occupational illnesses and injuries must be reported if they result in one or more of the following: (a) death to one or more workers; (b) one or more days away from work for the employee; (c) restricted motion or restrictions to the work an employee can do; (d) loss of consciousness to one or more workers; (e) transfer of an employee to another job; (f) medical treatment needed beyond in-house first aid; and (g) appear in Appendix B of the OSH Act.

11. All records required by OSHA can be maintained using the following forms: OSHA Form 300, OSHA Form 301, and OSHA Form 300A.

12. Employers are required to post the following material at locations where employee information is normally displayed: OSHA Poster 2203, summaries of variance requests of all types, copies of all OSHA citations received for failure to meet standards, and the summary page of OSHA Form 300.

13. OSHA compliance officers are authorized to take the following action with regard to workplace inspections: (a) enter at reasonable times any site, location, or facility where work is taking place; (b) inspect at reasonable times any condition, facility, machine, equipment, materials, and so on; and

(c) question in private any employee or other person formally associated with the company.

14. OSHA is empowered to issue citations and/or set penalties. Citations are for (a) other than serious violations, (b) serious violations, (c) willful violations, (d) repeat violations, (e) failure to correct prior violations, and (f) de minimis violations.

15. Employees may appeal the following aspects of OSHA's decisions regarding their workplace: (a) the amount of time (abatement period) given an employer to correct a hazardous condition that has been cited, and (b) an employer's request for an extension of an abatement period.

16. Employers may petition for modification of abatement or contest a citation, an abatement period, and/or a penalty.

17. States are allowed to develop their own safety and health programs. As an incentive, OSHA will fund up to 50 percent of the cost of operating a state program for states with approved plans. States may develop comprehensive plans covering public and private sector employers or limit their plans to coverage of public employers only.

18. In addition to setting standards and inspecting for compliance, OSHA provides services to help employers meet the latest safety and health standards. Services are typically offered at no cost and are intended for smaller companies, particularly those with especially hazardous processes or materials. Services available from OSHA include consultation, voluntary inspection programs, and training/education. OSHA is not without its detractors. Criticisms of OSHA take many forms, depending on the perspective of the critic. Some characterize OSHA as an overbearing bureaucracy with little or no sensitivity to the needs of employers who are struggling to survive in a competitive marketplace. Others label OSHA as timid and claim it does not do enough. At different times and different places, both points of view have probably been at least partially accurate. Other federal agencies and organizations that play important roles with regard to workplace safety and health are the National Institute for Occupational Safety and Health, which is part of the Department of Health and Human Services, and the Occupational Safety and Health Review Commission, which is an independent board consisting of members appointed by the president and given quasi-judicial authority to handle contested OSHA citations.

19. A standard is an operational principle, criterion, or requirement—or a combination of these. A code is a set of standards, rules, or regulations relating to the specific area. Standards and codes play an important role in modern safety and health management and engineering. These written procedures detail the safe and healthy way to perform jobs, which, consequently, makes for a safer and healthier workplace.

20. Fundamental legal principles with which safety and health professionals should be familiar are (a) negligence, (b) liability, (c) care, (d) ability to pay, (e) damages, (f) proximate cause, (g) willful/reckless conduct, (h) tort, and (i) foreseeability.

KEY TERMS AND CONCEPTS

Abatement period

Ability to pay

Appeals process

Closing conference

Code

Consultation services

Damages

De minimis violations

Demonstration Program

Division of Biomedical and Behavioral Science (DBBS)

Division of Respiratory Disease Studies (DRDS)

Division of Surveillance, Hazard Evaluations, and Field Studies (DSHEFS)

Division of Training and Manpower Development (DTMD)

Employee responsibilities

Employee rights

Employer appeals

Employer responsibilities

Employer rights

Foreseeability

Inspection tour

Liability

Merit Program

National Advisory Committee on Occupational Safety and Health (NACOSH)

National Institute for Occupational Safety and Health (NIOSH)

Negligence

Notice of contest

Notice of proposed rule making

Occupational diseases

Occupational Safety and Health Act (OSH Act)

Occupational Safety and Health Administration (OSHA)

Occupational Safety and Health Review Commission (OSHRC)

Opening conference

OSHA Form 300

OSHA Form 300A

OSHA Form 301

OSHA Poster 2203

Other-than-serious violation

Permanent variance

Petition for Modification of Abatement (PMA)

Proposed penalty

Proximate cause

Record keeping

Repeat violations

Reporting

Serious violation

Standard

Star Program

State-level OSHA program

Temporary emergency standards

Temporary variance

Tort

Voluntary Protection Programs (VPPs)

Willful/reckless conduct

Willful violation

Workplace accidents

Workplace inspection

REVIEW QUESTIONS

1. Briefly explain the rationale for the OSH Act.
2. What is OSHA's mission or purpose?
3. List those who are exempted from coverage by OSHA.
4. Explain the difference between an OSHA standard and an OSHA regulation.
5. Explain how the following processes relating to OSHA standards are accomplished: passage of a new standard; request for a temporary variance; appealing a standard.
6. Briefly describe OSHA's latest record-keeping requirements.
7. What are OSHA's reporting requirements?
8. Explain what employers are required to do to keep employees informed.
9. Explain the various components of OSHA's enhanced enforcement policy.
10. Describe how a hypothetical OSHA workplace inspection would proceed from the first step to the last.
11. List and explain three different types of OSHA citations and the typical penalties that accompany them.
12. Describe the process for appealing an OSHA citation.
13. List and briefly explain OSHA's voluntary protection programs.
14. List five employer responsibilities.
15. List five employee rights.
16. Describe the purpose and organization of NIOSH.
17. Define the following legal terms as they relate to workplace safety: *negligence; liability; ability to pay; tort.*

ENDNOTES

1. U.S. Department of Labor, Occupational Safety and Health Administration, *All About OSHA*, OSHA 2056, rev. ed. (Washington, DC: U.S. Department of Labor, Occupational Safety and Health Administration, 2002), 1.
2. Ibid.
3. Ibid., 2.
4. Ibid., 5.
5. Ibid., 7.
6. Ibid., 9.
7. Retrieved from http://www.osha.gov/standards. 2003.
8. Ibid.
9. Ibid.
10. U.S. Department of Labor, *All About OSHA*, 12.
11. Ibid.
12. Ibid., 12–13.
13. Ibid., 17.
14. Ibid., 24–25.

15. *Facility Manager's ALERT*, October 1, 2002, 3.
16. Ibid.
17. *Facility Manager's ALERT*, March 17, 2003, 3.
18. Ibid.
19. U.S. Department of Labor, *All About OSHA*, 27.
20. Ibid., 28
21. S. Davis, "Contested Cases, Missing or Misplaced," *Occupational Hazards*, March 1991, 47.
22. Ibid.
23. Ibid., 28.
24. Ibid.
25. U.S. Department of Labor, *All About OSHA*, 28.
26. Ibid.
27. Ibid., 29.
28. Ibid., 32.
29. Ibid., 34–35.
30. Ibid., 35–36.
31. Ibid., 37.
32. Ibid., 39–40.
33. Ibid., 37.
34. U.S. Department of Health and Human Services, *Occupational Safety & Health for Fiscal Year 1988 Under Public Law 91–596* (Washington, DC: Public Health Service, Centers for Disease Control and Prevention, National Institute for Occupational Safety and Health).
35. U.S. Department of Health and Human Services, handout (Washington, DC: Public Health Service, Centers for Disease Control and Prevention, National Institute for Occupational Safety and Health, 1998), 1.
36. Ibid., 2.
37. Ibid., 3.
38. Ibid., 4.
39. For more in-depth coverage of OSHA's construction standards, refer to D. L. Goetsch, *Construction Safety and Health* (Upper Saddle River, NJ: Prentice Hall, 2003).

WORKERS' COMPENSATION

Major Topics

- Overview of Workers' Compensation
- Historical Perspective
- Workers' Compensation Legislation
- Modern Workers' Compensation
- Workers' Compensation Insurance
- Resolution of Workers' Compensation Disputes
- Injuries and Workers' Compensation
- Disabilities and Workers' Compensation
- Monetary Benefits of Workers' Compensation
- Medical Treatment and Rehabilitation
- Medical Management of Workplace Injuries
- Administration and Case Management
- Cost Allocation
- Problems with Workers' Compensation
- Spotting Workers' Compensation Fraud and Abuse
- Future of Workers' Compensation
- Cost-Reduction Strategies

OVERVIEW OF WORKERS' COMPENSATION

The concept of **workers' compensation** developed as a way to allow injured employees to be compensated appropriately without having to take their employer to court. The underlying rationale for workers' compensation had two aspects: (1) fairness to injured employees, especially those without the resources to undertake legal actions that are often long, drawn out, and expensive; and (2) reduction of costs to employers associated with workplace injuries (e.g., legal, image, and morale costs). Workers' compensation is

intended to be a no-fault approach to resolving workplace accidents by reha-
bilitating injured employees and minimizing their personal losses because of
their reduced ability to perform and compete in the labor market.[1] Since its
inception as a concept, workers' compensation has evolved into a system that
pays out approximately $70 million in benefits and medical costs annually.
The national average net cost of workers' compensation in the manufacturing
sector is almost $6 per $100 of payroll.

Workers' compensation represents a compromise between the needs of
employees and the needs of employers. Employees give up their right to seek
unlimited compensation for pain and suffering through legal action.
Employers award the prescribed compensation (typically through insurance
premiums) regardless of the employee's negligence. The theory is that in the
long run both employees and employers will benefit more than either would
through legal action. As you will see later in this chapter, although workers'
compensation has reduced the amount of legal action arising out of workplace
accidents, it has not completely eliminated legal actions.

Objectives of Workers' Compensation

Workers' compensation laws are not uniform from state to state. In fact, there
are extreme variations. However, regardless of the language contained in the
enabling legislation in a specific state, workers' compensation as a concept has
several widely accepted objectives:

- Replacement of income
- Rehabilitation of the injured employee
- Prevention of accidents
- Cost allocation[2]

The basic premises underlying these objectives are described in the following paragraphs.

Replacement of Income

Employees injured on the job lose income if they are unable to work. For this
reason, workers' compensation is intended to replace the lost income adequately and promptly. Adequate **income replacement** is viewed as replacement
of current and future income (minus taxes) at a ratio of two-thirds (in most
states). Workers' compensation benefits are required to continue even if the
employer goes out of business.

Rehabilitation of the Injured Employee

A basic premise of workers' compensation is that the injured worker will
return to work in every case possible, although not necessarily in the same job
or career field. For this reason, a major objective of workers' compensation is to
rehabilitate the injured employee. The **rehabilitation** program is to provide the
needed medical care at no cost to the injured employee until he or she is

pronounced fit to return to work. The program also provides vocational training or retraining as needed. Both components seek to motivate the employee to return to the labor force as soon as possible.

Accident Prevention

Preventing future accidents is a major objective of workers' compensation. The theory underlying this objective is that employers will invest in **accident prevention** programs to hold down compensation costs. The payoff to employers comes in the form of lower insurance premiums that result from fewer accidents.

Cost Allocation

The potential risks associated with different occupations vary. For example, working as a miner is generally considered more hazardous than working as an architect. The underlying principle of **cost allocation** is to spread the cost of workers' compensation appropriately and proportionately among industries ranging from the most to the least hazardous. The costs of accidents should be allocated in accordance with the accident history of the industry so that high-risk industries pay higher workers' compensation insurance premiums than do low-risk industries.[3]

Who Is Covered by Workers' Compensation?

Workers' compensation laws are written at the state level, and there are many variations among these laws. As a result, it is difficult to make generalizations. Complicating the issue further is the fact that workers' compensation laws are constantly being amended, revised, and rewritten. Additionally, some states make participation in a workers' compensation program voluntary; others excuse employers with fewer than a specified number of employees.

In spite of the differences among workers' compensation laws in the various states, approximately 80 percent of the employees in the United States are covered by workers' compensation. Those employees who are not covered or whose coverage is limited vary as the laws vary. However, they can be categorized in general terms as follows:

- Agricultural employees
- Domestic employees
- Casual employees
- Hazardous work employees
- Charitable or religious employees
- Employees of small organizations
- Railroad and maritime employees
- Contractors and subcontractors
- Minors
- Extraterritorial employees[4]

According to Hammer, coverage in these types of employment, to the extent there is coverage, varies from state to state as follows:

- *Agricultural employees* have limited coverage in 38 states, Puerto Rico, and the Virgin Islands. In 15 states, workers' compensation coverage for agricultural employees is voluntary. In these states, employers are allowed to provide coverage if they wish but are not required to do so.
- *Domestic employees* have coverage available in all 50 states and Puerto Rico. However, coverage tends to be limited and subject to minimum requirements regarding hours worked and earnings.
- *Casual employees* are employed in positions in which the work is occasional, incidental, and scattered at irregular intervals. Such employees are not typically afforded workers' compensation coverage.
- *Hazardous employment* is the only type afforded workers' compensation coverage in some states. To qualify, a particular type of employment must be on an approved list of hazardous or especially hazardous jobs. However, the trend in these states is to broaden the list of approved jobs.
- *Charitable or religious employees* are not afforded workers' compensation in most states when this work is irregular, temporary, or short-term.
- *Small organizations* that employ fewer than a stipulated number of employees do not fall under the umbrella of workers' compensation in 26 states.
- *Railroad and maritime workers* are not typically covered by workers' compensation. However, in most cases, they are covered by the Federal Employer's Liability Act. This act disallows the use of common law defenses by employers if sued by an employee for negligence.
- *Contractors and subcontractors* are those who agree to perform a job or service for an agreed amount of money in a nondirected, nonsupervised format. In essence, contract and subcontract employees are viewed as being self-employed. For this reason, they are not covered by workers' compensation.
- *Minors* are afforded regular workers' compensation coverage as long as they are legally employed. In some states, coverage is significantly higher for minors who are working illegally.
- *Extraterritorial employees* are those who work in one state but live in another. In these cases, the employee is usually on temporary duty. Such employees are typically afforded the workers' compensation coverage of their home state.[5]

HISTORICAL PERSPECTIVE

Before workers' compensation laws were enacted in the United States, injured employees had no way to obtain compensation for their injuries except to take their employer to court. Although common law did require employers to provide a safe and healthy work environment, injured employees bore the burden of proof that negligence in the form of unsafe conditions contributed to these injuries. According to the Society of Manufacturing Engineers, prior to passage of workers' compensation, employees often had to sue their employer to receive compensation for injuries resulting from a workplace accident or occupational disease, even when the following circumstances prevailed:

- The employee was disabled or died as the result of a **workplace accident** or **occupational disease**.
- The injury might have been expected to occur when the risks and hazards of the job were considered.
- **Worker negligence** on the part of a fellow worker or the injured employee clearly caused the injury.[6]

Proving that an injury was the result of employee negligence was typically too costly, too difficult, and too time consuming to be a realistic avenue of redress for most injured employees. According to Somers and Somers, a New York commission determined that it took from six months to six years for an injured worker's case to work its way through the legal system.[7] Typically, injured workers, having lost their ability to generate income, could barely afford to get by, much less pay medical expenses, legal fees, and court costs. Another inhibitor was the *fear factor.* Injured employees who hoped to return to work after recovering were often afraid to file suit because they feared retribution by their employer. Employers not only might refuse to give them their jobs back but also might **blackball** them with other employers. Add to this that fellow employees were often afraid to testify to the negligence of the employer, and it is easy to see why few injured workers elected to take their employers to court.

Even with all of these inhibitors, some injured employees still chose to seek redress through the courts in the days before workers' compensation. Those who did faced a difficult challenge because the laws at that time made it easy for employers to defend themselves successfully. All an employer had to do to win a decision denying the injured plaintiff compensation was show that at least one of the following conditions existed at the time of the accident:

1. *Contributory negligence was a factor in the accident.* **Contributory negligence** meant that the injured worker's own negligence contributed to the accident. Even if the employee's negligence was a very minor factor, it was usually enough to deny compensation in the days before workers' compensation.
2. *There was negligence on the part of a fellow worker.* As with contributory negligence, negligence by a fellow employee, no matter how minor a contributing factor it was, could be sufficient to deny compensation. This defense was known as "the fellow servant rule."
3. *There was assumption of risk on the part of the injured employee.* If an employee knew that the job involved risk, he or she could not expect to be compensated when the risks resulted in accidents and injuries.[8] This defense relied on a long-standing principle of tort law known as "assumption of risk."

Because the majority of workplace accidents involve at least some degree of negligence on the part of the injured worker or fellow employees, employers typically won these cases. Because it required little more than a verbal warning by the employer to establish grounds for assumption of risk, the odds against an injured employee being awarded compensation become clear.

In his book *American Social Science*, Gagilardo gives an example of a case that illustrates how difficult it was to win compensation in the days before workers' compensation.[9] He relates the example of an employee who contracted tuberculosis while working under clearly hazardous conditions for a candy-making company. She worked in a wet, drafty basement that admitted no sunlight. Dead rats floated in the overflow of a septic tank that covered the basement floor, and a powerful stench permeated the workplace. Clearly, these were conditions that could contribute to the employee contracting tuberculosis. However, she lost the case and was denied compensation. The ruling judge justified the verdict as follows:

> We think that the plaintiff, as a matter of law, assumed the risk attendant upon her remaining in the employment (*Wager v. White Star Candy Company*, 217 N.Y. Supp. 173).

Situations such as this eventually led to the enactment of workers' compensation laws in the United States.

WORKERS' COMPENSATION LEGISLATION

Today, all 50 states, the District of Columbia, Guam, and Puerto Rico have workers' compensation laws. However, these laws did not exist prior to 1948. Considering that Prussia passed a workers' compensation law in 1838, the United States was obviously slow to adopt the concept. In fact, the first workers' compensation law enacted in the United States did not pass until 1908, and it applied only to federal employees working in especially hazardous jobs. The driving force behind passage of this law was President Theodore Roosevelt, who as governor of New York had seen the results of workplace accidents firsthand. Montana was the first state to pass a compulsory workers' compensation law. However, it was short-lived. Ruling that the law was unconstitutional, the Montana courts overturned it.

In 1911, the New York Court of Appeals dealt proponents of workers' compensation a serious blow. The New York state legislature had passed a compulsory workers' compensation law in 1910. However, in the case of *Ives v. South Buffalo Railway Company* (201 N.Y. 271, 1911), the New York Court of Appeals declared the law unconstitutional based on the contention that it violated the due process clause in the Fourteenth Amendment to the U.S. Constitution.[10]

This ruling had a far-reaching impact. According to Hammer, "The prestige of the New York court influenced legislators in many of the other states to believe that any compulsory law also would be held unconstitutional."[11] However, even with such precedent-setting cases as *Ives* on the books, pressure for adequate workers' compensation grew as unsafe working conditions continued to result in injuries, diseases, and deaths. In fact, shortly after the New York Court of Appeals released its due process ruling, tragedy struck in a New York City textile factory.

On March 25, 1911, the building that housed the Triangle Shirtwaist Factory on its eighth floor caught fire and burned.[12] As a result of the fire, 149 of the company's 600 workers died, and another 70 were injured. Although the cause of the accident could not be determined, it was clear to investigators and survivors alike that unsafe conditions created by the management of the company prevented those who died or were injured from escaping the fire.

Exit passageways on each floor of the building were unusually narrow (20 inches wide), which made it difficult for employees to carry out bolts of material. A wider exit on each floor was kept locked to force employees to use the narrow exit. The two elevators were slow and able to accommodate only small groups at a time.

As the fire quickly spread, employees jammed into the narrow passageways, crushing each other against the walls and underfoot. With all exits blocked, panic-stricken employees began to jump out of windows and down the elevator shafts. When the pandemonium subsided and the fire was finally brought under control, the harsh realization of why so many had been trapped by the deadly smoke and flames quickly set in.

The owners were brought into court on charges of manslaughter. Although they were not convicted, the tragedy did focus nationwide attention on the need for a safe workplace and adequate workers' compensation. As a result, new, stricter fire codes were adopted in New York, and in spite of the state court's ruling in *Ives*, the state legislature passed a workers' compensation law.

The next several years saw a flurry of legislation in other states relating to workers' compensation. In response to demands from workers and the general public, several states passed limited or noncompulsory workers' compensation laws. Many such states held back out of fear of being overturned by the courts. Others, particularly Washington, publicly disagreed with the New York Court of Appeals and passed compulsory laws. The constitutionality debate continued until 1917 when the U.S. Supreme Court ruled that workers' compensation laws were acceptable.

MODERN WORKERS' COMPENSATION

Since 1948, all states have had workers' compensation laws. However, the controversy surrounding workers' compensation has not died. As medical costs and insurance premiums have skyrocketed, many small businesses have found it difficult to pay the premiums. Unrealistic workers' compensation rates are being cited more and more frequently as contributing to the demise of small business in America.

The problem has even developed into an economic development issue. Business and industrial firms are closing their doors in those states with the highest workers' compensation rates and moving to states with lower rates. States with lower rates are using this as part of their recruiting package to attract new businesses and industry. Where low-rate states border high-rate states, businesses are beginning to move their offices across the border to the low-rate state while still doing business in the high-rate state.

Critics are now saying that workers' compensation has gotten out of hand and is no longer fulfilling its intended purpose. To understand whether this is the case, one must begin with an examination of the purpose of workers' compensation. According to Hammer, the U.S. Chamber of Commerce identified the following six basic objectives of workers' compensation:

1. To provide an appropriate level of income and medical benefits to injured workers or to provide income to the worker's dependents regardless of fault.
2. To provide a vehicle for reducing the amount of personal injury litigation in the court system.
3. To relieve public and private charities of the financial strain created by workplace injuries that go uncompensated.
4. To eliminate time-consuming and expensive trials and appeals.
5. To promote employer interest and involvement in maintaining a safe work environment through the application of an experience-rating system.
6. To prevent accidents by encouraging frank, objective, and open investigations of the causes of accidents.[13]

Early proponents of workers' compensation envisioned a system in which both injured workers and their employers would win. Injured workers would receive prompt compensation, adequate medical benefits, and appropriate rehabilitation to allow them to reenter the workforce and be productive again. Employers would avoid time-consuming, expensive trials and appeals and would improve relations with employees and the public in general.

What proponents of workers' compensation did not anticipate were the following factors: (1) employees who would see workers' compensation as a way to ensure themselves a lifelong income without the necessity of work; (2) enormous increases in the costs of medical care with corresponding increases in workers' compensation insurance premiums; and (3) the radical differences among workers' compensation laws passed by the various states.

Not all employees abide by the spirit of workers' compensation (i.e., rehabilitation in a reasonable amount of time). Attempted abuse of the system was perhaps inevitable. Unfortunately, such attempts result in a return to what workers' compensation was enacted to eliminate: time-consuming, drawn-out, expensive legal battles and the inevitable appeals.

Proponents of workers' compensation reform can cite a long list of cases that illustrate their point. The city of Pittsfield, Massachusetts, was once overwhelmed by workers' compensation claims. One of the more remarkable cases concerned a city worker who was receiving workers' compensation benefits as the result of a back injury. While collecting benefits, he was a star player for a local softball team.[14] He eventually agreed to waive his right to compensation for a lump sum settlement of $12,000 plus $3,000 for his lawyer, and city officials considered themselves cheaply rid of him.[15]

Another Pittsfield city employee in the Department of Public Works was injured and began collecting workers' compensation at a rate of $295.50 per week. In addition to his job with the city, this worker owned a small diesel oil company. When his workers' compensation benefits were called into question because he owned a business that produced an income, the injured employee sold the business to his son.[16]

One of the favorite examples of opponents of workers' compensation is that of the "fat deputy sheriff."

> The deputy sheriff was already despondent over the breakup of the extramarital affair he was having with a married colleague, when his supervisor made an expensive mistake. He rated the already overstressed sheriff's job performance as substandard and told him he was too fat. The sheriff promptly filed a claim for workers' compensation benefits to cover job-related pressures stemming from his performance evaluation that contributed to his mental injury.[17]

Medical costs have skyrocketed in the United States since the 1960s. There are many reasons for this. During this same period, the costs associated with other basic human needs including food, clothing, transportation, shelter, and education have also increased markedly. Increases in medical costs can be explained, at least partially, as the normal cost of living increases experienced in other sectors of the economy. However, the costs associated with medical care have increased much faster and much more than the costs in these other areas. The unprecedented increases can be attributed to two factors: (1) technological developments that have resulted in extraordinary but costly advances in medical care; and (2) a proliferation of **litigation** that has driven the cost of malpractice insurance steadily up. Each of these factors has contributed to higher medical costs. For example, X-ray machines that cost thousands of dollars have been replaced by magnetic resonance imaging (MRI) systems that may cost millions. **Malpractice** suits that once might not even have gone to court now result in multimillion-dollar settlements. Such costs are, of course, passed on to whoever pays the medical bill—in this case, employers who must carry workers' compensation insurance. California's workers' compensation system, for example, costs employers $7 billion annually in insurance premiums. Between 1987 and 1998, the cost of workers' compensation premiums increased by more than 60 percent. This trend continues.

In addition to contributing individually to increased medical costs, technology and litigation have interacted in such a way as to increase costs even further. This interaction occurs as follows. First, an expensive new technology is developed that enhances the predictive and/or prescriptive capabilities of the medical profession. Second, expensive malpractice suits force medical practitioners to be increasingly cautious and, accordingly, to order even more tests than a patient's symptoms may suggest. Finally, the tests involve expensive new technologies, adding even more to the cost of medical care.

Early supporters of the concept did not anticipate the radical differences among workers' compensation laws in the various states. The laws themselves differ, as do their interpretations. The differences are primarily in the areas of benefits, penalties, and workers covered. These differences translate into differences in the rates charged for workers' compensation insurance. As a result, the same injury incurred under the same circumstances but in different states can yield radically different benefits for the employee.

The potential for abuse, steadily increasing medical costs that lead to higher insurance premiums, and differences among workers' compensation laws all contribute to the controversy that still surrounds this issue. As business and industry continue to protest that workers' compensation has gotten out of hand, it will continue to be a heated issue in state legislatures. As states try to strike the proper balance between meeting the needs of the workforce while simultaneously maintaining a positive environment for doing business, workers' compensation will be an issue with which they will have to deal.

WORKERS' COMPENSATION INSURANCE

The costs associated with workers' compensation must be borne by employers as part of their overhead. In addition, employers must also ensure that the costs will be paid even if they go out of business. The answer for most employers is workers' compensation insurance.

In most states, workers' compensation insurance is compulsory. Exceptions to this are New Jersey, South Carolina, Texas, and Wyoming. New Jersey allows 10 or more employers to form a group and self-insure. Texas requires workers'

Safety Fact

Back Injuries Cost More with Workers' Compensation

One of the most common workplace injuries is the back injury. Almost one-fourth of all workers' compensation claims involve back injuries. According to the California Workers' Compensation Institute (CWCI), medical care costs are 43 percent higher when part of a workers' compensation claim than when part of a group medical plan. Back injuries treated under workers' compensation use medical services at a higher rate than those treated under a group medical plan. Workers' compensation claimants have more total visits for medical treatment, more total procedures, more procedures per week, and more visits per week.

State	State Fund	Private Insurer	Individual Employer Self-Insurance	Group of Employer's Self-Insurance
Alabama	No	Yes	Yes	Yes
Arkansas	No	Yes	Yes	Yes
California	Competitive	Yes	Yes	No
Florida	No	Yes	Yes	Yes
Indiana	No	Yes	Yes	No
Kansas	No	Yes	Yes	Yes
Maryland	Competitive	Yes	Yes	Yes
Montana	Competitive	Yes	Yes	Yes
Nebraska	No	Yes	Yes	No
New Jersey	No	Yes	Yes	No*
New Mexico	Competitive	Yes	Yes	Yes
Vermont	No	Yes	Yes	No
Washington	Exclusive	No	Yes	Yes
Wisconsin	No	Yes	Yes	No

* Permits 10 or more employees licensed by the state to group self-insure.

Figure 7–1
Workers' compensation coverage methods allowed for selected states.
Source: U.S. Department of Labor, 2005.

compensation only for *carriers*, as defined in Title 25, Article 911–A, Section II, Texas state statutes. Wyoming requires workers' compensation only for employers involved in specifically identified *extrahazardous occupations*.

A common thread woven through all the various compensation laws is the requirement that employers carry workers' compensation insurance. There are three types: **state funds**, **private insurance**, and **self-insurance**. Figure 7–1 summarizes the methods of insurance coverage allowed in a representative sample of states. Regardless of the method of coverage chosen, rates can vary greatly from company to company and state to state. Rates are affected by a number of different factors including the following:

- Number of employees
- Types of work performed (risk involved)
- Accident experience of the employer
- Potential future losses
- Overhead and profits of the employer
- Quality of the employer's safety program
- Estimates by actuaries

Insurance companies use one of the following six methods in determining the premium rates of employers:

1. *Schedule rating.* Insurance companies establish baseline safety conditions and evaluate the employer's conditions against the baselines. Credits are awarded for conditions that are better than the baseline, and debits are assessed for conditions that are worse. Insurance rates are adjusted accordingly.
2. *Manual rating.* A manual of rates is developed that establishes rates for various occupations. Each occupation may have a different rate based on its perceived level of hazard. The overall rate for the employer is a pro-rata combination of all the individual rates.
3. *Experience rating.* Employers are classified by type. Premium rates are assigned based on predictions of average losses for a given type of employer. Rates are then adjusted either up or down according to the employer's actual experience over the past three years.
4. *Retrospective rating.* Employees pay an established rate for a set period. At the end of the period, the actual experience is assessed, and an appropriate monetary adjustment is made.
5. *Premium discounting.* Large employers receive discounts on their premiums based on their size. The theory behind this method is that it takes the same amount of time to service a small company's account as it does a large company's, but the large company produces significantly more income for the insurer. Premium discounts reward the larger company for its size.
6. *Combination method.* The insurer combines two or more of the other methods to arrive at premium rates.[18]

The trend nationwide for the past decade has been for premiums to increase markedly. For example, over the past 10 years, some states experienced increases of over 60 percent. This trend will ensure that workers' compensation remains a controversial issue in the state legislatures.

RESOLUTION OF WORKERS' COMPENSATION DISPUTES

One of the fundamental objectives of workers' compensation is to avoid costly, time-consuming litigation. Whether this objective is being accomplished is questionable. When an injured employee and the employer's insurance company disagree on some aspect of the compensation owed (e.g., weekly pay, length of benefits, degree of disability), the disagreement must be resolved. Most states have an arbitration board for this purpose. Neither the insurance company nor the injured employee is required to hire an attorney. However, many employees do. There are a number of reasons for this. Some don't feel they can adequately represent themselves. Others are fearful of the "big business running over the little guy" syndrome. In any case, workers' compensation litigation is still very common and expensive.

Allowable attorney fees are set by statute, administrative rule, or policy in most states. In some states, attorney fees can be added to the injured employee's award. In others, the fee is a percentage of the award.

INJURIES AND WORKERS' COMPENSATION

The original workers' compensation concept envisioned compensation for workers who were injured in on-the-job accidents. What constituted an accident varied from state to state. However, all original definitions had in common the characteristics of being *sudden* and *unexpected*. Over the years, the definition of an accident has undergone continual change. The major change has been a trend toward the elimination of the sudden characteristic. In many states, the gradual onset of a disease as a result of prolonged exposure to harmful substances or a harmful environment can now be considered an accident for workers' compensation purposes.

A **harmful environment** does not have to be limited to its physical components. Psychological factors (such as stress) can also be considered. In fact, the highest rate of growth in workers' compensation claims over the past two decades has been in the area of stress-related injuries.

The National Safety Council maintains statistical records of the numbers and types of injuries suffered in various industries in the United States. Industries are divided into the following categories: agriculture, mining, construction, manufacturing, transportation/public utilities, trade, services, and public sector. Injuries in these industrial sectors are classified according to the type of accident that caused them. Accident types include overexertion, being struck by or against an object, falls, bodily reactions, caught in or between objects, motor vehicle accident, coming in contact with radiation or other caustics, being rubbed or abraded, and coming in contact with temperature extremes.

Over 30 percent of all disabling work injuries are the result of overexertion when all industry categories are viewed in composite. The next most frequent cause of injuries is struck by/struck against objects at 24 percent. Falls account for just over 17 percent. The remainder are fairly evenly distributed among the other accident types just listed.[19]

AOE and COE Injuries

Workers' compensation benefits are owed only when the injury arises *out of employment (AOE)* or occurs in the *course of employment (COE)*. When employees are injured undertaking work prescribed in their job description, work assigned by a supervisor, or work normally expected of employees, they fall into the AOE category. Sometimes, however, different circumstances determine whether the same type of accident is considered to be AOE. For example, say a soldering technician burns her hand while repairing a printed circuit board that had been rejected by a quality control inspector. This injury would be classified as AOE. Now suppose the same technician brings a damaged printed circuit board from her home stereo to work and burns her hand while trying to repair it. This injury would not be covered because the accident did not arise from her employment. Determining whether an injury should be classified as AOE or COE is often a point of contention in workers' compensation litigation.

Who Is an Employee?

Another point of contention in workers' compensation cases is the definition of the term *employee*. This is an important definition because it is used to determine AOE and COE. A person who is on the company's payroll, receives benefits, and has a supervisor is clearly an **employee**. However, a person who accepts a service contract to perform a specific task or set of tasks and is not directly supervised by the company is not considered an employee. Although definitions vary from state to state, there are common characteristics. In all definitions, the workers must receive some form of remuneration for work done, and the employer must benefit from this work. Also, the employer must supervise and direct the work, both process and result. These factors—supervision and direction—are what set **independent contractors** apart from employees and exclude them from coverage. Employers who use independent contractors sometimes require the contractors to show proof of having their own workers' compensation insurance.

Unless an employer provides transportation, employees are not generally covered by workers' compensation when traveling to and from work. However, in some circumstances, they can be covered. Consider the following example:

> Mary works at home three days a week and in the office two days. Wednesday is typically one of her days to telecommute from home. Around ten in the morning, Mary got a call from her supervisor asking her to rush to the office for an unplanned but important meeting and to bring several files that are critical to the subject of the meeting. On the way, Mary had an accident and was injured. Since she was transporting job-specific files at the request of her supervisor, the case was considered work-related, and Mary was eligible for workers' compensation.[20]

Another example shows that Mary's case was special. Tom works in an office in a large city. Because he preferred to drive his own car to work instead of using the public transportation system, Tom had to expose himself to dangerous rush-hour traffic twice every workday. One day, Tom was injured in an accident while driving home from work. He filed a workers' compensation claim, but it was denied because driving to and from work in his own automobile was not considered work related.[21]

DISABILITIES AND WORKERS' COMPENSATION

Injuries that are compensable typically fall into one of four categories: (1) temporary partial disability, (2) temporary total disability, (3) permanent partial disability, and (4) permanent total disability (Figure 7–2). Determining the extent of disability is often a contentious issue. In fact, it accounts for more workers' compensation litigation than any other issue. Further, when a disability question is litigated, the case tends to be complicated because the evidence is typically subjective, and it requires hearing officers, judges, or juries to determine the future.

Temporary Disability

Temporary disability is the state that exists when it is probable that an injured worker, who is currently unable to work, will be able to resume gainful employment with no or only partial disability. Temporary disability assumes that the employee's condition will substantially improve. Determining whether an employee is temporarily disabled is not normally difficult. Competent professionals can usually determine the extent of the employee's injuries, prescribe the appropriate treatment, and establish a timeline for recovery. They can then determine if the employee will be able to return to work and when the return might take place.

There is an important point to remember when considering a temporary disability case: The ability to return to work relates only to work with the company that employed the worker at the time of the accident.

Temporary disability can be classified as either *temporary total disability* or *temporary partial disability*. A **temporary total disability** classification means the injured worker is incapable of any work for a period of time but is expected to recover fully. Most workers' compensation cases fall in this classification. A **temporary partial disability** means the injured worker is capable of light or part-time duties. Depending on the extent of the injury, temporary partial disabilities sometimes go unreported. This practice is allowable in some states. It helps employers hold down the cost of their workers' compensation

Figure 7–2
Types of disabilities.

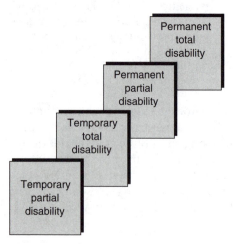

Safety Myth

Workers' Compensation Claims and Character

A leopard doesn't change its spots. A dishonest employee will always abuse the workers' compensation system with fraudulent claims. Right? Maybe not. A company in Oklahoma reduced its workers' compensation costs in just two years from $486,000 to $47,000. How did this company accomplish a reduction of this magnitude? With character training based on biblical principles. The program focuses on such character traits as honesty, attentiveness, gratitude, and dependability. At monthly training sessions, employees discuss one character trait and how it applies on the job.

premium. This is similar to not reporting a minor fender bender to your automobile insurance agent.

Most states prescribe in law the benefits owed in temporary total disability cases. Factors prescribed typically include a set percentage of an employee's wage that must be paid and a maximum period during which benefits can be collected. Figure 7–3 shows this information for a geographically distributed

State	Percentage of Employee's Wage	Maximum Period
Alabama	66⅔	Duration of disability
Arkansas	66⅔	450 weeks
California	66⅔	Duration of disability
Florida	66⅔	104 weeks
Indiana	66⅔	500 weeks
Kansas	66⅔	Duration of disability
Maryland	66⅔	Duration of disability
Montana	66⅔	Duration of disability*
Nebraska	66⅔	Duration of disability
New Jersey	70	400 weeks
New Mexico	66⅔	Duration of disability
Vermont	66⅔	Duration of disability
Washington	60–75	Duration of temporary disability
Wisconsin	60–75	Duration of disability

* Or until worker is released to preinjury job or similar employment.

Figure 7–3
Temporary total disability benefits for selected states.
Source: U.S. Department of Labor, 2005.

selection of states. Because workers' compensation legislation changes continually, this figure is provided only as an illustration of how benefits are prescribed in the laws of the various states. Actual rates are subject to change.

Permanent Partial Disability

Permanent partial disability is the condition that exists when an injured employee is not expected to recover fully. In such cases, the employee will be able to work again but not at full capacity. Often employees who are partially disabled must be retrained for another occupation.

Permanent partial disabilities can be classified as *schedule* or *nonscheduled* disabilities. *Schedule disabilities* are typically the result of nonambiguous injuries such as the loss of a critical but duplicated body part (e.g., arm, ear, hand, finger, or toe). Because such injuries are relatively straightforward, the amount of compensation that they generate and the period of time that it will be paid can be set forth in a standard schedule. A compilation of information from such schedules for a geographically distributed list of states is shown in Figure 7–4. Workers' compensation legislation changes continually; therefore this figure is provided only as an example. Actual rates are subject to change continually.

State	Percentage of Employee's Wage	Maximum Period
Alabama	66⅔	300 weeks
Arkansas	66⅔	450 weeks
California	66⅔	619.25 weeks
Florida	*	364 weeks
Indiana	66⅔	500 weeks
Kansas	66⅔	415 weeks
Maryland	66⅔	Duration of disability
Montana	66⅔	350 weeks
Nebraska	66⅔	300 weeks
New Jersey	70	600 weeks
New Mexico	66⅔	500 weeks
Vermont	66⅔	330 weeks
Washington	—	—
Wisconsin	66⅔	1,000 weeks

* Shall not exceed 66⅔.

Figure 7–4
Permanent partial disability benefits for selected states in 2005. Based on actual wages lost and not subject to minimums.
Source: U.S. Department of Labor.

Nonschedule injuries are less straightforward and must be dealt with on a case-by-case basis. Disabilities in this category tend to be the result of head injuries, the effects of which can be more difficult to determine. The amount of compensation awarded and the period over which it is awarded must be determined by studying the evidence. Awards are typically made based on a determination of disability percentage. For example, if it is determined that an employee has a 25 percent disability, the employee might be entitled to 25 percent of the income he or she could have earned before the injury with normal career progression factored in.

Four approaches to handling permanent partial disability cases have evolved. Three are based on specific theories, and the fourth is based on a combination of two or more of these theories. The three theories are (1) whole-person theory, (2) wage-loss theory, and (3) loss of wage-earning capacity theory.

Whole-Person Theory

The **whole-person theory** is the simplest and most straightforward of the theories for dealing with permanent partial disability cases. Once it has been determined that an injured worker's capabilities have been permanently impaired to some extent, this theory is applied like a subtraction problem. What the worker can do after recuperating from the injury is determined and subtracted from what he or she could do before the accident. Factors such as age, education, and occupation are not considered.

Wage-Loss Theory

The **wage-loss theory** requires a determination of how much the employee could have earned had the injury not occurred. The wages actually being earned are subtracted from what could have been earned, and the employee is awarded a percentage of the difference. No consideration is given to the extent or degree of disability. The only consideration is loss of actual wages.

Loss of Wage-Earning Capacity Theory

The most complex of the theories for handling permanent partial disability cases is the **loss of wage-earning capacity theory**, because it is based not just on what the employee earned at the time of the accident, but also on what he or she might have earned in the future. Making such a determination is obviously a subjective undertaking. Factors considered include past job performance, education, age, gender, and advancement potential at the time of the accident, among others. Once future earning capacity has been determined, the extent to which it has been impaired is estimated, and the employee is awarded a percentage of the difference. Some states prescribe maximum amounts of compensation and maximum periods within which it can be collected. For example, in Figure 7–4, Alabama sets 300 weeks as the maximum period for collecting on a nonschedule injury. Maryland, on the other hand, awards compensation for the duration of the disability. Scheduled disabilities are typically compensated for the duration in all states.

The use of schedules has reduced the amount of litigation and controversy surrounding permanent partial disability cases. This is the good news aspect of schedules. The bad news aspect is that they may be inherently unfair. For example, a surgeon who loses his hand would receive the same compensation as a laborer with the same injury if the loss of a hand is scheduled.

Permanent Total Disability

A **permanent total disability** exists when an injured employee's disability is such that he or she cannot compete in the job market. This does not necessarily mean that the employee is helpless. Rather, it means an inability to compete reasonably. Handling permanent total disability cases is similar to handling permanent partial disability cases except that certain injuries simplify the process. In most states, permanent total disability can be assumed if certain specified injuries have been sustained (i.e., loss of both eyes or both arms). In some states, compensation is awarded for life. In others, a time period is specified. Figure 7–5 shows the maximum period that compensation can be collected for a geographically distributed list of states. Notice in this Figure 7–5 that the time periods range from 401 weeks (Texas) to life (California and Wisconsin).

Figure 7–5
Duration of permanent total disability benefits for selected states.
Source: U.S. Department of Labor, 2005.

State	Maximum Paid
Alabama	Duration of disability
Arkansas	Duration of disability
California	Life
Florida	Duration of disability
Indiana	500 weeks
Kansas	Duration of disability
Maryland	Duration of disability
Montana	Duration of disability
Nebraska	Duration of disability
New Jersey	450 weeks (life in some cases)
New Mexico	Life
Vermont	Duration of disability*
Washington	Life
Wisconsin	Life

* Minimum of 330 weeks.

MONETARY BENEFITS OF WORKERS' COMPENSATION

The **monetary benefits** accruing from workers' compensation vary markedly from state to state. The actual amounts are of less importance than the differences among them. Of course, the amounts set forth in schedules change frequently. However, for the purpose of comparison, consider that at one time the loss of a hand in Pennsylvania resulted in an award of $116,245. The same injury in Colorado brought only $8,736.

When trying to determine a scheduled award for a specific injury, it is best to locate the latest schedule for the state in question. One way to do this is to contact the following agency:

U.S. Department of Labor
Employment Standards Administration
Office of State Liaison and Legislative Analysis
Division of State Workers' Compensation Programs
200 Constitution Ave. NW
Washington, DC 20210
http://www.dol.gov

Death and Burial Benefits

Workers' compensation benefits accrue to the families and dependents of workers who are fatally injured. Typically, the remaining spouse receives benefits for life or until remarriage. However, in some cases, a time period is specified. Dependents typically receive benefits until they reach the legal age of maturity unless they have a condition or circumstances that make them unable to support themselves even after attaining that age. Figure 7–6 contains the death benefits accruing to surviving spouses and children for a geographically distributed list of states. Because the actual amounts of benefits are subject to change, these are provided for illustration and comparison only.

Further expenses are provided in addition to death benefits in all states except Oklahoma. As is the case with all types of workers' compensation, the amount of burial benefits varies from state to state and is subject to change. Figure 7–7 contains the maximum burial benefits for a geographically distributed list of states.

MEDICAL TREATMENT AND REHABILITATION

All workers' compensation laws provide for payment of the medical costs associated with injuries. Most states provide full coverage, but some limit the amount and duration of coverage. For example, in Arkansas, employer liability ceases six months after an injury occurs in those cases in which the employee is able to continue working or six months after he or she returns to work in cases where there is a period of recuperation. In either case, the employer's maximum financial liability is $10,000.[22] In Ohio, medical benefits for silicosis,

State	Percentage of Employee's Wage		Maximum Period
	Spouse Only	Spouse and Children	
Alabama	50	66⅔	500 weeks
Arkansas	35	66⅔	Widow/widowerhood; children until 18 or married
California	66⅔	66⅔	—
Florida	50	66⅔	Widow/widowerhood; children until 18
Indiana	66⅔	66⅔	500 weeks
Kansas	66⅔	66⅔	Widow/widowerhood; children until 18
Maryland	66⅔	66⅔	Widow/widowerhood; children until 18
Montana	66⅔	66⅔	Surviving spouse— 10 years; children until 18
Nebraska	66⅔	75	Widow/widowerhood; children until 18
New Jersey	50	70	Widow/widowerhood; children until 18
New Mexico	66⅔	66⅔	700 weeks
Vermont	66⅔	76⅔	Widow/widowerhood until 62; children until 18
Washington	60	70	Widow/widowerhood; children until 18
Wisconsin	66⅔	—	300 weeks

Figure 7–6
Death benefits for surviving spouses and children for selected states.

asbestosis, and coal miner's pneumoconiosis are paid only in the cases of temporary total or permanent total disability.[23]

The laws also specify who is allowed or required to select a physician for the injured employee. The options can be summarized as follows:

- *Employee selects the physician of choice.* This option is available in Alaska, Arizona, Delaware, Hawaii, Illinois, Kentucky, Louisiana, Maine, Massachusetts, Mississippi, Nebraska, New Hampshire, North Dakota, Ohio, Oklahoma, Oregon, Rhode Island, Texas, the Virgin Islands, Washington, West Virginia, Wisconsin, and Wyoming.

Figure 7–7
Maximum burial allowances for selected states.
Note: Figures are provided for comparison and illustration only. They are subject to change.
Source: U.S. Department of Labor, 2005.

State	Maximum Allowance
Alabama	$3,000
Arkansas	6,000
California	5,000
Florida	2,500
Indiana	6,000
Kansas	5,000
Maryland	5,000
Montana	1,400
Nebraska	6,000
New Jersey	3,500
New Mexico	3,000
Vermont	5,500
Washington	*
Wisconsin	6,000
* 200% of state's average monthly wage	

- *Employee selects the physician from a list provided by the state agency.* This option applies in Connecticut, Nevada, New York, and the District of Columbia.
- *Employee selects the physician from a list provided by the employer.* This option applies in Georgia, Tennessee, and Virginia.
- *Employer selects the physician.* This option applies in Alabama, Florida, Idaho, Indiana, Iowa, Maryland, Montana, New Jersey, New Mexico, North Carolina, South Carolina, and South Dakota.
- *Employer selects the physician, but the selection may be changed by the state agency.* This option applies in Arkansas, Colorado, Kansas, Minnesota, Missouri, Utah, and Vermont.
- *Employer selects the physician, but after a specified period of time, the employee may choose another.* This option applies only in Puerto Rico.

Rehabilitation and Workers' Compensation

Occasionally an injured worker will need rehabilitation before he or she can return to work. There are two types of rehabilitation: medical and vocational. Both are available to workers whose ability to make a living is inhibited by physical and/or mental work-related problems.

Medical rehabilitation consists of providing whatever treatment is required to restore to the extent possible any lost ability to function normally. This may include such services as physical therapy or the provision of prosthetic devices. **Vocational rehabilitation** involves providing the education and training needed to prepare the worker for a new occupation. Whether the rehabilitation services are medical or vocational in nature or both, the goal is to restore the injured worker's capabilities to the level that existed before the accident.

MEDICAL MANAGEMENT OF WORKPLACE INJURIES

Out-of-control workers' compensation cases in the 1990s led to the concept of medical management of workplace injuries.[24] Through better management of workers' compensation claims, more than 30 states have merged the concepts of workers' compensation and managed care. The goals of these state-level efforts are to (1) speed up the processing of workers' compensation claims; (2) reduce costs; (3) reduce fraud and abuse; and (4) improve medical management of workplace injuries.

Workers' compensation and managed care have been merged through the creation of *Health Partnership Programs (HPPs)*. HPPs are partnerships between employers and their state's Bureau of Workers' Compensation (BWC). Employers who choose to participate (some states mandate participation) are required to have a *managed care organization (MCO)* that provides medical management of workplace injuries and illnesses.

Ohio's HPP is an example of how this concept typically works. According to Clairmonte Cappelle,

> When a workplace injury occurs or an illness manifests itself, the employee reports it to the employer and seeks initial medical treatment. At this early stage, the employer or health care provider informs the MCO, which files a first report of injury electronically with the state and begins medical management of the case. The MCO prefers that an injured worker stay within its health care provider network for care. However, injured workers may choose their own doctors and hospitals from the list of 100,000 BWC-certified providers. Except for emergency situations, workers who select non-BWC-certified health care providers will not have their workers' compensation medical costs covered by the state. Once a claim is filed, MCOs work with employers, injured workers, health care providers, third-party administrators and BWC. This includes authorizing certain medical procedures, processing providers' bills for payment by BWC and driving the return-to-work process.[25]

The HPPs are effective because an MCO coordinates the paperwork generated by the injured employee, the employer, health care providers, and the BWC. As the Ohio example illustrates, the results of this coordination are clear. The average time required to process an injury report prior to implementation of the HPP was more than 66 days. The HPP reduced this to approximately 33 days, effectively cutting reporting time in half. Whether managed medical care will be able to continue this positive trend remains to be seen, but the states that have implemented HPPs have learned the following lessons:

- It is better to mandate HPPs than to make them optional.
- Cost containment is only part of the goal. Managed care programs must also include criteria such as lost wages, ability to return to work, and administrative costs to the employer.
- Employees want choice in selecting health care providers.
- Smart return-to-work programs are critical.

ADMINISTRATION AND CASE MANAGEMENT

Even though OSHA specifies what constitutes a recordable accident, it is not uncommon for minor injuries to go unreported. The employee may be given the rest of the day off or treated with first aid and returned to work. This is done to avoid time-consuming paperwork and to hold down the cost of workers' compensation insurance. However, if an accident results in a serious injury, several agencies must be notified. What constitutes a serious injury, like many workers' compensation issues, can differ from state to state. However, as a rule, an injury is serious if it requires over 24 hours of active medical treatment (this does not include passive treatment such as observation). Of course, a fatality, a major disfigurement, or the loss of a limb or digit is also considered serious and must be reported.

At a minimum, the company's insurer, the state agency, and the state's federal counterpart must be notified. Individual states may require that additional agencies be notified. All establish a time frame within which notification must be made. Once the notice of injury has been filed, there is typically a short period before the victim or dependents can begin to receive compensation unless inpatient hospital care is required. However, when payments do begin to flow, they are typically retroactive to the date of the injury.

State statutes also provide a maximum time period that can elapse before a compensation claim is filed. The notice of injury does not satisfy the requirement of filing a claim notice. The two are separate processes. The statute of limitations on **claim notices** varies from state to state. However, most limit the period to no more than a year except in cases of work-related diseases in which the exact date of onset cannot be determined.

All such activities—filing injury notices, filing claim notices, arriving at settlements, and handling disputes—fall under the collective heading of administration and case management. Most states have a designated agency that is responsible for administration and case management. In addition, some states have independent boards that conduct hearings and hear appeals when disputes arise.

Once a workers' compensation claim is filed, an appropriate settlement must be reached. Three approaches can be used to settle a claim: (1) direct settlement, (2) agreement settlement, and (3) public hearing. The first two are used in uncontested cases, the third in contested cases.

1. *Direct settlement.* The employer or its insurance company begins making what it thinks are the prescribed payments. The insurer also sets the period over which payments will be made. Both factors are subject to review by the designated state agency. This approach is used in Arkansas, Michigan, Mississippi, New Hampshire, Wisconsin, and the District of Columbia.
2. *Agreement settlement.* The injured employee and the employer or its insurance company work out an agreement on how much compensation will be paid and for how long. Such an agreement must be reached before compensation payments begin. Typically, the agreement is reviewed by the

designated state administrative agency. In cases where the agency disapproves the agreement, the worker continues to collect compensation at the agreed-on rate until a new agreement can be reached. If this is not possible, the case becomes a contested case.

3. *Public hearing.* If an injured worker feels he or she has been inadequately compensated or unfairly treated, a hearing can be requested. Such cases are known as *contested cases*. The hearing commission reviews the facts surrounding the case and renders a judgment concerning the amount and duration of compensation. Should the employee disagree with the decision rendered, civil action through the courts is an option.

COST ALLOCATION

Workers' compensation is a costly concept. From the outset, one of the basic principles has been cost allocation. **Cost allocation** is the process of spreading the cost of workers' compensation across an industry so that no individual company is overly burdened. The cost of workers' compensation includes the costs of premiums, benefits, and administration. These costs have risen steadily over the years.

When workers' compensation costs by industry are examined, there are significant differences. For example, the cost of workers' compensation for a bank is less than one-half of 1 percent of gross payroll. For a ceramics manufacturer, the percentage might be as high as 3 or 4 percent.

Cost allocation is based on the experience rating of the industry. In addition to being the fairest method (theoretically) of allocating costs, this approach is also supposed to give employers an incentive to initiate safety programs. Opinions vary as to the fairness and effectiveness of this approach. Arguments against it include the following: (1) small firms do not have a sufficient number of employees to produce a reliable and accurate picture of the experience rating; (2) firms that are too small to produce an experience rating are rated by class of industry, thereby negating the incentive figure; (3) premium rates are more directly sensitive to experience levels in larger firms but are less so in smaller companies; and (4) in order to hold down experience ratings, employers may put their efforts into fighting claims rather than preventing accidents. Not much hard research has been conducted to determine the real effects of cost allocation. Such research is badly needed to determine if the theoretical construct of workers' compensation is, in reality, valid.

PROBLEMS WITH WORKERS' COMPENSATION

There are serious problems with workers' compensation in the United States. On one hand, there is evidence of abuse of the system. On the other hand, many injured workers who are legitimately collecting benefits suffer a substantial loss of income. Complaints about workers' compensation are common from all parties involved with the system (employers, employees, and insurance companies).

Earlier in this chapter, the example of the overweight deputy sheriff who applied for benefits because he was distraught over the breakup of his extramarital love affair and a poor performance evaluation was cited as an example of abuse. This individual is just one of thousands who are claiming that job stress has disabled them to the point that workers' compensation is justified. In 1980, there were so few stress claims they were not even recorded as a separate category. Today, they represent a major and costly category.

Stress claims are more burdensome than physical claims because they are typically reviewed in an adversarial environment. This leads to the involvement of expert medical witnesses and attorneys. As a result, even though the benefits awarded for stress-related injuries are typically less than those awarded for physical injuries, the cost of stress claims is often higher because of the litigation.

In addition to abuse, a steadily increasing caseload is also a problem. For example, California saw its caseload increase by 40,000 in one year.[26] As this trend continues, the cost of workers' compensation will increase on a parallel track.

Although the cost of workers' compensation is increasing steadily, the amount of compensation going to injured workers is often disturbingly low. In a given year, if workers' compensation payments in the United States amount to around $27 billion (which is typical), $17 billion of this goes to benefits. Almost $10 billion is taken up by medical costs.[27] The amount of wages paid to injured workers in most states is 66 percent. This phenomenon is not new.

> Almost one-half million families each year are faced with getting by on a drastically reduced income because of a disabling injury suffered by the principal income earner. On-the-job accidents are supposed to be covered by workers' compensation and all states have compensation systems. However, the injured worker rarely receives an income that comes close to what he or she was earning before the accident.[28]

The most fundamental problem with workers' compensation is that it is not fulfilling its objectives. Lost income is not being adequately replaced, the number of accidents has not decreased, and the effectiveness of cost allocation is questionable. Clearly, the final chapter on workers' compensation has not yet been written.

SPOTTING WORKERS' COMPENSATION FRAUD AND ABUSE

There is evidence of waste, fraud, and abuse of the system in all states that have passed workers' compensation laws. However, the public outcry against fraudulent claims is making states much less tolerant of, and much more attentive to, abuse. For example, the Ohio legislature passed a statute that allows criminal charges to be brought against employees, physicians, and lawyers who give false information in a workers' compensation case. This is a positive trend. However, even these measures will not completely eliminate abuse.

For this reason, it is important for organizations to know how to spot employees who are trying to abuse the system by filing fraudulent workers' compensation claims. Following are some factors that should cause employers to view claims with suspicion. However, just because one or more of these factors are present does not mean that an employee is attempting to abuse the system. Rather, these are simply factors that should raise cautionary flags:

- The person filing the claim is never home or available by telephone or has an unlisted telephone number.
- The injury in question coincides with a layoff, termination, or plant closing.
- The person filing the claim is active in sports.
- The person filing the claim has another job.
- The person filing the claim is in line for early retirement.
- The rehabilitation report contains evidence that the person filing the claim is maintaining an active lifestyle.
- No organic basis exists for disability. The person filing the claim appears to have made a full recovery.
- The person filing the claim receives all mail at a post office box and will not divulge a home address.
- The person filing the claim is known to have skills such as carpentry, plumbing, or electrical that could be used to work on a cash basis while feigning a disability.
- There are no witnesses to the accident in question.
- The person filing the claim has relocated out of state or out of the country.
- Demands for compensation are excessive.
- The person filing the claim has a history of filing.
- Doctors' reports are contradictory in nature.
- A soft tissue injury is claimed to have produced a long-term disability.
- The injury in question occurred during hunting season.[29]

These factors can help organizations spot employees who may be trying to abuse the workers' compensation system. It is important to do so because legitimate users of the system are hurt, as are their employers, by abusers of the system. If one or more of these factors are present, employers should investigate the claim carefully before proceeding.

FUTURE OF WORKERS' COMPENSATION

The future of workers' compensation will be characterized by an ongoing need for reform as well as higher premiums, higher deductibles, and less coverage in insurance policies. The key to reforming workers' compensation is finding a way to allocate more of the cost to benefits and medical treatment and less to administration and litigation. Key elements in any reform effort are as follows:

- Stabilizing workers' compensation costs over the long term
- Streamlining administration of the system
- Reducing the costs associated with the resolution of medical issues
- Limiting stress-related claims
- Limiting vocational rehabilitation benefits

Safety Fact

Workers' Compensation Scams in Florida

Hidden Payroll

The owner of a Citrus County roofing and painting business was arrested on charges of evading $832,000 in workers' compensation premiums. These premiums are based on payroll, and investigators charged that the roofer disguised about $1.5 million in payroll costs as subcontractor expenses.

Forklifted Foot

A Jacksonville warehouse manager was arrested on workers' compensation fraud charges after collecting $166,836 from a total disability claim. Although he claimed a forklift ran over his foot causing him to be unable to walk without assistance, investigators obtained a videotape showing him walking his daughter down the aisle at her wedding.

Slip 'n' Fall

A Gainesville woman was arrested on two counts of grand theft and two counts of filing false and fraudulent insurance claims. The charges stem from a series of 11 slip and fall accidents during a four-year period. Investigators allege the woman staged accidents in grocery and department stores to file claims for various back, head, arm, and leg injuries.

Drugstore Deal

The owner of a North Florida drugstore was arrested following a 62-count federal indictment charging that he submitted phony prescription claims for more than $1 million in Medicaid and state insurance reimbursements.

Insurance Mill

A suspended surgeon who owned an orthopedic clinic in Sunrise was arrested based on allegations of involvement in billing insurers for treatments not rendered, altering medical tests, and soliciting patients through runners.

Chiropractic Claims

A Pompano Beach chiropractor was arrested in an alleged scheme to bill insurers for patient visits and treatments never provided.

Source: Florida Department of Labor, Division of Workers' Compensation, 2006.

- Increasing benefits paid for temporary and permanent disabilities
- Reducing the amount that insurers may charge for overhead
- Providing more public input into the setting of rates
- Taking litigation out of the process to the extent possible
- Improving medical treatment management
- Improving overall case management
- Streamlining the claim notification and processing system
- Requiring more sufficient justification from insurance carriers concerning their rates

Controlling Rising Workers' Compensation Costs

The first and best line of defense against escalating workers' compensation costs is a safe and healthy workplace.[30] However, in today's litigious business environment, even this may not be enough. In order to be an effective member of the team responsible for keeping workers' compensation costs in check, safety and health professionals need to cooperate with risk management professionals in implementing the following strategies:

1. Establish an effective safety and health program, and document it clearly and comprehensively for the workers' compensation underwriters.
2. Review workers' compensation claims to ensure that they are accurate before they are submitted to underwriters.
3. Analyze concentrations of risk by location and have comprehensive, up-to-date plans on hand for preventing and responding to catastrophic events.
4. Advise risk management professionals on potential hazards and related risks so they can make informed decisions concerning levels of coverage and deductibles.
5. Communicate frequently with risk management personnel—you and they are on the same team.
6. Develop strategies for dealing with terrorism—many workers' compensation underwriters exclude terrorism from their coverage.[31]

COST-REDUCTION STRATEGIES

Safety and health professionals are responsible for helping their organizations hold down workers' compensation costs. Of course, the best way to accomplish this goal is to maintain a safe and healthy workplace, thereby preventing the injuries that drive up the costs. This section presents numerous other strategies that have proven effective in reducing workers' compensation costs after injuries have occurred, which happen in even the safest environments.

General Strategies

Regardless of the type of organization, there are several rules of thumb that can help reduce workers' compensation claims. These general strategies are as follows:

1. *Stay in touch with the injured employee.* Let injured employees know that they have not been forgotten and that they are not isolated. Answer all their questions and try to maintain their loyalty to the organization.
2. *Have a return-to-work program and use it.* The sooner an injured employee returns to work, even with a reduced workload, the lower workers' compensation costs will be. Reduced costs can, in turn, lower the organization's insurance premium. When using the return-to-work strategy, be cautious. Communicate with the employee and his or her medical treatment team. Have a clear understanding of the tasks that can be done and those that should be avoided, such as how much weight the employee can safely lift.

Colledge and Johnson recommend using the "SPICE" model for improving the effectiveness of return-to-work programs. It consists of the following components:

- Simplicity
- Proximity
- Immediacy
- Centrality
- Expectancy[32]

Simplicity means that the medical professionals who treat injured employees should work closely with safety professionals to prevent "system"-induced complications. Such complications occur when employees become convinced their injuries are more serious than they really are because of ominous-sounding diagnostic terminology and complicated tests and treatments. Medical professionals and safety personnel should work together to keep the terminology simple and to explain tests and treatments in easily understood lay terms.

Proximity means keeping the injured employee as close to the job as possible. Employees who are physically separated from their place of employment and their fellow employees also become mentally separated. Within a short time, what used to be "us" can become "them." Giving as much injury care as possible at the work site, providing light-duty assignments, and communicating regularly with employees whose injuries preclude on-site assignments or

Discussion Case

What Is Your Opinion?

Dr. Maxine McGuire, professor of safety and health at Orange Beach Community College, made the following statement to her class: "Workers compensation laws were passed to get the courts and the lawyers out of the process when employees are injured. The laws have not worked. There are more fraud, abuse, and lawsuits than ever." Is workers' compensation working as a concept? What is your opinion?

treatment will keep employees connected and maintain the advantages of proximity.

Immediacy means that the faster an employee's injury claims can be handled, the less likely he or she will be to develop psychosocial issues that can complicate the recovery process. The longer it takes to process a claim, conduct a diagnosis, and begin treatments the more likely the employee is going to worry about the injury and to become accustomed to being off work. Immediate diagnosis, processing, and treatment can decrease the amount of time that elapses before the employee can begin a return-to-work program.

Centrality means getting the employee, his or her family, the medical professionals handling the case, insurance personnel, and the employer to agree on a common vision for successfully returning the injured party to work as soon as possible. It is important for injured employees and their families to know that everyone involved has the same goal and that everyone is working in good faith to achieve that goal.

Expectancy means creating the expectation that getting the employee well and back to work is the goal of all parties involved. It is achieved by communicating this message clearly to all parties and reinforcing it by establishing short-term goals and timelines for actually being back on the job. Achievement of each respective short-term goal should move the employee one step closer to recovery and return to work.

3. *Determine the cause of the accident.* The key to preventing future accidents and incidents is determining the cause of the accident in question, and the key to holding down workers' compensation costs is preventing accidents. Eliminating the root cause of every accident is fundamental to any cost-containment effort.

Specific Strategies

In addition to the general strategies just presented, there are numerous specific cost-containment strategies that have proven to be effective. These specific strategies are presented in this section.

1. *Cultivate job satisfaction.* According to Jon Gice, increasing job satisfaction is just as important as eliminating physical hazards in the workplace.[33] High stress, aggression, alienation, and social maladjustment—all factors associated with, and aggravated by, a lack of job satisfaction—can make employees less attentive while working. An inattentive employee is an accident waiting to happen. Gice recommends the following strategies for improving job satisfaction:
   - Recognize and reward employees.
   - Communicate frequently and openly with employees about job-related problems.
   - Give employees as much control over their work as possible.
   - Encourage employees to talk freely among themselves.

- Practice conflict management.
- Provide adequate staffing and expense budgets.
- Encourage employees to use employee assistance programs.[34]

2. *Make safety part of the culture.* The Ohio Division of Safety and Hygiene recommends the following steps for making safety part of the organizational culture as a way to reduce workers' compensation costs:
   - Ensure visible, active leadership, involvement, and commitment from senior management.
   - Involve employees at all levels in the safety program and recognize them for their efforts.
   - Provide comprehensive medical care, part of which is a return-to-work program.
   - Ensure effective communication throughout the organization.
   - Coordinate all safety and health processes.
   - Provide orientation and training for all employees.
   - Have written safe work practices and procedures.
   - Have a comprehensive written safety policy.
   - Keep comprehensive safety records and analyze the data contained in those records.[35]

3. *Have a systematic cost-reduction program.* To reduce costs, an organization should have a systematic program that can be applied continually and consistently. Alpha Meat Packing Company of South Gate, California, has had success using the following strategies:
   - Insert safety notes and reminders in employees' paycheck envelopes.
   - Call injured employees at home to reassure them that they will have a job when they return.
   - Keep supervisors trained on all applicable safety and health issues, procedures, rules, and so on.
   - Hold monthly meetings to review safety procedures, strategies, and techniques.
   - Reward employees who give suggestions for making the workplace safer.[36]

4. *Use integrated managed care.* Managed care is credited by many with reducing workers' compensation costs nationwide. Others claim that cost reduction has occurred because managed care dangerously restricts the types and amount of health care provided to injured employees. GE Aircraft Engines is an advocate of managed care that is fully integrated as follows:
   - A comprehensive safety and hazard-prevention program to keep employees safe and healthy
   - On-site medical clinics
   - Plant compensation teams that include nurses, rehabilitation technicians, and third-party adjudicators
   - Claims management
   - Companywide safety and health database
   - Return-to-work program[37]

SUMMARY

1. Workers' compensation was developed to allow injured employees to be compensated without the need for litigation. It has four main objectives: replacement of income, rehabilitation, accident prevention, and cost allocation.

2. Prior to the enactment of workers' compensation laws, employees' only recourse when injured was through the courts, and the prevailing laws favored employers.

3. Early workers' compensation laws were ruled unconstitutional. The constitutional debate continued until 1917 when the U.S. Supreme Court ruled that workers' compensation laws were acceptable.

4. All 50 states have workers' compensation laws, but they vary markedly. All laws are enacted to provide benefits, pay medical costs, provide for rehabilitation when necessary, decrease litigation, and encourage accident prevention.

5. There are three types of workers' compensation insurance: state funds, private insurers, and self-insurance. Six methods are used for determining insurance premium rates for employers: schedule rating, manual rating, experience rating, retrospective rating, premium discounting, and a combination of these.

6. Although an often-stated objective of workers' compensation is the reduction of costly litigation, many cases still go to court. This is particularly true in stress-related cases.

7. Workers' compensation applies when an injury can be categorized as arising out of employment (AOE) or occurring in the course of employment (COE).

8. The definition of an employee can vary from state to state. However, a key concept in distinguishing between an employee and an independent contractor is direction (supervision). Employees are provided direction by the employer; contractors are not.

9. Injuries that are compensable through workers' compensation fall into one of four categories: (a) temporary partial disability, (b) temporary total disability, (c) permanent partial disability, and (d) permanent total disability.

10. Three theoretical approaches to handling permanent partial disability cases are (a) the whole-person theory, (b) the wage-loss theory, and (c) the loss of wage-earning capacity theory.

11. Workers' compensation benefits accrue to the families and dependents of workers who are fatally injured. Typically, the remaining spouse receives benefits for life or until remarriage. Dependents typically receive benefits until they reach the legal age of maturity.

12. All workers' compensation laws provide for payment of medical expenses within specified time periods, but there are differences on how a physician may be selected. The options are (a) employee selects the physician of choice; (b) employee selects the physician from a list provided by the state

agency; (c) employee selects the physician from a list provided by the employer; (d) employer selects the physician; (e) employer selects the physician, but the selection may be changed by the state agency; or (f) employer selects the physician but after a specified period of time the employee may choose another.

13. The definition of a serious injury can vary from state to state. However, as a rule an injury is serious if it requires over 24 hours of active medical treatment (this does not include passive treatment such as observation).

14. Workers' compensation claims can be settled in one of three ways. The first two are uncontested; the third is for contested claims. These approaches are (a) direct settlement, (b) agreement settlement, or (c) public hearing.

15. The goals of medical management of workplace injuries are to (a) speed up the processing of claims, (b) reduce costs, (c) reduce fraud and abuse, and (d) improve medical management.

16. Cost allocation is the process of attempting to spread the cost of workers' compensation across an industry so that no individual company is overly burdened. The cost of workers' compensation includes the cost of premiums, benefits, and administration.

17. The problems being experienced with workers' compensation can be summarized as follows: Workers' compensation is not achieving its intended objectives. It has not succeeded in taking litigation out of the process. The cost of the system rises steadily, but the benefits to injured workers have decreased in real terms. There is evidence of abuse. Therefore, the future of workers' compensation can be summarized in one word—reform.

18. Cost-reduction strategies include staying in touch with injured employees and determining the causes of accidents.

KEY TERMS AND CONCEPTS

Accident prevention

Agreement settlement

Assumption of risk

Blackball

Claim notice

Contributory negligence

Cost allocation

Direct settlement

Employee

Experience rating

Extraterritorial employees

Fear factor

Harmful environment

Income replacement

Independent contractor

Injury arises out of employment (AOE)

Injury occurs in the course of employment (COE)

Litigation

Loss of wage-earning capacity theory

Malpractice

Manual rating

Medical management of workplace injuries

Medical rehabilitation

Monetary benefits

Occupational disease

Permanent partial disability

Permanent total disability

Premium discounting

Private insurance

Public hearing

Reform

Rehabilitation

Retrospective rating

Schedule disabilities

Schedule rating

Self-insurance

State funds

Stress claims

Temporary partial disability

Temporary total disability

Vocational rehabilitation

Wage-loss theory

Whole-person theory

Worker negligence

Workers' compensation

Workplace accident

REVIEW QUESTIONS

1. Explain the underlying rationale of workers' compensation as a concept.
2. List four objectives of workers' compensation.
3. List five types of employees who may not be covered by workers' compensation.
4. What is meant by the term *contributory negligence?*
5. What is meant by the term *assumption of risk?*
6. Explain the reasons for the unprecedented increases in medical costs in the United States.
7. What are the three types of workers' compensation insurance?
8. Insurance companies use one of six methods for determining the premium rates of employers. Select three and explain them.
9. How can one determine if an injury should be considered serious?
10. Explain the concepts of AOE and COE.
11. Distinguish between an employee and an independent contractor.
12. Define the following terms: *temporary disability* and *permanent disability*.
13. Explain the following theories of handling permanent partial disability cases: *whole-person, wage-loss,* and *loss of wage-earning capacity*.
14. Distinguish between medical and vocational rehabilitation.
15. What are the three approaches for settling workers' compensation claims?
16. Explain the concept of medical management of workplace injuries.
17. Explain the theory of cost allocation.
18. Summarize briefly the problems most widely associated with workers' compensation.
19. What types of actions are workers' compensation reform movements likely to recommend in the future?
20. Explain the most common workers' compensation cost-reduction strategies.

ENDNOTES

1. "Workers' Compensation Beginner's Guide." Retrieved from http://beginnersguide.com/small_business/workers_compensation/ April 4, 2006.
2. Ibid.
3. Ibid.
4. Ibid.
5. Ibid.
6. Ibid.
7. H. M. Somers and A. R. Somers, *Workmen's Compensation* (New York: Wiley, 1945), 29.
8. Ibid.
9. D. Gagilardo, *American Social Science* (New York: Harper & Row, 1949), 149.
10. "Workers' Compensation: Beginner's Guide."
11. Ibid.
12. Ibid.
13. Ibid.
14. D. Goetsch, *Workers' Comp Cases*, Report 2006 (Niceville, FL: The Development Institute, April 2006), 12–13.
15. Ibid.
16. Ibid.
17. "Workers' Comp Update," *Occupational Hazards* 60, no. 7: 93.
18. Ibid.
19. Society of Manufacturing Engineers, *Accident Facts* (Chicago: Society of Manufacturing Engineers, 1998), 36.
20. D. Goetsch, "Workers' Comp Cases," 14.
21. Ibid.
22. U.S. Department of Labor, *State Workers' Compensation Laws* (Washington, DC: U.S. Department of Labor, January 2005).
23. Ibid.
24. C. Cappelle, "Making a Strong Case for Managed Care," *Occupational Hazards* 61, no. 4: 67–71.
25. Ibid., 68.
26. National Safety Council, *Accident Facts* (Chicago: National Safety Council 2004), 43.
27. Ibid.
28. U.S. Department of Labor, *Federal Worker 2000* (Washington, DC: U.S. Department of Labor, March 2000), 2.
29. D. Goetsch, "Workers' Comp Fraud Detection Checklist," Report 2005–6 (Niceville, FL: The Development Institute, April 2005), 8.
30. Christine Fuge, "Tough Times Ahead for Workers' Compensation," *Occupational Hazards* 64, no. 12: 39–42.
31. Ibid., 40.
32. A. Colledge and H. Johnson, "The S.P.I.C.E. Model for Return To Work," *Occupational Health & Safety* 69, no. 2: 64–69.

33. Jon Gice, *The Relationship between Job Satisfaction and Workers Compensation Claims* (Malvern, PA: Chartered Property Casualty Underwriters Society, 720 Providence Rd., 19355–0709).
34. Ibid.
35. Ohio Division of Safety and Hygiene, "Ohio Prompts 10 Steps to Reduced Comp Costs," *Occupational Hazards*, August 1996, 59.
36. "Workers' Comp Update: Meatpacking Industry Cuts Comp Claims," *Occupational Hazards*, May 1996, 103.
37. Candace Goforth, "Workers' Comp: Is Managed Care the Answer?" *Occupational Hazards*, October 1996, 126.

ACCIDENT INVESTIGATION AND REPORTING

Major Topics

- Types of Accident Investigations
- When to Investigate
- What to Investigate
- Who Should Investigate
- Conducting the Investigation
- Interviewing Witnesses
- Reporting Accidents
- Ten Accident Investigation Mistakes to Avoid

Dan Hartshorn defines an accident as "any unplanned event that causes injury, illness, property damage or harmful disruption of work process."[1]

When an accident occurs, it is important that it be investigated thoroughly. The results of a comprehensive accident report can help safety and health professionals pinpoint the cause of the accident. This information can then be used to prevent future accidents, which is the primary purpose of accident investigation.

The Society of Manufacturing Engineers describes the importance of thoroughly investigating accidents:

> The primary reason for investigating an accident is not to identify a scapegoat, but to determine the cause of the accident. The investigation concentrates on gathering factual information about the details that led to the accident. If investigations are conducted properly, there is the added benefit of uncovering problems that did not directly lead to the accident. This information benefits the ongoing effort of reducing the likelihood of accidents. As problems are revealed during the investigation, action items and improvements that can prevent similar accidents from happening in the future will be easier to identify than at any time. [2]

This chapter gives prospective and practicing safety and health professionals the information they need to conduct thorough, effective accident investigations and prepare comprehensive accident reports.

TYPES OF ACCIDENT INVESTIGATIONS

There are *accident reports* and there are *accident-analysis reports*. An **accident report** is completed when the accident in question represents only a minor incident. It answers the following questions: *who*, *what*, *where*, and *when*. However, it does not answer the *why* question.[3] An accident report can be completed by a person with very little formal investigation and reporting training or experience. Supervisors often complete accident reports which, in turn, might be used later as part one of a more in-depth accident report. OSHA's Form 301 (see Chapter 6) can be used for accident reports.

An **accident-analysis report** is completed when the accident in question is serious. This level of report should answer the same questions as the regular accident report plus one more—*why*. Consequently it involves a formal accident analysis. The analysis is undertaken for the purpose of determining the root cause of the accident. Accident analysis requires special skills and should be undertaken only by an individual with those skills. There are two reasons for this. First, the accident analysis must identify the actual root cause or the company will expend resources treating only symptoms or, even worse, solving the wrong problem. Second, serious accidents are always accompanied by the potential for litigation. If there might be legal action as a result of an accident, it is important to have a professional conduct the investigation even if it means bringing in an outside consultant.

How can safety and health professionals determine when an accident report is sufficient and when an accident-analysis report is called for? Accident reports are called for when the accident in question is a minor incident that did *not* result in any of the following circumstances: death, loss of consciousness, medical treatment beyond first aid, more than one additional day of lost work beyond the day of the accident, or any kind of modifications to the injured employee's work duties beyond those that might occur on the day of the accident.

Accident-analysis reports are called for when any of the following circumstances result from the accident in question: death, loss of consciousness, professional medical treatment beyond first aid, one or more days of lost work over and above any time lost the day of the accident, or any modifications to the injured employee's work duties beyond those that might occur on the day of the injury.

WHEN TO INVESTIGATE

Of course, the first thing to do when an accident takes place is to implement **emergency procedures**. This involves bringing the situation under control and caring for the injured worker. As soon as all emergency procedures have been accomplished, the accident investigation should begin. Waiting too long to complete an investigation can harm the results. This is an important rule of thumb to remember. Another is that *all* accidents, no matter how small, should be investigated. Evidence suggests that the same factors that cause minor accidents cause major accidents.[4] Further, a near miss should be treated like an accident and investigated thoroughly.

There are several reasons why it is important to conduct investigations immediately. First, immediate investigations are more likely to produce accurate information. Conversely, the longer the time span between an accident and an investigation, the greater the likelihood of important facts becoming blurred as memories fade. Second, it is important to collect information before the accident scene is changed and before witnesses begin comparing notes. Human nature encourages people to change their stories to agree with those of other witnesses.[5] Finally, an immediate investigation is evidence of management's commitment to preventing future accidents. An immediate response shows that management cares.[6]

WHAT TO INVESTIGATE

The purpose of an **accident investigation** is to collect facts. It is not to find fault. It is important that safety and health professionals make this distinction known to all involved. **Fault finding** can cause reticence among witnesses who have valuable information to share. **Causes** of the accident should be the primary focus. The investigation should be guided by the following words: **who**, **what**, **when**, **where**, **why**, and **how**.

This does not mean that mistakes and breaches of precautionary procedures by workers go unnoted. Rather, when these things are noted, they are recorded as facts instead of faults. If fault must be assigned, that should come later, after all the facts are in. The distinction is a matter of emphasis. The National Safety Council summarizes this approach:

> As you investigate, don't put the emphasis on identifying who could be blamed for the accident. This approach can damage your credibility and generally reduce the amount and accuracy of information you receive from workers. This does not mean you ignore oversights or mistakes on the part of employees nor does it mean that personal responsibility should not be determined when appropriate. It means that the investigation should be concerned with only the facts. In order to do a quality job of investigating accidents you must be objective and analytical. [7]

In attempting to find the facts and identify causes, certain questions should be asked, regardless of the nature of the accident. The Society of Manufacturing Engineers recommends using the following questions when conducting accident investigations:

- What type of work was the injured person doing?
- Exactly what was the injured person doing or trying to do at the time of the accident?

- Was the injured person proficient in the task being performed at the time of the accident? Had the worker received proper training?
- Was the injured person authorized to use the equipment or perform the process involved in the accident?
- Were there other workers present at the time of the accident? If so, who are they, and what were they doing?
- Was the task in question being performed according to properly approved procedures?
- Was the proper equipment being used, including personal protective equipment?
- Was the injured employee new to the job?
- Was the process, equipment, or system involved new?
- Was the injured person being supervised at the time of the accident?
- Are there any established safety rules or procedures that were clearly not being followed?
- Where did the accident take place?
- What was the condition of the accident site at the time of the accident?
- Has a similar accident occurred before? If so, were corrective measures recommended? Were they implemented?
- Are there obvious solutions that would have prevented the accident? [8]

The answers to these questions should be carefully and copiously recorded. You may find it helpful to dictate your findings into a microcassette recorder. This approach allows you to focus more time and energy on investigating and less on taking written notes.

Regardless of how the findings are recorded, it is important to be thorough. What may seem like a minor unrelated fact at the moment could turn out to be a valuable fact later when all the evidence has been collected and is being analyzed.

Common Causes of Accidents

Hartshorn places many of the common causes of accidents in the following categories: personal beliefs and feelings, decision to work unsafely, mismatch or overload, systems failures, traps, unsafe conditions, and unsafe acts.[9] The common causes in each of these categories can help investigators determine the root cause of an accident.

> *Personal beliefs and feelings.* Causes in this category include the following: individual did not believe the accident would happen to him or her; individual was working too fast, showing off, or being a know-it-all; individual ignored the rules out of contempt for authority and rules in general; individual gave in to peer pressure; and individual had personal problems that clouded his or her judgment.
>
> *Decision to work unsafely.* Some people, for a variety of reasons, feel it is in their best interests or to their benefit to work unsafely. Hence, they make a conscious decision to do so.

Mismatch or overload. Causes in this category include the following: individual is in poor physical condition; individual is fatigued; individual has a high stress level; individual is mentally unfocused or distracted; the task required is too complex or difficult; the task required is boring; the physical environment is stressful (e.g., excessive noise, heat, dust, or other factors); the work in question is very demanding—even for an individual in good physical condition; and individual has a negative attitude (e.g., hostile, uncooperative, apathetic, etc.).

Systems failure. Causes in this category consist of the various errors management makes that are not grossly negligent or *serious and willful.* Common causes in this category include lack of a clear policy; lack of rules, regulations, and procedures; poor hiring procedures; inadequate monitoring and inspections; failure to correct known hazards; insufficient training for employees; rules that are in place but are not enforced; no rewarding or reinforcement of safe behavior; inadequate tools and equipment provided; production requirements set too high; inadequate communication to employees of safety concerns, statistics, and rules; poor safety management; no or insufficient job safety analysis; and insufficient management support for safety.

Traps. Poor design of workstations and processes can create *traps* that, in turn, lead to unsafe behavior. Common causes in this category include defective equipment; failure to provide, maintain, and replace proper personal protective equipment; failure to train employees in the proper use of their personal protective equipment; overly complicated and confusing controls; poorly laid out work area; mechanical lifting equipment that is inadequate for the jobs required of it; uncontrolled hazards that might lead to slips and falls; excessive reaching, bending, stooping, and twisting; excessive contact pressure, vibration, or force; awkward postures that result from poor workstation or tool design; excessive temperature extremes; insufficient lighting; and insufficient ventilation.

Unsafe conditions. Common causes in this category include the following: unsafe condition created by the person injured in the accident; unsafe condition created by a fellow employee; unsafe condition created by a third party; unsafe condition created by management; unsafe condition knowingly overlooked by management; and unsafe condition created by the elements (e.g., rain, sun, snow, ice, wind, darkness, etc.).

Unsafe acts. Common causes in this category include the following: individual chooses to ignore the rules; people are involved in horseplay or fighting; individual uses drugs or alcohol; individual uses unauthorized tools or equipment; individual chooses an improper work method; individual fails to ask for information or other resources needed to do the job safely; individual forgets a rule, regulation, or procedure; individual does not pay proper attention; and individual uses improper body mechanics.

WHO SHOULD INVESTIGATE

Who should conduct the accident investigation? Should it be the responsible supervisor? The safety and health professional? A higher-level manager? An outside specialist? There is no simple answer to this question, and there is disagreement among professional people of goodwill.

In some companies, the supervisor of the injured worker conducts the investigation. In others, a safety and health professional performs the job. Some companies form an investigative team; others bring in outside specialists. There are several reasons for the various approaches used. Factors considered in deciding how to approach accident investigations include:

- Size of the company
- Structure of the company's safety and health program
- Type of accident
- Seriousness of the accident
- Technical complexity
- Number of times that similar accidents have occurred
- Company's management philosophy
- Company's commitment to safety and health

After considering all the variables just listed, it is difficult to envision a scenario in which the safety and health professional would not be involved in conducting an accident investigation. If the accident in question is very minor, the injured employee's supervisor may conduct the investigation, but the safety and health professional should at least study the accident report and be consulted regarding recommendations for corrective action.

If the accident is so serious that it has widespread negative implications in the community and beyond, responsibility for the investigation may be given to a high-level manager or corporate executive. In such cases, the safety and health professional should assist in conducting the investigation. If a company prefers the team approach, the safety and health professional should be a member of the team and, in most cases, should chair it. Regardless of the approach preferred by a given company, the safety and health professional should play a leadership role in collecting and analyzing the facts and developing recommendations.

CONDUCTING THE INVESTIGATION

The questions in the previous section summarize what to look for when conducting accident investigations. Figure 8–1 lists five steps to follow in conducting an accident investigation. [10] These steps are explained in the following paragraphs.

Figure 8–1
Steps in conducting an accident
investigation.

Isolate the Accident Scene

You may have seen a crime scene that was sealed off by the police. The entire area surrounding such a scene is typically blocked off by barriers or heavy yellow tape. This is done to keep curious onlookers from removing, disturbing, or unknowingly destroying vital evidence. This same approach should be used when conducting an accident investigation. As soon as emergency procedures have been completed and the injured worker has been removed, the accident scene should be *isolated* until all pertinent evidence has been collected or observed and recorded. Further, nothing but the injured worker should be removed from the scene. If necessary, a security guard should be posted to maintain the integrity of the **accident scene**. The purpose of isolating the scene is to maintain as closely as possible the conditions that existed at the time of the accident.

Record All Evidence

It is important to make a permanent record of all *pertinent evidence* as quickly as possible. There are three reasons for this: (1) certain types of evidence may be perishable; (2) the longer an accident scene must be isolated, the more likely it is that evidence will be disturbed, knowingly or unknowingly; and (3) if the isolated scene contains a critical piece of equipment or a critical component in a larger process, pressure will quickly mount to get it back in operation. Evidence can be recorded in a variety of ways including written notes, sketches, photography, videotape, dictated observations, and diagrams. In deciding what to record, a good rule of thumb is *if in doubt, record it*. It is better to record too much than to skip evidence that may be needed later after the accident scene has been disturbed.

Photograph or Videotape the Scene

This step is actually an extension of the previous step. Modern photographic and videotaping technology has simplified the task of observing and recording evidence. Safety and health professionals should be proficient in the operation of a camera, even if it is just an instant camera, and a videotaping camera.

The advent of the digital camera has introduced a new meaning for the concept of "instant photographs." Using a digital camera in conjunction with a computer, photographs of accident scenes can be viewed immediately and transmitted instantly to numerous different locations. Digital camera equipment is especially useful when photographs of accident scenes in remote locations are needed.

Both still and video cameras should be on hand, loaded, and ready to use immediately should an accident occur. As with the previous step, a good rule of thumb in photographing and videotaping is *if in doubt, shoot it*. When recording evidence, it is better to have more shots than necessary than it is to risk missing a vital piece of evidence.

A problem with photographs is that, by themselves, they don't always reveal objects in their proper perspective. To overcome this shortcoming, the National Safety Council recommends the following technique:

> When photographing objects involved in the accident, be sure to identify and measure them to show the proper perspective. Place a ruler or coin next to the object when making a close-up photograph. This technique will help to demonstrate the object's size or perspective. [11]

Identify Witnesses

In **identifying witnesses**, it is important to compile a witness list. Names on the list should be recorded in three categories: (1) **primary witnesses**, (2) **secondary witnesses**, and (3) **tertiary witnesses** (Figure 8–2). When compiling the witness list, ask employees to provide names of all three types of witnesses.

Figure 8–2
Categories of accident witnesses.

- Primary witnesses are eyewitnesses to the accident.

- Secondary witnesses are witnesses who did not actually see the accident happen, but were in the vicinity and arrived on the scene immediately or very shortly after the accident.

- Tertiary witnesses are witnesses who were not present at the time of the accident nor afterward but may still have relevant evidence to present (e.g., an employee who had complained earlier about a problem with the machine involved in the accident).

Interview Witnesses

Every witness on the list should be interviewed, preferably in the following order: primary witnesses first, secondary next, and tertiary last. After all witnesses have been interviewed, it may be necessary to reinterview witnesses for clarification or corroboration. Interviewing witnesses is such a specialized process that the next major section is devoted to it.

INTERVIEWING WITNESSES

The techniques used for interviewing accident witnesses are designed to ensure that the information is objective, accurate, as untainted by the personal opinions and feelings of witnesses as possible, and able to be corroborated. For this reason, it is important to understand the *when*, *where*, and *how* of interviewing the accident witnesses.

When to Interview

Immediacy is important. Interviews should begin as soon as the witness list has been compiled and, once begun, should proceed expeditiously. There are two main reasons for this. First, a witness's recollections will be best right after the accident. The more time that elapses between the accident and the interview, the more blurred the witness's memory will become. Second, immediacy avoids the possibility of witnesses comparing notes and, as a result, changing their stories. This is just human nature, but it is a tendency that can undermine the value of testimony given and, in turn, the facts collected. Recommendations based on questionable facts are not likely to be valid. Also, witnesses should be interviewed individually and separately, preferably before they have talked to each other.

Where to Interview

The best place to interview is at the accident scene. If this is not possible, interviews should take place in a private setting elsewhere. It is important to ensure that all distractions are removed, interruptions are guarded against, and the witness is not accompanied by other witnesses. All persons interviewed should be allowed to relate their recollections without fear of contradiction or influence by other witnesses or employees. It is also important to select a neutral location in which witnesses will feel comfortable. Avoid the **"principal's office syndrome"** by selecting a location that is not likely to be intimidating to witnesses.

How to Interview

The key to getting at the facts is to put the witness at ease and to listen. Listen to what is said, how it is said, and what is not said. Ask questions that will get at the information listed earlier in this chapter, but phrase them in an **open-ended** format. For example, instead of asking "Did you see the victim pull the red lever?" phrase your question as follows: "Tell me what you saw."

Safety Fact

Where to Conduct Accident Interviews

To ensure that employees are willing to give accurate information, safety and health professionals should conduct accident interviews in the privacy of their office. Right? Not necessarily. Experience has shown that the best way to promote accuracy is to interview witnesses at the site of the accident. This puts the accident interview *in context* in a setting that will help stimulate the memory. To ensure privacy and confidentiality, interview witnesses one at a time, at the accident site.

Don't lead witnesses with your questions or influence them with gestures, facial expressions, tone of voice, or any other form of nonverbal communication. Interrupt only if absolutely necessary to seek clarification on a critical point. Remain nonjudgmental and objective.

The information being sought in an accident investigation can be summarized as *who, what, when, where, why,* and *how* (Figure 8–3). As information is given, it may be necessary to take notes. If you can keep your notetaking to a minimum during the interview, your chances of getting uninhibited information are increased. Notetaking can distract and even frighten a witness.

An effective technique is to listen during the interview and make mental notes of critical information. At the end of the interview, summarize what you have heard and have the witness verify your summary. After the witness leaves, develop your notes immediately.

A question that sometimes arises is, "Why not tape the interview?" Safety and health professionals disagree on the effectiveness and advisability of taping. Those who favor taping claim it allows the interviewer to concentrate on listening without having to worry about forgetting a key point or having to interrupt the witnesses to jot down critical information. It also preserves everything that is said for the record as well as the tone of voice in which it is said. A complete transcript of the interview also ensures that information is not taken out of context.

Those opposed to taping say that taping devices tend to inhibit witnesses so that they are not as forthcoming as they would be without taping. Taping also slows down the investigation while the taped interview is transcribed and the interviewer wades through voluminous testimony trying to separate critical information from irrelevant information.

In any case, if the interview is to be taped, the following rules of thumb should be applied:

- Use the smallest, most unobtrusive taping device available, such as a microcassette recorder.
- Inform the witness that the interview will be taped.
- Make sure the taping device is working properly and that the tape it contains can run long enough so that you don't have to interrupt the witness to change it.

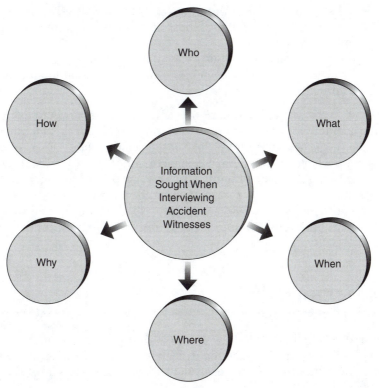

Figure 8–3
Questions to ask when interviewing witnesses.

Safety Fact

Management Is the Cause

One of the most difficult situations that a safety and health professional will face is when an investigation reveals that management—through action or inaction—is the root cause of an accident. How does the safety professional look his or her boss in the eye and say, "You are at fault"? Remember two things when facing such a situation. First, focus on the *condition* that led to the accident rather than whose action or inaction caused the condition. You want the condition corrected, and you don't want a higher manager wasting time and effort covering up. Second, be tactful. Remember, tact means "driving in the nail without breaking the board." By handling such situations in this way, you may achieve two things, both of which are good: (1) a hazardous condition will be corrected, and (2) you will win an important ally.

- Take time at the beginning of the interview to discuss unrelated matters long enough to put the witness at ease and overcome the presence of the taping device.
- Make sure that personnel are available to transcribe the tapes immediately.
- Read the transcripts as soon as they are available and highlight critical information.

An effective technique to use with eyewitnesses is to ask them to reenact the accident for you. Of course, the effectiveness of this technique is enhanced if the reenactment can take place at the accident site. However, even when this is not possible, an eyewitness reenactment can yield valuable information.

In using the **reenactment** technique, a word of caution is in order. If an eyewitness does exactly what the victim did, there may be another accident. Have the eyewitnesses explain what they are going to do before letting them do it. Then, have them *simulate* rather than actually perform the steps that led up to the accident.

REPORTING ACCIDENTS

An accident investigation should culminate in a comprehensive *accident report*. The purpose of the report is to record the findings of the accident investigation, the cause or causes of the accident, and recommendations for corrective action.

OSHA has established requirements for reporting and record keeping. According to OSHA document 2056,

> Employers of 11 or more employees must maintain records of occupational injuries and illnesses as they occur. Employers with 10 or fewer employees are exempt from keeping such records unless they are selected by the Bureau of Labor Statistics (BLS) to participate in the Annual Survey of Occupational Injuries and Illnesses.[12]

All injuries and illnesses are supposed to be recorded, regardless of severity, if they result in any of the outcomes shown in Figure 8–4. If an accident results

Figure 8–4
OSHA record-keeping requirements.

Injuries and illnesses must be recorded if they result in any of the following:

- Death
- One or more lost work days
- Restriction of motion or work
- Loss of consciousness
- Transfer to another job
- Medical treatment (more than first aid)

Discussion Case

What Is Your Opinion?

"Find out who is at fault and get rid of him," demanded the CEO. "I'm not going to have a careless employee running up our health care costs. It's tough enough trying to make a profit without some careless employee causing accidents. There will be a lawsuit, just you wait and see. We are going to be sued!" Gordon Jasperton, Clark Processing Company's safety director, bit his tongue and just listened. He had learned to let his boss vent before making a counterproposal. When the time seemed right, Jasperton said, "Sir, if we focus on finding an employee to blame, it's just going to make matters worse. I'll never get to the root of it that way." Whose approach is best in this case? What is your opinion?

in the death of an employee or hospitalization of five or more employees, a report must be submitted to the nearest OSHA office within 48 hours. This rule applies regardless of the size of the company. Reporting locally within an organization for insurance, legal, prevention, and management purposes and reporting for OSHA purposes can be two different tasks. OSHA's reporting requirements were covered in Chapter 6. Reporting procedures in this section pertain to local in-house reporting.

Accident report forms vary from company to company. However, the information contained in them is fairly standard. Regardless of the type of form used, an accident report should contain at least the information needed to meet the record-keeping requirements set forth by OSHA. This information includes at least the following, according to the National Safety Council:

- Case number of the accident
- Victim's department or unit
- Location and date of the accident or date that an illness was first diagnosed
- Victim's name, social security number, gender, age, home address, and telephone number
- Victim's normal job assignment and length of employment with the company
- Victim's employment status at the time of the accident (i.e., temporary, permanent, full-time, part-time)
- Case numbers and names of others injured in the accident
- Type of injury and body part(s) injured (e.g., burn to right hand; broken bone, lower right leg) and severity of injury (i.e., fatal, first aid only required, hospitalization required)
- Name, address, and telephone number of the physician called
- Name, address, and telephone number of the hospital to which the victim was taken
- Phase of the victim's work day when the accident occurred (e.g., beginning of shift, during break, end of shift, and so on)

- Description of the accident and how it took place, including a step-by-step sequence of events leading up to the accident
- Specific tasks and activities with which the victim was involved at the time of the accident (e.g., task: mixing cleaning solvent; activity: adding detergent to the mixture)
- Employee's posture or proximity related to his or her surroundings at the time of the accident (e.g., standing on a ladder; bent over at the waist inside the robot's work envelope)
- Supervision status at the time of the accident (i.e., unsupervised, directly supervised, indirectly supervised)
- Causes of the accident
- Corrective actions that have been taken so far
- Recommendations for additional corrective action[13]

In addition to these items, you may want to record such additional information as the list of witnesses; dates, times, and places of interviews; historical data relating to similar accidents; information about related corrective actions that were made previously but had not yet been followed up on; and any other information that might be relevant. Figure 8–5 is an example of an accident report form that meets the OSHA record-keeping specifications.

Why Some Accidents Are Not Reported

In spite of OSHA's reporting specifications, some accidents still go unreported. According to Cunningham and Kane,

> The majority of accidents are not being reported! Articles in the Wall Street Journal testify to this fact. Many firms failed to report OSHA recordable incidents, presumably either to avoid OSHA inspections that result from poor incident rates, or to achieve statistical goals. The saddest part of non-reporting of accidents is that they are not investigated to determine and eliminate the causes.[14]

There are several reasons why accidents go unreported. Be familiar with these reasons so that you can do your part to overcome them. Cunningham and Kane list the main reasons as follows:

1. *Red tape.* Some people see the paperwork involved in accident reporting as red tape and, therefore, don't report accidents just to avoid paperwork.
2. *Ignorance.* Not all managers and supervisors are as knowledgeable as they should be about the reasons for accident reporting. Many are not familiar with OSHA's reporting specifications.
3. *Embarrassment.* Occasionally, people do not report an accident because they are embarrassed by their part in it. A supervisor who did not properly supervise or a manager who has not provided the proper training for employees may be embarrassed to file a report.
4. *Record-spoiling.* Some accidents go unreported just to preserve a safety record, such as the record for days worked without an accident.

ACCIDENT REPORT FORM

Fairmont Manufacturing Company
1501 Industrial Park Road
Fort Walton Beach, FL 32548
904-725-4041

Victim-Related Information

Person completing report _____ Case no. _____

Victim's name _____

Gender _____ Age _____

Date of accident/illness _____

Victim's home address/telephone _____

Victim's assignment at the time of the accident and length of time in that assignment:

_____

_____

_____

Victim's normal job and length of time in that job: _____

_____

_____

Time of injury/illness and phase of victim's work day: _____

_____

_____

Severity of the injury (e.g., hospitalization required, first aid required, etc.): _____

_____

_____

Type of injury and body part(s) injured: _____

_____

Exact location of the accident (which facility, department, place within the department):

_____

Physician and hospital: _____

_____

Figure 8–5
Sample accident report form.
Note: Complete one form for each injured worker.

5. *Fear of repercussions.* Some accidents go unreported because the people involved are afraid of being found at fault, being labeled accident prone, and being subjected to other negative repercussions.

6. *No feedback.* Some accidents go unreported because those involved feel filing a report is a waste of time. This typically happens when

---

Accident-Related Information

Accident description with step-by-step sequence of events: _____

_____

_____

Task and specific activity at the time of the accident: _____

_____

_____

Posture/proximity of employee at the time of the accident: _____

_____

_____

Supervision status at the time of the accident: _____

_____

_____

Apparent causes including conditions, actions, events, and activities and other
contributing factors: _____

_____

_____

Recommendations for corrective action: _____

_____

_____

Case numbers and names of other persons injured in the accident: _____

_____

_____

Witnesses to the accident, and dates/places of their interviews:

_____   _____

_____   _____

_____   _____

_____   _____

_____   _____
Reporter name   Date

_____   _____
Employee name   Date

Figure 8–5 (*continued*)

management does not respond to recommendations made in earlier accident reports.[15]

Clearly, these reasons for not reporting accidents present safety and health professionals with a challenge. To overcome these inhibitors, it is necessary to develop a simple reporting system that will not be viewed as too much bureaucratic paperwork to have to do. Safety and health professionals must educate personnel at all levels concerning the purpose of accident reporting and why it is important. An important step is to communicate the fact that fault finding is not the purpose. Another important step is to follow up to ensure that recommendations are acted on or that employees are made aware of why they aren't. This helps ensure the integrity of the process.

Discipline and Accident Reporting

Fault finding is not the purpose of an accident investigation. However, an investigation sometimes reveals that an employee has violated or simply overlooked safety regulations. Should such violations be condoned? According to Kane and Cunningham,

> Many companies condone nonconformance to safety rules as long as no injury results. However, if the nonconformance results in an accident involving an injury, the disciplinary boom is promptly lowered. This inconsistency inevitably leads to resentment and failure to report accidents and a hiding of accident problems.[16]

There is a built-in dilemma here that modern safety and health professionals must be prepared to handle. On one hand, it is important that fault finding not be seen as the purpose of an accident investigation. Such a perception limits the amount of information that can be collected. On the other hand, if those workers whose behavior leads to accidents are not disciplined, the credibility of the safety program is undermined. Kane and Cunningham recommend the following procedures for handling this dilemma: *Never* discipline an employee because he or she had an accident. *Always* discipline employees for noncompliance with safety regulations.[17]

Such an approach applied with consistency will help maintain the integrity of both the accident investigation process and the overall safety program.

TEN ACCIDENT INVESTIGATION MISTAKES TO AVOID

The amount of information you collect and how you collect it will go a long way toward determining how effective your resultant corrective actions will be following a workplace accident. According to William R. Coffee Jr., safety and health professionals should avoid the following commonly made mistakes when investigating accidents.

- *Failing to investigate near misses.* A near miss is simply an accident that did not happen because of luck. Consequently, investigating near misses can reveal critical accident prevention information.
- *Taking ineffective corrective action.* Ineffective corrective action is often the result of a cursory accident investigation. When investigating, look for the root cause, not the symptoms. Corrective action based on symptoms will not prevent future accidents.

- *Allowing your biases to color the results of the investigation.* Look for facts and be objective when investigating an accident. Do not make assumptions or jump to conclusions. One of the best ways to eliminate bias in accident investigations is to use a standard, structured routine and to skip no steps in the routine.
- *Failing to investigate in a timely manner.* Accident investigations should begin as soon as possible after the accident. The longer you wait to begin, the more likely it is that evidence will be lost, corrupted, or compromised. For example, once people start talking to each other about what they saw, invariably their memories will be shaped by the opinions of their fellow workers and witnesses. People walking through an accident scene can compromise the integrity of the scene by unwittingly destroying evidence.
- *Failing to account for human nature when conducting interviews.* Often what those involved in an accident as well as witnesses to an accident will say during an interview will be shaped by their desire to escape blame, deflect blame to someone else, or protect a friend. This is why it is important to interview witnesses and others involved privately and individually, and to look for corroborating evidence to support (or refute) their input.
- *Failing to learn investigation techniques.* Before investigating an accident, safety and health professionals should complete specialized training or undertake self-study to learn investigation techniques such as those presented in this chapter. An unskilled investigator is not likely to conduct a valid investigation.
- *Allowing politics to enter into an investigation.* The goal of an investigation is and must be to identify the root cause so that appropriate corrective action can be taken. Personal likes, dislikes, favoritism, and office politics will corrupt an investigation from the outset.
- *Failing to conduct an in-depth investigation.* Everyone is in a hurry and investigating an accident was not on your agenda for the day. In addition, there is sometimes pressure from higher management to "get this thing behind us." Such pressures and circumstances can lead to a rushed investigation in which the goal is to get it over with, not to find the root cause of the accident. Surface-level investigations almost ensure that the same type of accident will happen again.
- *Allowing conflicting goals to enter into an investigation.* The ultimate goal of an accident investigation is to prevent future accidents and injuries. However, even when that is your goal, there may be other people who have different goals. Some may see the investigation as an opportunity to deflect blame, others may see it as an opportunity to protect the organization from litigation, and some may see it as a way to explain not meeting production quotas or performance standards. Safety and health professionals should be aware that other agendas may be in play every time an accident investigation is conducted. For this reason, objectivity, structure, and routine are critical.

- *Failing to account for the effects of uncooperative people.* One would think that employees and management personnel would automatically want to cooperate in accident investigation to ensure that similar accidents are prevented in the future. Unfortunately, this is not always the case. People will not always cooperate for a variety of reasons—all growing out of the concept of perceived self-interest. Further, the lack of cooperation will not always be overt. In fact, often it will be covert (e.g., a person you need to interview keeps putting you off or canceling meetings). Safety and health professionals need to understand that self-interest is one of the most powerful motivators of human beings and factor this into their planning for accident investigations.[18]

These 10 mistakes will probably never be completely eliminated from every accident investigation. However, if safety and health professionals are aware of them, they can at least ensure that such mistakes are minimized. The fewer of these mistakes that are made during an accident investigation, the better the quality of the investigation and the more likely that it will lead to effective corrective action.

SUMMARY

1. Accidents are investigated to identify causal factors that could lead to other accidents if not corrected, not to assign blame.
2. There are accident reports and accident-analysis reports. The latter attempt to determine *why* and should be completed by a professional.
3. It is important to begin an accident investigation as soon as possible after an accident occurs so that evidence and the memories of witnesses are still fresh.
4. Facts to be uncovered in an accident investigation can be summarized as who, what, when, where, why, and how.
5. Common causes of accidents fall into the following categories: personal beliefs and feelings, decision to work unsafely, mismatch or overload, systems failure, traps, unsafe conditions, and unsafe acts.
6. Who conducts the accident investigation can vary according to circumstances. However, regardless of how it is done, the safety and health professional should play an active role in the process.
7. Steps for conducting an accident investigation are as follows: (a) isolate the accident scene, (b) record all evidence, (c) photograph or videotape the accident scene, (d) identify witnesses, and (e) interview witnesses.
8. Witnesses to accidents fall into one of three categories: primary (eyewitnesses); secondary (were present at the scene, but did not see the accident); and tertiary (were not present but have information that may be relevant).
9. Interviews should take place at the accident site whenever possible. When this isn't practical, interviews should take place at a neutral location that is private and where the witness is comfortable.

10. The keys to getting at the facts in an interview are (a) put the witness at ease, (b) ask open-ended questions, and (c) listen. Interrupt only if absolutely necessary.
11. When possible, let eyewitnesses reenact the accident through simulation at the job site. Do not let them actually perform the tasks that led up to the accident.
12. The purpose of an accident report is to record the findings of the accident investigation, the cause or causes of the accident, and recommendations for corrective action. Report forms should meet the record-keeping specifications of OSHA.

KEY TERMS AND CONCEPTS

Accident-analysis report	Personal beliefs and feelings
Accident investigation	Primary witness
Accident report	Principal's office syndrome
Accident scene	Reenactment
Causes	Secondary witness
Decision to work unsafely	Systems failure
Emergency procedures	Tertiary witness
Evidence	Traps
Eyewitness	Unsafe acts
Fault finding	Unsafe conditions
How	What
Identify witnesses	When
Immediacy	Where
Isolate the scene	Who
Mismatch or overload	Why
Open-ended	

REVIEW QUESTIONS

1. Explain the rationale for investigating accidents.
2. When should an investigation be reported? Why?
3. Explain the difference between an accident report and an accident-analysis report.
4. List the categories of the most common causes of accidents.
5. What are the terms that should guide the conduct of an accident investigation?

6. What role should the safety and health professional play in the conduct of an accident investigation?
7. List and explain the steps for conducting an accident investigation.
8. Why is it important to record all pertinent evidence relating to an accident immediately after an accident has occurred?
9. What can you do when taking close-up photographs to put them in the proper perspective?
10. List and differentiate among the three categories of witnesses to an accident.
11. Briefly explain the *when* and *where* of interviewing witnesses.
12. Briefly explain the *how* of interviewing witnesses.
13. What is the purpose of an accident report?

ENDNOTES

1. D. Hartshorn, "Solving Accident Investigation Problems," *Occupational Hazards* 65, no. 1: 57.
2. Society of Manufacturing Engineers (SME), *Tool and Manufacturing Engineers Handbook* 6 (Dearborn, MI: Society of Manufacturing Engineers, 1998), 12–21.
3. Hartshorn, "Solving Problems," 57–58.
4. SME, *Tool and Manufacturing*, 12–21.
5. Ibid.
6. Ibid.
7. National Safety Council, *Supervisor's Safety Manual* (Chicago: National Safety Council, 1991), 69–70.
8. SME, *Tool and Manufacturing*, 12–21.
9. Hartshorn, "Solving Problems," 58–59.
10. National Safety Council, *Supervisor's Safety Manual*, 71.
11. Ibid.
12. U.S. Department of Labor, OSHA 2056, 2003 (Revised), 11.
13. National Safety Council, *Supervisor's Safety Manual*, 76–77.
14. J. Cunningham and A. Kane, "Accident Reporting—Part I: Key to Prevention," *Safety & Health* 139, no. 4, 70.
15. Ibid., 70–71.
16. A. Kane and J. Cunningham, "Accident Reporting—Part II: Consistent Discipline Is Vital," *Safety & Health* 139, no. 5, 78.
17. Ibid.
18. W. R. Coffee Jr., "Avoid These 10 Mistakes," *Occupational Health & Safety* 74, no. 5, 44–47.

PRODUCT SAFETY AND LIABILITY

9

Major Topics

- Product Liability and the Law
- Developing a Product Safety Program
- Evaluating the Product Safety Program
- Role of the Safety and Health Professional
- Quality Management and Product Safety
- Product Safety Program Record Keeping
- User Feedback Collection and Analysis
- Product Literature and Safety

The United States has become the most litigious industrialized country in the world. With approximately 70 percent of the world's attorneys doing business in the United States, a heavy volume of litigation is not surprising. One of the fastest growing areas of the law is **product liability**. Product manufacturers are being sued in large numbers by users, misusers, and even abusers of their products.

In the words of the National Safety Council:

> Injuries resulting from the use (or often misuse) of products are the basis for an increasing number of product liability lawsuits. These suits cost industry millions of dollars each year. The best way the manufacturer can prevent or defend such claims is by manufacturing a reasonably safe and reliable product, and, where necessary, by providing instructions for its proper use. The key to achieving a reasonably safe and reliable product and, at the same time, reducing the product liability exposure is to build in product safety.[1]

An example of a widely used consumer product that has risks associated with it is the microwave oven. Since they were first introduced, microwave ovens have been carefully scrutinized for potential radiation leak hazards. As a result, radiation leakage from microwave ovens is no longer of major concern as a public safety and health issue.

However, the number of burns associated with microwave ovens is on the rise. Scalding is the most common form of burn associated with microwave ovens.[2] Burns and scaldings together account for almost 5,000 microwave-related injuries every year.[3] Every one of these incidents represents a potential lawsuit against a microwave manufacturer. Clearly, these manufacturers are seeking ways to decrease their potential exposure to product safety litigation.

PRODUCT LIABILITY AND THE LAW

Prior to 1916, consumers and employees used products and machines at their own risk. If injured while using a manufacturer's product, a person had no legal recourse. Consequently, there was no vehicle through which damages could be sought. Finally, in 1916, the concept of negligent manufacture was established in the law. Since then, the concepts of breach of warranty and strict liability in tort have also been added to the body of law relating to product safety and product liability.

Lawsuit Process

What happens when an individual or organization decides to bring a product liability case against a manufacturer? Before dealing with this question, it is important to understand the nature of product liability law. Typically, product liability lawsuits fall within the realm of civil law. This means that such suits do not involve criminal charges. Rather, they involve one party seeking redress from another party in the form of monetary damages.

Figure 9–1 illustrates the various steps that take place when a product liability lawsuit is filed. The **discovery period** (step 3) can be lengthy. The collection of evidence can involve taking depositions from expert witnesses and examining the product in question. Product examination may involve running a variety of tests and observing simulations. The length of time involved in completing this step often leads both parties to seek out-of-court settlements.[4]

---

1. An injured party decides to seek redress against the manufacturer of a product and engages an attorney.
2. The attorney files a complaint in state or federal court.
3. The discovery period is undertaken during which evidence is collected, depositions are taken, and the product is examined.
4. A trial date is set.
5. The trial takes place, or a settlement is reached out of court.

Figure 9–1
Product liability lawsuit process.

History of Product Liability Law

Product liability law is relatively new. In fact, until 1960, manufacturers were not held liable unless they produced flagrantly dangerous products. The concept of nonliability, established in the courts of England in 1842, persisted in the law in this country until the turn of the century. Since then, four landmark cases have established what is still the foundation of product liability law in the United States. These cases are discussed in the following paragraphs.[5]

MacPherson v. Buick Motor Company

This case established the concept of **negligent manufacture,** which means that the maker of a product can be held liable for its performance from a safety and health perspective. While driving a Buick Motor Company car, MacPherson lost control and was injured in the resulting accident. The cause of the accident was a defective wheel. The courts ruled in MacPherson's favor.

Henningson v. Bloomfield Motors, Inc.

This case established the concept of **breach of warranty,** broadened the manufacturer's liability to include people without contractual agreements with the manufacturer, and limited the protection that manufacturers could derive from using disclaimers with their products. In this case, Henningson had purchased a new car for his wife, who was subsequently involved in an accident while driving it. The cause of the accident was a defective steering system. Two issues made this case different from other negligent manufacture cases: (1) the car was damaged to such an extent that negligent manufacture could not be proven; and (2) Mrs. Henningson had no contractual agreement with the manufacturer because Mr. Henningson had actually made the purchase. The court ruled that an implied warranty existed for the performance of the car and that it extended to persons other than those with an actual contractual agreement with the manufacturer.

Greenman v. Yuba Products, Inc.

This case established the concepts of **strict liability in tort** and **negligent design**. Greenman sued after being injured while using a lathe in the prescribed manner. His attorneys argued that the lathe's design was defective and that a breach of warranty had occurred. In finding in Greenman's favor, the court broadened the concept of negligent manufacture to include design of the product.

Van der Mark v. Ford Motor Company

This case confirmed the concepts established in *Greenman v. Yuba Products, Inc.* Van der Mark sued when his new car was destroyed because of a defective braking system. In finding for Van der Mark, the Supreme Court overturned a lower court decision against him, thereby confirming the concept of negligent design.

Strict Liability in Tort

Paragraph 402A of the Second Restatement of Torts (American Law Institute) reads as follows:

1. One who sells any product in a defective condition unreasonably dangerous to the user or consumer or to his property is subject to liability for physical harm thereby caused to the ultimate consumer or user, or to his property if:
   a. the seller is engaged in the business of selling such a product, and
   b. it is expected to and does reach the user or consumer without substantial change in the condition in which it is sold.
2. The rule stated in Subsection (1) applies although:
   a. the seller has exercised all possible care in preparation and sale of his product, and
   b. the user or consumer has not bought the product from or entered into any contractual relation with the seller.[6]

This definition of strict liability has become the standard and is used in the courts of most states.

Tort law also includes the concept of **duty to warn**. This is why there are warning labels on so many consumer products, particularly those made for children. Paragraph 388 of the Second Restatement of Torts established the following three criteria for determining whether a duty to warn exists:

1. The potential for an accident when the product is used without a warning, provided the use to which it is put is reasonably predictable
2. Probable seriousness of injuries if an accident does occur
3. Potential positive effectiveness and feasibility of a warning[7]

Statutory Product Liability Laws

In addition to the common and tort law concepts discussed in the previous section, a number of statutory laws relate to product liability. Those that have the most significant impact on product liability are as follows:

- Consumer Product Safety Act
- Flammable Fabrics Act
- Food, Drug, and Cosmetics Act
- Hazardous Substances Act
- Mine Safety and Health Act
- National Traffic and Motor Vehicle Safety Act
- Occupational Health and Safety Act
- Poison Prevention Packaging Act
- Refrigerator Safety Act

- Toxic Substances Control Act
- Workers' Compensation Act

Of these various statutes, the **Consumer Product Safety Act** has the most direct application to product liability. This is the statute with which Ralph Nader is closely associated. The act has four basic purposes:

- To protect the public from the risk of injuries incurred while using consumer products
- To help consumers make objective evaluations of the risks associated with using consumer products
- To encourage uniformity in standards and regulations and to minimize conflicts among regulations at the various levels of government
- To encourage research into the causes of product-related injuries, health problems, and deaths and how these things can be prevented[8]

The act is administered by the Consumer Product Safety Commission (CPSC). Since its inception in 1972, the CPSC has been successful in facilitating changes to widely used consumer products that have enhanced their safety. These changes include safety covers on matchbooks and childproof caps on medicines and other potentially dangerous substances. The CPSC also collects information on the safety of consumer products and maintains a database through the National Electronic Information Surveillance System (NEISS). NEISS is a nationally coordinated computer system containing hospital records of injuries resulting from the use of consumer products.[9]

Applying Product Liability Laws

To recover damages in a product liability case, a person must satisfy the burden of proof by meeting certain criteria, listed in Figure 9–2. Several important legal concepts are used to apply these criteria, including the following: patent defect, latent defect, prudent man concept, reasonable risk, and unreasonable risk.

- ■ The product is defective or unreasonably unsafe as produced by the manufacturer.
- ■ The condition that causes the product to be defective or unreasonably unsafe existed when the product left the care of the original manufacturer.
- ■ The defective or unreasonably unsafe condition actually caused the injury, damage, or loss that is the subject of the case.
- ■ The nature of the complainant's injury, damage, or loss is related to the defective or unreasonably unsafe condition.

Figure 9–2
Burden-of-proof criteria for product liability cases.

A **patent defect** is one that occurs in all items in a manufactured batch. For example, if a manufacturer produces a batch of 10,000 copies of a given product and the same defect occurs in all copies, it is classified as a patent defect. Patent defects are the kind that sometimes result in product recalls. **Latent defects** occur in only one or a limited number of copies in a batch.

A key concept in product liability cases is the concept of unreasonable risk. In determining whether a risk is reasonable or unreasonable, the **prudent man concept** is applied. A **reasonable risk** exists when consumers (1) understand risk, (2) evaluate the level of risk, (3) know how to deal with the risk, and (4) accept the risk based on reasonable risk–benefit considerations. In other words, they behave prudently. An **unreasonable risk** exists when (1) consumers are not aware that a risk exists, (2) consumers are not able to judge adequately the degree of risk even when they are aware of it, (3) consumers are not able to deal with the risk, and (4) risk could be eliminated at a cost that would not price the product out of the market.[10]

Community Right-to-Know Act

An important piece of federal legislation relating to product liability is Title III of the Superfund Amendments and Reauthorization Act (SARA). The law is commonly referred to as the **Community Right-to-Know Act**. The act gives people the right to obtain information about hazardous chemicals being used in their communities. It applies to all companies that make, transport, store, distribute, or use chemicals.

The act has four main components: (1) emergency planning, (2) emergency notification, (3) reporting requirements, and (4) toxic chemical release reporting. These four components are explained in the following paragraphs.

Emergency Planning

This component of the act requires the establishment of **local emergency planning committees (LEPCs)** and **state emergency response commissions (SERCs)**. LEPCs are required to do the following: (1) develop an emergency response plan for the local community, (2) provide information through public hearings, and (3) designate a community coordinator. SERCs must (1) supervise all LEPCs in the state, and (2) review the emergency response plans for all LEPCs.

Emergency Notification

Companies must report immediately the release of hazardous chemicals that exceed limits specified by the LEPC and SERC. When this occurs, companies must report on the (1) chemical name or names; (2) date, time, place, and amount of the release; (3) potential hazards to safety and health; (4) recommended precautions to limit the potential hazards of the release; and (5) name of a contact person who can provide additional information.

Reporting Requirements

Companies are required to maintain accurate, up-to-date information on chemicals that they produce, store, use, transport, and so on. They are required to provide material safety data sheets (MSDSs) or other communication devices that contain at least the following information: name, formula, and other technical information about the chemical; potential safety and health hazards; handling precautions; and emergency procedures. Finally, companies are required to give LEPCs and SERCs information about how much of a given chemical is typically present on their site or sites and where it is stored.

Toxic Chemical Release Reporting

Companies must report their annual level of toxic chemical emissions to the Environmental Protection Agency and to state-level regulatory agencies.

For information about the act, SERCs, and LEPCs, prospective and practicing safety and health professionals should use the Environmental Protection Agency's toll-free hotline telephone number: 1-800-535-0202.

DEVELOPING A PRODUCT SAFETY PROGRAM

The purpose of a **product safety program** is to limit as much as possible a company's exposure to product liability litigation and related problems. The key to limiting liability exposure is to develop and maintain a comprehensive product safety program. At a minimum, the program should have three functional components: product safety coordinator, product safety committee, and product safety auditor.

Product Safety Coordinator

A successful product safety program must involve all departments within the company (design, manufacturing, marketing, sales, installation or service, accounting, and so on). Therefore, it is important to have a **product safety coordinator** whose role is to coordinate and facilitate this involvement.

The factor that contributes most to the success or failure of a product safety program is the coordinator's level of authority. The higher the level of authority, the more likely the program is to succeed. Giving coordination responsibility to a person who lacks executive-level access and decision-making powers can contribute to the failure of a product safety program. This is why it is sometimes advantageous to add product safety responsibilities to those of an existing executive-level decision maker who (1) has line authority over a major component of the company, (2) has access to other executive-level managers, and (3) has authority equal to fellow executives.

According to the National Safety Council, the product safety coordinator should have the authority to undertake the following actions:

- Assist in setting product safety program policy
- Recommend product recalls, field modifications, product redesign, and special analysis
- Conduct complaint, incident, or accident analysis
- Coordinate all program documents
- Facilitate communication among all parties involved in the program
- Develop a base of product safety and liability information for use by all parties involved in the program
- Establish and maintain relationships with agencies and organizations that have missions relating to product safety and liability
- Conduct product safety program audits[11]

Product Safety Committee

The product safety coordinator is just that—a coordinator. Effective product safety management is the responsibility of all departments. For this reason, many companies find it advantageous to form a **product safety committee** with a representative from all major departments. This approach gives the product safety coordinator a broad base of expertise to call upon and encourages broad-based support among all departments.

Product Safety Auditor

Auditing is an important component of the product safety program, because the effectiveness of the program is determined through audits. According to the NSC,

> The program auditor's main duty is to evaluate the adequacy of the organization's product safety program activities in relation to actual and potential exposures. This evaluation determines what the organization should do to prevent product-related losses by comparing what is being done against what should be done.[12]

The **product safety auditor** is responsible for evaluating the overall organization and individual departments within it. Specific duties of the product safety auditor include the following:

- Identify evidence of a lack of commitment on management's part
- Observe the action taken when a product deficiency is identified (i.e., does management move immediately to determine the cause and correct the problem?)
- Bring deficiencies to the attention of management and make corresponding recommendations
- Review documentation of actions taken to correct product deficiencies (Were design changes made? Were manufacturing changes made? Was a product recall initiated?)

The NSC summarizes the duties of the product safety program auditor as follows: "To interview all key management personnel, to observe the actual manufacturing operation, and to investigate, question, and verify performance."[13] Obviously, the auditor needs a certain amount of autonomy to be able to perform these duties. Does an internal employee have the autonomy needed to do the job? Does the program coordinator? These are important questions to ask when establishing a product safety program.

Opinions vary on these questions. Some think that the product safety coordinator should also conduct product safety program audits. Others think that this is an inappropriate mix of duties with built-in conflict of interest. Some think that an external person or agency should conduct audits to provide complete objectivity. Others think an internal person who is familiar with the company, its processes, and its personnel can perform a better audit. In reality, what works best depends on the company, the individual situation, and the personnel involved. If management is truly committed to product safety, it won't matter whether the auditor is internal or external, nor will it matter whether the auditor is the product safety program coordinator or a different individual. The key to the success of a product safety program is *commitment*.

EVALUATING THE PRODUCT SAFETY PROGRAM

The job of the product safety auditor is to evaluate the program continually to identify and recommend corrections to the causes of product liability problems. Figure 9–3 lists some of the more common causes of product liability exposure. The product safety auditor should constantly search for these and other causes.

Figure 9–3
Common causes of product liability exposure.

- Insufficient research during product development
- Faulty product design
- Insufficient testing of product prototypes
- Faulty manufacturing
- Insufficient quality control
- Poorly written instructions
- Insufficient or unclear warnings
- Unethical representation of the product (recommending it for use in situations outside the scope of its design and intended use)

Preparing to Evaluate the Program

Before beginning an evaluation, the auditor should allow time to prepare for the task. Proper preparation can improve both the level of cooperation and the quality of information obtained. The following preparatory activities can improve the quality of the audit:

- Meet with top management officials and review the purpose of the audit
- Work with all managers involved to coordinate schedules so that audit activities occur at mutually acceptable times
- Let managers know how the audit will affect their organizations and what will be expected of them
- Review all documentation relating to the product safety program (e.g., minutes of meetings, memoranda, and letters)
- Review product-related literature (e.g., related processes, procedures, standards, and technical data)
- Review copies of all warning labels and other printed precautionary material relating to the product[14]

Conducting the Evaluation

Having accomplished the preparation activities just described, the auditor is ready to conduct the actual evaluation. Conducting the evaluation is a matter of going to all pertinent departments and looking for evidence of the types of factors set forth in Figure 9–3. The following departments should be audited at a minimum: engineering/design, manufacturing/production, marketing/sales, service, and purchasing.

ROLE OF THE SAFETY AND HEALTH PROFESSIONAL

There is no set rule for the role played by safety and health professionals in the operation of a product safety program. In some cases, the safety and health professional is a member of the product safety committee, typically in large companies. In other cases, the safety and health professional doubles as the product safety program coordinator, more often in smaller companies. In yet other cases, the safety and health professional may be designated as the product safety auditor, which may happen in both large and small companies.

Regardless of how their role is structured, safety and health professionals can make several contributions to the program. According to the NSC, these contributions include the following:

- Safety professionals, because of their knowledge of plant operations and because of their general expertise in safety-related matters, can evaluate and offer comments on the company's product safety program.

Discussion Case

What Is Your Opinion?

"Product liability is not my job," said Mark Conners, safety director for Richfield Toys, Inc. "The safety of the toys we manufacture is a design problem. Leave it to the engineers." "Engineers have a role to play, of course," said Amanda Garner, CEO of Richfield Toys. "But they are more function-oriented than safety-oriented. We need all our products to be scrutinized from a safety perspective." Does the safety director have a role to play here? If so, what is it? If not, why? What is your opinion?

- Safety professionals, because of their experience in safety training, can evaluate product safety-related training programs developed as part of the product safety program.
- Safety professionals, because of their knowledge of accident investigation techniques, can assist in conducting product-related accident investigations.
- Safety professionals can provide product safety surveillance in production areas. This can help to prevent errors and product-related accidents. All product safety-related complaints or problems uncovered as a result of surveillance should be discussed with the production manager and formally documented, with copies sent to the product safety program coordinator.
- Because of their past experience in developing and implementing employee safety programs, safety professionals are aware not only of potential product hazards, but also of ways in which customers misuse products. Therefore, a knowledgeable safety professional should be used as a consultant to the product safety program coordinator and the auditor.[15]

QUALITY MANAGEMENT AND PRODUCT SAFETY

It is widely accepted that the best way to limit liability exposure is to produce a quality product. But what is a **quality product?** For the purposes of this book, the following definition is used:

> *A quality product is one that meets or exceeds customer standards and expectations.*

An inherent expectation is that the product, if used properly, will cause no health or safety problems.

An approach to management with a large number of proponents is **quality management (QM)**. QM can be an effective way to ensure that a company's

- Requires a secure work environment
- Requires leadership at all levels
- Defines quality in terms of customer standards and expectations
- Focuses on the continual improvement of products and processes
- Requires long-term commitment of both management and employees
- Uses the teamwork approach in producing products
- Emphasizes continual training
- Requires total employee involvement

Figure 9–4
Major elements of quality management (QM).

products consistently meet or exceed customer standards and expectations, thereby reducing the company's exposure to product liability. Managing a company using QM means expecting a total and willing commitment to quality by all personnel at all levels. With QM, every employee is (1) responsible for quality and its continual improvement, and (2) empowered to make decisions/ recommendations to improve quality continually. With QM, quality is the overriding factor in all decisions at all levels.

A key element in this concept is the empowerment of employees. Empowered employees are able to think creatively, act independently, and pursue innovative solutions to problems. This approach is easy enough to talk about, but not so easy to implement. Figure 9–4 lists some of the main elements of QM. These key points make it clear that QM is not just another superficial quick-fix program puffed up by catchy slogans but lacking in substance. QM requires a complete change in a company's culture and management philosophy.

Another key element of QM is the elimination of the **vacuum mentality**. Workers often think that they work in a vacuum. They don't realize that their work affects that of other employees and vice versa. Products are produced through a sequentially arranged series of processes. Every task performed in one step affects all those that follow. This is why another key element of QM is **teamwork**.

In a QM setting, teamwork is engendered through five key strategies (Figure 9–5):

- **Involvement** of all personnel who must implement decisions in making those decisions.
- **Empowerment** of all personnel to take the action necessary to bring about product and process improvements in their areas of responsibility and to recommend action outside of their areas.
- **Communication**, both vertically and horizontally, on a continual basis. Communication must be a two-way activity. This means that supervisors and managers must *listen*.

Figure 9–5
Team-building strategies.

- **Reinforcement** of teamwork-oriented behavior and product and process improvements. Any rewards given, regardless of whether they involve money, recognition, or any other reinforcer, should be given to teams, not individuals.
- **Respect** for the dignity and worth of all team members, regardless of status. This is critical. A team is like a family in that every member is important.

Training and continual retraining are fundamental to QM. Employees must know how to do their jobs before they can be expected to do them well. Doing well means approaching all tasks in a manner that results in continual productivity and quality improvements while simultaneously ensuring a safe and healthy workplace. Therefore, training should be broad based. It should encompass specific job skills training, team-building training, and safety and health training.

PRODUCT SAFETY PROGRAM RECORD KEEPING

Despite a company's best efforts, it may still have to defend itself in a product safety lawsuit. When this happens, it is critical that the company is able to produce records to support its defense. In the words of the NSC,

> Complete, accurate records can be extremely convincing in a court of law. Consequently, records should be retained that are pertinent to all phases of a company's manufacturing, distributing, and importing activities, from the procurement of raw materials and components, through production and testing, to the marketing and distribution of the finished products.[16]

In addition to the litigation-related reasons, there are other compelling reasons for keeping accurate, up-to-date records. One key reason is compliance.

Safety Fact

Teams Must Have a Charter

A product safety team, like any other team, needs a charter. Teams need to know what their responsibilities are and what is expected of them. A team charter should contain at least the following elements:

- Clearly stated mission
- Broad goals
- Specific activities with schedules

Comprehensive product safety records are necessary to satisfy compliance requirements of many federal regulations. The most prominent of these are listed in Figure 9–6. Additional reasons for maintaining accurate product safety records are as follows:

- Comprehensive, up-to-date records are evidence of a company's commitment to produce quality products that pose no health or safety problems.
- Records document the amount of care required to produce, market, and distribute a high-quality product that is safe and reliable and poses no health risks.
- Records allow company officials to track products from the point of production, through the distribution system, and to customers.
- Records give company officials a database of information needed for determining and comparing insurance costs, identifying sources of supply, and determining the costs of field modifications or product recalls.[17]

An issue that has traditionally caused problems for product safety program personnel is record storage and retention. In most companies, either corporate policy or the requirements of outside agencies dictate how long records of various types should be retained. This is not the case with product safety records. Because there is no way to know when the records may be needed as evidence

---

- Consumer Product Safety Act (PL 92-573)
- Federal Hazardous Substances Act (15 USC 1261)
- Federal Food, Drug, and Cosmetic Act (21 USC 321)
- Poison Prevention Packaging Act (PL 91-601)
- Occupational Safety and Health Act (PL 91-596)
- Child Protection and Toy Act (PL 91-113)
- Magnuson-Moss Warranty—Federal Trade Commission Improvement Act (PL 93-637)

Figure 9–6
Federal regulations requiring product safety record keeping.

in a lawsuit, the only way to be sure that they are retained long enough is to retain them forever. Fortunately, record storage is no longer the problem that it was in the past thanks to the continuing advances in computer technology.

USER FEEDBACK COLLECTION AND ANALYSIS

The collection and analysis of feedback from users of the company's products is critical to the success of a product safety program. **User feedback** can come in the form of compliments, testimonials, complaints, problems, or accident reports. Such feedback can help identify modifications that should be made in a product's design, problems with manufacturing processes, the need for a product recall, and potential lawsuits.

Customer feedback can be collected in many different ways. Companies can publish a toll-free "hotline" telephone number that customers can call. Customers can be tracked and periodically surveyed. The Mazda Corporation does both. After purchasing a new Mazda vehicle, the customer receives a wallet-sized card containing a toll-free hotline to call with complaints or problems of any kind. The customer also receives periodic surveys and telephone calls from Mazda representatives.

Regardless of the collection methods used, it is important to have one central location into which feedback flows. It is also important that the product safety program coordinator have open access to this information and that it is shared regularly with the product safety committee for analysis.

The NSC says the following about consumer feedback collection and analysis:

> Based on information acquired from complaints, incidents, and accident reports from customers, distributors or dealers, and state (or provincial) or federal agencies, a company must be capable of determining immediately if a substantial hazard exists. Techniques for making the determination include: on-site investigation of the complaint, incident, or accident; hazard and failure analyses of the unit involved in the complaint, incident, or accident by the company or an independent laboratory; analysis of other units of the same product batch; evaluation of in-house test, or other records. If a company finds that a substantial product hazard does exist, the appropriate product recall or field-modification plan should be implemented.[18]

PRODUCT LITERATURE AND SAFETY

A key component in a product safety management program is a company's **product literature**. Such literature includes assembly instructions, warning labels, technical manuals, and operating instructions. It is not uncommon for product liability lawsuits to include charges relating to poorly written product literature. In fact, from a legal perspective, the quality of product literature is as important as the quality of the product itself.

Darzinskis describes the purpose of product literature:

> [The] proper role of safety instructions and warnings is to tell users about hazards that cannot be removed by design or controlled by guards and safety devices. A secondary role is to disclose a product's intended use, if this is not self-evident, as well as to recount the product's safety features and guards with the hope that understanding these features will encourage users not to remove, bypass, or alter them.[19]

Figure 9–7 contains a list of subjects that may be included in product literature. This list should be viewed as a menu from which subjects are selected. Once the subjects to be covered in a given piece of product literature have been identified, the literature can be developed. Darzinskis recommends the strategies discussed in the following paragraphs for producing high-quality, effective product literature.[20]

1. *Minimize and simplify narrative text.* Long, rambling paragraphs tend to turn off the hurried reader. Therefore, the use of narrative writing should be minimized. The text should consist of short, simple words and sentences written at a reading level below that of the intended audience.

2. *Use illustrations whenever possible.* The old saying that a picture is worth a thousand words applies when developing product literature. Tables, charts, graphs, flowcharts, photographs, line drawings, and so on should be used instead of narrative text whenever possible. Illustrations supplemented by the written word can be an effective approach.

- General information
- Safety information
- How to remove shipping crates from around the product
- Product components
- Handling instructions and precautions
- Assembly instructions
- Installation instructions
- Tools, equipment, special clothing, and other materials needed in conjunction with the product
- Instructions for proper use
- Routine servicing instructions
- Maintenance information
- Troubleshooting procedures
- Overhaul requirements and instructions
- Disposal procedures

Figure 9–7
Potential subjects for product literature.

3. *Consider the eye appeal of the layout.* The **eye appeal** of product literature can be enhanced by applying the following tips: (a) do not justify (line up) right-hand margins of text material; (b) use boldface print or underlining for short passages only; (c) use a typeface that has large lowercase letters as compared with its uppercase letters; (d) avoid the use of reverse print except for short passages; (e) use blue, black, green, or purple ink for single-color printing on packages.

4. *Maximize drawing power.* It is important to produce product literature that compels the reader to read it. The use of headlines, widely recognized symbols, and color increases the **drawing power** of the literature. Other strategies include increasing the amount of white space and using boxes around high-priority messages.

In addition to producing literature that is readable, compelling, and appealing to the eye, it is important to produce literature that is accurate. Product literature should never exaggerate or mislead users. The best design is rendered useless if the content of the readable, compelling, appealing information is incorrect.

SUMMARY

1. Prior to 1916, consumers and employees used products and machines at their own risk. In 1916, the concept of negligent manufacture was established in the law. Since then, the concepts of breach of warranty and strict liability in tort have also been established.
2. Typically, product liability lawsuits fall within the realm of civil rather than criminal law. Redress comes in the form of a monetary award or settlement.
3. The concept of strict liability in tort makes manufacturers of products liable for physical harm that results from their use as long as the product is not altered by a party other than the manufacturer while it is en route to the customer.
4. Tort law gives manufacturers a duty to warn consumers of potential hazards associated with products.
5. The most significant product liability statutes include the following: Consumer Product Safety Act; Flammable Fabrics Act; Food, Drug, and Cosmetics Act; Hazardous Substances Act; Mine Safety and Health Act; National Traffic and Motor Vehicle Safety Act; Occupational Health and Safety Act; Poison Prevention Packaging Act; Refrigerator Safety Act; Toxic Substances Control Act; and Workers' Compensation Act.
6. The Consumer Product Safety Act is designed to protect the public from injuries, help consumers evaluate the risk associated with products, encourage uniform standards, and encourage research.
7. Important legal concepts associated with product liability law are patent defect, latent defect, prudent man concept, reasonable risk, and unreasonable risk.

8. Title III of the Superfund Amendments and Reauthorization Act (SARA) is more commonly known as the Community Right-to-Know Act. The act has four components: emergency planning, emergency notification, reporting requirements, and toxic chemicals release reporting.

9. The three fundamental components of a product safety program are the product safety coordinator, product safety committee, and product safety auditor.

10. The role of the safety and health professional in the operation of a product safety program can vary depending on the size of the company and local circumstances. In some cases, a safety and health professional may serve as the program coordinator; in others, as the program auditor, and in others, as neither. In yet other cases, the safety and health professional may serve as an ex officio in-house consultant to the process.

11. Quality management (QM) is an approach to management that makes quality the responsibility of all employees while at the same time empowering them to make decisions and recommendations to improve quality. The goal is to meet or exceed customer expectations by building in quality. Key concepts in QM are involvement, empowerment, communication, reinforcement, and respect.

12. It is important to collect customer feedback and use it to make product improvements. It can be collected by mail surveys, telephone contact, and a variety of other methods. Regardless of how it is collected, user feedback should be shared with the product safety coordinator and committee.

13. From a legal perspective, the quality of product literature can be as important as the quality of the actual product. The purpose of product literature is to define the product's intended use and protect consumers from hazards that cannot be eliminated in the design or manufacturing processes.

KEY TERMS AND CONCEPTS

Breach of warranty

Communication

Community Right-to-Know Act

Consumer Product Safety Act

Discovery period

Drawing power

Duty to warn

Empowerment

Eye appeal

Involvement

Latent defect

Local emergency planning
committees (LEPCs)

Negligent design

Negligent manufacture

Patent defect

Product liability

Product literature

Product safety auditor

Product safety committee

Product safety coordinator

Product safety program

Prudent man concept

Quality management (QM)

Quality product

Reasonable risk

Reinforcement

Respect

State emergency response commissions (SERCs)

Strict liability in tort

Teamwork

Unreasonable risk

User feedback

Vacuum mentality

REVIEW QUESTIONS

1. Explain the best way for a manufacturer to limit its exposure to product liability lawsuits.
2. What are the steps to follow in filing a product liability lawsuit?
3. What is the discovery period in a product liability lawsuit?
4. Briefly explain the concept of strict liability in tort.
5. List the three criteria governing a company's duty to warn.
6. What are two purposes of the Consumer Product Safety Act?
7. Define the following concepts associated with product liability law: *patent defect*, *latent defect*, and *prudent man concept*.
8. Explain the reporting requirements component of the Community Right-to-Know Act.
9. List and briefly explain the three components of a comprehensive product safety program.
10. How may assigning product safety coordination duties to an executive of the company affect the success of the program?
11. What role do you think safety and health professionals should play in a product safety program?
12. Define the term *quality management*. Explain the five key strategies of QM.
13. Why is record keeping so important with product safety?
14. List four strategies for producing high-quality, effective product literature.

ENDNOTES

1. National Safety Council, *Accident Prevention Manual for Industrial Operations* (Chicago: National Safety Council), 425.
2. J. Castelli, "Are Microwave Ovens Safer Than Ever?" *Family Safety and Health* 50, no. 2, 28.
3. Ibid.
4. D. A. Colling, *Industrial Safety: Management and Technology* (Upper Saddle River, NJ: Prentice Hall), 263.
5. Ibid., 264.
6. American Law Institute, Second Restatement of Torts, 1965, paragraph 402A.
7. Ibid., paragraph 388.

8. Colling, *Industrial Safety*, 267.
9. Ibid.
10. Ibid., 266.
11. Ibid., 426.
12. National Safety Council, *Accident Prevention Manual*, 426.
13. Ibid.
14. Ibid., 427.
15. Ibid.
16. Ibid., 428.
17. Ibid., 434.
18. Ibid., 436.
19. K. Darzinskis, "How to Put Safety into Product Literature," *Safety & Health* 143, no. 6, 74.
20. Ibid., 76–78.

PART

3

THE HUMAN ELEMENT

10   Ergonomic Hazards: Musculoskeletal Disorders (MSDs) and Cumulative
     Trauma Disorders (CTDs)   255
11   Stress and Safety   296
12   Safety and Health Training   310
13   Violence in the Workplace   352

ERGONOMIC HAZARDS: MUSCULOSKELETAL DISORDERS (MSDs) AND CUMULATIVE TRAUMA DISORDERS (CTDs)

10

Major Topics

- Ergonomics Defined
- Human Factors and Ergonomic Hazards
- Factors Associated with Physical Stress
- Ergonomics: A Political Football
- OSHA's Voluntary Ergonomics Guidelines
- Worksite Analysis Program for Ergonomics
- Hazard Prevention and Control
- Medical Management Program
- Training and Education
- Common Indicators of Problems
- Identifying Specific Ergonomic Problems
- Ergonomic Problem-Solving Strategies
- Economics of Ergonomics
- Cumulative Trauma Disorders (CTDs)
- Participatory Ergonomics

The history of workplace development in the Western world is characterized by jobs and technologies designed to improve processes and productivity. All too often in the past, little or no concern was given to the impact of the job process or technology on workers. As a result, work processes and machines have sometimes been unnecessarily dangerous. Another result has been that new technologies have sometimes failed to live up to expectations. This is because, even in the age of high technology, human involvement in work processes is still the key to the most significant and enduring productivity improvements. If a machine or system is uncomfortable, difficult, overly complicated, or dangerous to use, human workers will not be able to derive its full benefit.

The proliferation of uncomfortable and dangerous workplace conditions, whether created by job design or unfriendly technologies, is now widely recognized as harmful to productivity, quality, and worker safety and health. The advent of the science of ergonomics is making the workplace more physically friendly. This, in turn, is making the workplace a safer and healthier place.

ERGONOMICS DEFINED

Minimizing the amount of physical stress in the workplace requires continuous study of the ways in which people and technology interact. The insight learned from this study must then be used to improve the interaction. This is a description of the science of **ergonomics**. For the purpose of this book, ergonomics is defined as follows:

> Ergonomics is a multidisciplinary science that seeks to conform the workplace and all of its physiological aspects to the worker. Ergonomics involves the following:
>
> Using special design and evaluation techniques to make tasks, objects, and environments more compatible with human abilities and limitations.
>
> Seeking to improve productivity and quality by reducing workplace stressors, reducing the risk of injuries and illnesses, and increasing efficiency.

The word *ergonomics* is derived from the Greek language. *Ergon* is Greek for *work*; *nomos* means *laws*. Therefore, in a literal sense, ergonomics means work laws. In practice, it consists of the scientific principles (laws) applied in minimizing the physical stress associated with the workplace (work). Figure 10–1 summarizes some of the widely accepted benefits of ergonomics.

There are benefits to be derived from ergonomics (as listed in Figure 10–1). There are also problems, both financial and health related, that can result from giving too little attention to ergonomics. The matter is complicated further because health problems tend to multiply a company's financial problems. Consequently, modern safety and health professionals need to be well versed in ergonomics.

Figure 10–1
Benefits of ergonomics.

- Improved health and safety for workers
- Higher morale throughout the workplace
- Improved quality
- Improved productivity
- Improved competitiveness
- Decreased absenteeism and turnover
- Fewer workplace injuries/health problems

HUMAN FACTORS AND ERGONOMIC HAZARDS

When the topic of ergonomics is discussed, the term *human factors* will usually find its way into the conversation.[1] But what is meant by the term? Tillman and Tillman define it as follows: Consumers are demanding safe and effective products. However, not all people have control over products they use. Therefore, all products must be carefully designed. For example, if a child car seat fails because it does not fit the child or is difficult to install, everyone will lose: the child, the parent, the designer, and the manufacturer. Human factors is a profession to help ensure that equipment and systems are safe and easy to operate by human beings. A human factors researcher gathers and analyzes data on human beings (how they work, their size, their capabilities and limitations). A human factors engineer works with designers as a team to incorporate data into designs to make sure people can operate and maintain the product or system. Human factors professionals then determine the skills needed to operate or maintain a finished product. Human factors is difficult to define because it is a compilation of many sciences dealing with both humans and machines. Some of the disciplines human factors experts are trained in include the following: psychology, anthropology, engineering, biology, medicine, education, and physiology.[2]

Human Factors Defined

Tillman and Tillman further define human factors as a "science combining research and application of human data."[3] The concept can also be viewed as a science that bridges research about human beings and the application of that research in designing products and systems for human beings.

Human Factors in Action

Tillman and Tillman provide several examples of how the science of human factors fits into the systems design process. Perhaps the best way to get a feel for the concept of human factors is to consider several of these examples:

1. *Predesign analysis.* In this stage of the design process, human factors professionals conduct research to answer such questions as: What is the best way for humans to interact with computers? What factors contribute to fatigue and stress in an office environment? How can designers overcome these factors?
2. *Preliminary design.* In this stage, human factors professionals study machine and human capabilities to determine which tasks should be undertaken manually and which should be automated.
3. *Detail design and development.* In design and development, human factors professionals define the environment required for operator safety, enhanced operator performance, and the reduction or prevention of operator stress and fatigue.
4. *Test and evaluation.* In this stage of the process, human factors professionals test actual humans in using the prototype equipment or system.[4]

Human Factors and Safety

Human factors can play an important role in both product safety and workplace safety (where many products are used). What follows is how the science of human factors can help reduce both product and workplace hazards:

1. *Hazard elimination by design.* Human error is frequently the root cause or a contributing cause in accidents on the job. Intelligent design can reduce human errors by providing controls that are simple to understand and operate and by proving human–machine interaction that is not boring or overly demanding physically.
2. *Provision and location of safety devices.* The design and location of safety devices such as emergency cutoff switches can reduce human error on the job, correspondingly reducing the chances of an accident.
3. *Provision of warning devices.* The color, location, and wording of warning devices; the pitch and volume of warning signals; and the design of caution markings on gauges and video displays are all important factors in reducing the likelihood of human error that might lead to an accident. The science of human factors can help determine the appropriate way to apply all of these factors in a given setting.
4. *Establishment of procedures/provision of training.* When hazards cannot be realistically designed out of a system, administrative procedures for hazard reduction must be established and training relating to those procedures must be provided. Human factors professionals can help establish appropriate administrative procedures and help develop the necessary training.[5]

FACTORS ASSOCIATED WITH PHYSICAL STRESS

Eight variables that can influence the amount of *physical stress* experienced on the job are as follows:

Sitting versus standing
Stationary versus moveable/mobile
Large demand for strength/power versus small demand for
    strength/power
Good horizontal work area versus bad horizontal work area
Good vertical work area versus bad vertical work area
Nonrepetitive motion versus repetitive motion
Low surface versus high surface contact
No negative environmental factors versus negative environmental
    factors

The following paragraphs summarize the extent of this influence and the forms that it can take.[6]

Sitting versus Standing

Generally speaking, sitting is less stressful than standing. Standing for extended periods, particularly in one place, can produce unsafe levels of stress on the back, legs, and feet. Although less so than standing, sitting can be stressful unless the appropriate precautions are taken. These precautions include proper posture, a supportive back rest, and frequent standing/stretching movement.

Stationary versus Mobile

Stationary jobs are those done primarily at one workstation. Of course, even these jobs involve movement at the primary workstation and occasional movement to other areas. Mobile jobs, on the other hand, require continual movement from one station to another. The potential for physical stress increases with stationary jobs when workers fail to take such precautions as periodically standing/stretching/moving. The potential for physical stress increases with mobile jobs when workers carry materials as they move from station to station.

Large versus Small Demand for Strength/Power

In classifying jobs by these two criteria, it is important to understand that repeatedly moving small amounts of weight over a period of time can have a cumulative effect equal to the amount of stress generated by moving a few heavy weights. Regardless of whether the stress results from lifting a few heavy objects or repeated lifting of lighter objects, jobs that demand larger amounts of strength/power are generally more stressful than those requiring less.

Good versus Bad Horizontal Work Area

A good **horizontal work area** is one that is designed and positioned so that it does not require the worker to bend forward or to twist the body from side to side. Horizontal work areas that do require these movements are bad. Bad horizontal work surfaces increase the likelihood of physical stress.

Safety Fact

High Cost of Pain

When employees are in pain, they respond by taking sick leave from work. Sick leave due to unspecified pain costs employers more than $3 billion in lost workdays. Approximately 17 million employees in the United States take an average of three days of sick leave per year to deal with unspecified pain (e.g., headaches, neck pain, back pain, and menstrual pain). Often the pain is not work related.

Good versus Bad Vertical Work Area

Good **vertical work areas** are designed and positioned so that workers are not required to lift their hands above their shoulders or bend down in order to perform any task. Vertical work areas that do require these movements are bad. Bad vertical work areas increase the likelihood of physical stress.

Nonrepetitive versus Repetitive Motion

Repetitive motion jobs involve short-cycle motion that is repeated continually. Nonrepetitive jobs involve a variety of tasks that are not, or only infrequently, repeated. Repetition can lead to monotony and boredom. When this happens, the potential for physical stress increases.

Low versus High Surface Contact

Surface stress can result from contact with hard surfaces such as tools, machines, and equipment. High-surface-contact jobs tend to be more stressful in a physical sense than are low-surface-contact jobs.

Absence versus Presence of Environmental Factors

Generally, the more **environmental factors** with which a worker has to contend on the job, the more stressful the job. For example, personal protective equipment, although conducive to reducing environmental hazards, can increase the amount of physical stress associated with the job.

ERGONOMICS: A POLITICAL FOOTBALL

OSHA's attempts to develop an ergonomics standard have been the subject of bitter disputes between labor and management in the private sector. These disputes, in turn, have found their way into the political arenas of both the U.S. Congress and the executive branch. OSHA established its voluntary ergonomics guidelines in 1989. Soon thereafter, organized labor began its campaign to have the guidelines made mandatory through passage of an ergonomics standard. During the Clinton administration (1993–2000), Congress steadfastly refused to approve OSHA's proposed ergonomics standard.

Consequently, just before leaving office, President Clinton signed an executive order putting the elements of the proposed ergonomics standard in place as a rule. However, with the election of George W. Bush as president, Congress used the Congressional Review Act (CRA) to overturn former President Clinton's executive order. Organized labor and other supporters of OSHA vowed to make congressional opponents of an ergonomics standard and President Bush pay for overturning the executive order that had in essence made the voluntary guidelines mandatory. However, the most ardent opponents in Congress of an ergonomics standard all won their elections and were returned to Congress, now even more determined to squelch any attempts by supporters to pass a standard.

OSHA has responded to all of this in two ways. First, OSHA continues to develop voluntary guidelines beyond those that apply to industry in general for businesses in specific industrial classifications. Second, OSHA has begun to claim that it will use its general duty clause to enforce ergonomic safety. However, there is little evidence that this is actually happening. After vowing to make enforcement through the general duty clause part of its four-pronged attack on musculoskeletal hazards, OSHA is, in reality, focusing most of its efforts on the other three prongs: voluntary guidelines, outreach and assistance, and continuing research.

OSHA'S VOLUNTARY ERGONOMICS GUIDELINES

OSHA first published guidelines for general safety and health program management in 1989. OSHA's **ergonomics guidelines** are voluntary and are designed to provide employers with the information and guidance needed to meet their obligations under the OSH Act regarding ergonomics.

OSHA followed these guidelines with another set designed specifically for the meat-packing industry. In June 2002, OSHA announced voluntary guidelines for retail grocery stores and the poultry-processing industry. These specific guidelines represent a model for guidelines that are likely to be developed for other specific industries. Meat packing and poultry processing were singled out because of the high incidence of **cumulative trauma disorders (CTDs)** associated with meat packing and processing. CTDs are injuries that result from an accumulation of repetitive motion stress. For example, using scissors continually over time can cause a CTD in the hand and wrist.

OSHA's Ergonomics Standard (Voluntary Guidelines)

OSHA's voluntary guidelines were well received by employees and labor organizations with one major caveat: the "voluntary" nature of the guidelines. Many people interested in occupational safety and health advocated a mandatory standard for ergonomics. This led to discussions concerning the development by OSHA of a new ergonomics standard. The standard has not yet been developed.

However, OSHA's proposed ergonomics standard failed to win approval from Congress. Consequently, after much political wrangling, OSHA officials decided to convert the proposed ergonomics standard into voluntary ergonomics guidelines. OSHA's current plan for reducing ergonomic hazards in the workplace has four elements:

1. Voluntary guidelines for specific industries.
2. Enforcement of the guidelines under the general duty clause of the OSH Act 5(a)(1). This is a controversial element of the four-part plan because some employers think that OSHA is using the general duty clause to make the "voluntary" guidelines mandatory against the stated will of Congress.
3. Compliance assistance to help employers reduce ergonomic hazards.
4. Research into ergonomic issues to help identify gaps in the body of knowledge surrounding this topic.

Enforcement by OSHA

Although the ergonomics guidelines developed by OSHA are voluntary, at least for the foreseeable future, the agency does claim that it will use the general duty clause of the OSH Act to enforce the guidelines in certain situations. OSHA's criteria for applying the general duty clause are as follows:

1. Is there currently an ergonomic hazard that is causing injuries?
2. Does the employer in question know about the hazard (or should the employer know)?
3. Are the injuries caused by the ergonomic hazard resulting in serious physical harm?
4. Are there feasible alternatives available to the employer for reducing, abating, or minimizing the hazard?

Although the use of the general duty clause to enforce voluntary ergonomics guidelines is controversial and has been questioned by many employers in theory, in practice this is an approach that is rarely used by OSHA.

Rationale for the Voluntary Guidelines

Musculoskeletal disorders (MSDs) and cumulative trauma disorders (CTDs) account for more than 30 percent of all occupational injuries and illnesses in the United States. Such injuries and illnesses result in more than 600,000 lost workdays annually and account for approximately one of every three dollars spent on workers' compensation every year. A conservative estimate of these costs is more than $20 billion annually. Organizations that are losing valuable employee time to CTDs and MSDs are well advised to implement the type of ergonomics program set forth in OSHA's standard on a voluntary, self-enforced basis.

Application of the Voluntary Guidelines

OSHA's ergonomics guidelines are geared toward manufacturing and materials handling in the general industry sector. They do not apply to construction, maritime operations, agriculture, or employers that operate a railroad, although they can be applied to other jobs in which the type of work that is fundamental and necessary to perform the job result in MSDs. The ergonomics guidelines apply to more than 1.5 million employees nationwide. Examples of jobs to which the guidelines apply are as follows:

- Patient-handling jobs (nurse assistants and orderlies)
- Shipping, receiving, and delivery (package sorting, handling, delivery, etc.)
- Baggage handlers
- Warehouse work (manual tasks)
- Beverage and water handling and delivery
- Grocery/retail store stocking and bagging
- Garbage and trash collecting
- Assembly-line work
- Piecework assembly
- Product inspection (involving manual tasks such as weighing objectives)
- Meat, poultry, and fish processing
- Machine loading, unloading, and operation
- Textile manufacturing
- Food preparation assembly-line work
- Commercial banking
- Cabinet making
- Tire making

Proposed Requirements of the Voluntary Guidelines

Organizations that fall into the general industry classifications of manufacturing and manual material handling are asked by the guidelines to implement a "basic ergonomics program." This amounts to assigning responsibility for ergonomics to one individual and informing employees about the risks of MSD-related injuries, symptoms of such injuries, and why early reporting of symptoms is important. In addition, the basic program requires employers to establish a system that employees can use to report symptoms of MSD injuries.

The so-called full ergonomics program set forth in the guidelines is not required unless and until an employee's job is determined, by use of a basic screening tool (such as the one shown in Figure 10–2) to have met the "action trigger." The full program consists of the following components:

- Management leadership and employee participation
- Employee participation
- Training
- Record keeping
- Job hazard analysis and control
- Work restriction protection
- MSD management
- Program evaluation

Workstation Checklist

for

Video Display Terminals

The workstation should be designed to ensure the following conditions:

A. *Head* and *neck* should be about upright.

B. *Head, neck,* and *trunk* should face forward.

C. *Trunk* should be about perpendicular to floor.

D. *Shoulders* and *upper arms* should be about perpendicular to floor and relaxed.

E. *Upper arms* and *elbows* should be close to body.

F. *Forearms, wrists,* and *hands* should be straight and parallel to floor.

G. Wrists and hands should be straight.

H. *Thighs* should be about parallel to floor, and lower legs should be about perpendicular to floor.

I. *Feet* should rest flat on floor or be supported by a stable footrest.

J. *VDT tasks* should be organized in a way that allows employee to vary them with other work activities or to take short breaks or recovery pauses while at the VDT workstation.

SEATING (The Chair)

1. *Backrest* should provide support for employee's lower back.

2. *Seat width* and *depth* should accommodate specific employee.

3. *Seat front* should not press against the back of employee's knees and lower legs.

4. *Seat* should have cushioning and be rounded with a "waterfall" front.

5. *Armrests* should support both forearms while the employee performs VDT tasks and not interfere with movement.

KEYBOARD/INPUT DEVICE

The keyboard/input device should be designed for doing VDT tasks so that. . . .

6. *Keyboard/input device platforms* are stable and large enough to hold keyboard and input device.

7. *Input device* (mouse or trackball) is located right next to keyboard so it can be operated without reaching.

8. *Input device* is easy to activate, and shape/size fits hand of specific employee (not too big/small).

9. *Wrists* and *hands* do not rest on sharp or hard edge.

Figure 10–2
Basic screening tool for VDT workstations.

MONITOR

The monitor should be designed for VDT tasks so that. . . .

10. *Top line* of screen is at or below eye level so the employee is able to read it without bending head or neck down/back.

11. *Employees with bifocals/trifocals* are able to read the screen without bending their heads or necks backward.

12. *Monitor distance* allows employees to read the screen without leaning their heads, necks, or trunks forward/backward.

13. *Monitor position* is directly in front of the employee so he or she does not have to twist the head or neck.

14. *No glare* (e.g., from windows, lights) is present on the screen that may cause employees to assume an awkward position to read screen.

WORK AREA

The work area is designed for doing VDT tasks so that. . . .

15. *Thighs* have clearance space between chair and the VDT table/keyboard platform.

16. *Legs* and *feet* have clearance space under the VDT table so the employee is able to get close enough to the keyboard/input device.

ACCESSORIES

17. *Document holder,* if provided, should be stable and large enough to hold documents that are used.

18. *Document holder,* if provided, should be placed at about the same height and distance as the monitor screen so there is little head movement while the employee looks from document to screen.

19. *Wrist rest,* if provided, should be padded and free of sharp and square edges.

20. *Wrist rest,* if provided, should allow employees to keep forearms, wrists, and hands straight and parallel to ground when using the keyboard/input device.

21. *Telephone* can be used with head upright and shoulders relaxed if the employee does VDT tasks at the same time.

GENERAL

22. Workstation and equipment should have sufficient adjustability so that the employee is able to be in a safe working posture and to make occasional changes in posture while performing VDT tasks.

23. VDT workstation, equipment, and accessories should be maintained in serviceable condition and function properly.

Note to Employers:
Workstation must meet ALL of criteria A–J and all but two of criteria 1–23.

Figure 10–2 (*continued*)

Source: Adapted from Appendix D-2 of OSHA Standard 1910.900.

WORKSITE ANALYSIS PROGRAM FOR ERGONOMICS

Although complex analyses are best performed by a professional ergonomist, the "ergonomics team"—or any qualified person—can use this program to conduct a *worksite analysis* and identify stressors in the workplace. The purpose of the following information is to give a starting point for finding and eliminating those tools, techniques, and conditions that may be the source of ergonomic problems.[7]

In addition to analyzing current workplace conditions, planned changes to existing and new facilities, processes, materials, and equipment should be analyzed to ensure that changes made to enhance production will also reduce or eliminate ergonomic risk factors. As emphasized before, this program should be adapted to each individual workplace.

The discussion of the recommended program for worksite analysis is divided into four main parts: (1) gathering information from available sources; (2) conducting baseline screening surveys to determine which jobs need closer analysis; (3) performing ergonomic job hazard analyses of those workstations with identified risk factors; and (4) after implementing control measures, conducting periodic surveys and follow-up studies to evaluate changes.

Information Sources

Records Analysis and Tracking

The essential first step in worksite analysis is **records analysis and tracking** to develop the information necessary to identify ergonomic hazards in the workplace. Existing medical, safety, and insurance records, including OSHA 300 logs, should be analyzed for evidence of injuries or disorders associated with CTDs. Health care providers should participate in this process to ensure confidentiality of patient records.

Safety Fact

Keys to a Successful Ergonomics Program

Regardless of the type of organization, the keys to having a successful ergonomics program are as follows:

- Commitment on the part of top management
- Written program
- Employee involvement
- Continuous monitoring of the program
- Adjusting as necessary based on the results of monitoring

Safety Myth

Monday Morning Syndrome Debunked

More workplace injury claims are filed on Monday than on any other day of the week. More than 23 percent of all workers' compensation claims are filed on Monday. This has led to a belief that employees are injuring themselves during non-work-related activities over the weekend and filing workers' compensation claims on Monday. The popular belief is that employees who don't carry personal medical insurance use this ploy to gain medical coverage under workers' compensation. However, this may not be the case. The high incidence of workers' compensation claims filed on Monday can also be attributed to such factors as the following:

- Lack of activity over the weekend, coupled with a sudden return to physical work on Monday.
- Effect of cold weather and age on muscles (Monday morning strains and sprains).

Source: The Ergonomics Report, April 7, 2006. Retrieved from http://www.ergonomicsreport.com/publish.

Incidence Rates

Incidence rates for upper extremity disorders and/or back injuries should be calculated by counting the incidence of CTDs and reporting the number for each 100 full-time workers per year by facility:

$$\text{Incidence rate} = \frac{(\text{Number of new cases/yr}) \times (200{,}000 \text{ work hrs}) \text{ per facility*}}{\text{Number of hours worked/facility/yr}}$$

Screening Surveys

The second step in worksite analysis is to conduct baseline screening surveys. Detailed baseline screening surveys identify jobs that put employees at risk of developing CTDs. If the job places employees at risk, an effective program will then require the ergonomic job hazard analysis.**

Checklist

The survey is performed with an ergonomic checklist. This checklist should include components such as posture, materials handling, and upper extremity factors. (The checklist should be tailored to the specific needs and conditions of the workplace.)

---

*Adapted from OSHA 3123.
**The same method should be applied to departments, production lines, or job types within the facility.

Ergonomic Risk Factors

Identification of **ergonomic hazards** is based on ergonomic risk factors: conditions of a job process, workstation, or work method that contribute to the risk of developing CTDs. Not all these risk factors will be present in every CTD-producing job, nor is the existence of one of these factors necessarily sufficient to cause a CTD.

CTD Risk Factors

Some of the risk factors for CTDs of the upper extremities include the following:

- Repetitive and/or prolonged activities
- Forceful exertions, usually with the hands (including pinch grips)
- Prolonged static postures
- Awkward postures of the upper body, including reaching above the shoulders or behind the back, and twisting the wrists and other joints to perform tasks
- Continued physical contact with work surfaces (soft tissue compression)
- Excessive vibration from power tools
- Cold temperatures
- Inappropriate or inadequate tool design
- High wrist acceleration
- Fatigue (inadequate recovery time)
- Use of gloves

Back Disorder Risk Factors

Risk factors for back disorders include:

- Bad body mechanics such as continued bending over at the waist, continued lifting from below the knees or above the shoulders, and twisting at the waist, especially while lifting
- Lifting or moving objects of excessive weight or asymmetric size
- Prolonged sitting, especially with poor posture
- Lack of adjustable chairs, footrests, body supports, and work surfaces at workstations
- Poor grips on handles
- Slippery footing

Multiple Risk Factors

Jobs, operations, or workstations that have **multiple risk factors** have a higher probability of causing CTDs. The combined effect of several risk factors in the development of CTDs is sometimes referred to as *multiple causation*.

Ergonomic Job Hazard Analyses

At this point, the employer has identified—through the information sources and screening surveys just discussed—jobs that place employees at risk of developing CTDs. As an essential third step in the worksite analysis, an effective ergonomics program requires a job hazard analysis for each job so identified.

The job hazard analysis should be routinely performed by a qualified safety and health professional, preferably an ergonomist, for jobs that put workers at risk of developing CTDs. This type of analysis helps verify lower risk factors at light-duty or restricted activity work positions and to determine if risk factors for a work position have been reduced or eliminated to the extent feasible.

Workstation Analysis

An adequate **workstation analysis** would be expected to identify all risk factors present in each studied job or workstation. For upper extremities, three measures of repetitiveness are the total hand manipulations per cycle, the cycle time, and the total manipulations or cycles per work shift. Force measurements may be noted as an estimated average effort and a peak force. They may be recorded as light, moderate, or heavy.

Tools should be checked for excessive vibration. The tools, personal protective equipment, and dimensions and adjustability of the workstation should be noted for each job hazard analysis. Finally, hand, arm, and shoulder postures and movements should be assessed for levels of risk.

Lifting Hazards

For manual materials handling, the maximum weight-lifting values should be calculated.

DVD/Videotape Method

The use of DVD/videotape, where feasible, is suggested as a method for analysis of the work process. Slow-motion videotape or equivalent visual records of workers performing their routine job tasks should be analyzed to determine the demands of the task on the worker and how each worker actually performs each task.

Periodic Ergonomic Surveys

The fourth step in worksite analysis is to conduct periodic review. Periodic surveys should be conducted to identify previously unnoticed factors or failures or deficiencies in work practices or engineering controls. The periodic review process should also include feedback, follow-up, and trend analysis.

Feedback and Follow-Up

A reliable system should be provided for employees to notify management about conditions that appear to be hazardous and to utilize their insight and experience to determine work practice and engineering controls. This may be initiated by an ergonomics questionnaire and maintained through an active safety and health committee or by employee participation with the ergonomics team. Reports of ergonomic hazards or signs and symptoms of potential CTDs should be investigated by ergonomics screening surveys and appropriate ergonomic hazard analysis in order to identify risk factors and controls.

Trend Analysis

Trends of injuries and illnesses related to actual or potential CTDs should be calculated, using several years of data where possible. Trends should be calculated for several departments, process units, job titles, or workstations. These trends may also be used to determine which work positions are most hazardous and need to be analyzed by the qualified person.

Using standardized job descriptions, incidence rates may be calculated for work positions in successive years to identify trends. Using trend information can help determine the priority of screening surveys and/or ergonomic hazard analyses.

HAZARD PREVENTION AND CONTROL

Engineering solutions, where feasible, are the preferred method for ergonomic **hazard prevention and control**. The focus of an ergonomics program is to make the job fit the person—not to make the person fit the job. This is accomplished by redesigning the workstation, work methods, or tool to reduce the demands of the job, including high force, repetitive motion, and awkward postures. A program with this goal requires research into currently available controls and technology. It should also include provisions for utilizing new technologies as they become available and for in-house research and testing. Following are some examples of engineering controls that have proven to be effective and achievable.

Workstation Design

Workstations should be designed to accommodate the persons who actually use them; it is not sufficient to design for the average or typical worker. Workstations should be easily adjustable and should be either designed or selected to fit a specific task, so that they are comfortable for the workers who use them. The work space should be large enough to allow for the full range of required movements, especially where knives, saws, hooks, and similar tools are used.

Design of Work Methods

Traditional work method analysis considers static postures and repetition rates. This should be supplemented by addressing the force levels and the hand and arm postures involved. The tasks should be altered to reduce these and the other stresses associated with CTDs. The results of such analyses should be shared with the health care providers to assist in compiling lists of light-duty and high-risk jobs.

Tool Design and Handles

Tools should be selected and designed to minimize the risks of upper extremity CTDs and back injuries. In any tool design, a variety of sizes should be available. Examples of criteria for selecting tools include the following:

- Designing tools to be used by either hand, or providing tools for both left- and right-handed workers.
- Using tools with triggers that depress easily and are activated by two or more fingers.
- Using handles and grips that distribute the pressure over the fleshy part of the palm, so that the tool does not dig into the palm.
- Designing and selecting tools for minimum weight; counterbalancing tools heavier than one or two pounds.
- Selecting pneumatic and power tools that exhibit minimal vibration and maintaining them in accordance with manufacturer's specifications or with an adequate vibration-monitoring program. Wrapping handles and grips with insulation material (other than wraps provided by the manufacturer for this purpose) is normally *not* recommended, as it may interfere with a proper grip and increase stress.

MEDICAL MANAGEMENT PROGRAM

An effective **medical management program** for cumulative trauma disorders is essential to the success of an employer's ergonomic program in industries with a high incidence of CTDs.[8] It is not the purpose of these guidelines to dictate medical practice for an employer's **health care providers**. Rather, they describe the elements of a medical management program for CTDs to ensure early identification, evaluation, and treatment of signs and symptoms; to prevent their recurrence; and to aid in their prevention. Medical management of CTDs is a developing field, and health care providers should monitor developments on the subject. These guidelines represent the best information currently available.

A physician or occupational health nurse (OHN) with training in the prevention and treatment of CTDs should supervise the program. Each work shift should have access to health care providers in order to facilitate treatment, surveillance activities, and recording of information. Where such personnel are not employed full time, the part-time employment of appropriately trained health care providers is recommended.

In an effective ergonomics program, health care providers should be part of the ergonomics team interacting and exchanging information routinely to prevent and treat CTDs properly. The major components of a medical management program for the prevention and treatment of CTDs are trained first-level health care providers, health surveillance, employee training and education, early reporting of symptoms, appropriate medical care, accurate record keeping, and quantitative evaluation of CTD trends throughout the plant.

Trained and Available Health Care Providers

Appropriately trained health care providers should be available at all times and on an ongoing basis as part of the ergonomics program. In an effective medical management program, first-level health care providers should be knowledgeable in the prevention, early recognition, evaluation, treatment, and rehabilitation of CTDs, as well as in the principles of ergonomics, physical assessment of employees, and OSHA record-keeping requirements.

Periodic Workplace Walk-Through

In an effective program, health care providers should conduct periodic, systematic workplace walk-throughs to remain knowledgeable of operations and work practices, to identify potential light-duty jobs, and to maintain close contact with employees. Health care providers should also be involved in identifying risk factors for CTDs in the workplace as part of the ergonomics team.

These walk-through surveys should be conducted every month or whenever a particular job task changes. A record should be kept documenting the date of the walk-through, areas visited, risk factors recognized, and action initiated to correct identified problems. Follow-up should be initiated to correct problems identified and should be documented to ensure that corrective action is taken when indicated.

Symptoms Survey

Those responsible for the medical management program should develop a standardized measurement to determine the extent of work-related disorder symptoms in each area of the plant. This measurement will help determine which jobs are exhibiting problems and to measure progress of the ergonomics program.

Institute a Survey

A **symptoms survey** of employees should be conducted to measure employee awareness of work-related disorders and to report the location, frequency, and duration of discomfort. Body diagrams should be used to facilitate gathering this information. Surveys normally should not include employee's personal identifiers to encourage employee participation.

The survey is one method for identifying areas or jobs where potential CTD problems exist. The major strength of the survey approach is in collecting data on the number of workers who may be experiencing some form of CTD.

Reported pain symptoms by several workers on a specific job would indicate the need for further investigation of that job.

Conduct the Survey Annually

Conducting the survey annually should help detect any major change in the prevalence, incidence, and/or location of reported symptoms.

Keep List of Light-Duty Jobs

The ergonomist or other qualified person should analyze the physical procedures used in the performance of each job, including lifting requirements, postures, hand grips, and frequency of repetitive motion.

The ergonomist and health care providers should develop a list of jobs with the lowest ergonomic risk. For such jobs, the ergonomic risk should be described. This information will assist health care providers in recommending assignments to light- or restricted-duty jobs. The light-duty job should, therefore, not increase ergonomic stress on the same muscle-tendon groups. Health care providers should likewise develop a list of known high-risk jobs. Supervisors should periodically review and update the lists.

Health Surveillance

Baseline

The purpose of baseline health surveillance is to establish a base against which changes in health care status can be evaluated, not to prevent people from performing work. Prior to assignment, all new and transferred workers who are to be assigned to positions involving exposure of a particular body part to ergonomic stress should receive a baseline health surveillance.

Conditioning Period Follow-Up

New and transferred employees should be given the opportunity during a four- to six-week break-in period to condition their muscle-tendon groups prior to working at full capacity. Unfortunately, this is not always possible; however, when it is possible, conditioning should be done. Health care providers should perform a follow-up assessment of these workers after the break-in period (or after one month, if the break-in period is longer than a month) to determine if conditioning of the muscle-tendon groups has been successful; whether any reported soreness or stiffness is transient and consistent with normal adaptation to the job or whether it indicates the onset of a CTD; and if problems are identified, what appropriate action and further follow-up are required.

Periodic Health Surveillance

Periodic health surveillance—every two to three years—should be conducted on all workers who are assigned to positions involving exposure of a particular body part to ergonomic stress. The content of this assessment should be similar to that outlined for the baseline. The worker's medical and occupational history should be updated.

Employee Training and Education

Health care providers should participate in the training and education of all employees, including supervisors and other plant management personnel, on the different types of CTDs and means of prevention, causes, early symptoms, and treatment of CTDs. This information should be reinforced during workplace walk-throughs and the individual health surveillance appointments. All new employees should be given such education during orientation. This demonstration of concern and the distribution of information should facilitate the early recognition of CTDs prior to the development of more severe and disabling conditions and increase the likelihood of compliance with prevention and treatment.

Encourage Early Report of Symptoms

Employees should be encouraged by health care providers and supervisors to report early signs and symptoms of CTDs to the in-plant health facility. This allows for timely and appropriate evaluation and treatment without fear of discrimination or reprisal by employers. It is important to avoid any potential disincentives for employee reporting, such as limits on the number of times that an employee may visit the health unit.

Protocols for Health Care Providers

Health care providers should use written protocols for health surveillance and the evaluation, treatment, and follow-up of workers with signs or symptoms of CTDs. A qualified health care provider should prepare the protocols. These protocols should be available in the plant health facility. Additionally, the protocols should be reviewed and updated annually or as state-of-the-art evaluation and treatment of these conditions changes.

Evaluation, Treatment, and Follow-Up of CTDs

If CTDs are recognized and treated appropriately early in their development, a more serious condition can likely be prevented. Therefore, a good medical management program that seeks to identify and treat these disorders early is important.

OSHA Record-Keeping Forms

The Occupational Safety and Health Act and record-keeping regulations in Title 29, Code of Federal Regulations (CFR) 1904 provide specific recording requirements that comprise the framework of the occupational safety and health recording system. The Bureau of Labor Statistics (BLS) has issued guidelines that provide official agency interpretations concerning the record keeping and reporting of occupational injuries and illnesses. These guidelines—U.S. Department of Labor, BLS: *Record-Keeping Guidelines for Occupational Injuries and Illnesses*, September 1986 (or later editions as published)—provide supplemental instructions for the OSHA record-keeping forms and should be available in every plant health care facility. Because health care providers often provide information for OSHA logs, they should be aware of record-keeping requirements and participate in fulfilling them.

Monitor Trends

Health care providers should periodically (e.g., quarterly) review health care facility sign-in logs, OSHA Form 300, and individual employee medical records to monitor trends for CTDs in the plant. This ongoing analysis should be made in addition to the symptoms survey to monitor trends continuously and to substantiate the information obtained in the annual symptoms survey. The analysis should be done by department, job title, work area, and so on.

The information gathered from the annual symptoms survey will help identify areas or jobs where potential CTD problems exist. This information may be shared with anyone in the plant because employees' personal identifiers are not solicited. The analysis of medical records (e.g., sign-in logs and individual employee medical records) may reveal areas or jobs of concern, but it may also identify individual workers who require further follow-up. The information gathered while analyzing medical records is confidential; thus, care must be exercised to protect the individual employee's privacy. The information gained from the CTD trend analysis and symptoms survey will help determine the effectiveness of the various programs initiated to decrease CTDs in the plant.

TRAINING AND EDUCATION

The fourth major program element for an effective ergonomics program is training and education.[9] The purpose of training and education is to ensure that employees are sufficiently informed about the ergonomic hazards to which they may be exposed and thus able to participate actively in their own protection.

Training and education allow managers, supervisors, and employees to understand the hazards associated with a job or process, their prevention and control, and their medical consequences. A training program should include all affected employees, engineers and maintenance personnel, supervisors, and health care providers.

The program should be designed and implemented by qualified persons. Appropriate special training should be provided for personnel responsible for administering the program. The program should be presented in language and at a level of understanding appropriate for the individuals being trained. It should provide an overview of the potential risk of illnesses and injuries, their causes and early symptoms, the means of prevention, and treatment.

The program should also include a means for adequately evaluating its effectiveness. This may be achieved by using employee interviews, testing, and observing work practices to determine if those who received the training understand the material and the work practices to be followed.

COMMON INDICATORS OF PROBLEMS

Does my company have ergonomic problems? Are injuries and illnesses occurring because too little attention is paid to ergonomic factors? These are questions that modern safety and health professionals should ask themselves. But how you answer such questions? According to the National Institute for Occupational Safety and Health (NIOSH), the factors discussed in the following paragraphs can be examined to determine if ergonomic problems exist in a given company.[10]

Apparent Trends in Accidents and Injuries

By examining accident reports, record-keeping documents such as OSHA Form 300 (Log of Work-Related Illnesses and Injuries), first-aid logs, insurance forms, and other available records of illnesses or injuries, safety and health professionals can identify trends if they exist. A pattern or a high incidence rate of a specific type of injury typically indicates that an ergonomic problem exists.

Incidence of Cumulative Trauma Disorders

Factors associated with CTDs include a high level of repetitive work, greater than normal levels of hand force, awkward posture, high levels of vibration, high levels of mechanical stress, extreme temperatures, and repeated hand-grasping or pinch-gripping. By observing the workplace and people at work, safety and health professionals can determine the amount of exposure that employees have to these factors and the potential for ergonomics-related problems.

Absenteeism and High Turnover Rates

High absentee rates and high turnover rates can be indicators of ergonomic problems. People who are uncomfortable on the job to the point of physical stress are more likely to miss work or leave for less stressful conditions.

Employee Complaints

A high incidence of **employee complaints** about physical stress or poor workplace design can indicate the presence of ergonomic problems.

Employee-Generated Changes

Employees tend to adapt the workplace to their needs. The presence of many workplace adaptations, particularly those intended to decrease physical stress, can indicate the presence of ergonomic problems. Have employees added padding, modified personal protective equipment, brought in extra lighting, or made other modifications? Such **employee-generated changes** may be evidence of ergonomic problems.

Poor Quality

Poor quality, although not necessarily caused by ergonomic problems, can be the result of such problems. Poor quality is at least an indicator that there may be ergonomic problems. Certainly, poor quality is an indicator of a need for closer inspection.

Manual Material Handling

The incidence of musculoskeletal injuries is typically higher in situations that involve a lot of **manual material handling**. Musculoskeletal injuries increase significantly when the job involves one or more of the following: lifting large objects, lifting bulky objects, lifting objects from the floor, and lifting frequently. When such conditions exist, the company has ergonomic problems.

IDENTIFYING SPECIFIC ERGONOMIC PROBLEMS

A task analysis of the job in question can identify specific ergonomic problems. Figure 10–3 lists the types of problems that can be identified by a thorough task analysis.[11]

- Tasks that involve potentially hazardous movements
- Tasks that involve frequent manual lifting
- Tasks that involve excessive wasted motion or energy
- Tasks that are part of a poor operations flow
- Tasks that require unnatural or uncomfortable posture
- Tasks with high potential for psychological stress
- Tasks with a high fatigue factor
- Tasks that could or should be automated
- Tasks that involve or lead to quality control problems

Figure 10–3
Problems that can be pinpointed by a task analysis.

General Observation

General observation of a worker or workers performing the task(s) in question can be an effective task analysis technique. The effectiveness is usually enhanced if the workers are not aware that they are being observed. When observing employees at work, be especially attentive to tasks requiring manual material handling and repetitive movements.

Questionnaires and Interviews

Questionnaires and **interviews** can be used for identifying ergonomic problems. Questionnaires are easier to distribute, tabulate, and analyze, but interviews generally provide more in-depth information.

Videotaping and Photography

Videotaping technology has simplified the process of task analysis considerably. Videotaping records the work being observed as it is done, it is silent so it is not intrusive, and such capabilities as freeze and playback enhance the observer's analysis capabilities significantly. Photography can also enhance the observer's analysis capabilities by recording each motion or step involved in performing a task. If photography is used, be aware that flashes can be disruptive. High-speed film will allow you to make photographs without using a flash.

Drawing or Sketching

Making a neat sketch of a workstation or a drawing showing workflow can help identify problems. Before using a drawing or sketch as part of a task analysis, make sure that it is accurate. Ask an employee who is familiar with the area or process sketched to check the drawing.

Measuring the Work Environment

Measurements can help identify specific ergonomic problems. How far must a worker carry the material manually? How high does a worker have to lift an object? How much does an object weigh? How often is a given motion repeated? Answers to these and similar questions can enhance the effectiveness of the analysis process.

Understanding the Ergonomics of Aging

When identifying specific ergonomic problems in the workplace, don't overlook the special challenges presented by aging workers. A good ergonomics program adapts the job to the person. Because nearly 30 percent of the workforce is 45 years of age or older, organizations must be prepared to adapt workstations to employees whose physical needs are different from those of their younger counterparts.

In adapting workstations and processes for employees who are 45 or older, keep in mind the following rules of thumb:

- Nerve conduction velocity, hand-grip strength, muscle mass, range of motion, and flexibility all begin to diminish about age 45.
- Weight and mass tend to increase through about the early fifties.
- Height begins to diminish beginning around age 30.
- Lower back pain is more common in people 45 years of age and older.
- Visual acuity at close range diminishes with age.[12]

These rules of thumb mean that safety and health professionals cannot take a "one-size-fits-all" approach to ergonomics. Adaptations for older workers must be individualized and should take aging factors into account.

ERGONOMIC PROBLEM-SOLVING STRATEGIES

The factors that influence stress were explained earlier in this chapter. These factors can be combined in different ways, and the ways in which they are combined determine the type and amount of stress experienced. For the purpose of recommending ergonomic problem-solving strategies, the following combinations can be used:

- Seated repetitive work with light parts
- Seated work with larger parts
- Seated control work
- Standing work
- Standing for heavy lifting and/or carrying work in one place or in motion
- Work with hands above chest height
- Work with hand tools
- Work with VDTs[13]

Seated Repetitive Work with Light Parts

This type of work can produce more physical stress than you may suspect. Back, neck, shoulder, and lower leg pain are commonly associated with this type of work. Some of the problems associated with seated repetitive work are the result of the nature of the job. The fixed work position and repetitive motion can contribute to ergonomic problems. To solve these problems, it may be necessary to modify both the job and the workstation. Improvement strategies include the following:

- Include other work tasks to break the monotony of repetition.
- Use job rotation, with workers rotating from one or more different jobs.
- Adjust the height of the work surface and/or position.
- Use an adjustable chair equipped with hand, wrist, or arm supports as appropriate.

Figure 10–4
Adjust the angle of work to reduce stress.
Source: From a drawing by Robin J. Miller.

- Make sure that there is sufficient legroom (height, width, and depth).
- Use ergonomic devices to adjust the height and angle of work (see Figure 10–4).[14]

Seated Work with Larger Parts

This type of work, which involves interacting with objects that may be too large to manipulate manually, is associated with assembly and welding jobs. Problems associated with this type of work are typically related to posture, illumination, reach, and lifting. Ergonomic strategies for improving work conditions include the following:

- Use technology to lift and position the work for easy access that does not require bending, twisting, and reaching.
- Use supplemental lighting at the worksite.
- Use adjustable chairs and work surfaces as appropriate (see Figure 10–5).[15]

Seated Control Work

This type of work involves sitting in one location and using wheels, levers, knobs, handles, and buttons to control a process, system, or piece of equipment. The physical stress associated with seated control work is typically the

Figure 10–5
Use adjustable work surfaces to reduce stress.
Source: From a drawing by Robin J. Miller.

result of excessive vibration or bending and twisting to achieve better visibility. Ergonomic strategies for improving work conditions include:

- Use an adjustable swivel chair with inflatable back and seat support.
- Provide comfortable and convenient locations for control devices.
- Use control devices that meet two standards: finger control systems that do not require more than five newtons (1.1 pounds); hand levers that do not exceed 20 newtons (4.5 pounds).
- Position the control seat so that a clear line of sight exists between the work and the person controlling it.
- Provide a ladder if the workstation is more than 14 inches above ground.[16]

Standing Work

This category includes most jobs that are performed while standing. Such jobs do not involve a great deal of repetitive motion but do involve handling medium to heavy materials. An example is a machine operator's job (lathe, mill, drill, punch, saw, and so on). Physical stress associated with this type of work includes leg, arm, and back strains. Occasionally, side strains occur when bending and twisting are necessary. Ergonomic strategies for improving work conditions include the following:

- Use adjustable machines and work surfaces to ensure the optimum height and position. When the machine height cannot be adjusted, portable platforms can serve the same purpose (Figure 10–6).
- When purchasing new machines, make sure there is a recess at the bottom for feet. This will allow the operator to stand close to the machine without

Figure 10–6
Adjust machines or work
surfaces to reduce stress.
Source: From a drawing by Robin J.
Miller.

having to bend over. Also, look for machines that have easily accessible
controls that fall within a comfortable reach zone for operators.

- Provide ample free space around machines for moving material in and out
and to allow for ease of movement in servicing machines.[17]

Standing for Heavy Lifting and Carrying

This type of work involves heavy lifting and moving material while standing.
Lifting and moving may be a relatively small part of the job but are required
somewhat regularly. The physical stress most commonly associated with this
type of work is back and muscle strains resulting from improper lifting. Falls
can also be a problem. Ergonomic strategies for improving work conditions
include the following:

- Eliminate manual lifting to the extent possible using various lifting and
hoisting technologies.
- Where manual lifting is necessary, train workers in proper lifting
techniques.
- Provide sufficient room around all objects to allow lifting without
twisting.
- Supply the appropriate personal protection equipment such as sure-grip
shoes and gloves.
- Keep floors around materials to be lifted clean and dry to prevent slips.
- Do not allow manual carrying of heavy objects upstairs. Stairs increase the
physical stress of carrying and, in turn, the potential for injury.[18]

Work with Hands above the Chest

This type of work can be done in either a standing or sitting position. It may or may not involve material handling. Physical stress associated with this type of work includes neck, upper body, and heart strain. Of these, the most potentially dangerous is heart strain. Prolonged work with the arms above the shoulder level requires the heart to work harder to pump blood to the elevated areas. Ergonomic strategies for improving work conditions include:

- Eliminate manual lifting to the extent possible by raising the work floor using lifts and various other technologies.
- Use extension arms or poles when the work floor cannot be raised.
- When purchasing new machines, look for machines with controls that are easily accessible below the horizontal plane of a worker's shoulders.[19]

Work with Hand Tools

All the types of work presented in this section may involve the use of hand tools to some extent. However, because hand tools introduce a variety of potential hazards that are indigenous to their use, they are best examined as a separate work type. Physical stress associated most commonly with the use of hand tools includes carpal tunnel syndrome (CTS) and muscle strains of the lower arm, hands, and wrist. Ergonomic strategies for improving the work conditions focus primarily on improving hand positions during the use of tools, enhancing the worker's grip on tools, and minimizing the amount of twisting involved. Following are some of these strategies:

- Select tools that are designed to keep hands in the rest position (palm down, gently curved, thumb outstretched, and knuckle of the index finger higher than that of the little finger).
- Reduce stress on the hand by selecting tools that have thick, rather than thin, handles (a good range for the diameter is 0.8 to 1.2 inches).
- Select tools that have enhanced gripping surfaces on handles such as knurling, filing, or other enhancements.
- To the extent possible, eliminate twisting by selecting tools designed so that the direction of movement or function is the same as the direction in which force is applied or by using technology (e.g., power screwdriver).
- For tools that do not involve twisting, select handles that have an oval-shaped cross section.
- Select tools with handles made of hard, nonpermeable materials that will not absorb toxic liquids that could be harmful to the skin.[20]

Work with VDTs

The video display terminal, primarily because of the all-pervasive integration of personal computers in the workplace, is now the most widely used piece of office equipment. This fact coupled with the ergonomic hazards associated with VDTs have created a whole new range of concerns for safety and health professionals. Using ergonomics to design a workspace will make it easier, safer, more comfortable, and more efficient to use. Following are some strategies that can be used to reduce the hazards associated with VDTs:

- *Arrange the keyboard properly.* It should be located in front of the user, not to the side. Body posture and the angle formed by the arms are critical factors (see Figure 10–7).
- *Adjust the height of the desk.* Taller employees often have trouble working at average height desks. Raising the desk with wooden blocks can solve this problem.
- *Adjust the tilt of the keyboard.* The rear portion of the keyboard should be lower than the front.
- *Encourage employees to use a soft touch on the keyboard and when clicking a mouse.* A hard touch increases the likelihood of injury.
- *Encourage employees to avoid wrist resting.* Resting the wrist on any type of edge can increase pressure on the wrist.
- *Place the mouse within easy reach.* Extending the arm to its full reach increases the likelihood of injury.

Head—Positioned directly over shoulders without straining, about an arm's length from the screen

Neck and Shoulders—Relaxed, without slouching

Back—Upright or inclined slightly forward from the hips; maintain the same spinal curve the back has when standing

Elbows—Relaxed at a 70°–135° angle

Wrist—Relaxed and straight, not angled up or down

Knees—Placed slightly lower than hips

Chair—Sloped slightly forward so knees are in the correct position

0°–60°

70°–135°

Screen—Positioned at eye level or slightly lower

Keyboard—Placed flat, close to elbow-level

Work Surface—Positioned at a height to allow the elbows to be relaxed at a 70°–135° angle; chair may need to be raised to accommodate this

Feet—Planted firmly on the floor. A footrest may be needed for shorter people or high work surfaces

Figure 10–7
Ergonomics of VDTs. The left of the diagram highlights optimal postures and positions for the computer user.

- *Remove dust from the mouse ball cavity.* Dust can collect, making it difficult to move the mouse. Blowing out accumulated dust once a week will keep the mouse easy to manipulate.
- *Locate the VDT at a proper height and distance.* The VDT's height should be such that the top line on the screen is slightly below eye level. The optimum distance between the VDT and user will vary from employee to employee, but it will usually be between 16 and 32 inches.
- *Minimize glare.* Glare from a VDT can cause employees to adopt harmful postures. Changing the location of the VDT, using a screen hood, and closing or adjusting blinds and shades can minimize glare.
- *Reduce lighting levels.* Reducing the lighting level in the area immediately around the VDT can eliminate vision strain.
- *Dust the VDT screen.* VDT screens are magnets to dust. Built-up dust can make the screen difficult to read, contributing to eye strain.
- *Eliminate telephone cradling.* Cradling a telephone receiver between an uplifted shoulder and the neck while typing can cause a painful disorder called cervical radiculopathy (compression of the cervical vertebrae in the neck). Employees who need to talk on the telephone while typing should wear a headset.
- *Require typing breaks.* Continuous typing for extended periods should be avoided. Repetitive strain injuries are cumulative. Breaking up the repetitive motion in question (typing and clicking) can help prevent the accumulation of strain.[21]

ECONOMICS OF ERGONOMICS

Perhaps the most underresearched subject relating to ergonomics is the cost-effectiveness of safety and health. There are two main reasons for this: (1) such research is often more complex and extensive than the safety and health measures that have been undertaken; and (2) many decision makers think such studies are irrelevant because promoting safety and health is the right thing to do, regardless of costs. As with safety and health in general, there are no in-depth studies available that conclusively pin down the cost benefits of specific ergonomic measures.

According to Bedal,

> Does the application of ergonomic principles make good business sense? The good news is most industry experts believe ergonomics does make good business sense. But the bad news is that if you're looking for studies to prove ergonomics is worth the investment, you'll be hard pressed to find them. Very few true cost-benefit analyses on applying ergonomic principles in an industrial setting have been done.[22]

There is disagreement among well-informed professionals on whether ergonomic improvements should be expected to meet the test of cost–benefit analysis. However, there is growing support for research that will produce

Discussion Case

What Is Your Opinion?

"We need to bring in an ergonomics consultant. I'm concerned that all of a sudden we are going to begin seeing CTD and back injuries resulting from some of our new processes and workstations. What do you think?" Mark Patron, safety director for Manufacturing Technologies Corporation, thought over the question before answering. "We don't need a consultant. The risk factors for CTD and back problems are well known in the business. Let me check a couple of reference books and get back to you." Patron sees ergonomic issues as a normal part of his job. His boss thinks an ergonomics expert is needed. What is your opinion?

reliable data. This point of view is beginning to characterize the outlook of safety and health professionals in government and academe.[23] Consequently, it is important to understand the problems in attempting to undertake or participate in cost–benefit studies.

Inhibitors of Cost–Benefit Studies

A number of factors inhibit hard research into the economics of ergonomics. Bedal summarizes the most problematic of these as follows:

* Record-keeping systems in industry are not sufficient to support such studies. As a result, a base of comparison from which to work has not been established.
* Industry does not track injuries and illnesses in ways that provide the controls necessary for true hard research. There are no control groups against which to compare groups of injured workers.
* It is difficult and sometimes impossible to determine what improvements can be attributed directly to specific ergonomic strategies and what improvements should be attributed to other factors.
* Undertaking hard research studies requires a commitment of both time and money. A longitudinal study of the effects of a given ergonomic improvement may take three to five years to produce reliable data.
* Follow-up evaluations of injuries that summarize the direct and indirect costs do not exist to the extent necessary to contribute to hard research.[24]

These inhibitors cannot be overcome unless industrial firms are willing to invest enormous sums of money and time. Although the data produced by such investments would be valuable, they are generally not perceived as adding to a company's bottom line. Therefore, industry is likely to continue to rely on academe and governmental agencies for research. In today's intensely competitive international marketplace, industry is likely to invest all available funds in efforts that are perceived as improving productivity and quality. Consequently, hard research into the economics of ergonomics may remain a low priority for some time to come.

CUMULATIVE TRAUMA DISORDERS (CTDS)

The personal computer has become an all-pervasive and universal work tool. Jobs from the shop floor to the executive office now involve frequent, repetitive computer use. This means that people in the workplace are typing and clicking at an unprecedented pace. Frequent and, for some, constant computer use have led to an explosion of injuries that until now were seen mostly in the meat-packing industry. Collectively, these injuries are known as **cumulative trauma disorders (CTDs)**.

Definition

CTD is an umbrella term that covers a number of injuries caused by forceful or awkward movements repeated frequently over time. Other aggravating factors include poor posture, an improperly designed workstation, poor tool design, and job stress. CTDs occur to the muscles, nerves, and tendons of the hands, arms, shoulders, and neck.

Classifications of CTDs

For years, CTDs have been incorrectly referred to as *carpal tunnel syndrome*, which is actually one type of CTD. This is like referring to all trees as oaks. Figure 10–8 is a checklist of the most common CTDs organized into four broad classifications.

Muscle and Tendon Disorders

Tendons connect muscles to bones. They can accommodate very little in the way of stretching and are prone to injury if overused. Overworking a tendon can cause small tears in it. These tears can become inflamed and cause intense pain. This condition is known as **tendinitis**.

Myofacial muscle damage can also be caused by overexertion. It manifests itself in soreness that persists even when resting. Muscles may burn and be sensitive to the touch. When sore muscles become inflamed and swell, the symptoms are aggravated even further by nerve compression. Tendons that curve around bones are encased in protective coverings called *sheathes*. Sheathes contain a lubricating substance known as synovial fluid. When tendons rub against the sheath too frequently, friction is produced. The body responds by producing additional synovial fluid. Excess build-up of this fluid

Figure 10–8
CTD checklist—types of injury by classification.

Muscle and Tendon Disorders

✓ Tendinitis

✓ Muscle damage (myofacial)

✓ Tenosynovitis

✓ Stenosing tenosynovitis

  • DeQuervain's disease

  • Trigger finger (flexor tenosynovitis)

✓ Shoulder tendinitis

✓ Bicipital tendinitis

✓ Rotator cuff tendinitis

✓ Forearm tendinitis

  • Flexor carpi radialis tendinitis

  • Extensor tendinitis

  • Flexor tendinitis

✓ Epicondylitis

✓ Ganglion cysts

Cervical Radiculopathy

Tunnel Syndromes

✓ Carpal tunnel syndrome

✓ Radial tunnel syndrome

✓ Sulcus ulnaris syndrome

✓ Cubital tunnel syndrome

✓ Guyon's canal syndrome

Nerve and Circulation Disorders

✓ Thoracic outlet syndrome

✓ Raynaud's disease

can cause swelling that, in turn, causes pressure on the surrounding nerves, causing a condition known as **tenosynovitis.**

Chronic tenosynovitis is known as *stenosing tenosynovitis* of which there are two types: DeQuervain's disease and flexor tenosynovitis (trigger finger). DeQuervain's disease affects the tendon at the junction of the wrist and thumb. It causes pain when the thumb is moved or when the wrist is twisted. Flexor tenosynovitis involves the locking of a digit in a bent position, hence the term *trigger finger.* However, it can occur in any finger.

Shoulder tendinitis is of two types: bicipital and rotator cuff tendinitis. Bicipital tendinitis occurs at the shoulder joint where the bicep muscle attaches. The rotator cuff is a group of muscles and tendons in the shoulder that move the arm away from the body and turn it in and out. Pitchers in baseball and quarterbacks in football often experience rotator cuff tendinitis.

Forearm tendinitis is of three types: flexor carpi radialis tendinitis, extensor tendinitis, and flexor tendinitis. Flexor carpi radialis tendinitis causes pain in the wrist at the base of the thumb. Extensor tendinitis causes pain in the muscles in the top of the hand, making it difficult to straighten the hands. Flexor tendinitis causes pain in the fingers, making them difficult to bend.

Epicondylitis and *ganglion cysts* are two muscle and tendon disorders. Epicondylitis (lateral) affects the outside of the elbow, whereas epicondylitis (medial) affects the inside. The common term for this disorder is "tennis elbow." Ganglion cysts grow on the tendon, tendon sheath, or synovial lining, typically on top of the hand, on the nail bed, above the wrist, or on the inside of the wrist.

Cervical Radiculopathy

This disorder is most commonly associated with holding a telephone receiver on an upraised shoulder while typing. This widely practiced act can cause compression of the cervical discs in the neck, making it painful to turn the head. Putting the body in an unnatural posture while using the hands is always dangerous.

Tunnel Syndromes

Tunnels are conduits for nerves that are formed by ligaments and other soft tissues. Damage to the soft tissues can cause swelling that compresses the nerves that pass through the tunnel. These nerves are the median, radial, and ulnar nerves that pass through a tunnel in the forearm and wrist. Pain experienced with tunnel injuries can be constant and intense. In addition to pain, people with a tunnel injury might experience numbness, tingling, and a loss of gripping power. The most common *tunnel syndromes* are carpal tunnel syndrome, radial tunnel syndrome, sulcus ulnaris syndrome, cubital tunnel syndrome, and Guyon's canal syndrome.

Nerve and Circulation Disorders

When friction or inflammation cause swelling, both nerves and arteries can be compressed, restricting the flow of blood to muscles. This can cause a disorder known as *thoracic outlet syndrome*. The symptoms of this disorder are pain in the entire arm, numbness, coldness, and weakness in the arm, hand, and fingers.

If the blood vessels in the hands are constricted, *Raynaud's disease* can result. Symptoms include painful sensitivity, tingling, numbness, coldness, and paleness in the fingers. It can affect one or both hands. This disorder is also known as *vibration syndrome* because it is associated with vibrating tools.

Safety Fact

CTDs: Work Related or Non-Work Related?

CTDs present a difficult challenge to safety and health professionals on two levels. First, identifying the source of an employee's injury can be difficult because of the interplay of non-work-related activities such as hobbies, playing musical instruments, medications, previous surgeries, hypertension, and other disorders. Second, treatment and rehabilitation can be hindered by outside activities. Consequently, prevention programs should be broad enough to take outside activities into account.

Preventing CTDs

The best way to prevent CTDs is proper work design. In addition, it also helps make employees aware of the hazards that can cause it. These hazards include poor posture at the workstation, inappropriate positioning of the hands and arms, a heavy hand on a keyboard or mouse, and any other act that repeatedly puts the body in an unnatural posture while using the hands. Ergonomically sound workstations can help prevent CTDs, especially when they can be modified to fit the individual employee. However, even the best ergonomic design cannot prevent a heavy hand on the keyboard or mouse. Consequently, ergonomics is only part of the answer. Following are some preventive strategies that can be applied in any organization:

1. *Teach employees the warning signs.* CTDs occur cumulatively over time; they sneak up on people. Employees should be aware of the following warning signs: weakness in the hands or forearms, tingling, numbness, heaviness in the hands, clumsiness, stiffness, lack of control over the fingers, cold hands, and tenderness to the touch.
2. *Teach employees how to stretch.* Employees whose jobs involve repetitive motion work such as typing may help prevent CTDs by using stretching exercises. Limbering up the hands and forearms each day before starting work and again after long breaks such as the lunch hour may help eliminate the stress on muscles and tendons that can lead to CTDs. The term "may" is used because the jury is still out as to the efficacy of stretching. There is no consensus in the ergonomics community on this preventive measure.
3. *Teach employees to start slowly.* Long-distance runners typically start slowly, letting their bodies adjust and their breathing find its rhythm. They pick up the pace steadily, until eventually settling in at a competitive pace. This approach is an excellent example of how employees in CTD-prone jobs should work. Teach employees to limber up, then begin slowly and increase their pace gradually.
4. *Avoid the use of wrist splints.* Teach employees to position their hands properly without using wrist splints. Splints can cause the muscles that they support to atrophy, thereby actually increasing the likelihood of problems.
5. *Start an exercise group.* Exercises that strengthen the hands and forearms coupled with exercises that gently stretch hand and forearm muscles may

be a preventive measure. Exercises that strengthen the back can help improve posture, and good posture helps prevent CTDs. As with stretching, the jury is still out on the efficacy of exercise as a preventive measure.

6. *Select tools wisely.* CTS and other cumulative trauma disorders (CTDs) are most frequently associated with the repetitive use of VDTs and hand tools. Selecting and using hand tools properly can help prevent CTDs. Figure 10–9

Use Anthropometric Data

Anthropometric data has to do with human body dimensions. Such data can be used to determine the proper handle length, grip span, tool weight, and trigger length when selecting tools.

Reduce Repetition

Repetition is a hazard that can and should be reduced using such strategies as the following:

- Limit overtime.
- Change the process.
- Provide mechanical assists.
- Require breaks.
- Encourage stretching and strengthening exercises.
- Automate where possible.
- Rotate employees regularly.
- Distribute work among more employees.

Reduce the Force Required

The more force required, the more potential for damage to soft tissue. Required force can be reduced using the following strategies:

- Use power tools wherever possible.
- Use the power grip instead of the pinch grip.
- Spread the force over the widest possible area.
- Eliminate slippery, hard, and sharp gripping surfaces.
- Use jigs and fixtures to eliminate the pinch grip.

Minimize Awkward Postures

Awkward postures contribute to CTDs. The following strategies can reduce posture hazards:

- Keep the wrist in a neutral position.
- Keep elbows close to the body (90°–110° where bent).
- Avoid work that requires overhead reaching.
- Minimize forearm rotation.

Figure 10–9
Checklist for safe selection and use of hand tools.

is a checklist for the proper selection and use of hand tools. Note that ergonomically designed hand tools will not overcome poor job design. Good job design and proper tool selection together are the best strategy.

PARTICIPATORY ERGONOMICS

Participatory ergonomics is the involvement of people at work in planning for and controlling the ergonomic aspects of their work environment. There is one caveat to be observed here: Before people are involved in planning and controlling the ergonomic aspects of their work environment, they must first be equipped with prerequisite knowledge. This means that employees will typically require training before they can become involved in participatory ergonomics.[25]

Participatory ergonomics is an approach to intervention that combines the best of two worlds: outside expertise and inside experience. When undertaking an ergonomics intervention in the workplace, the most common approach is to form a participatory ergonomics (or PE) team. This team should consist of the outside expert(s) in the field of ergonomics, an internal safety and health professional, management personnel, and employees. Before beginning the intervention process, the team is trained—typically by the outside expert(s) in the field of ergonomics. The goal of the PE team is to design ergonomic interventions that are tailored specifically to the workplace in question.

Nine of ten studies reviewed by the Institute for Work & Health showed that PE interventions can have a positive effect on safety and health in the workplace. These studies showed that PE interventions had a positive effect on the musculoskeletal symptoms of workers and on reducing workplace injuries as well as workers' compensation claims.[26]

Making PE Interventions Effective

The Institute for Work & Health found that there are several factors that, when taken together, serve to make PE interventions effective. These factors are:

- Acceptance of the PE team members by other employees and unions (where applicable)
- Acceptance of the PE team members by management
- Availability of an ergonomics expert to lead the team and provide the prerequisite training
- Sufficient support from management in the form of resources

When these factors are present, an organization can reduce injuries and workers' compensation claims by applying the concept of participatory ergonomics when making workplace interventions.

SUMMARY

1. Ergonomics is a multidisciplinary science that seeks to conform the workplace and all of its physiological aspects to the worker.
2. The word *ergonomics* is derived from the Greek language. *Ergon* means work and *nomos* means laws.
3. Human factors is a science that acts as a bridge between research about human beings and the application of that research in designing products and systems for human beings.
4. Common factors that can influence the amount of physical stress associated with a given job include the following: sitting, standing, demand for strength/power, work area, type of motion, amount of surface contact, and environmental factors.
5. OSHA's voluntary ergonomics guidelines have the following program elements: worksite analysis, hazard prevention and control, medical management, and training and education. A mandatory standard has been proposed that, when approved by Congress, will apply to manufacturing and material handling jobs.
6. Common indicators of the existence of ergonomic problems include the following: trends in accidents and injuries, incidence of cumulative trauma disorders (CTDs), absenteeism, high turnover rates, employee complaints, employee-generated changes, poor quality, and high incidence of manual material handling.
7. Procedures for identifying specific ergonomic problems include the following: general observation, questionnaires, interviews, videotaping, photography, drawing, sketching, and measuring.
8. In devising ergonomic problem-solving strategies, work might be divided into the following categories: seated repetitive work with light parts, seated work with larger parts, seated control work, standing work, standing work with heavy lifting or carrying work from place to place, work with hands above chest height, and work with hand tools.
9. There are two reasons why little hard research has been conducted concerning the economics of ergonomics: (1) the cost of such research is often more than the ergonomic improvements that were made; and (2) the feeling that ergonomic improvements should not have to be justified on the basis of cost-effectiveness.

KEY TERMS AND CONCEPTS

Cervical radiculopathy	Environmental factors
Cumulative trauma disorders (CTDs)	Ergonomic hazards
Employee complaints	Ergonomic risk factors
Employee-generated changes	Ergonomics

General duty clause
General observation
Hazard prevention and control
Health care providers
High wrist acceleration
Horizontal work area
Incidence rates
Interviews
Manual material handling
Medical management program
Multiple risk factors
OSHA's voluntary ergonomics
   guidelines
Participatory ergonomics

Physical stress
Questionnaires
Records analysis and tracking
Repetitive motion
Repetitive strain injury
Symptoms survey
Tendinitis
Tenosynovitis
Tunnel syndromes
Vertical work area
Worksite analysis
Workstation analysis

REVIEW QUESTIONS

1. Define the term *ergonomics*. Explain its origins.
2. Explain how the following opposing factors can influence the amount of physical stress associated with a job: sitting versus standing, large versus small demand for strength/power, nonrepetitive versus repetitive motion.
3. Explain the concept of human factors and how it relates to ergonomics.
4. Name the elements of OSHA's voluntary ergonomics guidelines.
5. What are the four main parts of OSHA's recommended worksite analysis program for ergonomics?
6. List five risk factors associated with CTDs.
7. Briefly explain the steps in conducting an ergonomic job hazard analysis.
8. Briefly explain the components of a hazard prevention and control program.
9. Who should participate in an ergonomics training program?
10. List and briefly explain three common indicators of the existence of ergonomic problems.
11. Describe three approaches that can be used to pinpoint specific ergonomic problems.
12. Describe an ergonomic problem-solving strategy for each of the following types of work: seated repetitive work with light parts, work with hands above the chest height, and work with hand tools.
13. Explain why so little hard research has been done concerning the economics of ergonomics.
14. Define the term *CTD*. Describe the most common types.

ENDNOTES

1. P. Tillman and B. Tillman, *Human Factors Essentials* (New York: McGraw-Hill, 1991), 3–22.
2. Ibid., 3–4.
3. Ibid., 9.
4. Ibid., 12–13.
5. Ibid., 21–22.
6. *Ergonomics Report*, April 7, 2006. Retrieved from http://www.ergonomics report.com/publish.
7. Adapted from OSHA 3123.
8. Ibid.
9. Ibid.
10. National Institute for Occupational Safety and Health (NIOSH), "Elements of Ergonomics Programs." Retrieved from http://www.cdc.gov/niosh.
11. Ibid., 12.
12. S. Minter, "Ergonomic Challenge: The Aging Workforce," *Occupational Hazards* 64, no. 9: 6.
13. NIOSH, 13–14.
14. Ibid., 14.
15. Ibid., 15.
16. Ibid., 16.
17. Ibid., 17.
18. Ibid., 18.
19. Ibid., 19.
20. Ibid., 20.
21. Roberta Carson, "Ergonomic Innovations: Free to a Good Company," *Occupational Hazards*, January 1996, 61–64.
22. T. Bedal, "The Economics of Ergonomics: What Are the Paybacks?", *Safety & Health* 142, no. 4: 34.
23. Ibid., 38.
24. Ibid., 39.
25. D. Cole, I. Rivilis, D. Van Eerd, K. Cullen, E. Irvin, and D. Kramer, "Effectiveness of Participatory Ergonomic Interventions: A Systematic Review," Institute for Work & Health, January 2005. Retrieved from http://www.iwh.on.ca.
26. Ibid.

STRESS AND SAFETY

Major Topics

- Workplace Stress Defined
- Sources of Workplace Stress
- Human Reactions to Workplace Stress
- Measurement of Workplace Stress
- Shift Work, Stress, and Safety
- Improving Safety by Reducing Workplace Stress
- Stress in Safety Managers
- Stress and Workers' Compensation

WORKPLACE STRESS DEFINED

Our emotions are affected in the workplace by social, occupational, environmental, and psychological factors that we perceive as threats. These perceived threats are external stimuli and **stress** is the reaction of the human body to these stimuli. The amount of stress felt depends as much on the individual's ability to deal with the external stimuli as the relative intensity of the stimuli. For example, the threat of falling out of a boat into the water would cause more stress for a person who cannot swim than it would for a person who is an accomplished swimmer.

In this example, the water is just as deep and the boat is just as far from shore for both individuals, but the threat and the amount of stress it causes are less for the one who can swim than for the nonswimmer. This is because of the good swimmer's ability to deal with the threatening situation. This same rule of thumb can be applied to all perceived threats in the workplace.

As they often do, management personnel and line employees often view stress from two different perspectives. Managers tend to view stress as an individual problem tied to the personality and emotional makeup of the employee.

Employees often view stress as a problem induced by poor supervision, unrealistic expectations, and other management shortcomings. Practically speaking, both managers and employees are partially right in their opposing perspectives. On one hand, the personality and emotional makeup of employees are both factors in how they respond to the stimuli that produce stress. On the other hand, management shortcomings are also factors.

However it is defined, stress is a serious problem in the modern workplace. Stress-related medical bills and the corresponding absentee rates cost employers over $150 billion annually. Almost 15 percent of all occupational disease claims are stress related.

Workplace stress involves the emotional state resulting from a perceived difference between the level of occupational demand and a person's ability to cope with this demand. Because preparations and emotions are involved, workplace stress is considered a subjective state. An environment that a worker finds stressful may generate feelings of tension, anger, fatigue, confusion, and anxiety.

Workplace stress is primarily a matter of person–workload fit. The status of the person–workload fit can influence the acceptance of the work and the level of acceptable performance of that work. The perception of workload may be affected by the worker's needs and his or her level of job satisfaction. The relation between job demands and the worker's ability to meet those demands further influence workplace stress. Because workplace stress may be felt differently by people in similar situations, it must be concluded that there are many causes of workplace stress.

SOURCES OF WORKPLACE STRESS

The sources of on-the-job stress may involve physical working conditions, work overload, **role ambiguity**, lack of **feedback**, personality, personal and family problems, or role conflict. Other sources of workplace stress are discussed in the following paragraphs.

- **Task complexity** relates to the number of different demands made on the worker. A job perceived as being too complex may cause feelings of inadequacy and result in emotional stress. Repetitive and monotonous work may lack complexity so that the worker becomes bored and dissatisfied with the job and possibly experiences some stress associated with the boredom.
- **Control** of employees over their job responsibilities can also be a source of workplace stress. Being responsible for work without being in control of it is a formula for stress. The more work can be organized and structured to allow for maximum control by those who do it, the less stressful the work will be. In addition, even when workers do feel as if they control the work they are responsible for, if they are monitored electronically their stress levels tend to rise. In the eyes of a line employee, electronic monitoring is

the same as having a supervisor standing over your shoulder constantly. Job control has also been tied to cardiovascular disease and heart attacks. Workers with high stress related to a lack of job control experience elevated levels of plasma fibrinogen—a blood-clotting factor that has been linked to cardiovascular events such as heart attacks. A study by Belgian scientists at Ghent University tied not just job stress, but also the specific type of stress associated with a lack of job control to plasma fibrinogen—a specific medical marker for cardiovascular problems.[1]

- A **feeling of responsibility** for the welfare or safety of family members may produce on-the-job stress. Being responsible for the welfare of his or her family may cause a worker to feel that options to take employment risks are limited. A worker may then perceive that he or she is "trapped in the job." Overly constrained employment options may lead to anxiety and stress. The feeling of being responsible for the safety of the general public has also been shown to be a stressor. Air traffic controllers are known to experience intense stress when their responsibility for public safety is tested by a near-accident event. A feeling of great responsibility associated with a job can transform a routine activity into a stress-inducing task.

- **Job security** involves the risk of unemployment. A worker who believes that his or her job is in jeopardy will experience anxiety and stress. The ready availability of other rewarding employment and a feeling that one's professional skills are needed reduce the stress associated with job security issues.

- An **organizational culture** leaves the employee feeling left out, out of the loop, and ill-informed. Organizations in which managers fail to communicate frequently and effectively with employees are creating high-stress environments for workers.

- **Work schedules** that are unpredictable, never-changing, and ever-changing can induce stress in employees. Employees have lives outside of their jobs. Consequently, the ability to predict their work schedules is important. When work schedules are unpredictable, stress increases. It also increases when work schedules are inflexible—they cannot be changed no matter what other obligation the employee might have. On the other hand, ever-changing work schedules such as those associated with shift work can also increase the level of employee stress. The big-picture issue with regard to work schedules is the level of control employees have over their lives. The less control, the more stress.

- **Home and family problems** can create added stress for workers. There was a time when employees were expected to "leave their problems at the front door" when they came to work. This, of course, is a practical impossibility. The demands of raising children, working out home and job schedules, dealing with the conflicting agendas inherent in dual-career households, and otherwise handling the everyday work and home conflicts that inevitably arise can markedly increase an employee's stress level.

- **Work relationships** can lead to on-the-job stress. People are social beings by nature. They like to get along with the people they spend a lot of time with. However, office politics, turf battles, and internal competition for recognition and rewards can be hard on work relationships. When employees do not get along with their fellow workers, stress levels increase.

- **Human resource management (HRM) issues** can be a source of workplace stress. People who work have a vested interest in their wages, salaries, working conditions, and benefits. If even one of these factors is a negative, employee stress levels can increase significantly. In addition to these factors, other stress-inducing HRM issues include being underemployed, failing to get promoted, and working in a position that is clearly not valued by management.

- **Workload demands** can stimulate stress when they are perceived as being overwhelming. These demands may involve time constraints and cognitive constraints such as speed of decision making and mandates for attention. Workload demands may also be physically overwhelming if the worker is poorly matched to the physical requirements of the job or is fatigued. Whenever the worker believes the workload to be too demanding, stress can result.

- **Psychological support** from managers and coworkers gives a feeling of acceptance and belonging and helps defuse stress. A lack of such support may increase the perception of a burdensome workload and result in stress.

- The lack of **environmental safety**, including the potential for workplace violence, can also be a cause of stress. Feeling that one is in danger can be a stressor. Workers need to feel safe from environmental hazards such as extreme temperatures, pressure, electricity, fire, explosives, toxic materials, ionizing radiation, noises, and dangerous machinery. To reduce the potential for stress due to environmental hazards, workers should feel that their managers are committed to safety and that their company has an effective safety program.

Common Causes of Stress in the Workplace

- The company was recently purchased by another company.
- Downsizing or layoffs have occurred in the past year.
- Employee benefits were significantly cut recently.
- Mandatory overtime is frequently required.
- Employees have little control over how they do their work.
- The consequences of making a mistake on the job are severe.
- Workloads vary greatly.
- Most work is machine paced or fast paced.
- Workers must react quickly and accurately to changing conditions.
- Personal conflicts on the job are common.
- Few opportunities for advancement are available.
- Workers cope with a great deal of bureaucracy in getting work done.
- Staffing, money, or technology is inadequate.
- Pay is below the going rate.
- Employees are rotated among shifts.[2]

It has long been known among safety and health professionals and practitioners of occupational medicine that stress can have a detrimental effect on attendance, productivity, employee retention, and morale. The sources of stress contained in this list are just some of the more common ones. There are many others. For example, technological developments have increased stress levels on the job.

The fact that cellular telephones can be obnoxiously intrusive and distracting has tended to increase stress levels at work. Continual change in the field of computer technology—change that forces workers to upgrade their skills almost as soon as they become comfortable with a given software package—has become a source of stress in the workplace. Stress has become such an all-pervasive source of safety and health–related problems in the workplace that professionals in the field must acknowledge it and work to reduce its harmful effects like any other workplace hazard.

HUMAN REACTIONS TO WORKPLACE STRESS

Human reactions to workplace stress may be grouped into the following categories: subjective or emotional (anxiety, aggression, guilt); behavioral (accident proneness, trembling); cognitive (inability to concentrate or make decisions); physiological (increased heart rate and blood pressure); and organizational (absenteeism and poor productivity). Continual or persistent stress has been linked to many physiological problems. Initially, the effects may be psychosomatic, but with continued stress, the symptoms show up as actual organic dysfunction. The most common forms of stress-related diseases are gastrointestinal, particularly gastric or duodenal ulcers. Research has linked some autoimmune diseases with increased long-term workplace stress.[3]

The human response to workplace stress can be compared to a rubber band being stretched. As the stress continues to be applied, the rubber band stretches until a limit is reached when the rubber band breaks. For humans, various physical and psychological changes are observed with the repetitive stimuli of stress. Until the limit is reached, the harmful effects can be reversed. With an increase in intensity or duration of the stress beyond the individual's limit, the effects on the human become pathological.[4]

There are three stages of the human stress response: (1) alarm, (2) resistance, and (3) exhaustion.[5] The alarm reaction occurs when the stress of a threat is sensed. The **stage of alarm** is characterized by pallor, sweating, and an increased heart rate. This stage is usually short. It prepares the body for whatever action is necessary.

When the stress is maintained, the **stage of resistance** initiates a greater physical response. The alarm symptoms dissipate, and the body develops an adaptation to the stress. The capacity for adaptation during this stage is limited.

Eventually, with sustained stress, the **stage of exhaustion** is reached. This stage is demonstrated by the body's failure to adapt to the continued stress. Psychosomatic diseases such as gastric ulcers, colitis, rashes, and autoimmune disorders may begin during this stage. The tendency to develop a specific stress-related disease may be partially predetermined by heredity, personal habits such as smoking, and personality.

From an evolutionary viewpoint, the adverse effects of stress on health may be considered a maladaptation of humans to stress. What does this tell us? Either we (1) learn to do away with all stress (unlikely); (2) avoid all stressful situations (equally unlikely); (3) learn to adapt to being sick because of stress (undesirable); or (4) learn to adapt to workplace stress (the optimal choice). The first step in learning to adapt to stress is understanding the amount of stress to which we are subjected.

MEASUREMENT OF WORKPLACE STRESS

Workplace stress can be seen as an individual's psychological reaction to the work environment. Although psychological response cannot be directly measured in physical terms, one method commonly employed uses a measurement of mental workload. Mental workload can be measured in one of three ways:

1. With **subjective ratings**, the workers are asked to rate their perceived level of workload. The perceived workload is then viewed as a direct reflection of workplace stress. The workers may be asked to rate their mood in relation to the work situation. The data gathered by this method are obviously subjective and state dependent. **State-dependent** data are directly related to the circumstances or state under which they are collected and, therefore, have a built-in state bias.
2. **Behavioral time sharing** techniques require the simultaneous performance of two tasks. One of the tasks is considered the primary or most important; the other is of secondary importance. The decrease in performance efficiency of the secondary task is considered an index of workload for behavioral time sharing or human multitasking. Workplace stress is thought to increase as behavioral time sharing increases.
3. **Psychophysiological techniques** require simultaneous measurement of heart rate and brain waves, which are then interpreted as indexes of mental workload and workplace stress.

Behavioral time sharing and psychophysiological techniques are related to theoretical models, making data easier to interpret. These two techniques also require sophisticated equipment and data collection methods.

Subjective ratings may be collected using questionnaires or survey instruments. These instruments may ask about the physical working conditions, the individual's health and mental well-being, and perceived overall satisfaction

Safety Fact

Workplace Stress and Health

The physical and mental health of working people is affected by their *psychological environment*. Autocratic, insensitive managers and supervisors can produce enough stress in employees to cause physical illness. When managers and supervisors show an interest in employees, empower them to participate, and provide positive reinforcement, fewer illnesses or cases of work-related depression are reported.

with the job. The data may then be compared to standardized scales developed by various researchers.

Psychosocial questionnaires evaluate workers' emotions about their jobs. Workers may be asked about job satisfaction, workload, pace, opportunities for advancement, management style, and organizational climate. Psychosocial questionnaires are another form of subjective rating and are also subject to state-dependent bias in the data. Regardless of the measurement method, because workplace stress is dependent on personal awareness, no direct means of measuring workplace stress are now available.

SHIFT WORK, STRESS, AND SAFETY

Shift work can require some employees to work when the majority of people are resting. In some cases, shift work requires rotating between two or three different starting times, which may vary by eight hours or more. Shift work has traditionally been required by the medical community, the transportation industry, utilities, security, and, increasingly, by retail sales.[6]

Basic physiological functions are scheduled by the biological clock called the **circadian rhythm**. Most children in the United States grow up on the day shift, going to school during the day and sleeping at night. After a life of being on the day shift, the body perceives a change in work shift as being stressful. If the person takes a job starting at midnight, his or her body will still expect to be sleeping at night and active during the day.

Many physical and psychological functions are affected by circadian rhythm. Blood pressure, heart rate, body temperature, and urine production are measurably slower at night. These same functions are normally faster during the day (active time).

Behavioral patterns also follow the circadian slower-at-night and more-active-during-the-day pattern. Sleep demand and sleep capacity for people age 14 or older are greatest at night and least during daylight hours. Alertness has been determined to be decreased at night.

Workers surveyed have consistently reported lower job satisfaction with rotating shifts. Day-shift workers with the same task definitions report higher job satisfaction and less stress than their second- or third-shift counterparts. Rotating shifts over several weeks can result in desensitization to the circadian rhythms. With this desensitization comes a measurable loss in productivity, increased numbers of accidents, and reported subjective discomfort. After returning to a predictable shift, workers regained their biological clock and circadian rhythm.

Not working the normal day-shift hours results in an increase in workplace stress, with rotating shifts being the most stressful. From a safety viewpoint, shift workers are subjected to more workplace stress in terms of weariness, irritability, depression, and a lack of interest in work. Shift work increases workplace stress and may lead to a less safe worker.

Reducing the Stress Associated with Shift Work

Shift work is and will probably always be a fact of life for employees in certain occupations. More than 15 million people in the United States have jobs that require shift work. To reduce the stress associated with shift work, safety and health professionals can apply the following strategies:

1. Encourage shift workers to exercise regularly. Regular exercise can have the double benefit of improving the quality of an individual's sleep and relieving pent-up stress.
2. Encourage shift workers to avoid caffeine, alcohol, or other drugs that can inhibit their ability to sleep.
3. If shift workers cannot sleep without some type of sleep aid, the food supplement melatonin or other natural sleep inducers should be recommended rather than sleeping pills that contain synthetic chemicals and can have side effects that might contribute to other stress-inducing effects.[7]

IMPROVING SAFETY BY REDUCING WORKPLACE STRESS

Not all sources of stress on the job can be eliminated, and employment screening is unlikely to identify all those who are sensitive to stress. People can learn to adapt to stress, however. **Training** can help people recognize and deal with stress effectively. Employees need to know what is expected of them at any given time and to receive recognition when it is deserved. Managers can reduce role ambiguity and stress caused by lack of feedback by providing frequent feedback.

Stress can result from low participation or lack of **job autonomy**. A manager can help employees realize their full potential by helping them match their career goals with the company's goals and giving them more control over their jobs.

Managers can help design jobs in ways that lead to worker satisfaction, thereby lessening work stress. **Physical stress** can be reduced by improving the work environment and establishing a sound safety and health program. Managers can also assist in the effort to provide varied and independent work with good possibilities for contact and collaboration with fellow workers and for personal development.

Organizational approaches to coping with work stress include avoiding a monotonous, mechanically **controlled pace**, standardized motion patterns, and constant repetition of **short-cycle operations**. Other stress-inducing work design features to avoid include jobs that do not make use of a worker's knowledge and initiative, that lack human contact, and that have authoritarian-type supervision.

There are also several individual approaches to coping with stress. One of the most important factors in dealing with stress is learning to recognize its symptoms and taking them seriously. Handling stress effectively should be a lifelong activity that gets easier with practice. Keeping a positive mental attitude can help defuse some otherwise stressful situations.

People can analyze stress-producing situations and decide what is worth worrying about. Individuals can effectively respond to a stressful workload by **delegating responsibility** instead of carrying the entire load. Relaxation techniques can also help reduce the effects of stress. Some common relaxation methods include meditation, biofeedback, music, and exercise.

The following strategies are recommended for reducing workplace stress.

- Management recognizes workplace stress and takes steps regularly to reduce this stress.
- Mental health benefits are provided in the employee's health insurance coverage.
- The employer has a formal employee communications program.
- Employees are given information on how to cope with stress.
- Workers have current, accurate, and clear **job descriptions**.
- Management and employees talk openly with one another.
- Employees are free to talk with each other during work.
- Employers offer exercise and other stress-reduction classes.

Discussion Case

What Is Your Opinion?

"We are beginning to see more and more stress-related problems. I'm afraid that if we don't deal with the issue, stress is going to cause even more serious injuries." "Nonsense. We all have stress. There has always been stress on the job. It goes with the territory. A few employees go to a seminar on workplace stress and, all of a sudden, everyone is complaining about stress." This discussion took place between the safety director and the CEO of Gulf Coast Electric Company. Is stress really a legitimate workplace hazard? What is your opinion?

- Employees are recognized and rewarded with nonmonetary prizes for their contributions.
- Work rules are published and are the same for everyone.
- Child care programs are available.
- Employees can work **flexible hours**.
- Perks are granted fairly based on a person's level in the organization.
- Workers have the training and **technology access** that they need.
- Employers encourage work and personal support groups.
- Workers have a place and time to relax during the workday.
- Elder care programs are available.
- Employees' work spaces are not crowded.
- Workers can put up personal items in their work areas.
- Management appreciates humor in the workplace.[8]

Additional ways to reduce stress in the workplace include the following:

- Match workload and pace to the training and abilities of employees.
- Make an effort to match work schedules with the personal lives of employees.
- Clearly define work roles.
- Before giving employees additional duties beyond their normal work roles, make sure they receive the necessary training.
- Enforce work rules equitably.
- Promote teamwork among employees and encourage it throughout the organization.
- Involve employees in making decisions that affect them.
- Inform employees in a timely manner of organizational changes that might affect them.[9]

There is no one clear answer to workplace stress. The suggestions given here are a good starting place for management and employees to begin the process of being aware of and dealing effectively with workplace stress.

STRESS IN SAFETY MANAGERS

Safety and health management can be a stressful profession. Robert Scherer of Wright State University identified the following four conditions that frequently trigger stress among modern safety and health professionals:

- Role overload
- Coping with regulatory breakdown
- Communication breakdown
- Competing loyalties[10]

According to Scherer, safety and health professionals are sometimes overloaded when corporate *downsizing* results and they are delegated more and more responsibilities. Trying to keep up with the ever-changing multitude of

Safety Myth

Managers Cannot Reduce Employee Stress

Of course, the workplace is stressful! This fact cannot be helped. Some people can handle stress, and some cannot. There is nothing managers can do. Right? Wrong. There is a lot that managers can do. Managers can provide training to help employees learn to recognize stress and deal with it. Managers can empower employees, giving them as much control as possible over their jobs. Managers can improve communication with employees. Managers can provide clear, understandable job descriptions that eliminate ambiguity. Managers can provide child care or child care assistance. Managers can provide exercise and stress reduction programs. There are many strategies that managers can employ to reduce stress on the job.

regulations is a stress-inducing challenge. Communication relating to safety and health is always a challenge. However, when economic forces focus an organization's attention on other matters, it can be even more difficult than usual to get the safety and health message across. This increased difficulty can lead to increased stress. Line managers who are more concerned with meeting production quotas than with the safety of their employees sometimes try to influence safety and health managers—their colleagues and sometimes their friends—to look the other way. This subjects safety and health managers to the pressures of competing loyalties.[11]

Scherer recommends that safety and health managers cope with these four common triggers of stress by applying the following strategies: (1) prioritize activities by focusing on those that present the most risk to the organization; (2) work closely with the organization's legal staff and subscribe to an online CD-ROM updating service; (3) formalize communication and hold regularly scheduled safety and health meetings for all operating employees; and (4) focus on the risks to the organization and refuse to take sides.[12]

STRESS AND WORKERS' COMPENSATION

There are serious problems with workers' compensation in the United States. On one hand, there is evidence of abuse of the system. On the other hand, many injured workers who are legitimately collecting benefits suffer a substantial loss of income. Complaints about workers' compensation are common from all parties involved with the system (employers, employees, and insurance companies).

Stress claims are more burdensome than physical claims because they are typically reviewed in an adversarial environment. This leads to the involvement of expert medical witnesses and attorneys. As a result, even though the benefits awarded for stress-related injuries are typically less than those awarded for physical injuries, the cost of stress claims is often higher because of litigation.

SUMMARY

1. Stress is the harmful physical and emotional response that occurs when the requirements of the job do not match the capabilities, resources, or needs of the worker.

2. Workplace stress involves a worker's feelings resulting from perceived differences between the demands of the job and the person's capacity to cope with these demands.

3. Sources of workplace stress include environmental conditions, work overload, role ambiguity, lack of feedback, personality, personal and family problems, and role conflict. Other sources of workplace stress are task complexity, lack of control over the job, public safety responsibility, job security, lack of psychological support, and environmental safety concerns.

4. A life insurance company poll of workers found that workplace stress may be caused by company reorganization, buyout or layoffs; mandatory overtime; varying workloads; work pace; lack of opportunity for advancement; bureaucracy; shortages of staff, money, or technology; low pay; or rotating shifts.

5. Human reaction to workplace stress may be grouped into five categories: (a) subjective, (b) behavioral, (c) cognitive, (d) physiological, and (e) organizational.

6. Psychosomatic reaction to stress may eventually lead to autoimmune disease.

7. Until an individual's limit is reached, the effects of stress may be reversed. After that limit, with continuing stress, the effects can become pathological.

8. Research has shown three stages of human reaction to stress: (a) alarm, (b) resistance, and (c) exhaustion.

9. The best policy regarding stress is to learn to adapt to it. Efforts to rid the workplace of all sources of stress are unlikely to succeed.

10. Mental workload can be measured in three ways: (a) subjective ratings that are state dependent, (b) behavioral time sharing, and (c) psychophysiological techniques.

11. Psychosocial questionnaires study how workers feel about their jobs.

12. Shift work occurs when the majority of people are at leisure. The circadian rhythm, or biological clock within the body, determines when a person will be comfortable either working or sleeping. Workers surveyed have reported lower job satisfaction with rotating shifts. Shift work may result in loss of productivity and an increased number of accidents.

13. All sources of stress on the job cannot be eliminated.

14. Managers can reduce workplace stress by reducing role ambiguity and increasing feedback and job autonomy. Managers can also lessen workers' stress by reducing exposure to physical hazards, varying the work pace, and eliminating monotonous or short-cycle operations.

15. Workplace stress reduction, according to a life insurance study, can be accomplished by providing employee mental health insurance benefits, improving employee–management communications, providing workers with information about how to deal with stress, providing job descriptions, talking with employees regularly, recognizing and rewarding contributions, having published work rules, and offering child care and elder care programs. Other stress reducers include permitting flexible work hours, granting perks fairly, giving adequate training and technology access, providing a place and time to relax, having uncrowded workplaces with space to put up personal items, and having a management that keeps a sense of humor.

16. Individuals can effectively respond to a stressful workload by delegating responsibility and learning how to relax. Relaxation methods include meditation, biofeedback, music, and exercise.

17. One of the most important factors in dealing with stress is learning to recognize its symptoms and taking the symptoms seriously.

KEY TERMS AND CONCEPTS

Behavioral time sharing

Circadian rhythm

Control

Controlled pace

Delegate responsibility

Employee communications

Environmental safety

Feedback

Feeling of responsibility

Flexible hours

Human reaction

Job autonomy

Job description

Job security

Physical stress

Psychological support

Psychophysiological techniques

Psychosocial questionnaires

Role ambiguity

Shift work

Short-cycle operations

Stage of alarm

Stage of exhaustion

Stage of resistance

State dependent

Stress

Subjective rating

Task complexity

Technology access

Training

Workload demands

Workplace stress

REVIEW QUESTIONS

1. Define *stress*.
2. How is workplace stress different from general stress?
3. List five sources of workplace stress and give an on-the-job example for each source.
4. Explain why lack of job autonomy may cause workplace stress.
5. Give five categories of human reaction to workplace stress.
6. How are psychosomatic reactions to stress and actual physiological illness related?
7. How are some autoimmune diseases and workplace stress related?
8. Explain three stages of human reaction to stress.
9. Explain three ways in which mental workload can be measured.
10. Discuss efforts to rid the workplace of all causes of workplace stress.
11. What type of data do psychosocial questionnaires provide? Discuss the bias in this type of data.
12. Discuss how shift work causes workplace stress. Give suggestions for minimizing workplace stress from shift work.
13. Give specific steps that can be taken by managers to help reduce workplace stress.
14. Discuss at least five methods to reduce workplace stress, according to the life insurance company research.
15. Explain how individuals can reduce workplace stress.

ENDNOTES

1. Occupational Health & Safety Online, "Study: Job Stress Linked to Increased Inflammation," September 2005. Retrieved from http://www.ohsonline.com/stevens/ohspub.nsf/d3d5b4f93862266e8625670c006dbc58/c16.
2. B. Farms, *Occupational Medicine Practical Guidelines: Evaluation and Management of Common Health Problems and Functional Recovery in Workers,* 2nd ed. (Elk Grove Village, IL: OEM Press, 2004), 492.
3. Ibid., 494.
4. Ibid.
5. Ibid.
6. D. Morshead, "Stress and Shiftwork," *Occupational Health & Safety* 71, no. 4: 36–37.
7. Ibid.
8. Farms, 498.
9. Ibid.
10. R. Scherer. "Stress Homes in on Our Safety's Ranks," *Occupational Hazards*, January 1994, 156–157.
11. Ibid., 157.
12. Ibid.

SAFETY AND HEALTH TRAINING

12

Major Topics

- Rationale for Safety and Health Training
- Education and Training Requirements
- Safety and Health Professionals as Trainers
- Preparing Safety and Health Instruction
- Presenting Safety and Health Instruction
- Applying Safety and Health Instruction
- Evaluating Safety and Health Instruction
- Training Supervisors
- Training New and Transferred Employees
- Job Safety Analysis as a Training Technique
- Training Opportunities Available
- Illiteracy and Safety
- English as a Second Language Training Issues
- OSHA Standards and Training

Education and training have been recognized as important components of organized safety programs for many years. In today's rapidly changing high-tech workplace, they are more important than ever. Modern safety and health professionals have a key role to play in ensuring that all employees at all levels receive the appropriate types and amounts of training. They must also be prepared to play an active role in preparing, presenting, arranging for the application of, and evaluating safety and health training. This chapter gives prospective and practicing professionals the information they need to play a positive role in providing effective safety and health education.

RATIONALE FOR SAFETY AND HEALTH TRAINING

Consider the following example. The management team at Will-Burt Company, a steel fabrication and parts manufacturer, was in financial trouble. According to Gregg LaBar,

> Low profits, too many product defects, and steep workers' compensation and medical costs were hurting the company. That's when new President and Chief Executive Officer Harry E. Featherstone stepped up and instituted an employee training and education program. . . . Naturally employees get their share of safety training, including lift truck operation, hazard communication, hearing protection, and lockout/tagout, as part of Will-Burt's overall commitment to education.[1]

The employee training program, including the safety components, helped turn Will-Burt's fortunes around. Today, the company is prosperous and competitive. In the words of Jack Bednarowski, director of human resources for Will-Burt, "Our overall attitude toward education, including safety training, is that we can't afford not to do it."[2]

Workers who have not been trained to perform their jobs safely are more likely to have accidents. According to the National Safety Council (NSC),

> Many studies have been made to determine why people fail to follow safety procedures or to take reasonable precautions on the job. Some of the reasons are that workers have:
>
> - Not been given specific instructions in the operation
> - Misunderstood the instructions
> - Not listened to the instructions
> - Considered the instructions either unimportant or unnecessary
> - Disregarded instructions

Any of the above lapses can result in an accident. To prevent such an occurrence, it is essential that safety training work be conducted efficiently.[3]

Besides the commonsense fact that a well-trained employee is more likely to be a safe employee, there are also **legal and ethical reasons** for providing safety and health training.

Legal and Ethical Reasons for Training

The Occupational Safety and Health Act, or OSH Act, mandates that employers provide safety and health training. The OSH Act requires

- Education and training programs for employees
- Establishment and maintenance of proper working conditions and precautions
- Provision of information about all hazards to which employees will be exposed on the job
- Provision of information about the symptoms of exposure to toxic chemicals and other substances that may be present in the workplace
- Provision of information about emergency treatment procedures

In addition, the OSH Act requires that employers make information available to workers concerning the results of medical or biological tests and that workers be given opportunities to observe when activities for monitoring regulated substances are undertaken.

However, the OSH Act did not specifically cover numerous toxic substances in its requirement for employee training. This broad-based exclusion represented a serious and dangerous loophole in the law. Since the passage of the OSH Act, this loophole has been closed. According to Spencer and Simonowitz,

> With the Federal standards promulgated in the early '80s called the 'Right-to-know' laws, composed of the "Access to Employee Exposure and Medical Records" (29 C.F.R. 1910.20) and the "Hazard Communication" (29 C.F.R. 1910.20), it is now the duty of employers to provide a wide variety of all information.[4]

As a result of the OSH Act and subsequent federal regulations, workers have a right to information about any aspect of the workplace that may affect their safety and health. In addition, the OSH Act and subsequent federal legislation clearly establish the employer's responsibility for providing employees with the information they need to work safely.

One of the specific responsibilities of employers set forth in the Hazard Communication regulation (29 CFR 1910.20) is the provision of **hazard communication** programs. These programs should include such components as warning labels, training, access to records, and the distribution of material safety data sheets.

The Occupational Safety and Health Administration (OSHA) makes grants available for companies to use for programs that will improve the understanding of **material safety data sheets (MSDSs)**. "The education program developed ... may also involve providing training to employees or a combination of both."[5] Information on grants for providing training relating to MSDSs is available from:

OSHA Office of Training and Education
Division of Training and Education Programs
1555 Times Drive
Des Plaines, IL 60018
http://www.osha.gov

Beyond the legal reasons for providing safety and health training, there are also ethical reasons. No amount of legislation can properly regulate every hazardous substance or every potentially hazardous situation. Chemicals and technologies are developed and put in place much more rapidly than is legislation. The federal *Registry of Toxic Effects of Chemical Substances* maintained by the National Institute for Occupational Safety and Health (NIOSH) typically adds several thousand new chemicals to its list each year.

Clearly, the only way to guarantee that employees are well informed about the safety and health aspects of their jobs is for companies to fulfill their moral obligations along these lines. According to Spencer and Simonowitz,

> If the sick person has rights to information and determination regarding his or her body, should the well, healthy employee or working person have any less right to

the same consideration when there are health implications of exposure on the job? If the sick person is entitled to know about procedures and prognosis, discomforts and inconveniences, risk, and experience of proposed treatment, surely working persons should have similar rights to information about the nature and toxicity of the substances with which they work, controls and their effectiveness, personal discomforts and inconvenience of both hazards and controls, morbidity and mortality data, and the relative risk.[6]

Who Is Responsible for Training?

The OSH Act requires safety and health training, but who is responsible for seeing that training is provided and who is responsible for providing it? Figure 12–1 illustrates the chain of responsibility for safety and health training. OSHA and the secretary of labor are responsible for both the direct provision of training and for ensuring that industrial firms provide training at the local level.

OSHA's original response to its training mandate was to develop and dispense educational materials. Local companies were to use these materials in the actual provision of training. This continues to be an important component of OSHA's overall training effort. However, OSHA officials soon learned that the printing and distributing of educational materials alone was not

Figure 12–1
Who is responsible for providing training?

Discussion Case

What Is Your Opinion?

"I don't need my employees wasting any more time in training," said the production supervisor. "I need them on the production line operating their machines." The safety director, who had heard this argument many times before, countered, "They won't be able to operate their machines from a hospital bed, and if they don't complete their safety training, the hospital is probably their next stop!"

Both of these managers are interested in the same thing—productivity. The production supervisor is interested in his unit's productivity today. The safety director is interested in the unit's long-term productivity. Who is right in this argument? What is your opinion?

sufficient. Consequently, OSHA added three additional components to its training effort:

- Monetary awards provided on a grant basis to companies, organizations, and educational institutions to finance the provision of safety and health training (the MSDS training grants referred to earlier in this chapter).
- Training requirements set forth in various standards developed by OSHA.
- Requirements set forth in the various OSHA regulations known collectively as the "right-to-know" regulations (referred to earlier in this chapter).

Financial awards are typically an effective way to generate activity in a given area. This has been the case with safety and health training, at least partially. The majority of the incentive funds awarded have gone to labor unions, whereas a relatively small percentage has gone to individual companies. This is partly because labor unions have been more aggressive in the development of safety and health programs and in seeking funds to cover the start-up costs of those programs. Corporate America has tended to rely primarily on local company programs for providing training.

A major inhibitor in the provision of safety and health training is *mistrust* between labor unions and management, a fact that safety and health professionals must be prepared to handle. Such professionals are normally part of a company's management team and, as a result, may be viewed with suspicion by workers. Safety and health professionals often feel as if they are caught in the middle with neither side appreciating their work.

When management and labor debate the safety and health issue, both sides usually make many charges and countercharges. In short, management claims labor is not sufficiently sensitive to the bottom line, and labor claims management is not sufficiently sensitive to the safety and health of workers.

The mistrust that is at the heart of this issue runs both ways. Therefore, safety and health professionals have a very important task in convincing both sides that providing a safe and healthy workplace is not only ethically right but also profitable in the long run.

Safety and health professionals should be able to articulate specifically how employers can meet their legal and ethical obligations concerning workforce training. Spencer and Simonowitz state what employers must do:

> To meet the legal and ethical requirements of educating the workforce, the information given must be sufficiently precise to answer questions about the substance, the kind and degree of exposure, the controls in use, and the degree of their effectiveness, and personal discomfort or inconvenience involved, morbidity and mortality data—both animal and human—and the relative risk. They must be informed not only about emergency procedures for the sudden acute exposure, but also of chronic illness associated with long-term low-level exposure and length of latency periods.[7]

To meet their legal and ethical obligations regarding safety and health training, companies must rely on their safety and health professionals who, in turn, must rely on first-line supervisors. Safety and health professionals and first-line supervisors should work closely in preparing, presenting, applying, and evaluating safety training. The division of labor between safety and health professionals and supervisors is not black and white. However, generally speaking, the following guidelines apply:

- Safety and health professionals train the supervisors and keep them up-to-date.
- Supervisors provide most of the training to workers.
- Safety and health professionals and supervisors work together to train workers jointly when appropriate.

In general, the less job-specific the information, the more likely it is provided by safety and health professionals. The more job-specific the information, the more likely it is provided by supervisors.

EDUCATION AND TRAINING REQUIREMENTS

It is well established that employers are obligated to provide safety and health training, but what are the actual training requirements? OSHA and the **Mine Safety and Health Administration (MSHA)** have established specific regulations that delineate training requirements by type of industry. For example, OSHA training requirements as set forth in Title 29—Labor, Code of Federal Regulations—cover the following industrial sectors:

General industry	Part 1910
Maritime industry	Part 1915–18
Construction industry	Part 1926
Agriculture industry	Part 1928

Regulations delineating the training requirements for mining workers are set forth in Subsection B of MSHA regulations. This subpart carries the title

"Training and Retraining Miners Working at Surface Mines and Surface Areas of Underground Mines." Because the training requirements mandated by OSHA and MSHA differ in accordance with the different industrial sectors for which they were developed, they are best examined separately. The following two subsections summarize the training requirements mandated by OSHA and MSHA.

OSHA Training Requirements

OSHA training requirements are set forth in broad occupational categories: general industry, maritime, construction, agriculture, and federal employees.[8] Figure 12–2 summarizes the general industry requirements.

Examples of training requirements included in 29 CFR Parts 1910 and 1926 follow. This is only a partial list, outlining some of the requirement types in the OSHA standards.

- *Personal protection equipment.* 29 CFR 1910.132(f)(2) requires that employees demonstrate that they know how to use personal protective equipment.
- *Confined spaces.* 29 CFR 1910.146(g)(1) requires that employees who work as entrants, attendants, or entry supervisors have the understanding, knowledge, and skills necessary for the safe performance of their assigned duties.
- *Respiratory protection.* 29 CFR 1910.134(k)(1) requires that each employee be able to demonstrate how to inspect, put on, remove, use, and check the seals of respirators.
- *Lockout/tagout.* 29 CFR 1910.147(c)(7)(i) requires that employees have the knowledge and skill required for the safe application, use, and removal of energy controls.
- *Laboratory safety.* 29 CFR 1910.1450(f)(4)(i)(C) requires that employees be trained in the specific procedures necessary to protect themselves from chemical hazards, including appropriate work practices, emergency procedures, and personal protective equipment to be used.
- *Ladders and stairways.* 29 CFR 1926.1060(a) requires that employees be able to recognize hazards related to ladders and stairways.

In addition to requiring training, OSHA also requires retraining in certain areas. The hearing conservation, respiratory protection, bloodborne pathogens, and lead standards all require annual training. The confined spaces, fall protection, and lockout/tagout standards require retraining whenever there is evidence that a trained employee no longer possesses the necessary knowledge and skills. Additional information about OSHA training requirements can be obtained by contacting OSHA at the following Web site or telephone number:

Occupational Safety and Health Administration
847-297-4913
http://www.osha.gov

Subpart F	**Powered Platforms, Manlifts, and Vehicle Mounted Work Platforms** Manlifts
Subpart G	**Occupational Health and Environmental Control** Ventilation, occupational noise exposure, ionizing radiation
Subpart H	**Hazardous Materials** Hydrogen, flammable and combustible agents, explosives and blasting agents, storage and handling of liquefied petroleum gases, storage and handling of anhydrous ammonia, hazardous waste operations and emergency response
Subpart I	**Personal Protective Equipment** Respiratory protection
Subpart J	**General Environmental Controls** Temporary labor camps, specifications for accident prevention signs and tags
Subpart K	**Medical and First Aid** Medical services and first aid
Subpart L	**Fire Protection** Fire brigade training, fixed dry chemical extinguishing systems, local fire alarm signaling systems
Subpart M	**Compressed Gas and Compressed Air Equipment** Safety relief devices for cargo and portable tanks storing compressed gases
Subpart N	**Materials Handling and Storage** Servicing of single piece and multi-piece rim wheels; powered industrial trucks; overhead and gantry cranes; crawler, locomotive, and truck cranes; derricks
Subpart O	**Machinery and Machine Guarding** Woodworking machinery requirements, mechanical power processes, forging machines
Subpart Q	**Welding, Cutting, and Brazing** Welding, cutting, and brazing
Subpart R	**Special Industries** Pulp, paper, and paperboard mills; laundry machinery and operations; sawmills; pulpwood logging; telecommunications
Subpart T	**Commercial Diving Operations** Qualifications of dive team
Subpart Z	**Toxic and Hazardous Substances** Asbestos, 4-nitrobiphenyl, alpha-naphthylamine, methyl chloromethyl ether, 3,3"-dichlorobenzidine (and its salts), bis-chloromethyl ether, beta-naphthylamine, benzidine, 4-aminodiphenyl, ethyleneimine, beta-propiolactone, 2-acetylamineofluorene, 4-dimethylaminoazobenzene, N-nitrosodiummethylamine, vinyl chloride, inorganic arsenic, lead, coke oven emissions, cotton dust, 1,2-dibromo-3-chloropropane, acrylonitrile (vinyl cyanide), ethylene oxide, hazard communication

Figure 12–2
General industry training requirements: 29 CFR Part 1910.

Training requirements in all the various areas discussed earlier in this section are similarly delineated in Part 1910. Safety and health professionals should be familiar with those that apply in their individual work settings.

MSHA Training Requirements

Subpart B of MSHA regulations contains **MSHA training requirements** for miners. These regulations define a miner as

> Any person working in a surface mine or surface area of an underground mine who is engaged in the extraction and production process or is regularly exposed to mine hazards, or who is a maintenance or service worker (whether employed by operator or contractor) working at the mine for frequent or extended periods.[9]

Because the regulations specify a different set of training requirements for inexperienced miners and newly employed but experienced miners, it is important to understand how MSHA defines *experienced miner*. Paragraph 48.22 of Subpart B defines this term as follows:

> A person currently employed as a miner; or a person who received training acceptable to MSHA from an appropriate state agency within the preceding one month; a person with 12 months experience working in surface operations during the preceding 3 years; a person who received new miner training . . . within the past 12 months.[10]

Paragraph 48.25 of Subpart B sets forth the following training requirements for new or inexperienced miners:

- A minimum of 24 hours of training is required. Typically, this training takes place before the miner begins work. However, with prior approval of MSHA, up to 16 hours of the training may be provided after the new miner begins work. This means that even with a waiver from MSHA, at least 8 hours of training must be provided before the miner begins work.
- The first eight hours of a new miner's training must include the following: (1) an introduction to the work environment (orientation); (2) recognition of workplace hazards; and (3) job-specific safety and health measures or concerns.
- All required training beyond the original eight hours must be completed within 60 days. The training program must include at least the following topics: statutory rights of miners and their representatives; authority and responsibility of supervisors; line authority of supervisors and miners' representatives; mine rules; hazard reporting procedures; self-rescue and respiratory devices; transportation controls and communication systems; introduction to the work environment; emergency evacuation and escape procedures; fire warning and firefighting procedures; ground control; personal health; hazard recognition; electrical hazards; MSHA-approved first aid; explosives; and job-specific safety and health procedures.
- Companies must have a training plan that specifies oral, written, or practical demonstration methods will be used to assess whether training has been completed successfully.[11]

Paragraph 48.26 of Subpart B of MSHA regulations lists the following training requirements for newly employed but experienced miners before they begin work: introduction to the work environment (orientation); mandatory safety and health standards, both general and job specific; authority and responsibility of supervisors and miners' representatives; emergency escape and evacuation procedures; fire warning and firefighting procedures; ground controls; and hazard recognition.[12]

In addition to these specific training requirements, MSHA requires eight hours per year of refresher training, comprehensive records of each miner's training, and compensation to miners for training time. Training is to be provided during work hours, and employees are to be paid their regular wages. If the training takes place at any location other than the normal work site, all expenses incurred by miners while participating in training (e.g., lodging, meals, and mileage) must be paid by the employer.[13] Additional information about MSHA training requirements can be obtained from the following address or Web site:

Director of Education and Training
MSHA
4015 Wilson Blvd.
Arlington, VA 22203
http://www.msha.gov

SAFETY AND HEALTH PROFESSIONALS AS TRAINERS

As mentioned earlier, determining whether training should be provided by supervisors or by safety and health professionals is not clear-cut. Generally speaking, supervisors are more likely to provide job- and task-specific training, whereas safety and health professionals are more likely to provide more generic training. Regardless of where this distinction is made, it is clear that today's safety and health professionals must be competent at developing, coordinating, and conducting training.

According to the NSC, persons conducting training should have the following characteristics: a thorough knowledge of the topics to be taught; a desire to teach; a positive, helpful, cooperative attitude; strong leadership abilities; a professional attitude and approach; and exemplary behavior that sets a positive example.[14]

In addition to having these characteristics, trainers should be knowledgeable about the fundamental principles of learning and the four-step teaching method. The principles of learning summarize much of what is known about how people learn best. It is important to conduct safety and health training in accordance with these principles. The four-step teaching method is a basic approach to conducting training that has proven effective over many years of use.

Principles of Learning

The **principles of learning** summarize what is known and widely accepted about how people learn. Trainers can do a better job of facilitating learning if they understand the following principles:

1. *People learn best when they are ready to learn.* You cannot *make* employees learn anything. You can only make them *want* to learn. Therefore, time spent motivating employees so that they want to learn about safety and health is time well spent. Explain why employees need to learn and how they will benefit personally from having done so.

2. *People learn more easily when what they are learning can be related to something they already know.* Build today's learning on what was learned yesterday and tomorrow's learning on what was learned today. Begin each new learning activity with a brief review of the one that preceded it.

3. *People learn best in a step-by-step manner.* This is an extension of the preceding principle. Learning should be organized into logically sequenced steps that proceed from the concrete to the abstract, from the simple to the complex, and from the known to the unknown.

4. *People learn by doing.* This is probably the most important principle for trainers to understand. Inexperienced trainers tend to confuse talking (i.e., lecturing or demonstrating) with teaching. Explanations can be part of the teaching process but are only useful if they are followed by application activities that require the learner to *do* something. To illustrate, consider the example of teaching an employee how to ride a bicycle. You can present a thorough lecture on the principles of pedaling and steering and give a comprehensive demonstration on how to do it. However, until the employee gets on and begins pedaling, he or she will not learn how to ride a bicycle.

5. *The more often people use what they are learning, the better they will remember and understand it.* How many things have you learned in your life that you can no longer remember? People forget what they do not use. Trainers should keep this principle in mind. It means that repetition and application should be built into the learning process.

6. *Success in learning tends to stimulate additional learning.* This principle is a restatement of a fundamental principle in management (i.e., success breeds success). Organize training into long enough segments to allow learners to see progress, but not so long that they become bored.

7. *People need immediate and continual feedback to know if they have learned.* Did you ever take a test and get the results back a week later? If so, that was probably a week later than you wanted them. People who are learning want to know immediately and continually how they are doing. Trainers should keep this principle in mind at all times. Feedback can be as simple as a nod, a pat on the back, or a comment such as "Good job!" It can also be more formal, such as a progress report or a graded activity. Regardless of its form, trainers should concentrate on giving **immediate and continual feedback**.

Four-Step Teaching Method

Regardless of the setting, teaching is a matter of helping people learn. One of the most effective approaches for facilitating learning is not new, innovative, gimmicky, or high tech in nature. It is known as the **four-step teaching method**, and it is an effective approach to use for safety and health training. A brief description of these steps follows:

- **Preparation** encompasses all tasks necessary to get participants prepared to learn, trainers prepared to teach, and facilities prepared to accommodate the process. Preparing participants means motivating them to want to learn. Personal preparation involves planning lessons and preparing all the necessary instructional materials. Preparing the facility involves arranging the room for both function and comfort, checking all equipment to ensure that it works properly, and making sure that all tools and other training aids are in place.
- **Presentation** is a matter of presenting the material that participants are to learn. It may involve giving a demonstration, presenting a lecture, conducting a question-and-answer session, helping participants interact with a computer or interactive video system, or assisting participants who are proceeding through self-paced materials. Recommends strategies for giving an effective presentation inculde: Begin dramatically, be brief, be organized, use humor, keep it simple, take charge, be sincere, consider conditions, and tell stories.[15]
- **Application** is a matter of opportunities for participants to use what they are learning. Application can range from simulation activities in which learners role-play to actual hands-on activities in which they use their new skills in a live format.
- **Evaluation** is a matter of determining the extent to which learning has taken place. In a training setting, evaluation does not need to be a complicated process. If the training objectives were written in measurable, observable terms, evaluation is simple. Employees were supposed to learn how to do X, Y, and Z safely. Have them do X, Y, and Z and then observe the results. In other words, have employees demonstrate proficiency in performing a task safely, and observe the results.

PREPARING SAFETY AND HEALTH INSTRUCTION

Preparing instruction involves the following steps: (1) preparing (planning) the instruction; (2) preparing the facility; and (3) preparing the learners. It is important to accomplish all three steps before attempting to present instruction. This section focuses on planning a safety and health course of instruction.

Planning Instruction

The instruction delivered by safety and health professionals is usually part of a course, workshop, or seminar. In any case, there must be a **course outline** that summarizes the major topics covered by the instruction. The outline should

---

Course Outline
Safety Engineering for Familiarity
(A Course for Supervisors)

Statement of Purpose

This course consists of 15 contact hours of instruction. Its purpose is to familiarize supervisors with the concept of safety engineering and the job of safety engineers so that supervisors and safety engineers can work together more effectively as team members.

Intended Outcomes

Upon completion of this course, supervisors should be able to do the following:

■ Demonstrate an understanding of safety engineering techniques.

■ Analyze workplace environmental hazards that safety engineers are concerned with.

■ Explain health and safety regulations that safety engineers are concerned with.

■ Explain the proper relationship between supervisors and safety engineers.

Figure 12–3
Sample course outline.

state the expected outcomes of the instruction in broad terms or, in other words, what the participant is supposed to be able to do after completing the course, workshop, or seminar. The outline should also have a brief statement of purpose. More specific instructional objectives are developed later when preparing lesson plans. Figure 12–3 is a sample outline for a short course on safety engineering for supervisory personnel.

Notice that the course outline contains just two components: a statement of purpose and a list of intended outcomes. Some instructors prefer to add additional components such as a list of equipment or training aids needed, but the components shown in Figure 12–3 are sufficient. A good course outline is a broadly stated snapshot of the scope and sequence of the course. Specific details are typically shown in the lesson plans that are developed next.

Lesson plans are an important part of the planning step. They are road maps or blueprints for the instruction that is to take place. In addition, they standardize instruction when more than one person may teach the same instruction to different groups. Standardization is particularly important for safety and health training. If even one member of a work team receives less training than the others, the potential for an accident is increased by his or her ignorance.

Lesson plans can vary in format according to the personal preferences of the trainer. However, all lesson plans should include the components discussed in the following paragraphs.

Lesson Title and Number

The lesson title should be as descriptive as possible of the content of the lesson. The number shows where the lesson fits into the sequence of lessons that make up the course.

Statement of Purpose

Like the statement of purpose in the course outline, the **statement of purpose** here consists of a concise description of the lesson's contents, where it fits into the course, and why it is included.

Learning Objectives

Learning objectives are specific statements of what the participant should know or be able to do as a result of completing the lesson. Objectives should be written in behavioral terms that can be measured or easily observed.

Training Aids List

This component serves as a handy checklist to help trainers quickly and conveniently ensure that all the **training aids** needed are present. The list should include every tool, handout, piece of equipment, video, chart, and so on needed to conduct the instruction for that lesson.

Instructional Approach

The **instructional approach** is a brief action plan for carrying out the instruction. It should begin with a short statement describing the instructional methodology to be used (lecture and discussion, demonstration, computer-assisted instruction, etc.). This statement is followed by a step-by-step summary of the trainer's major activities, such as: (1) deliver lecture on safety regulations; (2) distribute safety regulations handout, and so on.

Application Assignments

The **application assignments list** details the tasks that learners will be required to complete before they can apply what they are learning.

Evaluation Methodology

The **evaluation methodology** component explains how learning will be evaluated. Will there be a test? Will performance be observed? Will safety and health records be monitored for improvement? Such questions are answered in this section.

Figure 12–4 is an example of a lesson plan developed around one of the intended outcomes in Figure 12–3. This is the typical relationship between the course outline and the lesson plan (i.e., one intended outcome equals one lesson). Occasionally, an outcome may require more than one lesson for adequate coverage; at other times, several outcomes might be covered in one lesson.

**Lesson 1
Safety Engineering Techniques**

Statement of Purpose

This is the first lesson in a series of four that make up the course. The purpose of this lesson is to help supervisors understand the job of the safety engineer in the specific area of safety engineering techniques such as noise control, equipment guarding, and dust control.

Learning Objectives

Upon completion of this lesson, learners will be able to do the following:

■ Demonstrate how to apply noise control techniques to operating equipment.

■ Apply equipment guarding techniques to operating machines.

■ Apply dust control techniques in a shop setting.

Training Aids List

The following training aids are needed with this lesson: overhead projector, safety engineering techniques transparencies, dust control measures handout, and sample equipment guarding devices.

Instructional Approach

The lecture–discussion method supplemented with overhead transparencies, handouts, and live examples of equipment guards is used with this lesson. Instruction should proceed as follows: (1) lecture and discussion on noise control with overhead transparencies; (2) learners attach an equipment guarding device to a machine; and (3) learners examine their working area and make recommendations for dust control.

Evaluation

The instructor will observe the performance of learners as they undertake their application activities. Based on these observations, they will assign pass or fail assessments to each student.

Figure 12–4
Sample lesson plan.

PRESENTING SAFETY AND HEALTH INSTRUCTION

Educators hold that learners retain the following percentages from their instruction:

10 percent of what is read
20 percent of what is heard
30 percent of what is seen

50 percent of what is seen and heard

70 percent of what is seen and spoken

90 percent of what is said while doing what is talked about

Instruction can be presented in several different ways. The most widely used are the lecture–discussion format, group instruction, demonstration, conference, and multimedia methods. Regardless of the approach used, trainers should keep in mind the percentages just listed. The message is to get the learner actively engaged seeing, saying, listening, and, most importantly, *doing*.

Lecture–Discussion Method

The **lecture–discussion method** of teaching is the oldest, most familiar, most used, and probably most abused. A lecture is a planned, structured, and frequently illustrated (slides, charts, board, and so on) method of communicating information to a group of people. By itself, the lecture allows for only one-way communication. This serious deficiency is overcome by adding the discussion component. Discussion can be between the instructor and participants or among fellow learners. During discussion, the instructor's job is to keep the discussion on track and moving in the right direction.

The best justification for using the lecture–discussion method is that it is an effective way to communicate information to groups that are too large for individual interaction between instructor and participants. Another reason for using this method is that it allows the instructor to generate enthusiasm among participants about a topic.

The lecture–discussion method, if used properly, can be an effective teaching method. However, it does not work in every situation. Instructors or trainers need to know when to use this method and when to use another. Use the lecture–discussion method when

- The material to be presented deals strictly with data, theory, or information (no skills development)
- Participants need to be motivated before beginning a particular lesson
- The material to be presented is not available in print
- Sharing insight or experience in a particular area will enhance learning
- Information must be communicated to a large group in one session
- Interaction among participants is desired

Do *not* use the lecture–discussion method when

- The subject matter deals with skills development or how-to information
- The participant group is small enough to allow individual learner and teacher interaction
- There is no need for interaction among participants

This is probably the most overlooked step: *Participants must be thoroughly prepared prior to the session*. If they are not, the session will be all lecture and no discussion. Require participants to approach a lecture–discussion session having first prepared themselves. This will ensure that they are active, contributing learners rather than passive spectators. Prepare participants for a lecture–discussion session as follows:

1. Give them a written outline or overview of the lecture the day before so they can familiarize themselves with it.
2. At least a day ahead of time, pass out any handouts that will be used and ask the participants to read them, noting questions that they may have right on the handouts.
3. During the lecture, have group members raise their questions when they think of them. Use their questions to trigger open discussion and other questions.

There is a saying in teaching: "When giving a lecture, tell them what you are going to tell them, then tell them, then tell them what you told them." Although it is said with tongue in cheek, this is actually good advice. A well-planned, properly structured lecture contains three distinct components: the **opening**, in which you "tell them what you are going to tell them"; the **body**, in which you "tell them"; and the **closing**, in which you "tell them what you told them."

Discussion should be interspersed within the body of instruction in a workplace training setting. The best time to discuss an issue is when it is fresh on the minds of the participants. In training language, such instances are known as *Moms*, or moments of maximum opportunity.

The opening, body, and closing of a lecture all contain specific tasks that should be accomplished in order:

1. Opening
   a. Greet the class.
   b. State the title of the lecture.
   c. Explain the purpose of the lecture.
   d. List the objectives so that participants know exactly what they should be learning.
   e. Explain how the current lecture–discussion session relates to past topics studied.
   f. List and define any new terms that will be used during the session.
   g. Present a general overview of the content of the lecture–discussion session.
2. Body
   a. Present the information in the order listed in the participants' outline.
   b. Initiate discussion by raising specific questions, calling on participants for comments, or soliciting questions from them.
   c. Make frequent reference to all visual aids and supportive materials.
3. Closing
   a. Restate the title, purpose, and objectives.
   b. Briefly summarize major points.
   c. State your conclusions.

   d. Answer remaining questions.

   e. Make follow-up assignments to reinforce the lecture and discussion.

The elements just listed are the fundamental or tangible tasks that should be performed in all lecture–discussion sessions. You should also keep in mind a number of intangibles when conducting lecture–discussion sessions:

- Make sure that the classroom is arranged to accommodate a lecture and discussion.
- Be enthusiastic. Enthusiasm is contagious.
- Call on participants by name. They will appreciate the recognition and feel more at ease.
- Spread your attention evenly. This will make all participants feel that they are part of the lecture and discussion.
- Maintain eye contact with all learners in the session.
- Speak clearly, evenly, and slowly enough to be understood, but not in a monotone.
- Use facial expressions, body language, and movement to emphasize points.
- It is all right to use an outline or note cards to keep yourself on track, but never read a lecture to the participants.
- Use carefully prepared visual aids to reinforce major points.
- Do not dominate—facilitate. Participation is critical. Remember, this method is called lecture–*discussion*.

Demonstration Method

The **demonstration method** is the process in which the instructor shows participants how to perform certain skills or tasks. While demonstrating, the instructor also explains all operations step by step. The key to giving a good demonstration is preparation. The following checklist contains specific tasks for preparing for a demonstration:

1. Decide exactly what the purpose of the demonstration is, why it will be given, what participants should learn from it, what will be demonstrated and in what order, and how long the demonstration will last.
2. Gather all tools, equipment, and instructional aids. Make sure that everything is available and in working order. Never put yourself in a position of being forced to stop a demonstration unexpectedly because something does not work the way it should or because a necessary part of the demonstration is not on hand.
3. Set up the demonstration so that participants will easily be able to see what is going on and hear what you are saying.
4. Arrange all materials to be used in the demonstration so that they correspond with the order in which the various steps of the demonstration will be presented.
5. Practice the demonstration several times before giving it to work out any bugs.

Just as there are specific tasks to be performed in presenting a lecture, there are specific tasks to be performed in giving a demonstration:

1. Orient participants to the demonstration by explaining its purpose and objectives. Give them a brief overview of the content of the demonstration. Explain how the demonstration ties in with what they already know.
2. Present the demonstration in a slow, deliberate fashion so that participants can easily follow.
3. Pause between stages to determine if all class members are comprehending or if they have any questions that should be answered before continuing. Go back over any steps that participants did not seem to grasp.
4. Conclude the demonstration with a brief summary and question-and-answer session.

Participant activity in skills development is critical. Remember, no matter how well a demonstration is presented, it is really just showing. Showing is important, but participants learn best by doing. Therefore, it is vital to provide hands-on activities after a demonstration. An effective way to follow up demonstrations is to

1. Select several participants and ask them to repeat the demonstration you have just given.
2. Assign several practical application activities in which the participants are required to apply the skills demonstrated.
3. Observe participants on an individual basis as they attempt to perform the practical application activities. Give individual attention and assistance where needed. Be sure to correct misapplication immediately so that the wrong way does not become a habit.
4. Conduct performance evaluations so that skills development can be measured.

Conference Method

The **conference method** is particularly well suited for corporate training settings. It is less formal than a traditional classroom setting and requires that the trainer serve as a facilitator rather than a teacher. It is best used as a problem-solving teaching method. For example, a safety and health professional may use the conference method to make all supervisors aware of a new safety problem while simultaneously soliciting their input on how to solve the problem.

To be effective facilitators, safety and health professionals must become adept at defining the problem, soliciting input from participants, drawing out all of them, summarizing and repeating information, and building consensus. The conference teaching method, when effectively used, should result in both well-informed participants and a plan for solving the problem.

The following example is a scenario in which the conference method may be used. Annette Evans, safety and health manager for Shalimar Machine and Tool Company, has noticed an increase in minor cuts and abrasions on the hands of employees who are handling the new XRT composite material.

Through random observation, she has determined that employees are not properly applying the accident prevention techniques they are supposed to have learned for handling XRT. The employees do not appear to fully understand the techniques.

Evans decides to set up a training session for all supervisors of employees who handle XRT. She hopes to accomplish three things: (1) inform supervisors of the problem; (2) reacquaint supervisors with the appropriate accident prevention techniques for handlers of XRT; and (3) develop a plan to get handlers using the techniques properly. According to the NSC, Evans should proceed as follows:

1. State the problem.
2. Break the problem into segments to keep the discussion orderly.
3. Encourage free discussion.
4. Make sure that members have given adequate consideration to all of the significant points raised.
5. Record any conclusions that are reached.
6. State the final conclusion in such a way that it truly represents the findings of the group.[16]

If Evans follows these six steps, all supervisors of workers who handle the new XRT composite material will understand the problem. In addition, they will learn the proper application of the special accident prevention techniques that must be used when handling XRT. Finally, Evans will have solid recommendations that have the full support of the supervisors concerning how to ensure that XRT handlers properly apply the accident prevention techniques.

Other Presentation Methods

In addition to the presentation methods just discussed, several others can be used. The most widely used are simulation, videotapes and DVDs, and programmed instruction.

Simulation

Simulation, as the name implies, involves structuring a training activity that simulates a line situation. For example, if a safety and health professional is teaching a group of workers how to respond when a coworker is electrocuted, he or she may simulate that situation by having a worker role-play being electrocuted. Simulation can also be technology based. Computer simulation activities and those based on interactive laser disc and video technology are becoming more widely used. The military has used technology-based simulation for many years to train pilots.

Videotapes/DVDs

The use of videotapes and DVDs for presenting instruction has become common in corporate training settings. In essence, the videotape or DVD takes the place of a lecture or demonstration. The pause or stop functions on the player can be used to allow for discussion, questions and answers, and group interaction. The playback feature can be used for reviewing material or replaying portions of the tape/DVD that are not fully understood by participants. According to John Zeglin of NUS Training Corporation, the greatest advantage that videos and DVDs offer is consistency of presentation.[17] However, it is important to ensure that they meet the specific needs of individual companies. Magon suggests that the following features be checked when previewing videos and DVDs:

1. *Presentation.* What is the quality of the presentation? Look for both content and presentation style.
2. *Focus.* A good video or DVD should focus on one subject and give a comprehensive treatment in a time range of 12 to 25 minutes. Multiple subjects should be broken up into more than one video/DVD.
3. *Liveliness.* Is the video/DVD lively, upbeat, and interesting, or nothing more than a boring presentation by a "talking head"? A lively video/DVD will hold the learner's attention.
4. *Accuracy.* Is the content accurate? If not, none of the other features matter. Have subject matter experts preview videos for accuracy before incorporating them into a training program.[18]

Programmed Instruction

Programmed instruction is an approach used to individualize instruction. Traditionally, the programmed medium has been a workbook or text that presents information in segments that proceed as (1) information presentation; (2) information review; (3) questions, problems, or activities based on the information presented for the participant to work on; and (4) a self-test. Before proceeding to the next lesson, the learner must make a specified score on the self-test for the preceding lesson. Increasingly, programmed instruction is becoming computerized. This enhances the interactive nature of the instruction and, with good software, provides almost immediate feedback for the learner.

Interactive Video and CD-ROM

Interactive video training combines video/DVD and personal computer technologies to create an excellent approach to workplace training. Unlike many other media-based training methods, interactive video is not passive. Rather, it requires the learner to participate actively by making choices, selecting options, and participating in one-on-one simulations of workplace situations. According to Fisher,

> In operation, an employee sits privately at a TV monitor with (optional) headphones. Those being trained need only touch the TV screen for interaction, so no computer literacy is necessary. Employees became actively engaged in the learning process on a one-on-one basis. Dramatizations of workplace situations actually appear on the screen in live action.[19]

Fisher lists the following advantages of interactive video as a training methodology: It individualizes learning, increases retention, is self-paced, evaluates and records progress, reduces training time, simplifies learning, and is dependable, consistent, and flexible.[20]

CD-ROM technology is another excellent method for workplace training that is flexible, accessible, and particularly effective for use with employees who grew up in the age of electronic technology.

Online Training (E-Learning)

Online training provided over the Internet has begun to eclipse all the other forms of safety and health instruction. The e-learning business now exceeds $6 billion annually and continues to grow rapidly. The rise of online training in safety and health can be attributed to several factors including convenience, quality, computer proficiency of younger employees, and employee preference. However, as with all other commercially available forms of instruction, online training packages vary in quality. To ensure that you are purchasing quality instruction in the online format, apply the following strategies: (1) review before buying (e.g., make sure the material is up-to-date, accurate, comprehensive, and easy to use); (2) make sure the instructional design and the material are suited for the audience in question; (3) make sure that technical support is readily available; and (4) make sure that the training is not big on high-tech bells and whistles such as animation and graphics, but small on content and material.

APPLYING SAFETY AND HEALTH INSTRUCTION

One of the fundamental principles of learning states that *people learn by doing*. To the trainer, this means that learners must be given ample opportunity to apply what they are learning. If the topic of a training session is how to administer CPR, application involves having the learners actually practice on dummies. While the learners practice, the trainer observes, coaches, and corrects. Regardless of the nature of the material, participants should be given plenty of opportunities to apply what they are learning.

EVALUATING SAFETY AND HEALTH INSTRUCTION

Did the training provided satisfy the training objectives? Safety and health professionals need to know the answer to this question whenever training is given. However, this can be a difficult question to answer.

Evaluating training requires beginning with a clear statement of purpose. What is the overall purpose of the training? The objectives translate this purpose into specific, measurable terms.

The purpose of safety and health training is to improve the individual employee's ability to work safely and to improve the safety and health of the overall work environment.

To determine if training has improved performance, safety and health professionals need to know the answers to these questions: (1) Was the training provided valid? (2) Did the employees learn? (3) Has the learning made a difference? Valid training is training that is consistent with the training objectives.

Evaluating training for validity is a two-step process. The first step involves comparing the written documentation for the training, lesson plans, handouts, and so on with the training objectives contained in the course outline. If the training is valid in design and content, the written documentation will match the training objectives. The second step involves determining if the training provided is consistent with the documentation. Training that strays from the approved plan will not be valid. Participant evaluations of instruction conducted immediately after completion can provide information on consistency and the quality of instruction.

Determining if employees have learned is a matter of building evaluation into the training. Employees can be tested to determine if they have learned, but be sure that tests are based on the training objectives. If the training is valid and employees have learned, the training should make a difference in their performance. Performance on the job should improve. This means safety and health should be enhanced. Safety and health professionals can make determinations about performance using the same indicators that told them training was needed in the first place.

Safety Fact

Safety Training Is a Must

Too many business executives take the MBA (management by accounting) approach to their jobs. They will accept only dollars on the bottom line as proof of the value of an activity or function. Although the value of training can be shown in dollars and cents, the ubiquitous MBA mentality that persists in business and industry is hard to understand, and even harder to overcome.

The same executives who question the value of safety training have no problem understanding that professional football teams train all week to play just one game. Military units train constantly for the eventuality of a deployment. Training is fundamental to most professions and occupations. It should be fundamental to all.

Can employees perform tasks more safely than they could before the training? Is the accident rate down? Is the amount of time lost to accidents down? Are there fewer health-related complaints? These are the types of questions that safety and health professionals can ask to determine if training has improved performance. What follows is a checklist of questions for evaluating purchased training programs.

- Is the program recommended by someone you trust who has used it with good results?
- Does the program deal with the specific safety and health issues you are interested in?
- Are there specific behaviorally stated learning objectives for the program and, if so, are the objectives comprehensive?
- Does the program require actions that students may not know how to perform?
- Is the program appropriate to the intended audience?
- Is the content of the program up-to-date and accurate?
- Does the program reflect the appropriate philosophy and approach?
- Does the program require actual application of the content or is it just informational?
- Can the program be modified to adapt to different levels of learners?

TRAINING SUPERVISORS

Supervisors play a key role in the maintenance of a safe and healthy workplace. Consequently, safety and health professionals need to ensure that supervisors have had the training they need to be positive participants in the process. A *Safety & Health* survey revealed that only 53 percent of the companies responding provide safety and health training for their supervisors.[21] This is an unfortunate statistic because supervisors are the safety and health professional's link with employees. In the words of Peter Minetos, "Supervisors . . . are the ones who have to teach employees the safe way to conduct their jobs. With proper training, they can spot and eliminate risks that are waiting to create havoc for their workers."[22]

According to the NSC, the objectives of supervisor safety training are to

- Involve supervisors in the company's accident prevention programs
- Establish the supervisor as the key person in preventing accidents
- Get supervisors to understand their safety responsibilities
- Provide supervisors with information on causes of accidents and occupational health hazards and methods of prevention
- Give supervisors an opportunity to consider current problems of accident prevention and develop solutions based on their own and others' experience

- Help supervisors gain skill in accident prevention activities
- Help supervisors keep their own departments safe[23]

Responsibility for developing safety training programs for supervisors falls to the safety and health professional. Figure 12–5 is an outline for a basic safety training course for supervisors. This course was developed by the NSC and consists of 14 one-hour lessons.

**National Safety Council's
Basic Safety for Supervisors Course**

**Session 1
Loss Control for Supervisors**

Accidents and incidents, areas of responsibility, the cost of accidents, and a better approach to occupational safety and health.

**Session 2
Communications**

Elements of communication, methods of communication, and effective listening.

**Session 3
Human Relations**

Human relations concepts, leadership, workers with special problems, and the drug and alcohol problem.

**Session 4
Employee Involvement in Safety**

Promoting safe-worker attitudes, employee recognition, safety meetings, and off-the-job accident problems.

**Session 5
Safety Training**

New employee indoctrinations, job safety analysis (JSA), job instruction training (JIT), and other methods of instruction.

**Session 6
Industrial Hygiene and Noise Control**

General concepts, chemical agents, physical agents, temperature extremes, atmospheric pressures, ergonomics, biological stresses, threshold limit values (TLVs), and controls.

**Session 7
Accident Investigation**

Finding causes, emergency procedures, effective use of witnesses, and reports.

Figure 12–5
Sample safety training course outline.

Session 8
Safety Inspections

Formal inspections, inspection planning and checklists, inspecting work practices, frequency of inspections, recording hazards, and follow-up actions.

Session 9
Personal Protective Equipment

Controlling hazards, overcoming objections; protecting the head, eyes, and ears; respiratory protective equipment; safety belts and harnesses; protecting against radiation; safe work clothing; and protecting the hands, arms, feet, and legs.

Session 10
Materials Handling and Storage

Materials handling problems; materials handling equipment; ropes, chains, and slings; and material storage.

Session 11
Machine Safeguarding

Principles of guarding, safeguard design, safeguarding mechanisms, and safeguard types and maintenance.

Session 12
Hand Tools and Portable Power Tools

Safe working practices, use of hand tools, use of portable power tools, and maintenance and repair of tools.

Session 13
Electrical Safety

Electrical fundamentals review, branch circuits and grounding concepts, plug- and cord-connected equipment, branch circuit and equipment testing methods, ground fault circuit interrupters, hazardous locations, common electrical deficiencies, safeguards for home appliances, and safety program policy and procedures.

Session 14
Fire Safety

Basic principles, causes of fire, fire-safe housekeeping alarms, equipment, and evacuation; and reviewing the supervisor's fire job.

Figure 12–5 (*continued*)

TRAINING NEW AND TRANSFERRED EMPLOYEES

Perhaps the most important aspect of safety training is orientation for new and transferred employees. Consider the following quote from *Today's Supervisor:*

The confusion and stress that accompany the first day of any job are often the reasons that new employees are more than twice as likely to have accidents as experienced workers. A lack of experience, a strong desire to please and a hesitation to ask for help all cause one in eight new employees to be involved in some type of accident the first year on the job. The first month is the most critical.[24]

The early training provided for new and transferred employees should have at least the following components: orientation, job-specific procedures, and follow-up. These components are discussed in the following paragraphs.

Orientation

Orientation is critical. Too often, companies hand over new employees to an experienced employee who is supposed to "show them the ropes." This is a dangerous practice that can have the effect of minimizing, rather than emphasizing, the importance of safety. Orientation should be structured, and it should involve the new worker's supervisor as well as other personnel as appropriate. A good orientation program should teach the following at a minimum:

- Management is sincerely interested in preventing accidents.
- Accidents may occur, but it is possible to prevent them.
- Safeguarding equipment and the workplace has been done, and management is willing to go further as needs and methods are discovered.
- Each employee is expected to report to the supervisor any unsafe conditions encountered at work.
- The supervisor will give job instructions. No employee is expected to undertake a job before learning how to do it and being authorized to do it by a supervisor.
- The employee should contact the supervisor for guidance before undertaking a job that appears to be unsafe.
- If an employee suffers an injury, even a slight one, it must be reported at once.

In addition to these points, any safety rules that are a condition of employment, such as wearing eye protection or safety hats, should be understood and enforced from the first day of employment.[25]

Job-Specific Procedures

Before a new employee is allowed to begin work, he or she should be given instruction in safe operation procedures, the use of personal protective equipment, and any other procedures that promote a safe and healthy work environment.

Follow-Up

After a new employee has worked for three days to a week, a follow-up conference should be called. This conference should be led by the worker's supervisor. It should answer the following questions: Was the initial training effective?

Does the new employee have questions or concerns? Does the new employee have suggestions for improving the safety and health of his or her work environment? It can also be a good idea to ask the employee to verify that he or she has been adequately and properly oriented by signing a checklist such as the one in Figure 12–6.

Orientation Verification Checklist

Please review this checklist and indicate which topics were covered during your orientation by initialing the entry.

Reporting/Authority

☐ Your immediate supervisor
☐ Midmanagers, managers, and executives in your chain of command
☐ Safety and health manager
☐ Human resources manager

Job Requirements

☐ Job description
☐ Responsibility
☐ Authority
☐ Performance standards
☐ Evaluation system

Pay/Benefits

☐ Wages
☐ Pay days
☐ Payroll deduction options
☐ Medical benefits
☐ Life insurance
☐ Disability coverage
☐ Annuity opportunities
☐ Saving deductions
☐ Retirement

Rules/Regulations

☐ Company rules/regulations
☐ Department rules/regulations
☐ Unit rules/regulations
☐ General safety rules
☐ Job-specific safety rules
☐ Consequences of breaking rules

Tools/Equipment

☐ Check-out/check-in procedures
☐ Maintenance procedures
☐ Orientation to use
☐ Emergency shut-down
☐ Personal protective equipment

Work Hours

☐ Workdays
☐ Work hours
☐ Overtime compensation
☐ Holidays/down days
☐ Vacation
☐ Sick leave
☐ Nontraditional scheduling opportunities

Employee signature: _____ Date: _____

Figure 12–6
Sample employee orientation checklist.

JOB SAFETY ANALYSIS AS A TRAINING TECHNIQUE

Job safety analysis (JSA) is a process through which all the various steps in a job are identified and listed in order. Each step is then analyzed to identify any potential hazards associated with it. The final step involves developing procedures for reducing the hazard potential associated with each respective step. For example, say, in performing a job safety analysis for a solid ceramic molding process, you identified a step in which the technician could be seriously burned. With the setup and the potential hazard clearly identified, a procedure can be developed for reducing or completely eliminating the hazard.

Figure 12–7 is an example of a form that can be used to expedite the process. The form should be completed as described in the following paragraphs.

Break Down the Job into Steps

Go through the job slowly, taking note of the steps involved. Be especially cognizant of changes in direction, activity, or movement. Such changes typically signify the end of one step and the beginning of another. List all steps required to do the job even if a given step is not required every time.

Identify Potential Hazards

The most effective way to identify the potential hazards that may be associated with each step is to observe as another worker performs the job. While this is taking place, ask, "What could go wrong here? Is there danger of a back injury, burn, slip, being caught between objects, a fall, or muscle strain? Is there danger of exposure to dust, radiation, toxic fumes, or chemicals?" Record your concerns on the lines that correspond to the step(s) in question.

Develop Accident Prevention Procedures

Having identified potential hazards, ask these questions: How can this hazard be eliminated or reduced to the maximum extent possible? Will redesigning the job eliminate the hazard? Is personal protective equipment needed? Should worker–equipment interaction be changed ergonomically? Will better housekeeping solve the problem? State the accident prevention procedures decided on in behavioral (action) terms. Do not use vague generalities. Be specific about what must be done to prevent an accident (e.g., tighten the clamp firmly).

Using the JSA as a Training Technique

Conducting a JSA can be a valuable learning experience for both new and experienced employees. Not only does it help them understand their jobs better, it familiarizes them with potential hazards and involves them in developing

Kelton Electronics

Industrial Park Causeway
Fort Walton Beach, FL 32548
904-729-5218

Job Safety Analysis

Job title: _____

Analysis by: _____

Approved by: _____

Date: _____

Step-by-Step Sequence	Potential Hazard	Accident Prevention Procedure

Figure 12–7
Sample job safety analysis form.

accident prevention procedures. Workers are more likely to follow procedures that they had a voice in planning. Finally, the JSA process causes employees to think about safety and how it relates to their jobs.

TRAINING OPPORTUNITIES AVAILABLE

Numerous training opportunities are available for both practicing and prospective safety and health professionals. Training is available through colleges, universities, and technical schools, as well as from the following organizations:

Occupational Safety and Health Administration (OSHA)
National Institute for Occupational Safety and Health (NIOSH)
Mine Safety and Health Administration (MSHA)
National Safety Council (NSC)

OSHA, NIOSH, MSHA, and NSC
OSHA

OSHA maintains an Office of Training and Education and the OSHA Training Institute. Through these organizations, OSHA provides a comprehensive schedule of courses that changes every year, outreach training programs, training grants, and training materials. Information on training opportunities through OSHA is available at the following Web site:

http://www.osha.gov

NIOSH

NIOSH training grants are awarded to organizations that want to conduct occupational safety and health training and education. They are awarded based on criteria in specific announcements in the *Federal Register*. The grant program's objective is to provide training funds to eligible institutions or agencies to ensure that an adequate supply of qualified professional and paraprofessional occupational safety and health personnel is available, thus fulfilling the purposes of the Occupational Safety and Health Act. NIOSH awards Training Project Grants and Education and Research Center (formerly known as Educational Resource Center) Training Grants to support educational programs in the fields of industrial hygiene, occupational health nursing, occupational and industrial medicine, occupational safety, and other specialized training areas.

Through its 15 university-based Education and Research Centers (ERCs), NIOSH supports short-term *continuing education programs* for occupational safety and health responsibilities. In addition, academic degree programs and postgraduate research opportunities are offered in the areas of occupational medicine, occupational health nursing, industrial hygiene, and occupational safety, plus a number of special topics of interest to the health and safety field.

NIOSH offers a number of videotapes and DVDs on specific occupational safety and health problems such as dust control engineering, cumulative trauma disorders, and hand-arm vibration exposure. Additional information about these videotapes, DVDs, and training opportunities through NIOSH is available at the following Web site:

http://www.cdc.gov/niosh/training.html

MSHA

MSHA maintains the National Mine Health and Safety Academy, a comprehensive program of educational field services, and a program of state grants. MSHA's education and training programs are designed to promote safety and health in the nation's mining industry. The Directorate has two major components of operations: the Policy and Program Coordination Division and the National Mine Health and Safety Academy in Beckley, West Virginia. The Directorate plans, monitors, and evaluates all MSHA education and training programs addressing mine safety and health, and provides training at the Academy. The Directorate administers the State Grants Program, which provides funding to state mining agencies to supplement their mining health and safety programs. Additional information about training opportunities from MSHA is available at the following Web site:

http://www.msha.gov/PROGRAMS/EPD.htm

NSC

Prospective and practicing safety and health professionals should be familiar with NSC's training programs. Courses may be taken through the NSC, or customized on-site courses can be provided for companies that request such service.

Safety and health professionals can earn an Advanced Safety Certificate from the NSC by completing the following three courses within a span of five years: principles of occupational safety and health, safety training methods, and safety management techniques. The content of these courses is summarized as follows:

- *Principles of occupational safety and health.* This course is designed to give safety and health professionals the skills they need to develop and implement an effective safety program for their company. Major topics covered include safety organization, record keeping, communications for safety, accident investigation, electrical hazards, fire prevention and control, industrial hygiene, accident reporting, human factors engineering, safety inspections, new employee indoctrination, machine safeguarding, personal protective equipment, job safety analysis, job instruction training, federal legislation, materials handling, safety meetings, and office safety.

- *Safety training methods.* This course is designed to expose participants to proper techniques for planning, presenting, and evaluating instruction for adult learners. Major topics covered include methods of teaching, learning processes of adults, visual aid techniques, developing and using lesson plans, practice training, group involvement techniques, training needs of supervisors, computing training costs, course development, and extemporaneous speaking workshops.
- *Safety management techniques.* This course is designed to equip participants with communications and management skills. Major topics covered include safety management, meeting leadership, conference leading practice, managing change (communication and listening), human relations, safety management by objectives, problem solving, decision making, accident investigation, and safety program audits.

The courses offered by the institute vary each year. However, the following courses are typically available at a minimum: principles of occupational safety and health, safety training methods, safety management techniques, advanced safety concepts, practical aspects of industrial hygiene, safety in chemical operations, product safety management, OSHA hazard communication standard, motor fleet accident investigation, fundamentals of occupational ergonomics, back power master trainer, and defensive driving. Additional information about the NSC's safety training opportunities is available at the following Web site:[26]

http://www.nsc.org/osh/oshtrain.htm

ILLITERACY AND SAFETY

In recent years, industry has been forced to face a tragic and potentially devastating problem: *adult illiteracy*, which is having a major impact on the productivity of industry in the United States. It is estimated that more than 60 million people, or approximately one-third of the adult population in this country, are marginally to functionally illiterate.[27] The problem is compounded by non-English-speaking immigrants who may also be illiterate in their native language.

Illiteracy in the broadest sense is the inability to read, write, compute, solve problems, and communicate. Functionally illiterate people read at the fourth-grade level or below. Marginally literate people read at the fifth- through eighth-grade levels. Functionally literate people read at the ninth-grade level or above. This is the literacy continuum as it now exists. However, technological developments are having the effect of shifting these levels up the continuum. As a result, what constitutes functional or marginal literacy now will have to be redefined periodically.

People are sometimes shocked to learn that the number of illiterate adults is so high. There are several reasons why this number has been obscured in the past and why there is now a growing awareness of the adult illiteracy problem.

- Traditionally, the number of low-skilled jobs available has been sufficient to accommodate illiterate adults.
- Faulty research methods for collecting data on illiterate adults have obscured the reality of the situation.
- Reticence on the part of illiterate adults to admit that they have a problem and to seek help has further obscured the facts.[28]

The reasons we are now becoming more aware of the adult illiteracy problem are

- Basic skill requirements are being increased by technological advances and the need to compete in the international marketplace.
- Broader definitions of literacy that go beyond just reading and writing now include speaking, listening, and mathematics abilities.
- We are realizing that old views of what constitutes literacy no longer apply.[29]

Impact of Illiteracy on Industry

The basic skills necessary to be productive in a modern industrial setting are increasing steadily. At the same time, the high school dropout rate nationwide continues to increase as does the number of high school graduates who are illiterate in spite of their diploma. This means that although the number of high-skill jobs in modern industry is increasing, the number of people able to fill them is on the decline. The impact this will have on industry in the United States can be summarized as follows:

- Difficulty in filling high-skill jobs
- Lower levels of productivity and, as a result, a lower level of competitiveness
- Higher levels of waste
- Higher potential for damage to sophisticated technological systems
- Greater number of dissatisfied employees in the workplace
- Greater potential for safety and health problems in the workplace

The fact that illiteracy on the job increases the likelihood of accidents should be of concern to safety and health professionals. For example, an illiterate employee would not be able to read or understand a material safety data sheet (MSDS) explaining the safe handling procedures for a given chemical or toxic material.[30] Written notification of safety and health hazards and written instruction for minimizing those hazards are of little use if the workforce or any part of it is illiterate.[31]

OSHA and the Illiteracy Problem

Complaints that OSHA's material safety data sheets are too technical and too complex are common. In fact, OSHA awards grants aimed at simplifying the MSDSs. The American Subcontractors Association has noted a gap between the reading level of MSDSs and the average worker.[32] An MSDS is technical by its very nature and written at a collegiate level.[33]

OSHA responds to complaints relating to illiteracy in the workplace by taking the stand that it is not OSHA's problem. It accepts no responsibility because safety training does not have to be in written form. As the situation evolves, however, OSHA and other governmental agencies are becoming more sensitive to the issue of illiteracy. The Department of Labor has increased its efforts to deal with the problem by sponsoring research and demonstration projects designed to develop literacy training program models, test and demonstrate new and innovative instructional techniques, develop instruments for assessing literacy levels, and encourage the development of state-level policies that promote workforce literacy. However, with immigrants flooding the workforce, illiteracy will continue to be an issue.[34]

Industry's Role in Fighting Illiteracy

Industry in the United States has found it necessary to confront the illiteracy problem head on by providing remedial education for employees in the workplace. Some companies contract with private training firms, others provide the remediation themselves, and others form partnerships with community colleges or vocational schools.

The National Center for Education Statistics conducted a study of industry-based adult literacy training programs in several industrial firms.[35] According to this study, technology companies require math, verbal and written communication, and basic physics of their employees. Physics skills have not traditionally been viewed as required for functional literacy. However, employees must have these skills to function in this high-tech company. The approach used by some technology companies to provide literacy training can be one of the least expensive. They work with public community colleges or vocational schools to provide literacy training at little or, in some cases, no cost. Safety and health professionals should establish a working relationship with representatives of local community colleges and vocational schools.[36]

High-tech companies such as Rockwell International also define literacy more stringently than has been typical in the past. To function effectively at such companies, employees must be skilled in chemistry and physics. Rockwell hires its own certified teachers and provides its own literacy program.[37]

Polaroid is a high-tech company that takes an aggressive approach in defining functional literacy. The skills taught in Polaroid's program are also indicative of the trend toward higher levels of knowledge. Polaroid requires statistics, problem solving, and computer literacy. The need for these skills is technology driven. Statistics is needed in order to use statistical process control (SPC), which is becoming a widely used quality control method in automated manufacturing settings. Few employees in a modern industrial firm can function without using a computer on the job, hence the need for computer literacy training.

Problem-solving skills are becoming critical as companies implement quality management (QM) programs. Such programs involve all employees in identifying and correcting problems that negatively affect quality or do not add value to the company's products. This is the rationale for including problem-solving skills in the definition of functional literacy.

Role of Safety and Health Professionals in Literacy Training

Kutner, et al., made recommendations to assist in the planning and implementation of such programs.[38] Several of these recommendations have relevance for safety and health professionals:

- The definitions should be driven by the needs of the company.
- Companies should establish an environment in which employees feel comfortable having their literacy skills assessed and in seeking help to raise those skills.
- Whenever possible, companies should establish programs to raise the skills of existing employees rather than laying them off and hiring new employees.
- Whenever possible, companies should collaborate with educational institutions or education professionals in providing literacy training.

As industrial companies continue to enhance their technological capabilities and safety and health concerns increase correspondingly, the literacy levels of the workforce will also have to increase. Because the number of people in the labor force who are highly literate is not increasing, the need for workforce literacy training is a fact of life that safety and health professionals must confront.

ENGLISH AS A SECOND LANGUAGE TRAINING ISSUES

In an article for *Occupational Hazards*, Joseph Halcarz wrote America's blue-collar work force stands as a testament to diversity. . . . OSHA recognizes that more than 10 million Americans speak little or no English, and one in five Americans does not speak English at home. . . . By truly understanding and respecting the literacy, cultural and language differences of their workers, companies stand to gain financially while creating a safer work environment for employees.[39]

To enjoy the gains that come with a safe and healthy workplace, organizations that have employees who either speak only limited English or for whom English is a second language must take steps to accommodate their special needs in terms of both training and cultural perspectives. There is little question that a language barrier can magnify the potential for accidents and injuries in the workplace. Recent studies show that the on-the-job fatality rate for Hispanics is more than 10 percent higher than for other groups. In addition to increasing the likelihood of accidents, language problems can also result in lower productivity and higher turnover.[40]

Modifying Training Methods

Traditional training methods—videos/DVDs, lectures, handouts, and other *classroom*-oriented methods—will not work well with employees who do not speak English as their principal language. *Hands-on* training that requires more doing and less listening, reading, and writing is the better approach with limited-English-speaking audiences. In other words, when working with a diverse group of employees in which language might be a barrier, don't use training methods that rely heavily on understanding the language. Instead of *telling* employees how to work safely, *show* them. Set up demonstrations that replicate the actual situation in question and let employees do what they are supposed to do on the job rather than just watching a video/DVD of someone else doing it.

According to Halcarz, the best way to overcome language, culture, and literacy barriers is to provide safety training programs that offer the following attributes:

1. Bilingual training materials that address cultural issues
2. Training materials that recognize the dignity of all employees regardless of their background
3. Training materials that minimize literacy requirements[41]

Considering Cultural Concerns

We tend to think of cultural issues in terms of language and national origin, and many of them are. However, even people who speak the same language and come from the same country can have cultural differences. For example, a person who grew up in the South may not know about working in extremely cold weather. This is a cultural issue, and it should be considered when developing safety training programs. Conversely, an individual who grew up in a cold climate might not understand the heat-related problems of working in the subtropics.

An employee's attitude toward reporting an injury can have cultural implications. For example, an employee from one culture might be quick to report an injury, whereas an employee from another culture might consider filing a report as being a troublemaker or as a sign of weakness. Immigrants, for example, might be afraid to report an injury for fear of being reported to the Immigration and Naturalization Service.

An effective way to identify the various cultural barriers that might have a negative effect on an employee's ability to fully grasp the training provided is to arrange facilitated brainstorming sessions. In these sessions, a third-party facilitator helps employees discuss freely, openly, and without attribution any concerns they might have about the training they need and any barriers that might inhibit their progress in completing the training. A third-party facilitator, especially one who speaks the principal language of the employees in question, is more likely to bring out employee concerns than their supervisor or manager.

OSHA STANDARDS AND TRAINING

Training has always been a high priority with OSHA. To promote training and to assist companies in providing it, OSHA maintains a Division of Training and Education Programs Web site (http://www.osha.gov). However, in recent years the trend in developing and revising OSHA standards is toward being increasingly prescriptive about training. In other words, not only do OSHA standards require that training be provided but they are also beginning to prescribe what should be taught, how much training must be provided, and how often. OSHA's Powered Industrial Truck Standard (29 CFR 1910.178), typically referred to as the Forklift Standard, is an example of the newly prescriptive nature of OSHA standards as they relate to training.

OSHA's Forklift Standard: An Example of Prescriptive Training Requirements

Originally, the Forklift Standard required only that industrial truck operators be trained and that employers develop methods for training them. The standard now requires employers to develop a training program specifically geared toward the type of truck to be driven and the conditions in which it will be driven. In addition, employers must now evaluate the employee's performance in operating specific types of trucks under specific work conditions, and the employer must certify that the required training has taken place. The employer is responsible for implementing the required training and for ensuring that only employees who have completed the necessary training be allowed to operate the type of industrial truck in question under the conditions present.[42]

OSHA even specifies the type of training that is required. It must be a combination of formal instruction (defined as lectures, videos/DVDs, etc.) and practical instruction (defined as demonstrations and hands-on exercises). Further, the training must include an evaluation of the operator's performance in an actual work setting. Topics that must be covered in the training are listed in 29 CFR 1910.178.

The standard specifies who is allowed to provide the training, when and under what conditions refresher training is mandated, certification requirements, and retraining requirements. In other words, OSHA has developed a standard that is very prescriptive about training requirements. This is not a one-time undertaking relating only to powered industrial trucks. Rather, prescriptiveness as it relates to training is the trend in the development and revising of OSHA's standards.

SUMMARY

1. The rationale for safety and health training is that workers who know how to do their jobs properly are less likely to have accidents.
2. The OSH Act established a legal foundation for safety and health training. In addition to producing and distributing training materials, OSHA also promotes safety and health training by providing grants, publishing training requirements, and writing training-oriented regulations.
3. The division of labor between safety and health professionals and supervisors is not clear-cut. However, generally speaking, the following guidelines apply: Safety and health professionals help train supervisors and keep them up-to-date; supervisors provide most of the training to workers; and safety and health professionals and supervisors work together as appropriate to train workers.
4. The Mine Safety and Health Administration (MSHA) sets training requirements for miners in Subpart B of its regulations. Paragraph 48.25 contains the regulations that pertain to new or inexperienced miners.
5. Any person who is going to conduct safety and health training should have the following characteristics: thorough knowledge of the subject matter, desire to teach, positive attitude, leadership ability, professional approach, and exemplary behavior.
6. The principles of learning summarize what is known about how people learn. Trainers should be familiar with these principles.
7. The four-step teaching method is as follows: preparation, presentation, application, and evaluation.
8. Widely used presentation methods include lecture–discussion, demonstration, conference, simulation, videotapes/DVDs, programmed instruction, and online training.
9. People learn by doing. Therefore, it is critical that people be given plenty of opportunities to apply what they are learning.
10. In evaluating training, safety and health professionals need to know (a) if the training was valid, (b) if the employees learned, and (c) if the learning made a difference.
11. Supervisors play a key role in the maintenance of a safe and healthy workplace. Consequently, safety and health professionals need to ensure that supervisors have had the training they need to be positive participants in the process.
12. One of the most important aspects of a safety program is the orientation of new employees. It should consist of a general orientation, job-specific procedures, and follow-up.
13. Job safety analysis, or JSA, can be an excellent way to teach safety. It consists of three steps: (a) breaking down a job into a chronological sequence of steps, (b) identifying potential hazards for each step, and (c) developing accident procedures to eliminate or reduce potential hazards.

14. Employers who hire workers with limited English-speaking capabilities should use less classroom instruction and more hands-on activities when providing safety training.
15. OSHA standards are becoming more prescriptive concerning training requirements. OSHA's Forklift Standard is an excellent example of this trend.

KEY TERMS AND CONCEPTS

Application
Application assignments list
Conference method
Course outline
Cultural concerns
Demonstration method
Ethical reasons
Evaluation
Evaluation methodology
Four-step teaching method
Hazard communication
Immediate and continual feedback
Instructional approach
Interactive video/DVD training
Job safety analysis (JSA)
Labor–management mistrust
Learning objectives
Lecture–discussion method
Legal reasons

Lesson plan
Material safety data sheets (MSDSs)
Mine Safety and Health Administration (MSHA)
MSHA training requirements
Online training
Opening, body, and closing of a lecture
Orientation
OSHA training requirements
Planning instruction
Preparation
Presentation
Principles of learning
Programmed instruction
Simulation
Statement of purpose
Training aids

REVIEW QUESTIONS

1. Explain briefly the rationale for providing safety training.
2. Establish the legal framework for providing safety and health training.
3. What does the hazard communication regulation (29 CFR 1910) require of employers?
4. Beyond the legal requirements of the OSH Act, how does OSHA promote safety and health training?
5. Summarize the mistrust that sometimes exists between labor and management concerning safety and health training.

6. What document could you use as a guide to the OSHA training requirements?
7. How does MSHA define the term *experienced miner*?
8. Summarize the MSHA training requirements for newly employed but experienced miners.
9. What are the characteristics that persons who plan to conduct safety training should possess?
10. Explain why online safety and health training has become so popular.
11. List five principles of learning.
12. Explain briefly each of the steps in the four-step teaching method.
13. What are the essential components of a lesson plan?
14. Give three examples of when the lecture–discussion method is best used.
15. Give three examples of when the lecture–discussion method is not appropriate.
16. What are the three components of a lecture?
17. Give an example of when the demonstration method may be used.
18. Briefly summarize the National Safety Council's recommendations for using the conference teaching method.
19. List five questions that should be asked when evaluating training.
20. What are the objectives of supervisor safety training, according to the National Safety Council?
21. List the minimum content of a good orientation program.
22. Explain how to conduct a job safety analysis.
23. What is the safety training value of a job safety analysis?
24. Explain how an employer can accommodate the training needs of workers with limited English-speaking ability.

ENDNOTES

1. G. LaBar, "Worker Training: An Investment in Safety," *Occupational Hazards*, August 1991, 25.
2. Ibid.
3. P. M. Laing, ed., *Supervisor's Safety Manual*, 7th ed. (Chicago: National Safety Council, 1991), 35.
4. J. A. Spencer and J. A. Simonowitz, "Employee Education," in *Introduction to Occupational Health and Safety*, ed. Joseph LaDon (Chicago: National Safety Council, 1986), 276.
5. B. K. Morris, "Grants Available for Training Workers on MSDS," *Occupational Health & Safety Letter* 20, no. 26: 210.
6. Spencer and Simonowitz, "Employee Education," 277–278.
7. Ibid., 280.
8. Title 29, Code of Federal Regulations, Parts 1910 and 1926.
9. Paragraph 48.22, Subpart B, MSHA Regulations, as published in the *Federal Register* 43, no. 199.
10. Ibid.

11. Paragraph 48.25, Subpart B, MSHA Regulations, as published in the *Federal Register* 43, no. 199.

12. Ibid.

13. Ibid.

14. Laing, *Supervisor's Safety Manual*, 36.

15. S. Woo, "Creative Training," *Occupational Health & Safety* 75, no. 1: 34–38.

16. National Safety Council, *Accident Prevention Manual for Industrial Operations: Administration and Programs*, 9th ed. (Chicago: National Safety Council, 1988), 195.

17. K. M. Magon, "Videos Add New Dimension to Safety Programs," *Safety & Health* 143, no. 1: 34.

18. Ibid., 35.

19. R. Fisher, "Interactive Training, Wave of the Future," *Safety & Health* 138, no. 1: 34.

20. Ibid., 36–37.

21. P. Minetos, "Supervisors: Teach Them Well," *Safety & Health* 140, no. 4: 64.

22. Ibid.

23. National Safety Council, *Accident Prevention Manual*, 186.

24. K. Knowles, ed., "Start Safety Training Early," *Today's Supervisor* 55, no. 7: 4.

25. National Safety Council, *Accident Prevention Manual*, 199.

26. C. Piepho, "Safety Training Institute," *Safety & Health* 141, no. 5: 80–85.

27. M. Kutner, E. Greenberg, and J. Baer, "A First Look at the Literacy of America's Adults in the 21st Century," National Center for Education Statistics, December 2005. Retrieved from http://nces.ed.gov/pubsearch /pubsinfo.asp?pubid=2006470.

28. Ibid., 4.

29. Ibid., 5.

30. Ibid., 8.

31. Ibid., 14.

32. Ibid., 16.

33. Ibid., 17.

34. Ibid., 18.

35. Ibid., 23.

36. Ibid., 24.

37. Ibid., 26.

38. Ibid., 27.

39. Joseph E. Halcarz, "Safety Training for Diverse Worker Population," *Occupational Hazards* 65, no. 2: 26.

40. Ibid., 26–28.

41. Ibid., 28.

42. Retrieved from http://www.osha.gov/Training/PIT/geninddpc.html (accessed July 20, 2006, 1–3.)

VIOLENCE IN THE WORKPLACE

Major Topics

- Occupational Safety and Workplace Violence: The Relationship
- Workplace Violence: Definitions
- Workplace Violence: Cases
- Size of the Problem
- Legal Considerations
- Risk-Reduction Strategies
- Contributing Social and Cultural Factors
- OSHA's Voluntary Guidelines
- Conflict Resolution and Workplace Violence
- Do's and Don'ts for Supervisors
- Emergency Preparedness Plan

> America has been hard at work in the past 10 days and here is what happened: A Federal Express pilot took a claw hammer and attacked three others in the cockpit, forcing one of them to put the fully loaded DC-10 cargo plane through a series of violent rolls and nose dives in a melee that brought the whole crew back bleeding. A purchasing manager in suburban Chicago stabbed his boss to death because, police say, they couldn't agree on how to handle some paperwork. And a technician who quit because he had trouble working for a woman sneaked back inside the fiber optics laboratory, pulled out a 9-mil semiautomatic pistol and started firing at workers, who ducked or fled and curled up in closets and file cabinets. By the time he finished the job, two were dead, two were injured; he walked upstairs to an office and shot himself in the head.[1]

Workplace violence has emerged as a critical safety and health issue. According to the Bureau of Labor Statistics, homicide is the second leading cause of death to American workers, "accounting for 16 percent of the 6,588 fatal work injuries in the United States."[2] Although more than 80 percent of workplace homicide victims are men, workplace violence is not just a male problem. In fact, workplace homicide is the leading cause of death on the job for women in the United States.

Almost one million people are injured or killed in workplace-violence incidents every year in the United States, and the number of incidents is on the rise. In fact, according to the U.S. Department of Justice, the workplace is the most dangerous place to be in the United States.[3] Clearly, workplace violence is an issue of concern to safety and health professionals.

OCCUPATIONAL SAFETY AND WORKPLACE VIOLENCE: THE RELATIONSHIP

The prevention of workplace violence is a natural extension of the responsibilities of safety and health professionals. Hazard analysis, records analysis and tracking, trend monitoring, incident analysis, and prevention strategies based on administrative and engineering controls are all fundamental to both concepts. In addition, emergency response and employee training are key elements of both. Consequently, occupational safety and health professionals are well suited to add the prevention of workplace violence to their normal duties.

WORKPLACE VIOLENCE: DEFINITIONS

Safety and health professionals should be familiar with the language that has developed around the issue of workplace violence. This section contains the definitions of several concepts as they relate specifically to workplace violence.

- *Workplace violence.* Violent acts, behavior, or threats that occur in the workplace or are related to it. Such acts are harmful or potentially harmful to people, property, or organizational capabilities.
- *Occupational violent crime (OVC).* Intentional battery, rape, or homicide during the course of employment.[4]
- *Employee.* An individual with an employment-related relationship (present or past) with the victim of a workplace-violence incident.
- *Outsider.* An individual with no relationship of any kind with the victim of a workplace-violence incident or with the victim's employer.

Safety Fact

High-Risk Occupations and People

High-risk occupations in terms of workplace violence are taxicab drivers; retail workers; police and security officers; and finance, insurance, real estate, health care, and community service employees. Employees 65 years and older are more likely to be victims than are younger employees.

- *Employee-related outsider.* An individual with some type of personal relationship (past or present) with an employee, but who has no work-related relationship with the employee.
- *Customer.* An individual who receives products or services from the victim of a workplace-violence incident or from the victim's employer.

Each of these terms has other definitions. Those presented here reflect how the terms are used in the language that has evolved around workplace violence.

WORKPLACE VIOLENCE: CASES

This section contains numerous cases of workplace violence that occurred in the United States during a one-year period. These cases are provided to give students of occupational safety and health a better understanding of the types of violent incidents that occur frequently in today's workplace. The names of individuals involved have been changed, but the incidents are real.

1. *Jackson, Mississippi.* A 32-year-old firefighter shot his wife through the head and then proceeded to a firehouse. Using an assault rifle, the man shot six coworkers—all supervisors—killing four of them and seriously wounding another two. He then fled the scene and exchanged fire with a police officer, wounding the officer, before being shot in the head himself and critically injured. The president of the union representing the shooter described him as "a time bomb waiting to go off."

2. *Fort Lauderdale, Florida.* Shouting "Everyone is going to die," a maintenance employee who had been fired walked into a meeting of his former coworkers and began shooting, killing five people and injuring another. The employees were inside a temporary trailer office when the disgruntled former employee showed up with two handguns. Police say that he chased workers around the office and shot them methodically, pausing only to reload.

3. *Honolulu, Hawaii.* An employee returned to his former workplace after being fired from his job. He held five coworkers hostage—including his former boss—for up to six hours. During the incident, the disgruntled former employee shot and seriously wounded his former supervisor. He was eventually killed by police after he held a shotgun to the head of one coworker for several hours while negotiating with officers. Prior to being shot by police, the perpetrator threatened to kill the coworker and started a countdown to pulling the trigger of the gun. The countdown prompted the hostage to grab the barrel of the gun and gave police the opportunity to shoot. The perpetrator died of his wounds.

4. *Waterville, Maine.* A former patient of a mental institution was accused of beating and stabbing four nuns, killing two of them. Police said that they caught the perpetrator in the convent's chapel Saturday evening, standing over one of the nuns and beating her with a religious figurine. Officers said that

they had to pull him off the woman and that he had also beaten and stabbed three others in an adjacent part of the convent. Nuns at the convent of the Servants of the Blessed Sacrament said that the man had applied for a job but had been turned down. Nuns had just finished a prayer service when the perpetrator smashed the glass on a locked door, opened it, and walked inside. The perpetrator was described as an accomplished musician who had played trumpet in a local jazz combo and studied at the University of Maine at Augusta.

5. *Evensdale, Ohio.* A male in his early fifties returned to the offices from which he had recently been fired. Brandishing two pistols, he shot and killed three employees and wounded a fourth. A witness quoted the perpetrator as saying he was "going after someone who had screwed him over." After the murders, he surrendered quietly to authorities.

6. *Harlem, New York.* A man who was apparently involved in protesting the closure of an electronics store entered a clothing store with a gun and arson materials. He immediately set the store on fire and began shooting at employees. Eight individuals were killed (including the perpetrator), and another four were wounded—three seriously—by gunfire.

7. *Columbus, Ohio.* A man was fired from his job at a bank credit card center following charges of sexual harassment against coworkers. A year later, he forced his way into at least two homes of former coworkers, fatally shooting four individuals (including a four-month-old child) and wounding two others. He was captured by police attempting to flee the city in his automobile.

8. *San Jose, California.* A young accountant, on the job for just six weeks, shot and killed his female supervisor, then committed suicide with the same gun. This happened one day after he received his first performance counseling session. The killer was a 28-year-old Asian male. The victim was a 32-year-old white female.

9. *Palatine, Illinois.* A postal worker reported to work with a handgun and shot two coworkers who he claimed were his friends. The shooter was a 53-year-old white male who had worked at the same location for almost 20 years and had an exemplary service record.

10. *Los Angeles, California.* A city electrician with a 12-year work history shot and killed four of his supervisors when he learned he was facing possible dismissal for poor performance. After shooting his supervisors, the killer, 42 years old, quietly waited for police to arrive and arrest him. He was heard saying that he had specifically targeted the four murdered individuals because he "felt he was being picked on and singled out" by them.

11. *Industry, California.* A "quiet, unassuming" postal worker who had been on the job for 22 years shot and killed his supervisor. The perpetrator was 58 years old and was easily disarmed by coworkers after the shooting.

12. *Asheville, North Carolina.* A "classic loner" just fired from his job at a machine tool company returned the next day with a rifle and a pistol. The perpetrator, 47 years old, killed three workers and wounded another four before quietly surrendering to police.

13. *Littleton, Colorado.* A man distraught over marital problems opened fire in a crowded grocery store, killing three people before he was subdued by a bystander. The perpetrator was a 35-year-old white man. Among the victims was his wife, age 37, and the store manager, age 39. This type of incident is becoming more common. It is often referred to as *spillover* violence.

14. *Richmond, California.* After a Richmond Housing Authority employee was fired from his job, he went to his automobile, retrieved a handgun, and returned to shoot and kill a supervisor and a coworker. The perpetrator was 38 years old. Both victims were women, aged 47 and 24 years.

SIZE OF THE PROBLEM

Violence in the workplace no longer amounts to just isolated incidents that are simply aberrations. In fact, workplace violence should be considered a common hazard worthy of the attention of safety and health professionals. In a report on the subject, the U.S. Department of Justice revealed the following information:

- About one million individuals are the direct victims of some form of violent crime in the workplace every year. This represents approximately 15 percent of all violent crimes committed annually in America. Approximately 60 percent of these violent crimes were categorized as *simple assaults* by the Department of Justice.
- Of all workplace violent crimes reported, over 80 percent were committed by males; 40 percent were committed by complete strangers to the victims; 35 percent by casual acquaintances, 19 percent by individuals well known to the victims, and 1 percent by relatives of the victims.
- More than half of the incidents (56 percent) were not reported to police, although 26 percent were reported to at least one official in the workplace.
- In 62 percent of violent crimes, the perpetrator was not armed; in 30 percent of the incidents, the perpetrator was armed with a handgun.
- In 84 percent of the incidents, there were no reported injuries; 10 percent required medical intervention.
- More than 60 percent of violent incidents occurred in private companies, 30 percent in government agencies, and 8 percent to self-employed individuals.
- It is estimated that violent crime in the workplace caused 500,000 employees to miss 1,751,000 days of work annually, or an average of 3.5 days per incident. This missed work equates to approximately $55 million in lost wages.[5]

The Society for Human Resource Management (SHRM) periodically surveys its members on the issue of workplace violence. One such survey produced the following results:[6]

Regarding violent incidents in the workplace:

- 33 percent of all managers surveyed experienced at least one violent incident in the workplace.
- 54 percent of these managers reported between two and five acts of violence in the five years prior to the survey.

Regarding the type of violence experienced:

- 75 percent of the reported incidents were fistfights.
- 17 percent of the incidents were shootings.
- 8 percent of the incidents were stabbings.
- 6 percent of the incidents were sexual assaults.

Regarding the victims of the incidents:

- 54 percent of the incidents were employee against employee.
- 13 percent of the incidents were employee against a supervisor.
- 7 percent of the incidents were customer against worker(s).

Regarding the gender of the perpetrator:

- 80 percent of all violent acts were committed by males.

Regarding the injuries sustained by the victims:

- 22 percent of the incidents involved serious harm.
- 42 percent of the incidents required medical intervention.

Regarding the reasons for the violent incidents:

- 38 percent were attributed to personality conflicts.
- 15 percent were attributed to marital or family problems.
- 10 percent were attributed to drug or alcohol abuse.
- 7 percent were nonspecific as to attribution.
- 7 percent were attributed to firings or layoffs.

Regarding crisis management programs:

- 28 percent of the organizations had a crisis management program in place prior to the violent incident.
- 12 percent of the organizations implemented a crisis management program after the violent incident occurred.

Regarding the effect of a violent incident on the workplace:

- 41 percent of the organizations reported increased stress levels in the workplace after a violent incident.
- 20 percent reported higher levels of paranoia.
- 18 percent reported increased distrust among employees.

LEGAL CONSIDERATIONS

Most issues relating to safety and health have legal ramifications, and workplace violence is no exception. The legal aspects of the issue revolve around the competing rights of violent employees and their coworkers (Figure 13–1). These conflicting rights create potential liabilities for employers.

Rights of Violent Employees

It may seem odd to be concerned about the rights of employees who commit violent acts on the job. After all, logic suggests that in such situations the only concern would be the protection of other employees. However, even violent employees have rights. Remember, the first thing that law enforcement officers must do after taking criminals into custody is to read them their rights.

According to Stephen C. Yohay,

> Employee rights are granted by a number of sources, including individual employment contracts, collective bargaining agreements, statutes such as the various state, local, and federal civil rights statutes, and sometimes from policies and procedures utilized by the employer. To the extent possible under the circumstances, the employer should consider following the procedures outlined in applicable contracts or policies before taking any adverse employment action against a threatening or violent employee.[7]

This does not mean that an employer cannot take the immediate action necessary to prevent a violent act or the recurrence of such an act. In fact,

Figure 13–1
Conflicting rights.

Rights of Violent Employees	⋈	Rights of Coworkers

Safety Myth

Workplace Violence—An Isolated Problem?

Most working Americans are unconcerned about workplace violence. They see it as an isolated problem that happens in other organizations, but not theirs. Right? Not any more. A Time/CNN poll of the general population in the United States found that 37 percent of respondents viewed workplace violence as a growing problem. Of those responding to the poll, 18 percent had personally witnessed some form of violence at work. The same percentage of respondents (19 percent) said they feared for their own safety while at work.

failure to act prudently in this regard can subject an employer to charges of negligence. However, before taking long-term action that will adversely affect the violent individual's employment, employers should follow applicable laws, contracts, policies, and procedures. Failure to do so can serve to exacerbate an already difficult situation.

For example, in *International Union v. Micro Manufacturing, Inc.*, the employer was required to reinstate an employee who had been fired for assaulting and battering the owner of the company.[8] The reason? The violent employee had been terminated on the spot without a union representative present. This approach violated the terms of the union's collective bargaining agreement.

In addition to complying with all applicable laws, policies, and procedures, it is also important to apply these consistently when dealing with violent employees. Dealing with one violent employee one way while dealing with another in a different way puts the employer at a disadvantage legally. Consequently, it is important for employers to be prepared to deal both promptly and properly with violent employees. According to Yohay,

> The courts and the EEOC as a general matter do not require an employer to retain an employee who is violent or threatening to the other employees. However, . . . the violent and threatening employee does have some competing rights. Employers must check the law in their jurisdictions to ascertain exactly where the line is drawn regarding the rights of the threatening or violent employee.[9]

Employer Liability for Workplace Violence

Having to contend with the rights of both violent employees and their coworkers, employers often feel as if they are caught between a rock and a hard place. Fortunately, the situation is less bleak than it may first appear due primarily to the **exclusivity provision** of workers' compensation laws. This provision makes workers' compensation the employee's exclusive remedy for injuries that are work related. This means that even in cases of workplace violence, as long as the violence occurs within the scope of the victim's employment, the employer is protected from civil lawsuits and the excessive jury verdicts that have become so common.

The key to enjoying the protection of the exclusivity provision of workers' compensation laws lies in determining that violence-related injuries are within the scope of the victim's employment—a more difficult undertaking than one may expect. For example, if the violent act occurred at work but resulted from a non-work-related dispute, does the exclusivity provision apply? What if the dispute was work related, but the violent act occurred away from the workplace?

Making Work-Related Determinations

The National Institute for Occupational Safety and Health (NIOSH) developed the following guidelines for categorizing an injury as being work related:

- If the violent act occurred on the employer's premises, it is considered an on-the-job event if one of the following criteria apply:

    The victim was engaged in work activity, apprenticeship, or training.

    The victim was on break, in hallway, restrooms, cafeteria, or storage areas.

    The victim was in the employer's parking lots while working, arriving at, or leaving work.

- If the violent act occurred off the employer's premises, it is still considered an on-the-job event, if one of the following criteria apply:

    The victim was working for pay or compensation at the time, including working at home.

    The victim was working as a volunteer, emergency services worker, law enforcement officer, or firefighter.

    The victim was working in a profit-oriented family business, including farming.

    The victim was traveling on business, including to and from customer–business contacts.

    The victim was engaged in work activity in which the vehicle is part of the work environment (e.g., taxi driver, truck driver, and so on).[10]

RISK-REDUCTION STRATEGIES

Figure 13–2 is a checklist that can be used by employers to reduce the risk of workplace violence in their facilities. Most of these risk-reduction strategies grow out of the philosophy of *crime reduction through environmental design (CRTED)*.[11] CRTED has the following four major elements, to which the author has added a fifth (administrative controls).

- Natural surveillance
- Control of access
- Establishment of territoriality
- Activity support
- Administrative controls

The following explains how these elements can help avoid workplace violence:

1. *Natural surveillance.* This strategy involves designing, arranging, and operating the workplace in a way that minimizes secluded areas. Making all areas inside and outside the facility easily observable allows for natural surveillance.

2. *Control of access.* One of the most common occurrences of workplace violence involves an outsider entering the workplace and harming employees. The most effective way of stopping this type of incident is to control access to the workplace. Channeling the flow of outsiders to an access-control station,

✓ Identify high-risk areas and make them visible. Secluded areas invite violence.

✓ Install good lighting in parking lots and inside all buildings.

✓ Minimize the handling of cash by employees and the amount of cash available on the premises.

✓ Install silent alarms and surveillance cameras where appropriate.

✓ Control access to all buildings (employee badges, visitor check-in and check-out procedure, visitor passes, and so on).

✓ Discourage working alone, particularly late at night.

✓ Provide training in conflict resolution as part of a mandatory employee orientation.

✓ Conduct background checks before hiring new employees.

✓ Train employees how to handle themselves and respond when a violent act occurs on the job.

✓ Develop policies that establish ground rules for employee behavior and responses in threatening or violent situations.

✓ Nurture a positive, harmonious work environment.

✓ Encourage employees to report suspicious individuals and activities or potentially threatening situations.

✓ Deal with allegations of harassment or threatened violence promptly before the situation escalates.

✓ Take threats seriously and act appropriately.

✓ Adopt a *zero-tolerance* policy toward threatening or violent behavior.

✓ Establish a *violence hot line* so that employees can report potential problems anonymously.

✓ Establish a *threat-management team* with responsibility for preventing and responding to violence.

✓ Establish an *emergency response team* to deal with the immediate trauma of workplace violence.

Figure 13–2
Checklist for workplace-violence risk reduction.

requiring visitor's passes, issuing access badges to employees, and isolating pickup and delivery points can minimize the risk of violence perpetrated by outsiders.

3. *Establishment of territoriality.* This strategy involves giving employees control over the workplace. With this approach, employees move freely within their established territory but are restricted in other areas. Employees come to know everyone who works in their territory and can, as a result, immediately recognize anyone who shouldn't be there.

4. *Activity support.* Activity support involves organizing work flow and natural traffic patterns in ways that maximize the number of employees

Discussion Case

What Is Your Opinion?

A man walks into an office building and asks to see his wife. The man is well known to the other employees, one of whom escorts him to his wife's workstation. Suddenly, the man pulls a gun and shoots his wife and another employee who tries to intervene. Is this an on-the-job event? Is the employer at fault? What is your opinion about this incident involving an employee-related outsider?

conducting natural surveillance. The more employees observing the activity in the workplace, the better.

5. *Administrative controls.* Administrative controls consist of management practices that can reduce the risk of workplace violence. These practices include establishing policies, conducting background checks, and providing training for employees.

CONTRIBUTING SOCIAL AND CULTURAL FACTORS

Another way to reduce the risk of workplace violence is to ensure that managers understand the social and cultural factors that can lead to it. These factors fall into two broad categories: individual and environmental factors.

Individual Factors Associated with Violence

The factors explained in this section can be predictors of the potential for violence. Employees and individuals with one or more of the following factors may respond to anger, stress, or anxiety in a violent way.

1. *Record of violence.* Past violent behavior is typically an accurate predictor of future violent behavior. Consequently, thorough background checks should be a normal part of the employment process.

2. *Membership in a hate group.* Hate groups often promote violence against the subjects of their prejudice. Hate-group membership on the part of an employee should raise a red flag in the eyes of management.

3. *Psychotic behavior.* Individuals who incessantly talk to themselves, express fears concerning conspiracies against them, say that they hear voices, or become increasingly disheveled over time may be violence prone.

4. *Romantic obsessions.* Workplace violence is often the result of romantic entanglements or love interests gone awry. Employees who persist in making unwelcome advances may eventually respond to rejection with violence.

Safety Fact

Eight Steps for Preventing Workplace Violence

- Complete a risk-assessment survey of the entire workplace.
- Review existing security procedures.
- Develop and publish a policy statement that explains expectations, rules for behavior, roles, duties, and responsibilities.
- Develop work-site-specific prevention procedures.
- Train all managers, supervisors, and employees.
- Establish incident reporting and investigation procedures.
- Establish incident follow-up procedures (trauma plan, counseling services, and disciplinary guidelines).
- Monitor, evaluate, and adjust.

---

5. *Depression.* People who suffer from depression are prone to hurt either themselves or someone else. An employee who becomes increasingly withdrawn or overly stressed may be suffering from depression.

6. *Finger pointers.* Refusal to accept responsibility is a factor often exhibited by perpetrators of workplace violence. An employee's tendency to blame others for his or her own shortcomings should raise the caution flag.

7. *Unusual frustration levels.* The workplace has become a competitive, stressful, and sometimes frustrating place. When frustration reaches the boiling point, the emotional explosion that results can manifest itself in violence.

8. *Obsession with weapons.* Violence in the workplace often involves a weapon (gun, knife, or explosive device). A normal interest in guns used for hunting or target practice need not raise concerns. However, an employee whose interest in weapons is unusually intense and focused is cause for concern.

9. *Drug dependence.* It is common for perpetrators of workplace violence to be drug abusers. Consequently, drug dependence should cause concern not only for all the usual reasons but also for its association with violence on the job.

Environmental Factors Associated with Violence

The environment in which employees work can contribute to workplace violence. An environment that produces stress, anger, frustration, feelings of powerlessness, resentment, and feelings of inadequacy can increase the potential for violent behavior. The following factors can result in such an environment:

1. *Dictatorial management.* Dictatorial, overly authoritative management that shuts employees out of the decision-making process can cause them to

feel powerless, as if they have little or no control over their jobs. Some people respond to powerlessness by striking out violently—a response that gives them power, if only momentarily.

2. *Role ambiguity.* One of the principal causes of stress and frustration on the job is role ambiguity. Employees need to know for what they are responsible, how they will be held accountable, and how much authority they have. When these questions are not clear, employees become stressed and frustrated, factors often associated with workplace violence.

3. *Partial, inconsistent supervision.* Supervisors who play favorites engender resentment in employees who aren't the favorite. Supervisors who treat one employee differently than another or one group of employees differently from another group also cause resentment. Employees who feel that they are being treated unfairly or unequally may show their resentment in violent ways.

4. *Unattended hostility.* Supervisors who ignore hostile situations or threatening behavior are unwittingly giving them their tacit approval. An environment that accepts hostile behavior will have hostile behavior.

5. *No respect for privacy.* Supervisors and managers who go through the desks, files, tool boxes, and work areas of employees without first getting their permission can make them feel invaded or even violated. Violent behavior is a possible response to these feelings.

6. *Insufficient training.* Holding employees accountable for performance on the job without providing the training that they need to perform well can cause them to feel inadequate. People who feel inadequate can turn their frustration inward and become depressed or turn it outward and become violent.

The overriding message in this section is twofold. First, managers should establish and maintain a positive work environment that builds up employees rather than tearing them down. Second, managers should be aware of the individual factors that can contribute to violent behavior and respond promptly if employees show evidence of responding negatively to these factors.

OSHA'S VOLUNTARY GUIDELINES

The U.S. Department of Labor, working through the Occupational Safety and Health Administration (OSHA), has established *advisory guidelines* relating to workplace violence.[12] Two key points to understand about these guidelines are:

- The guidelines are *advisory* in nature and *informational* in content. The guidelines *do not* add to or enhance in any way the requirements of the general duty clause of the OSH Act.
- The guidelines were developed with night retail establishments in mind. Consequently, they have a service-oriented emphasis. However, much of the advice contained in the guidelines can be adapted for use in manufacturing, processing, and other settings.

Figure 13–3
Broadly applicable elements of OSHA's advisory guidelines on workplace violence.

✓	Management commitment and employee involvement
✓	Workplace analysis
✓	Hazard prevention and control
✓	Safety and health training
✓	Record keeping and evaluation

Figure 13–3 is a checklist of those elements of the specifications that have broader applications. Any management program relating to safety and health in the workplace should have at least these four elements.

Management Commitment and Employee Involvement

Management commitment and employee involvement are fundamental to developing and implementing any safety program, but they are especially important when trying to prevent workplace violence. The effectiveness of a workplace-violence prevention program may be a life-or-death proposition.

Figure 13–4 is a checklist that explains what management commitment means in practical terms. Figure 13–5 describes the practical application of

✓	Hands-on involvement of executive management in developing and implementing prevention strategies.
✓	Sincere, demonstrated concern for the protection of employees.
✓	Balanced commitment to both employees and customers.
✓	Inclusion of safety, health, and workplace-violence prevention in the job descriptions of all executives, managers, and supervisors.
✓	Inclusion of safety, health, and workplace-violence prevention criteria in the performance evaluations of all executives, managers, and supervisors.
✓	Assignment of responsibility for providing coordination and leadership for safety, health, and workplace-violence prevention to a management-level employee.
✓	Provision of the resources needed to prevent workplace violence effectively.
✓	Provision of or guaranteed access to appropriate medical counseling and trauma-related care for employees affected physically or emotionally by workplace violence.
✓	Implementation, as appropriate, of the violence-prevention recommendations of committees, task forces, and safety professionals.

Figure 13–4
Elements of management commitment to workplace-violence prevention.

✓   Staying informed concerning all aspects of the organization's safety, health, and workplace-violence program.

✓   Voluntarily complying—in both letter and spirit—with all applicable workplace-violence prevention strategies adopted by the organization.

✓   Making recommendations—through proper channels—concerning ways to prevent workplace violence and other hazardous conditions.

✓   Prompt reporting of all threatening or potentially threatening situations.

✓   Accurate and immediate reporting of all violent or threatening incidents.

✓   Voluntary participation on committees, task forces, and focus groups concerned with preventing workplace violence.

✓   Voluntary participation in seminars, workshops, and other educational programs relating to the prevention of workplace violence.

Figure 13–5
Elements of employee involvement in workplace-violence prevention.

employee involvement. Figure 13–6 is a checklist that can be used to ensure that violence prevention becomes a standard component of organizational plans and operational practices. These three checklists can be used by any type of organization to operationalize the concepts of management commitment and employee involvement.

Workplace Analysis

Workplace analysis is the same process used by safety and health professionals to identify potentially hazardous conditions unrelated to workplace violence. Work-site analysis should be ongoing and have at least four components (Figure 13–7). An effective way to conduct an ongoing program of workplace analysis is to establish a threat-assessment team with representatives from all departments and led by the organization's chief safety and health professional.

Records Monitoring and Tracking

The purpose of **records monitoring and tracking** is to identify and chart all incidents of violence and threatening behavior that have occurred within a given time frame. Records to analyze include the following: incident reports, police reports, employee evaluations, and letters of reprimand. Of course, individual employees' records should be analyzed in confidence by the human resources member of the team. The type of information that is pertinent includes the following:

- Where specifically did the incident occur?
- What time of day or night did the incident occur?
- Was the victim an employee? Customer? Outsider?
- Was the incident the result of a work-related grievance? Personal?

✓ Include the prevention of workplace violence in the safety and health component of the organization's strategic plan.

✓ Adopt, disseminate, and implement a *no-tolerance* policy concerning workplace violence.

✓ Adopt, disseminate, and implement a policy that protects employees from reprisals when they report violent, threatening, or potentially threatening situations.

✓ Establish procedures for reporting violent and threatening incidents.

✓ Establish procedures for making recommendations for preventing workplace violence.

✓ Establish procedures for monitoring reports of workplace violence so that trends can be identified and incidents predicted and prevented.

✓ Develop a comprehensive workplace-violence prevention program that contains operational procedures and standard practices.

✓ Develop a workplace-violence component to the organization's emergency response plan.

✓ Train all employees in the application of standard procedures relating to workplace violence.

✓ Conduct periodic emergency-response drills for employees.

Figure 13–6
Checklist for incorporating workplace-violence prevention in strategic plans and operational practices.

Figure 13–7
Four components of workplace analysis.

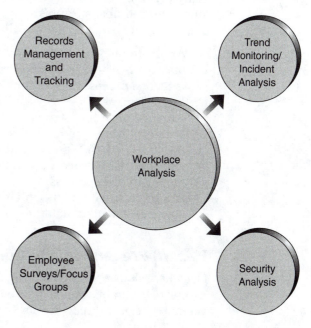

Trend Monitoring and Incident Analysis

Trend monitoring and incident analysis may prove helpful in determining patterns of violence. If there have been enough incidents to create one or more graphs, the team will want to determine if the graphs suggest a trend or trends. If the organization has experienced only isolated incidents, the team may want to monitor national trends. By analyzing both local and national incidents, the team can generate information that will be helpful in predicting and, thereby, preventing workplace violence. The team should look for trends in severity, frequency, and type of incidents.

Employee Surveys and Focus Groups

Employees are one of the best sources of information concerning workplace hazards. This is also true when it comes to identifying vulnerabilities to workplace violence. Employee input should be solicited periodically through written **employee surveys and focus groups** or one of these methods. Where are we vulnerable? What practices put our employees at risk? These are the types of questions that should be asked of employees. An effective strategy for use with focus groups is to give participants case studies of incidents that occurred in other organizations. Then ask such questions as, "Could this happen here? Why? Or why not? How can we prevent such incidents from occurring here?"

Security Analysis

Is the workplace secure, or could a disgruntled individual simply walk in and harm employees? It is important to ask this question. The team should periodically perform a **security analysis** of the workplace to identify conditions, situations, procedures, and practices that make employees vulnerable. The types of questions to ask include the following:

- Are there physical factors about the facility that make employees vulnerable (e.g., isolated, poorly lighted, infrequently trafficked, or unobservable)?
- Is there a process for handling disgruntled customers? Does it put employees at risk?
- Are the prevention strategies already implemented working?
- Is the training provided to employees having a positive effect? Is more training needed? Who needs the training? What kind of training is needed?
- Are there situations in which employees have substantial amounts of money in their possession, on- or off-site?
- Are there situations in which employees are responsible for highly valuable equipment or materials late at night or at isolated locations?

Hazard Prevention and Control

Once hazardous conditions have been identified, the strategies and procedures necessary to eliminate them must be put in place. The two broad categories of prevention strategies are engineering controls and administrative controls, just as they are with other safety and health hazards. In addition to these, organizations should adopt postincident response strategies as a way to prevent future incidents.

Engineering Controls

Engineering controls relating to the prevention of workplace violence serve the same purpose as engineering controls relating to other hazards. They either remove the hazard, or they create a barrier between it and employees. Engineering controls typically involve changes to the workplace. Examples of engineering controls include:

- Installing devices and mechanisms that give employees a complete view of their surroundings (e.g., mirrors, glass or clear plastic partitions, interior windows, and so on)
- Installing surveillance cameras and television screens that allow for monitoring of the workplace
- Installing adequate lighting, particularly in parking lots
- Pruning shrubbery and undergrowth outside and around the facility
- Installing fencing so that routes of egress and ingress to company property can be channeled and, as a result, better controlled
- Arranging outdoor sheds, storage facilities, recycling bins, and other outside facilities for maximum visibility

Administrative Controls

Whereas engineering controls involve making changes to the workplace, **administrative controls** involve making changes to how work is done. This amounts to changing work procedures and practices. Administrative controls fall into four categories (Figure 13–8).

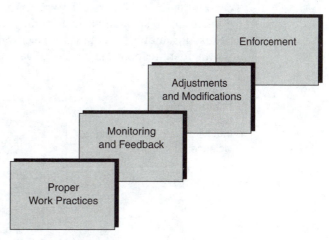

Figure 13–8
Categories of administrative controls.

- **Proper work practices** are those that minimize the vulnerability of employees. For example, if a driver has to make deliveries in a high-crime area, the company may employ a security guard to go along, change delivery schedules to daylight hours only, or both.
- **Monitoring and feedback** ensures that proper work practices are being used and that they are having the desired effect. For example, say, a company established a controlled access system in which visitors must check in at a central location and receive a visitor's pass. Is the system being used? Are all employees sticking to specified procedures? Has unauthorized access to the workplace been eliminated?
- **Adjustments and modifications** are made to violence prevention practices if it becomes clear from monitoring and feedback that they are not working or that improvements are needed.
- **Enforcement** involves applying meaningful sanctions when employees fail to follow the established and proper work practices. An employee who has been fully informed concerning a given administrative control, has received the training needed to practice it properly, but consciously decides not to follow the procedure should be disciplined appropriately.

Postincident Response

Postincident response relating to workplace violence is the same as postincident response relating to traumatic accidents. The first step is to provide immediate medical treatment for injured employees. The second step involves providing psychological treatment for traumatized employees. This step is even more important in cases of workplace violence than with accidents. Employees who are present when a violent incident occurs in the workplace, even if they don't witness it, can experience the symptoms of psychological trauma shown in Figure 13–9. Employees experiencing such symptoms or any others growing out of psychological trauma should be treated by professionals such as psychologists, psychiatrists, clinical nurse specialists, or certified social workers. In addition to one-on-one counseling, employees may also be enrolled in support groups. The final aspect of postincident response is the investigation, analysis, and report. In this step, safety

Figure 13–9
Symptoms of psychological trauma in cases of workplace violence.

✓	Fear of returning to work
✓	Problems in relationships with fellow employees and/or family members
✓	Feelings of incompetence
✓	Guilt feelings
✓	Feelings of powerlessness
✓	Fear of criticism by fellow employees, supervisors, and managers

professionals determine how the violent incident occurred and how future incidents may be prevented, just as postaccident investigations are handled.

Training and Education

Training and education are as fundamental to the prevention of workplace violence as they are to the prevention of workplace accidents and health-threatening incidents. A complete safety and health training program should include a comprehensive component covering all aspects of workplace violence (e.g., workplace analysis, hazard prevention, proper work practices, and emergency response). Such training should be provided on a mandatory basis for supervisors, managers, and employees.

Record Keeping and Evaluation

Maintaining accurate, comprehensive, up-to-date records is just as important when dealing with violent incidents as it is when dealing with accidents and nonviolent incidents. By evaluating records, safety personnel can determine how effective their violence prevention strategies are, where deficiencies exist, and what changes need to be made. Figure 13–10 shows the types of records that should be kept.

- *OSHA log of injury and illness.* OSHA regulations require inclusion in the Injury and Illness Log of any injury that requires more than first aid, is a

Figure 13–10
Types of records that should be kept.

lost-time injury, requires modified duty, or causes loss of consciousness. Of course, this applies only to establishments required to keep OSHA logs. Injuries caused by assaults which are otherwise recordable also must be included in the log. A fatality or catastrophe that results in the hospitalization of three or more employees must be reported to OSHA within eight hours. This includes those resulting from workplace violence and applies to all establishments.

- *Medical reports.* Medical reports of all work injuries should be maintained. These records should describe the type of assault (e.g., unprovoked sudden attack), who was assaulted, and all other circumstances surrounding the incident. The records should include a description of the environment or location, potential or actual cost, lost time, and the nature of injuries sustained.
- *Incidents of abuse.* Incidents of abuse, verbal attacks, aggressive behavior—which may be threatening to the employee but do not result in injury, such as pushing, shouting, or acts of aggression—should be evaluated routinely by the affected department.
- *Minutes of safety meetings.* Minutes of safety meetings, records of hazard analyses, and corrective actions recommended and taken should be documented.
- *Records of all training programs.* Records of all training programs, attendees, and qualifications of trainers should be maintained.

As part of its overall program, an employer should regularly evaluate its safety and security measures. Top management should review the program regularly, as well as each incident, to determine the program's effectiveness. Responsible parties (managers, supervisors, and employees) should collectively evaluate policies and procedures on a regular basis. Deficiencies should be identified, and corrective action taken. An evaluation program should involve the following activities:

- Establishing a uniform violence reporting system and regular review of reports
- Reviewing reports and minutes from staff meetings on safety and security issues
- Analyzing trends and rates of illness or injury or fatalities caused by violence relative to initial or *baseline* rates
- Measuring improvements based on lowering the frequency and severity of workplace violence
- Keeping up-to-date records of administrative and work practice changes to prevent workplace violence to evaluate their effectiveness
- Surveying employees before and after making job or workplace changes or installing security measures or new systems to determine their effectiveness
- Keeping abreast of new strategies available to deal with violence as they develop
- Surveying employees who experience hostile situations about the medical treatment they received initially and, again, several weeks afterward, and then several months later

- Complying with OSHA and state requirements for recording and reporting deaths, injuries, and illnesses
- Requesting periodic law enforcement or outside consultant review of the workplace for recommendations on improving employee safety

Management should share violence prevention evaluation reports with all employees. Any changes in the program should be discussed at regular meetings of the safety committee, union representatives, or other employee groups.

CONFLICT RESOLUTION AND WORKPLACE VIOLENCE

When developing a violence prevention program for an organization, the natural tendency is to focus on protecting employees from outsiders. This is important. However, increasingly with workplace violence, the problem is internal. All too often in the modern workplace, conflict between employees is turning violent. Consequently, a violence prevention program is not complete without the following elements: **conflict management** and **anger management**.[13]

Conflict Management Component

Disagreements on the job can generate counterproductive conflict. This is one of the reasons why managers in organizations should do what is necessary to manage conflict properly. However, it is important to distinguish between just conflict and counterproductive conflict. Not all conflict is bad. In fact, properly managed conflict that has the improvement of products, processes, people, and the work environment as its source is positive conflict.

Counterproductive conflict—the type associated with workplace violence—occurs when employees behave in ways that work against the interests of the overall organization and its employees. This type of conflict is often characterized by deceitfulness, vindictiveness, personal rancor, and anger. *Productive conflict* occurs when right-minded, well-meaning people disagree, without being disagreeable, concerning the best way to support the organization's mission. Conflict management has the following components:

- Establishing conflict guidelines
- Helping all employees develop conflict prevention and resolution skills
- Helping all employees develop anger management skills

Establishing Conflict Guidelines

Conflict guidelines establish ground rules for discussing and debating differing points of view, differing ideas, and differing opinions concerning how best to accomplish the organization's vision, mission, and broad objectives. Figure 13–11 is an example of an organization's conflict guidelines. Guidelines such as these should be developed with a broad base of employee involvement from all levels in the organization.

Conflict Guidelines
Micro Electronics Manufacturing (MEM)

Micro Electronics Manufacturing encourages discussion and debate among employees at all levels concerning better ways to improve continually the quality of our products, processes, people, and work environment. This type of interaction, if properly handled, will result in better ideas, policies, procedures, practices, and decisions. However, human nature is such that conflict can easily get out of hand, take on personal connotations, and become counterproductive. Consequently, in order to promote productive conflict, MEM has adopted the following guidelines. These guidelines are to be followed by all employees at all levels:

- The criteria to be applied when discussing/debating any point of contention is as follows: Which recommendation is most likely to move our company closer to accomplishing its mission?

- Disagree, but don't be disagreeable. If the debate becomes too hot, stop and give all parties an opportunity to cool down before continuing. Apply your conflict resolution skills and anger management skills. Remember, even when we disagree about how to get there, we are all trying to reach the same destination.

- Justify your point of view by tying it to our mission and require others to do the same.

- In any discussion of differing points of view, ask yourself the following question: "Am I just trying to win the debate for the sake of winning (ego), or is my point of view really the most valid?"

Figure 13–11
Sample conflict guidelines.

Develop Conflict Prevention and Resolution Skills

If managers are going to expect employees to disagree without being disagreeable, they are going to have to ensure that all employees are skilled in the art and science of conflict resolution. The first guideline in Figure 13–11 is an acknowledgment of human nature. It takes advanced human relation skills and constant effort to disagree without being disagreeable. Few people are born with this ability. Fortunately, it can be learned. The following strategies are based on a three-phase model developed by Tom Rusk and described in his book *The Power of Ethical Persuasion.*[14]

Explore the Other Person's Viewpoint

Allow the other person to present his or her point of view. The following strategies will help make this phase of the discussion more positive and productive.

1. Establish that your goal at this point is mutual understanding.
2. Elicit the other person's complete point of view.
3. Listen nonjudgmentally and do not interrupt.

Safety Fact

Aggressive Employees Threaten Productivity

No one wants to work with or around an aggressive person. As a result, aggressive employees cause tardiness, absenteeism, and turnover. All three of these factors are known to harm productivity. Consequently, it is important to deal with aggressive employees through counseling, aggression management training, or even termination.

4. Ask for clarification if necessary.
5. Paraphrase the other person's point of view and restate it to show that you understand.
6. Ask the other person to correct your understanding if it appears to be incomplete.

Explain Your Viewpoint

After you accurately and fully understand the other person's point of view, present your own. The following strategies will help make this phase of the discussion more positive and productive:

1. Ask for the same type of fair hearing for your point of view that you gave the other party.
2. Describe how the person's point of view affects you. Don't point the finger of blame or be defensive. Explain your reactions objectively, keeping the discussion on a professional level.
3. Explain your point of view accurately and completely.
4. Ask the other party to paraphrase and restate what you have said.
5. Correct the other party's understanding if necessary.
6. Review and compare the two positions (yours and that of the other party). Describe the fundamental differences between the two points of view and ask the other party to do the same.

Agree on a Resolution

Once both viewpoints have been explained and are understood, it is time to move to the resolution phase. This is the phase in which both parties attempt to come to an agreement. Agreeing to disagree—in an agreeable manner—is an acceptable solution. The following strategies will help make this phase of the discussion more positive and productive:

1. Reaffirm the mutual understanding of the situation.
2. Confirm that both parties are ready and willing to consider options for coming to an acceptable solution.

3. If it appears that differences cannot be resolved to the satisfaction of both
parties, try one or more of the following strategies:

Take time out to reflect and try again.

Agree to third-party arbitration or neutral mediation.

Agree to a compromise solution.

Take turns suggesting alternative solutions.

Yield (this time), once your position has been thoroughly stated and is
understood. The eventual result may vindicate your position.

Agree to disagree while still respecting each other.

Develop Anger Management Skills

It is difficult, if not impossible, to keep conflict positive when anger enters the
picture. If individuals in an organization are going to be encouraged to question, discuss, debate, and even disagree, they must know how to manage their
anger. Anger is an intense, emotional reaction to conflict in which self-control
may be lost. Anger is a major cause of workplace violence. Anger occurs when
people feel that one or more of their fundamental needs are being threatened.
These needs include the following:

- Need for approval
- Need to be valued
- Need to be appreciated
- Need to be in control
- Need for self-esteem

When one or more of these needs is threatened, a normal human response
is to become angry. An angry person can respond in one of four ways:

1. *Attacking.* With this response, the source of the threat is attacked. This
response often leads to violence, or at least verbal abuse.

2. *Retaliating.* With this response, you fight fire with fire, so to speak.
Whatever is given, you give back. For example, if someone calls your
suggestion ridiculous (threatens your need to be valued), you may retaliate by
calling his or her suggestion dumb. Retaliation can escalate into violence.

3. *Isolating.* This response is the opposite of venting. With the isolation
response, you internalize your anger, find a place where you can be alone, and
simmer. The childhood version of this response was to go to your room and
pout. For example, when someone fails to even acknowledge your suggestion
(threatens your need to be appreciated), you may swallow your anger, return to
your office, and boil over in private.

4. *Coping.* This is the only positive response to anger. Coping does not
mean that you don't become angry. Rather, it means that, even when you do,
you control your emotions instead of letting them control you. A person
who copes well with anger is a person who, in spite of his or her anger, stays in

control. The following strategies will help employees manage their anger by becoming better at coping:

> Avoid the use of anger-inducing words and phrases including the following: *but, you should, you made me, always, never, I can't, you can't,* and so on.
>
> Admit that others don't make you angry; you allow yourself to become angry. You are responsible for your emotions and your responses to them.
>
> Don't let pride get in the way of progress. You don't have to be right every time.
>
> Drop your defenses when dealing with people. Be open and honest.
>
> Relate to other people as equals. Regardless of position or rank, you are no better than they, and they are no better than you.
>
> Avoid the human tendency to rationalize your angry responses. You are responsible and accountable for your behavior.

If employees in an organization can learn to manage conflict properly and to deal with anger positively, the potential for workplace violence will be diminished substantially. Conflict and anger management will not prevent violent acts from outsiders. There are other methods for dealing with outsiders. However, properly managing conflict and anger can protect employees from each other.

DO'S AND DON'TS FOR SUPERVISORS

Supervisors can play a pivotal role in the prevention of workplace violence. Following are some rules-of-thumb that will enhance the effectiveness of supervisors in this regard:

- *Don't* try to diagnose the personal, emotional, or psychological problems of employees.
- *Don't* discuss an employee's drinking unless it occurs on the job. Restrict comments to performance.
- *Don't* preach to employees. Counsel employees about attendance, tardiness, and job performance, not about how they should live their lives.
- *Don't* cover up for employees or make excuses for inappropriate behavior. Misguided kindness may allow problems to escalate and get out of hand.
- *Don't* create jobs to get problem employees out of the way. Stockpiling an employee simply gives him or her more time to brood and to allow resentment to build.
- *Don't* ignore the warning signs explained earlier in this chapter. The problems that they represent will not simply go away. Sooner or later, they will have to be handled. Sooner is better.

- *Do* remember that chemical dependence and emotional problems tend to be progressive. Left untreated, they get worse, not better.
- *Do* refer problem employees to the employee assistance program or to other mental health service providers.
- *Do* make it clear to employees that job performance is the key issue. They are expected to do what is necessary to maintain and improve their performance.
- *Do* make it clear that inappropriate behavior will not be tolerated.[15]

EMERGENCY PREPAREDNESS PLAN

To be prepared for properly handling a violent incident in the workplace, employers should form a crisis management team.[16] The team should have only one mission—immediate response to violent acts on the job—and be chaired by a safety and health professional. Team members should receive special training and be updated regularly. The team's responsibilities should be as follows:

- Undergo trauma response training
- Handle media interaction
- Operate telephone and communication teams
- Develop and implement, as necessary, an emergency evacuation plan
- Establish a backup communication system
- Calm personnel after an incident
- Debrief witnesses after an incident
- Ensure that proper security procedures are established, kept up-to-date, and enforced
- Help employees deal with posttraumatic stress
- Keep employees informed about workplace violence as an issue, how to respond when it occurs, and how to help prevent it

SUMMARY

1. Preventing workplace violence is a natural extension of the responsibilities of safety and health professionals. Like the traditional responsibilities of such professionals, dealing with workplace violence involves such activities as hazard analysis, records analysis and tracking, trend monitoring, and incident analysis.
2. Key concepts relating to workplace violence include the following: occupational violent crime (OVC), employee, outsider, employee-related outsider, and customer. These terms have definitions relating specifically to workplace violence.
3. Approximately one million people are victims of workplace violence every year. These incidents result in more than 1.75 million lost days of work annually.

4. Almost 40 percent of all violent acts in the workplace are committed by males. The majority of violent incidents reported each year (75 percent) are fistfights.
5. When dealing with violent incidents on the job, it is important to remember that even the perpetrator has rights. Employee rights are protected by employment contracts, collective bargaining agreements, and various local, state, and federal civil rights statutes. To the extent possible, when dealing with a violent employee, follow the procedures stipulated in contracts, agreements, and statutes.
6. Although it is important that employers consider perpetrators' rights when dealing with workplace violence, it is equally important that they act prudently to prevent harm to other employees and customers.
7. The exclusivity provision of workers' compensation laws provide employers with some protection from liability in cases of workplace violence, provided that the incident is work related. When this is the case, workers' compensation is the injured employee's exclusive remedy.
8. A violent act can be considered an on-the-job incident, even if it is committed away from the workplace. Specific guidelines have been established by NIOSH for determining whether a violent act can be classified as an on-the-job incident.
9. The concept of crime reduction through environmental design (CRTED) has four major elements as follows: natural surveillance, control of access, establishment of territoriality, and activity support. The author has added another: administrative controls.
10. OSHA has produced voluntary advisory guidelines relating to workplace violence. Although the guidelines are aimed specifically at the night retail industry, they provide an excellent framework that can be used in other industries including manufacturing, transportation, and processing. The framework has the following broad elements: management commitment and employee involvement, workplace analysis and hazard prevention control, safety and health training, and record keeping and evaluation.

KEY TERMS AND CONCEPTS

Activity support

Adjustments and modifications

Administrative controls

Anger management

Conflict guidelines

Conflict management

Control of access

Customer

Employee

Employee-related outsider

Employee surveys and focus groups

Enforcement

Engineering controls

Establishment of territoriality

Exclusivity provision

Management commitment and employee involvement

Monitoring and feedback

Natural surveillance

Occupational violent crime (OVC)

Outsider

Postincident response

Proper work practices

Record keeping and evaluation

Records monitoring and tracking

Security analysis

Trend monitoring and incident analysis

Workplace analysis

Workplace violence

REVIEW QUESTIONS

1. Define the following terms as they relate to violence in the workplace: *occupational violent crime, employee,* and *outsider.*
2. Approximately how many people are direct victims of workplace violence annually?
3. Defend or refute the following statement: Employees who commit violent acts forfeit their rights and can be dealt with accordingly.
4. What is the exclusivity provision of workers' compensation laws? Why is this provision significant?
5. Defend or refute the following statement: A violent act that occurs away from the employer's premises cannot be considered work related.
6. Explain the concept of crime reduction through environmental design (CRTED).
7. Defend or refute the following statement: A manufacturer must comply with OSHA's guidelines on workplace violence.
8. What elements of OSHA's guidelines on workplace violence can be adapted for future use in most types of business and industrial firms?
9. What are the primary causes of conflict on the job?
10. Explain the four ways in which an angry person may respond in a work setting.

ENDNOTES

1. U.S. Department of Labor, Occupational Safety and Health Administration, *OSHA Workplace Violence Guidelines.* Retrieved June 2003 from http://www.osha.gov/workplaceviolence/viol. html.
2. Anastasia Toufexis, "Workers Who Fight Firing with Fire," *Time* 143, no. 17: 34.
3. U.S. Department of Justice, *Violence and Theft in the Workplace* (NCJ-148199) (Annapolis Junction, MD: Bureau of Justice Statistics Clearinghouse).

4. Lloyd G. Nigro, "Violence in the American Workplace: Challenges to the Public Employees," *Public Administration Review* 56, no. 4: 326.

5. U.S. Department of Justice, *Violence and Theft in the Workplace.*

6. Society for Human Resource Management, "Workplace Violence Survey," results published in *USA Today Magazine,* January 2006.

7. Stephen C. Yohay and Melissa L. Peppe, "Workplace Violence: Employer Responsibilities," *Occupational Hazards,* July 1996, 22.

8. *International Union v. Micro Manufacturing, Inc.,* 895 F. Supp. 170, 171 (E.D. Mich. 1995).

9. Yohay and Peppe, "Workplace Violence," 24.

10. National Institute for Occupational Safety and Health and U.S. Department of Health and Human Services, *Homicide in U.S. Workplaces: A Strategy for Prevention and Research* (Washington, DC: Centers for Disease Control and Prevention, NIOSH).

11. Janice L. Thomas, "A Response to Occupational Violent Crime," *Professional Safety,* 27–31, July 2006.

12. U.S. Department of Labor, Occupational Safety and Health Administration, *Guidelines for Workplace Violence Prevention Programs for Night Retail Establishments.* Retrieved July 2003 from http://www.osha.gov/workplace evidence.

13. Ibid.

14. Tom Rusk, *The Power of Ethical Persuasion* (New York: Penguin Books), xv–xvii.

15. Illinois State Police, *Do's and Don'ts for the Supervisor.* Retrieved from http://www.state.il.us/isp/viowkplc/vwpp6c.htm.

16. Illinois State Police, *Do's and Don'ts for the Supervisor.* Retrieved from http://www.state.il.us/isp/viowkplc/vwpp8.htm.

4

HAZARD ASSESSMENT, PREVENTION, AND CONTROL

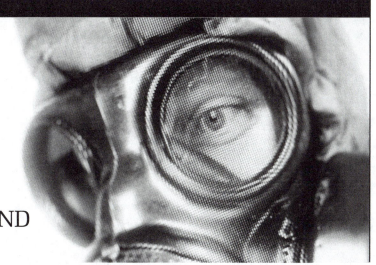

14 Mechanical Hazards and Machine Safeguarding   385

15 Falling, Impact, Acceleration, Lifting, and Vision Hazards   411

16 Hazards of Temperature Extremes   452

17 Pressure Hazards   470

18 Electrical Hazards   487

19 Fire Hazards and Life Safety   511

20 Industrial Hygiene and Confined Spaces   544

21 Radiation Hazards   604

22 Noise and Vibration Hazards   626

23 Computers, Automation, and Robots   659

24 Bloodborne Pathogens in the Workplace   681

MECHANICAL HAZARDS AND MACHINE SAFEGUARDING

Major Topics

- Common Mechanical Injuries
- Safeguarding Defined
- OSHA's Requirements for Machine Guarding
- Risk Assessment in Machine Operation
- Requirements for All Safeguards
- Point-of-Operation Guards
- Point-of-Operation Devices
- Machine Guarding Self-Assessment
- Feeding and Ejection Systems
- Robot Safeguards
- Control of Hazardous Energy (Lockout/Tagout Systems)
- General Precautions
- Basic Program Content
- Taking Corrective Action

Failure to provide proper machine guards and enforce their use can be costly for companies. A manufacturing firm in Syracuse, New York, learned this fact the hard way when it was cited by OSHA for failure to provide appropriate machine guards and require their proper use by employees. Because the company had been issued a similar citation earlier, it was fined $119,000. Mechanical hazards that are not properly guarded are implicated in thousands of workplace injuries every year.

As another example, a windshield manufacturing firm was fined more than $105,000 by OSHA when it allowed employees to be exposed to electrical hazards and knowingly failed to meet lockout/tagout standards. The small gains in productivity that might be obtained by willfully bypassing mechanical safeguards on machines can cost companies huge fines and even expensive medical bills.

Mechanical hazards are those associated with power-driven machines, whether automated or manually operated. Concerns about mechanical hazards date back to the Industrial Revolution and the earliest days of mechanization. Machines driven by steam, hydraulic, or electric power introduced new hazards into the workplace. In spite of advances in safeguarding technologies and techniques, mechanical hazards are still a major concern today. In addition, automated machines have introduced new concerns.

COMMON MECHANICAL INJURIES

In an industrial setting, people interact with machines that are designed to drill, cut, shear, punch, chip, staple, stitch, abrade, shape, stamp, and slit such materials as metals, composites, plastics, and elastomers. If appropriate safeguards are not in place or if workers fail to follow safety precautions, these machines can apply the same procedures to humans. When this happens, the types of **mechanical injuries** that result are typically the result of cutting, tearing, shearing, crushing, breaking, straining, or puncturing (see Figure 14–1). Information about each of these hazards is provided in the following paragraphs.

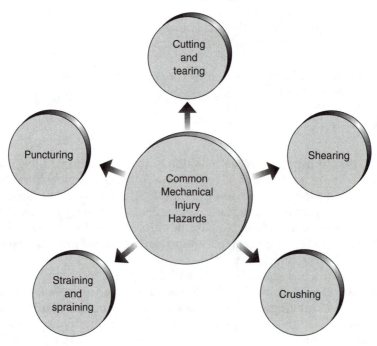

Figure 14–1
Some common mechanical hazards.

Cutting and Tearing

A cut occurs when a body part comes in contact with a sharp edge. The human body's outer layer consists of the following, starting from the outside: *epidermis*, the tough outer covering of the skin; *dermis,* the greatest part of the skin's thickness; *capillaries*, the tiny blood vessels that branch off the small arteries and veins in the dermis; *veins*, the blood vessels that collect blood from the capillaries and return it to the heart; and *arteries*, the larger vessels that carry blood from the heart to the capillaries in the skin. The seriousness of **cutting** or **tearing** the skin depends on how much damage is done to the skin, veins, arteries, muscles, and even bones.

Shearing

To understand what **shearing** is, think of a paper cutter. It shears the paper. Power-driven shears for severing paper, metal, plastic, elastomers, and composite materials are widely used in manufacturing. In times past, such machines often amputated fingers and hands. These tragedies typically occurred when operators reached under the shearing blade to make an adjustment or placed materials there and activated the blade before fully removing their hand. Safeguards against shearing accidents are explained later in this chapter.

Crushing

Injuries from **crushing** can be particularly debilitating, painful, and difficult to heal. They occur when a part of the body is caught between two hard surfaces that progressively move together, thereby crushing anything between them. Crushing hazards can be divided into two categories: *squeeze-point* types and *run-in points*.

Squeeze-point hazards exist where two hard surfaces, at least one of which must be in motion, push close enough together to crush any object that may be between them. The process can be slow, as in a manually operated vise, or fast, as with a metal-stamping machine.

Run-in point hazards exist where two objects, at least one of which is rotating, come progressively closer together. Any gap between them need not become completely closed. It need only be smaller than the object or body part lodged in it. Meshing gears and belt pulleys are examples of run-in point hazards (see Figures 14–2, 14–3, and 14–4).

Body parts can also be crushed in other ways—for example, a heavy object falling on a foot or a hammer hitting a finger. However, these are impact hazards, which are covered in Chapter 15.

Breaking

Machines used to deform engineering materials in a variety of ways can also cause broken bones. A break in a bone is known as a *fracture*. Fractures are classified as simple, compound, complete, and incomplete.

Figure 14–2
This nip point can pull hands,
feet, or articles of clothing.

Figure 14–3
Fingers might get caught
between the bench top
and the revolving wheel.

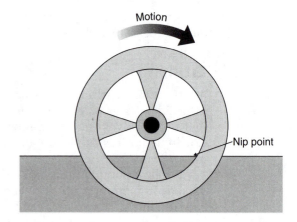

Figure 14–4
Nip points can catch fingers,
hands, hair, clothing, etc., with
dangerous results.

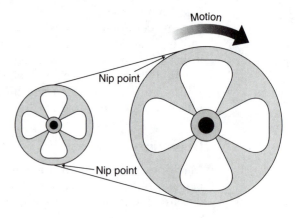

A simple fracture is a break in a bone that does not pierce the skin. A compound fracture is a break that has broken through the surrounding tissue and skin. A complete fracture divides the affected bone into two or more separate pieces. An incomplete fracture leaves the affected bone in one piece but cracked.

Fractures are also classified as transverse, oblique, and comminuted. A transverse fracture is a break straight across the bone. An oblique fracture is diagonal. A comminuted fracture exists when the bone is broken into a number of small pieces at the point of fracture.

Straining and Spraining

There are numerous situations in an industrial setting when **straining** of muscles or **spraining** of ligaments is possible. A strain results when muscles are overstretched or torn. A sprain is the result of torn ligaments in a joint. Strains and sprains can cause swelling and intense pain.

Puncturing

Punching machines that have sharp tools can puncture a body part if safety precautions are not observed or if appropriate safeguards are not in place. **Puncturing** results when an object penetrates straight into the body and pulls straight out, creating a wound in the shape of the penetrating object. The greatest hazard with puncture wounds is the potential for damage to internal organs.

SAFEGUARDING DEFINED

All the hazards explained in the previous section can be reduced by the application of appropriate safeguards. CFR 1910 Subpart O contains the OSHA standards for machinery and machine guarding (1910.211–1910.222). The National Safety Council defines **safeguarding** as follows:

> Machine safeguarding is to minimize the risk of accidents of machine–operator contact. The contact can be:
>
> 1. An individual making the contact with the machine—usually the moving part—because of inattention caused by fatigue, distraction, curiosity, or deliberate chance taking;
> 2. From the machine via flying metal chips, chemical and hot metal splashes, and circular saw kickbacks, to name a few;
> 3. Caused by the direct result of a machine malfunction, including mechanical and electrical failure.[1]

Safeguards can be broadly categorized as point-of-operation guards, point-of-operation devices, and feeding/ejection methods. The various types of safeguards in these categories are explained later in this chapter.

OSHA'S REQUIREMENTS FOR MACHINE GUARDING

The OSHA standard containing the general requirements for machine guarding is 29 CFR 1910.212. A more specific standard (29 CFR 1926.300) exists for the construction industry. This section focuses on 29 CFR 1910.212—OSHA's requirements for all industries. Those requirements are summarized as follows:

Types of guarding. One or more methods of machine guarding must be provided to protect people from such point of operation hazards as nip points, rotating parts, flying chips, and sparks. "Point of operation" refers to the area on the machine where work is performed on the material being processed. Examples of point-of-operation machine guards are barriers, two-hand switches and tripping devices, and electronic sensors.

General requirements for machine guards. Where possible, guards should be affixed to the machine in question. When this is not possible, guards should be secured in the most feasible location and method away from the machine. Guards must be affixed in such a way that they do not create a hazard themselves.

Guarding the point of operation. Any point of operation that might expose a person to injury must be guarded. Guarding devices must comply with all applicable standards. In the absence of applicable standards, the guard must be designed, constructed, and installed in such a way as to prevent the machine operator from having any part of his body (including clothing, hair, etc.) in the danger zone during the operating cycle of the machine.

Machines requiring point of operation guards. The following are examples of machines that require point of operation guards: guillotine cutters, shears, alligator shears, power presses, milling machines, power saws, jointers, portable power tools, forming rolls, and calendars.

Exposure of blades. Fans must be guarded in any case in which the periphery of the fan blades is less than seven feet above the floor or working level. Guards for fans shall have no openings that exceed one-half inch.

Anchoring fixed machinery. Machines that are designed to be fixed in one location must be securely anchored to prevent movement.[2]

RISK ASSESSMENT IN MACHINE OPERATION

Risk assessment in this context is the process of quantifying the level of risk associated with the operation of a given machine. It should be a structured and systematic process that answers the following four specific questions:

- How *severe* are potential injuries?
- How *frequently* are employees exposed to the potential hazards?
- What is the *possibility* of avoiding the hazard if it does occur?
- What is the *likelihood* of an injury should a safety control system fail?[3]

Safety Fact

Failure to Properly Guard Machines Can Be Expensive

Failure to properly guard machines can be costly to employers and tragic to employees. This is the lesson a manufacturing firm in Illinois learned when an employee working on an unguarded power press lost three fingers. OSHA subsequently cited the company for failure to use sensors and guards on dangerous parts of machines, periodically inspect machines to ensure they are properly guarded, and properly train and supervise machine operators.

Source: From "Worker Loses His Fingers; Company Pays the Price," *Facility Manager's Alert* 8, no. 175: 3.

The most widely used risk-assessment technique is the decision tree, coupled with codes representing these four questions and defined levels of risk. Figure 14–5 is an example of a risk-assessment decision tree. In this example, the codes and their associated levels of risk are as follows:

S = Severity

Question 1: Severity of potential injuries
S1   Slight injury (bruise, abrasion)

S2   Severe injury (amputation or death)

F = Frequency

Question 2: Frequency of exposure to potential hazards
F1   Infrequent exposure

F2   From frequent to continuous exposure

P = Possibility

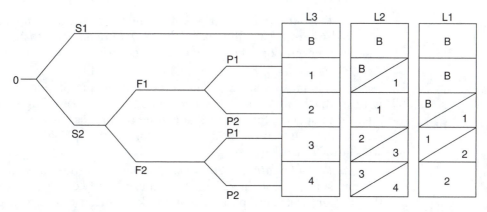

Figure 14–5
Risk-assessment decision tree.

Question 3: Possibility of avoiding the hazard if it does occur

P1   Possible

P2   Less possible to not impossible

L = Likelihood

Question 4: Likelihood that the hazard will occur

L1   Highly unlikely

L2   Unlikely

L3   Highly likely

Risk Levels

Associated risk factors ranging from lowest (B) to highest (4)
By applying the decision tree in Figure 14–5 or a similar device, the risk associated with the operation of a given machine can be quantified. This allows safety personnel to assign logical priorities for machine safeguarding and hazard prevention.

REQUIREMENTS FOR ALL SAFEGUARDS

The various machine motions present in modern industry involve mechanisms that rotate, reciprocate, or do both. This equipment includes tools, bits, chucks, blades, spokes, screws, gears, shafts, belts, and a variety of different types of stock. Safeguards can be devised to protect workers from harmful contact with such mechanisms while at the same time allowing work to progress at a productive rate. The National Safety Council has established the following requirements for safeguards.

1. *Prevent contact.* Safeguards should prevent human contact with any potentially harmful machine part. The prevention extends to machine operators and any other person who might come in contact with the hazard.

2. *Be secure and durable.* Safeguards should be attached so that they are secure. This means that workers cannot render them ineffective by tampering with or disabling them. This is critical because removing safeguards in an attempt to speed production is a common practice. Safeguards must also be durable enough to withstand the rigors of the workplace. Worn-out safeguards won't protect workers properly.

3. *Protect against falling objects.* Objects falling onto moving machine mechanisms increase the risk of accidents, property damage, and injury. Objects that fall on a moving part can be quickly hurled out, creating a dangerous projectile. Therefore, safeguards must do more than just prevent human contact. They must also shield the moving parts of machines from falling objects.

4. *Create no new hazard.* Safeguards should overcome the hazards in question without creating new ones. For example, a safeguard with a sharp edge, unfinished surface, or protruding bolts introduces new hazards while protecting against the old.

5. *Create no interference.* Safeguards can interfere with the progress of work if they are not properly designed. Such safeguards are likely to be disregarded or disabled by workers feeling the pressure of production deadlines.

6. *Allow safe maintenance.* Safeguards should be designed to allow the more frequently performed maintenance tasks (e.g., lubrication) to be accomplished without the removal of guards. For example, locating the oil reservoir outside the guard with a line running to the lubrication point will allow for daily maintenance without removing the guard.[4]

Design and construction of safeguards are highly specialized activities requiring a strong working knowledge of machines, production techniques, and safety. However, it is critical that all the factors explained in this section be considered and accommodated during the design process.

POINT-OF-OPERATION GUARDS

Guards are most effective when used at the point of operation, which is where hazards to humans exist. Point-of-operation hazards are those caused by the shearing, cutting, or bending motions of a machine. Pinch-point hazards result from guiding material into a machine or transferring motion (e.g., from gears, pressure rollers, or chains and sprockets). Single-purpose safeguards, because they guard against only one hazard, typically are permanently fixed and non-adjustable. Multiple-purpose safeguards, which guard against more than one hazard, typically are adjustable.[5]

Point-of-operation guards are of three types, each with its own advantages and limitations: fixed, interlocked, and adjustable (Figure 14–6).

Figure 14–6
Point-of-operation guards.

- **Fixed guards** provide a permanent barrier between workers and the point of operation. They offer the following advantages: They are suitable for many specific applications, can be constructed in-plant, require little maintenance and are suitable for high-production, repetitive operations. Limitations include the following: They sometimes limit visibility, are often limited to specific operations, and sometimes inhibit normal cleaning and maintenance.
- **Interlocked guards** shut down the machine when the guard is not securely in place or is disengaged. The main advantage of this type of guard is that it allows safe access to the machine for removing jams or conducting routine maintenance without the need for taking off the guard. There are also limitations. Interlocked guards require careful adjustment and maintenance and, in some cases, can be easily disengaged.
- **Adjustable guards** provide a barrier against a variety of different hazards associated with different production operations. They have the advantage of flexibility. However, they do not provide as dependable a barrier as other guards do, and they require frequent maintenance and careful adjustment.

Figures 14–7 through 14–12 show various guards used in modern manufacturing settings.

Figure 14–7
Series 12 PRO-TECH-TOR GATE GUARD used on an open-back power press.
Courtesy of PROTECH SYSTEMS.

Figure 14–8
Series 17 CHECKMATE RIVET
GUARD used on a foot-operated
riveting machine. Note that more
modern models have light beam
protection.
Courtesy of PROTECH SYSTEMS.

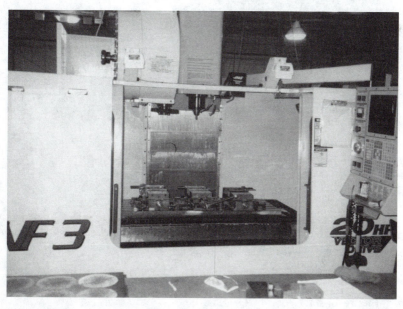

Figure 14–9
When the doors are opened, the milling tool stops automatically.

Figure 14–10
In order for this shearing machine to cut, both the foot pedal and the hand button must be engaged.

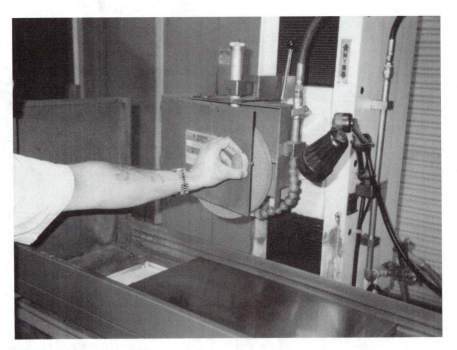

Figure 14–11
This door protects the operator in the event of an exploding or shattering grinding wheel.

Figure 14–12
The safety door on this drilling machine must be closed
or the drill will not operate.

POINT-OF-OPERATION DEVICES

A number of different **point-of-operation devices** can be used to protect workers. The most widely used are explained in the following paragraphs.

- **Photoelectric devices** are optical devices that shut down the machine whenever the light field is broken. These devices allow operators relatively free movement. They do have limitations, including the following: They do not protect against mechanical failure, they require frequent calibration, they can be used only with machines that can be stopped, and they do not protect workers from parts that might fly out of the point-of-operation area.

- **Radio-frequency devices** are capacitance devices that brake the machine if the capacitance field is interrupted by a worker's body or another object. These devices have the same limitations as photoelectric devices.

- **Electromechanical devices** are contact bars that allow only a specified amount of movement between the worker and the hazard. If the worker moves the contact bar beyond the specified point, the machine will not cycle. These devices have the limitation of requiring frequent maintenance and careful adjustment.

- **Pullback devices** pull the operator's hands out of the danger zone when the machine starts to cycle. These devices eliminate the need for auxiliary barriers. However, they also have limitations. They limit operator movement, must be adjusted for each individual operator, and require close supervision to ensure proper use.

- **Restraint devices** hold the operator back from the danger zone. They work well, with little risk of mechanical failure. However, they do limit the operator's movement, must be adjusted for each individual operator, and require close supervision to ensure proper use.

- **Safety trip devices** include trip wires, trip rods, and body bars. All these devices stop the machine when tripped. They have the advantage of simplicity. However, they are limited in that all controls must be activated manually. They protect only the operator and may require the machine to be fitted with special fixtures for holding work.
- **Two-hand controls** require the operator to use both hands concurrently to activate the machine (e.g., a paper cutter or metal-shearing machine). This ensures that hands cannot stray into the danger zone. Although these controls do an excellent job of protecting the operator, they do not protect onlookers or passers-by. In addition, some two-hand controls can be tampered with and made operable using only one hand.
- **Gates** provide a barrier between the danger zone and workers. Although they are effective at protecting operators from machine hazards, they can obscure the work, making it difficult for the operator to see.[6]

Figure 14–13 is a point-of-operation device that shuts down the machine when an operator's hand breaks a light beam. How quickly will the machine shut down? The stopping distance equation is as follows:

$$SD = \text{Stoptime [Seconds} \times \text{Hand speed constant (63''/second)]}$$

Figure 14–13
Series 25 EAGLE EYE
INFRA-RED LIGHT BARRIER.
A point-of-operation guarding
system on a roller press
machine.
Courtesy of PROTECH SYSTEMS.

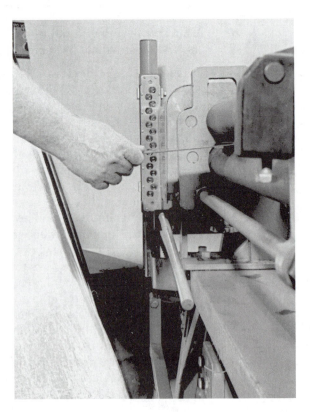

MACHINE GUARDING SELF-ASSESSMENT

One of the most effective ways to ensure that machines are properly guarded is to conduct periodic self-assessments. These self-assessments can be conducted by safety personnel, supervisors, or employees. Developing self-assessment criteria and encouraging supervisors and employees to use them daily is an excellent strategy for safety and health professionals. The following questions can be used for conducting machine guarding self-assessments:

1. Are all machines that might expose people to rotating parts, nip points, flying chips, sparks, flying particles, or other similar hazards properly guarded?
2. Are all mechanical power transmission belts and the nip points they create properly guarded?
3. Are all exposed power shafts located less than seven feet above the working level properly guarded?
4. Are all hand tools and other hand-operated equipment inspected regularly for hazardous conditions?
5. Is compressed air used to clean tools, machines, equipment, and parts reduced to less than 30 pounds per square inch (psi)?
6. Are power saws and similar types of equipment properly guarded?
7. Are the tool rests for grinding wheels set to within one-eighth or less of the grinding wheel?
8. Are hand tools regularly inspected on a systematic basis for burred ends, cracked handles, and other potentially hazardous conditions?
9. Are all compressed gas cylinders inspected regularly and systematically for obvious signs of defects, deep rusting, or leakage?
10. Do all employees who handle and store gas cylinders and valves know how to do so without causing damage?
11. Are all air receivers periodically and systematically inspected, including safety valves?
12. Are all safety valves tested regularly, systematically, and frequently?[7]

FEEDING AND EJECTION SYSTEMS

Feeding and ejection systems can be effective safeguards if properly designed and used. The various types of feeding and ejection systems available for use with modern industrial machines are summarized as follows:

- **Automatic feed** systems feed stock to the machine from rolls. Automatic feeds eliminate the need for operators to enter the danger zone. Such systems are limited in the types and variations of stock that they can feed. They also typically require an auxiliary barrier guard and frequent maintenance.
- **Semiautomatic feed** systems use a variety of approaches for feeding stock to the machine. Prominent among these are chutes, moveable dies, dial

feeds, plungers, and sliding bolsters. They have the same advantages and limitations as automatic feed systems.

- **Automatic ejection** systems eject the work pneumatically or mechanically. The advantage of either approach is that operators don't have to reach into the danger zone to retrieve workpieces. However, these systems are restricted to use with relatively small stock. Potential hazards include blown chips or debris and noise. Pneumatic ejectors can be quite loud.[8]
- **Semiautomatic ejection** systems eject the work using mechanisms that are activated by the operator. Consequently, the operator does not have to reach into the danger zone to retrieve workpieces. These systems do require auxiliary barriers and can be used with a limited variety of stock.

ROBOT SAFEGUARDS

Robots have become commonplace in modern industry. The safety and health concerns relating to robots are covered in Chapter 23. Only the guarding aspects of robot safety are covered in this section. The main hazards associated with robots are (1) entrapment of a worker between a robot and a solid surface, (2) impact with a moving robot arm, and (3) impact with objects ejected or dropped by the robot.

The best guard against these hazards is to erect a physical barrier around the entire perimeter of a robot's **work envelope** (the three-dimensional area established by the robot's full range of motion). This physical barrier should be able to withstand the force of the heaviest object that a robot could eject.

Various types of shutdown guards can also be used. A guard containing a sensing device that automatically shuts down the robot if any person or object enters its work envelope can be effective. Another approach is to put sensitized doors or gates in the perimeter barrier that automatically shut down the robot as soon as they are opened.

These types of safeguards are especially important because robots can be deceptive. A robot that is not moving at the moment may simply be at a stage between cycles. Without warning, it might make sudden and rapid movements that could endanger any person inside the work envelope.

CONTROL OF HAZARDOUS ENERGY (LOCKOUT/TAGOUT SYSTEMS)

OSHA's standard for the control of hazardous energy, often referred to as the "lockout/tagout" standard, is 29 CFR 1910.147. The purpose of this standard is to protect people in the workplace from hazardous energy while they are performing service or maintenance on machines, tools, and equipment. A key element of the standard is to prevent the accidental or inadvertent activation of a machine while it is being serviced or repaired. The lockout/tagout standard identifies the proper procedures for shutting down machines and equipment and locking or tagging it out so that accidental or inadvertent activation does not occur. The standard also calls for employee training and periodic inspections. The overall requirement of 29 CFR 1910.147 is that before service or maintenance are performed, the machines or equipment in question must be disconnected from their energy source, and the energy source must be either locked out or tagged out to prevent accidental or inadvertent activation.[9]

Lockout/Tagout Language

The following terms and phrases are frequently used in the language of lockout/tagout. Safety and health professionals should be knowledgeable of these terms:

- *Affected employee.* Employees who perform their jobs in areas in which the procedure in question is implemented and in which service or maintenance operations are performed. Affected employees do not implement energy control procedures unless they are *authorized.*
- *Authorized employee.* Employees who perform service or maintenance on a machine and use lockout/tagout procedures for their own protection.
- *Energized.* Machines, equipment, and tools are energized if they are connected to an energy source or when they still contain stored or residual energy even after being disconnected.
- *Capable of being locked out.* A device is considered to be capable of being locked out if it meets one of the following requirements: (1) it has a hasp to which a lock can be attached; (2) it has another appropriate integral part through which a lock can be attached; (3) it has a built-in locking mechanism; or (4) it can be locked without permanently dismantling, rebuilding, or replacing the energy-isolating device.
- *Energy-isolating device.* Any mechanical device that physically prevents the release or transmission of energy (e.g., circuit breakers, disconnect switches, blocks, etc.).
- *Energy source.* Any source of power that can activate a machine or piece of equipment (e.g., electrical, mechanical, hydraulic, pneumatic, chemical, thermal, etc.).
- *Energy control procedure.* A written document containing all the information an authorized person needs to know in order to properly control hazardous energy when shutting down a machine or equipment for maintenance or service.
- *Energy control program.* A systematic program for preventing the accidental or inadvertent energizing of machines or equipment during maintenance or servicing. This is sometimes called the organization's lockout/tagout program.
- *Lockout.* Placing a lockout device such as a padlock on an energy-isolating device to prevent the accidental or inadvertent energizing of a machine during maintenance or servicing.

- *Lockout device.* Any device (see Figure 14–14) that uses a positive means to keep an energy-isolation device in the *safe* position to prevent the - accidental or inadvertent energizing of a machine or piece of equipment.
- *Tagout.* Placing a tag (see Figure 14–15) on an energy-isolation device to warn people so that they do not accidentally or inadvertently energize a machine or piece of equipment.
- *Tagout device.* Any prominent warning device such as a tag that can be affixed to an energy-isolation device to prevent the accidental or inadvertent energizing of a machine or piece of equipment.

Figure 14–14
Lockout system.

Figure 14–15
Tagout system.

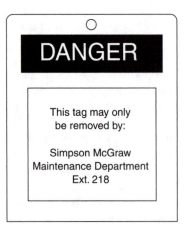

Provisions of the Standard

OSHA's standard for control of hazardous energy contains provisions in the following areas: energy control program, energy control procedure, energy-isolating devices, lockout/tagout devices, periodic inspections, application of controls and lockout/tagout devices, removal of locks and tags, testing or positioning of machines, outside personnel, group lockout or tagout, and shift or personnel changes. Those provisions are as follows:

- *Energy control program.* Organizations must establish energy control programs that have fully documented energy control procedures, provide employee training, and ensure periodic inspections.
- *Energy control procedure.* Organizations must develop, document, and use energy control procedures that contain at least the following elements: (a) a statement on how the procedure will be used; (b) procedural steps used to shut down, isolate, block, and secure machines or equipment; (c) steps designating the safe placement, removal, and transfer of lockout/tagout devices and who has responsibility for them; and (d) specific requirements for testing machines or equipment to verify the effectiveness of energy control measures.
- *Energy-isolating devices.* Organizations must have appropriate energy-isolating devices for preventing the accidental or inadvertent release of energy on all machines and equipment. The preferred type of device is one that can be locked out. However, when this type of device is not feasible, tagout procedures may be used. When a tagout system is used, employees must receive training on the following limitations of tags: (a) tags are just warning devices and do not provide the safety margin of locks; (b) tags may be removed only by the person who affixes them and should never be bypassed, ignored, or otherwise overcome; (c) tags must be legible and understandable by all employees (this can mean providing labels in more than one language); (d) tags and their means of attachment must be made of material that is durable enough to withstand the environment to which they will be subjected; (e) tags can evoke a false sense of security; and (f) tags must be attached securely enough that they will not come off during servicing or maintenance operations.
- *Requirements for lockout/tagout devices.* Lockout and tagout devices must meet the following requirements: (a) durable enough to withstand the environment to which they will be exposed; (b) standardized in terms of color or size and print and format; (c) substantial enough to minimize the likelihood of accidental or unauthorized removal; and (d) identifiable in terms of the employee who affixed them and the warning message (e.g., Do Not Start, Do Not Close, Do Not Energize, Do Not Open, etc.).
- *Employee training.* Organizations must provide both initial training and retraining as necessary and certify that the necessary training has been given to all employees covered in the standard (e.g., authorized, affected, and other). Training for authorized employees must cover the following topics at a minimum: (a) details about the types and magnitude of hazardous energy

sources present in the workplace; and (b) methods, means, and procedures for isolating and controlling these sources. Training for affected employees (usually machine operators or users) and other employees must cover the following topics at a minimum: (a) how to recognize when the energy control procedure is implemented; and (b) the importance of never attempting to start up a locked out or tagged out machine. All training must ensure that employees understand the purpose, function, and restrictions of the energy control program, and that authorized employees have the knowledge and skills necessary to properly apply, use, and remove energy controls.

- *Periodic inspections.* Inspections must be performed at least annually to ensure that the energy control program is up-to-date and being properly implemented. In addition, the organization must certify that the periodic inspections have actually taken place.
- *Application of controls and lockout/tagout devices.* Controls and lockout/tagout devices must be applied properly. The appropriate procedure consists of the following steps that must be implemented in sequence: (a) prepare for shutdown; (b) shut down the machine or equipment in question; (c) affix the lockout or tagout device; (d) render safe any residual or stored energy that might remain in the machine; and (e) verify that the energy source has been effectively isolated and that the machine or equipment has been effectively deenergized.
- *Removal of locks or tags.* Before locks or tags are removed, the following procedures must be completed: (a) inspect the work area to ensure that nonessential items have been removed and that the machine is capable of operating properly; (b) check the area around the machine to ensure that all employees are safely back or removed from the area; (c) notify affected employees immediately after removing the energy control devices and before energizing the machine; and (d) ensure that energy control devices are removed by the individual who affixed them (if this is not possible, make sure that the person who does remove them follows the proper procedures).

Safety Fact

Machines That Are Not Properly Locked Out Can Kill

Failure to properly lock out machines can be fatal. It was in the case of a 17-year-old employee who was killed when the waste handling machine he was working on was inadvertently energized. His employer, a sheet-metal manufacturer, was cited by OSHA for failure to implement a lockout/tagout program, provide appropriate personal protective equipment, and train employees on the hazards of releasing stored energy. The company was fined $175,000, but the employee paid a much higher price.

Source: "Worker Killed by Machine That Wasn't Locked Out," *Facility Manager's Alert* 8, no. 184: 3.

- *Additional safety requirements.* OSHA allows for special circumstances as set forth in this subsection. When a machine must be energized in order to test or position it, energy control devices may be removed only as follows: (a) clear the machine of tools and materials; (b) remove employees from the area; (c) remove the devices as set forth in the standard; (d) energize the machine and conduct the test or proceed with the positioning procedure; and (e) deenergize the machine, isolate the energy source, and reapply the energy control devices. Organizations must ensure that outside personnel such as contractors are fully informed about energy control procedures. In group lockout or tagout situations, each individual employee performing maintenance or service tasks must be protected by his or her own personal energy control device. Organizations must have specific procedures for ensuring continuity in spite of personnel and shift changes.

Evaluating Lockout/Tagout Programs

Lockout/tagout violations are frequently cited by OSHA during on-site inspections. The following questions developed by Linda F. Johnson can be used to evaluate an organization's lockout/tagout program:

- Are all machinery or equipment capable of movement required to be deenergized or disengaged and blocked or locked out during cleaning, servicing, adjusting, or setting up operations?
- Where the power disconnect equipment does not disconnect the electrical control circuit, are the appropriate electrical enclosures identified?
- If the power disconnect for equipment does not disconnect the electrical control circuit, is a means provided to ensure that the control circuit can be disconnected and locked out?
- Is it required to lock out main power disconnects instead of locking out control circuits?
- Are all equipment control valve handles equipped with a means for locking out?
- Does the lockout procedure require that stored energy—whether it is mechanical, hydraulic, or air—be released or blocked before the equipment is locked out for repairs?
- Are appropriate employees provided with individually keyed personal safety locks?
- Are these employees required to keep personal control of their keys while they have safety locks in use?
- Is only the employee exposed to the hazard required to install or remove the safety lock?
- Are employees required to check the safety lockout by attempting a startup after making sure no one is exposed?
- After the safety is checked, does the employee again place the switch in the "off" position?
- Are employees instructed always to push the control circuit stop button before reenergizing the main power switch?

Discussion Case

What Is Your Opinion?

John Martin, director of manufacturing, is not happy with his colleague Pete Chang who is director of safety and health at Robbins Engineering Corporation. Chang has ordered two machines in the Manufacturing Department tagged out until point-of-operation guards are replaced. The machining supervisor, in an attempt to increase output, had his machinists remove the guards. As a result, there have been several minor injuries and a couple of more serious near misses. The issue is short-term productivity versus the safety and health of employees. What is your opinion on this issue?

- Are all employees who are working on locked-out equipment identified by their locks or accompanying tags?
- Are enough accident prevention signs, tags, and safety padlocks provided for any reasonably foreseeable repair emergency?
- When machine operations, configuration, or size require the operator to leave his or her control station to install tools or perform other operations, is he or she required to lock or tag out separately any parts of the machine that could move if accidentally activated?
- If the equipment or lines cannot be shut down, locked out, and tagged, is a safe job procedure established and rigidly followed?
- Have employees been trained not to start machinery or equipment if it is locked out or tagged out?
- Are all workers notified when the machinery or equipment they usually use is shut down and locked out for maintenance or servicing purposes?
- After maintenance is completed, is the machinery checked to ensure that nonessential items have been removed and the machine is operationally intact?
- Before the machinery is activated, are employees removed from possible danger?
- When the machinery is fully operational, are employees notified?[10]

GENERAL PRECAUTIONS

The types of safeguards explained in this chapter are critical. In addition to these specific safeguards, there are also a number of general precautions that apply across the board in settings where machines are used. Some of the more important general precautions are as follows:

- All operators should be trained in the safe operation and maintenance of their machines.

- All machine operators should be trained in the emergency procedures to take when accidents occur.
- All employees should know how to activate emergency shutdown controls. This means knowing where the controls are and how to activate them.
- Inspection, maintenance, adjustment, repair, and calibration of safeguards should be carried out regularly.
- Supervisors should ensure that safeguards are properly in place when machines are in use. Employees who disable or remove safeguards should be disciplined appropriately.
- Operator teams (two or more operators) of the same system should be trained in coordination techniques and proper use of devices that prevent premature activation by a team member.
- Operators should be trained and supervised to ensure that they dress properly for the job. Long hair, loose clothing, neckties, rings, watches, necklaces, chains, and earrings can become caught in equipment and, in turn, pull the employee into the hazard zone.
- Shortcuts that violate safety principles and practices should be avoided. The pressures of deadlines should never be the cause of unsafe work practices.
- Other employees who work around machines but do not operate them should be made aware of the emergency procedures to take when an accident occurs.

BASIC PROGRAM CONTENT

Machine safeguarding should be organized, systematic, and comprehensive. A company's safeguarding program should have at least the following elements:

- Safeguarding policy that is part of a broader companywide safety and health policy
- Machine hazard analysis
- Lockout/tagout (materials and procedures)
- Employee training
- Comprehensive documentation
- Periodic safeguarding audits (at least annually)

TAKING CORRECTIVE ACTION

What should be done when a mechanical hazard is observed? The only acceptable answer to this question is, *take immediate corrective action*. The specific action indicated will depend on what the problem is. Figure 14–16 shows selected examples of problems and corresponding corrective actions.

Problem	Action
Machine is operating without the safety guard.	Stop machine immediately and activate the safety guard.
Maintenance worker is cleaning a machine that is operating.	Stop machine immediately and lock or tag it out.
Visitor to the shop is wearing a necktie as he observes a lathe in operation.	Immediately pull the visitor back and have him remove the tie.
An operator is observed disabling a guard.	Stop the operator, secure the guard, and take disciplinary action.
A robot is operating without a protective barrier.	Stop the robot and erect a barrier immediately.
A machine guard has a sharp, ragged edge.	Stop the machine and eliminate the sharp edge and ragged burrs by rounding it off.

Figure 14–16
Selected examples of problems and corresponding actions.

These are only a few of the many different types of problems that require corresponding corrective action. Regardless of the type of problem, the key to responding is immediacy. As shown in the examples given earlier in this chapter, waiting to take corrective action can be fatal. It is important to note that it is often prudent to exceed some OSHA guarding requirements (i.e., the 7-foot rule).

SUMMARY

1. The most common mechanical injuries are cutting and tearing, shearing, crushing, breaking, straining, spraining, and puncturing.
2. Safeguarding involves devices or methods that minimize the risk of accidents resulting from machine–operator contact. The contact may result from an individual making contact with a machine, flying metal chips, chemical or hot metal splashes, stock kickbacks, or mechanical malfunction.
3. The OSHA standard containing the general requirements for machine guarding is 29 CFR 1910.212. This standard contains requirements in the following areas: types of guarding, general requirements, guarding the point of operation, machines requiring point-of-operation guards, exposure of fan blades, and anchoring fixed machinery.
4. All safeguards should have the following characteristics: prevent contact, be secure and durable, protect against falling objects, create no new hazards, create no interference, and allow safe maintenance.
5. Point-of-operation devices come in a variety of different types including the following: photoelectric, radio frequency, electromechanical, pullback, restraint, safety trip, two-hand controls, trips, and gates.
6. Feeding and ejection systems can be effective safeguards if properly designed and used. These systems come in two types: automatic and semiautomatic.

7. The main mechanical hazards associated with robots are as follows: (a) entrapment of a worker between a robot and a solid surface, (b) impact with a moving robot arm, and (c) impact with objects ejected or dropped by the robot. The best safeguard for a robot is a barrier around the perimeter of its work envelope. Sensitized doors or gates in the barrier can also decrease the hazard potential.

8. OSHA's standard for the control of hazardous energy (29 CFR 1910.147) is often referred to as the lockout/tagout standard. The overall purpose of the standard is to prevent injuries from the accidental or inadvertent energizing of machines or equipment while they are shut down for maintenance or servicing. Key concepts in this standard are as follows: affected employee, authorized employee, capable of being locked out, energy-isolating device, energy control procedure, energy control program, lockout, lockout device, tagout, and tagout device. Key provisions in the standard include energy control program, energy control procedure, energy-isolating devices, requirements for lockout/tagout, employee training, periodic inspections, application of controls and lockout/tagout devices, and removal of locks or tags.

9. When hazards or hazardous behaviors are observed, corrective action should be taken immediately. Waiting to act can be fatal.

KEY TERMS AND CONCEPTS

Adjustable guards	Lockout/tagout systems
Affected employee	Mechanical hazards
Authorized employee	Mechanical injuries
Automatic ejection	Photoelectric devices
Automatic feed	Point-of-operation devices
Breaking	Point-of-operation guards
Crushing	Pullback devices
Cutting	Puncturing
Electromechanical devices	Radio-frequency devices
Energized	Restraint devices
Energy control procedure	Risk assessment
Energy control program	Safeguarding
Energy source	Safety trip devices
Energy-isolating device	Semiautomatic ejection
Fixed guards	Semiautomatic feed
Gates	Shearing
Interlocked guards	Spraining
Lockout device	Straining

Tagout   Two-hand controls
Tagout device   Work envelope
Tearing

REVIEW QUESTIONS

1. List and briefly explain the common types of mechanical injury hazards.
2. Explain the concept of safeguarding.
3. Summarize OSHA's requirements for machine guarding.
4. What are the requirements all safeguards should meet?
5. Describe the three types of point-of-operation guards.
6. Describe four types of point-of-operation devices.
7. What are the relative advantages and disadvantages of feeding and ejection systems?
8. Describe the primary hazards associated with robots.
9. Explain how to guard against the hazards associated with robots.
10. What is a lockout system?
11. What is a tagout system?
12. What impact may a lockout/tagout system have if carefully followed nationwide?
13. Summarize the main provisions of OSHA's lockout/tagout standard.
14. Explain the concept of risk assessment as it relates to machine operation.
15. Explain how to evaluate lockout/tagout programs.

ENDNOTES

1. National Safety Council, *Guards: Safeguarding Concepts Illustrated*, 7th ed. (Chicago: National Safety Council, 2002), 1.
2. Retrieved from http://www.osha.gov/pls/oshaweb.
3. EN 954, Part I, "Safety of Machinery—Principles of Safety Related to Control Systems," *European Union*, 1997.
4. National Safety Council, *Guards: Safeguarding Concepts Illustrated*, 2–3.
5. Ibid., 36.
6. Ibid., 38–39.
7. Retrieved from http://online.misu.nodak.edu/19577/BADM309checklist.htm.
8. Ibid., 44.
9. Retrieved from http://www.osha.gov/SLTC/smallbusiness/sec11.html.
10. L. Johnson, "The 'Red Flags' of LOTC," *Occupational Health & Safety* 68, no. 3:55.

FALLING, IMPACT, ACCELERATION, LIFTING, AND VISION HAZARDS

Major Topics

- Causes of Falls
- Kinds of Falls
- Walking and Slipping
- Slip and Fall Prevention Programs
- OSHA Fall Protection Standards
- Ladder Safety
- Impact and Acceleration Hazards (head, eyes, face, and feet protection)
- Lifting Hazards
- Standing Hazards
- Hand Protection
- Personal Protective Equipment
- Forklift Safety (Powered Industrial Trucks)

Some of the most common accidents in the workplace happen as the result of slipping, falling, and improper lifting. Impact from a falling object is also a common cause of accidents. This chapter provides the information needed by modern safety and health professionals to prevent such accidents. It also provides specific information about head, hand, back, eye, face, and foot protection as well as forklift safety.

CAUSES OF FALLS

More than 16 percent of all disabling work-related injuries are the result of falls. Clearly, falls are a major concern of safety and health professionals. The primary causes of falls are as follows:

- A foreign object on the walking surface
- A design flaw in the walking surface

- Slippery surfaces
- An individual's impaired physical condition[1]

A **foreign object** is any object that is out of place or in a position to trip someone or to cause a slip. There is an almost limitless number of **design flaws** that may cause a fall. A poorly designed floor covering, a ladder that does not seat properly, or a catwalk that gives way are all examples of design flaws that may cause falls. Slippery surfaces are particularly prevalent in industrial plants where numerous different lubricants and cleaning solvents are used.

Automobile accidents are often caused when a driver's attention is temporarily drawn away from the road by a visual distraction. This is also true in the workplace. Anything that distracts workers visually can cause a fall. When a person's physical condition is impaired for any reason, the potential for falls increases. This is a particularly common problem among aging workers. Understanding these causes is the first step in developing fall prevention techniques.

KINDS OF FALLS

Falling from ladders and other elevated situations is covered later in this chapter. This section deals with the more common surface falls. Such falls can be divided into the following four categories:

- **Trip and fall** accidents occur when workers encounter an unseen foreign object in their path. When the employee's foot strikes the object, he or she trips and falls.
- **Stump and fall** accidents occur when a worker's foot suddenly meets a sticky surface or a defect in the walking surface. Expecting to continue at the established pace, the worker falls when his or her foot is unable to respond properly.
- **Step and fall** accidents occur when a person's foot encounters an unexpected step down (e.g., a hole in the floor or a floorboard that gives way). This can also happen when an employee thinks he or she has reached the bottom of the stairs when, in reality, there is one more step.
- **Slip and fall** accidents occur when the worker's center of gravity is suddenly thrown out of balance (e.g., an oily spot causes a foot to shoot out from under the worker). This is the most common type of fall.

WALKING AND SLIPPING

Judging by the number of injuries that occur each year as the result of slipping, it is clear that walking can be hazardous to a worker's health. This is, in fact, the case when walking on an unstable platform. A stable platform for walking is any surface with a high degree of traction that is free of obstructions. It follows that an unstable platform is one lacking traction, one on which there are obstructions, or both.

Measuring Surface Traction

In order to understand *surface traction*, you must have a basis for comparison. An effective way for comparing the relative traction of a given surface is to use the **coefficient of friction**, which is a numerical comparison of the resistance of one surface (shoe or boot) against another surface (the floor).

Figure 15–1 is a continuum showing coefficients of friction ratings from very slippery to good traction. Surfaces with a coefficient of friction of 0.2 or less are very slippery and very hazardous. At the other end of the continuum, surfaces with a coefficient of friction of 0.4 or higher have good traction.

To gain a feel for what different coefficients actually mean, consider the following: (1) ice has a coefficient of friction of 0.10; (2) concrete has a coefficient of 0.43; (3) linoleum has a coefficient of 0.33; and (4) waxed white oak has a coefficient of 0.24. Compare these coefficients with Figure 15–1 to determine the degree of hazard and how the surfaces compare.

Factors That Decrease Traction

Good housekeeping can be a major factor in reducing slip and fall hazards. Water, oil, soap, coolant, and cleaning solvents left on a floor can decrease traction and turn an otherwise safe surface into a danger zone. For example, the friction coefficient of concrete (0.43) is reduced by almost 15 percent if the concrete is wet. Rubber-soled shoes can decrease slipping hazards somewhat, but changing the type of shoe is not enough to ensure safety. Additional precautions are needed.

General Strategies for Preventing Slips

Modern safety and health professionals are concerned with preventing slips and falls. Slip prevention should be a part of the company's larger safety and health program. Here are some strategies that can be used to help prevent slipping:

Figure 15–1
Coefficients of friction and relative traction ratings.

1. *Choose the right material from the outset.* Where the walking surface is to be newly constructed or an existing surface is to be replaced, safety and health professionals should encourage the selection of surface materials that have the highest possible coefficient of friction. Getting it right from the start is the best way to prevent slipping accidents.

2. *Retrofit an existing surface.* If it is too disruptive or too expensive to replace a slippery surface completely, **retrofit** it with friction enhancement devices or materials. Such devices or materials include runners, skid strips, carpet, grooves, abrasive coatings, grills, and textured coverings.

3. *Practice good housekeeping.* Regardless of the type of surface, keep it clean and dry. Spilled water, grease, oil, solvents, and other liquids should be removed immediately. When the surface is wet intentionally, as when cleaning or mopping, rope off the area and erect warning signs.

4. *Require nonskid footwear.* Employees who work in areas where slipping is likely to be a problem should be required to wear shoes with special nonskid soles. This is no different from requiring steel-toed boots to protect against falling objects. **Nonskid footwear** should be a normal part of a worker's personal protective equipment.

5. *Inspect surfaces frequently.* Employees who are working to meet production deadlines may be so distracted that they don't notice a wet surface, or they may notice it but feel too rushed to do anything about it. Consequently, safety and health professionals should conduct frequent inspections and act immediately when a hazard is identified.[2]

Effective strategies for preventing slips and falls include the following:

1. Review and analyze accident statistics to determine where slip and fall accidents are happening and why; then take the appropriate corrective measures.
2. Monitor the condition of walking surfaces continually and make appropriate preventive corrections immediately.
3. Make sure that ramps and sloped floors have high-friction surfaces.
4. Use safety mats, nonslip flooring, and slip-resistant safety shoes.
5. Make sure that stairs have handrails.
6. Make sure that visibility is good in potentially hazardous areas. Add extra lighting if necessary. Also make sure that the color of paint in these areas is bright and helpful in calling attention to potential hazards.
7. Make sure that spills are cleaned up immediately and that the underlying cause of the spill is corrected.
8. Make sure that employees who work in potentially hazardous areas select and wear the right slip-resistant footwear.
9. Use appropriate technologies such as **vertical incidence tribometers (VITs)** to measure the slip resistance of floors and take appropriate action based on the results.
10. Conduct periodic audits of walking surfaces throughout the facilities in question, document carefully the findings, and take appropriate action in a systematic way.[3]

Specific Strategies for Preventing Slips

In addition to the general strategies already explained there are a number of specific strategies that will be helpful for preventing slips and falls. These specific strategies are presented in this section.[4]

Building Lobbies

The lobbies to buildings often have slick, highly polished floor surfaces designed to project a certain image. Unfortunately, such floors represent a real hazard for slipping and falling. The hazard level is often increased when outside moisture from rain, sleet, or snow is brought in by pedestrian traffic. The following prevention strategies will help decrease the hazard level: (1) use large welcome mats that are wide enough to allow several "drying steps" to be taken before reaching the slick floor; (2) provide umbrella holders so that dripping umbrellas are not brought onto the slick floor; (3) monitor the floor surface continually and dry any moisture that makes its way onto the floor immediately; and (4) substitute nonslip surfaces for the slick, highly polished flooring.

Restrooms

Because of the nature of their use, restrooms are likely to have tiled flooring. Certain types of tiles become slippery when water from the sinks, toilets, or urinals splash onto the floor or overflow. The hazards in this situation are multiplied when soap is added in restrooms. The following prevention strategies will help decrease the hazard level in restrooms: (1) monitor restrooms continually and clean up spills immediately; (2) use "wet floors" warning signs; (3) block off any wet areas until they dry; and (4) conduct periodic inspections of public restrooms on a systematic basis.

Kitchens

Some business and industrial firms have commercial kitchens for providing meals to their personnel. When this is the case, the most common hazard is polymerized grease. Grease on almost any floor surface, but especially on tile, concrete, and linoleum floors, creates a serious slip and fall hazard. The following strategies will help decrease the level of hazard in commercial kitchens: (1) use a nonslip floor surface; (2) require kitchen employees to wear slip-resistant footwear; and (3) frequently dry clean the floors after-hours using a method other than wet mopping (which typically just distributes the grease).

Processing Areas

Companies that process materials typically experience high hazard levels in the processing areas. Processing often requires the use of fluids that can spill over and onto the floor. For example, companies that process meat have to contend with blood, fats, and meat juices on the floor. Companies that process chemicals must contend with spillage of those chemicals. The following strategies will help decrease the hazard level in processing areas: (1) use nonslip flooring; (2) monitor floor surfaces continually and take immediate action to clean up spills; (3) require processing employees to wear slip-resistant footwear; and (4) inspect and clean floor surfaces on a regular basis.

SLIP AND FALL PREVENTION PROGRAMS

Every year slips, trips, and falls cause more than one million workplace injuries and approximately 16,000 deaths. A company's overall safety and health program should include a slip and fall prevention component. Such a component should have the following elements:

1. *A policy statement/commitment.* Statement to convey management's commitment. Areas that should be included in the policy statement are management's intent, scope of activity, responsibility, accountability, the safety professional's role, authority, and standards.

2. *Review and acceptance of walkways.* Establish the criteria that will be used for reviewing all walking surfaces and determining if they are acceptable. For example, a criterion may be a minimum coefficient of friction value. Regardless of the criteria, the methodology that will be used for applying them to the review and acceptance of walkways should also be explained.

3. *Reconditioning and retrofitting.* Include recommendations and timetables for reconditioning or retrofitting existing walking surfaces that do not meet review and acceptance criteria.

4. *Maintenance standards and procedures.* State the maintenance standards for walking surfaces (e.g., how often surfaces should be cleaned, resurfaced, replaced, and so on). In addition, this section should contain procedures for meeting the standards.

5. *Inspections, audits, tests, and records.* Provide a comprehensive list of inspections, audits, and tests (including the types of tests) that will be done, how frequently, and where. Maintain records of the results.

6. *Employee footwear program.* Specify the type of footwear required of employees who work on different types of walking surfaces.

7. *Defense methods for legal claims.* Outline the company's legal defenses so that aggressive action can be taken immediately should a lawsuit be filed against the company. In such cases, it is important to be able to show that the company has not been negligent (e.g., the company has a slip and fall prevention program that is in effect).

8. *Measurement of results.* Contain the following two parts: (a) an explanation of how the program will be evaluated and how often (e.g., comparison of yearly, quarterly, or monthly slip and fall data); (b) records of the results of these evaluations.[5]

Safety Fact

Language of Fall Protection

Fall protection has a language of its own. In order to understand OSHA regulations and other fall protection guidelines, it is necessary to know the following terms:

- *Anchorage.* A secure point of attachment for lifelines, lanyards, or deceleration devices.
- *Body belt.* A strap with means both for securing it about the waist and for attaching it to a lanyard, lifeline, or deceleration device.
- *Body harness.* Straps that may be secured about the person in a manner that distributes the fall-arrest forces over at least the thighs, pelvis, waist, chest, and shoulders, with a means for attaching the harness to other components of a personal fall arrest system.
- *Connector.* A device used to couple (connect) parts of a personal fall arrest system or positioning device system together.
- *Hole.* A void or gap 2 inches (5.1 centimeters) or more in the least dimension in a floor, roof, or other walking/working surface.
- *Lanyard.* A flexible line of rope, wire rope, or strap that generally has a connector at each end for connecting the body belt or body harness to a deceleration device, lifeline, or anchorage.
- *Lifeline.* A component consisting of a flexible line for connection to an anchorage at one end to hang vertically (vertical lifeline) or for connection to anchorages at both ends to stretch horizontally (horizontal lifeline). Serves as a means for connecting other components of a personal fall arrest system to the anchorage.
- *Low-slope roof.* A roof having a slope less than or equal to 4 in 12 (vertical to horizontal).
- *Opening.* A gap or void 30 inches (76 centimeters) or more high and 18 inches (46 centimeters) or more wide, in a wall or partition, through which employees can fall to a lower level.
- *Personal fall arrest system.* A system including but not limited to an anchorage, connectors, and a body belt or body harness used to arrest an employee in a fall from a working level. The use of a body belt for fall arrest is prohibited.
- *Platform.* A working space for persons, elevated above the surrounding floor or ground.
- *Positioning device system.* A body belt or body harness system rigged to allow an employee to be supported on an elevated vertical surface, such as a wall, and work with both hands free while leaning backward.
- *Rope grab.* A deceleration device that travels on a lifeline and automatically, by friction, engages the lifeline and locks to arrest a fall.
- *Self-retracting lifeline/lanyard.* A deceleration device containing a drum-wound line that can be slowly extracted from, or retracted onto, the drum under minimal tension during normal employee movement and which, after onset of a fall, automatically locks the drum and arrests the fall.

OSHA FALL PROTECTION STANDARDS

The OSH Act mentions fall protection in several places. Although the General Industry Standards are silent on fall protection, the problem is covered in the following subparts:

Subpart D   Walking/working surfaces

Subpart F   Powered platforms, manlifts, and vehicle-mounted work platforms

Subpart R   Special industries

In addition to these OSHA standards, the American National Standards Institute (ANSI) publishes a Fall Protection Standard (ANSI Z359.1: *Safety Requirements for Personal Fall Arrest Systems, Subsystems, and Components*). The most comprehensive and most controversial fall protection standard is OSHA's Fall Protection Standard for the construction industry (Subpart M of 29 CFR 1926).

OSHA's Fall Protection Standard for Construction

OSHA's current Fall Protection Standard sets the *trigger height* at 6 feet. This means that any construction employee working higher than 6 feet off the ground must use a fall protection device such as a safety harness and line (see Figure 15–2).

This trigger height means that virtually every small residential builder and roofing contractor is subject to the standard. Because most residential builders and roofing contractors are small, Subpart M of 29 CFR 1926 is a source of much controversy.

OSHA officials argue that the 6-foot trigger height saves up to 80 lives per year and prevents more than 56,000 injuries. The rationale is that 6 percent of all lost-time fall injuries in the construction industry are caused by falls from less than 10 feet. Opponents counter that the cost of complying with the standard is almost $300 million annually. Commercial contractors, whose employees typically work much higher than the 6- to 16-foot range, are not concerned about the height controversy.

Items from OSHA Regulation 1926 that apply specifically to fall protection in scaffolding work are as follows:

- 1926.451(g)(2) reads: "The employer shall have a competent person determine the feasibility and safety of providing fall protection for employees erecting or dismantling supported scaffolds. Employers are required to provide fall protection for employees erecting or dismantling supported scaffolds where the installation and use of such protection is feasible and does not create a greater hazard."
- 1926.502(d)(15) reads: "Anchorages used for attachment of personal fall arrest equipment shall be independent of any anchorage being used to

Figure 15–2
Personal fall arrest harness.
Courtesy of Dalloz Safety (Miller Brand Products).

support or suspend platforms and capable of supporting at least 5,000 pounds (22.2 kilograms) per employee attached, or shall be designed, installed and used as follows:

   (i) as part of a complete personal fall arrest system which maintains a safety factor of at least two; and

  (ii) under the supervision of a qualified person."

- 1926.451(d)(16) "Scaffold."
    - (i) limit maximum arresting force on an employee to 1500 pounds (4 kg) when used with a body belt;
    - (ii) limit maximum arresting force on an employee to 1,800 pounds (8 kg) when used with a body harness;
    - (iii) be rigged such that an employee can neither free-fall more than 6 feet (1.8 m), nor contact any lower level;
    - (iv) bring an employee to a complete stop and limit maximum deceleration distance an employee travels to 3.5 feet (1.07 m); and
    - (v) have sufficient strength to withstand twice the potential impact energy of an employee free-falling a distance of 6 feet (1.8 m), or the free fall distance permitted by the system, whichever is less.[6]

OSHA's Recommendations for Effective Fall Protection

Because slip and fall accidents account for approximately one million workplace injuries every year, organizations obviously need to have a strong fall prevention program in place. But what does it take to have an effective fall prevention program? OSHA recommends the following strategies:

- *Have a plan.* An organization should develop a written plan that is part of its larger safety and health plan. The fall protection plan should contain a statement of commitment from both management and employees, rules and regulations relating to fall protection, and an explanation of the training program and training requirements.
- *Establish proper fall protection requirements.* Require the use of fall protection equipment any time an employee works more than 4 feet above the floor in general industry, 6 feet or more in construction, and 10 feet or more when on scaffolding.
- *Provide proper fall protection equipment and procedures and require their use.* Organizations should determine what types of fall protection equipment and procedures are needed, provide them to employees, and require their proper use. As examples, this might include personal fall arrest systems, guardrails, safety nets, positioning devices, warning lines, controlled access zones, and safety monitoring.
- *Ensure fall protection device replacement.* Organizations should replace fall protection devices on a regular schedule even if there are no significant signs of wear.
- *Ensure proper use and type of equipment.* Ensure that the fall protection equipment provided to employees is the proper type for the situation in question and that employees inspect it before putting it on, that it fits properly, and that it is properly attached to anchorage points.
- *Provide training.* Provide fall protection training for supervisors and employees including how to recognize fall-related hazards and how to properly use all applicable fall protection equipment.[7]

LADDER SAFETY

Jobs that involve the use of ladders (see Figure 15–3) introduce their own set of safety problems, one of which is an increased potential for falls. The National Safety Council recommends that ladders be inspected before every use and that employees who use them follow a set of standard rules.[8]

Inspecting Ladders

Taking a few moments to look over a ladder carefully before using it can prevent a fall. The NSC recommends the following when inspecting a ladder:

- See if the ladder has the manufacturer's instruction label on it.
- Determine whether the ladder is strong enough.
- Read the label specifications about weight capacity and applications.
- Look for the following conditions: cracks on side rails; loose rungs, rails, or braces; or damaged connections between rungs and rails.
- Check for heat damage and corrosion.
- Check wooden ladders for moisture that may cause them to conduct electricity.
- Check metal ladders for burrs and sharp edges.
- Check fiberglass ladders for signs of *blooming*, deterioration of exposed fiberglass.[9]

Do's and Don'ts of Ladder Use

Many accidents involving ladders result from improper use. Following a simple set of rules for the proper use of ladders can reduce the risk of falls and other ladder-related accidents. The NSC recommends the following do's and don'ts of ladder use:

- Check for slipperiness on shoes and ladder rungs.
- Secure the ladder firmly at the top and bottom.
- Set the ladder's base on a firm, level surface.
- Apply the **four-to-one ratio** (base one foot away from the wall for every 4 feet between the base and the support point).
- Face the ladder when climbing up or down.
- Barricade the base of the ladder when working near an entrance.
- Don't lean a ladder against a fragile, slippery, or unstable surface.
- Don't lean too far to either side while working (stop and move the ladder).
- Don't rig a makeshift ladder; use the real thing.
- Don't allow more than one person at a time on a ladder.
- Don't allow your waist to go any higher than the last rung when reaching upward on a ladder.
- Don't separate the individual sections of extension ladders and use them individually.

Figure 15–3
Portable ladder/stair.
Courtesy of Lapeyre Stair, Inc.

- Don't carry tools in your hands while climbing a ladder.
- Don't place a ladder on a box, table, or bench to make it reach higher.[10]

OSHA standards for walking and working surfaces and ladder safety are set forth in 29 CFR Part 1910 (Subpart D). The standards contained in Subpart D are as follows:

1910.21	Definitions
1910.22	General requirements
1910.23	Guarding floor and wall openings and holes
1910.24	Fixed industrial stairs
1910.25	Portable wood ladders
1910.26	Portable metal ladders
1910.27	Fixed ladders
1910.28	Safety requirements for scaffolding
1910.29	Manually propelled mobile ladder stands and scaffolds (towers)
1910.30	Other working surfaces
1910.31	Sources of standards
1910.32	Standards organizations

WHAT TO DO AFTER A FALL

If, in spite of your best efforts, a fall occurs on the job, what employees do in the immediate aftermath can mean the difference between life and death for the victim. First, make sure your organization has a fall rescue plan in place that includes the following: (1) training for all personal in how to carry out a rescue—what to do and what not to do; (2) proper equipment on-site and readily available; (3) coordination with local emergency authorities; and (4) assigned responsibilities. Then, in addition, make sure all employees who work at heights or with others who work at heights understand the following basics:[11]

- *Never work alone.* These should always be two or more people working in close proximity when working at heights. It is important to have someone available to intiate the fall rescue plan.
- *Keep legs moving.* When a worker is dangling from his fall arrest gear, it is important for him to keep his legs moving—not frantically, but just rhythmically and regularly. This will help prevent the venous pooling of blood that can lead to shock. If it is possible, the suspended worker should try to move into an upright position.
- *Raise the worker to a seated position.* Once the suspended worker has been brought to the ground, the tendency is to lie him down in a horizontal position. This is a mistake because it can suddenly release pooled blood that can strain the heart and cause death. Instead, move the victim into a seated position.

MONITOR FALL PROTECTION EQUIPMENT AND KNOW WHY IT FAILS

Although more people than ever are using fall protection equipment on the job, the number of injuries and deaths from falls continues to increase. There are several reasons for this, including poor training, deterioration of equipment over time, and selection of the wrong equipment for the job. However, the reason that should concern safety and health professionals most is failure due to lack of monitoring. In many cases, equipment failures could have been prevented by a systematic monitoring process in which all equipment is inspected before being used. Consequently, it is important for safety and health professionals to understand why fall protection equipment fails, and to ensure that equipment is properly monitored to detect potential failure points before it is used.[12]

When inspecting fall protection equipment, look for the following types of potential problems: weld splatter; webbing cuts and abrasions; broken stitching; frayed or burned webbing; chemical damage; discoloration; deformed hardware; loose, distorted, or broken grommets; knotted webbing; and malfunctioning snap hooks. Just making sure that employees know how to monitor their equipment and that they follow through and do it—every time—will save lives.

IMPACT AND ACCELERATION HAZARDS

An employee working on a catwalk drops a wrench. The falling wrench accelerates over the 20-foot drop and strikes an employee below. Had the victim not been wearing a hard hat he might have sustained serious injuries from the impact. A robot loses its grip on a part, slinging it across the plant and striking an employee. The impact from the part breaks one of the employee's ribs. These are examples of accidents involving **acceleration** and **impact**. So is any type of fall because, having fallen, a person's rate of fall accelerates (increases) until striking a surface (impact). Motor vehicle accidents are also acceleration and impact instances.

Because falls were covered in the previous section, this section will focus on hazards relating to the acceleration and impact of objects. Approximately 25 percent of the workplace accidents that occur each year as the result of acceleration and impact involve objects that become projectiles.

Protection from Falling or Accelerating Objects

Objects that fall, are slung from a machine, or otherwise become projectiles pose a serious hazard to the heads, faces, feet, and eyes of workers. Consequently, protecting workers from projectiles requires the use of appropriate personal protective equipment and strict adherence to safety rules by all employees.

Head Protection

Approximately 120,000 people sustain head injuries on the job each year.[13] Falling objects are involved in many of these accidents. These injuries occur in spite of the fact that many of the victims were wearing hard hats. Such statistics have been the driving force behind the development of tougher, more durable hard hats.

Originally introduced in 1919, the hard hats first used for head protection in an industrial setting were inspired by the helmets worn by soldiers in World War I. Such early versions were made of varnished resin-impregnated canvas. As material technology evolved, hard hats were made of vulcanized fiber, then aluminum, and then fiberglass. Today's hard hats are typically made from the thermoplastic material polyethylene, using the injection-molding process.[14] Basic hard hat design has not changed radically since before World War II. They are designed to provide limited protection from impact primarily to the top of the head, and thereby reduce the amount of impact transmitted to the head, neck, and spine.[15]

It is important not only to wear hard hats, but to wear them properly (e.g. never wear them backwards to make a fashion statement). Some companies adhere double-stick tape or flat magnets to the upper visor area to minimize the amount of dust or iron filings that fall into workers' eyes.

The American National Standards Institute (ANSI) standard for hard hats is Z89–1986. OSHA subsequently adopted this standard as its hard hat standard (29 CFR 1010.135).

> This standard calls for testing hard hats for impact attenuation and penetration resistance as well as electrical insulation. Specifically, hard hats are tested to withstand a 40-foot-pound impact, which is equivalent to a two-pound hammer falling about 20 feet. Hard hats are also designed to limit penetration of sharp objects that may hit the top of the hard-hat shell and to provide some lateral penetration protection.[16]

Safety Fact

Employee Killed by Falling Object during Windstorm

Safety and health professionals should take the weather into account when considering potential hazards in the workplace. Weather conditions can increase the probability of accidents and injuries. This is what happened on a job site in Lynbrook, New York. During a windstorm, an object was blown off a high surface. As it fell, the object's rate of speed accelerated, making the object lethal. It struck an employee working below in the head, killing him. The employee was not wearing any type of head protection at the time of the accident. His employer was cited by OSHA for failure to provide employees with personal protective equipment and fined $10,000.

Source: "Worker Killed by Falling Object during Windstorm," *Facility Manager's Alert* 8, no. 183: 3.

Safety Fact

Lateral Protection and Hard Hats

Conventional hard hats are designed to deflect a downward vertical blow. However, every year, there are head injuries on the job from lateral blows. Consequently, the American National Standards Institute (ANSI) has added a new type of hard hat to its protective headgear standard (ANSI Z89.1). Under this standard, conventional hard hats are classified as Type I. Hard hats that provide lateral protection are classified as Type II. Expect lateral protection to become the workplace standard eventually because 70 percent of the employees injured each year while wearing hard hats received blows to the unprotected areas of the head.

Hard hats can help reduce the risk associated with falling or projected objects, but only if they are worn. The use of hard hats in industrial settings in which falling objects are likely has been mandated by federal law since 1971.[17] In addition to making the use of hard hats mandatory when appropriate and supervising to ensure compliance, Feuerstein recommends the use of incentives.[18] According to Feuerstein,

> It would seem that the sweetest offer a head-injury prevention program makes is a work environment free of injuries from falling objects. But sometimes this ultimate reward is too abstract to excite employees. They need to be led into safety for its own sake by concrete incentives, such as intra-department competition, monetary rewards for good suggestions, points toward prizes, and peer recognition for the most improved behavior.[19]

Resources expended promoting the use of hard hats are resources wisely invested. "Work accidents resulting in head injuries cost employers and workers an estimated $2.5 billion per year in workers' compensation insurance, medical expenses and accident investigation as well as associated costs due to lost time on the job and substitute workers. That is an average cost of $22,500 for each worker who received a head injury."[20]

Eye and Face Protection

Eye and face protection are critical in the modern workplace. Eye injuries are a common and costly phenomenon.

Every day, an estimated 1,000 eye injuries occur in American workplaces, according to the Bureau of Labor Statistics (BLS). The result? More than $300 million a year in lost production time, medical expenses, and workers' compensation. Why is this continuing to happen?

First, too many people are not wearing eye protection. BLS found three out of every five workers with eye injuries were not wearing eye protection. Second, they were not wearing the right kind of eye protection. Flying particles, according to the bureau, cause most eye injuries. Almost 70 percent of injuries resulted from flying or falling objects or sparks striking the eye. About 20 percent of the

Discussion Case

What Is Your Opinion?

Two safety and health professionals are debating an issue over lunch. "Hard hats don't really do any good because employees won't wear them. It's like dieting. The best diet in the world won't work if you don't follow it." "I don't agree," said the second safety professional. "Hard hats have prevented thousands of head injuries. If employees don't wear them, it's because managers don't enforce their own rules." Join this debate. What is your opinion?

injuries were caused by contact with chemicals. Nearly half of the accidents occurred in manufacturing; just over 20 percent were in construction.[21]

Eye and face protection typically consist of safety glasses, safety goggles, or face shields. The ANSI standard for face and eye protective devices is Z87.1–2003. OSHA has also adopted this standard. It requires that nonprescription eye and face protective devices pass two impact tests: a high-mass, low-speed test and a low-mass, high-speed test. Figure 15–4 summarizes the

High-Mass Impact Test—Purpose

This test is intended to ensure the level of mechanical integrity of a protective device and a level of protection from relatively heavy, pointed objects traveling at low speeds. Frames shall be capable of resisting impact from a 500-gram (17.6-ounce) missile with a 30-degree conical heat-treated tip and a 1-mm (.039-inch) radius dropped from a height of 130 cm (51.2 inches). No parts or fragments shall be ejected from the protector that could contact an eye.

High-Velocity Impact Test—Purpose

This test is intended to ensure a level of protection from high-velocity, low-mass projectiles. Frames shall be capable of resisting impact from 6.35-mm (¼-inch) steel balls weighing 1.06 grams (.04 ounce) at 150 feet per second (fps) from 0 degrees to 90 degrees for frames; 250 fps for goggles, 300 fps for face shields.

Impact Test—Drop Ball

A 25.4-mm (1-inch) steel ball, weighing 68 grams (24 ounces), free fall from 127 cm (50 inches).

Lens Thickness

Thickness is 3.0 mm (.118 inch) except lenses that withstand high velocity impact, then 2.0-mm (.079-inch) thickness is acceptable.

Impact Test—Penetration

Lens shall be capable of resisting penetration from a Singer needle on a holder weighing 1.56 ounces dropped freely from 50 inches.

Figure 15–4
ANSI Standard Z87.1–2003.

purpose of the tests and their individual requirements concerning impact and penetration. Figures 15–5, 15–6, and 15–7 show examples of the types of devices available for eye and face protection.

The high-mass impact test determines the level of protection provided by face and eye protective devices from relatively heavy, pointed objects that are moving at low speeds. The high-velocity impact test determines the level of protection provided from low-mass objects moving at high velocity.

Figure 15–5
Safety glasses that wrap around for lateral protection.
Courtesy of ELVEX Corporation, Bethel, CT.

Figure 15–6
Eye safety shield combined with ear and head protection.
Courtesy of ELVEX Corporation, Bethel, CT.

Figure 15–7
Face shield (Huntsman® Model
K Facesaver® with 8154L
window attached).
Courtesy of Kedman Company,
Huntsman Products Division.

Assessing the Workplace for Eye Hazards

The type of eye protection needed in a given setting depends on the type of work done in that setting and the corresponding hazards. Before establishing a vision protection program, it is necessary to assess the workplace. OSHA recommends using the following questions in making a workplace assessment:

- Do employees perform tasks that may produce airborne dust or flying particles?
- Do employees work near others who perform tasks that may produce airborne dust or flying particles?
- Do employees handle hazardous liquid chemicals or blood?
- Do employees work near others who handle hazardous liquid chemicals or blood?
- Do employees work in conditions in which their lens may become fogged?
- Do employees work in situations that may expose their eyes to chemical or physical irritants?
- Do employees work in situations that may expose their eyes to intense light or lasers?[22]

Based on the answers to these questions, a vision protection program can be developed to protect employees. That program should meet certain requirements as recommended by OSHA. These requirements are summarized in the next section.

Safety Fact

Selecting the Right Face Protection Device

Hazard-Protection Devices

- Flying particles, chips, sand, and so on—face shield
- Hot sparks—face shield
- Heat—reflective face shield
- Molten metal—face shield
- Chemical splash—face shield
- Ultraviolet light and infrared heat—welding helmet or welding shield with shaded lens

---

Requirements When Choosing Vision Protection Devices

There are many different types of eye protection devices available that vary in terms of function, style, fit, lens, and other options. OSHA recommends applying the following criteria when selecting vision protection devices:

- Select only those that meet the standards set forth in ANSI Z87.1–2003.
- Select devices that protect against the specific hazard(s) identified in that assessment.
- Select devices that are as comfortable as possible to wear.
- Select devices that do not restrict vision in any way.
- Select devices the have fogging prevention capabilities built in.
- Select devices that are durable, easy to clean, and easy to disinfect.
- Select devices that do not interfere with the functioning of other personal protective equipment.[23]

Training

Once the workplace has been assessed and eye protection devices have been selected, it is important to provide employees with training in the proper use of the devices. This accomplishes the following: First, it ensures that the eye protection devices are used properly. Second, it shows employees that they have a critical role to play in the protection of their eyes. OSHA recommends training that covers the following topics:

- Why it is important to use the eye protection devices
- How the devices protect the eyes
- Limitations of the devices
- When the devices should be used
- How the devices are properly worn
- How straps are adjusted for both effectiveness and comfort
- How the employee can identify signs of wear that may lessen the effectiveness of the devices
- How the devices are cleaned and disinfected and how often[24]

First Aid for Eye Injuries

Even with proper eye protection, there is still the risk that an employee may sustain an injury. Even the best vision protection program is not perfect. When this happens, the following guidelines for first aid apply:

- Be gentle with the employee. Don't add to the injury with rough treatment.
- Do not attempt to remove objects embedded in the eyeball.
- Rinse the eyes with a copious amount of water for 15 to 30 minutes to remove the chemicals. Call for professional help. Cover both eyes after the rinsing has been completed.
- Never press on an injured eye or put any pressure on it (as when covering the eyes).
- Do not allow the employee to rub his or her eyes.[25]

Contact Lenses in a Chemical Enviornment

For years it was commonly thought among safety and health professionals that workers should not wear contact lenses in a chemical environment. In fact, until 2003 NIOSH recommended that workers in chemical environments not wear contact lenses. However, over time much has been learned about this issue. Contact lenses may, in fact, be worn in chemical environments and NIOSH has published an "intelligence bulletin" (number 59) explaining how to safely wear contact lenses in chemical environments.[26]

There are still environments in which contacts lens should not be worn. These are environments in which certain chemicals such as the following are present:

- 1,2-dibromo-3-chloropropane (DBCP)
- 4,4'-methylene dianiline (MDA)
- Ethyl alcohol
- Ethylene oxide
- Isopropyl alcohol
- Methylene chloride

This list is neither exhaustive nor comprehensive. Rather, it contains a partial list of the types of chemicals that can make a work environment especially hazardous for contact lens wearers. Before allowing employees who wear contact lenses to work in a chemical environment, it is best to conduct a comprehensive hazard assessment in which the types of chemicals that will be present are identified. NIOSH's Intelligence Bulletin 59: "Contact Lens Use in a Chemical Environment" is a good source of information when conducting hazard assessments of chemical environments. Once the hazard assessment has been completed, provided none of the "contact-lens-prohibited" chemicals are present in the work environment, workers who wear contact lenses should use the same eye protection recommended for other workers.

Foot Protection

The OSHA regulations for foot protection are found in 29 CFR 1910.132 and 126. Foot and toe injuries account for almost 20 percent of all disabling workplace injuries in the United States.[27] There are over 180,000 foot and toe injuries in the workplace each year.[28] The major kinds of injuries to the foot and toes are from the following:

- Falls or impact from sharp or heavy objects (this type accounts for 60 percent of all injuries)
- Compression when rolled over by or pressed between heavy objects
- Punctures through the sole of the foot
- Conductivity of electricity or heat
- Electrocution from contact with an energized, conducting material
- Slips on unstable walking surfaces
- Hot liquid or metal splashed into shoes or boots
- Temperature extremes[29]

The key to protecting workers' feet and toes is to match the protective measure with the hazard. This involves the following steps: (1) identify the various types of hazards present in the workplace, (2) identify the types of footwear available to counter the hazards, and (3) require that proper footwear be worn. Shoes selected should meet all applicable ANSI standards and have a corresponding ANSI rating. For example, "a typical ANSI rating is Z41PT83M1–75C–25. This rating means that the footwear meets the 1983 ANSI standard and the steel toe cap will withstand 75 foot pounds of impact and 2,500 pounds of compression."[30]

Modern safety boots are available that provide comprehensive foot and toe protection. The best safety boots provide all of the following types of protection:

- *Steel* toe for impact protection
- *Rubber or vinyl* for chemical protection
- *Puncture-resistant* soles for protection against sharp objects
- *Slip-resistant* soles for protection against slippery surfaces
- *Electricity-resistant material* for protection from electric shock

Employers are not required to provide footwear for employees, but they are required (29 CFR 1910.132 and 136) to provide training on foot protection. The training must cover the following topics as a minimum:

- Conditions when protective footwear should be worn
- Type of footwear needed in a given situation
- Limitations of protective footwear
- Proper use of protective footwear

OSHA Regulations Relating to Footwear

Foot protection is a high priority with OSHA. This can be seen in the number of regulations OSHA has developed and now mandates relating to footwear. The most prominent of these and a summary of the respective requirements follow:

1. *29 CFR 1910.132(d):* Requires hazard assessment in the workplace.

Safety Fact

Footwear Terms

Safety and health professionals should know their terms to be better prepared to help workers select and use the right protective footwear. The following terms are widely used in the safety and health community:

- *Steel toe:* protective cap built into the shoe or boot to protect toes from impact hazards such as falling objects and compression hazards.
- *Metatarsal guard:* protective surface built into the tongue of the shoe or boot or an external surface worn over the upper instep area to protect the metatarsal bones of the foot.
- *Puncture-resistant guard:* protective surface built into the sole of the shoe or boot to protect the bottom of the foot from sharp objects that might pierce the sole.
- *Electrical hazard guard:* protective material used to construct the sole and heel of the shoe or boot that is resistant to electricity.
- *Slip resistant guard:* special material/surface used to construct the sole and heel of the shoe or boot or applied to them to reduce slip and fall hazards.

Source: J. Goodwin, "A Cure for Common Foot Hazards," *Occupational Health & Safety* 74, no. 7: 86.

---

2. *29 CFR 1910.136:* Lists the general requirements for foot protection in the workplace. Requires that employers ensure that employees use the appropriate foot protection when working in areas where there is a danger of foot injuries due to falling or rolling objects, or objects piercing the sole, and where such employees' feet are exposed to electrical hazards.
3. *29 CFR 1910.132(f)(a,iv,v):* Specifies the training mandated for employees and fitting of footwear.[31]

In addition to these regulations, the following additional regulations also deal with specific aspects of footwear and foot protection: 29 CFR 1910.94 (ventilation), 29 CFR 1910.156 (foot, hand, eye, face, head, and body protection for employees who serve in fire brigades), 29 CFR 1910.269 (foot protection when working in electric power generation, transmission, and distribution), and 29 CFR 1910.1029 (foot protection when working with coke ovens).

LIFTING HAZARDS

Back injuries that result from improper lifting are among the most common in an industrial setting. In fact, back injuries account for approximately $12 billion in workers' compensation costs annually. The following statistics concerning workplace back injuries illustrate the scope and seriousness of this problem:

- Lower back injuries account for 20 to 25 percent of all workers' compensation claims.

- Thirty-three to 40 percent of all workers' compensation costs are related to lower back injuries.
- Each year, there are approximately 46,000 back injuries in the workplace.
- Back injuries cause 100 million lost workdays each year.
- Approximately 80 percent of the population will experience lower back pain at some point in their lives.[32]

Typical cause of back injuries in the workplace include: improper lifting, reaching, sitting, and bending. **Lifting hazards** such as poor posture, ergonomic factors, and personal lifestyles also contribute to back problems. Consequently, a company's overall safety and health program should have a back safety/lifting component.

Back Safety/Lifting Program

Prevention is critical in back safety. Consequently, safety and health professionals need to know how to establish back safety programs that overcome the hazards of lifting and other activities. Dr. Alex Kaliokin recommends the following six-step program:

1. *Display poster illustrations.* Posters that illustrate proper lifting, reaching, sitting, and bending techniques should be displayed strategically throughout the workplace. This is as important in offices as in the plant. Clerical and office personnel actually sustain a higher proportion of back injuries than employees in general. Sitting too long without standing, stretching, and walking can put as much pressure on the back as lifting.

2. *Preemployment screening.* Preemployment screening can identify people who already have back problems when they apply. This is important because more than 40 percent of back injuries occur in the first year of employment and the majority of these injuries are related to preexisting back problems.

3. *Regular safety inspections.* Periodic inspections of the workplace can identify potential problem areas so that corrective action can be taken immediately. Occasionally bringing a workers' compensation consultant in to assist with an inspection can help identify hazards that company personnel may miss.

4. *Education and training.* Education and training designed to help employees understand how to lift, bend, reach, stand, walk, and sit safely can be the most effective preventive measure undertaken. Companies that provide back safety training report a significant decrease in back injuries.

5. *Use external services.* A variety of external health care agencies can help companies extend their programs. Identify local health care providing agencies and organizations, what services they can provide, and a contact person in each. Maintaining a positive relationship with these external service contact people can increase the services available to employers.

6. *Map out the prevention program.* The first five steps should be written down and incorporated in the company's overall safety and health program. The written plan should be reviewed periodically and updated as needed.[33]

In spite of a company's best efforts, back injuries will still occur. Consequently, safety and health professionals should be familiar with the treatment and therapy that injured employees are likely to receive. Treatment for reconditioning addresses five goals: restoring function, reducing pain, minimizing deficits in strength, reducing lost time, and returning the body to preinjury fitness levels.[34]

A concept that is gaining acceptance in bridging the gap between treatment or therapy and a safe return to work is known as **work hardening**.[35] Work hardening and its objectives are explained as follows: In specially designed "work centers," various work stations, exercise equipment, and aggressive protocols are used for work reconditioning. The objectives are:

- A return to maximum physical abilities as soon as possible
- Improvement of general body fitness
- Reducing the likelihood of reinjury
- Work simulation that duplicates real work conditions[36]

The work centers referred to above replicate in as much detail as possible the injured employee's actual work environment. In addition to undergoing carefully controlled and monitored therapy in the work center, the employee is encouraged to use exercise equipment. Employees who undergo work center therapy should have already completed a program of acute physical therapy and pain management, and they should be medically stable.[37]

Health and safety managers can help facilitate the fastest possible safe resumption of duties by injured employees by identifying local health care providers that use the work-hardening approach. Such services and local providers of them should be made known to higher management so that the company can take advantage of them.

Proper Lifting Techniques

One of the most effective ways to prevent back injuries is to teach employees proper lifting techniques. Following are lifting techniques that should be taught as part of an organization's safety program.

Plan Ahead

- Determine if you can lift the load. Is it too heavy or too awkward?
- Decide if you need assistance.
- Check your route to see whether it has obstructions and slippery surfaces.

Lift with Your Legs, Not Your Back

- Bend at your knees, keeping your back straight.
- Position your feet close to the object.

- Center your body over the load.
- Lift straight up smoothly; don't jerk.
- Keep your torso straight; don't twist while lifting or after the load is lifted.
- Set the load down slowly and smoothly with a straight back and bent knees; don't let go until the object is on the floor.

Push, Don't Pull

- Pushing puts less strain on your back; don't pull objects.
- Use rollers under the object whenever possible.

NIOSH and the Ergonomic Guidelines for Manual Material Handling (EGMMH)

The National Institute for Occupational Safety and Health (NIOSH) originally developed guidelines for lifting and lowering in 1981. The guidelines include a formula for calculating the recommended weight limit (RWL) for a given lifting job. The 1981 formula was simple and easy to use because it considered only a few factors that affect a lifting task. In 1993, the NIOSH guidelines were revised. The formula now takes into account nonsymmetrical lifting and lifting of items that don't have handles. Another important aspect of the guidelines was the new **multitask analysis strategy**. This strategy gives safety professionals a method for considering a variety of related lifting variables and how they interact. This is a much more complicated method than the original, but it is also much more accurate and realistic.

The multitask analysis strategy is particularly useful when dealing with tasks in which the lifting variables change throughout the task. For example, consider the task of a stacking job in which each successive item takes a different vertical location in the stack. The ergonomics of the task change with each successive item added to the stack (e.g., reach span required, height of lift) as do the corresponding hazards.

Because of its complexity, the NIOSH *lifting equation* is now more difficult to use than the original version. Safety professionals should know the formulas to understand and identify the various risk factors that contribute to back injuries. Software that allows safety professionals simply to plug selected values into the formulas is becoming readily available and is recommended. These values may be easily collected using nothing more than a stopwatch and a tape measure.

To apply the new lifting equation, safety professionals need to understand the types of information that they must collect and either turn it over to mathematicians or plug it into a computer program. This information is as follows:

LC: Load constant (always use 51 pounds or 23 kilograms)

HM: Horizontal line measured from the midpoint between the ankles forward to the midpoint between the hands, at both the origin and destination of lift

VM: Vertical line from the floor to the hands (also measured at the origin and destination of the lift)

DM: Vertical distance between the origin and destination of the lift

AM: Turning or twisting angle of asymmetry

FM: Average frequency rate of lifting measured in lifts per minute

CM: Coupling value (Does the item to be lifted have a good, fair, or poor grasping mechanism?)

More recently, NIOSH teamed with the California Occupational Safety and Health Administration (CAL/OSHA) and the Ergonomic Assist and Systems Equipment (EASE) Council of the Material Handling Industry of America to develop the **Ergonomic Guidelines for Manual Material Handling (EGMMH).** The guidelines were finalized in 2006 and were developed primarily for smaller companies that cannot afford to hire a staff of safety and health professionals. They are divided into four parts: (1) the ergonomic process; (2) a matrix of common material handling tasks; (3) several chapters on how to approach common material handling tasks; and (4) a resource index.

The guidelines provide more than one safe way to approach the various material handling tasks covered, and they incorporate photographs and pictures to illustrate these approaches. Organizations or safety and health professionals who are interested in obtaining a copy of the EGMMH should visit the NIOSH Web site:

http://www.cde.goc/niosh

STANDING HAZARDS

Many jobs require that workers stand or walk for prolonged periods. Prolonged walking and/or standing can cause lower back pain, sore feet, varicose veins, and a variety of other related problems. The following hazard mitigation strategies can help minimize these problems for workers who must stand and walk for prolonged periods.

Antifatigue Mats

Antifatigue mats provide cushioning between the feet and hard working surfaces such as concrete floors (see Figure 15–8). This cushioning effect can reduce muscle fatigue and lower back pain. However, too much cushioning can be just as bad as too little. Consequently, it is important to test mats on a trial basis before buying a large quantity. Mats that become slippery when wet should be avoided. In areas where chemicals are used, be sure to select mats that will hold up to the degrading effects of chemicals.

Figure 15–8
Padded mat.
Courtesy of Tennessee Mat Company, Ohio.

Shoe Inserts

When antifatigue mats are not feasible because employees must move from area to area and, correspondingly, from surface to surface, shoe inserts may be the answer. Such inserts are worn inside the shoe and provide the same type of cushioning the mats provide. Shoe inserts can help reduce lower back, foot, and leg pain. It is important to ensure proper fit. If inserts make an employee's shoes too tight, they will do more harm than good. In such cases, employees may need to wear a slightly larger shoe size.

Foot Rails

Foot rails added to work stations can help relieve the hazards of prolonged standing. Foot rails allow employees to elevate one foot at a time four or five inches. The elevated foot rounds out the lower back, thereby relieving some of the pressure on the spinal column. Placement of a rail is important. It should not be placed in a position that inhibits movement or becomes a tripping hazard.

Workplace Design

A well-designed workstation can help relieve the hazards of prolonged standing. The key is to design workstations so that employees can move about while they work and can adjust the height of the workstation to match their physical needs.

Sit/Stand Chairs

Sit/stand chairs are higher-than-normal chairs that allow employees who typically stand while working to take quick mini-breaks and return to work without the hazards associated with getting out of lower chairs. They have the advantage of giving the employee's feet, legs, and back an occasional rest without introducing the hazards associated with lower chairs.

Proper Footwear

Proper footwear is critical for employees who stand for prolonged periods. Well-fitting, comfortable shoes that grip the work surface and allow free movement of the toes are best.

HAND PROTECTION

In the United States there are more than 500,000 hand injuries every year. Hand injuries are both serious and costly for employers and for employees. Section 138 of OSHA standard 29 CFR 1910.132 covers personal protective equipment for hands. This standard requires employers to base the selection of hand protection (gloves) on a comprehensive assessment of the tasks performed for a given job, hazards present, and the duration of exposure to the hazards. The assessment must be documented in writing.[38]

With the assessment completed, employers are required to review specification information from manufacturers of safety gloves and select the gloves that are best suited for the individual situation. Selecting just the right gloves for the job has historically been one of employers' greatest difficulties in complying with the standard. It is not a simple task. For example, take the issue of fit. A poorly fitted set of gloves cannot offer the degree of protection that a responsible employer or employee wants. Yet, because manufacturers have not developed a consistent set of metrics for sizing gloves, the only way to determine whether a pair fits properly is for the employee to try them on.

Fit is just one of the problems faced when selecting gloves. Other critical features include the protection capability, comfort, and tactile sensitivity of the gloves. Often, greater comfort and tactile sensitivity can mean less protection. Correspondingly, greater protection can mean less comfort and tactile sensitivity.

In an attempt to simplify the process of selecting the right gloves, the American National Standards Institute (ANSI) and the Industrial Safety Equipment Association (ISEA) developed a joint hand-protection standard, ANSI/ISEA 105–1999. This standard simplifies glove selection by (1) defining characteristics of protection in a variety of critical areas including cuts, puncture resistance, abrasion, protection from cold and heat, chemical resistance (including both permeation and degradation), viral penetration, dexterity, liquid-tightness, and flame/heat resistance; and (2) standardizing the tests used to measure all of these various characteristics.

Common Glove Materials

Depending on the individual hazards available in a given situation, the right gloves for the application may be made of a variety of different materials (see Figures 15–9, 15–10, and 15–11). The most widely used materials in manufacturing gloves are as follows:

- *Leather.* Offers comfort, excellent abrasion resistance, and minimum cut resistance.
- *Cotton.* Offers comfort, minimal abrasion resistance, and minimum cut resistance.
- *Aramids.* Offer comfort, good abrasion resistance, excellent cut resistance, and excellent heat resistance.
- *Polyethylene.* Offers comfort, excellent abrasion resistance, and minimal cut resistance. Gloves made of this material should not be subjected to high temperatures.
- *Stainless steel cord (wrapped in synthetic fiber).* Offers comfort, good abrasion resistance, and optimal cut resistance.
- *Chain link or metal mesh.* Offers very little comfort, but maximum abrasion and cut resistance.

Figure 15–9
Cut-resistant work gloves.
Courtesy of Best Manufacturing Company.

- *Butyl rubber.* Offers little comfort, but has excellent resistance to heat, ozone, tearing, and certain chemicals including alcohols, aldehydes, ketones, esters, nitriles, gases, amides, acids, and nitro compounds.
- *Nitrile-based material.* Offers greater comfort and protection. Consequently, there is increased use of this type of material for the substrate coating of glasses.
- *Viton rubber.* Offers little comfort, but performs well with chemicals that butyl rubber cannot protect against, including aliphatics, halogenated, and aromatics. Like butyl gloves, viton gloves also perform well in handling alcohols, gases, and acids.[39]

Figure 15–10
Synthetic fiber gloves reinforced with stainless steel.
Courtesy of Best Manufacturing Company.

PERSONAL PROTECTIVE EQUIPMENT

Personal protective equipment (PPE) is a critical component in the safety program of most organizations. Head, hand, back, eye, face, foot, skin, and breathing protection all involve the use of PPE. Unfortunately, it can be difficult to convince employees to wear PPE properly or, sometimes, to wear it at all. Employees often balk at the perceived "inconvenience" of PPE. They don't like the way it looks or how it feels. They think it is cumbersome in which to work in or time consuming to put on and take off. Sometimes, they just forget.

Making employees comfortable with PPE is a serious and sometimes difficult challenge for safety and health professionals. The following strategies can be used to meet this challenge:

• *Make maximum use of engineering and administrative controls.* PPE should be the last line of defense in protecting employees from hazardous conditions. Before adopting PPE, organizations should first use every

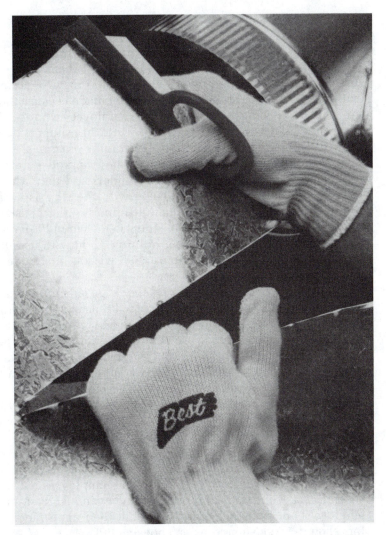

Figure 15–11
Natural rubber gloves.
Courtesy of Best Manufacturing Company.

engineering and administrative control available to minimize potential hazards. If employees see that the organization is doing its part by applying these controls, they will be less reluctant to do their part in properly using PPE.

- *Ensure the optimum choice of PPE by using risk assessment.* Employees know when there is a mismatch between the hazards they face and the PPE they are provided. OSHA requires that PPE be selected on the basis of a comprehensive risk assessment. This approach helps employers make the optimum choice when selecting PPE. Employees who know that the PPE provides adequate protection from hazards will be more likely to use it.

- *Involve employees in all aspects of the PPE program.* Always involve employees when making decisions that affect them. This is good policy for two reasons: (1) Employees may be able to provide input that will improve the quality of the decisions being made because they understand the work tasks being discussed; and (2) employees who are involved in the decision making are more likely to buy into and support that decision.

- *Provide comprehensive education and training programs.* Employees need to understand why PPE is important and how to properly use it. Employers should never assume that employees understand the "why" or "how" of PPE. Training programs should begin with the why aspects, cover them thoroughly, and give employees ample opportunities to ask questions and voice concerns. Once employees understand why PPE is important, they should be given comprehensive training on how to use it properly. No employee should be expected to use PPE without first understanding why and how they should use it.

- *Reinforce the proper use of PPE and challenge its improper use.* Employers should never fall into the trap of taking PPE use for granted. Proper behavior relating to PPE should be reinforced by supervisors and managers. Correspondingly, improper use should be challenged. Employers should use PPE properly themselves and reward employees who follow suit. Rewards need not be formal. Publicly complimenting an employee can be reward enough. Correspondingly, when an employee is seen failing to use the required PPE or using it improperly, that employee should be corrected. However, whereas compliments are given publicly, correction should be done in private.

- *Be sensitive to fit, comfort, and style issues.* Ill-fitting PPE poses a double problem: (1) it may not provide the necessary protection because of the improper fit; and (2) if it does not fit well, employees may be reluctant to wear it because it is uncomfortable. Style can also be a problem in that employees are often self-conscious about their appearance. All these factors should be considered when choosing PPE.

- *Work to make PPE a normal part of the uniform.* By applying these strategies, employers can make PPE a normal part of the uniform. When this happens, using PPE will become standard operating procedure, and its proper use will cease to be an issue.[40]

FORKLIFT SAFETY (POWERED INDUSTRIAL TRUCKS)

Powered industrial truck or forklift safety is included here because forklift-related injuries often result from impact or acceleration hazards. OSHA issues its standards for forklift safety in 29 CFR 1910.138 under the heading "Powered Industrial Truck." These regulations apply to forklifts, platform lift trucks, and motorized hand trucks. The latest edition of OSHA's forklift standard is based primarily on a standard produced by the American Society of Mechanical Engineers titled, "Safety for Low Lift and High Lift Trucks" (ASME B56.1).

Forklifts are different from cars and trucks in several ways. Employees who drive forklifts should understand how they are different. The primary differences are as follows:

- Forklifts are typically steered by the rear wheels.
- An empty forklift can be more difficult to steer than one with a load.
- Forklifts are frequently driven in reverse.
- Forklifts have three-point suspension so that the center of gravity can move from the rear of the vehicle closer to the front when it is loaded.

Because of these differences, it is important to ensure that only properly trained employees drive forklifts and that these employees follow some basic rules of accident prevention. The rules fall into four categories: (1) general, (2) lifting, (3) traveling, and (4) placing.

General Rules

The rules in this section are general and apply to all phases of forklift operation:

- Keep arms, hands, and legs inside the vehicle at all times.
- Face in the direction of travel at all times.
- If the load blocks your view, drive backward.
- Allow plenty of room for braking—at least three vehicle lengths.
- Make sure there is sufficient overhead clearance before moving a load.

Rules for Picking Up a Load

- Make sure the load is within the capacity of the forklift.
- Make sure forks are positioned properly.
- Make sure the load is properly balanced.
- Make sure the load is secure.
- Raise the load to the proper height.

Safety Fact

Staying Up-to-Date on Personal Protective Equipment

One of the challenges faced by safety and health professionals is keeping up-to-date with the latest developments in the area of personal protective equipment (PPE). New materials are developed and improvements are made continually. The International Safety Equipment Association (ISEA) can be a helpful resource for keeping current. The ISEA helps safety and health professionals select the right PPE for the job in question. The association's Web site address is http://www.safetyequipment.org.

- Run the forks all the way into the pallet, and tilt the mast back to stabilize the load before moving.
- Back out and stop completely before lowering the load.

Rules for Traveling with a Load

- Always give pedestrians the right-of-way.
- Never allow passengers on the forklift.
- Keep the forks low while moving.
- Keep the load tilted back slightly while moving.
- Drive slowly; a forklift is not a car.
- Slow down at all intersections; stop and sound the horn at blind intersections.
- Drive up and back down ramps and inclines.
- Never lift or lower the load when traveling.
- Keep to the right just as you do when driving a car.
- Watch for oil, grease, and wet spots, which could inhibit traction.
- Cross railroad tracks at a skewed angle, never at a right angle.
- Watch for edges on loading docks and other changes in elevation.
- Maintain at least four seconds of spacing between your forklift and the one in front of you.

Rules for Placing a Load

- Stop the forklift completely before raising or lowering the load.
- Move slowly and cautiously with the load raised.
- Never walk or stand under a raised load or allow anyone else to do so.
- Be certain the forks have cleared the pallet before turning and before changing height.
- Stack the load square and straight.
- Check behind and on both sides before backing up.

OSHA's Training Requirements

OSHA estimates that 155,000 injuries and 100 fatalities occur every year as a result of accidents involving powered industrial trucks. Because of this, OSHA revised its training requirements for operators of powered industrial trucks (29 CFR 1910.178).[41] OSHA estimates that employers can save as much as $135 million annually by following these training requirements. Of this sum, $83 million could be saved in direct costs such as medical treatment, and $52 million could be saved in indirect costs.

Initial Training Program Content

Before they are allowed to operate a powered industrial truck (forklift), employees are required by OSHA's 29 CFR 1910.178 to complete initial training in the following topics:

1. Operating instruction, warnings, and precautions for the types of trucks the operator will be authorized to operate.
2. Differences between the truck and the automobile.

3. Truck controls and instrumentation: where they are located, what they do, and how they work.
4. Engine or motor operation.
5. Steering and maneuvering.
6. Visibility (including restrictions due to loading).
7. Fork and attachment adaptation, operation, and use limitations.
8. Vehicle capacity.
9. Vehicle stability.
10. Any vehicle inspection and maintenance the operator will be required to perform.
11. Refueling and/or charging and recharging of batteries.
12. Operating limitations.
13. Any other operating instructions, warnings, or precautions listed in the operator's manual for the types of vehicle the employee is being trained to operate.
14. Workplace-related topics including the following: surface conditions where the vehicle will be operated; composition of loads to be carried and load stability; load manipulation, stacking, and unstacking; pedestrian traffic in areas where the vehicle will be operated; narrow aisles and other restricted places where the vehicle will be operated; hazardous (classified) locations where the vehicle will be operated; ramps and other sloped surfaces that affect the vehicle's stability; closed environments and other areas where insufficient ventilation or poor vehicle maintenance could cause a buildup of carbon monoxide or diesel exhaust; other unique or potentially hazardous environmental conditions in the workplace that could affect safe operation.

Refresher Training Program Content

Refresher training, including an evaluation of the effectiveness of that training, must be conducted to ensure that operators of powered industrial trucks have the up-to-date knowledge and skills to operate them safely. Refresher training

Safety Fact

Ambient Noise Can Increase Forklift Hazards

Part of safely operating forklifts is ensuring that pedestrians in the vicinity can hear the truck and thereby avoid it. If pedestrians in the workplace cannot hear a forklift, the results can be fatal as was the case at Revere Copper Products, Inc. in Rome, New York. After an employee was killed by a forklift, the company's safety investigation revealed that ambient noise was making it difficult for pedestrians to hear forklifts in Revere's mill. To correct this problem, Revere's safety personnel installed high-intensity strobe lights on the forklift mast uprights. Because the flashing strobe lights of an approaching truck reflect off the surroundings, pedestrians can see a truck approaching even if they cannot hear it.

Source: "Tip of the Week," *Occupational Health & Safety*. Retrieved from http://stevenspub-ea-14478352@process.request.com.

is required whenever any one of the following conditions exists: the operator has been observed operating a vehicle in an unsafe manner; the operator has been involved in an accident or a near-hit accident; the operator has received an evaluation that indicates he or she is not operating the powered truck safely; the operator is assigned to drive a different type of truck; or any condition in the workplace changes in such a manner that it could affect the safe operation of the truck. In addition to these requirements, employers are required to conduct an evaluation of the performance of all industrial powered truck operators at least once every three years.

SUMMARY

1. The primary causes of falls are a foreign object on the walking surface, a design flaw in the walking surface, a slippery surface, and a person's impaired physical condition.
2. Most falls fit into one of four categories: trip and fall, stump and fall, step and fall, and slip and fall.
3. The coefficient of friction between surfaces is an effective method for comparing the traction of a walking surface. A coefficient of friction of 0.20 or less means the surface is very slippery and very hazardous. A coefficient of 0.40 or higher means there is good traction. Coefficients of friction decrease when a surface is wet.
4. Strategies for preventing slips include the following: choose the right material from the outset, retrofit an existing surface, practice good housekeeping, require nonskid footwear, and inspect surfaces frequently.
5. A slip and fall prevention program should have the following components: policy statement/commitment; review and acceptance of walkways; reconditioning and retrofitting; maintenance standards and procedures; inspection, audits, tests, and records; employee footwear; legal defenses; and measurement of results. Special attention should be given to building lobbies, restrooms, kitchens, and processing areas.
6. OSHA's recommendations for effective fall protection include the following: have a plan, establish requirements, provide equipment and procedures, ensure proper use and type of equipment, and provide training.
7. Do's and don'ts of ladder safety include checking for slipperiness, allowing only one person on the ladder at a time, securing the base and top on a level surface, applying the four-to-one ratio, facing the ladder when climbing, avoiding leaning, and always holding on with one hand.
8. OSHA regulations that apply specifically to fall protection in scaffolding work are as follows: 1926.451(g)(2), 1926.502(d)(15), and 1926.451(d)(16).
9. Protection from impact injuries from falling or projected objects includes personal protective equipment to protect the head (hard hats), eyes and face (goggles or shields), and feet (footwear).

10. Back/lifting safety programs should have the following components: poster illustrations, preemployment screening, regular safety inspections, education and training, external services, and a written map of the program.
11. Standing hazards can be minimized by using the following: antifatigue mats, shoe inserts, foot rails, improved workplace design, and sit/stand chairs.
12. The most widely used materials in manufacturing gloves for hand protection are leather, cotton, aramids, polyethylene, stainless-steel cord, chain link, metal mesh, butyl rubber, nitrile-based material, and viton rubber.
13. Personal protective equipment should be the last line of defense protecting employees from hazards. Before using PPE, every feasible engineering and administrative control should be employed.
14. OSHA standard 29 CFR 1910.178 requires initial training before an employee is allowed to operate a forklift, refresher training under specified conditions, and an evaluation of forklift operator performance at least every three years.

KEY TERMS AND CONCEPTS

Acceleration	Metatarsal guard
Anchorage	Multitask analysis strategy
Back safety/lifting	Nonskid footwear
Body belt/body harness	OSHA Fall Protection Standard
Coefficient of friction	Personal fall arrest system
Connector	Preemployment screening
Design flaw	Puncture-resistant guard
Ergonomic Guidelines for Manual Material Handling (EGMMH)	Reconditioning
Electrical hazard guard	Refresher training program content
External services	Retrofit
Eye and face protection	Review and acceptance
Foot protection	Self-retracting lifeline/lanyard
Foreign object	Slip and fall
Four-to-one ratio	Slip-resistant guard
Good housekeeping	Slippery surface
Head protection	Steel toe
Impact	Step and fall
Initial training program content	Stump and fall
Ladder safety	Surface traction
Lanyard/lifeline	Trip and fall
Lifting hazards	Vertical incidence tribometer (VIT)
	Work hardening

REVIEW QUESTIONS

1. List the primary causes of falls.
2. Explain briefly the most common kinds of falls.
3. Explain how surface traction is measured.
4. List and briefly explain five strategies for preventing slips.
5. Describe the various components of a slip and fall prevention program.
6. Explain the "trigger height" controversy over OSHA's Fall Protection Standard for construction.
7. What are the requirements for personal fall arrest systems in OSHA regulation 1926.451(d)(16)?
8. What is a lanyard?
9. Explain OSHA's recommendations for effective fall protection, and what action to take when a worker is dangling from his fall arrest gear.
10. Explain how to assess the workplace for eye hazards.
11. What topics should be covered in eye protection training?
12. What should an employee look for when inspecting a ladder?
13. Briefly summarize the evolution of hard hats in this country.
14. List six major kinds of injuries to the foot and toes that occur in the workplace each year, and explain the most widely used footwear terms.
15. What are the typical causes of back injuries in the United States?
16. Describe the six-step back safety/lifting program, and explain what the EGMMH are.
17. List four ways to minimize standing hazards.
18. Explain the strategies for proper lifting that should be taught as part of the safety program.
19. What are the critical factors to consider when selecting gloves?
20. Explain how to get employees more comfortable with using PPE.
21. Explain the training requirement contained in OSHA's standard 29 CFR 1910.178 (powered industrial trucks).

ENDNOTES

1. J. Rhodes, "A Risk Manager's Roadmap," *Occupational Health & Safety Online*, May 11, 2006, 1–4. Retrieved from http://www.stevenspublishing.com/stevens/ohspub.nsf/pubhome/of108dOc21adf8doc21adf8136257.
2. Ibid., 2.
3. Ibid., 2–3.
4. Ibid., 3–4.
5. T. Christensen and S. Rupard, "After the Fall: Why Fall Protection Isn't Always Enough," *Occupational Health & Safety* 75, no 3: 48–49.
6. OSHA Regulation 24 CFR 1926.
7. James L. Nash, "OSHA's Fight against Fatal Falls," *Occupational Hazards* 65, no. 4: 37.

8. National Safety Council, "Ladder Safety Tips." Retrieved from http://www.nsc.org/pubs/sh/clip1099.htm.
9. Ibid.
10. Ibid.
11. T. Christensen and S. Rupard, 49.
12. T. Cox, "Is Your Fall Protection Equipment a Silent Hazard?" *Occupational Health & Safety* 75, no. 5: 69–70.
13. Retrieved from http://public.ansi.org/ansionline/portal/search.
14. Ibid.
15. Ibid.
16. Ibid.
17. P. Feuerstein, "Head Protection Looks Up," *Safety & Health* 144, no. 3: 38.
18. Ibid., 39.
19. Ibid.
20. Bross, "Advances Lead to Tougher, More Durable Hard Hat," *Occupational Health & Safety* 74, no. 10: iv.
21. A. Chambers, "Safety Goggles at a Glance," *Occupational Health & Safety* 71, no. 10: 58.
22. J. Hensel, "Setting Up a Vision Program," *Occupational Health & Safety* 68, no. 10: 36.
23. Ibid.
24. Ibid.
25. Ibid.
26. B. Weissman, "Contact Lenses in a Chemical Enviornment," *Occupational Health & Safety* 74, no. 10: 56–58.
27. J. Goodwin, "A Cure for Common Foot Hazards," *Occupational Health & Safety* 74, no. 7: 84.
28. Ibid.
29. Ibid., 84–86.
30. Ibid., 86.
31. Retrieved from http://www.osha.gov/pls/oshaweb.
32. Retrieved from http://www.nsc.org/ergorisk/.
33. A. Kaliokin, "Six Steps Can Help Prevent Back Injuries and Reduce Compensation Costs," *Safety & Health* 138, no. 4: 50.
34. B. Urborg, "How to Comply with OSHA's Ergonomic Standard." Retrieved from" http://www.nsc.org/news/nr11601.htm.
35. Ibid.
36. Ibid.
37. Ibid.
38. 29 CFR 1910.132, Section 38.
39. T. Busshart, "A Cut Above," *Occupational Safety & Health* 67, no. 1: 36.
40. T. Andrews, "Getting Employees Comfortable with PPE," *Occupational Hazards* 62, no. 1: 35–38.
41. 29 CFR 1910.178—Powered Industrial Trucks.

HAZARDS OF TEMPERATURE EXTREMES

Major Topics

■ Thermal Comfort
■ Heat Stress and Strain
■ Cold Stress
■ Burns and Their Effects
■ Chemical Burns

Part of providing a safe and healthy workplace is appropriately controlling the temperature, humidity, and air distribution in work areas. A work environment in which the temperature is not properly controlled can be uncomfortable. Extremes of either heat or cold can be more than uncomfortable—they can be dangerous. Heat stress, cold stress, and burns are major concerns of modern safety and health professionals. This chapter provides the information that professionals need to know to overcome the hazards associated with extreme temperatures.

THERMAL COMFORT

Thermal comfort in the workplace is a function of a number of different factors.[1] Temperature, humidity, air distribution, personal preference, and acclimatization are all determinants of comfort in the workplace. However, determining optimum conditions is not a simple process.

To understand fully the hazards posed by temperature extremes, safety and health professionals must be familiar with several basic concepts related to thermal energy. The most important of these are summarized here:

- **Conduction** is the transfer of heat between two bodies that are touching, or from one location to another within a body. For example, if an employee touches a workpiece that has just been welded and is still hot, heat will be conducted from the workpiece to the hand. Of course, the result of this heat transfer is a burn.

- **Convection** is the transfer of heat from one location to another by way of a moving medium (a gas or a liquid). Convection ovens use this principle to transfer heat from an electrode by way of gases in the air to whatever is being baked.
- **Metabolic heat** is produced within a body as a result of activity that burns energy. All humans produce metabolic heat. This is why a room that is comfortable when occupied by just a few people may become uncomfortable when it is crowded. Unless the thermostat is lowered to compensate, the metabolic heat of a crowd will cause the temperature of a room to rise to an uncomfortable level.
- **Environmental heat** is produced by external sources. Gas or electric heating systems produce environmental heat as do sources of electricity and a number of industrial processes.
- **Radiant heat** is the result of electromagnetic nonionizing energy that is transmitted through space without the movement of matter within that space.

HEAT STRESS AND STRAIN

The key question that must be answered by safety and health professionals concerning employees whose work may subject them to heat stress is as follows:

> "What are the conditions to which most adequately hydrated, unmedicated, healthy employees may be exposed without experiencing heat strain or any other adverse effects?"

The American Conference of Governmental Industrial Hygienists (ACGIH) publishes a comprehensive manual to help safety and health professionals answer this question for the specific situations and conditions that they face. This manual, titled *TLVs and BEIs: Threshold Limit Values for Chemical Substances and Physical Agents and Biological Exposure Indices*, provides reliable guidance and should be in every safety and health professional's library. In addition to using the information contained in this manual, all safety and health professionals should have a comprehensive heat stress management program in place and apply sound professional judgment.[2]

Heat Stress Defined

Heat stress is the net heat load to which a worker may be exposed from the combined contributions of metabolic effect of work, environmental factors (i.e., air temperature, humidity, air movement, and radiant heat exchange) and clothing requirements. A mild or moderate heat stress may cause discomfort and may adversely affect performance and safety, but it is not harmful to health.

As the heat stress approaches human tolerance limits, the risk of heat-related disorders increases.[3]

What follows are some widely used heat stress–related terms safety and health professionals should be familiar with:

Heat exhaustion. This is physical state in which the worker's skin becomes clammy and moist and his or her body temperature is still normal or slightly higher than normal. Heat exhaustion results from loss through sweating of fluid and salt that are not properly replaced during exertion.

Heat cramps. Heat cramps are muscle cramps that can occur when workers exert themselves sufficiently to lose fluids and salt through sweating, but replace only the fluids by drinking large amounts of water containing no salt.

Heat syncope or fainting. Workers who exert themselves in a hot environment will sometimes faint. This is especially the case with workers who are not accustomed to working in such an environment.

Heat rash. Workers who exert themselves in a hot environment in which sweat does not evaporate can develop a prickly rash known as heat rash. Before air conditioning was widely used in the hot and humid summer months in the southeastern United States, children often developed heat rash. Periodic rest breaks in a cool environment that allows sweat to evaporate will prevent heat rash.

Work tolerance time (WTT). WTT is a formula safety and health professionals can use to determine what steps can be taken to allow a worker to safely perform his or her required tasks in the environment in question for the time required. The formula takes into account such factors as temperature, humidity, level of energy that will be expended in performing the task, rest periods, and personal protective equipment (PPE).

Moisture vapor transfer rate (MVTR). The MVTR is a measure of the ability of the fabric used in making personal protective equipment (PPE) to dissipate heat. The best MVTR occurs on an unclothed body. Even the lightest cotton fabric is less capable of dissipating heat when the unclothed body is used for baseline comparisons. The MVTR of impermeable fabric is zero (because the fabric does not allow the skin to "breathe"). The higher the MVTR the better in hot environments.

Heat Strain Defined

Heat strain is the overall physiological response resulting from heat stress. The physiological adjustments are dedicated to dissipating excess heat from the body. **Acclimatization** is a gradual physiological adaptation that improves an individual's ability to tolerate heat stress.

Recognizing Heat Strain

Safety and health professionals, supervisors, and coworkers should know how to recognize heat strain. The following factors are signs of excessive heat strain. Exposure to heat stress should be stopped immediately for any employee experiencing any of these symptoms:

- A sustained rapid heart rate (180 beats per minutes minus the employee's age in years). For example, a 40-year-old employee has a sustained heart rate of 150 beats per minutes. This is a problem because the heart rate exceeds 140 (180 minus 40) beats per minute.
- Core body temperature is greater than 38.5°C.
- Recovery rate one minute after a peak work effort is greater than 110 beats per minute.
- Sudden and severe fatigue, nausea, dizziness, or light-headedness.

These symptoms can be assessed on the spot in real-time. In addition, other symptoms can be monitored only over time. Employees are at greater risk of excessive heat strain if they experience any of the following:

- Profuse sweating that continues for hours.
- Weight loss of more than 1.5 percent of body weight during one work shift.
- Urinary sodium excretion of less than 50 moles (24-hour period).

Clothing

Heat is best removed from the body when there is free movement of cool dry air over the skin's surface. This promotes the evaporation of sweat from the skin, which is the body's principal cooling mechanism. Clothing impedes this

Safety Fact

Symptoms of Heat Exhaustion

Employees need to be able to observe and recognize the symptoms of heat exhaustion in themselves and their coworkers. The following are observable symptoms of heat exhaustion:

- Fatigue
- Nausea and/or vomiting
- Headache
- Light-headedness
- Clammy, moist skin
- Pale or flushed complexion
- Fainting when trying to stand
- Rapid pulse

process, some types more than others. Encapsulating suits and clothing that is impermeable or highly resistant to the flow of air and water vapor multiply the potential for heat strain.

When assessing heat stress hazards in the workplace, safety and health professionals should consider the added effect of clothing. For example, the *wet bulb globe temperature* (*WBGT*) of working conditions should be increased by 3.5°C for employees wearing cloth overalls. This factor increases to 5°C with double cloth overalls.

Because the WBGT is influenced by air temperature, radiant heat, and humidity, it can be helpful in establishing a threshold for making judgments about working conditions. WBGT values can be calculated using the following formula:

Exposed to Direct Sunlight

$$WBGT = 0.7 \ T_{nwb} \text{ to } 0.2 \ T_g + 0.1 \ T_{db}$$

T_{nwb} = Natural wet bulb temperature
T_g = Globe temperature
T_{db} = Dry bulb (air) temperature

Not Exposed to Direct Sunlight

$$WBGT = 0.7 \ T_{nwb} + 0.3 \ T_g$$

These formulas for WBGT give safety and health professionals a beginning point for making judgments. The WBGT must be adjusted for clothing, work demands, and the employee's acclimatization state. The key is to ensure that employees never experience a core body temperature of 38°C or higher. Figures 16–1 and 16–2 provide screening criteria for heat stress exposure. Once the WBGT has been calculated and adjusted for clothing, these figures may be used for factoring in work demands and acclimatization. To use Figures 16–1 and 16–2, apply the following example:

Several acclimatized employees have a job to do that has a work demand of 75 percent work and 25 percent rest. The WBGT has been computed as 26. The work is considered "heavy." Because the employees will wear long-sleeved shirts and long trousers made of woven material an additional 3.5° must be added: 26 + 3.5 = 29.5°C. Using the proper column and row of Figure 16–1, a WBGT of 26 can be determined. Because the calculated and adjusted WBGT is 29.5, there is a problem. In order to work in these conditions, the employees should adjust the work demand to 25 percent work and 75 percent rest.

Heat Stress Management

Safety and health professionals should continually emphasize the importance of paying attention to recognizable symptoms of heat stress. In addition, they should ensure that a comprehensive heat stress management program is in place. Such a program should consist of both general and specific controls.

Screening Criteria (°C) Acclimatized Employees				
Work Demands	Light Work	Moderate Work	Heavy Work	Very Heavy Work
100% Work	29.5	27.5	26.0	—
75% Work 25% Rest	30.5	28.5	27.5	—
50% Work 50% Rest	31.5	29.5	28.5	27.5
25% Work 75% Rest	32.5	31.0	30.0	29.5

Figure 16–1
Criteria for determining the allowable work periods for acclimatized employees.
Source: American Conference of Governmental Industrial Hygienists (ACGIH).

Screening Criteria (°C) Not-Acclimatized Employees				
Work Demands	Light Work	Moderate Work	Heavy Work	Very Heavy Work
100% Work	27.5	25.0	22.5	—
75% Work 25% Rest	29.0	26.5	24.5	—
50% Work 50% Rest	30.0	28.0	26.5	25.0
25% Work 75% Rest	31.0	29.0	28.0	26.5

Figure 16–2
Criteria for determining the allowable work periods for employees who are not acclimatized.
Source: American Conference of Governmental Industrial Hygienists (ACGIH).

General Controls

The ACGIH recommends the following general controls:

- Provide accurate verbal and written instructions, training programs, and other information about heat stress and strain.
- Encourage drinking small volumes (approximately 1 cup) of cool water about every 20 minutes.

Safety Fact

Heat Stress Guidelines and Materials Are Available from NIOSH and OSHA

NIOSH developed criteria for a proposed federal standard regulating the exposure of workers to heat stress. The criteria are available online at:

> http://www.cdc.gov/niosh/topics/heatstress

OSHA provides a variety of helpful heat stress–related materials that can be used for training, assessing heat-related hazards, or creating a site-specific heat stress prevention program. These materials are available online at:

> http://www.osha.gov/SLTC/heatstress/recognition.html

- Permit self-limitation of exposure. Encourage coworker observation to detect signs and symptoms of heat strain in others.
- Counsel and monitor those employees who take medications that may compromise normal cardiovascular, blood pressure, body temperature regulation, renal, or sweat gland functions, as well as those who abuse or who are recovering from the abuse of alcohol and other intoxicants.
- Encourage healthy lifestyles, ideal body weight, and electrolyte balance.
- Adjust expectations of those returning to work after absence from heat stress situations and encourage consumption of salty foods (with approval of the employee's physician if on a salt-restricted diet).
- Consider replacement medical screening to identify those susceptible to systemic heat injury.[4]

Specific Controls

The ACGIH recommends the following specific controls:

- Establish engineering controls that reduce the metabolic rate, provide general air movement, reduce process heat and water-vapor release, and shield radiant heat sources, among others.
- Consider administrative controls that set acceptable exposure times, allow sufficient recovery, and limit physiological strain.
- Consider personal protection that has been demonstrated to be effective for the specific work practices and conditions at the location.[5]

COLD STRESS

Excessive exposure to cold can lead to hypothermia, which can be fatal. The goal of safety and health professionals in protecting employees from acute cold stress is to prevent the deep body temperature from falling below 36°C (98.6°F) and to prevent cold injuries to body extremities, especially the hands, feet, and head. A fatal exposure to cold typically results from failure to remove the employee from a cold air environment or immersion in cold water.[6]

Excessive exposure to cold stress, even when not fatal, can result in impaired judgment, reduced alertness, and poor decision making. Acute cold stress can cause reduced muscular function, decreased tactile sensitivity, reduced blood flow, and thickening of the synovial fluid. Chronic cold stress can lead to reduced functioning of the peripheral nervous system. All these factors increase the likelihood of accidents and injuries. Figure 16–3 shows the effects of allowing the core body temperature to fall to selected levels.

Whether employees are exposed to cold air or are immersed in cold water, wind can magnify the level of cold stress. This phenomenon is often referred to as *windchill*. Figure 16–4 shows the effect of wind on selected temperatures. To read this chart, locate the actual temperature (50, 40, 30 . . . 0). Then, find the applicable wind speed. Reading across that row to the right, find the equivalent temperature. For example, if employees are working in an environment that is 30°F and has a wind speed of 15 miles per hour (mph), the equivalent temperature is 9°F.

Preventing Cold Stress

When the equivalent air temperature reaches −32°C (−25.6°F), continuous exposure of exposed skin should not be allowed. At equivalent air temperatures of 2°C (35.6°F), employees who are immersed in water or whose clothing gets wet should be treated for hypothermia immediately. Figure 16–5 shows selected threshold limit values (TLVs) for employees who work in environments with temperature below freezing.

Effects of Reducing the Core Body Temperature

Core Temperature

°C	°F	Body's Response
37.6	99.6	Normal rectal temperature
36.0	96.8	Metabolic rate increases
35.0	95.0	Pronounced shivering
33.0	91.4	Severe hypothermia
30.0	86.0	Progressive loss of consciousness begins
24.0	75.2	Pulmonary edema
20.0	68.0	Cardiac standstill

Figure 16–3
The body's response to reducing its core temperature.

Cooling Effect of Wind						
	Actual Temperature (°F) and Equivalent Temperatures (°F)					
Wind Speed (in mph)	50	40	30	20	10	0
5	48	37	27	16	6	−5
15	36	22	9	−5	−18	−32
25	30	16	0	−15	−29	−44
35	27	11	−4	−20	−35	−51

Figure 16–4
Effect of wind on the actual temperature.
Source: U.S. Army Research Institute of Environmental Medicine.

TLVs for a Four-Hour Shift							
Air Temperature		No Wind		5 mph Wind		10 mph Wind	
C°	F°	Max. Work Time	No. of Breaks	Max. Work Time	No. of Breaks	Max. Work Time	No. of Breaks
−32 to −34	−25 to −29	75 Min.	2	55 Min.	3	40 Min.	4
−35 to −37	−30 to −34	55 Min.	3	40 Min.	4	30 Min.	5
−38 to −39	−35 to −39	40 Min.	4	30 Min.	5	Nonemergency Work Prohibited	

Note: This applies to workers properly dressed in dry clothing.

Figure 16–5
Partial table for determining TLVs in selected circumstances.
Source: American Conference of Governmental Industrial Hygienists (ACGIH).

To use Figure 16–5, locate the applicable temperature in the leftmost column. Reading to the right, locate the applicable wind speed. For example, employees working a four-hour shift in an environment with an air temperature of −32°C and a 5-mph wind should be exposed no longer than 55 minutes at a time and should warm up at least three times during the shift.

When work is to be performed in an environment with an air temperature of 4°C (39.2°F) or less, total body protective clothing is advisable. What follows are several strategies that can be used to decrease the hazards of cold stress:

• When working in a setting in which wind is a factor, reduce the effect of the wind by (1) erecting a windscreen; or (2) wearing wind-breaking clothing.

- When working in a setting in which clothing may get wet, apply one or more of the following strategies: (1) with light work, wear an outer layer of impermeable clothing; (2) with heavier work, wear an outer layer that is water repellent, but not impermeable (change outerwear as it becomes wet); (3) select outer garments that are ventilated to prevent internal wetting from sweat; (4) if clothing gets wet before going into the cold environment, change first; (5) change socks daily or more often to keep them dry; and (6) use vapor barrier boots to help keep the feet dry.
- When working in a cold setting, use auxiliary heat applied directly to the hands and feet.
- When working in a cold setting, use facial protection to prevent cold stress to the face and lungs.
- If adequate protective clothing that is appropriate for the conditions in question is not available, the work should be modified or suspended until conditions change or the clothing is available.

When work is to be performed in an environment with an air temperature of −12°C (10.4°F), the following additional strategies should be applied:

- Employees should be under continuous observation using either direct supervision or the buddy system.
- The work rate should be paced to avoid sweating. When heavy work is necessary, employees should take frequent warming breaks in heated shelters. If clothing becomes wet—internally or externally—it should be changed during a break.
- Do not allow new employees to work full time in these conditions until they have several days to become accustomed to the conditions and the necessary protective clothing.
- When determining the required work level for employees (light, heavy, or very heavy), consider the weight and bulkiness of protective clothing.
- Organize work in cold environments to minimize long periods of sitting or standing still. Never use unprotected metal chairs or seats.
- Before allowing employees to work in a cold environment, make sure they have been trained in safety and health procedures. Figure 16–6 is a checklist of topics that should be covered as a minimum during employee training.
- When work in a refrigerated room is required, the air velocity should be minimized and maintained at one meter per second (200 feet per minute) or less.
- When work outdoors in snow is required, employees should be provided special safety goggles that protect the eyes from ultraviolet light, glare, and blowing ice crystals.
- Employees who suffer from diseases or take medications that inhibit normal body functions or that reduce normal body tolerances should be prohibited from working in environments where temperatures are at −1°C (30.2°F) or less.
- Employees who are routinely exposed to the following conditions should be medically certified as being suitable for work in such conditions: (1) air

---

Training Checklist for Employees Who Work in a Cold Environment

✓ Proper warming procedures

✓ Applicable first-aid procedures

✓ Protective clothing requirements and proper use of protective clothing

✓ Proper eating habits

✓ Proper drinking habits (e.g., avoid caffeine and other stimulants)

✓ Recognizing of cold stress and strain (e.g., impending frostbite, impending hypothermia, and so on)

✓ Safe work processes

Figure 16–6
Checklist for training employees who will work in a cold environment.

temperatures of less than $-24°C$ ($-11.2°F$) with wind speeds less than 5 mph; and (2) air temperatures of less than $-18°C$ ($0°F$) with wind speeds greater than 5 mph.

BURNS AND THEIR EFFECTS

One of the most common hazards associated with heat in the workplace is the burn. Burns can be especially dangerous because they disrupt the normal functioning of the skin, which is the body's largest organ and the most important in terms of protecting other organs. It is necessary first to understand the composition of, and purpose served by, the skin to understand the hazards that burns can represent.

Human Skin

Human skin is the tough, continuous outer covering of the body. It consists of the following two main layers: (1) the outer layer, which is known as the **epidermis;** and (2) the inner layer, which is known as the **dermis, cutis,** or **corium**. The dermis is connected to the underlying subcutaneous tissue.

The skin serves several important purposes including the following: protection of body tissue, sensation, secretion, excretion, and respiration (see Figure 16–7). Protection from fluid loss, water penetration, ultraviolet radiation, and infestation by microorganisms is a major function of the skin. The sensory functions of touching, sensing cold, feeling pain, and sensing heat involve the skin.

The skin helps regulate body heat through the sweating process. It excretes sweat that takes with it electrolytes and certain toxins. This helps keep the

Figure 16–7
Functions of the human skin.

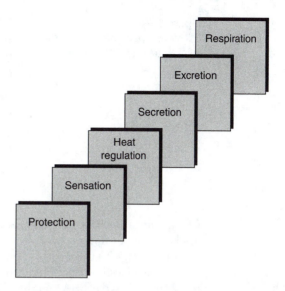

body's fluid level in balance. By giving off minute amounts of carbon dioxide and absorbing small amounts of oxygen, the skin also aids slightly in respiration.

What makes burns particularly dangerous is that they can disrupt any or all of these functions depending on their severity. The deeper the penetration, the more severe the burn.

Severity of Burns

The severity of a burn depends on several factors. The most important of these is the depth to which the burn penetrates. Other determining factors include location of the burn, age of the victim, and amount of burned area.

The most widely used method of classifying burns is by degree (i.e., first-, second-, or third-degree burns). Modern safety and health professionals should be familiar with these classifications and what they mean.

First-degree burns are minor and result only in a mild inflammation of the skin, known as *erythema*. Sunburn is a common form of first-degree burn. It is easily recognizable as a redness of the skin that makes the skin sensitive and moderately painful to the touch.

Second-degree burns are easily recognizable from the blisters that form on the skin. If a second-degree burn is superficial, the skin will heal with little or no scarring. A deeper second-degree burn will form a thin layer of coagulated, dead cells that feels leathery to the touch. A temperature of approximately 98.9°C can cause a second-degree burn in as little as 15 seconds of contact.

Third-degree burns are very dangerous and can be fatal depending on the amount of body surface affected. A third-degree burn penetrates through both the epidermis and the dermis. A deep third-degree burn will penetrate body tissue. Third-degree burns can be caused by both moist and dry hazards. Moist hazards include steam and hot liquids; these cause burns that appear white.

Figure 16–8
Estimating percentage of body surface area (BSA) burned.

Right arm	9% of BSA
Left arm	9% of BSA
Head/neck	9% of BSA
Right leg	18% of BSA
Left leg	18% of BSA
Back	18% of BSA
Chest/stomach	18% of BSA
Perineum	1% of BSA

Dry hazards include fire and hot objects or surfaces; these cause burns that appear black and charred.

In addition to the depth of penetration of a burn, the amount of surface area covered is also a critical concern. This amount is expressed as a percentage of **body surface area (BSA)**. Figure 16–8 shows how the percentage of BSA can be estimated. Burns covering over 75 percent of BSA are usually fatal.

Using the first-, second-, and third-degree burn classifications in conjunction with BSA percentages, burns can be classified further as minor, moderate, or critical. According to Mertz, these classifications can be summarized as described in the following paragraphs.[7]

Minor Burns

All first-degree burns are considered **minor**. Second-degree burns covering less than 15 percent of the body are considered minor. Third-degree burns can be considered minor provided they cover only 2 percent or less of BSA.

Moderate Burns

Second-degree burns that penetrate the epidermis and cover 15 percent or more of BSA are considered **moderate**. Second-degree burns that penetrate the dermis and cover from 15 to 30 percent of BSA are considered moderate. Third-degree burns can be considered moderate provided they cover less than 10 percent of BSA and are not on the hands, face, or feet.

Critical Burns

Second-degree burns covering more than 30 percent of BSA or third-degree burns covering over 10 percent of BSA are considered **critical**. Even small-area third-degree burns to the hands, face, or feet are considered critical because of the greater potential for infection to these areas by their nature. In addition, burns that are complicated by other injuries (fractures, soft tissue damage, and so on) are considered critical.

Safety Fact

Burn-Related Accidents Can Be Costly

Burn-related accidents on the job can be costly. The first and highest cost is the human suffering of the victims. The second is the damaged morale of other employees who were not physically injured, but may be emotionally injured. The third cost is in fines assessed against the organization by regulatory agencies. For example, an oil company in OSHA Region 7 was fined $135,000 when an explosion at the company's bulk storage plant left a truck driver with burns over 90 percent of his body. The company was cited by OSHA for both willful and serious violations including the following: (1) exposing workers to hazards by failing to segregate areas for tank off-loading and storage tanks containing flammable liquids; (2) failing to regularly inspect and maintain tank pressure vents, overfill alarms, and emergency vents; (3) failing to establish an emergency response plan; and (4) using valves and fittings made of low-melting materials without protecting them from exposure to fire.

CHEMICAL BURNS

Chemicals are widely used in modern industry even by companies that do not produce them as part of their product base. Many of the chemicals produced, handled, stored, transported or otherwise used in industry can cause burns similar to those caused by heat (i.e., first-, second-, and third-degree burns). The hazards of chemical burns are very similar to those of thermal burns.

Chemical burns, like thermal burns, destroy body tissue; the extent of destruction depends on the severity of the burn. However, chemical burns continue to destroy body tissue until the chemicals are washed away completely.

The National Safety Council describes the physiological process in cases of chemical burns:

> Many concentrated chemical solutions have an affinity for water. When they come in contact with body tissue, they withdraw water from it so rapidly that the original chemical composition of the tissue (and hence the tissue itself) is destroyed. In fact, a strong caustic may dissolve even dehydrated animal tissue. The more concentrated the solution, the more rapid is the destruction.[8]

The severity of the burn produced by a given chemical depends on the following factors:

- Corrosive capability of the chemical
- Concentration of the chemical
- Temperature of the chemical or the solution in which it is dissolved
- Duration of contact with the chemical[9]

Chemical	Potential Harmful Effect
Acetic acid	Tissue damage
Liquid bromide	Corrosive effect on the respiratory system and tissue damage
Formaldehyde	Tissue hardening
Lime	Dermatitis and eye burns
Methylbromide	Blisters
Nitric/sulfuric acid mixture	Severe burns and tissue damage
Oxalic acid	Ulceration and tissue damage
White phosphorus	Ignites in air causing thermal burns
Silver nitrate	Corrosive/caustic effect on the skin
Sodium (metal)	Ignites with moisture causing thermal burns
Trichloracetic acid	Tissue damage

Figure 16–9
Harmful effects of selected widely used chemicals.

Effects of Chemical Burns

Different chemicals have different effects on the human body. The harmful effects of selected widely used chemicals are summarized in Figure 16–9.[10] These are only a few of the many chemicals widely used in industry today. All serve an important purpose; however, all carry the potential for serious injury.

The primary hazardous effects of chemical burns are infection, loss of body fluids, and shock, and are summarized in the following paragraphs.[11]

Infection

The risk of **infection** is high with chemical burns—as is it with heat-induced burns—because the body's primary defense against infection-causing microorganisms (the skin) is penetrated. This is why it is so important to keep burns clean. Infection in a burn wound can cause **septicemia** (blood poisoning).

Fluid Loss

Body **fluid loss** in second- and third-degree burns can be serious. With second-degree burns, the blisters that form on the skin often fill with fluid that seeps out of damaged tissue under the blister. With third-degree burns, fluids are lost internally and, as a result, can cause the same complications as a hemorrhage. If these fluids are not replaced properly, the burns can be fatal.

Shock

Shock is a depression of the nervous system. It can be caused by both physical and psychological trauma. In cases of serious burns, it may be caused by the intense pain that can occur when skin is burned away, leaving sensitive nerve endings exposed. Shock from burns can come in the following two forms: (1) primary shock, which is the first stage and results from physical pain or psychological trauma; and (2) secondary shock, which comes later and is caused by a loss of fluids and plasma proteins as a result of the burns.

First Aid for Chemical Burns

There is a definite course of action that should be taken when chemical burns occur, and the need for immediacy cannot be overemphasized. According to the NSC, the proper response in cases of chemical burns is to "wash off the chemical by flooding the burned areas with copious amounts of water as quickly as possible. This is the only method for limiting the severity of the burn, and the loss of even a few seconds can be vital."[12]

In the case of chemical burns to the eyes, the continuous flooding should continue for at least 15 minutes. The eyelids should be held open to ensure that chemicals are not trapped under them.

According to the NSC,

> The Committee on Industrial Ophthalmology, Council of Industrial Health of the American Medical Association has noted the tremendous saving of eyesight among industrial employees brought about by immediate and thorough flushing of harmful chemicals from the eyes by copious amounts of water. It is the belief of the committee that this is the most effective and practical emergency first aid treatment of eyes injured by chemicals.[13]

Clothing is another consideration when an employee comes in contact with a caustic chemical. If chemicals have saturated the employee's clothes, they must be removed quickly. The best approach is to remove the clothes while flooding the body or the affected area. If necessary for quick removal, clothing should be ripped or cut off.

The critical need to apply water immediately in cases of chemical burns means that water must be readily available. Health and safety professionals should ensure that special eye wash and shower facilities are available wherever employees handle chemicals.

SUMMARY

1. Important thermal-related terms include *conduction, convection, metabolic heat, environmental heat*, and *radiant heat*.
2. Heat stress is the net load to which a worker may be exposed from the combined contributions of metabolic cost of work, environmental factors, and clothing requirements. Key heat stress concepts are heat exhaustion, heat cramps, heat syncope, and heat rash.
3. The goal in protecting employees from cold stress is to prevent the deep body temperature from falling below 36°C (98.6°F).

4. Cold stress can be prevented by applying the following strategies: medical screening and supervision, orientation and training, proper work practices, and engineering and administrative controls.

5. Wind or air movement causes the body to sense coldness beyond what the thermometer registers. This phenomenon is known as the windchill factor. This should be considered when planning work schedules.

6. The most common form of cold stress is hypothermia.

7. The most widely used method of classifying burns is by degree: first-, second-, and third-degree burns. The amount of surface area covered by burns is expressed as a percentage of body surface area (BSA). Burns are also classified as minor, moderate, and critical.

8. The severity of chemical burns depends on the corrosive capability of the chemical, concentration and temperature of the chemical, and the duration of contact.

9. The primary hazards associated with chemical burns beyond the damage to the body tissues are infection, fluid loss, and shock. The most important first aid for chemical burns is immediate and continual flushing with water.

KEY TERMS AND CONCEPTS

Acclimatization	Heat rash
Body surface area (BSA)	Heat strain
Chemical burn	Heat stress
Cold stress	Heat stress management
Conduction	Heat syncope
Convection	Hypothermia
Corium	Infection
Critical burn	Metabolic heat
Cutis	Minor burn
Dermis	Moderate burn
Electrolytes	Moisture vapor transfer rate (MVTR)
Environmental heat	Radiant heat
Epidermis	Second-degree burn
First-degree burn	Shock
Fluid loss	Third-degree burn
Heat cramps	Work tolerance time (WTT)
Heat exhaustion	

REVIEW QUESTIONS

1. Define the following thermal comfort–related terms: *conduction, convection*, and *metabolic heat.*
2. What is heat stress?
3. What is heat strain?
4. Define the following terms: *heat exhaustion, heat cramps, heat syncope, heat rash, work tolerance time*, and *moisture vapor transfer rate.*
5. What are the symptoms of heat exhaustion?
6. How can heat strain be recognized?
7. How does clothing affect the cooling process?
8. What factors influence the WBGT?
9. Describe the various general controls in heat stress management.
10. How can cold stress be prevented?
11. How does wind movement affect the way the body perceives temperature?
12. Describe the symptoms of cold stress and hypothermia.
13. Describe the various components of a cold stress prevention program.
14. Describe the various purposes served by the skin.
15. Describe and differentiate among first-, second-, and third-degree burns.
16. Describe and differentiate among minor, moderate, and critical burns.
17. List the factors that determine the severity of a chemical burn.
18. Explain the hazards of chemical burns besides tissue damage.
19. What should you do if an employee accidentally splashes a caustic chemical on himself or herself?

ENDNOTES

1. American Conference of Governmental Industrial Hygienists (ACGIH), 2006, *TLVs and BEIs* (Cincinnati, OH: 2006), 172–181.
2. Ibid., 180–188.
3. Ibid., 181.
4. Ibid., 187.
5. Ibid.
6. Ibid., 171–179.
7. Patricia M. Mertz, *Burn Study*, an unpublished paper, University of Miami School of Medicine, Department of Dermatology and Cutaneous Surgery, Miami, Florida, October 2000, 2.
8. National Safety Council, "Chemical Burns," Data Sheet 1–523 Rev. 87 (Chicago: National Safety Council): 1.
9. Ibid.
10. Ibid., 3–4.
11. Ibid., 2.
12. Ibid.
13. Ibid.

PRESSURE HAZARDS

Major Topics

- Pressure Hazards Defined
- Sources of Pressure Hazards
- Boilers and Pressure Hazards
- High-Temperature Water Hazards
- Hazards of Unfired Pressure Vessels
- Hazards of High-Pressure Systems
- Cracking Hazards in Pressure Vessels
- Nondestructive Testing of Pressure Vessels
- Pressure Dangers to Humans
- Decompression Procedures
- Measurement of Pressure Hazards
- Reduction of Pressure Hazards

PRESSURE HAZARDS DEFINED

Pressure is defined in physics as the force exerted against an opposing fluid or thrust distributed over a surface. This may be expressed in force or weight per unit of area, such as psi (pounds per square inch). A **hazard** is a condition with the potential of causing injury to personnel, damage to equipment or structures, loss of material, or lessening of the ability to perform a prescribed function. Thus, a **pressure hazard** is a hazard caused by a dangerous condition involving pressure. Critical injury and damage can occur with relatively little pressure. OSHA defines high-pressure cylinders as those designated with a service pressure of 900 pounds psi or greater.

We perceive pressure in relation to the earth's atmosphere. Approximately 21 percent of the atmosphere is oxygen, with most of the other 79 percent being nitrogen. In addition to oxygen and nitrogen, the atmosphere contains trace amounts of several inert gases: argon, neon, krypton, xenon, and helium.

At sea level, the earth's atmosphere averages 1,013 H (hydrogen) 10 N/m^2, or 1.013 millibars, or 760 mm Hg (29.92 inches), or 14.7 psi, depending on the measuring scale used.[1] The international system of measurement utilizes newtons per square meter (N/m^2). However, in human physiology studies, the typical unit is millimeters of mercury (mm Hg).

Atmospheric pressure is usually measured using a **barometer.** As the altitude above sea level increases, atmospheric pressure decreases in a nonlinear fashion. For example, at 5,486 meters (18,000 feet) above sea level, the barometric pressure is equal to 390 mm Hg. Half of this pressure, around 195 mm Hg, can be found at 2,010 meters (23,000 feet) above sea level.

Boyle's law states that the product of a given pressure and volume is constant with a constant temperature:

$$P_1V_1 = P_2V_2, \quad \text{when } T \text{ is constant}$$

Air moves in and out of the lungs because of a pressure gradient or difference in pressure. When atmospheric pressure is greater than pressure within the lungs, air flows down this pressure gradient from the outside into the lungs. This is called **inspiration,** inhalation, or breathing in, and occurs with greater lung volume than at rest. When pressure in the lungs is greater than atmospheric pressure, air moves down a pressure gradient outward from the lungs to the outside. **Expiration** occurs when air leaves the lungs and the lung volume is less than the relaxed volume, increasing pressure within the lungs.

Gas exchange occurs between air in the lung alveoli and gas in solution in blood. The pressure gradients causing this gas exchange are called partial pressures. **Dalton's law of partial pressures** states that, in a mixture of theoretically ideal gases, the pressure exerted by the mixture is the sum of the pressures exerted by each component gas of the mixture:

$$P_A = P_O + P_N + P_{\text{else}}$$

Air entering the lungs immediately becomes saturated with water vapor. Water vapor, although it is a gas, does not conform to Dalton's law. The partial pressure of water vapor in a mixture of gases is not dependent on its fractional concentration in that mixture. Water vapor partial pressure, instead, is dependent on its temperature. From this exception to Dalton's law comes the fact that at the normal body temperature of 37°C (98.6°F), water vapor maintains a partial pressure of 47 mm Hg as long as that temperature is maintained. With this brief explanation of how pressure is involved in human breathing, we now focus on the various sources of pressure hazards.

SOURCES OF PRESSURE HAZARDS

There are many sources of pressure hazard—some natural, most created by humans. Because the human body is comprised of approximately 85 percent liquid, which is virtually incompressible, increasing pressure does not create problems by itself. Problems can result from air being trapped or expanded within body cavities.

When sinus passages are blocked so that air cannot pass easily from the sinuses to the nose, expansion of the air in these sinuses can lead to problems. The same complications can occur with air trapped in the middle ear's eustachian tube. As Boyle's law states, gas volume increases as pressure decreases. Expansion of the air in blocked sinus passages or the middle ear occurs with a rapid increase in altitude or rapid ascent underwater. This can cause pain and, if not eventually relieved, disease. Under extreme circumstances of rapid ascent from underwater diving or high-altitude decompression, lungs can rupture.

Nitrogen absorption into the body tissues can become excessive during underwater diving and breathing of nitrogen-enriched air. Nitrogen permeation of tissues occurs in proportion to the partial ppressure of nitrogen taken in. If the nitrogen is permeating tissues faster than the person can breathe it out, bubbles of gas may form in the tissues.

Decompression sickness can result from the decompression that accompanies a rapid rise from sea level to at least 5,486 meters (18,000 feet) or a rapid ascent from around 40 to 20 meters (132 to 66 feet) underwater. Several factors influence the onset of decompression sickness:

- *A history* of previous decompression sickness increases the probability of another attack.
- *Age* is a component. Being over 30 increases the chances of an attack.
- *Physical fitness* plays a role. People in better condition have a reduced chance of the sickness. Previously broken bones and joint injuries are often the sites of pain.
- *Exercise* during the exposure to decompression increases the likelihood and brings on an earlier onset of symptoms.
- *Low temperature* increases the probability of the sickness.
- *Speed of decompression* also influences the sickness. A rapid rate of decompression increases the possibility and severity of symptoms.
- *Length of exposure* of the person to the pressure is proportionately related to the intensity of symptoms. The longer the exposure, the greater the chances of decompression sickness.

A reduction in partial pressure can result from reduced available oxygen and cause a problem in breathing known as *hypoxia.* Too much oxygen or oxygen breathed under pressure that is too high is called **hyperoxia.** Another partial pressure hazard, *nitrogen narcosis,* results from a higher-than-normal level of nitrogen pressure.

When breathed under pressure, nitrogen causes a reduction of cerebral and neural activity. Breathing nitrogen at great depths underwater can cause a feeling of euphoria and loss of reality. At depths greater than 30 meters (100 feet), nitrogen narcosis can occur even when breathing normal air. The effects may become pathogenic at depths greater than 60 meters (200 feet), with motor skills threatened at depths greater than 91 meters (300 feet). Cognitive processes deteriorate quickly after reaching a depth of 99 meters (325 feet). Decompression procedures are covered later in this chapter.

BOILERS AND PRESSURE HAZARDS

A boiler is a closed vessel in which water is heated to form steam, hot water, or high-temperature water under pressure.[2] Potential safety hazards associated with boilers and other pressurized vessels include the following:

- Design, construction, or installation errors
- Poor or insufficient training of operators
- Human error
- Mechanical breakdown or failure
- Failure or blockage of control or safety devices
- Insufficient or improper inspections
- Improper application of equipment
- Insufficient preventive maintenance[3]

Through years of experience, a great deal has been learned about how to prevent accidents associated with boilers. OSHA recommends the following daily, weekly, monthly, and yearly accident prevention measures:

1. *Daily check.* Check the water to make sure that it is at the proper level. Vent the furnace thoroughly before starting the fire. Warm up the boiler using a small fire. When the boiler is operating, check it frequently.
2. *Weekly check.* At least once every week, test the low-water automatic shutdown control and record the results of the test on a tag that is clearly visible.
3. *Monthly check.* At least once every month, test the safety valve and record the results of the test on a tag that is clearly visible.
4. *Yearly check.* The low-level automatic shutdown control mechanism should be either replaced or completely overhauled and rebuilt. Arrange to have the vendor or a third-party expert test all combustion safeguards, including fuel pressure switches, limit switches, motor starter interlocks, and shutoff valves.[4]

HIGH-TEMPERATURE WATER HAZARDS

High-temperature water (HTW) is exactly what its name implies—water that has been heated to a very high temperature, but not high enough to produce steam.[5] In some cases, HTW can be used as an economical substitute for steam (e.g., in industrial heating systems). It has the added advantage of releasing less energy (pressure) than steam does.

In spite of this, there are hazards associated with HTW. Human contact with HTW can result in extremely serious burns and even death. The two most prominent sources of hazards associated with HTW are operator error and improper design. Proper training and careful supervision are the best guards against operator error.

Design of HTW systems is a highly specialized process that should be undertaken only by experienced engineers. Mechanical forces such as water hammer, thermal expansion, thermal shock, or faulty materials cause system failures more often than do thermodynamic forces. Therefore, it is important to allow for such causes when designing an HTW system.

The best designs are simple and operator-friendly. Designing too many automatic controls into an HTW system can create more problems than it solves by turning operators into mere attendants who are unable to respond properly to emergencies.

HAZARDS OF UNFIRED PRESSURE VESSELS

Not all pressure vessels are fired. Unfired pressure vessels include compressed air tanks, steam-jacketed kettles, digesters, and vulcanizers, as well as others that can create heat internally by various means rather than by external fire.[6] The various means of creating internal heat include (1) chemical action within the vessel, and (2) application of some heating medium (electricity, steam, hot oil, and so on) to the contents of the vessel. The potential hazards associated with unfired pressure vessels include hazardous interaction between the material of the vessel and the materials that will be processed in it; inability of the filled vessel to carry the weight of its contents and the corresponding internal pressure; inability of the vessel to withstand the pressure introduced into it plus pressure caused by chemical reactions that occur during processing; and inability of the vessel to withstand any vacuum that may be created accidentally or intentionally.

The most effective preventive measure for overcoming these potential hazards is proper design. Specifications for the design and construction of unfired pressure vessels include requirements in the following areas: working pressure range, working temperature range, type of materials to be processed, stress relief, welding or joining measures, and radiography. Designs that meet the specifications set forth for unfired pressure vessels in such codes as the ASME (American Society of Mechanical Engineers) Code (Section VIII) will overcome most predictable hazards.

Beyond proper design, the same types of precautions taken when operating fired pressure vessels can be used when operating unfired vessels. These include continual inspection, proper housekeeping, periodic testing, visual observation (for detecting cracks), and the use of appropriate safety devices.

HAZARDS OF HIGH-PRESSURE SYSTEMS

The hazards most commonly associated with high-pressure systems are leaks, pulsation, vibration, release of high-pressure gases, and whiplash from broken high-pressure pipe, tubing, or hose.[7] Strategies for reducing these hazards include limiting vibration through the use of vibration dampening (use of anchored pipe supports); decreasing the potential for leaks by limiting the number of joints in the system; using pressure gauges; placing shields or barricades around the system; using remote control and monitoring; and restricting access.

CRACKING HAZARDS IN PRESSURE VESSELS

One of the most serious hazards in pressure vessels is the potential for cracking.[8] Cracking can lead to either a complete rupture or to leaks. The consequences of a complete rupture include (1) blast effects due to the sudden expansion of the contents of the vessel, and (2) possible injuries and damage from fragmentation. The consequences of a leak include (1) suffocation or poisoning of employees depending on the contents of the vessel, (2) explosion and fire, and (3) chemical and thermal burns from contact with the contents of the vessel.

Pressure vessels are used in many different applications to contain many different types of substances ranging from water to extremely toxic chemicals. Leakage or rupture may occur in welded seams, bolted joints, or at nozzles. Figure 17–1 shows a diagram of a typical pressure vessel showing the potential

Figure 17–1
Diagram of a typical pressure vessel showing potential points for leakage or rupture.

points of leakage and rupture. The types of vessels that are most susceptible to leakage and rupture, primarily because of the processes they are part of or their contents, are as follows:

Deaerator Vessels

Deaeration is the process of removing noncondensible gases, primarily oxygen, from the water used in steam generation. Deaerator vessels are used in such applications as power generation, pulp and paper processing, chemical processing, and petroleum refining. The most common failures associated with deaerator vessels are (1) cracks caused by water hammer at welded joints that were not postweld heat treated, and (2) cracks caused by corrosion fatigue.

Amine Vessels

The *amine process* removes hydrogen sulfide from petroleum gases such as propane and butane. It can also be used for removing carbon dioxide in some processes. Amine vessels are used in petroleum refineries, gas treatment facilities, and chemical plants. The most common failures associated with amine vessels are cracks in stressed or unrelieved welds.

Wet Hydrogen Sulfide Vessels

Any fluid that contains water and hydrogen sulfide is considered wet hydrogen sulfide. Many of the vessels used to contain wet hydrogen sulfide are made of steel. Hydrogen is generated when steel is exposed to such a mixture. Dissolved hydrogen can cause cracking, blistering, and embrittlement, particularly in high-strength steels. Consequently, low-strength steels are recommended for wet hydrogen sulfide vessels.

Ammonia Vessels

Vessels for the containment of ammonia are widely used in commercial refrigeration systems and chemical processes. Such ammonia vessels are typically constructed as spheres of carbon steel. The water and oxygen content in ammonia can cause carbon steel to crack, particularly near welds.

Pulp Digester Vessels

The process used to digest pulp in the manufacture of paper involves the use of a weak water solution of sodium hydroxide and sodium sulfide in a temperature range of 110°C to 140°C (230°F to 285°F). The most common failure in pulp digester vessels is cracking along welded seams due primarily to caustic stress corrosion.

NONDESTRUCTIVE TESTING OF PRESSURE VESSELS

To prevent leakage or rupture, it is necessary to examine pressure vessels periodically. There are five widely used nondestructive methods for testing: (1) visual examination, (2) liquid penetration test, (3) magnetic particle test, (4) gamma and X-ray radiography, and (5) ultrasonic test. Visual, liquid penetration, and magnetic particle tests can detect only those defects that are either on the surface or near it. Radiographic and ultrasonic tests can detect problems within the material. Consequently, the visual, liquid penetration, and magnetic particle tests are referred to as *surface tests.* Radiography and ultrasonic are called *volumetric tests.*

Visual Examination

A visual examination consists of taking a thorough look at the vessel to detect signs of corrosion, erosion, or hydrogen blistering. In order to conduct a dependable visual examination of a pressure vessel, it is necessary to have a clean surface and good lighting.

Liquid Penetration Test

This test involves placing a specially formulated liquid penetrant over an area and letting it seep in. When the penetrant is removed from the surface, some of it remains entrapped in the area of discontinuity. A developing agent is then applied, which draws out the entrapped penetrant and magnifies the discontinuity. The process can be enhanced by adding fluorescent chemicals to the penetrant to aid in the detection of problems.

Magnetic Particle Test

This test is based on the fact that discontinuities in or near the surface of a pressure vessel disturb magnetic flux lines that are induced in a ferromagnetic material. Disturbances are detected by applying fine particles of ferromagnetic material to the surface of the vessel. The necessary magnetic field is produced most frequently using the "prod" technique in which electric current is run through an area by applying opposing "prods" (contact probes). A drawback of this test is that corners and surface irregularities in the vessel material can produce the same disturbances as defects. Consequently, special care is needed when using this test in a region with corners or welded joints. Because this test works only with ferromagnetic material, its use is limited to vessels made of carbon and low-alloy steels.

Radiography Test

This test amounts to making an X-ray negative of a given portion of the vessel. The process works in the same way as those used by physicians and dentists.

Irregularities such as holes, voids, or discontinuities produce a greater exposure (darker area) on the X-ray negative.

Ultrasonic Test

This test is similar to radar and other uses of electromagnetic and acoustic waves for detecting foreign objects. Short signals are induced into the material. Waves that are reflected back from discontinuities are detected by one or more transducers. Ultrasonic testing requires an electronic system for generating a signal, a transducer system for converting the electrical into mechanical vibrations and vice versa, and an electronic system for amplifying, processing, and displaying the return signal.

PRESSURE DANGERS TO HUMANS

The term *anoxia* refers to the rare case of a total lack of oxygen. Hypoxia, a condition that occurs when the available oxygen is reduced, can occur while ascending to a high altitude or when oxygen in air has been replaced with another gas, which may happen in some industrial situations.

Altitude sickness is a form of hypoxia associated with high altitudes. Ascent to an altitude of 10,000 feet above sea level can result in a feeling of malaise, shortness of breath, and fatigue. A person ascending to 14,000 to 15,000 feet may experience euphoria, along with a reduction in powers of reason, judgment, and memory. Altitude sickness includes a loss of **useful consciousness** at 20,000 to 25,000 feet. After approximately five minutes at this altitude, a person may lose consciousness. The loss of consciousness comes at approximately one minute or less at 30,000 feet. Over 38,000 feet, most people lose consciousness within 30 seconds and may fall into a coma and possibly die.

Hyperoxia, or an increased concentration of oxygen in air, is not a common situation. Hyperbaric chambers or improperly calibrated scuba equipment can create conditions that may lead to convulsions if pure oxygen is breathed for greater than three hours. Breathing air at a depth of around 300 feet can be toxic and is equivalent to breathing pure oxygen at a depth of 66 feet.

At high pressures of oxygen, around 2,000 to 5,000 mm Hg, dangerous cerebral problems such as dizziness, twitching, vision deterioration, and nausea may occur. Continued exposure to these high pressures will result in confusion, convulsion, and eventual death.

Changes in total pressure can induce *trapped gas effects.* With a decrease in pressure, trapped gases will increase in volume (according to Boyle's law). Trapped gases in the body include air pockets in the ears, sinuses, and chest. Divers refer to the trapped gas phenomenon as the *squeeze.* Jet travel causes the most commonly occurring instance of trapped gas effects. Takeoff and landing may cause relatively sudden shifts in pressure, which may lead to discomfort and pain. With very rapid ascent or descent, injury can develop.

Lung rupture can be caused by a swift return to the surface from diving or decompression during high-altitude flight. This event is rare and happens only if the person is holding his or her breath during the decompression.

Evolved gas effects are associated with the absorption of nitrogen into body tissues. When breathed, nitrogen can be absorbed into all body tissues in concentrations proportional to the partial pressure of nitrogen in air. When a person is ascending in altitude, on the ground, in flight, or under water, nitrogen must be exhaled at a rate equal to or exceeding the absorption rate to avoid evolved gas effects.

If the nitrogen in body tissues such as blood is being absorbed faster than it is being exhaled, bubbles of gas may form in the blood and other tissues. Gas bubbles in the tissues may cause decompression sickness, which can be painful and occasionally fatal. Early symptoms of this disorder occur in body bends or joints such as elbows, knees, and shoulders. The common name for decompression sickness is the **bends.**

When the formation of gas bubbles is due to rapid ambient pressure reduction, it is called **dysbarism.**[9] The major causes of dysbarism are (1) the release of gas from the blood, and (2) the attempted expansion of trapped gas in body tissues. The sickness may occur with the decompression associated with rapidly moving from sea level (considered zero) to approximately 20,000 feet above sea level. Dysbarism is most often associated with underwater diving or working in pressurized containers (such as airplanes). Obese and older people seem to be more susceptible to dysbarism and decompression sickness.

Dysbarism manifests itself in a variety of symptoms. The **creeps** are caused by bubble formation in the skin, which causes an itchy, crawling, rashy feeling in the skin. Coughing and choking, resulting from bubbles in the respiratory system, are called the **chokes.** Bubbles occurring in the brain, although rare, may cause tingling and numbing, severe headaches, spasticity of muscles, and in some cases, blindness and paralysis. Dysbarism of the brain is rare. Rapid pressure change may also cause pain in the teeth and sinuses.[10]

Aseptic necrosis of bone is a delayed effect of decompression sickness. Blood in the capillaries supplying the bone marrow may become blocked with gas bubbles, which can cause a collection of platelets and blood cells to build up in a bone cavity. The marrow generation of blood cells can be damaged as well as the maintenance of healthy bone cells. Some bone areas may become calcified with severe complications when the bone is involved in a joint.

DECOMPRESSION PROCEDURES

Employees who work in an environment that is under pressure must undergo decompression procedures before returning to a normal atmosphere.[11] Such procedures are planned based on the amount of pressure to which the employee is subjected and for how long. In 29 CFR 1926 (Subpart S, Appendix A), OSHA

Partial Table (Two-Stage Decompression)							
		Decompression Data					
Working Chamber Pressure (psig)	Working Period (hours)	Stage No.	Pressure Reduction (psig)		Time in Stage (minutes)	Pressure Reduction Rate (min/pound)	Total Time Decompress (minutes)
			From	To			
20	3	1	20	4	3	0.20	—
		2	4	0	12	3.00	15
	4	1	20	4	3	0.20	—
		2	4	0	40	10.00	43
	5	1	20	4	3	0.20	—
		2	4	0	60	15.00	63

Figure 17–2
Portion of a table for planning a two-stage decompression.
(*Note:* Do not interpolate. Always use the next higher value for conditions that fall between numbers in the table.)

provides tables that can be used for planning appropriate decompression procedures for employees. Figure 17–2 is an example of a portion of such a table.

In most cases, decompression will need to occur in two stages. Figure 17–2 shows a portion of a table to be used for planning two-stage decompressions. The following example demonstrates how to use such a table:

An employee will be working for four hours in an environment with a working chamber pressure of 20 pounds per square inch gauge (psig). Locate 20 psig and 4 working period hours in the table in Figure 17–2. Stage 1 of the decompression will require a reduction in pressure from 20 psig to 4 psig over a period of 3 minutes at the uniform rate of 0.20. Stage 2 of the decompression will require a reduction in pressure from 4 psig to 0 psig over a period of 40 minutes at the uniform rate of 10 minutes per pound. The total time for the decompression procedure is 43 minutes.

Decompression procedures are designed to prevent the various effects of decompression sickness that were explained in the previous section. For a complete set of decompression tables refer to the following Web address:

http://www.osha.gov

MEASUREMENT OF PRESSURE HAZARDS

Confirming the point of pressurized gas leakage can be difficult. After a gas has leaked out to a level of equilibrium with its surrounding air, the symptoms of the leak may disappear. There are several methods of detecting pressure hazards:

- *Sounds* can be used to signal a pressurized gas leak. Gas discharge may be indicated by a whistling noise, particularly with highly pressurized gases escaping through small openings. Workers should not use their fingers to probe for gas leaks as highly pressurized gases may cut through tissue, including bone.
- *Cloth streamers* may be tied to the gas vessel to help indicate leaks. Soap solutions may be smeared over the vessel surface so that bubbles are formed when gas escapes. A stream of bubbles indicates gas release.
- *Scents* may be added to gases that do not naturally have an odor. The odor sometimes smelled in homes that cook or heat with natural gas is not the gas but a scent added to it.
- *Leak detectors* that measure pressure, current flow, or radioactivity may be useful for some types of gases.
- *Corrosion* may be the long-term effect of escaping gases. Metal cracking, surface roughening, and general weakening of materials may result from corrosion.

There are many potential causes of gas leaks. The most common of these are as follows:

- *Contamination* by dirt can prevent the proper closing of gas valves, threads, gaskets, and other closures used to control gas flow.
- *Overpressurization* can overstress the gas vessel, permitting gas release. The container closure may distort and separate from gaskets, leading to cracking.
- *Excessive temperatures* applied to dissimilar metals that are joined may cause unequal thermal expansion, loosening the metal-to-metal joint and allowing gas to escape. Materials may crack because of excessive cold, which may also result in gas escape. Thermometers are often used to indicate the possibility of gas release.
- *Operator errors* may lead to hazardous gas release from improper closure of valves, inappropriate opening of valves, or overfilling of vessels. Proper training and supervision can reduce operator errors.

Destructive as well as nondestructive methods may be used to detect pressure leaks and incorrect pressure levels. **Nondestructive testing** methods do not harm the material being tested. Nondestructive methods may include mixing dye penetrants and magnetic or radioactive particles with the gas and then measuring the flow of the gas. Ultrasonic and X-ray waves are another form of nondestructive testing and are often used to characterize materials and detect cracks or other leakage points.

Destructive testing methods destroy the material being checked. Proof pressures generate stresses to the gas container, typically 1.5 to 1.667 times the maximum expected operating pressure for that container. Strain measurements may also be collected to indicate permanent weakening changes to the container material that remain after the pressure is released. **Proof pressure tests** often call for the pressure to be applied for a specified time and released. Stress and strain tests are then applied to the material. Proof pressure tests may or may not result in the destruction of the container being tested.

REDUCTION OF PRESSURE HAZARDS

The reduction of pressure hazards often requires better maintenance and inspection of equipment that measures or uses high-pressure gases. Proper storage of pressurized containers reduces many pressure hazards. Pressurized vessels should be stored in locations away from cold or heat sources, including the sun. Cryogenic compounds (those that have been cooled to unusually low temperatures) may boil and burst the container when not kept at the proper temperatures. The whipping action of pressurized flexible hoses can also be dangerous. Hoses should be firmly clamped at the ends when pressurized.

Gas compression can occur in sealed containers exposed to heat. For this reason, aerosol cans must never be thrown into or exposed to a fire. Aerosol cans may explode violently when exposed to heat, although most commercially available aerosols are contained in low-melting point metals that melt before pressure can build up.

Pressure should be released before working on equipment. Gauges can be checked before any work on the pressurized system is begun. When steam equipment is shut down, liquid may condense within the system. This liquid or dirt in the system may become a propellant, which may strike bends in the system, causing loud noises and possible damage.

Water hammer is a shock effect caused by liquid flow suddenly stopping.[12] The shock effect can produce loud noises. The momentum of the liquid is conducted back upstream in a shock wave. Pipe fittings and valves may be damaged by the shock wave. Reduction of this hazard involves using air chambers in the system and avoiding the use of quick-closing valves.

Negative pressures or **vacuums** are caused by pressures below atmospheric level. Negative pressures may result from hurricanes and tornadoes. Vacuums may cause collapse of closed containers. Building code specifications usually allow for a pressure differential. Vessel wall thickness must be designed to sustain the load imposed by the differential in pressure caused by negative pressure. Figure 17–3 describes several methods to reduce the hazards associated with pressurized containers.

Discussion Case

What Is Your Opinion?

While visiting a friend, Mary Carpenter—safety director for a small manufacturer of pressurized metal containers—saw something that really bothered her. While cleaning up his yard, her friend threw all his trash into a fire contained in a metal drum. Carpenter noticed two aerosol cans being thrown in the fire and quickly warned her friend of the danger of explosion. He laughed and shrugged off her warning, saying "There is no danger. The can will melt before it explodes." Who is right in this situation? What is your opinion?

- Install valves so that failure of a valve does not result in a hazard.
- Do not store pressurized containers near heat or sources of ignition.
- Train and test personnel dealing with pressurized vessels. Only tested personnel should be permitted to install, operate, maintain, calibrate, or repair pressurized systems. Personnel working on pressure systems should wear safety face shields or goggles.
- Examine valves periodically to ensure that they are capable of withstanding working pressures.
- Operate pressure systems only under the conditions for which they were designed.
- Relieve all pressure from the system before performing any work.
- Label pressure system components to indicate inspection status as well as acceptable pressures and flow direction.
- Connect pressure relief devices to pressure lines.
- Do not use pressure systems and hoses at pressure exceeding the manufacturer's recommendations.
- Keep pressure systems clean.
- Keep pressurized hoses as short as possible.
- Avoid banging, dropping, or striking pressurized containers.
- Secure pressurized cylinders by a chain to prevent toppling.
- Store acetylene containers upright.
- Examine labels before using pressurized systems to ensure correct matching of gases and uses.
- Use dead man's switches on high-pressure hose wands.

Figure 17–3
Reduction of pressure hazards.

SUMMARY

1. Pressure is defined in physics as the force exerted against an opposing fluid or solid.
2. Pressure is perceived in relation to the earth's atmosphere.
3. Barometers are used to measure atmospheric pressure.
4. Boyle's law states that the product of a given pressure and volume is constant with constant temperature:

$$P_1V_1 = P_2V_2, \quad \text{when } T \text{ is constant}$$

5. Inspiration is breathing air into the lungs.
6. Expiration occurs when air leaves the lungs.

7. Dalton's law of partial pressures states that in a mixture of ideal gases, the pressure exerted by the mixture is the sum of the pressures exerted by each component gas of the mixture:

$$P_A = P_O + P_N + P_{else}$$

8. Water vapor, although a gas, does not conform to Dalton's law.
9. Increasing pressure on the body does not create problems by itself.
10. Decompression sickness can occur from the decompression involved with a rapid rise from sea level to 18,000 feet or a rapid ascent from around 132 to 66 feet underwater.
11. Under extreme circumstances of rapid ascent from underwater diving or high-altitude decompression, lung rupture can occur.
12. Factors involved with decompression sickness include previous exposure history, age, physical fitness, exercise, low temperatures, speed of decompression, and length of exposure.
13. The bends are an example of decompression sickness.
14. Aseptic necrosis of bone can be a delayed effect of decompression sickness.
15. Hypoxia is a reduction of available oxygen.
16. Excessive nitrogen absorption into body tissues can occur from breathing nitrogen-enriched air and is called nitrogen narcosis.
17. Evolved gas effects are associated with the absorption of nitrogen into body tissues.
18. Altitude sickness is a form of hypoxia.
19. Altitude sickness may involve a loss of useful consciousness.
20. Hyperoxia is an increased concentration of oxygen in air and is not common.
21. Trapped gas effects can result from changes in total pressure.
22. Dysbarism is the rapid formation of gas bubbles in the tissue due to rapid ambient pressure reduction.
23. The creeps are caused by bubble formation in the skin.
24. Formation of bubbles in the respiratory tract is called the chokes.
25. Pressure vessels are of many types including deaerator, amine, wet hydrogen sulfide, ammonia, and pulp digester vessels.
26. Nondestructive testing of pressure vessels can be accomplished by visual examination, the liquid penetration test, magnetic particle test, radiography test, and ultrasonic testing.
27. Several methods are used to detect pressure hazards: sounds, cloth streamers, soap solutions, scents, leak detectors, visual checks for corrosion or contamination.
28. Detection of pressure hazards includes destructive and nondestructive testing.
29. Proof pressures can be used to test container strength to contain a pressurized gas.
30. Pressurized cylinders and other vessels should be stored away from cold or heat sources, including the sun.
31. Aerosol cans should not be discarded in fires or by any method using heat.

32. Water hammer is a series of loud noises caused by pressurized liquid flow suddenly stopping.
33. Negative pressures or vacuums are caused by pressures below atmospheric level.

KEY TERMS AND CONCEPTS

Altitude sickness

Amine vessels

Ammonia vessels

Aseptic necrosis

Barometer

Bends

Boyle's law

Chokes

Creeps

Dalton's law of partial pressures

Deaerator vessels

Decompression sickness

Destructive testing

Dysbarism

Evolved gas effects

Expiration

Hazard

Hyperoxia

Hypoxia

Inspiration

Liquid penetration test

Magnetic particle test

Negative pressures

Nitrogen narcosis

Nondestructive testing

Pressure

Pressure hazard

Proof pressure

Pulp digester vessels

Radiography test

Trapped gas effects

Ultrasonic testing

Useful consciousness

Vacuums

Visual examination

Water hammer

Wet hydrogen sulfide vessels

REVIEW QUESTIONS

1. Against which references is pressure measured? How are these references measured?
2. Define *inspiration* and *expiration*.
3. Explain Dalton's law of partial pressures.
4. Does water vapor conform to Dalton's law?
5. Briefly discuss decompression sickness.
6. What do length of exposure, the bends, the chokes, and aseptic necrosis of bone have in common?

7. Define *hypoxia* and *hyperoxia.*

8. Explain nitrogen narcosis.

9. Discuss altitude sickness.

10. What is the relationship between trapped gas effects and dysbarism?

11. What is the difference between destructive and nondestructive testing?

12. Briefly explain proof pressures.

13. What causes vacuums?

14. Explain three ways to conduct nondestructive testing of pressure vessels.

15. What is the total decompression time for an employee who works for four hours under pressure of 20 psig?

ENDNOTES

1. Occupational Safety and Health Administration, "Pressure Vessel Guidelines," *OSHA Technical Manual* (Washington, DC: Occupational Safety and Health Administration, 2000), sec. IV, chap. 3, 30.

2. Ibid., 31.

3. Ibid.

4. Ibid.

5. Ibid.

6. Ibid.

7. Ibid., 38.

8. Ibid., 39.

9. Ibid., 39.

10. Ibid.

11. OSHA Regulations 29 CFR 1926 (Subpart S, Appendix A), Decompression Tables.

12. Ibid.

ELECTRICAL HAZARDS

Major Topics

- Electrical Hazards Defined
- Sources of Electrical Hazards
- Electrical Hazards to Humans
- Detection of Electrical Hazards
- Reduction of Electrical Hazards
- OSHA's Electrical Standards
- Electrical Safety Program
- Electrical Hazards Self-Assessment
- Prevention of Arc Flash Injuries (NFPA 70E)

Consider the following scenario: A textile mill in Massachusetts was fined $66,375 when an employee contacts OSHA and complains about unsafe conditions at the mill. The Region 1 Office of OSHA conducted an investigation in response to the complaint that uncovered the following willful violation: allowing employees to perform live electrical work without safe work procedures or appropriate personal protective equipment.[1] In addition, the investigation uncovered several serious violations including (1) storage of flammable materials near emergency exits, (2) improper storage of oxygen and acetylene cylinders, (3) failure to post load ratings, and (4) an exposed live electrical source and unsuitable electrical outlets for wet or damp locations.[2]

ELECTRICAL HAZARDS DEFINED

Electricity is the flow of negatively charged particles called *electrons* through an electrically conductive material. Electrons orbit the nucleus of an atom, which is located approximately in the atom's center. The negative charge of the

Figure 18–1
Basic structure of an atom.

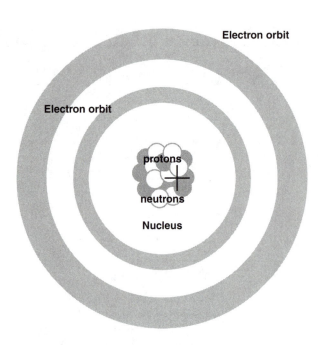

electrons is neutralized by particles called **neutrons**, which act as temporary energy repositories for the interactions between positively charged particles called **protons** and **electrons**.

Figure 18–1 shows the basic structure of an atom, with the positively charged nucleus in the center. The electrons are shown as energy bands of orbiting negatively charged particles. Each ring of electrons contains a particular quantity of negative charges. The basic characteristics of a material are determined by the number of electron rings and the number of electrons in the outer rings of its atoms. A *positive charge* is present when an atom (or group of atoms) in a material has too many electrons in its outer shell. In all other cases, the atom or material carries a *negative charge*.

Electrons that are freed from an atom and are directed by external forces to travel in a specific direction produce *electrical current*, also called *electricity*. **Conductors** are substances that have many free electrons at room temperature and can pass electricity. *Insulators* do not have a large number of free electrons at room temperature and do not conduct electricity. Substances that are neither conductors nor insulators can be called **semiconductors**.

Electrical current passing through the human body causes a shock. The quantity and path of this current determines the level of damage to the body. The path of this flow of electrons is from a negative source to a positive point, because opposite charges attract one another.

When a surplus or deficiency of electrons on the surface of a material exists, **static electricity** is produced. This type of electricity is called "static" because there is no positive material nearby to attract the electrons and cause them to move. Friction is not required to produce static electricity, although it can increase the charge of existing static electricity. When two surfaces of opposite static electricity charges are brought into close range, a discharge, or spark, will occur. The spark from static electricity is often the first clue that such static exists. A common example is the sparks that come from rustling woolen blankets in dry heated indoor air.

The *potential difference* between two points in a circuit is measured by **voltage**. The higher the voltage, the more likely it is that electricity will flow between the negative and positive points.

Pure conductors offer little **resistance** to the flow of electrons. Insulators, on the other hand, have very high resistance to electricity. Semiconductors have a medium-range resistance to electricity. The higher the resistance, the lower the flow of electrons. Resistance is measured in **ohms**.

Electrical current is produced by the flow of electrons. The unit of measurement for current is **amperes** (or amps). One amp is a current flow of 6.28×10^{18} electrons per second. Current is usually designated by I. **Ohm's law** describes the relationship among volts, ohms, and amps. One ohm is the resistance of a conductor that has a current of one amp under the potential of one volt. Ohm's law is stated as

$$V = IR$$

where

V = potential difference in volts
I = current flow in amps
R = resistance to current flow in ohms

Power is measured in wattage (or *watts*) and can be determined from Ohm's law:

$$W = VI \quad \text{or} \quad W = I^2R$$

where

W = power in watts

Most industrial and domestic use of electricity is supplied by **alternating current** (or **AC current**). In the United States, standard AC circuits cycle 60 times per second. The number of cycles per second is known as **frequency** and is measured in **hertz**. Because voltage cycles in AC current, an **effective current** for AC circuits is computed, which is slightly less than the peak current during a cycle.

A **direct current** (or **DC current**) has been found to generate as much heat as an AC current that has a peak current 41.4 percent higher than the DC. The ratio of effective current to peak current can be determined by

(Effective current)/(Peak current) = (100%)/(100% + 41.4%) = 0.707 or 70.7%

Effective voltages are computed using the same ratios as effective current. A domestic 110-volt circuit has an effective voltage of 110 volts, with peaks of voltage over 150 volts.

The path of electrical current must make a complete loop for the current to flow. This loop includes the source of electrical power, a conductor to act as the path, a device to use the current (called a **load**), and a path to the ground. The earth maintains a relatively stable electrical charge and is a good conductor. The earth is considered to have **zero potential** because of its massive size. Any electrical conductor pushed into the earth is said to have zero potential. The earth is used as a giant common conductor back to the source of power.

Electrocution occurs when a person makes contact with a conductor carrying a current and simultaneously contacts the ground or another object that includes a conductive path to the ground. This person completes the circuit loop by providing a load for the circuit and thereby enables the current to pass through his or her body. People can be protected from this danger by insulating the conductors, insulating the people, or isolating the danger from the people.

The National Electrical Code (NEC) is published by the National Fire Protection Association (NFPA). This code specifies industrial and domestic electrical safety precautions. The NEC categorizes industrial locations and gases relative to their degree of fire hazard and describes in detail the safety requirements for industrial and home wiring. The NEC has been adopted by many jurisdictions as the local electrical code. The National Board of Fire Underwriters sponsors Underwriters Laboratories (UL). The UL determines whether equipment and materials for electrical systems are safe in the various NEC location categories. The UL provides labels for equipment that it approves as safe within the tested constraints.

Typical 110-volt circuit wiring has a **hot wire** carrying current, a **neutral wire**, and a **ground wire**. The neutral wire may be called a **grounded conductor**, with the ground wire being called a **grounding conductor**. Neutral wires usually have white insulation, hot wires have red or black insulation, and ground wires have green insulation or are bare. Figure 18–2 shows a typical three-wire circuit.

The hot wire carries an effective voltage of 110 volts with respect to the ground, whereas the neutral wire carries nearly zero voltage. If the hot wire makes contact with an unintended conductor, such as a metal equipment case,

Figure 18–2
Typical three-wire circuit.

the current can bypass the load and go directly to the ground. With the load skipped, the ground wire is a low-resistance path to the earth and carries the highest current possible for that circuit.

A **short circuit** is a circuit in which the load has been removed or bypassed. The ground wire in a standard three-wire circuit provides a direct path to the ground, bypassing the load. Short circuits can be another source of electrical hazard if a human is the conductor to the ground, thereby bypassing the load.

SOURCES OF ELECTRICAL HAZARDS

Short circuits are one of many potential electrical hazards that can cause electrical shock. Another hazard is water, which considerably decreases the resistance of materials, including humans. The resistance of wet skin can be as low as 450 ohms, whereas dry skin may have an average resistance of 600,000 ohms. According to Ohm's law, the higher the resistance, the lower the current flow. When the current flow is reduced, the probability of electrical shock is also reduced.

The major causes of electrical shock are

- Contact with a bare wire carrying current. The bare wire may have deteriorated insulation or be normally bare.
- Working with electrical equipment that lacks the UL label for safety inspection.
- Electrical equipment that has not been properly grounded. Failure of the equipment can lead to short circuits.
- Working with electrical equipment on damp floors or other sources of wetness.
- Static electricity discharge.
- Using metal ladders to work on electrical equipment. These ladders can provide a direct line from the power source to the ground, again causing a shock.
- Working on electrical equipment without ensuring that the power has been shut off.
- Lightning strikes.

Figure 18–3 depicts some of these electrical shock hazards.

Electrostatic Hazards

Electrostatic hazards may cause minor shocks. Shocks from static electricity may result from a single discharge or multiple discharges of static. Sources of electrostatic discharge include the following:

- Briskly rubbing a nonconductive material over a stationary surface. One common example of this is scuffing shoes across a wool or nylon carpet. Multilayered clothing may also cause static sparks.[3]

Figure 18–3
Electrical shock hazards.

- Moving large sheets of plastic, which may discharge sparks.
- The explosion of organic and metallic dusts, which have occurred from static buildup in farm grain silos and mine shafts.
- Conveyor belts. Depending on their constituent material, they can rub the materials being transported and cause static sparks.
- Vehicle tires rolling across a road surface.
- Friction between a flowing liquid and a solid surface.[4]

The rate of discharge of electrical charges increases with lower humidity. Electrostatic sparks are often greater during cold, dry winter days. Adding humidity to the air is not commonly used to combat static discharge, however, because higher humidity may result in an uncomfortable working environment and adversely affect equipment.[5]

Arcs and Sparks Hazards

With close proximity of conductors or contact of conductors to complete a circuit, an electric **arc** can jump the air gap between the conductors and ignite combustible gases or dusts. When the electric arc is a discharge of static electricity, it may be called a **spark**. A spark or arc may involve relatively little or a great deal of power and is usually discharged into a small space.

Combustible and Explosive Materials

High currents through contaminated liquids may cause the contaminants to expand rapidly and explode. This situation is particularly dangerous with contaminated oil-filled circuit breakers or transformers. A poor match between current or polarity and capacitors can cause an explosion. In each of these cases, the conductor is not capable of carrying a current of such high magnitude. Overheating from high currents can also lead to short circuits, which in turn may generate fires or explosions.

Lightning Hazards

Lightning is static charges from clouds following the path of least resistance to the earth, involving very high voltage and current. If this path to the earth involves humans, serious disability may result, including electrocution. Lightning may also damage airplanes from intracloud and cloud-to-cloud flashes. Electrical equipment and building structures are commonly subject to lightning hazards. Lightning tends to strike the tallest object on the earth below the clouds. A tree is a common natural path for lightning.

Improper Wiring

Improper wiring permits equipment to operate normally but can result in hazardous conditions. The section of this chapter on detection of electrical hazards discusses tests to identify unsafe wiring practices. One common mistake is to "jump" the ground wire to the neutral wire. In this case, the ground wire is actually connected to the neutral wire. Equipment usually operates in a customary way, but the hazard occurs when low voltages are generated on exposed parts of the equipment, such as the housing. If the neutral circuit becomes corroded or loose, the voltage on the ground wire increases to a dangerous level.

Improper wiring (or miswiring) can cause other hazards. When the ground is connected improperly, the situation is referred to as **open ground**. Usually the equipment with this miswiring will operate normally. If a short occurs in the equipment circuitry without proper grounding, anyone touching that equipment may be severely shocked.

With **reversed polarity**, the hot and neutral wires have been reversed. A worker who is not aware that the black lead (hot) and white lead (neutral) have been reversed could be injured or cause further confusion by connecting the circuit to another apparatus. If a short between the on/off switch and the load occurred, the equipment may run indefinitely, regardless of the switch position. In a reversed polarity light bulb socket, the screw threads become conductors.[6]

Temporary wiring installations sometimes remain in place for years until an accident occurs. Flexible wiring should rarely be substituted for fixed wiring in permanent buildings. A loose knot should be tied in a flexible cord when the plug is installed or replaced. The knot can prevent a pull on the cord from being transmitted to electrical connections such as the plug.

Insulation Failure

The degradation of insulation can cause a bare wire and resulting shock to anyone coming in contact with that wire. Most insulation failure is caused by environments toxic to insulation. These environments include:

- Direct sunlight or other sources of ultraviolet light, which can induce gradual breakdown of plastic insulation material.
- Sparks or arcs from discharging static electricity, which can result in burned-through holes in insulation.
- Repeated exposure to elevated temperatures, which can produce slow but progressive degradation of insulation material.
- Abrasive surfaces, which can result in erosion of the material strength of the insulation.
- Substance incompatibility with the atmosphere around the insulation and the insulation material, which can induce chemical reactions. Such reactions may include oxidation or dehydration of the insulation and eventual breakdown.
- Animals such as rodents or insects chewing or eating the insulation material, leading to exposure of the circuit. Insects can also pack an enclosed area with their bodies so tightly that a short circuit occurs. This is a common occurrence with electrical systems near water, such as pump housings and television satellite dishes.
- Moisture and humidity being absorbed by the insulation material, which may result in the moisture on the insulation carrying a current.

Equipment Failure

There are several ways in which equipment failure can cause electrical shocks. Electrical equipment designers attempt to create devices that are explosion-proof, dust-ignition-proof, and spark-proof. Following are some of the more common types of equipment failure:

- Wet insulation can become a conductor and cause an electrical shock.
- Portable tool defects can result in the device's housing carrying an electric current. Workers do not expect tool housings to be charged and may be shocked when they touch a charged tool housing.
- Broken power lines carry great amperage and voltage and can cause severe disability.
- When equipment is not properly grounded or insulated, an unshielded worker may receive a substantial electrical shock.

		Division	
Class	**Group**	**I**	**II**
I. Flammable vapors and gases	A. Acetylene B. Hydrogen C. Ether D. Hydrocarbon fuels and solvents	Normally explosive; flammable paint spray areas	Not normally in explosive concentration; adjacent to paint spray area
II. Combustible dusts	E. Metal dusts F. Carbon dusts G. Flour, starch, grain, plastic, or chemical dusts	Conductive or ignitable dusts may be present; grain mills or processors	Not normally in ignitable concentration; grain storage areas
III. Ignitable fibers	Textiles, woodworking	Handled or used in manufacturing; cotton gins	Stored or handled in storage, not in manufacturing; excelsior storage

Figure 18–4
Hazardous electrical equipment location categories.

Hazardous Locations for Electrical Equipment

The NEC classifies hazardous locations for electrical equipment. There are three basic classes: Class I for flammable vapors and gases, Class II for combustible dusts, and Class III for ignitable fibers. There are also two divisions of hazard categories. Division I has more stringent requirements for electrical installation than Division II does. Figure 18–4 gives examples for each location category.[7]

ELECTRICAL HAZARDS TO HUMANS

The greatest danger to humans suffering electrical shock results from current flow. The voltage determines whether a particular person's natural resistance to current flow will be overcome. Skin resistance can vary between 450 ohms and 600,000 ohms, depending on skin moisture.[8] Some levels of current "**freeze**" a person to the conductor; the person cannot voluntarily release his or her grasp. **Let-go current** is the highest current level at which a person in contact with the conductor can release the grasp of the conductor. Figure 18–5 shows the relationship between amperage dosage and danger with a typical domestic 60-cycle AC current.

The severity of injury with electrical shock depends on the dosage of current, as shown in Figure 18–5, but also on the path taken through the body by

Dose in Current in Milliamps	Effect on Human Body
Less than 1	No sensation, no perceptible effect.
1	Shock perceptible, reflex action to jump away. No direct danger from shock but sudden motion may cause accident.
More than 3	Painful shock.
6	Let-go current for women.*
9	Let-go current for men.*
10–15	Local muscle contractions. Freezing to the conductor for 2.5% of the population.
30–50	Local muscle contractions. Freezing to the conductor for 50% of the population.
50–100	Prolonged contact may cause collapse and unconsciousness. Death may occur after three minutes of contact due to paralysis of the respiratory muscles.
100–200	Contact of more than a quarter of a second may cause ventricular fibrillation of the heart and death. AC currents continuing for more than one heart cycle may cause fibrillation.
Over 200	Clamps and stops the heart as long as the current flows. Heart beating and circulation may resume when current ceases. High current can produce respiratory paralysis, which can be reversed with immediate resuscitation. Severe burns to the skin and internal organs. May result in irreparable body damage.

*Difference between men and women is based on the relative body mass of the "average"-sized man and woman (60-cycle AC current).

Figure 18–5
Current effects on the human body (60-cycle AC current).

the current. The path is influenced by the resistance of various parts of the body at the time of contact with the conductor. The skin is the major form of resistance to current flow. Current paths through the heart, brain, or trunk are generally much more injurious than paths through extremities.

DETECTION OF ELECTRICAL HAZARDS

Several items of test equipment can be used to verify electrical equipment safety. A **circuit tester** is an inexpensive piece of test equipment with two wire leads capped by probes and connected to a small bulb. Most circuit testers test

| | Lights | | |
Situation	1	2	3
Correct wiring	On	On	Off
Ground jumped to neutral	On	On	Off
Neutral and ground reversed	On	On	Off
Reversed polarity	On	Off	On
Open ground	On	Off	Off

Figure 18–6
Receptacle wiring tester indicator lights.

at least a 110- to 220-volt range. This simple tester can ensure that power has been turned off before electrical maintenance begins. The tester may also be used to determine whether housings and other equipment parts are carrying a current. When one of the leads makes contact with a hot wire and the other lead connects to a grounded conductor, the bulb lights.

A **receptacle wiring tester** is a device with two standard plug probes for insertion into an ordinary 110-volt outlet and a probe for the ground. Indicator lights show an improperly wired receptacle (outlet). However, there are several types of miswiring that are not disclosed by using this tester, including the ground wire to neutral wire mistake. Figure 18–6 shows the meaning of lit indicator lights on the receptacle wiring tester.

A continuity tester may be used to determine whether a conductor is properly grounded or has a break in the circuit. Continuity is checked on circuits that are disconnected from a power source. Continuity testers often have an alligator clip on one end of a wire and a bulb and probe on the other end of the same wire. One terminal of the tester can be connected to the equipment housing; the other terminal is connected to a known ground. If the bulb does not light, the equipment is shown to be improperly grounded. With a circuit, the bulb lights when a current is capable of passing through the complete circuit. The unlit bulb of a continuity tester indicates a break in the circuit.

Infrared thermal imaging is another technique that can be used for detecting electrical hazards.

REDUCTION OF ELECTRICAL HAZARDS

Grounding of electrical equipment is the primary method of reducing electrical hazards. The purpose of grounding is to safeguard people from electrical shocks, reduce the probability of a fire, and protect equipment from damage.

Grounding ensures a path to the earth for the flow of excess current. Grounding also eliminates the possibility of a person being shocked by contact with a charged capacitor. The actual mechanism of grounding was discussed at the beginning of this chapter.

Electrical system grounding is achieved when one conductor of the circuit is connected to the earth. Power surges and voltage changes are attenuated and usually eliminated with proper system grounding. **Bonding** is used to connect two pieces of equipment by a conductor. Bonding can reduce potential differences between the equipment and thus reduce the possibility of sparking. Grounding, in contrast, provides a conducting path between the equipment and the earth. Bonding and grounding together are used for entire electrical systems.

Separate equipment grounding involves connecting all metal frames of the equipment in a permanent and continuous manner. If an insulation failure occurs, the current should return to the system ground at the power supply for the circuit. The equipment ground wiring will be the path for the circuit current, enabling circuit breakers and fuses to operate properly. The exposed metal parts of the equipment shown in Figure 18–7 must be grounded or provided with double insulation.[9]

A **ground fault circuit interrupter (GFCI)**, also called a *ground fault interrupter (GFI)*, can detect the flow of current to the ground and open the circuit, thereby interrupting the flow of current. When the current flow in the hot wire is greater than the current in the neutral wire, a **ground fault** has occurred. The GFI provides a safety measure for a person who becomes part of the ground fault circuit. The GFI cannot interrupt current passing between two circuits or between the hot and neutral wires of a three-wire circuit. To ensure safety, equipment must be grounded as well as protected by a GFI. A GFI should be replaced periodically based on the manufacturer's recommendations.

There are several options for reducing the hazards associated with static electricity. The primary hazard of static electricity is the transfer of charges to surfaces with lower potential. Bonding and grounding are two means of controlling static discharge. **Humidification** is another mechanism for reducing electrical static; it was discussed in the section on sources of electrical hazards. Raising

Safety Fact

Workplace Deaths from Electrocution

Workplace deaths from electrocution represent a serious and ongoing problem. Almost 450 traumatic work-related deaths per year are caused by electrocution. This is approximately 6 percent of all work-related deaths annually. The workplace hazards most frequently associated with electrocution are internal wiring, buried electrical cables, and overhead power lines. Accidental contact with these hazards by cranes, booms, hoists, riggings, scaffolds, ladders, trucks, and vehicles are the primary causes of electrocution on the job.

Portable electric tools such as drills and saws.

Communication receivers and transmitters.

Electrical equipment in damp locations.

Television antenna towers.

Electrical equipment in flammable liquid storage areas.

Electrical equipment operated with over 150 volts.

Figure 18–7
Equipment requiring grounding or double insulation.

the humidity above 65 percent reduces charge accumulation.[10] However, when the relative humidity exceeds 65 percent, biological agents can begin to grow in heating, ventilation, and air conditioning (HVAC) ducts and unventilated areas.

Antistatic materials have also been used effectively to reduce electrical static hazards. Such materials either increase the surface conductivity of the charged material or absorb moisture, which reduces resistance and the tendency to accumulate charges.

Ionizers and electrostatic neutralizers ionize the air surrounding a charged surface to provide a conductive path for the flow of charges. Radioactive neutralizers include a radioactive element that emits positive particles to neutralize collected negative electrical charges. Workers need to be safely isolated from the radioactive particle emitter.

Safety Fact

Handling Equipment Exposed to Water

Additional electrical hazards are introduced into the workplace by fires or natural disasters every year. Electrical equipment that is exposed to water by flood, firefighting, tropical storms, hurricanes, or any other calamity must be handled carefully. Such equipment can be extremely dangerous if powered up without proper reconditioning. The National Electrical Manufacturers Association provides guidelines for the proper handling of electrical equipment that has been exposed to water. NEMA can be contacted at the following address:

National Electrical Manufacturers Association
300 N. 17th St., Suite 1847
Rosslyn, VA 22209
Telephone: 703-841-3268
FAX: 703-841-3368
http://www.nema.org/

Fuses consist of a metal strip or wire that melts if a current above a specific value is conducted through the metal. Melting the metal causes the circuit to open at the fuse, thereby stopping the flow of current. Some fuses are designed to include a time lag before melting to allow higher currents during startup of the system or as an occasional event.

Magnetic circuit breakers use a solenoid (a type of coil) to surround a metal strip that connects to a tripping device. When the allowable current is exceeded, the magnetic force of the solenoid retracts the metal strip, opening the circuit. Thermal circuit breakers rely on excess current to produce heat and bending in a sensitive metal strip. Once bent, the metal strip opens the circuit. Circuit breakers differ from fuses in that they are usually easier to reset after tripping and often provide a lower time lag or none at all before being activated.

Double insulation is another means of increasing electrical equipment safety. Most double-insulated tools have plastic nonconductive housings in addition to standard insulation around conductive materials.

There are numerous methods of reducing the risk of electrocution by lightning. Figure 18–8 lists the major precautions to take.[11]

Another means of protecting workers is isolating the hazard from the workers or vice versa. **Interlocks** automatically break the circuit when an unsafe situation is detected. Interlocks may be used around high-voltage areas

---

- ■ Place lightning rods so that the upper end is higher than nearby structures.
- ■ Avoid standing in high places or near tall objects. Be aware that trees in an open field may be the tallest object nearby.
- ■ Do not work with flammable liquids or gases during electrical storms.
- ■ Ensure proper grounding of all electrical equipment.
- ■ If inside an automobile, remain inside the automobile.
- ■ If in a small boat, lie down in the bottom of the boat.
- ■ If in a metal building, stay in the building and do not touch the walls of the building.
- ■ Wear rubber clothing if outdoors.
- ■ Do not work touching or near conducting materials, especially those in contact with the earth such as fences.
- ■ Avoid using the telephone during an electrical storm.
- ■ Do not use electrical equipment during the storm.
- ■ Avoid standing near open doors or windows where lightning may enter the building directly.

Figure 18–8
Lightning hazard control.

- Ensure that power has been disconnected from the system before working with it. Test the system for deenergization. Capacitors can store current after power has been shut off.
- Allow only fully authorized and trained people to work on electrical systems.
- Do not wear conductive material such as metal jewelry when working with electricity.
- Screw bulbs securely into their sockets. Ensure that bulbs are matched to the circuit by the correct voltage rating.
- Periodically inspect insulation.
- If working on a hot circuit, use the buddy system and wear protective clothing.
- Do not use a fuse with a greater capacity than was prescribed for the circuit.
- Verify circuit voltages before performing work.
- Do not use water to put out an electrical fire.
- Check the entire length of electrical cord before using it.
- Use only explosion-proof devices and nonsparking switches in flammable liquid storage areas.
- Enclose uninsulated conductors in protective areas.
- Discharge capacitors before working on the equipment.
- Use fuses and circuit breakers for protection against excessive current.
- Provide lightning protection on all structures.
- Train people working with electrical equipment on a routine basis in first aid and cardiopulmonary resuscitation (CPR).

Figure 18–9
Summary of safety precautions for electrical hazards.

to keep personnel from entering the area. Elevator doors typically have interlocks to ensure that the elevator does not move when the doors are open. Warning devices to alert personnel about detected hazards may include lights, colored indicators, on/off blinkers, audible signals, or labels.

It is better to design safety into the equipment and system than to rely on human behavior such as reading and following labels. Figure 18–9 summarizes the many methods of reducing electrical hazards.

OSHA'S ELECTRICAL STANDARDS

OSHA's standards relating to electricity are found in 29 CFR 1910 (Subpart S). They are extracted from the National Electrical Code. This code should be referred to when more detail is needed than appears in OSHA's excerpts. Subpart S is divided into the following two categories of standards: (1) Design of Electrical Systems, and (2) Safety-Related Work Practices. The standards in each of these categories are as follows:

Design of Electrical Systems

1910.302	Electric utilization systems
1910.303	General requirements
1910.304	Wiring design and protection
1910.305	Wiring methods, components, and equipment for general use
1910.306	Specific-purpose equipment and installations
1910.307	Hazardous (classified) locations
1910.308	Special systems

Safety-Related Work Practices

1910.331	Scope
1910.332	Training
1910.333	Selection and use of work practices
1910.334	Use of equipment
1910.335	Safeguards for personal protection

ELECTRICAL SAFETY PROGRAM

With electrocution accounting for approximately 6 percent of all workplace deaths in the United States every year, it is important that employers have instituted an effective electrical safety program. The National Institute for Occupational Safety and Health (NIOSH) recommends the following strategies for establishing such a program:

- Develop and implement a comprehensive safety program and, when necessary, revise existing programs to address thoroughly the area of electrical safety in the workplace.
- Ensure compliance with existing OSHA regulations, Subpart S of 29 CFR 1910.302 through 1910.399 of the General Industry Safety and Health Standards, and Subpart K of 29 CFR 1926.402 through 1926.408 of the OSHA Construction Safety and Health Standards.
- Provide all workers with adequate training in the identification and control of the hazards associated with electrical energy in their workplace.
- Provide additional specialized electrical safety training to those working with or around exposed components of electric circuits. This training should include, but not be limited to, training in basic electrical theory,

proper safe work procedures, hazard awareness and identification, proper use of personal protective equipment (PPE), proper lockout/tagout procedures, first aid including CPR, and proper rescue procedures. Provide periodic retraining as necessary.

- Develop and implement procedures to control hazardous electrical energy that include lockout and tagout procedures. Ensure that workers follow these procedures.
- Provide testing or detection equipment for those who work directly with electrical energy that ensure their safety during performance of their assigned tasks.
- Ensure compliance with the National Electrical Code and the National Electrical Safety Code.
- Conduct safety meetings regularly.
- Conduct scheduled and unscheduled safety inspections at work sites.
- Actively encourage all workers to participate in workplace safety.
- In a construction setting, conduct a job site survey before starting any work to identify all electrical hazards, implement appropriate control measures, and provide training to employees specific to all identified hazards.
- Ensure that proper personal protective equipment is available and worn by workers where required (including fall protection equipment).
- Conduct job hazard analyses of all tasks that may expose workers to the hazards associated with electrical energy and implement control measures that will adequately insulate and isolate workers from electrical energy.
- Identify potential electrical hazards and appropriate safety interventions during the planning phase of construction or maintenance projects. This planning should address the project from start to finish to ensure that workers have the safest possible work environment.[12]

ELECTRICAL HAZARDS SELF-ASSESSMENT

Even the best safety professional cannot be everywhere at once. Consequently, one of the best strategies for safety personnel is to enlist the assistance of supervisors. After all, helping ensure a safe and healthy work environment is part of every supervisor's job description, or at least it should be. To help prevent accidents and injuries from electrical hazards, safety personnel should consider developing checklists supervisors can use to undertake periodic self-assessments in their areas of responsibility. What follows are the types of questions that should be contained in such checklists:

1. Are all electricians in your company up-to-date with the latest requirements of the National Electrical Code (NEC)?
2. Does your company specify compliance with the NEC as part of its contracts for electrical work with outside personnel?
3. Do all electrical installations located in the presence of hazardous dust or vapors meet the NEC requirements for hazardous locations?

4. Are all electrical cords properly strung (i.e., so that they do not hang on pipes, nails, hooks, etc.)?
5. Is all conduit, BX cable, and so on properly attached to supports and tightly connected to junction boxes and outlet boxes?
6. Are all electrical cords free of fraying?
7. Are rubber cords free of grease, oil, chemicals, and other potentially damaging materials?
8. Are all metallic cables and conduit systems properly grounded?
9. Are all portable electric tools and appliances grounded or double insulated?
10. Are all ground connections clean and tightly made?
11. Are all fuses and circuit breakers the proper size and type for the load on each circuit?
12. Are all fuses free of "jumping" (i.e., with pennies or metal strips)?
13. Are all electrical switches free of evidence of overheating?
14. Are all switches properly mounted in clean, tightly closed metal boxes?
15. Are all electrical switches properly marked to show their purpose?
16. Are all electric motors kept clean and free of excessive grease, oil, or potentially damaging materials?
17. Are all electric motors properly maintained and provided with the necessary level of overcurrent protection?
18. Are bearings in all electrical motors in good condition?
19. Are all portable lights equipped with the proper guards?
20. Are all lamps kept free of any and all potentially combustible materials?
21. Is the organization's overall electrical system periodically checked by a person competent in the application of the NEC?[13]

Safety Tip

Do's and Don'ts of Extension Cords

Approximately 2,500 workplace injuries annually are tied to improper use of extension cords. The following tips can help prevent accidents when extension cords are used in the workplace:

1. Never use an extension cord for an extended period—more than a few weeks—even if it appears to be in perfect condition.
2. Never cover extension cords by rugs or carpet in an attempt to prevent tripping—this can hide shorts and bare spots in the cord that might set the rug on fire.
3. Never just unplug an extension cord that feels hot in order to let it cool down—get rid of it (extension cords that get hot have probably reached the end of their period of safe usage).
4. Avoid using extruded-type extension cords that have only one layer of insulation.

Source: National Fire Protection Association, NFPA 70E–2004.

PREVENTION OF ARC FLASH INJURIES

Arc flash injuries occur in the workplace every day in this country. Many of these injuries lead to severe burns and even death, which is doubly tragic because the accidents could have been prevented. An **arc flash** is an electrical short circuit that travels through the air rather than flowing through conductors, bus bars, and other types of equipment. The uncontrolled energy released by an arc flash can produce high levels of heat and pressure. It can also cause equipment to explode, sending dangerous shrapnel flying through the air.[14]

Arc flashes are sometimes produced by electrical equipment malfunctions, but a more common cause is accidental human contact with an electrical circuit or conductor. For example, a person working near a piece of energized electrical equipment might accidentally drop a tool that then makes contact with an electrical circuit or conductor. The result is an arc flash that can injure or even kill the worker, not to mention the equipment damage.

Arc flashes become even more hazardous when workers are wearing flammable clothing instead of appropriate personal protective equipment (PPE). Arc flashes can produce sufficient heat to easily ignite clothing, cause severe burns, and even damage hearing (hearing damage is caused by the high level of pressure that can be released by an arc flash). The best and most obvious way to prevent arc flash injuries is to deenergize the electrical equipment in question and lock or tag it out before beginning maintenance or service work on it.

However, this is not always possible. Some maintenance and service functions such as troubleshooting require that the equipment being worked on be energized. When this is the case, it is important to consult the National Fire Protection Association's *Handbook for Electrical Safety in the Workplace* (NFPA 70E) and proceed as follows:

- Perform a flash hazard analysis in accordance with NFPA 70E, Article 130.3, or use Table 130.7(C)(9)(a) (Hazard/Risk Category Classifications) to identify the hazard/risk category of the job tasks that must be performed on the energized equipment.
- Establish a flash protection boundary around the equipment in question in accordance with NFPA 70E, Article 130.3(A).
- Select the PPE that will be worn by the worker(s) who will perform the tasks in question on the energized equipment from Table 130.7(C)(10) (Protective Clothing and Personal Protective Equipment Matrix) based on the level of risk identified for these tasks in Table 130.7(C)(9)(a) in the first step above.

For example, if you determine that a worker must perform tasks on a piece of energized electrical equipment and these tasks are rated in "Hazard/Risk Category 3," Table 130.7(C)(10) would require the following PPE:

1. Cotton underwear.
2. Fire-resistant pants and shirt.
3. Fire-resistant coverall.[15]

Safety Fact

Arc Flash Injury Prevention Procedures

An excellent source of guidelines and procedures for preventing arc flash accidents and injuries is the *Handbook for Electrical Safety in the Workplace* (NFPA 70E) produced by the National Fire Protection Association. This handbook is available at the following Web site:
http://www.nfpa.org

SUMMARY

1. Electricity is the flow of negatively charged particles through an electrically conductive material.
2. Atoms have a centrally located nucleus that consists of protons and neutrons. Electrons orbit the nucleus.
3. Conductors are substances that have many free electrons at room temperature and can pass electricity.
4. Insulators do not have a large number of free electrons at room temperature and do not conduct electricity.
5. When a surplus or deficiency of electrons on the surface of a material exists, static electricity is produced.
6. Resistance is measured in ohms.
7. Current flow is measured in amperes or amps.
8. Ohm's law is $V = IR$, where V = volts, I = amps, and R = ohms.
9. Power is measured in watts.
10. Watts (W) are calculated by $W = VI$ or $W = I^2R$.
11. Frequency is measured in hertz.
12. A load is a device that uses electrical current.
13. The NEC specifies industrial and domestic electrical safety precautions.
14. The UL determines whether equipment and materials for electrical systems are safe in the various NEC location categories.
15. Common 110-volt circuits include a hot wire, a neutral wire, and a ground wire.
16. A short circuit is one in which the load has been removed or is bypassed.
17. Sources of electrical hazards include contact with a bare wire, deteriorated insulation, equipment lacking the UL label, improper grounding of equipment, short circuits, dampness, static electricity discharge, metal ladders, power sources remaining on during electrical maintenance, and lightning strikes.
18. Electrostatic hazards may cause minor shocks.
19. A spark or arc involves little power and is discharged into a small space.

20. If the conductor is not capable of carrying a particular amperage of current, the material surrounding the conductor may become overheated and explode or burst into flame.
21. Lightning is a collection of static charges from clouds following the path of least resistance to the earth.
22. Lightning tends to strike the tallest object on the earth.
23. Jumping the ground wire to the neutral wire is unsafe wiring.
24. Open grounds are those with improperly connected ground wires.
25. Reversing the hot and neutral wires results in reversed polarity and an unsafe situation.
26. Flexible wiring should not be substituted for fixed wiring in permanent buildings.
27. Most insulation failure is caused by an environment toxic to insulation.
28. Electrical equipment designers attempt to create devices that are explosion-proof, dust-ignition-proof, and spark-proof.
29. The NEC classifies hazardous locations for electrical equipment.
30. The greatest danger to humans with electrical shock is current flow and the path through the body that the current takes.
31. Above a particular amperage of current, people freeze to conductors and are unable to let go of the conductor.
32. Circuit testers can ensure that power has been turned off.
33. A receptacle wiring tester indicates improperly wired outlets.
34. A continuity tester may be used to check whether a conductor is properly grounded or has a break in the circuit.
35. Grounding ensures a path to the earth for the flow of excess current.
36. Bonding and grounding increase the safety for entire electrical systems.
37. A GFI can detect current flow to the ground and open the circuit.
38. A ground fault occurs when the current in the hot wire is greater than the current in the neutral wire.
39. Antistatic materials, ionizers, and radioactive neutralizers reduce electrical static buildup.
40. Fuses and circuit breakers open the circuit with excess amperage.
41. Double insulation increases electrical safety.
42. Lightning hazard control includes using lightning rods, avoiding tall objects and flammable materials, not touching conductive materials, not using the telephone, not touching the walls of metal buildings, and not standing near open doors and windows during electrical storms.
43. Interlocks automatically open the circuit when an unsafe condition is detected.
44. It is better to design in safety for the electrical system than to deal effectively with accidents.
45. With electrocution accounting for approximately 6 percent of all workplace deaths in the United States every year, it is important that employers have instituted an effective electrical safety program.
46. An arc flash is an electrical short circuit that travels through the air. An arc flash can be caused by equipment malfunctions or accidental human contact.

KEY TERMS AND CONCEPTS

Alternating current (AC)

Amperes (amps)

Antistatic materials

Arc

Arc flash

Bonding

Circuit tester

Conductors

Continuity tester

Direct current (DC)

Double insulation

Effective current

Electrical current

Electrical hazards

Electrical safety program

Electrical system grounding

Electricity

Electrons

Electrostatic hazards

Electrostatic neutralizers

Equipment failure

Freeze

Frequency

Fuses

Ground fault

Ground fault circuit interrupter (GFCI)

Ground fault interrupter (GFI)

Ground wire

Grounded conductor

Grounding conductor

Hazardous locations

Hertz

Hot wire

Humidification

Improper wiring

Insulation failure

Insulators

Interlocks

Ionizers

Isolating the hazard

Jump the ground wire

Let-go current

Lightning

Load

Magnetic circuit breaker

National Board of Fire Underwriters

National Electrical Code (NEC)

Negative charge

Neutral wire

Neutrons

NFPA 70E

Ohms

Ohm's law

Open ground

Positive charge

Potential difference

Power

Protons

Radioactive neutralizers

Receptacle wiring tester

Resistance

Reversed polarity

Semiconductor

Separate equipment grounding

Shock

Short circuit

Spark

Static electricity

Thermal circuit breaker

Underwriters Laboratories (UL)   Watts

Voltage   Zero potential

Warning devices

REVIEW QUESTIONS

1. Define *zero potential*. Explain the relationship between zero potential and grounding.
2. Explain what each of the following terms measure: *volt, amp, ohm, hertz,* and *watt*.
3. Briefly discuss the difference between DC and AC using the concept of effective current.
4. What is the relationship between the NEC and UL?
5. Briefly state the relationship among potential difference, lightning, and grounding.
6. Define *open ground*.
7. List at least five lightning hazard control measures.
8. Explain the relationship between circuit load and short circuits.
9. How do ionizers, radioactive neutralizers, and antistatic materials work? Why does humidification work?
10. Discuss how bonding and grounding work together to increase electrical safety.
11. How do continuity testers, circuit testers, and receptacle wiring testers operate?
12. Explain freeze and let-go current.
13. Describe the structure of an atom.
14. Discuss the proper wiring of a three-wire circuit.
15. Why is jumping the ground wire a hazard?
16. Explain reversed polarity.
17. Why are warning devices less effective than designed-in safety precautions?
18. Explain five strategies for establishing an effective electrical safety program.
19. Explain why it is important for safety personnel to help employees and supervisors conduct self-assessments.
20. Explain how to reduce arc flash hazards.

ENDNOTES

1. "Massachusetts Mill Cited for Safety and Health Violations," *Occupational Health & Safety News* 20, no. 9: 10.
2. Ibid.
3. National Fire Protection Association, *Handbook for Electrical Safety in the Workplace* (Quincy, MA: National Fire Protection Association, 2004), 57.

4. Ibid., 59.

5. Ibid., 60.

6. Ibid., 61.

7. National Fire Protection Association, *Handbook for Electrical Safety*, 103.

8. Ibid., 104.

9. Ibid., 105.

10. National Fire Protection Association, *Handbook for Electrical Safety*, 109.

11. Ibid., 110.

12. National Institute for Occupational Safety and Health, *Worker Deaths by Electrocution: A Summary of Surveillance Findings and Investigative Case Reports*, 2006, 13–16. Retrieved from http://www.cdc.gov/niosh/homepage .html.

13. Retrieved from http://online.migu.nodak.edu/19577/BADM309checklist .htm.

14. W. Wallace, "NFPA 70E: Performing the Electrical Flash Hazard Analysis," *Occupational Safety & Health* 74, no. 8: 38–44.

15. National Fire Protection Association, *Handbook for Electrical Safety*, Table 130.7(C)(9)(a).

FIRE HAZARDS AND LIFE SAFETY

Major Topics

- Fire Hazards Defined
- Sources of Fire Hazards
- Fire Dangers to Humans
- Detection of Fire Hazards
- Reduction of Fire Hazards
- Development of Fire Safety Standards
- OSHA Fire Standards
- Life Safety
- Flame-Resistant Clothing
- Fire Safety Programs
- Explosive Hazards
- OSHA's Firefighting Options
- Self-Assessment in Fire Protection

The assistant manager of a fuel storage plant was killed while trying to repair a broken-down piece of equipment. According to the *NFPA Journal*,

> The victim had been called to investigate a strong odor of gasoline that had been detected by a gasoline tank driver. . . . It was suspected that the facility's vapor-recovery system, which captures gasoline vapors displaced from tank trucks being filled with product, had malfunctioned. Soon after the manager entered the area, a violent explosion occurred. The victim's severely burned body and repair tools were found near the damaged equipment.[1]

In another recent example, facility damage was held to a minimum and human injury was avoided when a sprinkler system suppressed a fire that broke out on the second floor of a polyurethane foam manufacturing plant in North Carolina. The *NFPA Journal* stated,

> The fire broke out just after 4:00 P.M., an hour after all employees but one had left for the day. The ballast of a ceiling-mounted fluorescent light fixture short-circuited

and cracked open, allowing burning ballast material to drop to the floor and the conveyor below. The burning material ignited scrap urethane.[2]

Fortunately, in this case, the company had a properly functioning automatic fire suppression system, and tragedy was avoided.

Two employees were killed and another received second- and third-degree burns over 45 percent of his body when an explosion and fire occurred at a chemical plant in Ohio. According to the *NFPA Journal*,

> The employees inserted a long metal rod into the hopper and attempted to dislodge the blockage. As the men moved the rod around, an explosion occurred inside the hopper.[3]

What all these tragedies have in common is that they did not have to happen. These incidents could have been prevented. The damage and injuries that can result from fire can be both physically and psychologically damaging. The resulting trauma can affect even those employees who are not physically injured. Therefore, modern safety and health professionals should be familiar with fire hazards and their prevention.

FIRE HAZARDS DEFINED

Fire hazards are conditions that favor fire development or growth. Three elements are required to start and sustain fire: (1) oxygen, (2) fuel, and (3) heat. Because oxygen is naturally present in most earth environments, fire hazards usually involve the mishandling of fuel or heat.

Fire, or **combustion**, is a chemical reaction between oxygen and a combustible fuel. Combustion is the process by which fire converts fuel and oxygen into energy, usually in the form of heat. By-products of combustion include light and smoke. For the reaction to start, a source of ignition, such as a spark or open flame, or a sufficiently high temperature is needed. Given a sufficiently high temperature, almost every substance will burn. The **ignition temperature** or **combustion point** is the temperature at which a given fuel can burst into flame.

Fire is a chain reaction. For combustion to continue, there must be a constant source of fuel, oxygen, and heat (see Figure 19–1). The flaming mode is represented by the tetrahedron on the left (heat, oxidizing agent, and reducing agent) that results from a chemical chain reaction. The smoldering mode is represented by the triangle on the right. **Exothermic** chemical reactions create heat. Combustion and fire are exothermic reactions and can often generate large quantities of heat. **Endothermic** reactions consume more heat than they generate. An ongoing fire usually provides its own sources of heat. It is important to remember that cooling is one of the principal ways to control a fire or put it out.

All chemical reactions involve forming and breaking chemical bonds between atoms. In the process of combustion, materials are broken down into basic elements. Loose atoms form bonds with each other to create molecules of substances that were not originally present.

Figure 19–1
Fire tetrahedron (L) and fire triangle (R).

Carbon is found in almost every flammable substance. When a substance burns, the carbon is released and then combines with the oxygen that *must* be present to form either **carbon dioxide** or **carbon monoxide**.

Carbon dioxide is produced when there is more oxygen than the fire needs. It is not toxic, but it can be produced in such volumes that it seriously reduces the concentration of oxygen in the air surrounding the fire site. Carbon monoxide—a colorless, odorless, deadly gas—is the result of incomplete combustion of a fuel. It is produced when there is insufficient oxygen to burn the fuel present efficiently. In general, most fires have insufficient oxygen and therefore produce large quantities of carbon monoxide. It is important in any intentional industrial fire that the fuel be consumed as completely as possible. This will reduce ash and minimize smoke and gases, including carbon monoxide.

Hydrogen, found in most fuels, combines with oxygen to form water. Synthetic polymers, found in plastics and vinyls, often form deadly fumes when they are consumed by fire, or when they melt or disintegrate from being near fire or high heat. Burning, melting, or disintegrating plastic at a fire site should be presumed to be releasing toxic fumes.

Liquids and solids, such as oil and wood, do not burn directly but must first be converted into a flammable **vapor** by heat. Hold a match to a sheet of paper, and the paper will burst into flames. Look closely at the paper, and you will see that the paper is not burning. The flames reside in a vapor area just above the surface of the sheet.

Vapors will burn only at a specific range of mixtures of oxygen and fuel, determined by the composition of the fuel. At the optimum mixture, a fire burns, generates heat and some light, and produces no other by-products. In an unintentional fire, the mixture is constantly changing as more or less oxygen is brought into the flames and more or less heat is generated, producing more or fewer vapors and flammable gases.

Remove the fire's access to fuel or remove the oxygen, and the fire dies. Although a spark, flame, or heat may start a fire, the heat that a fire produces is necessary to sustain it. Therefore, a fire may be extinguished by removing the

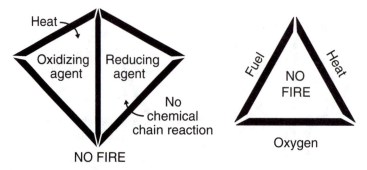

Figure 19–2
The broken tetrahedron and broken triangle.

fuel source, starving it of oxygen, or cooling it below the combustion point. Even in an oxygen-rich, combustible environment, such as a hospital oxygen tent, fire can be avoided by controlling heat and eliminating sparks and open flames (see Figure 19–2). The broken lines in the tetrahedron and the triangle indicate that the necessary elements are removed.

An **explosion** is a very rapid, contained fire. When the gases produced exceed the pressure capacity of the vessel, a rupture or explosion must result. The simplest example is a firecracker. The fuse, which usually contains its own source of oxygen, burns into the center of a firecracker. The surrounding powder ignites, and the heat produced vaporizes the balance of the explosive material and ignites it. The tightly wrapped paper of the firecracker cannot contain the expanding gases. The firecracker explodes, in much less time than was required to read about it.

Heat always flows from a higher temperature to a lower temperature, never from a lower temperature to a higher temperature without an outside force being applied. Fires generate heat, which is necessary to sustain the fire. Excess heat is then transferred to surrounding objects, which may ignite, explode, or decompose. **Heat transfer** is accomplished by three means, usually simultaneously: (1) conduction, (2) radiation, and (3) convection.

Conduction is direct thermal energy transfer. On a molecular level, materials near a source of heat absorb the heat, raising their kinetic energy. **Kinetic energy** is the energy resulting from a moving object. Energy in the form of heat is transferred from one molecule to the next. Materials conduct heat at varying rates. Metals are very good conductors of heat. Concrete and plastics are poor conductors, hence good insulators. Nevertheless, a heat buildup on one side of a wall will transfer to the other side of the wall by conduction.

Radiation is electromagnetic wave transfer of heat to a solid. Waves travel in all directions from the fire and may be reflected off a surface, as well as absorbed by it. Absorbed heat may raise the temperature beyond a material's combustion point, and then a fire erupts. Heat may also be conducted through a vessel to its contents, which will expand and may explode. An example is the spread of fire through an oil tank field. A fire in one tank can spread to nearby tanks through radiated heat, raising the temperature and pressure of the other tank contents.

Convection is heat transfer through the movement of hot gases. The gases may be the direct products of fire, the results of chemical reaction, or additional gases brought to the fire by the movement of air and heated at the fire surfaces by conduction. Convection determines the general direction of the spread of a fire. Convection causes fires to rise as heat rises and move in the direction of the prevailing air currents.

All three forms of heat transfer are present at a campfire. A metal poker left in a fire gets red hot at the flame end. Heat is conducted up the handle, which gets progressively hotter until the opposite end of the poker is too hot to touch. People around the fire are warmed principally by radiation, but only on the side facing the fire. People farther away from the fire will be warmer on the side facing the fire than the backs of people closer to the fire. Marshmallows toasted above the flames are heated by convection (see Figure 19–3).

Spontaneous combustion is rare, but it can happen. Organic compounds decompose through natural chemical processes. As they degrade, they release methane gas (natural gas), an excellent fuel. The degradation process—a chemical reaction—produces heat. In a forest, the concentrations of decomposing matter are relatively minimal, and both the gas and the heat vent naturally.

A classic example of spontaneous combustion is a pile of oil-soaked rags. A container of oil seldom ignites spontaneously. A collection of clean fabrics seldom bursts into flames. Rags soaking completely within oil are usually safe. One oil-soaked rag is unlikely to cause a problem. However, in a pile of oil-soaked rags—especially in a closed container—the chemistry is quite different.

The fibers of the rags expose a large surface area of oil to oxidation. The porous nature of rags allows additional oxygen to be absorbed, replacing the oxygen already consumed. When the temperature rises sufficiently, the surfaces of the oil on the rags vaporize.

Hypergolic reactions occur when mixing fuels. Oxidizers produce just such a rapid heat buildup, causing immediate combustion at room temperature with no apparent source of ignition. Although the term *hypergolic* originated with rocket propellants, the phenomenon has been around for a long time.

Figure 19–3
Campfire with convection heat.

Convection

Pyrophor hypergolic fuels are those that self-ignite in the presence of oxygen found at normal atmospheric concentrations. One example is white phosphorus, which is kept underwater. If it starts to dry out, the phosphorus erupts in flames.

SOURCES OF FIRE HAZARDS

Almost everything in an industrial environment can burn. Metal furniture, machines, plaster, and concrete block walls are usually painted. Most paints and lacquers will easily catch fire. Oxygen is almost always present. Therefore, the principal method of fire suppression is passive—the absence of sufficient heat. Within our environment, various conditions elevate the risk of fire and so are termed *fire hazards*.

For identification, fires are classified according to their properties, which relate to the nature of the fuel. The properties of the fuel directly correspond to the best means of combating a fire (see Figure 19–4).

Without a source of fuel, there is no fire hazard. However, almost everything in our environment can be a fuel. Fuels occur as solids, liquids, vapors, and gases.

Solid fuels include wood, building decorations and furnishings such as fabric curtains and wall coverings, and synthetics used in furniture. What would an office be without paper? What would most factories be without cardboard and packing materials such as Styrofoam molds and panels, shredded or crumpled papers, bubble wrap, and shrink wrap? All these materials easily burn.

Few solid fuels are, or can be made, fireproof. Even fire walls do not stop fires, although they are defined by their ability to slow the spread of fire. Wood and textiles can be treated with fire- or flame-retardant chemicals to reduce their flammability.

Class A fires	Solid materials such as wood, plastics, textiles, and their products: paper, housing, clothing.
Class B fires	Flammable liquids and gases.
Class C fires	Electrical (referring to live electricity situations, not including fires in other materials started by electricity).
Class D fires	Combustible, easily oxidized metals such as aluminum, magnesium, titanium, and zirconium.
Special categories	Extremely active oxidizers or mixtures, flammables containing oxygen, nitric acid, hydrogen peroxide, and solid missile propellants.

Figure 19–4
Classes of fire.
Source: National Fire Protection Association, *Fire Protection Handbook*, 19th ed. (Quincy, MA: National Fire Protection Association, 2003).

Flammable Liquids

Class I–A	Flash point below 73°F, boiling point below 100°F.
Class I–B	Flash point below 73°F, boiling point at or above 100°F.
Class I–C	Flash point at or above 73°F, but below 100°F.

Combustible Liquids

Class II	Flash point at or above 100°F, but below 140°F.
Class III–A	Flash point at or above 140°F, but below 200°F.
Class III–B	Flash point at or above 200°F.

Figure 19–5
Classes of flammable and combustible liquids.
Source: Fire Protection Handbook, 19th ed. (Quincy, MA: National Fire Protection Association, 2003).

Solid fuels are involved in most industrial fires, but mishandling flammable liquids and flammable gases is a major cause of industrial fires. Two often-confused terms applied to flammable liquids are *flash point* and *fire point*. The **flash point** is the lowest temperature for a given fuel at which vapors are produced in sufficient concentrations to flash in the presence of a source of ignition. The **fire point** is the minimum temperature at which the vapors continue to burn, given a source of ignition. The **auto-ignition temperature** is the lowest point at which the vapors of a liquid or solid self-ignite *without* a source of ignition.

Flammable liquids have a flash point below 37.7°C (100°F). **Combustible liquids** have a flash point at or higher than that. Both flammable and combustible liquids are further divided into the three classifications shown in Figure 19–5.

As the temperature of any flammable liquid increases, the amount of vapor generated on the surface also increases. Safe handling, therefore, requires both a knowledge of the properties of the liquid and an awareness of ambient temperatures in the work or storage place. The **explosive range**, or **flammable range**, defines the concentrations of a vapor or gas in air that can ignite from a source. The auto-ignition temperature is the lowest temperature at which liquids spontaneously ignite.

Most flammable liquids are lighter than water. If the flammable liquid is lighter than water, water cannot be used to put out the fire.[4] The application of water floats the fuel and spreads a gasoline fire. Crude oil fires burn even while floating on fresh or sea water.

Unlike solids (which have a definite shape and location) and unlike liquids (which have a definite volume and are heavier than air), gases have no shape. Gases expand to fill the volume of the container in which they are enclosed, and they are frequently lighter than air. Released into air, gas concentrations are difficult to monitor due to the changing factors of air, current direction, and temperature. Gases may stratify in layers of differing concentrations but often collect near the top of whatever container in which they are enclosed.

Concentrations found to be safe when sampled at workbench level may be close to, or exceed, flammability limits if sampled just above head height.

The products of combustion are gases, flame (light), heat, and smoke. Smoke is a combination of gases, air, and suspended particles, which are the products of incomplete combustion. Many of the gases present in smoke and at a fire site are toxic to humans. Other, usually nontoxic, gases may replace the oxygen normally present in air. Most fatalities associated with fire are from breathing toxic gases and smoke and from being suffocated because of oxygen deprivation. Gases that may be produced by a fire include acrolein, ammonia, carbon monoxide, carbon dioxide, hydrogen bromide, hydrogen cyanide, hydrogen chloride, hydrogen sulfide, sulfur dioxide, and nitrogen dioxide. Released gases are capable of traveling across a room and randomly finding a spark, flame, or adequate heat source, **flashing back** to the source of the gas.

The National Fire Protection Association (NFPA) has devised the NFPA 704 system for quick identification of hazards presented when substances burn (see Figure 19–6). The NFPA's red, blue, yellow, and white diamond is used on product labels, shipping cartons, and buildings. Ratings within each category are 0 to 4, where zero represents no hazard; 4, the most severe hazard level. The colors refer to a specific category of hazard:

Red   = flammability

Blue   = health

Yellow = reactivity

White   = special information

Although we do not think of electricity as burning, natural and generated electricity play a large role in causing fires. Lightning strikes cause many fires every year. In the presence of a flammable gas or liquid mixture, one spark can produce a fire.

Electrical lines and equipment can cause fires either by a short circuit that provides an ignition spark, by arcs, or by resistances generating a heat buildup. Electrical switches and relays commonly arc as contact is made or broken.

Figure 19–6
Identification of fire hazards.

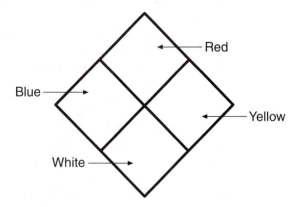

Flammability has a red background and is the top quarter of the diamond.

0   No hazard. Materials are stable during a fire and do not react with water.
1   Slight hazard. Flash point well above normal ambient temperature.
2   Moderate hazard. Flash point is slightly above normal ambient temperature.
3   Extreme fire hazard. Gases or liquids that can ignite at normal temperature.
4   Extremely flammable gases or liquids with very low flash points.

Health has a blue background and is the left quarter of the diamond.

0   No threat to health.
1   Slight health hazards. Respirator is recommended.
2   Moderate health hazard. Respirator and eye protection required.
3   Extremely dangerous to health. Protective clothing and equipment is required.
4   Imminent danger to health. Breathing or skin absorption may cause death. A fully encapsulating suit is required.

Reactive has a yellow background and is the right quarter of the diamond.

0   No hazard. Material is stable in a fire and does not react with water.
1   Slight hazard. Materials can become unstable at higher temperatures or react with water to produce a slight amount of heat.
2   Moderate or greater hazard. Materials may undergo violent chemical reaction, but will not explode. Materials react violently with water directly or form explosive mixtures with water.
3   Extreme hazard. Materials may explode given an ignition source or have violent reactions with water.
4   Constant extreme hazard. Materials may polymerize, decompose, explode, or undergo other hazardous reactions on their own. Area should be evacuated in event of a fire.

Special information has a white background and is the bottom quarter of the diamond.

This area is used to note any special hazards presented by the material.

Another source of ignition is heat in the form of hot surfaces. It is easy to see the flame hazard present when cooking oil is poured on a very hot grill. The wooden broom handle leaning up against the side of a hot oven may not be as obvious a hazard. Irons used in textile manufacturing and dry-cleaning plants also pose a heat hazard.

Space heaters frequently have hot sides, tops, backs, and bottoms, in addition to the heat-generating face. Hotplates, coffee pots, and coffee makers often create heated surfaces. Many types of electric lighting generate heat, which is transferred to the lamp housing.

Engines produce heat, especially in their exhaust systems. Compressors produce heat through friction, which is transferred to their housings. Boilers produce hot surfaces, as do steam lines and equipment using steam as power. Radiators, pipes, flues, and chimneys all have hot surfaces. Metal stock that has been cut by a blade heats up as the blade does. Surfaces exposed to direct sunlight become hot surfaces and transmit their heat by conduction to their other side. Heated surfaces are a potential source of fire.

FIRE DANGERS TO HUMANS

Direct contact with flame is obviously dangerous to humans. Flesh burns, as do muscles and internal organs. The fact that we are 80 percent water, by some estimations, does not mitigate the fact that virtually all the other 20 percent burns. Nevertheless, burns are not the major cause of death in a fire.

NFPA statistics show that most people die in fires from suffocating or breathing smoke and toxic fumes. Carbon dioxide can lead to suffocation because it can be produced in large volumes, depleting oxygen from the air. Many fire extinguishers use carbon dioxide because of its ability to starve the fire of oxygen while simultaneously cooling the fire. The number one killer in fires is carbon monoxide, which is produced in virtually all fires involving organic compounds. Carbon monoxide is produced in large volumes and can quickly reach lethal dosage concentrations.

Figure 19–7 shows the major chemical products of combustion. Other gases may be produced under some conditions. Not all these gases are present at any particular fire site. Many of these compounds will further react with other substances often present at a fire. For example, sulfur dioxide will combine with water to produce sulfuric acid. Oxides of nitrogen may combine with water to produce nitric acid. Sulfuric acid and nitric acid can cause serious acid burns.

Product	Fuels	Pathology
Acrolein	Cellulose, fatty substances, woods, and paints	Highly toxic irritant to eyes and respiratory system
Ammonia (NH_3)	Wool, silk, nylon, melamine, refrigerants, hydrogen–nitrogen compounds	Somewhat toxic irritant to eyes and respiratory system
Carbon dioxide (CO_2)	All carbon and organic compounds	Not toxic, but depletes available oxygen
Carbon monoxide (CO)	All carbon and organic compounds	Can be deadly
Hydrogen chloride (HCl)	Wool, silk, nylon, paper, polyurethane, rubber, leather, plastic, wood	Quickly lethal asphyxiant
Hydrogen sulfide (H_2S)	Sulfur-containing compounds, rubber, crude oil	Highly toxic gas; strong odor of rotten eggs, but quickly destroys sense of smell
Nitrogen dioxide (NO_2)	Cellulose nitrate, celluloid, textiles, other nitrogen oxides	Lung irritant, causing death or damage
Sulfur dioxide (SO_2)	Sulfur and sulfur-containing compounds	Toxic irritant

Figure 19–7
Major chemical products of combustion.

DETECTION OF FIRE HAZARDS

Many automatic fire detection systems are used in industry today. Many systems can warn of the presence of smoke, radiation, elevated temperature, or increased light intensity. **Thermal expansion detectors** use a heat-sensitive metal link that melts at a predetermined temperature to make contact and ultimately sound an alarm. Heat-sensitive insulation can be used, which melts at a predetermined temperature, thereby initiating a short circuit and activating the alarm.

Photoelectric fire sensors detect changes in infrared energy that is radiated by smoke, often by the smoke particles obscuring the photoelectric beam. A relay is open under acceptable conditions and closed to complete the alarm circuit when smoke interferes.

Ionization or **radiation sensors** use the tendency of a radioactive substance to ionize when exposed to smoke. The substance becomes electrically conductive with the smoke exposure and permits the alarm circuit to be completed.

Ultraviolet or **infrared detectors** sound an alarm when the radiation from fire flames is detected. When rapid changes in radiation intensities are detected, a fire alarm signal is given.

OSHA has mandated the monthly and annual inspection and recording of the condition of fire extinguishers in industrial settings. A hydrostatic test to determine the integrity of the fire extinguisher metal shell is recommended according to the type of fire extinguisher. The hydrostatic test measures the capability of the shell to contain internal pressures and the pressure shifts expected to be encountered during a fire.

REDUCTION OF FIRE HAZARDS

The best way to reduce fires is to prevent them. A major cause of industrial fires is hot, poorly insulated machinery and processes. One means of reducing a fire hazard is the isolation of the three triangle elements: fuel, oxygen, and heat. In the case of fluids, closing a valve may stop the fuel element.

Fires may also be prevented by the proper storage of flammable liquids. Liquids should be stored as follows:

- In flame-resistant buildings that are isolated from places where people work. Proper drainage and venting should be provided for such buildings.
- In tanks below ground level.
- On the first floor of multistory buildings.

Substituting less-flammable materials is another effective technique for fire reduction. A catalyst or fire inhibitor can be employed to create an endothermic energy state that eventually smothers the fire. Several ignition sources can be eliminated or isolated from fuels:

- Prohibit smoking near any possible fuels.
- Store fuels away from areas where electrical sparks from equipment, wiring, or lightning may occur.
- Keep fuels separate from areas where there are open flames. These may include welding torches, heating elements, or furnaces.
- Isolate fuels from tools or equipment that may produce mechanical or static sparks.

Other strategies for reducing the risk of fires are as follows:

- Clean up spills of flammable liquids as soon as they occur. Properly dispose of the materials used in the cleanup.
- Keep work areas free from extra supplies of flammable materials (e.g., paper, rags, boxes, and so on). Have only what is needed on hand with the remaining inventory properly stored.
- Run electrical cords along walls rather than across aisles or in other trafficked areas. Cords that are walked on can become frayed and dangerous.
- Turn off the power and completely deenergize equipment before conducting maintenance procedures.
- Don't use spark- or friction-prone tools near combustible materials.
- Routinely test fire extinguishers.

Fire Extinguishing Systems

In larger or isolated industrial facilities, an employee fire brigade may be created. (See OSHA requirements in next section.) **Standpipe and hose systems** provide the hose and pressurized water for firefighting. Hoses for these systems usually vary from 1 inch to 2.5 inches in diameter.[5]

Automatic sprinkler systems are an example of a fixed extinguishing system because the sprinklers are fixed in position. Water is the most common fluid released from the sprinklers. Sprinkler supply pipes may be kept filled with water in heated buildings; in warmer climates, valves are used to fill the pipes with water when the sprinklers are activated. When a predetermined heat threshold is breached, water flows to the heads and is released from the sprinklers.

Portable fire extinguishers are classified by the types of fire that they can most effectively reduce. Figure 19–8 describes the four major fire extinguisher classifications. Blocking or shielding the spread of fire can include covering the fire with an inert foam, inert powder, nonflammable gas, or water with a thickening agent added. The fire may suffocate under such a covering. Flooding a liquid fuel with nonflammable liquid can dilute this fire element. Figures 19–9, 19–10, and 19–11 are photographs of effective fire prevention equipment.

Fire Class	Extinguisher Contents	Mechanism	Disadvantages
A	Foam, water, dry chemical	Chain-breaking cooling, smothering, and diluting	Freezes if not kept heated.
B	Dry chemical, bromotrifluoromethane, and other halogenated compounds, foam, CO_2, dry chemical	Chain-breaking smothering, cooling, and shielding	Halogenated compounds are toxic.
C	Bromotrifluoromethane, CO_2, dry chemical	Chain-breaking smothering, cooling, and shielding	Halogenated compounds are toxic; fires may ignite after CO_2 dissipates.
D	Specialized powders such as graphite, sand, limestone, soda ash, sodium chloride	Cooling, smothering	Expensive cover of powder may be broken with resultant reignition.

Figure 19–8
Fire extinguisher characteristics.

Figure 19–9
Fireproof storage cabinet.
Courtesy of JUSTRITE®.

Figure 19–10
Fire-protective drums on a poly
spill pallet.
Courtesy of JUSTRITE®.

Figure 19–11
Fire-protective drums stored in an outside storage cabinet.
Courtesy of JUSTRITE®.

OSHA Regulations for Fire Brigades

Fire brigade regulations are covered in 29 CFR 1910.156 (Subpart L, Appendix A). Relevant requirements from the regulations are as follows:

1. *Scope.* Employers are not required to form a fire brigade. However, if an employer does decide to organize a fire brigade, the requirements of this section apply.

2. *Prefire planning.* Have prefire planning conducted by the local fire department or the workplace fire brigade so that they may become familiar with the workplace and process hazards. Involvement with the local fire department or fire prevention bureau is encouraged to facilitate coordination and cooperation between members of the fire brigade and those who may be called upon for assistance during a fire emergency.

3. *Organizational statement.* The organizational statement should contain the following information: a description of the duties that the fire brigade members are expected to perform; the line authority of each fire brigade officer; the number of the fire brigade officers and number of training instructors; and a list and description of the types of awards or recognition that brigade members may be eligible to receive.

4. *Physical capability.* The physical capability requirement applies only to those fire brigade members who perform interior structural firefighting. Employees who cannot meet the physical capability requirement may still be members of the fire brigade as long as such employees do not perform interior structural firefighting. It is suggested that fire brigade members who are unable to perform interior structural firefighting be assigned less stressful and physically demanding fire brigade duties (e.g., certain types of training, record keeping, fire prevention inspection and maintenance, and fire pump operations).

"Physically capable" can be defined as being able to perform those duties specified in the training requirements of Section 1910.156(c). Physical capability can also be determined by physical performance tests or by a physical examination when the examining physician is aware of the duties that the fire brigade member is expected to perform.

5. *Training and education.* Training and education must be commensurate with those functions that the fire brigade is expected to perform (i.e., those functions specified in the organizational statement). Such a performance requirement provides the necessary flexibility to design a training program that meets the needs of individual fire brigades.

At a minimum, hands-on training is required to be conducted annually for all fire brigade members. However, for those fire brigade members who are expected to perform interior structural firefighting, some type of training or education session must be provided at least quarterly.

6. *Firefighting equipment.* It is important to remove from service and replace any firefighting equipment that is damaged or unserviceable. This prevents fire brigade members from using unsafe equipment by mistake.

Firefighting equipment, except portable fire extinguishers and respirators, must be inspected at least annually. Portable fire extinguishers and respirators are required to be inspected at least monthly.

7. *Protective clothing.* Paragraph (e) of 1910.156 does not require all fire brigade members to wear protective clothing. It is not the intention of the standards to require employers to provide a full ensemble of protective clothing for every fire brigade member without consideration given to the types of hazardous environments to which the fire brigade member may be exposed. It is

the intention of the standards to require adequate protection for those fire brigade members who may be exposed to fires in an advanced stage, smoke, toxic gases, and high temperatures. Therefore, the protective clothing requirements apply only to those fire brigade members who perform interior structural firefighting operations.

Additionally, the protective clothing requirements do not apply to the protective clothing worn during outside firefighting operations (brush and forest fires, crash crew operations) or other special firefighting activities. It is important that the protective clothing to be worn during these types of firefighting operations reflect the hazards that may be encountered by fire brigade members.

8. *Respiratory protective devices.* Respiratory protection is required to be worn by fire brigade members while working inside buildings or confined spaces where toxic products of combustion or an oxygen deficiency are likely to be present; respirators are also to be worn during emergency situations involving toxic substances. When fire brigade members respond to emergency situations, they may be exposed to unknown contaminants in unknown concentrations. Therefore, it is imperative that fire brigade members wear proper respiratory protective devices during these situations. Additionally, there are many instances where toxic products of combustion are still present during mop-up and overhaul operations. Therefore, fire brigade members should continue to wear respirators during these types of operations.[6]

Disaster Preparations

Training employees may be the most successful lifesaving preparation for a fire disaster. Company fire brigade members should be trained and tested at least quarterly. Disaster preparation initially requires management commitment and planning and continued response and recovery practice by the fire brigade on a regular basis. Also necessary are regular, but less frequent, fire drills for all personnel.

Disaster preparations also include the integration of company planning with community plans. Community disaster relief agencies such as the police, fire department, Red Cross, and hospitals should be consulted and informed of company disaster preparation plans.

Preventing Office Fires

The shop floor is not the only part of the plant where fire hazards exist. Offices are also susceptible to fires. According to Vogel, "Every year about 7,000 fires occur in office buildings, which cause injuries, deaths, and millions of dollars in fire damages."[7] The following strategies are helpful in preventing office fires:

- Confine smoking to designated areas that are equipped with nontip ashtrays and fire-resistant furnishings.

- Periodically check electrical circuits and connections. Replace frayed or worn cords immediately.
- Make sure that extension cords and other accessories are UL-approved and used only as recommended.
- Make sure there is plenty of air space left around copying machines and other office machines that can overheat.
- Locate heat-producing appliances away from the wall or anything else that can ignite.
- Frequently inspect personal appliances such as hotplates, coffee pots, and cup warmers. Assign responsibility for turning off such appliances every day to a specific person.
- Keep aisles, stairwells, and exits clear of paper, boxes, and other combustible materials.[8]

DEVELOPMENT OF FIRE SAFETY STANDARDS

The purpose of modern fire safety standards is the protection of life and the prevention of property damage. However, the impetus for developing standards has always been and continues to be the occurrence of major disasters. Typically, standards are developed after a major tragedy occurs in which property is damaged on a large scale and lives are lost. Public shock turns into an outcry for action. A flurry of political activity follows, and agencies and organizations that develop standards are called on to develop new standards.

According to Arthur E. Cote,

> There are approximately 89,000 standards in the United States. Of these, 50,000 of them actually are procurement specifications developed and used by the government. The remaining 39,000 are voluntary standards developed in the U.S. Most codes and standards in the fire protection field are developed by three organizations: the National Fire Protection Association (NFPA), the American Society for Testing Materials (ASTM), and Underwriter's Laboratories (UL). In addition, two model code organizations—the International Conference of Building Officials and Code Administration (BOCA), and the Southern Building Code Congress International (SBCC)—develop model building and fire prevention codes.[9]

The trend in fire safety standards is toward performance-based standards and away from the traditional specification-based approach. An example of each type of standard will help illustrate the difference. A specification-based standard may require that brick, concrete, or steel material be used in a given type of building. A performance-based standard may specify that materials used have a one-, two-, or four-hour fire resistance rating.[10] Advances in the testing of engineering materials will help overcome most of the barriers to full development and implementation of performance-based standards.

Cote summarizes his views on the future of fire safety standards as follows:

Codes and standards will survive in the 21st century. They may, however, be considerably different from the codes and standards we now have. They certainly will be based more on standardized fire tests, models, data, and related science and engineering than on consensus judgment. How much more will depend on the extent to which there is widespread acceptance of the anticipated breakthroughs in fire science and in related modeling and calculation methods.[11]

OSHA FIRE STANDARDS

OSHA standards for fire protection appear in 29 CFR 1910.156 (Subpart L). This subpart contains the standards for fire brigades, fixed fire suppression equipment, and other fire protection systems. Employers are not required to form fire brigades, but those who choose to do so must meet a number of specific requirements. There are other fire-related requirements in other subparts. For example, fire exits, emergency action plans, and means of egress are covered in Subpart E. The standards in Subpart L are as follows:

Fire Protection

1910.155	Scope, application, and definitions
1910.156	Fire brigades

Portable Fire Suppression Equipment

1910.157	Portable fire extinguishers
1910.158	Standpipe and hose systems

Fixed Fire Suppression Equipment

1910.159	Automatic sprinkler systems
1910.160	Fixed extinguishing systems, general
1910.161	Fixed extinguishing systems, dry chemical
1910.162	Fixed extinguishing systems, gaseous agent
1910.163	Fixed extinguishing systems, water spray and foam

Other Fire Protection Systems

1910.164	Fire detection systems
1910.165	Employee alarm systems

Figure 19–12 summarizes fire prevention and suppression strategies.

- Use the least-flammable materials whenever possible.
- Analyze the company to determine types of potential fires and provide appropriate sprinklers and/or extinguishers.
- Develop a database of the flammability of materials available in the company.
- Store containers or flammable materials away from sources of heat or sparks and away from humans.
- Do not permit smoking near flammable materials.
- Include a venting mechanism in storage containers and locate them near a drain.
- Minimize fuel storage container size to reduce the size of the fire that may involve those fuels.
- Isolate fuels from sources of heat.
- Include a smoke detection system and portable fire extinguishers in the facility. Extinguishers should be easily available to every workstation.
- Make sure sources of heat have controlling mechanisms and are near fire detection equipment.
- Check fire extinguishing equipment regularly.
- Perform periodic inspections for fire hazards and reappraisal of fire hazards.
- Train plant personnel in basic fire prevention, which should include periodic fire drills.
- Make sure fire brigade personnel are well trained, tested, and regularly practice fire control.
- Stress cleanliness and an organized method of disposal of flammable materials.

Figure 19–12
Fire prevention and suppression summary.

LIFE SAFETY

Life safety involves protecting the vehicles, vessels, and lives of people in buildings and structures from fire. The primary reference source for life safety is the *Life Safety Code*, published by the National Fire Protection Association. The code applies to new and existing buildings. It addresses the construction, protection, and occupancy features necessary to minimize the hazards of fire, smoke, fumes, and panic. A major part of the code is devoted to the minimum requirements for design of egress necessary to ensure that occupants can quickly evacuate a building or structure.

Basic Requirements

In this section, the term *structure* refers to a structure or building.

- Every structure, new and existing, that is to be occupied by people must have a **means of egress** and other fire protection safeguards that (1) ensure that occupants can promptly evacuate or be adequately protected without evacuating; and (2) provide sufficient backup safeguards to ensure that human life is not endangered if one system fails.

- Every structure must be constructed or renovated, maintained, and operated in such a way that occupants are (1) protected from fire, smoke, or fumes; (2) protected from fire-related panic; (3) protected long enough to allow a reasonable amount of time for evacuation; and (4) protected long enough to defend themselves without evacuating.

- In providing structures with means of egress and other fire protection safeguards, the following factors must be considered: (1) character of the occupancy, (2) capabilities of occupants, (3) number of occupants, (4) available fire protection, (5) height of the structure, (6) type of construction, and (7) any other applicable concerns.

- No lock or other device may be allowed to obstruct egress in any part of a structure at any time that it is occupied. The only exceptions to this requirement are mental health detention and correctional facilities. In these, the following criteria are required: (1) responsible personnel must be available to act in the case of fire or a similar emergency; and (2) procedures must be in place to ensure that occupants are evacuated in the event of an emergency.

- All exits in structures must satisfy the following criteria: (1) be clearly visible or marked in such a way that an unimpaired individual can readily discern the route of escape; (2) all routes to a place of safety must be arranged or clearly marked; (3) any doorway and passageway that may be mistaken as a route to safety must be arranged or clearly marked in such a way as to prevent confusion in an emergency; and (4) all appropriate steps must be taken to ensure that occupants do not mistakenly enter a dead-end passageway.

- Egress routes and facilities must be included in the lighting design wherever artificial illumination is required in a structure.

- Fire alarm systems must be provided in any facility that is large enough or so arranged that a fire itself may not adequately warn occupants of the danger. Fire alarms should alert occupants to initiate appropriate emergency procedures.

- In any structure or portion of a structure in which a single means of egress may be blocked or overcrowded in an emergency situation, at least two means of egress must be provided. The two means of egress must be arranged in such a way as to minimize the possibility of both becoming impassable in the same emergency situation.

- All stairs, ramps, and other means of moving from floor to floor must be enclosed (or otherwise protected) to afford occupants protection when used as a means of egress in an emergency situation. These means of vertical movement should also serve to inhibit the spread of fire, fumes, and smoke from floor to floor.

- Compliance with the requirements summarized herein does not eliminate or reduce the need to take other precautions to protect occupants from fire hazards, nor does it permit the acceptance of any condition that could be hazardous under normal occupancy conditions.[12]

The information in this section is a summary of the broad fundamental requirements of the *Life Safety Code* of the NFPA. More specific requirements relating to means of egress and features of fire protection are explained in the following sections.

Means of Egress

This section explains some of the more important issues in the *Life Safety Code* relating to means of egress. Students and practitioners who need more detailed information are encouraged to refer to the *Life Safety Code*.

1. *Doors.* Doors that serve as exits must be designed, constructed, and maintained in such a way that the means of egress is direct and obvious. Windows that could be mistaken for doors in an emergency situation must be made inaccessible to occupants.

2. *Capacity of means of egress.* The means of egress must have a capacity sufficient to accommodate the occupant load of the structure calculated in accordance with the requirements of the *Life Safety Code*.

3. *Number of means of egress.* Any component of a structure must have a minimum of two means of egress (with exceptions as set forth in the code). The minimum number of means of egress from any story or any part of a story is three for occupancy loads of 500 to 1,000 and four for occupancy loads of more than 1,000.

4. *Arrangement of means of egress.* All exits must be easily accessible at all times in terms of both location and arrangement.

5. *Measurement of travel distance to exits.* The travel distance to at least one exit must be measured on the walking surface along a natural path of travel beginning at the most remote occupied space and ending at the center of the exit. Distances must comply with the code.

6. *Discharge from exits.* All exits from a structure must terminate at a public way or at yards, courts, or open spaces that lead to the exterior of the structure.

Safety Myth

Fire Ratings for Doors, Walls, and Floors

A three-hour fire wall will provide protection for three hours. Right? Not necessarily. In fact, any relationship between a fire rating and the reality of fire resistance may be little more than coincidental. The problem is that the factors used to determine fire ratings are outdated. They are based on materials that were in use many years ago and no longer have any relevance. The materials used for constructing and furnishing buildings today are radically different from those upon which fire ratings are based. Consider fire ratings overvalued, and plan for less time than they allow.

7. *Illumination of means of egress.* All means of egress shall be illuminated continuously during times when the structure is occupied. Artificial lighting must be used as required to maintain the necessary level of illumination. Illumination must be arranged in such a way that no area is left in darkness by a single lighting failure.

8. *Emergency lighting.* Emergency lighting for all means of egress must be provided in accordance with the code. In cases where maintaining the required illumination depends on changing from one source of power to another, there shall be no appreciable interruption of lighting.

9. *Marking of means of egress.* Exits must be marked by readily visible, approved signs in all cases where the means of egress is not obviously apparent to occupants. No point in the exit access corridor shall be more than 100 feet from the nearest sign.

10. *Special provisions for high hazard areas.* If an area contains contents that are classified as highly hazardous, occupants must be able to exit by traveling no more than 75 feet. At least two means of egress must be provided, and there shall be no dead-end corridors.[13]

The requirements summarized in this section relate to the fundamental specifications of the *Life Safety Code* relating to means of egress. For more detailed information concerning general requirements, means of egress, and other factors such as fire protection and fire protection equipment, refer to the actual code.

FLAME-RESISTANT CLOTHING

For employees who work in jobs in which flames or electric arcs may occur, wearing flame-resistant clothing can be a lifesaver.[14] Electric arcs are the result of electricity passing through ionized air. Although electric arcs last for only a few seconds, during that time they can produce extremely high levels of heat and flash flame.

OSHA's standards relating to flame-resistant clothing are found in CFR 1910.269, paragraph 1. Key elements of paragraph 1 explain the employer's responsibilities regarding personal protective equipment and flame-resistant clothing.

Apparel

CFR 1910.269, paragraph 1(6) reads as follows:

(i) When work is performed within reaching distance of exposed energized parts of equipment, the employer shall ensure that each employee removes or renders nonconductive all exposed conductive articles, such as key or watch chains, rings, or wrist watches or bands, unless such articles do not increase the hazards associated with contact with the energized parts.

(ii) The employer shall train each employee who is exposed to the hazards or flames or electric arcs in the hazards involved.

(iii) The employer shall ensure that each employee who is exposed to the hazards of flames or electric arcs does not wear clothing that, when exposed to flames or electric arcs, could increase the extent of injury that would be sustained by the employee.

Note: Clothing made from the following types of fabrics, either alone or in blends, is prohibited by this paragraph, unless the employer can demonstrate that the fabric has been treated to withstand the conditions that may be encountered or that the clothing is worn in such a manner as to eliminate the hazard involved: acetate, nylon, polyester, rayon.

(iv) Fuse Handling. When fuses must be installed or removed with one or both terminals energized at more than 300 volts or with exposed parts energized at more than 50 volts, the employer shall ensure that tools or gloves rated for the voltage are used. When expulsion-type fuses are installed with one or both terminals energized at more than 300 volts, the employer shall ensure that each employee wears eye protection meeting the requirements of Subpart I of this Part, uses a tool rated for the voltage, and is clear of the exhaust path of the fuse barrel.[15]

FIRE SAFETY PROGRAMS

Organizations that are interested in protecting their employees from fire hazards should remember the Boy Scouts' motto: Be prepared. The best way to be prepared is to establish a comprehensive fire safety program that encompasses all the functional activities required for being prepared.[16] A comprehensive fire safety program should have at least the following components: assessment, planning, awareness/prevention, and response.

An effective way to develop, implement, and maintain a comprehensive fire safety program is to establish a cross-functional fire safety committee.

"Cross-functional" means that it should have members from all the organization's various functional units. It should also have at least one executive-level manager to ensure and demonstrate that level of support. This approach has several advantages including the following: (1) it focuses the eyes and ears of a broad cross section of the workforce on fire safety, (2) it ensures a broad base of input, and (3) it ensures executive level commitment. This committee should be staffed and chaired by the organization's highest ranking safety and health professional.

Assessment

Assessment of the workplace for fire hazards should be continuous and ongoing. Although the organization's safety and health professional will have primary responsibility for this, committee members also need to be involved and involve the departments that they represent. Members of the safety committee should be trained in the fundamentals of fire hazard assessment by the safety and health professional. They should then pass on this knowledge to employees in their departments, units, and teams. In this way, all employees are involved in continually looking for fire hazards and communicating their concerns to the safety committee.

Planning

OSHA requires that an organization's emergency fire safety plan have at least the following components:

- Emergency escape procedures and routes
- Critical "shutdown" procedures
- Employee headcount procedures
- Rescue and medical procedures
- Procedures for reporting fires and emergencies
- Important contact personnel for additional information

Once the plan is in place, it should be reviewed at least annually and updated as necessary.

Awareness and Prevention

After the fire safety committee has completed the emergency plan and upper management has approved it, employees must become acquainted with it. All employees should receive awareness training so that they understand their role in carrying out the emergency plan. The fire safety committee should evaluate the training program periodically, using guidelines such as the following:

- Do all employees know the role they play in implementing the emergency plan?
- How are employees with disabilities provided for?

- Do all employees understand the escape plans? Evacuation procedures?
- Do all new and temporary employees receive training?
- Are all employees informed when the plan is revised?
- Is a comprehensive drill undertaken at least once each year?
- Are all employees familiar with the sound of the alarm system?
- Is the alarm system checked periodically?
- Are sufficient fire detection devices in place? Are they tested periodically?
- Do all employees know the most likely causes of fires?

Response

Accidents can happen in even the safest organizations. Therefore, it is very important that employees understand the emergency plan and periodically practice responding. Just knowing what the plan says is not sufficient. People do not always think clearly in an emergency situation. They will, however, do what they have learned to do through practice. Consequently, one of the fire safety committee's most important responsibilities is to arrange periodic drills so that employees automatically respond properly.

EXPLOSIVE HAZARDS

Many chemical and toxic substances used in modern organizations are flammable or combustible. Consequently, under certain conditions, they can explode. Working in these conditions involves hazards that require special precautions for handling, storing, transporting, and using such substances.

Explosives-Related Concepts

Safety relating to explosive materials is a highly specialized field. This section discusses terms and concepts used in this field with which modern safety and health professionals should be familiar.

- A **flammable substance** is any substance with a flash point below 37.8°C (100°F) and a vapor pressure of less than 40 psi at that temperature. Such liquids are also known as *Class I liquids*. They tend to be compositions of hydrogen and carbon such as crude oil and its numerous by-products.
- A **combustible substance** is any substance with a flash point of 37.8°C (100°F) or higher. Such liquids are known as *Class II liquids*. They also tend to be compositions of hydrogen and carbon such as crude oil and its numerous by-products.
- The *flash point* is the lowest temperature at which a substance gives off sufficient vapors to combine with air to form an ignitable mixture. Ignition can be precipitated by a spark.

- The *auto-ignition temperature* is the lowest temperature at which a vapor-producing substance or a flammable gas ignites even without the presence of a spark or a flame. This is sometimes known as *spontaneous ignition*.
- In most cases, a certain amount of oxygen must be present in a vapor–air mixture for an explosion to occur. The amount that must be present for a given substance is the **oxygen limit** for that substance.
- **Volatility** is the evaporation (vaporization) capability of a given substance. The greater the tendency of a substance to vaporize, the more volatile it is.[17]

Common Uses of Flammable and Combustible Substances

Flammable and combustible substances are widely used in modern organizations. Therefore, the hazards associated with them are not limited to the industries producing such materials and substances. The National Safety Council lists the following as common uses of flammable and combustible substances in modern industry and specific related precautions that should be taken with each.

- *Dip tanks.* Dipping operations involving flammable or combustible substances should take place in a stand-alone, one-story building constructed of noncombustible materials. The building should be (1) well ventilated, (2) clearly marked as a hazardous area, (3) free of ignition sources, and (4) large. The dip tank itself should be covered and should contain an automatic fire extinguishing system.
- *Japanning and drying ovens.* Ovens used to evaporate varnish, japan enamel, and any other combustible substance should be (1) well ventilated, (2) equipped with an automatic fire protection system, and (3) have a shutdown system that activates automatically in the event of a fire or explosion.
- *Oil burners.* Selecting the proper type of fuel for use in an oil burner is the best precaution to prevent the accumulation and potential ignition of soot. The safest fuel to use in an oil burner is one that meets the following criteria: (1) flash point higher than 37.8°C (100°F), (2) hydrocarbon based, and (3) acid- and grit-free. In addition, the supply tank should be located outside the building housing the oil burner and should be underground. The top of the storage tank should be lower than all pipes entering it. Finally, the oil burner should have an automatic system for preventing the discharge of unburned oil into a hot firebox.
- *Cleaning solvents.* Many of the solvents used to clean metal parts are combustible. The primary precaution when cleaning metal parts is selecting substances that are not easily ignited. Additional precautions include ventilation and the selection of a cleaning area that is free of ignition sources.
- *Internal combustion engines.* Internal combustion engines are widely used in modern industry for powering such equipment as forklifts and lift trucks. Because they are typically fueled with gasoline or diesel fuel, there are fire and explosive hazards associated with this operation. Precautions include (1) proper maintenance, (2) good housekeeping, (3) shutdown of

engines and cooling of exhaust pipes before filling fuel tanks, and (4) a well-ventilated area for filling fuel tanks.

- *Spray-painting booths.* The hazard associated with spray-painting booths is that an explosive mixture of paint vapor and air can occur. To prevent such occurrences, proper ventilation is critical. Regular cleaning of the booth to remove accumulated spray deposits is also important. Paint booths should be equipped with automatic fire protection systems.[18]

The preceding lists only some of the many ways that potentially explosive materials are used in modern organizations.

Other Health Hazards of Explosive Materials

The health hazards associated with explosions and fires are well known. The potential for serious injury or death from the force of a blast or from burns is very high. However, there are other hazards associated with explosive and combustible materials. These include skin irritation, intoxication, and suffocation.

Irritation can occur when the skin comes in contact with hazardous substances. The degree of irritation can range from minor to severe, depending on the type of substance, its concentration, and the duration of contact. Intoxication can occur when an employee breathes the vapors of combustible substances. This can cause impaired judgment, performance, and reaction time, which can, in turn, result in an accident. Finally, the vapors from combustible materials can accumulate in confined spaces. When this happens, the air becomes contaminated and is both toxic and explosive. In such cases, the hazard of suffocation must be added to those associated with explosives.[19]

OSHA'S FIREFIGHTING OPTIONS

Even with the best fire prevention program, and even with the best engineering controls in place, it may still be necessary to manually fight fires at some point. Some companies prefer to have their employees evacuate the premises in the event of a fire. However, for some companies the potential for fires is so much a part of daily operations that they prefer to equip their employees to fight fires. Companies that either allow or expect employees to help fight fires should follow the guidelines set forth by OSHA for manual firefighting.[20]

There are three options available to companies that wish to have their employees participate in firefighting:

1. All employees are involved.
2. Only designated employees are involved.
3. Only employees who are part of an established fire brigade are involved.

Each of these options has its own set of requirements.

Option 1: All Employees Fight Fires

With this option, all employees are allowed to fight fires. However, first they are required to

1. Have and understand an emergency action plan provided by the company
2. Have and understand a fire prevention plan provided by the company
3. Complete annual training and refresher training concerning their duties in fighting fires and in the proper use of fire extinguishers

Option 2: Designated Employees Fight Fires

With this option, only selected employees are allowed to fight fires. First they are required to

1. Have and understand an emergency action plan provided by the company
2. Have and understand a fire prevention plan provided by the company
3. Complete annual training and refresher training concerning their duties fighting fires and in how to properly use fire extinguishers

Option 3: Fire Brigades Fight Fires

With this option, only those employees who are part of an established fire brigade are allowed to fight fires. Fire brigades are divided into two types—**incipient** and **interior structural**. An incipient fire brigade is used to control only small fires. It requires no special protective clothing or equipment. An interior structural fire brigade may fight any type of fire provided it has been issued the appropriate protective clothing and equipment.

The requirements for each type of fire brigade are different. Employees who are part of an incipient fire brigade are required to

1. Have and understand an emergency action plan provided by the company
2. Have and understand a fire protection plan provided by the company
3. Have and understand an organizational statement that establishes the scope, organizational structure, training, equipment, and functions of the fire brigade
4. Have and understand standard operating procedures for the fire brigade to follow during emergencies
5. Complete annual training and refresher training that is hands-on in nature

The requirements of an interior structural fire brigade are the same as those for an incipient fire brigade through the standard operating procedures. However, there are also additional requirements including

- Satisfactory completion of medical examinations that verify their fitness to participate
- Special protective clothing and equipment of the type used by local fire departments including self-contained breathing equipment
- Quarterly, as opposed to annual, training and retraining that is hands-on in nature

SELF-ASSESSMENT IN FIRE PROTECTION

Safety and health personnel cannot be everywhere at the same time. Consequently, it is wise to enlist the assistance of supervisors and employees in fire protection. An excellent way to do this is to provide them with a self-assessment checklist that will guide them in scanning their areas of responsibility for fire hazards. Such checklists should contain at least the following questions:

1. Are portable fire extinguishers properly mounted, readily accessible, and available in adequate number and type?
2. Are fire extinguishers inspected monthly for both operability and general condition with appropriate notation made on their respective tags?
3. Are fire extinguishers recharged regularly and are the dates noted on their tags?
4. Are interior standpipes and valves inspected regularly?
5. Is the fire alarm system tested regularly?
6. Are employees trained in the proper use of fire extinguishers?
7. Are employees trained concerning under what conditions they should help fight fires and under what conditions they should evacuate?
8. Are the nearest fire hydrants flushed annually?
9. Are the nearest fire hydrants maintained regularly?
10. Are avenues and ingress and egress clearly marked?
11. Are all avenues of ingress and egress kept free of clutter and other types of obstructions?
12. Are fire doors and shutters in good working condition?
13. Are fusible links in place and readily accessible?
14. Is the local fire department familiar with the facility and any specific hazards?
15. Is the automatic sprinkler system in good working order, maintained on a regular basis, given the proper overhead clearance, and protected from inadvertent contact damage?[21]

SUMMARY

1. Fire hazards are conditions that favor fire development or growth.
2. The elements required to start and sustain a fire in the smoldering mode are oxygen, fuel, and heat.
3. The elements required to start and sustain a fire in the flaming mode are heat, an oxidizing agent, a reducing agent, and a chemical chain reactive.
4. The product of combustion is energy in the form of heat.
5. By-products of combustion include light and smoke.
6. For a fire to start, there must be either a source of ignition or a sufficiently high temperature for the fuel.
7. Fire is an exothermic chemical reaction. Exothermic reactions generate heat. Endothermic reactions consume more heat than they generate.

8. Chemical reactions in a fire break down materials into basic elements.
9. Loose atoms bond with each other to create substances that were not originally present.
10. Cooling is one of the principal ways to control a fire or extinguish it.
11. Carbon is found in almost every flammable substance.
12. In a fire, released carbon atoms combine with oxygen to form either carbon dioxide or carbon monoxide.
13. Carbon dioxide can deplete oxygen concentrations in the air near the fire.
14. Carbon monoxide is a colorless, odorless, deadly gas.
15. Hydrogen, found in most fuels, combines with oxygen to form water.
16. Synthetic polymers in plastics and vinyls often form deadly toxic fumes when they are consumed, melted, or disintegrated in the presence of fire or high heat.
17. Liquid and solid fuels are first converted to a vapor before they burn.
18. The trend with regard to safety standards is away from the traditional specifications-based approach to a performance-based approach.
19. Removing the fuel, starving the fire of oxygen, or cooling it below the combustion point may extinguish a fire.
20. An explosion is a very rapid, contained fire.
21. Heat always travels from a higher temperature to a lower one.
22. Excess heat is transferred to other objects by conduction, radiation, or convection.
23. Conduction is direct thermal energy heat transfer, molecule to molecule, through solids and liquids.
24. Radiation is electromagnetic wave heat transfer through air in a straight line to surrounding solids.
25. Convection is heat transfer through the movement of hot gases.
26. Spontaneous combustion is rare, but not impossible.
27. Almost everything in the industrial environment can burn.
28. Fires are classified according to their properties, which relate to the fuels.
29. Class A fires involve solid fuels.
30. Class B fires involve flammable liquids and gases.
31. Class C fires involve live electricity.
32. Class D fires involve combustible metals.
33. Special categories include extremely active oxidizers and flammables containing oxygen.
34. All common packing materials burn easily.
35. Fire walls are defined by their ability to slow the spread of fire.
36. Wood and textiles can be treated to reduce their flammability.
37. The flash point is the lowest temperature at which vapors are produced in sufficient concentration to burn, given a source of ignition.
38. The fire point is the lowest temperature at which vapors will continue to burn, given a source of ignition.
39. The auto-ignition temperature is the lowest point at which the vapors of a liquid or solid self-ignites without a source of ignition.
40. Flammable liquids have a flash point below 37.8°C (100°F).

41. Combustible liquids have a flash point at or above 37.8°C (100°F).
42. Flammable and combustible liquids are each divided into three classifications.
43. Most flammable liquids are lighter than water; therefore, water cannot be used to put out these fires.
44. Most gases are lighter than air.
45. Many of the gases present in smoke and at a fire site are toxic to humans.
46. Most fatalities from fire result from breathing toxic gases and smoke, or from suffocating because of a lack of oxygen.
47. A red, blue, yellow, and white diamond label is used to identify hazards present when a substance burns.
48. Natural and generated electricity play a major role in causing fires.
49. Heat, in the form of hot surfaces, can be a source of ignition.
50. Automatic fire detection systems employ different means of detecting a fire.
51. The best way to reduce fires is to prevent their occurrence.
52. OSHA regulations for fire brigades include requirements for prefire planning, an organizational statement, physical capability, training and education, fire-fighting equipment, protective clothing, and respiratory protective devices.
53. A comprehensive fire safety program should have at least the following elements: assessment, planning, awareness and prevention, and response.
54. OSHA provides specific requirements for manual firefighting in three optional approaches: all employees, designated employees, and fire brigades.

KEY TERMS AND CONCEPTS

Auto-ignition temperature	Fire
Automatic sprinkler systems	Fire brigade
Carbon	Fire point
Carbon dioxide	Fixed extinguishing system
Carbon monoxide	Flame-resistant clothing
Combustible liquids	Flammable gases
Combustible substance	Flammable liquids
Combustion	Flammable range
Combustion point	Flammable substance
Conduction	Flash point
Convection	Flashing back
Endothermic	Fuel
Exothermic	Heat
Explosion	Heat transfer
Explosive range	Hydrogen

Hypergolic

Ignition temperature

Incipient fire brigade

Infrared detectors

Interior structural fire brigade

Ionization sensors

Isolation

Kinetic energy

Life Safety Code

Means of egress

Oxygen

Oxygen limit

Photoelectric fire sensors

Polymers

Products of combustion

Pyrophor hypergolic fuels

Radiation

Radiation sensors

Smoke

Source of ignition

Spontaneous combustion

Standpipe and hose systems

Thermal energy

Thermal expansion detectors

Ultraviolet detectors

Vapor

Volatility

REVIEW QUESTIONS

1. What are the three elements of the fire triangle?
2. Fire is a chemical reaction. What is going on?
3. Where is carbon found?
4. Compare and contrast carbon monoxide and carbon dioxide.
5. How is combustion of liquids and solids different from gases?
6. What are the three methods of heat transfer? Describe each.
7. What can happen to a pile of oil-soaked rags in a closed container? Describe the process.
8. In which direction does a fire normally travel?
9. Name something in this room that will not burn.
10. What are the classes of fires?
11. What property do almost all packing materials share?
12. What are the differences among flash point, fire point, and auto-ignition temperature?
13. Which are more stable: combustible liquids or flammable liquids?
14. Which way do gases usually travel?
15. Describe the NFPA hazards identification system.
16. In what four ways can electricity cause a fire?
17. What are the leading causes of fire-related deaths?
18. What are some of the toxic chemicals often produced by fires?
19. What are some of the systems utilized by smoke detectors?
20. What is the most successful life-saving preparation for a fire disaster?
21. Explain the best ways to prevent an office fire.
22. What is the trend with regard to future fire safety standards?

23. Define the term *life safety*.
24. What types of fabrics are prohibited in environments that are flame or arc prone?
25. Explain briefly OSHA's regulations for fire brigades.
26. Summarize the key components of a fire safety program.
27. Describe the precautions that should be taken for dip tanks, oil burners, and spray-painting booths.
28. Explain all three of OSHA's manual firefighting options.

ENDNOTES

1. "Worker Killed in Vapor Cloud Ignition at Tank Storage Facility," *NFPA Journal* 85, no. 3: 29.
2. Ibid., 29–30.
3. Ibid., 31.
4. National Fire Protection Association, *Fire Protection Handbook*, 19th ed. (Quincy, MA: 2003), 112.
5. Ibid., 132.
6. 29 CFR 1910.156 (Subpart L, Appendix A).
7. C. Vogel, "Fires Can Raze Office Buildings," *Safety & Health* 144, no. 3: 27.
8. Ibid., 26–27.
9. A. E. Cote, "Will Fire Safety Standards Survive in the 21st Century?" *NFPA Journal* 85, no. 5: 37.
10. Ibid., 42.
11. Ibid.
12. National Fire Protection Association, *Life Safety Code* (NFPA 101) (Quincy, MA: 2006), 101–119.
13. Ibid., 101–26 through 101–50.
14. National Fire Protection Association, *Life Safety Code*, 151.
15. CFR 1910.269, paragraph 1(6).
16. A. Burke, "Before the Fire," *Occupational Health & Safety* 69, no. 2: 50–52.
17. National Fire Protection Association, *Life Safety Code*, 162.
18. Ibid.
19. Ibid.
20. C. Schroll, "Manual Fire Control," *Occupational Health & Safety* 71, no. 2: 27–28.
21. Retrieved from http://online.misu.nodak.edu/19577/BADM309checklist. htm.

INDUSTRIAL HYGIENE AND CONFINED SPACES

Major Topics

- Overview of Industrial Hygiene
- Industrial Hygiene Standards
- OSH Act and Industrial Hygiene
- Hazards in the Workplace
- Toxic Substances Defined
- Entry Points for Toxic Agents
- Effects of Toxic Substances
- Relationship of Doses and Responses
- Airborne Contaminants
- Effects of Airborne Toxics
- Effects of Carcinogens
- Asbestos Hazards
- Indoor Air Quality and "Sick-Building" Syndrome
- Toxic Mold and Indoor Air Quality
- Threshold Limit Values
- Hazard Recognition and Evaluation
- Prevention and Control
- NIOSH and Industrial Hygiene
- NIOSH Guidelines for Respirators
- Standards and Regulations
- General Safety Precautions
- Confined Space Hazards
- OSHA Confined Space Standard
- OSHA Standards for Toxic and Hazardous Materials
- OSHA's Hazard Communication Standard

Industrial hygiene is an area of specialization within the broader field of industrial safety and health. This chapter provides prospective and practicing safety and health professionals with the information they need to know about this area of specialization.

OVERVIEW OF INDUSTRIAL HYGIENE

Industrial hygiene is a safety and health profession that is concerned with predicting, recognizing, assessing, controlling, and preventing environmental stressors in the workplace that can cause sickness or serious discomfort to workers. An environmental stressor is any factor in the workplace that can cause enough discomfort to result in lost time or illness. Common stressors include gases, fumes, vapors, dusts, mists, noise, and radiation.

The Code of Ethics of the American Academy of Industrial Hygiene describes the responsibilities of industrial hygienists:

- To ensure the health of employees
- To maintain an objective approach in recognizing, assessing, controlling, and preventing health hazards regardless of outside pressure and influence
- To help employees understand the precautions that they should take to avoid health problems
- To respect employers' honesty in matters relating to industrial hygiene
- To make the health of employees a higher priority than obligations to the employer[1]

Role of the Safety and Health Professional

The role of modern safety and health professionals vis-à-vis industrial hygiene often depends on the size of the company employing them. Large companies often employ professionals who specialize in industrial hygiene. These specialists have titles such as occupational physician, industrial hygienist, industrial toxicologist, and health physicist. In smaller companies, safety and health professionals often have responsibility for all safety and health matters, including industrial hygiene.

In companies that employ specialists, their recommendations are used by safety and health professionals to develop, implement, monitor, and evaluate the overall safety and health program. If specialists are not employed, safety and health professionals are responsible for seeking the advice and assistance necessary to predict, recognize, assess, control, and overcome environmental stressors that may cause sickness or serious discomfort to employees.

INDUSTRIAL HYGIENE STANDARDS

Industrial hygiene is a broad field encompassing many different areas of specialization. As such, there are a variety of different organizations and agencies that develop standards of practice relating to some aspect of the field. Safety and health professionals need to know how to contact these organizations in order to stay current with their latest standards. Contact information for the most prominent standards-developing organizations in the broad field of industrial hygiene are as follows:

Air-Conditioning and Refrigeration Institute: http://www.ari.org
Air Movement and Control Association International, Inc.:
    http://www.amca.org
American Conference of Governmental Industrial Hygienists:
    http://www.acgih.org
American Industrial Hygiene Association: http://www.aiha.org
American National Standards Institute: http://www.ansi.org
American Society of Heating, Refrigerating and Air-Conditioning
    Engineers: http://www.ashrae.org
American Society of Safety Engineers: http://www.asse.org
American Society for Testing Materials: http://www.astm.org
International Organization for Standardization: http://www.iso.ch
National Air Duct Cleaners Association: http://www.nadca.com
National Environmental Balancing Bureau: http://www.nebb.org
National Fire Protection Association: http://www.nfpa.org
National Institute for Occupational Safety and Health:
    http://www.cdc.gov/niosh
Occupational Safety and Health Administration: http://www.osha.gov
Underwriters Laboratories: http://www.ul.com
U.S. Department of Energy: http://tis.eh.doe.gov
U.S. Environmental Protection Agency: http://www.epa.gov

OSH ACT AND INDUSTRIAL HYGIENE

The principal federal legislation relating to industrial hygiene is the Occupational Safety and Health Act of 1970 (OSH Act) as amended. The OSH Act sets forth the following requirements relating to industrial hygiene:

- Use of warning labels and other means to make employees aware of potential hazards, symptoms of exposure, precautions, and emergency treatment
- Prescription of appropriate personal protective equipment and other technological preventive measures (29 CFR 1910.133 and 1910.134 [Subpart I])
- Provision of medical tests to determine the effect on employees of exposure to environmental stressors
- Maintenance of accurate records of employee exposures to environmental stressors that are required to be measured or monitored
- Accessibility of monitoring tests and measurement activities to employees
- Availability of monitoring tests and measurement activities records to employees on request
- Notification of employees who have been exposed to environmental stressors at a level beyond the recommended threshold and corrective action being taken

OSHA Process Safety Standard

The OSHA Process Safety Standard is found in 29 CFR 1910.119. Its purpose is to prevent *catastrophic* accidents caused by major releases of highly hazardous chemicals. To comply with this standard, companies must have written operating procedures, mechanical integrity programs, and formal incident investigation procedures. Other key elements are as follows:

1. *Coverage.* Although the process safety standard is typically associated with large chemical and petrochemical processing plants, its coverage is actually much broader than this. Any company is covered that uses the threshold amount of a chemical listed in the standard—or 10,000 pounds or more of a flammable material at one site in one location.

2. *Employee participation.* Section (c) of the standard requires that employees be involved in all aspects of the process safety management program. In addition, employees must be given access to information developed as part of the program.

3. *Process safety information (PSI).* Section (d) of the standard requires organizations to establish and maintain process safety information files. Information included in the files includes chemical, process, and equipment data.

4. *Process hazard analyses (PHAs).* Section (e) of the standard requires that companies conduct process hazard analyses for all processes covered by the standard. Like any other hazard analysis, the PHAs are supposed to identify potential problems so that prompt corrective action or preventive measures can be taken.

5. *Standard operating procedures (SOPs).* Section (f) of the standard requires employers to establish and maintain written standard operating procedures for using chemicals safely. The requirement applies to handling, processing, transporting, and storing chemicals.

6. *Requirements for contractors.* Section (h) of the standard describes the special requirements imposed on companies that contract portions of their work to other companies. Complying with the standard is a matter of making sure that contractors comply. The following requirements are imposed by Section (h):

- Screen contractors before issuing a contract to ensure that they have a comprehensive safety and health program.
- Orient contractors concerning the chemicals with which they may be required to work or be around, the emergency action plan, and other pertinent information.
- Evaluate contractors periodically to ensure that their safety performance is acceptable.
- Maintain an OSHA injury and illness log for the contractor that is separate from, and in addition to, that of the host company.

OSHA Regulation for Chemical Spills

OSHA issues a special regulation dealing with chemical spills. The standard (29 CFR 1910.120) is called the Hazardous Waste Operations and Emergency Response Standard, or *HAZWOPER*. **HAZWOPER** gives organizations two options for responding to a chemical spill. The first is to evacuate all employees in the event of a spill and to call in professional emergency response personnel. Employers who use this option must have an emergency action plan (EAP) in place in accordance with 29 CFR 1010.38(a). The second option is to respond internally. Employers using this must have an emergency response plan that is in accordance with 29 CFR 1010.120.

1. *Emergency action plans (EAPs).* An **emergency action plan** should have at least the following elements: alarm systems, evacuation plan, a mechanism or procedure for emergency shutdown of the equipment, and a procedure for notifying emergency response personnel.

2. *Emergency response plan.* Companies that opt to respond internally to chemical spills must have an **emergency response plan** that includes the provision of comprehensive training for employees. OSHA Standard 29 CFR 1910.120 specifies the type and amount of training required, ranging from awareness to in-depth technical training for employees who will actually deal with the spill. It is important to note that OSHA forbids the involvement of untrained employees in responding to a spill. The following topics are typical of those covered in up-to-date HAZWOPER courses.

- Summary of key federal laws
- Overview of impacting regulations
- Classification and categorization of hazardous waste
  Definition of hazardous waste
  Characteristics
  Toxic characteristic leaching procedure (TCLP)
  Lists of hazardous wastes
- Hazardous waste operations
  Definitions
  Levels of response
- Penalties for noncompliance
  Civil penalty policy
- Responses to spills
  Groundwater contamination
  Sudden releases
  Clean-up levels
  Risk assessment
  Remedial action

- Emergency response
  Workplan
  Site evaluation and control
  Site specific safety and health plan
  Information and training program
  Personal protective equipment
  Monitoring
  Medical surveillance
  Decontamination procedures
  Emergency response
  Other provisions
- Contingency plans
  Alarm systems
  Action plan
- Personal protective equipment
  Developing a PPE program
  Respiratory equipment
  Protective clothing
  Donning PPE
  Doffing PPE
- Material safety data sheets
  Introduction
  Preparing MSDSs
  MSDS information
  Hazardous ingredients
  Physical/chemical characteristics
  Fire and explosion hazard data
  Reactivity data
  Health hazard data
  Precautions for safe handling and use
  Control measures
- Site control
  Site maps
  Site preparation
  Work zones
  Buddy system
  Site security
  Communications
  Safe work practices

- Hazardous waste containers
  Emergency control
  Equipment
  Tools
  Safety
- Decontamination
  Types
  Decontamination plan
  Prevention of contamination
  Planning
  Emergencies
  Physical injury
  Heat stress
  Chemical exposure
  Medical treatment area
  Decontamination of equipment
  Decontamination procedures
  Sanitation of PPE
  Disposal of contaminated materials

In a given year, industrial facilities in the United States release almost 6 billion pounds of toxic substances into the environment. The Environmental Protection Agency (EPA) began measuring such emissions in 1987. Since that time, the trend in toxic emissions has been downward. Even with the downward trend, however, in a typical year, toxic substances will be released into the environment in the following amounts (according to the EPA):

- 2.3 billion pounds released into the air
- 1.1 billion pounds injected into underground wells
- 910 million pounds placed in treatment and disposal facilities
- 550 million pounds placed in municipal wastewater treatment plants
- 440 million pounds placed in landfills
- 100 million pounds released into the water[2]

People are exposed to a variety of substances every day in the home and at work—paints, paint remover, detergent, cleaning solvents, antifreeze, and motor oil, to name just a few. Most substances with which we interact are not dangerous in small amounts or limited exposure. However, high levels of exposure to certain substances in high concentrations can be dangerous. Levels of exposure and concentration, as well as how we interact with substances, help determine how hazardous substances are. In addition, some substances used frequently in certain industrial settings can explode. Following is some information that modern safety and health professionals need to know about toxic and explosive substances.

HAZARDS IN THE WORKPLACE

The environmental stressors on which industrial hygiene focuses can be divided into the following broad categories: chemical, physical, biological, and ergonomic hazards. Typical **chemical hazards** include mists, vapors, gases, dusts, and fumes. Chemical hazards are either inhaled or absorbed through the skin or both. **Physical hazards** include noise, vibration, extremes of temperature, and excessive radiation (electromagnetic or ionizing). **Biological hazards** come from molds, fungi, bacteria, and insects. Bacteria may be introduced into the workplace through sewage, food waste, water, or insect droppings. **Ergonomic hazards** are related to the design and condition of the workplace. Poorly designed workstations and tools are ergonomic hazards. Conditions that put workers in awkward positions and impair their visibility are also hazards.

Material Safety Data Sheets

Employees should be warned of chemical hazards by labels on containers or material safety data sheets (MSDSs). MSDSs are special sheets that summarize all pertinent information about a specific chemical. The hazard communication standard of the OSH Act requires that chemical suppliers provide users with an MSDS for each chemical covered by the standard.

An MSDS should contain the following information as appropriate: manufacturer's name, address, and telephone number; a list of hazardous ingredients; physical and chemical characteristics; fire and explosion hazard information; reactivity information; health hazard information; safety precautions for handling; and recommended control procedures.

An MSDS must contain specified information in eight categories:

- *Section I: General information.* This section contains directory information about the manufacturer of the substance, including the manufacturer's name and address, telephone number of an emergency contact person, a nonemergency telephone number for information, and a dated signature of the person who developed or revised the MSDS.
- *Section II: Hazardous ingredients.* This section should contain the common name, chemical name, and Chemical Abstracts Service (CAS) number for the substance. Chemical names are the scientific designations given in accordance with the nomenclature system of the International Union of Pure and Applied Chemistry. The CAS number is the unique number for a given chemical that is assigned by the Chemical Abstracts Service.
- *Section III: Physical and chemical characteristics.* Data relating to the vaporization characteristics of the substance are contained in this section.
- *Section IV: Fire and explosive hazard data.* Data relating to the fire and explosion hazards of the substance are contained in this section. Special firefighting procedures are also included in this section.

- *Section V: Reactivity data.* Information concerning the stability of the substance as well as the potential for hazardous decomposition or polymerization of the substance is contained in this section.
- *Section VI: Health hazards.* This section contains a list of the symptoms that may be suffered as a result of overexposure to the substance. Emergency first-aid procedures are also explained in this section.
- *Section VII: Safe handling and use.* This section explains special handling, storage, spill, and disposal methods and precautions relating to the substance.
- *Section VIII: Control measures.* The types of ventilation, personal protective equipment, and special hygienic practices recommended for the substance are explained in this section.[3]

Environmental Stressors

Noise is sound that is unwanted or that exceeds safe limits. It can cause problems ranging from annoyance to hearing loss. Acceptable levels of noise have been established by OSHA, NIOSH, and the EPA. OSHA mandates that an employee's exposure level be limited to 90 decibels (dB) calculated as an eight-hour, time-weighted average. Applying the OSHA recommendation, 14 percent of workers are employed in situations that expose them to excess noise levels.[4] Chapter 22 contains the latest data from NIOSH and OSHA on noise-level standards.

Temperature control is the most basic way to eliminate environmental hazards:

> People function efficiently only in a very narrow body temperature range, a "core" temperature measured deep inside the body, not on the skin or at body extremities. Fluctuations in core temperatures exceeding about 2°F below, or 3°F above, the normal core temperature of 37.6°C (99.6°F), which is 37°C mouth temperature (98.6°F mouth temperature), impair performance markedly. If this five-degree range is exceeded, a health hazard exists.[5]

Radiation hazards are increasingly prevalent in the age of high technology. In the category of **ionizing radiation,** safety and health professionals are concerned with five kinds of **radiation** (alpha, beta, X-ray, gamma, and neutron). Of these, alpha radiation is the least penetrating, making shielding simple, whereas the others are more difficult to shield against. Meters and other instruments are available to measure radiation levels in the workplace. The greatest risk for nonionizing radiation in the modern workplace comes from lasers. Shielding requirements for lasers are described in the construction regulations of the OSH Act.[6]

Extremes of pressure also represent a potential hazard in the workplace. According to Olishefski,

> One of the most common troubles encountered by workers under compressed air is pain and congestion in the ears from inability to ventilate the middle ear properly

during compression and decompression. As a result, many workers subjected to increased air pressures suffer from temporary hearing loss; some have permanent hearing loss. This damage is believed to be caused by obstruction of the eustachian tubes, which prevents proper equalization of pressure from the throat to the middle ear. The effects of reduced pressure on the worker are much the same as the effects of decompression from a high pressure. If pressure is reduced too rapidly, decompression sickness and ear disturbances similar to the diver's conditions can result.[7]

Biological hazards from various biological organisms can lead to disease in workers. The now-famous outbreak of what has come to be known as Legionnaire's disease is an example of what can result from biological hazards. This disease first surfaced at a convention where numerous participants became sick and soon died. The cause was eventually traced back to bacteria that grew in the cooling/air-moving systems serving the convention center. That bacteria has since been named *Legionnella*.

Ergonomic hazards are conditions that require unnatural postures and unnatural movement. The human body can endure limited amounts of unnatural postures or motions. However, repeated exposure to such conditions can lead to physical stress and injury. Design of tools, workstations, and jobs can lead to or prevent ergonomic hazards.

TOXIC SUBSTANCES DEFINED

A **toxic substance** is one that has a negative effect on the health of a person or animal. Toxic effects are a function of several factors, including the following: (1) properties of the substance, (2) amount of the dose, (3) level of exposure, (4) route of entry, and (5) resistance of the individual to the substance. The issue of toxic substances can be summarized as follows:

> When a toxic chemical acts on the human body, the nature and extent of the injurious response depends upon the dose received—that is, the amount of the chemical that actually enters the body or system and the time interval during which this dose was administered. Response can vary widely and might be as little as a cough or mild respiratory irritation or as serious as unconsciousness and death.[8]

ENTRY POINTS FOR TOXIC AGENTS

The development of preventive measures to protect against the hazards associated with industrial hygiene requires first knowing how toxic agents enter the body. A toxic substance must first enter the bloodstream to cause health problems. The most common routes of entry for toxic agents are inhalation, absorption, injection, and ingestion (see Figure 20–1). These routes are explained in the following paragraphs.

Figure 20–1
Common routes of entry
of toxic substances.

Inhalation

The route of entry about which safety and health professionals should be most concerned is **inhalation.** Airborne toxic substances such as gases, vapors, dust, smoke, fumes, aerosols, and mists can be inhaled and pass through the nose, throat, bronchial tubes, and lungs to enter the bloodstream. The amount of a toxic substance that can be inhaled depends on the following factors: (1) concentration of the substance, (2) duration of exposure, and (3) breathing volume.

According to Olishefski,

> Inhalation, as a route of entry, is particularly important because of the rapidity with which a toxic material can be absorbed in the lungs, pass into the bloodstream, and reach the brain. Inhalation is the major route of entry for hazardous chemicals in the work environment.[9]

Absorption

The second most common route of entry in an industrial setting is **absorption,** or passage through the skin and into the bloodstream. The human skin is a protective barrier against many hazards. However, certain toxic agents can penetrate the barrier through absorption. Of course, unprotected cuts, sores, and abrasions facilitate the process, but even healthy skin will absorb certain chemicals. Humans are especially susceptible to absorbing such chemicals as

Figure 20–2
Substances can be absorbed
through the skin.

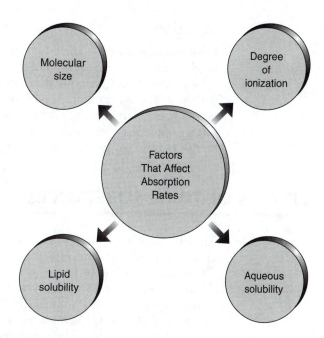

organic lead compounds, nitro compounds, organic phosphate pesticides, TNT, cyanides, aromatic amines, amides, and phenols.[10]

With many substances, the rate of absorption and, in turn, the hazard levels increase in a warm environment. The extent to which a substance can be absorbed through the skin depends on the factors shown in Figure 20–2. Another factor is body site. Different parts of the body have different absorption capabilities. For example, the forearms have a lower absorption potential than do the scalp and forehead.

Ingestion

Ingestion, not a major concern in an industrial setting, is entry through the mouth.[11] An ingested substance is swallowed. It moves through the stomach into the intestines and from there into the bloodstream. Toxic agents sometimes enter the body by ingestion when they are accidentally consumed by workers eating lunch or a snack. Airborne contaminants can also rest on food or the hands and, as a result, be ingested during a meal or snack. The possibility of ingesting toxic agents makes it critical to confine eating and drinking to sanitary areas away from the work site and to make sure that workers practice good personal hygiene such as washing their hands thoroughly before eating or drinking.

As it moves through the gastrointestinal tract, the toxic substance's strength may be diluted. In addition, depending on the amount and toxicity of the substance, the liver may be able to convert it to a nontoxic substance. The liver can, at least, decrease the level of toxicity and pass along the substance to the kidneys, where some of the substance is eliminated in the urine.

Injection

Injection involves the introduction of a substance into the body by way of a needle and syringe, compressed air, high-pressure hydraulic leaks, or any other capable medium. Consequently, this is not often a route of entry for a toxic substance in the workplace. Injection is sometimes used for introducing toxic substances in experiments involving animals. However, this approach can produce misleading research results because the needle bypasses some of the body's natural protective mechanisms.

EFFECTS OF TOXIC SUBSTANCES

The effects of toxic substances vary widely, as do the substances themselves. However, all the various effects and exposure times can be categorized as being either acute or chronic.

Acute effects and exposures involve a sudden dose of a highly concentrated substance. They are usually the result of an accident (a spill or damage to a pipe) that results in an immediate health problem ranging from irritation to death. Acute effects and exposures are (1) sudden, (2) severe, (3) typically involve just one incident, and (4) cause immediate health problems. Acute effects and exposures are not the result of an accumulation over time.

Chronic effects and exposures involve limited continual exposure over time. Consequently, the associated health problems develop slowly. The characteristics of chronic effects and exposures are (1) continual exposure over time, (2) limited concentrations of toxic substances, (3) progressive accumulation

Blood	Kidneys	Heart	Brain
Benzene	Mercury	Aniline	Lead
Carbon monoxide	Chloroform		Mercury
Arsenic			Benzene
Aniline			Manganese
Toluene			Acetaldehyde

Eyes	Skin	Lungs	Liver
Cresol	Nickel	Asbestos	Chloroform
Acrolein	Phenol	Chromium	Carbon tetrachloride
Benzyl chloride	Trichloroethylene	Hydrogen sulfide	Toluene
Butyl alcohol		Mica	
		Nitrogen dioxide	

Figure 20–3
Selected toxic substances and the organs that they endanger most.

of toxic substances in the body and progressive worsening of associated health problems, and (4) little or no awareness of exposures on the part of affected workers.

When a toxic substance enters the body, it eventually affects one or more body organs. Part of the liver's function is to collect such substances, convert them to nontoxics, and send them to the kidneys for elimination in the urine. However, when the dose is more than the liver can handle, toxics move on to other organs, producing a variety of different effects. The organs that are affected by toxic substances are the blood, kidneys, heart, brain, central nervous system, skin, liver, lungs, and eyes. Figure 20–3 lists some of the more widely used toxic substances and the organs that they endanger most.

RELATIONSHIP OF DOSES AND RESPONSES

Safety and health professionals are interested in predictability when it comes to toxic substances. How much of a given substance is too much? What effect will a given dose of a given substance produce? These types of questions concern dose–response relationships. A **dose** of a toxic substance can be expressed in a number of different ways depending on the characteristics of the substance; for example, amount per unit of body weight, amount per body surface area, or amount per unit of volume of air breathed. The dose–response relationship may be expressed mathematically as follows:[12]

$$(C) \times (T) = K$$

where

> C = concentration
> T = duration (time) of exposure
> K = constant

Note that in this relationship, C times T is *approximately* equal to K. The relationship is not exact.

Three important concepts to understand relating to doses are dose threshold, lethal dose, and lethal concentration. These concepts are explained in the following paragraphs.

Dose Threshold

The **dose threshold** is the minimum dose required to produce a measurable effect. Of course, the threshold is different for different substances. In animal tests, thresholds are established using such methods as (1) observing pathological changes in body tissues, (2) observing growth rates (are they normal or retarded?), (3) measuring the level of food intake (has there been a loss of appetite?), and (4) weighing organs to establish body weight to organ weight ratios.

Lethal Dose

A **lethal dose** of a given substance is a dose that is highly likely to cause death. Such doses are established through experiments on animals. When lethal doses of a given substance are established, they are typically accompanied by information that is of value to medical professionals and industrial hygienists. Such information includes the type of animal used in establishing the lethal dose, how the dose was administered to the animal, and the duration of the administered dose. Lethal doses do not apply to inhaled substances. With these substances, the concept of lethal concentration is applied.

Lethal Concentration

A **lethal concentration** of an inhaled substance is the concentration that is highly likely to result in death. With inhaled substances, the duration of exposure is critical because the amount inhaled increases with every unprotected breath.

AIRBORNE CONTAMINANTS

It is important to understand the different types of airborne contaminants that may be present in the workplace.[13] Each type of contaminant has a specific definition that must be understood in order to develop effective safety and health measures to protect against it. The most common types of airborne contaminants are dusts, fumes, smoke, aerosols, mists, gases, and vapors (Figure 20–4).

- *Dusts.* **Dusts** are various types of solid particles that are produced when a given type of organic or inorganic material is scraped, sawed, ground, drilled, handled, heated, crushed, or otherwise deformed. The degree of hazard represented by dust depends on the toxicity of the parent material and the size and level of concentration of the particles.

Figure 20–4
Common airborne contaminants.

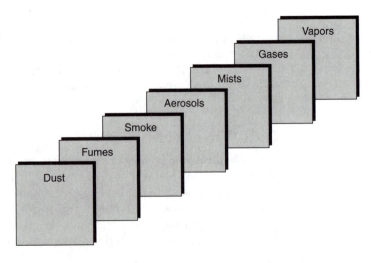

- *Fumes.* The most common causes of fumes in the workplace are such manufacturing processes as welding, heat treating, and metalizing, all of which involve the interaction of intense heat with a parent material. The heat volatilizes portions of the parent material, which then condenses as it comes in contact with cool air. The result of this reaction is the formation of tiny particles that can be inhaled.
- *Smoke.* Smoke is the result of the incomplete combustion of carbonaceous materials. Because combustion is incomplete, tiny soot or carbon particles remain and can be inhaled.
- *Aerosols.* **Aerosols** are liquid or solid particles that are so small they can remain suspended in air long enough to be transported over a distance. They can be inhaled.
- *Mists.* **Mists** are tiny liquid droplets suspended in air. Mists are formed in two ways: (1) when vapors return to a liquid state through condensation; and (2) when the application of sudden force or pressure turns a liquid into particles.
- *Gases.* Unlike other airborne contaminants that take the form of either tiny particles or droplets, gases are actually formless fluids. Gases become particularly hazardous when they fill a confined, unventilated space. The most common sources of gases in an industrial setting are from welding and the exhaust from internal combustion engines.
- *Vapors.* Certain materials that are solid or liquid at room temperature and at normal levels of pressure turn to **vapors** when heated or exposed to abnormal pressure. Evaporation is the most common process by which a liquid is transformed into a vapor.

In protecting workers from the hazards of airborne contaminants, it is important to know the permissible levels of exposure for a given contaminant and to monitor continually the level of contaminants using accepted measurement practices and technologies. The topic of exposure thresholds is covered later in this chapter.

EFFECTS OF AIRBORNE TOXICS

Airborne toxic substances are also classified according to the type of effect they have on the body. The primary classifications are shown in Figure 20–5 and explained in the paragraphs that follow. With all airborne contaminants, concentration and duration of exposure are critical concerns.

Irritants

Irritants are substances that cause irritation to the skin, eyes, and the inner lining of the nose, mouth, throat, and upper respiratory tract. However, they produce no irreversible damage:

> Irritants can be subdivided into primary and secondary irritants. A primary irritant is a material that exerts little systemic toxic action, either because the products formed

Figure 20–5
Airborne toxic substances.

on the tissues of the respiratory tract are nontoxic or because the irritant action is far in excess of any systemic toxic action. A secondary irritant produces irritant action on mucous membranes, but this effect is overshadowed by systemic effects resulting from absorption. Normally, irritation is a completely reversible phenomenon.[14]

Asphyxiants

Asphyxiants are substances that can disrupt breathing so severely that suffocation results. Asphyxiants may be simple or chemical in nature. A simple asphyxiant is an inert gas that dilutes oxygen in the air to the point that the body cannot take in enough air to satisfy its needs for oxygen. Common simple asphyxiants include carbon dioxide, ethane, helium, hydrogen, methane, and nitrogen. Chemical asphyxiants, by chemical action, interfere with the passage of oxygen into the blood or the movement of oxygen from the lungs to body tissues. Either way, the end result is suffocation due to insufficient or no oxygenation. Common chemical asphyxiants include carbon monoxide, hydrogen cyanide, and hydrogen sulfide.

Safety Fact

Hazardous Airborne Emissions Can Be Costly

Failure to properly monitor, measure, and report hazardous airborne emissions cost a manufacturing company in Danbury, Connecticut, $281,000 in fines levied by the Environmental Protection Agency. Degreasing agents used by the company contained the chemical trichloroethylene (TCE). The company failed to calculate TCE emissions, submit written reports about them, or post warnings about the chemical near machines. If inhaled for even a short period, TCE can cause lung irritation, headaches, dizziness, and poor coordination, none of which are conditions conducive to safe machine operation.

Source: From "Firm Hit for Solvent Emissions," *Facility Manager's Alert* 8, no. 174: 3.

Narcotics and Anesthetics

Narcotics and **anesthetics** are similar in that carefully controlled dosages can inhibit the normal operation of the central nervous system without causing serious or irreversible effects. This makes them particularly valuable in a medical setting. Dentists and physicians use narcotics and anesthetics to control pain before and after surgery. However, if the concentration of the dose is too high, narcotics and anesthetics can cause unconsciousness and even death. When this happens, death is the result of asphyxiation. Widely used narcotics and anesthetics include acetone, methyl-ethyl-ketone, acetylene hydrocarbons, ether, and chloroform.

EFFECTS OF CARCINOGENS

A **carcinogen** is any substance that can cause a malignant tumor or a **neoplastic growth.** A *neoplasm* is cancerous tissue or tissue that may become cancerous. Other terms used synonymously for carcinogen are *tumorigen, oncogen,* and *blastomogen:*

> It is well established that exposure to some chemicals can produce cancer in laboratory animals and man. There are a number of factors that have been related to the incidence of cancer—the genetic pattern of the host, viruses, radiation including sunshine, and hormone imbalance, along with exposure to certain chemicals. Other factors such as cocarcinogens and tumor accelerators are involved. It is also possible that some combination of factors must be present to induce cancers. There is pretty good clinical evidence that some cancers are virus-related. It may be that a given chemical in some way inactivates a virus, activates one, or acts as a cofactor.[15]

Medical researchers are not sure exactly how certain chemicals cause cancer. However, there are a number of toxic substances that are either known, or are strongly suspected, to be carcinogens. These include coal tar, pitch, creosote oil, anthracene oil, soot, lamp black, lignite, asphalt, bitumen waxes, paraffin oils, arsenic, chromium, nickel compounds, beryllium, cobalt, benzene, and various paints, dyes, tints, pesticides, and enamels.[16]

ASBESTOS HAZARDS

The EPA estimates that approximately 75 percent of the commercial buildings in use today contain asbestos in some form.[17] Asbestos was once thought to be a miracle material because of its many useful characteristics, including fire resistance, heat resistance, mechanical strength, and flexibility. As a result, asbestos was widely used in commercial and industrial construction between 1900 and the mid-1970s (see Figure 20–6).[18]

Figure 20–6
Asbestos use, 1900 to present.

In the mid-1970s, medical research clearly tied asbestos to respiratory cancer, scarring of the lungs (now known as *asbestosis*), and cancer of the chest or abdominal lining (*mesothelioma*).[19] Its use has been severely restricted in the United States since 1989.

The following quote on **friable asbestos** shows why asbestos is still a concern even though its further use has been banned:

> When asbestos becomes friable (crumbly), it can release fibers into the air that are dangerous when inhaled. As asbestos-containing material (ACM) ages, it becomes less viable and more friable. Asbestos can be released into the air if it is disturbed during renovation or as a result of vandalism.[20]

OSHA has established an exposure threshold known as the permissible exposure limit (PEL) for asbestos. The PEL for asbestos is 0.1 fiber per cubic centimeter of air for an eight-hour time-weighted average.[21] The addresses for OSHA and other sources of further information about asbestos are given in Figure 20–7.

Asbestos Removal and Containment

When an industrial facility is found to contain asbestos, safety and health professionals are faced with the question of whether to remove it or contain it. Before making this decision, the following factors should be considered:

- Is there evidence that the *asbestos-containing material (ACM)* is deteriorating? What is the potential for future deterioration?

Asbestos Action Program
EPA
Mail Code TS-799
401 M St. SW
Washington, DC 20460
202-382-3949

Cancer Information Service
National Cancer Institute
Bldg. 31, Room 10A24
9000 Rockville Pike
Bethesda, MD 20892
800-4CANCER

NIOSH Publications Office
4676 Columbia Pkwy.
Cincinnati, OH 45226
513-533-8287

OSHA Publications Office
Room N–3101
200 Constitution Ave. NW
Washington, DC 20210
202-523-9649

Asbestos Abatement Council of AWCI
1600 Cameron St.
Alexandria, VA 22314
703-684-2924

Asbestos Information Association of
   North America
1745 Jefferson Davis Hwy., Suite 509
Arlington, VA 22202
703-979-1150

The National Asbestos Council
1777 Northeast Expressway, Suite 150
Atlanta, GA 30329
404-633-2622

Figure 20–7
Sources of information about asbestos in the workplace.

- Is there evidence of physical damage to the ACM? What is the potential for future damage?
- Is there evidence of water damage to the ACM or spoilage? What is the potential for future damage or spoilage?[22]

Several approaches can be used for dealing with asbestos in the workplace. The most widely used are removal, enclosure, and encapsulation. These methods are explained in the following paragraphs.

Removal

Asbestos removal is also known as *asbestos abatement.* The following procedures are recommended for removal of asbestos: (1) the area in question must be completely enclosed in walls of tough plastic; (2) the enclosed area must be ventilated by *high-efficiency particle absolute (HEPA)* filtered negative air machines (these machines work somewhat like a vacuum cleaner in eliminating asbestos particles from the enclosed area); (3) the ACM must be covered with a special liquid solution to cut down on the release of asbestos fibers; and (4) the ACM must be placed in leakproof containers for disposal.[23]

Safety Fact

Respirable Silica Exposure and Lung Disease

Employees exposed to respirable silica are at risk of developing the lung disease silicosis. Particularly hazardous activities are hammering, drilling, chipping, crushing, loading, and dumping rock. Employees involved in abrasive blasting are also at risk. Symptoms of silicosis include coughing, difficulty in breathing, fever, and night sweats.

Enclosure

Enclosure of an area containing ACMs involves completely encapsulating the area in airtight walls.[24] The following procedures are recommended for enclosing asbestos: (1) use HEPA-filtered negative air machines in conjunction with drills or any other tools that may penetrate or otherwise disturb ACMs; (2) construct the enclosing walls of impact-resistant and airtight materials; (3) post signs indicating the presence of ACMs within the enclosed area; and (4) note the enclosed area on the plans of the building.

Encapsulation

Encapsulation of asbestos involves spraying the ACMs with a special sealant that binds them together, thereby preventing the release of fibers.[25] The sealant should harden into a tough, impact-resistant skin. This approach is generally used only on acoustical plaster and similar materials.

Personal Protective Equipment for Asbestos Removal

It is important to use the proper types of personal protective clothing and respiratory devices. Clothing should be disposable and should cover all parts of the body.[26] Respirators used when handling asbestos should be "high-efficiency cartridge filter type (half-and-full-face types); any powered-air purifying respirator; any type C continuous-flow supplied-air, pressure-demand respirator, equipped with an auxiliary positive pressure self-contained breathing apparatus."[27]

Medical Records and Examinations

It is important that employees who handle ACMs undergo periodic medical monitoring. Medical records on such employees should be kept current and maintained for at least 20 years.[28] They should contain a complete medical history on the employee. These records must be made available on request to employees, past employees, health care professionals, employee representatives, and OSHA personnel.

Medical examinations, conducted at least annually, should also be required for employees who handle ACMs. These examinations should include front and back chest X-rays that are at least 7 inches by 14 inches. The examination should also test pulmonary function, including forced vital capacity and forced expiratory volume at one second.

INDOOR AIR QUALITY AND "SICK-BUILDING" SYNDROME

A key concept relating to indoor air quality is what has come to be called "sick-building syndrome." In reality, a "sick" building is one that makes people sick because it has become infested with mold, mildew, spores, and other airborne microorganisms. Although much is still unknown about sick-building syndrome, the EPA estimates that as many as 30 percent of the buildings in the United States have air quality problems.

Poor **indoor air quality (IAQ)** can cause a variety of health problems ranging from the temporary to the long term. Health problems commonly associated with poor indoor air quality include allergic reactions, respiratory problems, eye irritation, sinusitis, bronchitis, and pneumonia. Often, the cause of poor indoor air quality can be slipshod maintenance such as failure to keep fans, ductwork, and filters clean. Other contributors are the particles and gases that can be released by office equipment, carpets, paints, cleaning solvents, and office supplies.

One of the keys to preventing sick-building syndrome is air exchange. Important factors in a building's ability to eliminate contaminated air and bring in fresh air are:

- Ventilation
- Air infiltration rates
- Airflow rates in ducts
- Airflow patterns
- Fume exhaust

The most accurate methods available for measuring these factors fall under the broad heading of **tracer gas techniques.** A tracer gas is any gas or vapor not normally found in a building. The best have the following characteristics:

- Nontoxic
- Nonallergic at the levels used
- Chemically inert
- Odorless and tasteless
- Nonflammable and nonexplosive
- Easily transported
- Easily dispersed as an atmospheric gas
- Easily and economically measured with a high degree of reliability

The most widely used tracer gases are sulfur hexafluoride, halogenated refrigerants, and perfluorocarbons. To perform a tracer gas test, the following materials and equipment are needed:

- A suitable tracer gas
- A device for measuring tracer gas concentrations
- An air-sampling system
- A tracer gas injection system
- A data acquisition and control system

There are several different types of tracer gas tests including tracer decay, constant concentration, buildup/decay, CO_2 measurement, and reentrainment/recirculation. Regardless of the type of test used, the testing process involves the following steps:

1. Inject the tracer gas into the building.
2. Measure the concentration of the tracer gas in different parts of the building at different times over a certain period.

The data collected during a tracer gas test can give safety and health professionals the following types of information:

- Total air exchange rate for the building
- Air change rate due to the operation of the building's HVAC system
- Air change rate due to air infiltration and leakage
- Percentage of outside air supplied by the building's HVAC system
- Effectiveness of the ventilation system in removing contaminants
- The distribution of the ventilation air throughout the building

With this type of information available, safety and health personnel can determine if there are pockets where contaminated air is trapped; if the ventilation and air infiltration rates are sufficient; if airflow rates through ducts are sufficient; if airflow patterns are what they should be; and if fume hoods are performing as they should. This type of information is needed to detect and prevent indoor air quality problems.

ANSI's Indoor Air Quality/HVAC Standard

The American National Standards Institute (ANSI) developed a standard specifically addressing indoor air quality (IAQ). The standard, ANSI Z9.8, carries the following title: "Fundamentals Governing the Management, Operation, Testing, and Maintenance of HVAC Systems for Maintaining Acceptable Indoor Air Quality In Employee Occupancies Through Dilution Ventilation." Key components of the standard are summarized in the following paragraphs:

General coverage. ANSI Z9.8 is very specifically written to apply primarily to office space. It applies specifically to employee occupancies in nonindustrial spaces including general office spaces, commercial operations, and office spaces within industrial facilities.

Safety Fact

Indoor Environmental Quality Information

Information about indoor environmental quality is available from the Environmental Protection Agency (EPA). The EPA's information hotline is open from 9:00 A.M. to 5:00 P.M., Eastern Standard Time, Monday–Friday. The hotline number is 1-800-438-4318 (http://www.epa.gov).

Another good source of information is the American Society of Heating, Refrigerating and Air-Conditioning Engineers at 1-800-527-4723 (http://www.ashrae.org).

Application flexibility. The requirements set forth in the standard are minimums. Employers may use demonstrably equal or better approaches. Where the provisions of ANSI Z9.8 conflict with other standards, the more stringent standard is to take precedence. If employers deviate from the standard, they must justify the deviation(s) in writing.

Acceptable air quality. The standard does not prescribe or define what is or is not "acceptable" air quality. Employers are required to determine and define "acceptable." Employers may use threshold limit values (TLVs) or PELs as guidelines or they may establish ceiling limits (e.g., if more than 2 percent of employees complain about air quality problems the ceiling limit has been reached).

Tobacco smoke. The standard requires employers to evaluate smoking in the workplace and apply whatever management controls are appropriate. This gives employers a great deal of latitude. The attitude of the committee that developed the standard (ANSI Z9.8) seems to be that most employers already have smoking controls established.

TOXIC MOLD AND INDOOR AIR QUALITY

Toxic mold has surfaced as an issue relating to IAQ. The issue is complicated by at least two factors. First, there are thousands of types of molds, but only a few are toxic. Second, different people have different levels of sensitivity to mold. On OSHA's list of the 13 most critical indoor air quality hazards, mold is the last entry. On the other hand, in those limited instances in which molds are toxic, they can cause coughing, atypical asthma, nasal congestion, sinusitis, rhinitis, skin rashes, and fatigue. In severe cases, toxic molds can be deadly.

OSHA recommends a three-step process for investigating the possibility of the presence of toxic mold in the workplace. Step one involves interviewing employees about symptoms. Step two involves conducting an on-site review of the workplace in question. Step three, which is undertaken only if the symptoms and physical evidence warrant it, involves an on-site environmental evaluation in which a specialist takes samples.

The principal causal factor in most cases of toxic mold is inadequate ventilation. Consequently, an effective approach for preventing the accumulation of toxic mold in the workplace is to apply the following steps: (1) check outdoor intakes and make sure they are not near trash storage areas, standing water, exhausts, or anything else that might contribute to the growth of mold; (2) make sure the drip pans are sloped sufficiently to prevent the accumulation of standing water; and (3) check ductwork regularly to ensure that the lining is dry and clean.

Toxic Mold Assessment and Remediation

Modern safety and health professionals must be prepared to deal with moisture and mold issues. It is important to investigate periodically to identify sources of moisture and mold. The following procedures may be used to guide investigations:

- *Look* for mold in likely locations such as around pipes, drains, windows, and dark, poorly ventilated areas.
- *Listen* to the feedback and comments of employees who might complain about allergies that could be affected by mold or about any aspect of indoor air quality.
- *Smell* the air in the work environment. You can tell if the air is damp, stale, or musty. If it is, there is a moisture problem, even if it is hidden under floors or behind walls.
- *Train* employees how to be "mold investigators" by showing them how to look, listen, and smell in their work environment.
- *Inventory* the principal areas of moisture and mold risk in your facility and monitor these high-risk areas continually.[29]

When mold is found, it is important to act. Mold remediation, in general terms, proceeds as follows: (1) stop the moisture intrusion, (2) contain and isolate the moisture that is already present, (3) dry and filter the affected area, (4) remove anything in the affected area that cannot be dried, (5) kill existing bacteria with disinfectants and sanitizing agents, (6) clean and then reclean the area, and (7) take whatever steps are necessary to prevent any further moisture intrusion.

According to Alfred Draper III, an industrial hygienist who specializes in mold-related restoration and remediation, mold remediation projects can be divided into four classes ranging from low impact to high. Class I remediation is just good housekeeping (e.g., minimizing dust, using drop cloths, and cleaning up with HEPA-filtered vacuums). Class II requires the use of EPA-registered disinfectants, containing construction waste, and limiting access to work areas.

Class III projects are where remediation begins to be more challenging. For Class III projects, Draper recommends the following procedures:

- Remove or isolate the affected area's HVAC system to prevent ductwork contamination.

- Seal off the affected area using hard critical barriers (e.g., plywood, sheetrock, etc.).
- Require all personnel who must enter the affected area to wear the proper personal protective equipment (PPE) including disposable full-body coverings, gloves, and half-mask HEPA filter respirators.
- Contain all remediated waste in tightly covered containers before it is shipped.
- Clean and decontaminate all equipment prior to making the final wipedown and cleanup of the affected area.

Class IV projects are the most challenging and demanding of moisture or mold remediations. For Class IV, Draper recommends all the Class III procedures plus the following:

- Cover all equipment, structures, and surfaces not being cleaned.
- Require all remediation personnel to wear the PPE which should include disposable full-body coverings, gloves, rubber boots, and full-face respirators.
- Construct a decontamination facility and require all personnel to pass through it when leaving the affected area.
- Monitor exterior air throughout the duration of the remediation project.
- Evaluate high-risk personnel who work near the affected area, at least during the more hazardous aspects of the remediation.
- Thoroughly "air wash" any materials in the affected room that cannot be disposed of.
- Use negative pressure throughout the cleaning process for local exhaust at in the affected area.
- Use air scrubbing throughout the remediation project to remove any mold spores that might be introduced into the air.
- Have all air ducts professionally cleaned.
- Collect cultured and noncultured samples that can be tested to ensure that the remediation process was effective.

THRESHOLD LIMIT VALUES

How much exposure to a toxic substance is too much? How much is acceptable? Guidelines that answer these questions for safety and health professionals are developed and issued annually by the American Conference of Governmental Industrial Hygienists (ACGIH). The guidelines are known as **threshold limit values (TLVs).** The ACGIH describes threshold limit values as follows:

> Threshold limit values refer to airborne concentrations of substances and represent conditions under which it is believed that nearly all workers may be repeatedly exposed day after day without adverse effect. Because of wide variation in individual susceptibility, however, a small percentage of workers may experience discomfort

Safety Fact

Asking ACGIH to Develop a TLV

What if you are a safety professional who believes that a given substance or agent should have a threshold limit value (TLV)? You may write to the Science Group of the American Conference of Governmental Industrial Hygienists (ACGIH) and ask that a TLV be assigned. The request should contain documentation of why a TLV is needed, production volumes, uses, toxicity data, unique health hazards, and any other scientific data that may be relevant.

> ACGIH
> 1330 Kemper Meadow Dr.
> Suite 600
> Cincinnati, OH 45240
> 513-742-2020
> http://www.acgih.org

from some substances at concentrations at or below the threshold limit; a smaller percentage may be affected more seriously by aggravation of a preexisting condition or by development of an occupational illness.

Threshold limits are based on the best available information from industrial experience, from experimental human and animal studies, and, when possible, from a combination of the three. The basis on which the values are established may differ from substance to substance; protection against impairment of health may be a guiding factor for some, whereas reasonable freedom from irritation, narcosis, nuisance, or other forms of stress may form the basis for others.[30]

ACGIH's Classifications of TLVs and BEIs

The ACGIH develops threshold limit values (TLVs) and **biological exposure indices (BEIs)** to help safety and health professionals control certain chemical, biological, and physical health hazards in the workplace.[31] TLVs and BEIs are not legal standards and are not intended to be; rather, they are guidelines. However, their impact is increasingly felt, and in a positive way. As government organizations and agencies continue to find that political considerations make it difficult to promulgate legally authorized standards in a timely manner, the ability of the ACGIH to produce TLV guidelines that are updated annually makes its guidelines more and more valuable. Key concepts about TLVs and BEIs that should be understood by safety and health professionals are as follows:

Threshold limit value–time-weighted average (TLV-TWA). The time-weighted average for a conventional eight-hour workday and a 40-hour workweek for a given substance to which it is believed that nearly all workers may be repeatedly exposed on a daily basis without suffering ill effects. For example, the TLV-TWA for liquefied petroleum gas is 1,000 parts per million (ppm).

Threshold limit value–short-term exposure limit (TLV-STEL). The concentration of a given substance to which it is believed that workers may be exposed continuously for short periods without suffering ill effects. A STEL is defined as a 15-minute TWA exposure that should not be exceeded at any time during the workday period. Also, exposures above the TLV-TWA up to the STEL should not exceed 15 minutes and should not occur more than four times in a day (with at least 60 minutes between exposures). For example, the TLV-STEL for isopropyl ether is 310 ppm.

Threshold limit value–ceiling (TLV-C). The concentration of a given substance that should not be exceeded at any point during an exposure period.

Biological exposure indices. The levels of determinants that are expected to be present in specimens taken from healthy workers who have been exposed to selected substances to the same extent as other workers with inhalation exposure to the substance at the TLV. For example, the BEI for acetone in the urine of a worker is 50 milligrams per liter (mg/L).

Physical agents. Substances or factors that can introduce added stress on the human body so that the effects of a given substance at the TLV might be magnified. Physical factors include acoustics (noise), ergonomic conditions, ionizing radiation, lasers, nonionizing radiation, subfrequency and static electric fields, and thermal stress (cold and heat). For example, the TLV for noise of 94 dB (average) is one hour per day.

Calculating a TWA

Time-weighted averages (TWAs) can be calculated for exposures to given substances.[32] Olishefski gives the following formula for calculating the TWA for an eight-hour day:

$$TWA = \frac{CaTa + CbTb + \ldots CnTn}{8}$$

where

Ta = time of the first exposure period during the eight-hour shift
Ca = concentration of the substance in question in period a
Tb = another time period during the same shift
Cb = concentration of the substance in question in period b
Tn = nth or final time period in the eight-hour shift
Cn = concentration during period n

HAZARD RECOGNITION AND EVALUATION

The degree and nature of the hazard must be understood before effective hazard control procedures can be developed. This involves recognizing that a hazard exists and then making judgments about its magnitude with regard to chemical, physical, biological, and ergonomic stresses.

Questions that can be used for recognizing hazards in the workplace are as follows:

- What is produced?
- What raw materials are used in the process?
- What additional materials are used in the process?
- What equipment is used?
- What operational procedures are involved?
- What dust control procedures are involved?
- How are accidental spills cleaned up?
- How are waste by-products disposed?
- Is there adequate ventilation?
- Are processes equipped with exhaust devices?
- How does the facility layout contribute to employee exposure?
- Are properly working personal protective devices available?
- Are safe operating procedures recorded, made available, monitored, and enforced?[33]

Olishefski recommends that all processes be subjected to the following hazard recognition procedures:

- Determine the exposure threshold for each hazardous substance identified when applying the questions just listed, including airborne contaminants.
- Determine the level of exposure to each hazardous substance.
- Determine which employees are exposed to each hazardous material, how frequently, and for how long.
- Calculate the TWAs to the exposure thresholds identified earlier.[34]

For hazard evaluation, the following considerations are important: the nature of the material or substance involved, the intensity of the exposure, and the duration of the exposure. Key factors to consider are how much exposure is required to produce injury or illness; the likelihood that enough exposure to produce injury or illness will take place; the rate of generation of airborne contaminants; the total duration of exposure; and the prevention and control measures used.[35]

For example, the textile industry was once ranked fifth on a list of 43 industries in terms of preventing workplace injuries and fatalities. The industry improved this ranking to first, a feat attributed to hazard evaluation and recognition efforts within the industry adopted to comply with the OSHA cotton-dust standard (1910.1043).[36]

The textile industry has become a leader in the area of industrial safety and health. Even so, there are still problems. According to *Safety & Health,*

Although the textile industry has stopped the use of cancer-associated dyes, as many as 50,000 current and former industry workers have had significant major exposure during their work lives and are at high risk for developing bladder cancer.[37]

One of the ways in which textile companies are dealing with their industrial hygiene problems is by adding more safety and health professionals to their staffs. According to Bone, there will be "a growth of safety positions at the plant level. Each plant is going to need one safety-and-health professional to keep up with regulations."[38]

Honeywell (GE) includes industrial hygiene as a major component in its overall safety and health program. A corporatewide environmental auditing program for recognizing and evaluating hazards is the mainstay of Honeywell's industrial hygiene effort.

As with all effective programs of this nature, Honeywell's program has the involvement and support of the board of directors and higher management. The program came into being in response to what came to be known as the "Kepone tragedy," an incident in which employees at a company in Hopewell, Virginia, suffered Kepone poisoning. The tragic results of this incident motivated Honeywell to establish a rigorous environmental auditing component to its overall industrial hygiene program. Honeywell now conducts approximately 50 environmental audits annually in such areas as occupational health, medical programs, solid waste disposal, hazardous waste disposal, safety/loss prevention, and product safety. The results of the audits are used to develop prevention and control strategies.

PREVENTION AND CONTROL

Most prevention and control strategies can be placed in one of the following four categories: (1) engineering controls, (2) ventilation, (3) personal protective equipment, and (4) administrative controls.[39] Examples of strategies in each category are given in the following paragraphs.

Engineering Controls

The category of **engineering controls** includes such strategies as replacing a toxic material with one that is less hazardous or redesigning a process to make it less stressful or to reduce exposure to hazardous materials or conditions. Other engineering controls are isolating a hazardous process to reduce the number of people exposed to it and introducing moisture to reduce dust.[40]

For example, exhaust ventilation, which involves trapping and removing contaminated air, is an engineering control. This type of ventilation is typically used with such processes as abrasive blasting, grinding, polishing, buffing, and spray painting or finishing. It is also used in conjunction with open-surface tanks. Dilution ventilation involves simultaneously removing and adding air to dilute a contaminant to acceptable levels.[41]

Personal Protection from Hazards

When the work environment cannot be made safe by any other method, **personal protective equipment (PPE)** is used as a last resort. PPE imposes a barrier between the worker and the hazard but does nothing to reduce or eliminate the hazard. Typical equipment includes safety goggles, face shields, gloves, boots, earmuffs, earplugs, full-body clothing, barrier creams, and respirators.[42]

Occasionally, in spite of an employee's best efforts in wearing PPE, his or her eyes or skin will be accidentally exposed to a contaminant. When this happens, it is critical to wash away or dilute the contaminant as quickly as possible. Specially designed eyewash and emergency wash stations such as those shown in Figures 20–8, 20–9, and 20–10 should be readily available and accessible in any work setting where contaminants may be present.

Figure 20–8
Haws Model 700 BT Pedestal OMNI-FLO eye-face station.
Courtesy of Haws Drinking Faucet Company.

Figure 20–9
Haws Model 700BT wall-mounted eyewash station.
Courtesy of Haws Drinking Faucet Company.

Figure 20–10
Haws Model 8730 emergency
polar shower booth.
Courtesy of Haws Drinking
Faucet Company.

Administrative Controls

Administrative controls involve limiting the exposure of employees to hazardous conditions using such strategies as the following: rotating schedules, required breaks, work shifts, and other schedule-oriented strategies.[43]

Additional Strategies

The type of prevention and control strategies used will depend on the evaluation of the specific hazards present in the workplace. The following list of generic strategies apply regardless of the setting:

- Practicing **good housekeeping,** including workplace cleanliness, waste disposal, adequate washing and eating facilities, healthful drinking water, and control of insects and rodents
- Using special control methods for specific hazards, such as reduction of exposure time, film badges and similar monitoring devices, and continuous sampling with preset alarms
- Setting up medical programs to detect intake of toxic materials
- Providing training and education to supplement engineering controls[44]

Self-Protection Strategies

One of the best ways to protect employees from workplace hazards is to teach them to protect themselves. Modern safety and health professionals should ensure that all employees are familiar with the following rules of self-protection:

1. *Know the hazards in your workplace.* Take the time to identify all hazardous materials and conditions in your workplace and know the safe exposure levels for each.

Safety Fact

General Requirements for Eyewash Stations

ANSI Standard Z358.1 states that emergency eyewash stations must deliver tepid water or solution for at least 15 minutes at a minimum flow rate of 0.4 gallon per minute. Other general requirements are as follows:

- Must be able to flush both eyes simultaneously
- Must have freeze protection when the possibility of freezing exists
- Must have no sharp projects in the operating area of the station
- Must have nozzles protected from contaminants
- Must not corrode in the presence of flushing fluid

Safety Fact

Selecting Chemical Protective Clothing

The National Institute for Occupational Safety and Health (NIOSH) provides a guide for evaluating and selecting chemical protective clothing (CPC). Guidelines are contained in NIOSH Publication No. 90–109. The process is as follows:

- Evaluate the workplace.
- Obtain samples of CPC.
- Test the candidate CPC.
- Choose the best performing CPC.
- Monitor the CPC's use and performance in the workplace.

The entire set of guidelines is available from NIOSH at the following Web site: http://www.cdc.gov/niosh.

2. *Know the possible effects of hazards in your workplace.* Typical effects of workplace hazards include respiratory damage, skin disease and irritation, injury to the reproductive system, and damage to the blood, lungs, central nervous system, eyesight, and hearing.

3. *Use personal protective equipment properly.* Proper use of personal protective equipment means choosing the right equipment, getting a proper fit, correctly cleaning and storing equipment, and inspecting equipment regularly for wear and damage.

4. *Understand and obey safety rules.* Read warning labels before using any contained substance, handle materials properly, read and obey signs, and do only authorized work.

5. *Practice good personal hygiene.* Wash thoroughly after exposure to a hazardous substance, shower after work, wash before eating, and separate potentially contaminated work clothes from others before washing them.[45]

NIOSH AND INDUSTRIAL HYGIENE

The National Institute for Occupational Safety and Health (NIOSH) is part of the Department of Health and Human Services (DHHS). This agency is important to industrial hygiene professionals. The main focus of the agency's research is on toxicity levels and human tolerance levels of hazardous substances. NIOSH prepares recommendations for OSHA standards dealing with hazardous substances, and NIOSH studies are made available to employers.

The areas of research of NIOSH's four major divisions—Biomedical and Behavioral Science; Respiratory Disease Studies; Surveillance, Hazard Evaluations, and Field Studies; and Training and Manpower Development—were discussed in Chapter 6. The results of these divisions' studies and their continually updated lists of toxic materials and recommended tolerance levels are extremely helpful to industrial hygienists concerned with keeping the workplace safe.

NIOSH GUIDELINES FOR RESPIRATORS

The respirator is one of the most important types of personal protective equipment available to individuals who work in hazardous environments (see Figures 20–11 and 20–12). Because the performance of a respirator can mean the difference between life and death, NIOSH publishes strict guidelines regulating the manufacture of respirators. The standard with which manufacturers must comply is 42 CFR 84. In addition, safety and health professionals must ensure that employees are provided respirators that meet all of the specifications set forth in 42 CFR 84.

There are two types of respirators: air filtering and air supplying. Air-filtering respirators filter toxic particulates out of the air. To comply with 42 CFR 84, an air-filtering respirator must protect its wearer from the most penetrating aerosol size of particle, which is 0.3 micron aerodynamic mass in median diameter. The particulate filters used in respirators are divided into three classes, each class having three levels of efficiency as follows:

Class N respirators may be used only in environments that contain no oil-based particulates. They may be used in atmospheres that contain solid or nonoil contaminants.

	Class N (Not Oil Resistant)	Class R (Oil Resistant)	Class P (Oil Proof)
Efficiency	95%	95%	95%
Efficiency	99%	99%	99%
Efficiency	99.97%	99.97%	99.97%

Class R respirators may be used in atmospheres containing any contaminant. However, the filters in Class R respirators must be changed after each shift if oil-based contaminants are present. Class P respirators may be used in any atmosphere containing any particulate contaminant.

If there is any question about the viability of an air-filtering respirator in a given setting, employees should use air-supplying respirators. This type of respirator works in much the same way as an air tank for a scuba diver. Air from the atmosphere is completely blocked out, and fresh air is provided via a self-contained breathing apparatus.

In addition to the NIOSH regulations regarding respirators (42 CFR 84), there are regulations published by OSHA in 29 CFR 1010.134. Key provisions of the OSHA regulations are as follows:

- Respirators, when they are required, must be provided by the employer.
- Medical evaluations must be provided for respirator users.

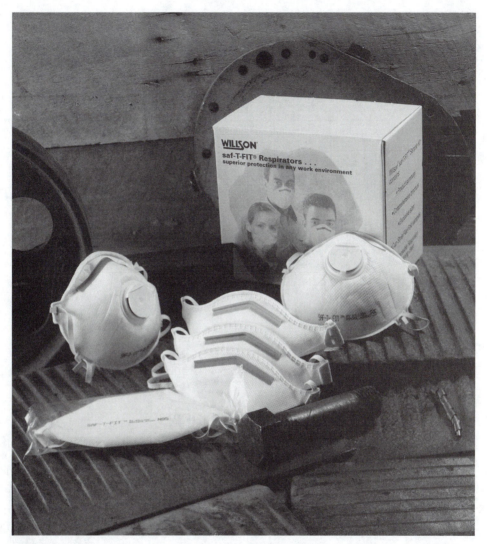

Figure 20–11
Breathing protection devices.
Courtesy of Dalloz Safety (Willson Safety Equipment).

- Fit testing according to standards must be conducted.
- Respirators must be used in reasonably foreseeable emergency situations.
- Respirators must be properly cleaned and maintained.
- Adequate air quality, quantity, and flow for atmosphere-supplied respirators must be ensured.
- Training and evaluation programs must be provided to ensure effectiveness.

Figure 20–12
Respirator.
Courtesy of Dalloz Safety (Willson Safety Equipment).

Air Safety Program Elements

Companies with facilities in which fumes, dust, gases, vapors, or other potentially harmful particulates are present should have an air safety program as part of their overall safety and health program. The program should have at least the following elements:

- Accurate hazard identification and analysis procedures to determine what types of particulates are present and in what concentration

Safety Myth

Contact Lenses and Respirators

Firefighters and other personnel who wear respirators should not wear contact lenses, Right? Not necessarily. OSHA Standard 29 CFR 1910.134(e)(5)(ii) forbids the practice of wearing contact lenses with a respirator, but the standard is based on old information relating to hard contact lenses. The problem with hard contacts is their tendency to pop out. However, most contact lens wearers now use the soft variety that stay in place better. Employees who wear soft contact lenses can use respirators.

- Standard operating procedures (in writing) for all elements of the air safety program
- Respirators that are appropriate in terms of the types of hazards present and that are included in 42 CFR 84
- Training including fit testing, limitations, use, and maintenance of respirators
- Standard procedures for routine maintenance and storage of respirators

STANDARDS AND REGULATIONS

Standards and regulations relating to toxic substances come from a number of different sources. Prominent among these are OSHA and the EPA. Important standards and regulations in this area include the OSHA Chemical Process Standard, the EPA Clean Air Act, Title III of the Superfund Amendments and Reauthorization Act of 1986 (SARA), the Hazardous Materials Transportation Uniform Safety Act of 1990, the Toxic Substances Control Act and the HAZ-MAT Act of 1975. These and other standards and regulations are explained in this section.

OSHA Chemical Process Guidelines

OSHA developed its Chemical Process Guidelines (Process Safety Management of Highly Hazardous Chemicals—Compliance Guidelines and Enforcement Procedures) in response to incidents in which fires and explosions resulted from accidental chemical releases. The guidelines require chemical procedures to analyze their processes to identify potentially hazardous situations and to assess the extent of the hazard. Having completed their analysis, the results must be used in their emergency response plans and to take action to minimize the hazards identified. Specific additional requirements include the following:

- Compiling process safety information
- Maintaining safe operating procedures
- Training and educating employees
- Maintaining equipment
- Conducting accident investigations
- Developing emergency response plans
- Conducting safety compliance audits[46]

EPA Clean Air Act

President George H. W. Bush signed the Clean Air Act amendments into law in November 1990, thereby renewing and extending the original Clean Air Act of 1970.

A key element of the Clean Air Act is its requirement that companies use the **maximum achievable control technology (MACT).** Such technologies represent the current state of the art in pollution control on an ever-changing, ever-improving basis. This means that as technological improvements occur, companies will have to upgrade their pollution control systems to stay in compliance.

The standard also focuses on preventing accidental releases of toxic substances. In addition to identifying potential hazards and taking steps to control them, companies are required to develop plans for minimizing the damage that results when, in spite of controls, releases occur.

The Clean Air Act should enhance a downward trend in air pollution that was already occurring when it was renewed in 1990. That trend saw the amount of toxics released into the air being reduced at a rate of approximately 650 million pounds per year. Of course, this is far short of the 56 billion pounds per year reduction required by the Clean Air Act Amendments of 1990, but it does represent movement in the right direction.

Superfund Amendments and Reauthorization Act

Title III of the **Superfund Amendments and Reauthorization Act (SARA)** is also known as the Emergency Planning and Community Right-to-Know Act. This law is designed to allow individuals to obtain information about hazardous chemicals in their communities so that they can protect themselves in case of an emergency. It applies to all companies that use, make, transport, or store chemicals.

Safety and health professionals involved in developing emergency response plans for their companies should be familiar with SARA and its requirements relating to emergency planning. The major components of the Emergency Planning and Community Right-to-Know Act are discussed in the following paragraphs.

Emergency Planning (Sections 301–303)

Communities are required to form local **emergency planning** committees (LEPCs), and states are required to form state emergency response commissions (SERCs). LEPCs must develop emergency response plans for their local communities, host public forums, select a planning coordinator for the community, and work with the coordinator in developing local plans. SERCs must oversee LEPCs and review their emergency response plans. Plans for individual companies in a given community should be part of that community's larger plan. Local emergency response professionals should use their community's plan as the basis for simulating emergencies and practicing their response.

Emergency Notification (Section 304)

Chemical spills or releases of toxic substances that exceed established allowable limits must be reported to appropriate LEPCs and SERCs. Immediate **emergency notification** may be verbal, provided a written notification is filed promptly thereafter. The report must contain at least the following information: (1) names of the substances released, (2) where the release occurred, (3) when the release occurred, (4) the estimated amount of the release, (5) known hazards to people and property, (6) recommended precautions, and (7) name of a contact person in the company.

Information Requirements (Section 311)

Local companies are required to keep their LEPCs and SERCs—and, through them, the public—informed about the hazardous substances that they store, handle, transport, and use. These information requirements include keeping comprehensive, up-to-date records of the substances on file and readily available; providing copies of MSDSs for all hazardous substances; recording general storage locations for all hazardous substances; estimating the amount of each hazardous substance on hand on a given day; and estimating the average annual amount of hazardous substances kept on hand.

Toxic Chemical Release Reporting (Section 313)

Local companies must report the total amount of toxic substances released into the environment as either emissions or hazardous waste. Toxic chemical release reports go to the EPA and the state-level environmental agency. Section 313 applies to companies that meet the following criteria:

- Employ 10 or more full-time personnel
- Produce or process more than 25,000 pounds of a given toxic substance or use more than 10,000 pounds of a given toxic substance in any capacity
- Conduct 50 percent or more of their business in areas defined by Standard Industrial Classification (SIC) Codes 20–39

Section 313 of SARA requires companies that produce, store, use, or transport chemicals to estimate and report their releases of toxic substances.

Hazardous Materials Transportation and Uniform Safety Act

Transportation of toxic substances always involves a certain amount of hazard. According to the NSC, in a given year,

> hazardous materials incidents caused 165 injuries, 17 deaths and more than $21 million in damages. Most of these happened on the nation's highways where more than 60 percent of the materials are transported. . . . Human error caused most of the incidents. Package failure caused another 30 percent of incidents and vehicle crashes caused 8 percent.[47]

The Hazardous Materials Transportation and Uniform Safety Act of 1975 was passed in response to statistics such as these. The basic provisions of the act are shown in Figure 20–13. In addition to these provisions, the act makes companies that transport hazardous materials partially liable for damages

- Training relating directly to specific functions for private manufacturers, shippers, and carriers.
- Training for employees of public agencies.
- National registration with registration fees used to help pay for the costs of emergency response training.
- A national permit system for motor carriers.

Figure 20–13
Basic provisions of the Hazardous Materials Transportation and Uniform Safety Act.

when an accident occurs and the carrier does not have a satisfactory rating from the Department of Transportation.

The following steps can be used for minimizing the risks associated with transporting hazardous materials:

- Hire personnel who are certified and experienced in dealing with hazardous materials. Then make sure to brief them properly on the hazardous materials that they will be handling and keep them up-to-date on the latest federal regulations.
- Establish a training program that covers at least the following topics: safety equipment, identification of hazardous shipments, routine handling procedures, and emergency procedures.
- At each location where hazardous materials will be loaded or unloaded, name one person who is responsible for making periodic inspections, instructing new employees, meeting with representatives of companies that ship major hazardous materials, maintaining comprehensive records, filing all necessary reports, and contacting the shipper and cosigner when there is a problem.
- Identify all hazardous materials and post a diamond-shaped placard on the outside of all vehicles used to transport hazardous materials.[48]

GENERAL SAFETY PRECAUTIONS

Following are a number of general safety precautions that apply in any settings where explosive and combustible materials are present.

1. *Prohibit smoking.* Smoking should be prohibited in any areas of a plant where explosive and combustible materials are present. Eliminating potential sources of ignition is a standard safety precaution in settings where explosions and fire are possible. In the past, such areas were marked off as restricted, and No Smoking signs were posted. It is becoming common practice to prohibit smoking on the premises altogether or to restrict smoking to designated areas that are well removed from hazard areas.

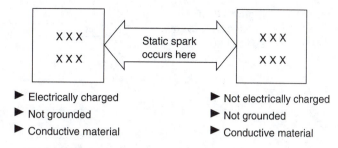

Figure 20–14
How static electricity occurs.

X X X
X X X

Static spark occurs here

X X X
X X X

► Electrically charged
► Not grounded
► Conductive material

► Not electrically charged
► Not grounded
► Conductive material

2. *Eliminate static electricity.* Static electricity occurs when dissimilar materials come into contact and then separate. If these materials are combustible or are near other materials that are combustible, an explosion can occur. Therefore, it is important to eliminate static electricity. The potential for the occurrence of static electricity can be reduced substantially by the processes of grounding and bonding. **Bonding** involves eliminating the difference in static charge potential between materials. *Grounding* involves eliminating the difference in static charge potential between a material and the ground. Figures 20–14, 20–15, and 20–16 illustrate these concepts.

3. *Use spark-resistant tools.* Sparks from tools used in settings where explosive and combustible materials are present represent a threat that must be addressed. In such settings, spark-resistant tools should be used to the maximum extent possible. Wooden, leather-faced, and rubber-covered tools can help prevent sparks that might ignite volatile materials.[49]

CONFINED SPACE HAZARDS

A **confined space** is any area with limited means of entry and exit that is large enough for a person to fit into but is not designed for occupancy. Examples of confined spaces include vaults, vats, silos, ship compartments,

Figure 20–15
Bonding prevents static spark.

Bonding Wire

A

+++
+++

B

+++
+++

► A and B are connected by a bonding wire

► Electrical charge is shared by A and B via the bonding wire eliminating charge differences between them

Figure 20–16
Grounding prevents static spark.

▶ A and B are connected by a bonding wire and to the ground by a ground wire

▶ Charge from A to B runs into the ground

Figure 20–17
Flammable/explosive range for vapors and gases.

train compartments, sewers, and tunnels. What makes confined spaces hazardous, beyond those factors that define the concept, is their potential to trap toxic and explosive vapors and gases.

OSHA's Confined Space Standard (29 CFR 1910.146, paragraph (b)) defines a hazardous atmosphere as one "that may expose employees to the risk of death, incapacitation, impairment of ability to self-rescue, injury, or acute illness."[50] The definition goes on to describe a hazardous atmosphere as a "flammable gas, vapor, or mist in excess of 10 percent of its lower flammable limit."[51]

Safety and health professionals should be cautious in assuming that 10 percent of a lower flammable limit (LFL) for a given substance constitutes an acceptable atmosphere. Even 10 percent of the LFL may exceed the TLV for gases and vapors. Consequently, safety and health professionals should encourage the use of the most sensitive instruments available to detect airborne contaminants.

Certain substances have upper and lower flammable limits (sometimes referred to as "explosive limits," a less-accurate label because some will flame without exploding). The LFL is the lowest concentration of a gas or a vapor that can generate a flame when in the presence of a sufficient ignition source. The upper flammable limit (UFL) is the highest concentration that can propagate a flame. The range between these two extremes is the explosive/flammable range (Figure 20–17).

Using a concentration of 10 percent of the LFL can be dangerous for two reasons: (1) the gas sensor used may be inaccurate; and (2) the threshold limit value (TLV) for some substances is exceeded at the LFL. For example, the TLV for acetone is 500 ppm. The 10 percent LFL for acetone is 2,500 ppm, or five times the TLV.

Safety and health professionals are well advised to take the following precautions when dealing with confined spaces that may have a toxic environment. First, use the most sensitive detection instrument available. Detector tubes, portable gas chromatographs, and infrared analyzers are all more accurate than combustible gas sensors are. Second, compare the 10 percent LFL for any substance in question with its TLV, and let the TLV take precedence.

To ensure that a confined space is safe, the following questions should be asked and answered in the affirmative before allowing entry. A negative response to even one of these questions means that entry into the confined space in question is not yet safe.

- Are access and exit equipment such as ladders and steps in good working condition?
- Has the confined space been properly purged of the toxic vapors and other toxic substances?
- Are all lines that transport potentially hazardous substances into or through the confined space turned off and properly capped?
- Are all moving equipment and moving parts of equipment in the confined space shut down and locked out?
- Has proper ventilation (natural or mechanical) been provided?
- Has the atmosphere inside the confined space been checked by appropriately sensitive detection devices?
- Have provisions been made to monitor continually the atmosphere inside the confined space during work?

In addition to the toxic and explosive hazards associated with confined spaces, there are often physical hazards. For example, tunnels often contain pipes that can trip an employee or that can leak and cause a fall. Empty liquid or gas storage vessels may contain mechanical equipment or pipes that must be carefully maneuvered around, often in the dark.

OSHA CONFINED SPACE STANDARD

The OSHA standard relating to confined spaces is found in 29 CFR 1910.146. This standard mandates that entry permits be required before employees are allowed to enter a potentially hazardous confined space. This means that an employee must have a written permit to enter a confined space. Before the permit is issued, a supervisor, safety or health professional, or some other designated individual should do the following:

1. *Shut down equipment/power.* Any equipment, steam, gas, power, or water in the confined space should be shut off and locked or tagged to prevent its accidental activation.

2. *Test the atmosphere.* Test for the presence of airborne contaminants and to determine the oxygen level in the confined space. Fresh, normal air contains 20.8 percent oxygen. OSHA specifies the minimum and maximum safe levels of oxygen as 19.5 and 23.5 percent, respectively. Atmospheric tests indicate whether a respirator is required and, if so, what type, classification, and level. Figures 20–18, 20–19, 20–20, and 20–21 are examples of devices used for checking the atmosphere.

3. *Ventilate the space.* Spaces containing airborne contaminants should be purged to remove them. Such areas should also be ventilated to keep contaminants from building up again while an employee is working in the space.

4. *Have rescue personnel stand by.* Never allow an employee to enter a confined space without having rescue personnel standing by in the immediate vicinity. These personnel should be fully trained and properly equipped. It is not uncommon for an untrained, improperly equipped employee to be injured or killed trying to rescue a colleague who gets into trouble in a confined space.

Figure 20–18
SAFE T NET 2000 detection device.
Courtesy of Gas Tech, Inc.

Figure 20–19
GT Land Surveyor detection device.
Courtesy of Gas Tech, Inc.

Figure 20–20
STM 2100 detection device.
Courtesy of Gas Tech, Inc.

5. *Maintain communication.* An employee outside the confined space should stay in constant communication with the employee inside. Communication can be visual, verbal, or electronic (radio, telephone) depending on the distance between the employee inside and the entry point.

6. *Use a lifeline.* A lifeline attached to a full-body harness and a block and tackle will ensure that the employee who is inside can be pulled out should he or she lose consciousness. The apparatus should be rigged so that one employee working alone can pull an unconscious employee out of the confined space.

Ventilation of Confined Spaces

Before allowing employees to enter a confined space, it is important to make the space as safe as possible. One of the most effective strategies for doing so is ventilation. Because confined spaces vary in size, shape, function, and hazard potential, there must be a number of different methods for ventilating them.

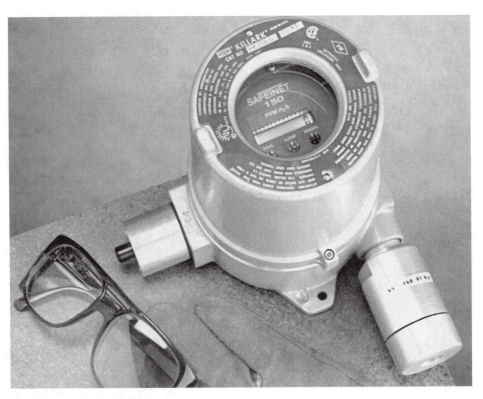

Figure 20–21
SAFE T NET 150 detection device.
Courtesy of Gas Tech, Inc.

Before ventilating a confined space, it should be *purged*. Purging is the process of initially clearing the space of contaminants. Once the area has been purged, ventilation can begin. Ventilation is the process of continually moving fresh air through a space. Ventilation, when properly done, will accomplish the following:

- Dilute and replace airborne contaminants that may still be present in the confined space.
- Ensure an adequate supply of oxygen (between 19.5 and 23.5 percent).
- Exhaust contaminants produced by work performed in the confined space (e.g., welding, painting, and so on).

Ventilation and Local Exhaust

Providing ventilation in a confined space can maintain a comfortable temperature, it can remove odors, and it can dilute contaminants. However, never depend solely on general ventilation to remove toxic contaminants from the air. To eliminate the hazards posed by toxic contaminants such as solvent vapors and welding fumes, it is necessary to exhaust the confined space aggressively. The combination of initial purging, local exhaust, and ventilation is the ideal approach. If contaminant concentrations remain too high even with this approach, employees should wear an appropriate respirator.

Rescue Preparation

The time to think about getting injured employees out of a confined space is well before they enter the space in the first place. Every year, employees are killed trying to save an injured colleague inside a confined space. In an attempt to save injured colleagues, well-meaning employees, who are neither properly trained nor adequately equipped, often fall victim to the toxic atmosphere themselves and die. This is a tragic circumstance made even more so because it is unnecessary and avoidable.

With the right amount of planning and training, employees can be quickly and effectively rescued from confined spaces. Planning should answer the following questions:

- What types of injuries or incidents may occur in a given space?
- What types of hazards may be present in the space?
- What precautions should be taken by rescue personnel entering the space (e.g., lifelines, hoist, respirator, and so on)?
- How much maneuvering room is in the confined space?
- What if the victim needs first aid before he or she can be moved?

All these questions should be answered in the organization's emergency action plan. In addition, all members of the rescue team should have received the training necessary to respond quickly, safely, and effectively. An effective

Discussion Case

What Is Your Opinion?

"You'll be fine as long as you wear your respirator," said the supervisor to the employee as he entered the empty vessel. "There might be some residual toxic gas left over, but there won't be much. If you don't breathe it, the gas can't hurt you. Get in, do the inspection, and get out. It shouldn't take five minutes." Is this supervisor giving the employee accurate advice? What is your opinion?

response is one that is appropriate to the magnitude of the incident and is carried out safely.

OSHA's Confined Space Rescue Requirements

OSHA's Confined Space Standard (29 CFR 1910.146) sets forth the following procedures for effecting the rescue of a worker from a confined space:

1. Preplan a confined space incident.
2. Assess the incident.
3. Conduct monitoring of the confined space environment.
4. Control any hazards present.
5. Prepare for entry—then enter the confined space.
6. Package and remove the victim from the confined space.
7. Remove entrants (rescuers) from the confined space.
8. Secure the confined space.[52]

CONFINED SPACE MANAGEMENT POLICY

Organizations that expose workers to confined spaces in the course of doing their jobs should adopt a comprehensive confined space management policy and enforce it carefully and consistently. The policy should cover at least the following areas of concern: (1) administrative controls, (2) training for all applicable personnel, (3) permitting procedures, and (4) work-team requirements.[53]

Administrative Controls

Organizations that expose workers to confined spaces should include the following administrative controls in their confined space management policy:

1. Identification of all confined spaces and related hazards with signs and placards posted to warn employees appropriately

2. Establishment of an entry-permitting process for controlling and restricting entry into confined spaces
3. Selection of confined space work teams including posting of the names of team members at all sites in question
4. Training for all members of confined space work teams
5. Provision of the proper personal protective equipment to all members of confined space work teams as well as training in its proper use
6. Communication of all requirement and cautions to all members of confined space work teams before their work begins
7. Provision of stand-alone rescue equipment as well as periodic rescue drills
8. Establishment of safe exposure levels inside the confined spaces in question including posting of these levels at the sites
9. Monitoring and testing of atmospheric conditions inside confined spaces
10. Issuance of evacuation orders when safe exposure levels are exceeded in confined spaces
11. Maintenance of all applicable records for the confined space management program

Training for Applicable Personnel

All personnel who will be assigned to a confined space work team as well as their supervisors should receive periodic training that covers at least the following areas of concern:

1. Entry permitting
2. Hazard awareness
3. Selection and proper use of monitoring equipment
4. Selection and proper use of PPE
5. Selection and proper use of stand-alone rescue equipment as well as all applicable rescue procedures
6. Communication procedures
7. Performance auditing
8. Documentation and recordkeeping

Permitting Procedures

Organizations that expose workers to confined spaces should have a comprehensive permitting procedure that is adhered to without exception. What follows is a checklist that should be completed before issuing a permit allowing any member of a confined space work team to begin work:

- Has an entry supervisor been designated to ensure compliance with all applicable procedures?
- Has confined space monitoring been completed and documented?
- Have all applicable PPE requirements been established for the confined space in question?
- Are all entrants wearing the proper PPE and have they been trained in its proper use?

- Have all concerns of work-team members have satisfactorily resolved?
- Has an attendant who will monitor from outside the confined space been designated and properly trained?
- Is the attendant properly situated to carry out his responsibilities?
- Does the attendant know how and when to order an evacuation of the confined space?
- Has access to the confined space in question been properly controlled?
- Have communication and evacuation procedures been discussed and practiced?
- Has an emergency rescue team been established, properly trained, and made ready to act?
- Is stand-alone rescue equipment available outside the entrance to the confined space?
- Has the emergency rescue team established its procedures and practiced them?
- Has an expiration date and time for work in the confined space been established and made known to all members of the work team?
- Have those who are authorized to enter the confined space been designated and are there names known by all members of the work team?
- Has the work that is authorized to be done in the confined space been outlined and communicated to all members of the work team?
- Have procedures for monitoring the confined space during work been established and practiced?

When the answer is "Yes" to all the questions on the checklist, the entry supervisor is allowed to issue the necessary permits for entry into the confined space in question. The checklist should be completed every time work must be done in a confined space. Attempting to save time by skipping steps can be a fatal mistake.

Safety Fact

ANSI Standard for Confined Spaces

The American National Standards Institute (ANSI) established its own standard for work conducted in confined spaces. ANSI Z117.I–2003, Safety Requirements for Confined Spaces, covers confined space requirements including identification and evaluation of confined spaces, emergency response and rescue, permit and nonpermit requirements, atmospheric testing, isolation and decontamination, lockout/tagout safeguarding procedures, personal protection equipment, and warning signs. The standard is available from ANSI at the following Web site:

   http://www.asse.org/Z117

Work-Team Requirements

Organizations that expose workers to confined spaces should establish work teams that consist of the following: (1) entry supervisor who issues the work permit, (2) attendant who monitors from outside the confined space and enforces the work permit, (3) entrants who actually enter the confined space to perform the work, (4) monitor who regularly monitors the atmosphere in the confined space and records the results, (5) rescue team, and (6) safety and health professional to coordinate the team and all its activities.

OSHA STANDARDS FOR TOXIC AND HAZARDOUS MATERIALS

The OSHA standards for hazardous materials are contained in 29 CFR (Subpart H). Nine of the standards apply to specific materials. Four of the standards have broader applications. The standards applying to specific materials are as follows:

Hazardous Materials (Specific Standards)

1910.101	Compressed gases
1910.102	Acetylene
1910.103	Hydrogen
1910.104	Oxygen
1910.105	Nitrous oxide
1910.108	Dip tanks
1910.109	Explosives and blasting agents (An amendment to the OSH Act passed in 1992 now requires that manufacturers of explosives and pyrotechnics must observe the requirements of the Process Safety Management Standards in 1910.119 in addition to this standard.)
1910.110	Liquefied petroleum gases
1910.111	Anhydrous ammonia

In addition to these specific standards, Subpart H contains four standards that have broad applications. Standard 1910.106 parallels the National Fire Protection Association's NFPA 30: *Flammable and Combustible Liquids Code.* Standard 1910.107 regulates processes in which paint is applied by compressed air, electrostatic steam, or other continuous or intermittent processes. Standard 1910.119 regulates process safety management relating to 125 specific chemicals. Standard 1910.120 regulates both hazardous waste operations and spills or accidental releases.

Toxic and hazardous substances are covered in 29 CFR (Subpart Z). The standards in this subpart establish permissible exposure limits (PELs) for over 450 toxic and hazardous substances. Each standard deals with a specific substance or substances. The standards contained in Subpart Z begin with 1910.1000 and run through 1910.1500.

OSHA'S HAZARD COMMUNICATION STANDARD

Any organization that uses hazardous materials in the workplace is required to fully inform employees and on-site contractors of the hazards and to provide training concerning the safe handling, storage, and use of the materials. OSHA's Hazard Communication Standard is 29 CFR 1910.1200.[54] Organizations that use any of the following types of substances should be especially attentive to complying with this standard: acids, cleaning solvents and compounds, flammable gases and liquids, paints, lacquers, and enamels, insecticides, lubricants and oils, fumigants, fungicides and herbicides, and adhesives. This list is not exhaustive, but it does provide a good start concerning what substances to watch for.

Written Hazard Communication Program

Organizations that use hazardous materials are required by 29 CFR 1910.1200 to develop a written program and to communicate applicable elements of it to employees and on-site contractors. An organization's written hazard communication program should contain at least the following elements:

1. Purpose and objectives
2. Inventory (list) of hazardous chemicals and substances
3. Labels and other forms for warning (prevention procedures)
4. Material safety data sheets (procedures and sheets)
5. Hazards of nonroutine tasks (methods of informing)
6. On-site contractors (methods of informing)
7. Training (procedures and elements)

Employee Right-to-Know

The purpose of 29 CFR 1910.1200 is to prevent accidents and injuries by making sure that employees and on-site contractors know about any hazardous substances they might come in contact with in the course of their work, and how to use those substances safely. Those portions of an organization's written hazard communication program that must be shared with employees and on-site contractors are as follows:

1. *Hazardous product inventory.* A comprehensive list of all hazardous substances found in the workplace in question.

Figure 20–22
Employees must be
notified of this right to
know.

ATTENTION ALL EMPLOYEES!
As an employee, or on-site contractor, you have the right to be informed of all hazardous chemical substances to which you may be exposed in the workplace.
29 CFR 1910.1200

2. *Material safety data sheets file.* A master file of data sheets for all hazardous substances in the inventory must be maintained both at the main offices of the organization and any branch sites.

3. *Proper labeling.* All containers—original, immediate-use, and storage—must be properly labeled. The label must contain the name of the hazardous substance, the manufacturer, emergency procedures, and instructions for the safe use of the substance in question. "Immediate-use" containers (those filled by an employee for use during a shift) must have a label that contains the name of the substance and the nature of the hazard (e.g., caustic, corrosive, flammable, etc.).

4. *Emergency plan.* Such plans must contain all the necessary actions to be taken in case of an incident (e.g., fire, spill, accidental release, reaction, etc.). The plan must also list all equipment and materials required to properly respond to an incident, an indication of their locations, and a description of the procedures for using the equipment and materials.

5. *Employee training program.* Training provided to employees and onsite contractors must cover at least the following: (a) discussion of material safety data sheets for all substances in the inventory; (b) right-to-know requirements so that employees and on-site contractors understand what they have a right to know (see Figure 20–22); (c) explanation of labeling requirements; (d) proper handling and use procedures; (e) emergency procedures; and (f) completion of a standardized form verifying completion of the training.

SUMMARY

1. The responsibilities of industrial hygienists include the following: ensuring the health of employees; objectively recognizing, assessing, controlling, and preventing health hazards; helping employees understand precautions; and making the health of employees a high priority.

2. The OSH Act established the following requirements relating to industrial hygiene: use of warning labels, use of personal protective equipment, medical testing, records maintenance, accessibility of information about monitoring activities open to employees, availability of such records to employees, and notification of exposure to environmental stressors.

3. The most prominent hazards in the workplace are chemical, physical, biological, and ergonomic.

4. The main routes of entry for toxic agents are inhalation, absorption, and ingestion.

5. The most common types of airborne contaminants are dusts, fumes, smoke, aerosols, mists, gases, and vapors.

6. Asbestos, once thought to be a miracle material, is now known to be an extremely hazardous substance. It has been tied to respiratory cancer, scarring of the lungs, and cancer of the chest or abdominal lining. When identified in the workplace, asbestos should be handled by removal, enclosure, or encapsulation.

7. The three most important concepts to understand concerning exposure thresholds are time-weighted average (TWA), short-term exposure limit, and exposure ceiling.

8. The American National Standards Institute (ANSI) developed its own indoor air quality standard (ANSI Z9.8). Key concepts in the standard are application flexibility, acceptable air quality, and tobacco smoke.

9. Hazard recognition procedures include the following: determine the exposure threshold for each hazardous substance in the workplace, determine the level of exposure to each, determine which employees are exposed and for how long, and calculate the TWAs.

10. General prevention and control strategies include the following: substitution, process changes, isolation, moisture to reduce dust, exhaust methods, ventilation, personal protective equipment, good housekeeping, special control methods, medical programs, and education and training.

11. The National Institute for Occupational Safety and Health (NIOSH) is part of the Department of Health and Human Services (DHHS). Its two broad functions are research and education in the areas of toxic materials and human tolerance levels.

12. A toxic substance is one that has a negative effect on the health of a person or animal. The effect produced by a toxic substance depends on its properties, the amount of the dose, the level of exposure, and the individual's resistance.

13. The route of entry of a toxic substance is an important consideration. Common routes of entry include ingestion, injection, absorption, and inhalation.

14. The dose threshold is the minimum dose of a toxic substance required to produce a measurable effect. A lethal dose is one that is highly likely to cause death. A lethal concentration of an inhaled substance is the concentration that is likely to cause death.

15. Exposures to toxic substances are either acute or chronic. Acute exposure involves sudden exposure to high concentrations of the substance in question. Chronic exposure involves limited but continual exposure to the substance in question.

16. Airborne contaminants are classified according to the type of effect that they have on the body. There are irritants, asphyxiants, and narcotics and anesthetics.

17. A carcinogen is any substance that can cause a malignant tumor or a neoplastic growth. Other terms used synonymously for carcinogen are *tumorigen, oncogen,* and *blastomogen.*

18. Pertinent standards relating to toxic substances include the OSHA Chemical Process Standard, the EPA Clean Air Act, the Superfund Amendments and Reauthorization Act, and the Hazardous Materials Transportation and Uniform Safety Act.

19. OSHA's Confined Space Standard defines a hazardous atmosphere as one "that may expose employees to the risk of death, incapacitation, impairment of ability to self-rescue, injury, or acute illness." A confined space management policy should have the following elements: administration, controls, training, permitting, and work-team requirements.

20. Threshold limit value (TLV) refers to airborne concentrations of substances and represents conditions under which it is believed that nearly all workers may be repeatedly exposed day after day, without adverse effect. TLVs are expressed as time-weighted average, short-term exposure limit, and ceiling.

21. Material safety data sheets are an excellent source of help for safety and health professionals concerned about the potential hazards of a given toxic substance. Information in an MSDS is presented in eight sections: general information, hazardous ingredients, physical and chemical characteristics, fire and explosive hazard data, reactivity data, health hazards, safe handling and use, and control measures.

22. Important concepts relating to explosive materials include flammable substance, combustible substance, flash point, auto-ignition temperature, oxygen limits, and volatility.

23. OSHA's Hazard Communication Standard (29 CFR 1910.1200) requires organizations to fully inform employees and on-site contractors of the presence of hazardous substances in the workplace and to provide safe-use training.

KEY TERMS AND CONCEPTS

Absorption	Airborne contaminants
Acceptable air quality	Air-filtering respirators
Acute effects and exposures	Air safety program
Aerosols	Air-supplying respirators

Anesthetics

Application flexibility

Asbestos-containing material (ACM)

Asbestos removal

Asphyxiants

Biological exposure index (BEI)

Biological hazards

Bonding

Carcinogen

Ceiling

Chemical hazards

Chemical Process Standard

Chronic effects and exposures

Clean Air Act

Confined space

Dose

Dose threshold

Dusts

Emergency action plan

Emergency notification

Emergency planning

Emergency response plan

Encapsulation

Enclosure

Engineering controls

Ergonomic hazards

Exposure ceiling

Exposure thresholds

Friable asbestos

Good housekeeping

Grounding

Hazard recognition

HAZWOPER

High-efficiency particle absolute (HEPA)

Indoor air quality (IAQ)

Industrial hygiene

Ingestion

Inhalation

Injection

Ionizing radiation

Irritants

Isolating or enclosing

Lethal concentration

Lethal dose

Local exhaust

Material safety data sheet (MSDS)

Maximum achievable control technology (MACT)

Mists

Narcotics

Neoplastic growth

Noise

Nonionizing radiation

OSHA Process Safety Standard

OSHA's Hazard Communication Standard

Personal hygiene

Personal protective equipment (PPE)

Physical hazards

Radiation

Respirator

Route of entry

Self-protection strategies

Short-term exposure limit

Sick-building syndrome

Static electricity

Substituting

Superfund Amendments and Reauthorization Act (SARA)

Temperature control

Threshold limit value (TLV)

Time-weighted average (TWA)

Toxic substance

Tracer gas techniques   Volatility

Vapors   Wet methods

Ventilation

REVIEW QUESTIONS

1. Define the term *industrial hygiene.*
2. Briefly explain the responsibilities of the modern industrial hygienist.
3. What is the role of the safety and health professional regarding industrial hygiene?
4. List five OSHA requirements relating to industrial hygiene.
5. Briefly explain the typical categories of hazards in the workplace.
6. What are the most common routes of entry for toxic agents?
7. Describe the following types of airborne contaminants: dusts, fumes, smoke, mists, and gases.
8. What factors should be considered in deciding whether to remove or contain asbestos?
9. Summarize the various elements of ANSI Z9.8 (indoor air quality/HVAC).
10. Explain the following ways of dealing with asbestos in the workplace: removal, enclosure, and encapsulation.
11. What types of medical examinations should be required of employees who handle ACMs?
12. Briefly explain the following concepts relating to exposure thresholds: time-weighted average, short-term exposure limit, and exposure ceiling.
13. List the most important considerations when evaluating hazards in the workplace.
14. List five generic prevention and control strategies that can be used in any workplace.
15. Give an example of a prevention/control strategy in each of the following categories: engineering controls, ventilation, and personal protective equipment.
16. Explain five self-protection strategies that employees can use in the workplace.
17. How does NIOSH relate to industrial hygiene?
18. What is a toxic substance?
19. List the factors that determine the effect that a toxic substance will have.
20. Describe the most common routes of entry for toxic substances.
21. Explain the mathematical expression of the dose–response relationship.
22. Define the following terms: *dose threshold, lethal dose,* and *lethal concentration.*
23. Differentiate between acute and chronic effects and exposures.
24. List and describe the various classifications of airborne toxics.
25. What is a carcinogen?

26. Describe the basic provisions of the following standards: OSHA Chemical Process Standard, EPA Clean Air Act, and SARA.
27. What is a threshold limit value?
28. Define the following terms: *time-weighted average* and *ceiling.*
29. Explain the three NIOSH categories of respirators.
30. What is "sick-building" syndrome?
31. Explain the major tenets of the OSHA Confined Space Standard.
32. Explain the various elements that should be included in an organization's confined space management policy.
33. Summarize the requirements of OSHA's Hazard Communication Standard.

ENDNOTES

1. J. B. Olishefski, "Overview of Industrial Hygiene," in *Fundamentals of Industrial Hygiene,* 3rd ed. (Chicago: National Safety Council), 5.
2. Retrieved from http://www.epa.gov/air/oagps/peg-caa/pegcaain.html.
3. Olishefski, "Overview of Industrial Hygiene," 160–161.
4. Ibid., 164–165.
5. Ibid., 10.
6. Ibid., 12–14.
7. Ibid., 14.
8. Ibid., 152.
9. Ibid., 17.
10. Ibid.
11. Ibid., 14.
12. Ibid., 148.
13. Ibid., 18–19.
14. Ibid., 22–23.
15. Ibid., 368.
16. Ibid., 369.
17. Centers for Disease Control and Prevention, National Institute for Occupational Safety and Health, *Selected Topics: Asbestos, May 25,* 2006, 4–7. Retrieved from http://www.cdc.gov/niosh/asbestos.html.
18. Ibid.
19. Ibid.
20. Ibid.
21. Ibid.
22. Ibid., 3–4.
23. Ibid., 48.
24. Ibid.
25. Ibid.
26. Ibid., 49.
27. Ibid.
28. Ibid.

29. A. Draper, "I Think It's Mold (Now What?)," *Occupational Health & Safety* 74, no. 5: 67–69.

30. American Conference of Governmental Industrial Hygienists, *Threshold Limit Values* (Cincinnati, OH: American Conference of Governmental Industrial Hygienists, 2006), 3.

31. Ibid., v–vi, 45, 44, 96, and 112.

32. Olishefski, "Overview of Industrial Hygiene," 21.

33. Ibid.

34. Ibid.

35. Ibid., 22.

36. J. Bone, "Textile Industry Weaves a Safety Future," *Safety & Health* 144, no. 3: 17.

37. Ibid., 52.

38. Ibid., 53.

39. Olishefski, "Overview of Industrial Hygiene," 25.

40. Ibid.

41. Ibid., 26.

42. Ibid.

43. Ibid., 27.

44. Olishefski, "Overview of Industrial Hygiene," 24.

45. National Institute for Occupational Safety and Health, Publication No. 90–109. Retrieved June 2006 from http://www.cdc.gov/NIOSH.

46. Retrieved from http://www.osha.gov/pls/oshaweb/owadisp.show.

47. National Safety Council, Accident Facts, 2006, 19.

48. Hazardous Materials Transportation and Uniform Safety Act. Retrieved June 2006 from http://www.eh.doe.gov/oepa/laws/hmta.htm.

49. National Institute for Occupational Safety and Health, Publication 90–109.

50. 29 CFR 1910.146, paragraph (b).

51. Ibid.

52. 29 CFR 1910.146.

53. S. V. Magyar, Jr., "Confined Space Entry," *Occupational Health & Safety* 75, no. 2: 38–50.

54. Retrieved from http://online.misu.nokak.edu/19577/hazMater.htm.

RADIATION HAZARDS

Major Topics

- Ionizing Radiation: Terms and Concepts
- Exposure of Employees to Radiation
- Precautions and Personal Monitoring
- Caution Signs and Labels
- Evacuation Warning Signal
- Instructing and Informing Personnel
- Storage and Disposal of Radioactive Material
- Notification of Incidents
- Reports and Records of Overexposure
- Notice to Employees
- Nonionizing Radiation
- Electromagnetic Fields in the Workplace
- OSHA Standards for Health and Environmental Controls

The widow of a construction worker who helped build the British Nuclear Fuels (BNF) Sellafield plant was awarded $286,500 when it was determined that his death from chronic myeloid leukemia was the result of overexposure to radiation. Sellafield was constructed for the purpose of separating uranium from used fuel rods. Working at the plant for approximately nine months, the victim received a total cumulative dose of almost 52 millisieverts of radiation, which exceeded the established limit for an entire 12-month period. BNF compensated the victim's wife and the families of 20 additional workers who died from causes related to radiation.

Radiation hazards in the workplace fall into one of two categories: ionizing or nonionizing. This chapter provides prospective and practicing safety and health professionals with the information they need concerning radiation hazards in both categories.

IONIZING RADIATION: TERMS AND CONCEPTS

An *ion* is an electrically charged atom (or group of atoms) that becomes charged when a neutral atom (or group of atoms) loses or gains one or more electrons as a result of a chemical reaction. If an electron is lost during this process, a positively charged ion is produced; if an electron is gained, a negatively charged ion is produced. To *ionize* is to become electrically charged or to change into ions. Therefore, **ionizing radiation** is radiation that becomes electrically charged or changed into ions. Types of ionizing radiation, as shown in Figure 21–1, include alpha particles, beta particles, neutrons, X-radiation, gamma radiation, highspeed electrons, and high-speed protons.

To understand the hazards associated with radiation, safety and health professionals need to understand the basic terms and concepts summarized in the following paragraphs, adapted from CFR (Code of Federal Regulations) 1910.1096.

- **Radiation** consists of energetic nuclear particles and includes alpha rays, beta rays, gamma rays, X-rays, neutrons, high-speed electrons, and high-speed protons.
- **Radioactive material** is material that emits corpuscular or electromagnetic emanations as the result of spontaneous nuclear disintegration.
- A **restricted area** is any area to which access is restricted in an attempt to protect employees from exposure to radiation or radioactive materials.

Figure 21–1
Types of ionizing radiation.

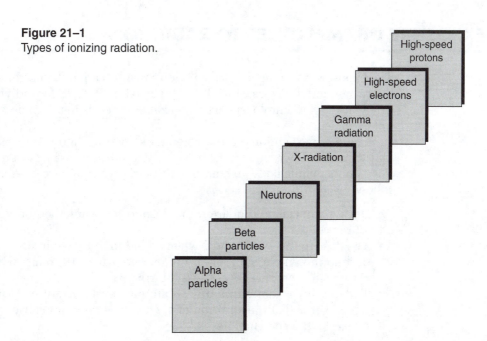

- An **unrestricted area** is any area to which access is not controlled because there is no radioactivity hazard present.
- A **dose** is the amount of ionizing radiation absorbed per unit of mass by part of the body or the whole body.
- **Rad** is a measure of the dose of ionizing radiation absorbed by body tissues stated in terms of the amount of energy absorbed per unit of mass of tissue. One rad equals the absorption of 100 ergs per gram of tissue.
- **Rem** is a measure of the dose of ionizing radiation to body tissue stated in terms of its estimated biological effect relative to a dose of 1 roentgen (r) of X-rays.
- **Air dose** means that an instrument measures the air at or near the surface of the body where the highest dosage occurs to determine the level of the dose.
- **Personal monitoring devices** are devices worn or carried by an individual to measure radiation doses received. Widely used devices include film badges, pocket chambers, pocket dosimeters, and film rings.
- A **radiation area** is any accessible area in which radiation hazards exist that could deliver doses as follows: (1) within one hour, a major portion of the body could receive more than 5 millirems; or (2) within five consecutive days, a major portion of the body could receive more than 100 millirems.
- A **high-radiation area** is any accessible area in which radiation hazards exist that could deliver a dose in excess of 100 millirems within one hour.

EXPOSURE OF EMPLOYEES TO RADIATION

The exposure of employees to radiation must be carefully controlled and accurately monitored. Figure 21–2 shows the maximum doses for individuals in one calendar quarter. Employers are responsible for ensuring that these dosages are not exceeded.

There are exceptions to the amounts shown in Figure 21–2. According to OSHA, an employer may permit an individual in a restricted area to receive doses to the whole body greater than those shown in Figure 21–2 as long as the following conditions are met:

- During any calendar quarter, the dose to the whole body does not exceed 1¼ rems.
- The dose to the whole body, when added to the accumulated occupational dose to the whole body, shall not exceed $5(N - 18)$ rems, where N is the employee's age in years at the last birthday.
- The employer maintains up-to-date past and current exposure records, which show that the addition of such a dose does not cause the employee to exceed the specified doses.[1]

Figure 21–2
Ionizing radiation exposure limits of humans.

Body/Body Region	Rems per Calendar Quarter
Whole body	1.25
Head and trunk	1.25
Blood-forming organs	1.25
Lens of eyes	1.25
Gonads	1.25
Hands and forearms	18.75
Feet and ankles	18.75
Skin of whole body	7.50

Employers must ensure even more careful controls with individuals under 18 years of age. Such individuals may receive only doses that do not exceed 10 percent of those specified in Figure 21–2 in any calendar quarter.

OSHA is not the only agency that regulates radiation exposure. The *Nuclear Regulatory Commission (NRC)* is also a leading agency in this area. The NRC's regulations specify that the total internal and external dose for employees may not exceed 5 rems per year. This same revision established a total exposure limit of 0.6 rem over the entire course of a pregnancy for female employees. According to the NRC, the average radiation exposure of nuclear plant workers is less than 400 millirems annually.[2]

PRECAUTIONS AND PERSONAL MONITORING

Personal monitoring precautions are important for employees of companies that produce, use, release, dispose of, or store radioactive materials or any other source of ionizing radiation. Accordingly, OSHA requires the following precautions:

- Employers must conduct comprehensive surveys to identify and evaluate radiation hazards present in the workplace from any and all sources.
- Employers must provide appropriate personal monitoring devices such as film badges, pocket chambers, pocket dosimeters, and film rings.
- Employers must require the use of appropriate personal monitoring devices by the following: (1) any employee who enters a restricted area where he or she is likely to receive a dose greater than 25 percent of the total limit of exposure specified for a calendar quarter; (2) any employee 18 years of age or less who enters a restricted area where he or she is likely to receive a dose greater than 5 percent of the total limit of exposure specified for a calendar quarter; and (3) any employee who enters a highradiation area.[3]

CAUTION SIGNS AND LABELS

Caution signs and **labels** have always been an important part of safety and health programs. This is particularly true in companies where radiation hazards exist. The universal color scheme for caution signs and labels warning of radiation hazards is purple or magenta superimposed on a yellow background.

Both OSHA and the NRC require caution signs in radiation areas, highradiation areas, airborne radiation areas, areas containing radioactive materials, and containers in which radioactive materials are stored or transported.[4]

Figure 21–3 shows the universal symbol for radiation. Along with the appropriate warning words, this symbol should be used on signs and labels. Figure 21–4 shows a warning sign and label that may be used in various radioactive settings. On containers, labels should also include the (1) quantity of radioactive material, (2) kinds of radioactive materials, and (3) date on which the contents were measured.[5]

Figure 21–3
Universal radiation symbol.

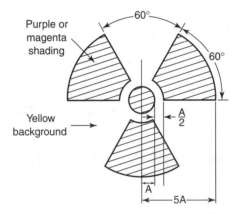

Figure 21–4
Sample warning sign.

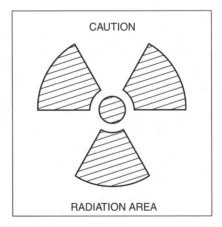

EVACUATION WARNING SIGNAL

Companies that produce, use, store, or transport radioactive materials are required to have a signal-generating system that can warn of the need for evacuation.[6] OSHA describes the **evacuation warning signal** system as follows:

> The signal shall be a mid-frequency complex sound wave amplitude modulated at a subsonic frequency. The complex sound wave in free space shall have a fundamental frequency (f1) between 450 and 500 hertz (Hz) modulated at a subsonic rate between 4 and 5 hertz. The signal generator shall not be less than 75 decibels at every location where an individual may be present whose immediate, rapid, and complete evacuation is essential.[7]

In addition to this basic requirement, OSHA also stipulates the following:

- A sufficient number of signal generators must be installed to cover all personnel who may need to be evacuated.
- The signal shall be unique, unduplicated, and instantly recognizable in the plant where it is located.
- The signal must be long enough in duration to ensure that all potentially affected employees are able to hear it.
- The signal generator must respond automatically without the need for human activation, and it must be fitted with backup power.[8]

INSTRUCTING AND INFORMING PERSONNEL

It is critical that companies involved in producing, using, storing, handling, or transporting radioactive materials keep employees informed concerning radiation hazards and the appropriate precautions for minimizing them. Consequently, OSHA has established specific requirements along these lines. They are summarized as follows:

- All employees must be informed of existing radiation hazards and where they exist; the extent of the hazards; and how to protect themselves from the hazards (precautions and personal protective equipment).
- All employees must be advised of any reports of radiation exposure requested by other employees.
- All employees must have ready access to 29 CFR 1910.1096(i) (Instruction of Personnel, Posting) and any related company operating procedures.[9]

These requirements apply to all companies that do not have superseding requirements (i.e., companies regulated by the Atomic Energy Commission and companies in states with their own approved state-level OSHA plans).

Instruction and information are important in all safety and health programs. They are especially important in settings in which radiation hazards exist. Employees in these settings must be knowledgeable about radiation hazards and how to minimize them. Periodic updating instruction for experienced workers is as important as initial instruction for new employees and should not be overlooked. Often, it is the overly comfortable, experienced worker who overlooks a precaution and thereby causes an accident.

STORAGE AND DISPOSAL OF RADIOACTIVE MATERIAL

Radioactive materials that are stored in restricted areas must be appropriately labeled, as described earlier in this chapter. Radioactive materials that are stored in unrestricted areas "shall be secured against unauthorized removal from the place of storage."[10] This requirement precludes the handling and transport, intentional or inadvertent, of radioactive materials by persons who are not qualified to move them safely.

A danger inherent in storing radioactive materials in unrestricted areas is that an employee, such as a maintenance worker, may unwittingly attempt to move the container and damage it in the process. This could release doses that exceed prescribed acceptable limits.

The disposal of radioactive material is also a regulated activity. There are only three acceptable ways to dispose of radioactive waste: (1) transfer to an authorized recipient; (2) transfer in a manner approved by the Atomic Energy Commission; or (3) transfer in a manner approved by any state that has an agreement with the Atomic Energy Commission pursuant to Section 27(b) 42 U.S.C. 2021(b) of the Atomic Energy Act.[11] States having such agreements are listed in Figure 21–5.

Figure 21–5
States having agreements with the Atomic Energy Commission.

The following states have agreements with the Atomic Energy Commission to dispose of radioactive waste pursuant to 27(b) 42 U.S.C. 2021(b) of the Atomic Energy Act:

Alabama	Mississippi
Arizona	New Hampshire
Arkansas	New York
Colorado	North Carolina
Florida	North Dakota
Georgia	Oregon
Idaho	South Dakota
Kansas	Tennessee
Kentucky	Texas
Louisiana	Washington
Maryland	

NOTIFICATION OF INCIDENTS

A radiation-related incident must be reported if employees meet a specific set of requirements. An **incident** is defined by OSHA as follows:

> Exposure of the whole body of any individual to 25 rems or more of radiation; exposure of the skin of the whole body of any individual to 150 rems or more of radiation; or exposure of the feet, ankles, hands, or forearms of any individual to 375 rems or more of radiation.[12]
>
> The release of radioactive material in concentrations which, if averaged over a period of 24 hours, would exceed 5,000 times the limit specified.[13]

If an incident meeting one of these criteria occurs, the employer must notify the proper authorities immediately. Companies regulated by the Atomic Energy Commission are to notify the commission. Companies in states that have agreements with the Atomic Energy Commission (Figure 21–5) are to notify the state designee. All other companies are to notify the U.S. assistant secretary of labor.[14] Telephone or telegraph notifications are sufficient to satisfy the immediacy requirement.

The notification requirements are eased to 24 hours in cases where whole-body exposure is between 5 and 24 rems; exposure of the skin of the whole body is between 30 and 149 rems; or exposure of the feet, ankles, hands, or forearms is between 75 and 374 rems.[15]

REPORTS AND RECORDS OF OVEREXPOSURE

In addition to the immediate and 24-hour notification requirements explained in the previous section, employers are required to follow up with a written report within 30 days.Written reports are required when an employee is exposed as set forth in the previous section, or when radioactive materials are on hand in concentrations greater than specified limits. Each report should contain the following material, as applicable: extent of exposure of employees to radiation or radioactive materials; levels of radiation and concentration of radiation involved; cause of the exposure; levels of concentrations; and corrective action taken.[16]

Whenever a report is filed concerning the overexposure of an employee, the report should also be given to that employee. The following note should be placed prominently on the report or in a cover letter: "You should preserve this report for future reference."[17]

Records of the doses of radiation received by all monitored employees must be maintained and kept up-to-date. Records should contain cumulative doses for each monitored employee. Figure 21–6 is an example of a cumulative exposure report form of the kind that can be used to satisfy reporting requirements. Notice that the maximum quarterly dose per body or body region is indicated for that region. Such records must be shared with monitored employees

	Rems in 1st Quarter	Rems in 2nd Quarter	Rems in 3rd Quarter	Rems in 4th Quarter	Total
Employee _____					
Dates Covered _____					
Whole body; head and trunk; blood-forming organs; lens of eyes; or gonads (1¼ rems max/quarter)					
Hands and forearms; feet and ankles (18 rems max/quarter)					
Skin of whole body (7½ rems max/quarter)					

Figure 21–6
Cumulative radiation exposure record.

at least annually.[18] A better approach is to advise monitored employees of their cumulative doses as soon as the new cumulative amount is recorded for that period.

Cumulative radiation records must be made available to former employees on request. Upon receiving a request from a former employee, employers must provide the information requested within 30 days.[19] According to OSHA,

> Such report shall be furnished within 30 days from the time the request is made, and shall cover each calendar quarter of the individual's employment involving exposure to radiation or such lesser period as may be requested by the employee. The report shall also include the results of any calculations and analysis of radioactive material deposited in the body of the employee. The report shall be in writing and contain the following statement: "You should preserve this report for future reference."[20]

NOTICE TO EMPLOYEES

The U.S. Nuclear Regulatory Commission (NRC) publishes NRC Form 3 to inform employees of the standards for protection against radiation.[21] Form 3 is presented in a question-and-answer format. Selected questions and answers follow:

- *What is the Nuclear Regulatory Commission?* The Nuclear Regulatory Commission is an independent federal regulatory agency responsible for licensing and inspecting nuclear power plants and other commercial uses of radioactive materials.

- *What does the NRC do?* The NRC's primary responsibility is to ensure that workers and the public are protected from unnecessary or excessive exposure to radiation and that nuclear facilities, including power plants, are constructed to high-quality standards and operated in a safe manner. The NRC does this by establishing requirements in Title 10 of the Code of Federal Regulations (10 CFR) and in licenses issued to nuclear users.

- *What responsibility does my employer have?* Any company that conducts activities licensed by the NRC must comply with the NRC's requirements. If a company violates NRC requirements, it can be fined or have its license modified, suspended, or revoked. Your employer must tell you which NRC radiation requirements apply to your work and must post NRC Notices of Violation involving radiological working conditions.

- *What is my responsibility?* For your own protection and the protection of your co-workers, you should know how NRC requirements relate to your work and should obey them. If you observe violations of the requirements or have a safety concern, you should report them.

- *What if I cause a violation?* If you engaged in deliberate misconduct that may cause a violation of the NRC requirements or would have caused a violation if it had not been detected, or deliberately provided inaccurate or incomplete information to either the NRC or to your employer, you may be subject to enforcement action. If you report such a violation, the NRC will consider the circumstances surrounding your reporting in determining the appropriate enforcement action, if any.

- *How do I report violations and safety concerns?* If you believe that violations of NRC rules or the terms of the license have occurred, or if you have a safety concern, you should report them immediately to your supervisor. You may report violations or safety concerns directly to the NRC. However, the NRC encourages you to raise your concerns with the licensee because it is the licensee who has the primary responsibility for, and is most able to ensure, safe operation of nuclear facilities. If you choose to report your concern directly to the NRC, you may report this to an NRC inspector, or call or write to the NRC Regional Office serving your area. If you send your concern in writing, it will assist the NRC in protecting your identity if you clearly state in the beginning of your letter that you have a safety concern or that you are submitting an allegation. The NRC's toll-free Safety Hotline for reporting safety concerns is listed below. The addresses for the NRC Regional Offices and the toll-free telephone numbers are also listed in the Safety Fact.

- *What if I work with radioactive material or in the vicinity of a radioactive source?* If you work with radioactive materials or near a radiation source, the amount of radiation exposure that you are permitted to receive may be limited by NRC regulations. The limits on your exposure are contained in Sections 20.1201, 20.1207, and 20.1208 of Title 10 of the Code of Federal Regulations (10 CFR 20) depending on the part of the regulations to which your employer is subject. Although these are the maximum allowable limits, your employer should also keep your radiation exposure as far below those limits as "reasonably achievable."

Safety Fact

NRC: Regional Offices

Region	Address	Telephone
I	U.S. Nuclear Regulatory Commission, Region I 475 Allendale Rd. King of Prussia, PA 19406-1415 http://www.nrc.gov	800-432-1156
II	U.S. Nuclear Regulatory Commission, Region II Atlanta Federal Center 61 Forsyth St. SW, Suite 23T85 Atlanta, GA 30303-3415	800-577-8510
III	U.S. Nuclear Regulatory Commission, Region III 801 Warrenville Rd. Lisle, IL 60532-4351	800-522-3025
IV	U.S. Nuclear Regulatory Commission, Region IV 611 Ryan Plaza Dr., Suite 400 Arlington, TX 76011-8064	800-952-9677

- *May I get a record of my radiation exposure?* Yes. Your employer is required to advise you of your dose annually if you are exposed to radiation for which monitoring was required by NRC. In addition, you may request a written report of your exposure when you leave your job.
- *How are violations of NRC requirements identified?* NRC conducts regular inspections at licensed facilities to assure compliance with NRC requirements. In addition, your employer and site contractors conduct their own inspections to assure compliance. All inspectors are protected by federal law. Interference with them may result in criminal prosecution for a federal offense.
- *May I talk with an NRC inspector?* Yes. NRC inspectors want to talk to you if you are worried about radiation safety or have other safety concerns about licensed activities, such as the quality of construction or operations at your facility. Your employer may not prevent you from talking with an inspector. The NRC will make all reasonable efforts to protect your identity where appropriate and possible.
- *May I request an inspection?* Yes. If you believe that your employer has not corrected violations involving radiological working conditions, you may request an inspection. Your request should be addressed to the nearest NRC Regional Office and must describe the alleged violation in detail. It must be signed by you or your representative.
- *How do I contact the NRC?* Talk to an NRC inspector on-site, or call or write to the nearest NRC Regional Office in your geographical area. If you

call the NRC's toll-free Safety Hotline during normal business hours, your call will automatically be directed to the NRC Regional Office for your geographical area. If you call after normal business hours, your call will be directed to the NRC's Headquarters Operations Center, which is staffed 24 hours a day.

- *Can I be fired for raising a safety concern?* Federal law prohibits an employer from firing or otherwise discriminating against you for bringing safety concerns to the attention of your employer or the NRC. You may not be fired or discriminated against because you (1) ask the NRC to enforce its rules against your employer; (2) refuse to engage in activities which violate NRC requirements; (3) provide information or are about to provide information to the NRC or your employer about violations of requirements or safety concerns; or (4) are about to ask for, or testify, help, or take part in an NRC, congressional, or any federal or state proceeding.

NONIONIZING RADIATION

Nonionizing radiation is that radiation on the electromagnetic spectrum that has a frequency (hertz, cycles per second) of 1015 or less and a wavelength in meters of 3 Hz 10^{-7} or less. This encompasses visible, ultraviolet, infrared, microwave, radio, and AC power frequencies. Radiation at these frequency levels does not have sufficient energy to shatter atoms and ionize them.[22] However, such radiation can cause blisters and blindness. In addition, there is mounting evidence of a link between nonionizing radiation and cancer. The warning symbol for radio frequency radiation is shown in Figure 21–7.

The greatest concerns about nonionizing radiation relate to the following sources: visible radiation, ultraviolet radiation, infrared radiation, radio frequency and microwave radiation, extremely low frequency radiation, lasers, and video display terminals. The main concerns in each of these areas are explained in the following paragraphs. The following section deals specifically with electromagnetic radiation (EMR) from power lines and other sources.

Discussion Case

What Is Your Opinion?

"We are moving out of this house today!" said the wife to her husband. "What are you talking about? You love this house." "Yes, I do," said the wife. "But I love our kids more, and I don't intend to stay here and see them die of cancer." "Cancer? What do you mean?" asked the husband, feeling increasingly lost. "It's the power lines that run along our back lot line," said the wife. "I read an article today that says power lines cause cancer." "I've heard that," said the husband. "But I don't believe it. That's just a myth." Who is correct in this discussion? What is your opinion?

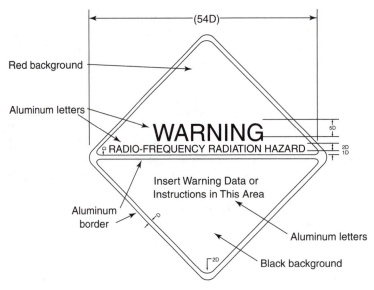

Figure 21–7
Warning symbol for radio frequency radiation.

1. **Visible radiation** comes from light sources that create distortion. This can be a hazard to employees whose jobs require color perception. For example, 8 percent of the male population is red color-blind and cannot properly perceive red warning signs.[23]

2. The most common source of **ultraviolet radiation** is the sun. Potential problems from ultraviolet radiation include sunburn, skin cancer, and cataracts. Precautionary measures include special sunglasses treated to block out ultraviolet rays and protective clothing. Other sources of ultraviolet radiation include lasers, welding arcs, and ultraviolet lamps.[24]

3. **Infrared radiation** creates heat. Consequently, the problems associated with this kind of nonionizing radiation involve heat stress and dry skin and eyes. Primary sources of infrared radiation are high-temperature processes such as the production of glass and steel.[25]

4. *Radio frequency (RF)* and *microwave (MW) radiation* are electromagnetic radiation in the frequency range of 3 kilohertz (kHz) to 300 gigahertz (GHz). Usually MW radiation is considered a subset of RF radiation, although an alternative convention treats RF and MW radiation as two spectral regions. Microwaves occupy the spectral region between 300 GHz and 300 MHz, whereas RF or radio waves include 300 MHz to 3 kHz. RF and MW radiation are nonionizing in that there is insufficient energy (less than 10 electron volts [eV]) to ionize biologically important atoms. The primary health effects of RF and MW energy are considered to be thermal. The absorption of RF and MW energy varies with frequency. Microwave frequencies produce a skin effect—you can literally sense your skin starting to feel warm. RF radiation may penetrate the

body and be absorbed in deep body organs without the skin effect that can warn an individual of danger.

Research has revealed other nonthermal effects. All the standards of Western countries have, so far, based their exposure limits solely on preventing thermal problems. In the meantime, research continues. Use of RF and MW radiation includes aeronautical radios, citizen's (CB) radios, cellular phones, processing and cooking of foods, heat sealers, vinyl welders, high-frequency welders, induction heaters, flow solder machines, communications transmitters, radar transmitters, ion implant equipment, microwave drying equipment, sputtering equipment, glue curing, and power amplifiers used in EMC and metrology (calibration).[26]

5. *Extremely low frequency (ELF) radiation* includes alternating current (AC) fields and nonionizing radiation from 1 Hz to 300 Hz. Because ELF frequencies are low (wavelengths are on the order of 1,000 km), static electromagnetic fields are created. ELF fields are considered as separate, independent, nonradiating electric and magnetic fields. Electric and magnetic fields (EMFs) at 60 Hz are produced by power lines, electrical wiring, and electrical equipment. Electric fields are produced by voltage and increase in strength as the voltage increases. Electric field strength is measured in units of volts per meter (V/m). Magnetic fields are from the flow of current through wires or electrical devices and increase in strength as the current increases. Magnetic fields are measured in units of gauss (G) or tesla (T).

Electrical equipment usually must be turned on for a magnetic field to be produced. Electric fields are present even when equipment is turned off, as long as it is plugged in. Current research has focused on potential health effects of magnetic fields. Some inconclusive epidemiological studies have suggested increased cancer risk associated with estimates of magnetic field exposure. No similar associations have been reported for electric fields. Exposure to EMFs depends on the strength of the magnetic field sources, the distance from those sources, and the time spent in the magnetic field. The American Conference of Governmental Industrial Hygienists (ACGIH) has established occupational threshold limit values (TLVs) for static magnetic fields, sub-radio-frequency (30 kHz and below) magnetic fields, and sub-radio-frequency (30 kHz) and below static electric fields.[27]

6. **Lasers** are being used increasingly in modern industry. The hazards of lasers consist of a thermal threat to the eyes and the threat of electrocution from power sources. In addition, the smoke created by lasers in some processes can be toxic.[28]

7. **Video display terminals (VDTs)** are widely used in the modern workplace. They emit various kinds of nonionizing radiation. Typically, the levels are well below established standards. However, concerns persist about the long-term effects of prolonged and continual exposure to VDT-based radiation.[29] The National Institute for Occupational Safety and Health (NIOSH) monitors this issue and looked into the potential impact of VDT use on pregnant users. NIOSH found that "women who work with VDTs have no greater risk of miscarriage than those who do not."[30]

ELECTROMAGNETIC FIELDS IN THE WORKPLACE

The first studies of the potential effects on worker health of occupational exposure to electric and magnetic fields were published in the literature of the former Soviet Union in the mid-1960s. In these studies, electric switchyard workers reported a variety of subjective complaints, including problems with their cardiovascular, digestive, and central nervous systems.[31] Since then, numerous studies of the effects of **electromagnetic fields (EMFs)** on humans have been done. Although much of the research has been inconclusive in establishing a clear cause-and-effect relationship between EMFs and health problems, the case for a clear link between EMFs and a variety of health problems is strong. Consequently, safety and health professionals must be prepared to take appropriate precautions in the workplace.

According to Savitz, "associations reported between electric occupations and leukemia and brain cancer seem too consistent to be attributable to chance."[32] Occupations with a higher-than-normal incidence of leukemia and brain cancer are as follows:

- Telephone operators
- Electrical manufacturing workers
- Power plant workers
- Telecommunication workers
- Electrical engineers
- Electrical line workers
- Power station operators
- Electricians
- Cable splicers

The health problems most frequently associated with EMF exposure are brain cancer, acute myeloid leukemia, leukemia, and lymphatic leukemia.

EMFs and Health Problems

Studies have shown that some employees exposed to high magnetic fields have increased cancer rates. However, such associations do not necessarily show that EMF exposures cause cancer (any more than the springtime association of robins and daffodils shows that one causes the other). Scientists have looked carefully at all the EMF evidence, but they still disagree on the health effects of EMFs, except to say that better information is needed.[33]

Many studies report small increases in the rate of leukemia or brain cancer in groups of people living or working in high magnetic fields. Other studies have found no such increases. The most important data come from six recent studies of workers wearing EMF monitors to measure magnetic fields. All but one study found significantly higher cancer rates for men with average workday exposures above 4 milligauss. However, the results of these studies disagree in important ways, such as the type of cancer associated with EMF exposures. Consequently, scientists cannot be sure whether the increased risks are caused by EMFs or by other factors. A few preliminary studies have also associated workplace EMFs with breast cancer, and one study has reported a possible link between occupational EMF exposure and Alzheimer's disease. The data from all these studies are too limited for scientists to draw conclusions.

A study involving almost 139,000 subjects of the possible link between exposure to low-frequency electromagnetic fields and suicide among electric utility workers has uncovered what appears to be a distinct association. Electricians working for five power companies faced twice the expected risk of suicide, whereas line workers faced one-and-a-half times the expected risk, according to the University of North Carolina at Chapel Hill study.[34] Suicides among power plant operators occurred at a rate slightly lower than expected, researchers found. Younger people appeared to be at greater risk of suicide than older ones. Higher, more recent exposures also seemed to boost the chances that workers may take their own lives.

Using employment files, death certificates, and other records, researchers found 536 suicides among current or former electric utility workers. Researchers also analyzed various job titles and duties and estimated occupational exposures to EMFs based on devices worn by randomly selected workers to monitor exposures. Researchers were unable to control past mental illness, addictions, or family disturbances such as divorce, nor were they able to explain the suggested link between EMFs and suicide. Consequently, the results are inconclusive and the debate continues.

An Appropriate Response for Safety Professionals

The research into the possible cause-and-effect relationship between EMFs and health problems is inconclusive. On the other hand, the anecdotal and circumstantial evidence strongly suggests a link. How, then, is the safety and health professional to respond? According to William E. Feero,

> The emergence of new electrical technologies has produced an extremely complex electric and magnetic field environment in which we must live and work. Although there presently exists no universally accepted human exposure guidelines, current biological research may someday produce such limits. It is in anticipation of these exposure limits that electric and magnetic field management techniques are being investigated. At low frequencies, two categories of field management exist: cancellation and shielding. The particular class of control used will depend on several functions of the field and its source, and will be determined finally on a case-by-case basis.[35]

Cancellation Approach

Cancellation is an attenuation technique in which the magnetic fields produced by sources of electricity are, in effect, canceled out. It works as follows: Phase currents flowing through a given conductor are canceled out or drastically reduced by phase currents flowing in the opposite conductors. The cancellation approach can be used for both single-phase and balanced multiphase systems.

According to Feero,

> Cancellation fields can be set up in some cases with very little cost penalty. In many cases, a principal source of magnetic fields is found to be the conductor systems leading to tools or power apparatus. In such relatively simple cases, these fields could be canceled via compaction of the conductor systems. For example, a low voltage device, either a 120 or 240V service, is typically supplied by a power cord and the fields any distance from the cord are rapidly attenuated. In the situation where the cords or leads have to be very close to the worker, the added precaution of twisting or interleaving of these leads with each other will further reduce the field. The net effect of twisting a pair of conductors is that each individual conductor appears to occupy the same space. Therefore, the fields are much more effectively canceled because the spacing between the conductors is reduced to nearly zero.
>
> Electrical apparatus that consume considerable energy would generally be supplied by three-phase cables rather than single-phase cables. If single-phase cables are used, care in bundling of the cables within cable trays and the routing of the cable trays can be a very effective control technology. In many instances where strong fields have been found near transformer vaults in buildings, the source of the magnetic field is the cable system leading to and from the transformer vault.
>
> The magnetic field produced directly from most apparatus exhibits the characteristics of loop current source fields. Therefore, a simple control technique may be to move the device, i.e., a compressor motor in a refrigeration unit, to the back of the unit's housing. If the device's function does not permit it to be moved (hand-held tools, for instance) then more sophisticated and possibly expensive techniques must be employed.[36]

Shielding Approach

Shielding is another approach available for decreasing exposure to EMFs. According to Feero,

> Shielding of magnetic fields requires either that the magnetic fields be diverted around the volume considered to be sensitive to the magnetic fields, or the magnetic fields be contained within the device that produces the fields. Effectively accomplishing shielding at either the source or the subject requires extreme care in

Safety Fact

TLVs and Personal Medical Devices

Routine exposures to static magnetic fields should not exceed 60 millitesla (mT) or the equivalent 600 gauss (G) for the whole body or 600 mT (6000 G) to the limbs on a daily time-weighted average basis. Wearers of cardiac pacemakers and other personal electronic medical devices should be exposed to no more than 0.5 mT (5 G) on a daily time-weighted average basis.

choosing the shielding material. The electrical properties of ferromagnetic materials are very complex functions of magnetic field frequencies and magnitudes. For strong magnetic fields, the highly non-linear saturation characteristics of ferromagnetic materials have been widely recognized and reasonably adjusted to achieve source shielding. However, only a few engineers and physicists are aware of the effects of coerciveness at very low magnetizing forces. Subject shielding invariably involves weak magnetic fields. Ferromagnetic materials that are normally considered to have very high permeability may exhibit quite low permeability if being used to attempt to shield milligauss field levels. Thus, the problem of dynamic range encountered when trying to apply cancellation techniques reappears in a different form when attempting to utilize shielding techniques.[37]

Both cancellation and shielding are highly technical approaches requiring specialized knowledge. Safety and health professionals who are not specialists in electromagnetic fields may find it necessary to consult with EMF experts before attempting to implement either approach.

OSHA STANDARDS FOR HEALTH AND ENVIRONMENTAL CONTROLS

OSHA's standards relating to radiation hazards are contained in 29 CFR 1910 (Subpart G). These standards are as follows:

1910.94	Ventilation
1910.95	Occupational noise exposure
1910.96	Ionizing radiation
1910.97	Nonionizing radiation
1910.98	Effective dates
1910.99	Sources of standards
1910.100	Standards organizations

Safety Fact

Reducing EMF Exposure in the Workplace

Employees are often exposed to electromagnetic fields in the workplace without even knowing it. It is important to inform employees about EMFs and the associated hazards and to take steps to reduce exposure levels. Strategies for reducing exposure include the following:

- Magnetic fields drop off significantly approximately three feet from the source. Increase the distance between employees and sources of EMF.
- Substitute low-EMF-producing equipment (power supplies) wherever possible.
- Reduce the amount of time that employees are exposed to EMF sources.

Safety Fact

Average Magnetic Field Exposures*

Type of Worker	Average Daily Median**	Exposures Range
Workers on the job:		
Clerical workers without computers	0.5	0.2–2.0
Clerical workers with computers	1.2	0.5–4.5
Machinists	1.9	0.6–27.6
Electric line workers	2.5	0.5–34.8
Electricians	5.4	0.8–34.0
Welders	8.2	1.7–96.0

*Magnetic fields are often measured in gauss or milligauss (one thousandth of a gauss = 1 milligauss).
**The median is the middle measurement: half the workers have average daily exposures above this point and half below.

Source: National Institute for Occupational Safety and Health (NIOSH).

SUMMARY

1. Widely used terms relating to ionizing radiation are *radiation, radioactive material, restricted area, unrestricted area, dose, rad, rem, air dose, personal monitoring devices, radiation area*, and *high-radiation area*.

2. Exposure of individuals to radiation must be carefully controlled and accurately monitored. Doses are typically measured in rems.

3. Employers must require the use of personal monitoring devices such as film badges, pocket chambers, pocket dosimeters, and film rings. Employers are also required to conduct comprehensive surveys to identify and evaluate radiation hazards present in the workplace.

4. Caution signs are required by both OSHA and the NRC in the following areas and situations: radiation areas, high-radiation areas, airborne radiation areas, areas containing radioactive materials, and containers in which radioactive materials are stored or transported.

5. Companies that produce, use, store, or transport radioactive materials are required to have a signal-generating system to warn workers to evacuate immediately. There must be a sufficient number of signal generators to cover all personnel who must be evacuated.

6. Employees must be informed and instructed regarding potential radiation hazards, precautions that they should take, and records of exposure.

7. Radioactive materials that are stored in restricted areas must be appropriately labeled. Materials stored in unrestricted areas must be secured against unauthorized removal.

8. Radiation incidents that involve exposure beyond prescribed limits must be reported immediately or within 24 hours, depending on the dose. Such incidents must be reported in writing within 30 days.

9. Reports of overexposure should contain the following information: extent of exposure, levels of radiation, concentration of radiation, cause of exposure, and corrective action taken.

10. The U.S. Nuclear Regulatory Commission (NRC) publishes Form 3 in a question-and-answer format to inform employees of the standards for protection against radiation.

11. Nonionizing radiation is radiation on the electromagnetic spectrum that lacks sufficient energy to ionize atoms. This encompasses visible, ultraviolet, infrared, radio, microwave, and AC power frequencies.

12. Electromagnetic fields, or EMFs, encompass radiation from power lines and a long list of electrical appliances. Concerns about, and evidence of, a link between EMFs and cancer exist.

KEY TERMS AND CONCEPTS

Air dose

Calendar quarter

Caution signs

Dose

Electromagnetic field (EMF)

Evacuation warning signal

Exposure

High-radiation area

Incident

Infrared radiation

Ionizing radiation

Labels

Lasers

Nonionizing radiation

Nuclear Regulatory Commission (NRC)

Personal monitoring devices

Rad

Radiation

Radiation area

Radioactive material

Rem

Restricted area

Ultraviolet radiation

Unrestricted area

Video display terminals (VDTs)

Visible radiation

REVIEW QUESTIONS

1. Define the following terms relating to ionizing radiation: *radiation, restricted area, dose, rem*, and *radiation area*.

2. What is the maximum dose of radiation allowed to the whole body during a calendar quarter for a 17-year-old individual?

3. Explain the requirements of employers regarding the use of personal monitoring devices.

4. List the situations in which caution signs are required of employers.
5. Describe, in OSHA's language, the required evacuation warning signal.
6. How should radioactive material be treated when stored in a nonrestricted area?
7. Describe the incidents that require immediate notification concerning radiation overexposure.
8. List the required contents in a written report of radiation overexposure.
9. Define the term *nonionizing radiation*.
10. What are the primary sources of concern regarding nonionizing radiation?
11. Describe the anecdotal and scientific evidence linking EMFs to cancer.
12. What does the Nuclear Regulatory Commission do?

ENDNOTES

1. 29 CFR 1910.1096(b)(2).
2. Nuclear Regulatory Commission, 10 CFR (Part 20).
3. 29 CFR 1910.1096(d).
4. 29 CFR 1910.1096(e).
5. 29 CFR 1910.1096(e)(6)(iv).
6. 29 CFR 1910.1096(g)(1)(2)(3).
7. 29 CFR 1910.1096(f)(i)(ii).
8. 29 CFR 1910.1096(f)(iii-vi)(2)(ii).
9. 29 CFR 1910.1096(i).
10. 29 CFR 1910.1096(j).
11. 29 CFR 1910.1096(k) and (i).
12. 29 CFR 1910.1096(l)(i).
13. 29 CFR 1910.1096(l)(ii).
14. 29 CFR 1910.1096(l)(1).
15. 29 CFR 1910.1096(l)(2).
16. 29 CFR 1910.1096(m)(1).
17. 29 CFR 1910.1096(m)(2).
18. 29 CFR 1910.1096(n)(1).
19. 29 CFR 1910.1096(o)(1).
20. Ibid.
21. U.S. Nuclear Regulatory Commission, *NRC Form 3* (Washington, DC: 2006).
22. American Conference of Governmental Industrial Hygienists, *TLVs and BEIs* (Cincinnati, OH: 2006), 147.
23. Ibid., 42.
24. Ibid., 43.
25. Ibid.
26. OSHA, "Technical Links." Retrieved June 2006 from http://www.osha-slc.gov.
27. Ibid.
28. Ibid.

29. Ibid.

30. NIOSH, "Women's Safety and Health Issues at Work." Retrieved April 2006 from http://www.edc.gov.do?action=search&subset=NIOSH&queryText= Radiation+from+video+display.

31. William E. Murray, Jr. and Robert M. Patterson, "Electric and Magnetic Fields: What Do We Know?" *American Industrial Hygiene Association Journal* 54, no. 4: 164.

32. David A. Savitz, "Overview of Epidemiologic Research on Electric and Magnetic Fields and Cancer," *American Industrial Hygiene Journal* 54, no. 4: 202.

33. NIOSH, "Fact Sheet," 3. Retrieved June 2006 from http/www.cdc.gov/ niosh/emf2.html.

34. Carolina News Service, "Largest Study Finds Evidence of Association Between EMFs and Exposed Worker Suicide" 1–2. Retrieved March 15, 2000, from http://www.unc.edu/news/newssery.

35. William E. Feero, "Electric and Magnetic Field Management," *American Industrial Hygiene Journal* 54, no. 4: 205.

36. Ibid., 207–208.

37. Ibid., 208–209.

NOISE AND VIBRATION HAZARDS

Major Topics

- Hearing Loss Prevention Terms
- Characteristics of Sound
- Hazard Levels and Risks
- Standards and Regulations
- Workers' Compensation and Noise Hazards
- Identifying and Assessing Hazardous Noise Conditions
- Noise Control Strategies
- Vibration Hazards
- Other Effects of Noise Hazards
- Corporate Policy
- Evaluating Hearing Loss Prevention Programs

The modern workplace can be noisy. This poses two safety- and health-related problems. First, there is the problem of distraction. Noise can distract workers and disrupt their concentration, which can lead to accidents. Second, there is the problem of hearing loss. Exposure to noise that exceeds prescribed levels can result in permanent hearing loss.

Modern safety and health professionals need to understand the hazards associated with noise and vibration, how to identify and assess these hazards, and how to prevent injuries related to them. This chapter provides the necessary information for prospective and practicing safety and health professionals to do so.

HEARING LOSS PREVENTION TERMS

There are certain terms common to hearing loss prevention that must be understood by safety and health professionals. You may find the definitions in this section helpful when trying to understand the content of this chapter.

- *Attenuation: Real-world baseline audiogram.* Estimated sound protection provided by hearing protective devices as worn in "real-world" environments.
- *Baseline audiogram.* A valid audiogram against which subsequent audiograms are compared to determine if hearing thresholds have changed. The baseline audiogram is preceded by a quiet period to obtain the best estimate of the person's hearing at that time.
- *Continuous noise.* Noise of a constant level measured over at least one second using the "slow" setting on a sound level meter. Note that an intermittent noise (e.g., on for over a second and then off for a period) is both variable *and* continuous.
- *Decibel (dB).* The unit used to express the intensity of sound. The decibel was named after Alexander Graham Bell. The decibel scale is a logarithmic scale in which 0 dB approximates the threshold of hearing in the mid-frequencies for young adults and in which the threshold of discomfort is between 85 and 95 dB and the threshold for pain is between 120 and 140 dB.
- *Dosimeter.* When applied to noise, the instrument that measures sound levels over a specified interval, stores the measures, and calculates the sound as a function of sound level and sound duration. It describes the results in terms of dose, time-weighted average, and other parameters such as peak level, equivalent sound level, sound exposure level, and so on.
- *Exchange rate.* The relationship between intensity and dose. OSHA uses a 5-dB exchange rate. Thus, if the intensity of an exposure increases by 5 dB, the dose doubles. This may also be referred to as the *doubling rate.* The U.S. Navy uses a 4-dB exchange rate; the U.S. Army and U.S. Air Force use a 3-dB exchange rate.
- *Hazardous noise.* Any sound for which any combination of frequency, intensity, or duration is capable of causing permanent hearing loss in a specified population.
- *Conductive and sensorineural loss.* Hearing loss is often characterized by the area of the auditory system responsible for the loss. For example, when injury or a medical condition affects the *outer ear* or *middle ear* (i.e., from the pinna, ear canal, and eardrum to the cavity behind the eardrum—which includes the ossicles), the resulting hearing loss is referred to as a conductive loss. When an injury or medical condition affects the *inner ear* or the auditory nerve that connects the inner ear to the brain (i.e., the cochlea and the VIIIth cranial nerve), the resulting hearing loss is referred to as a sensorineural loss. Thus, a welder's spark that damages the eardrum causes a conductive hearing loss. Because noise can damage the tiny hair cells located in the cochlea, it causes a sensorineural hearing loss.
- *Hearing threshold level (HTL).* The hearing level, above a reference value, at which a specified sound or tone is heard by an ear in a specified fraction of the trials. Hearing threshold levels have been established so that dB HTL reflects the best hearing of a group of persons.
- *Hertz (Hz).* The unit measurement for audio frequencies. The frequency range for human hearing lies between 20 Hz and approximately 20,000 Hz.

The sensitivity of the human ear drops off sharply below about 500 Hz and above 4,000 Hz.

- *Impulsive noise.* Generally used to characterize impact or impulse noise typified by a sound that rapidly rises to a sharp peak and then quickly fades. The sound may or may not have a "ringing" quality (such as striking a hammer on a metal plate or a gunshot in a reverberant room). Impulsive noise may be repetitive, or may be a single event (as with a sonic boom). *Note:* If impulses occur in very rapid succession (e.g., some jackhammers), the noise is not described as impulsive.
- *Material hearing impairment.* As defined by OSHA, a material hearing impairment is an average hearing threshold level of 25 dB HTL at the frequencies of 1,000, 2,000, and 3,000 Hz.
- *Noise.* Any unwanted sound.
- *Noise dose.* The noise exposure expressed as a percentage of the allowable daily exposure. For OSHA, a 100 percent dose equals an eight-hour exposure to a continuous 90-dBA noise; a 50 percent dose equals an eight-hour exposure to an 85-dBA noise or a four-hour exposure to a 90-dBA noise. If 85 dBA is the maximum permissible level, an eight-hour exposure to a continuous 85-dBA noise equals a 100 percent dose. If a 3-dB exchange rate is used in conjunction with an 85-dBA maximum permissible level, a 50 percent dose equals a two-hour exposure to 88 dBA or an eight-hour exposure to 82 dBA.
- *Noise-induced hearing loss.* A sensorineural hearing loss that is attributed to noise and for which no other etiology can be determined.
- *Standard threshold shift.* OSHA uses this term to describe a change in hearing threshold relative to the baseline audiogram of an average of 10 dB or more at 2,000, 3,000, and 4,000 Hz in either ear. Used by OSHA to trigger additional audiometric testing and related follow-up.
- *Significant threshold shift.* NIOSH uses this term to describe a change of 15 dB or more at any frequency, 400 through 6,000 Hz, from baseline levels that is present on a retest in the same ear and at the same frequency. NIOSH recommends a confirmation audiogram within 30 days with the confirmation audiogram preceded by a quiet period of at least 14 hours.
- *Time-weighted average (TWA).* A value, expressed in dBA, computed so that the resulting average is equivalent to an exposure resulting from a constant noise level over an eight-hour period.

CHARACTERISTICS OF SOUND

Sound is any change in pressure that can be detected by the ear. Typically, sound is a change in air pressure. However, it can also be a change in water pressure or any other pressure-sensitive medium. **Noise** is unwanted sound. Consequently, the difference between noise and sound is in the perception of the person hearing it (e.g., loud rock music may be considered "sound" by a rock fan but "noise" by a shift worker trying to sleep).

What we think of as sound, the eardrum senses as fluctuations in atmospheric pressure. The eardrum responds to these fluctuations in atmospheric pressure by vibrating. These vibrations are carried to the brain in the form of neural sensations and interpreted as sound.

Sound and vibration are very similar. Sound typically relates to a sensation that is perceived by the inner ear as hearing. **Vibration,** on the other hand, is inaudible and is perceived through the sense of touch. Sound can occur in any medium that has both mass and elasticity (air, water, and so on). It occurs as elastic waves that cross over (above and below) a line representing normal atmospheric pressure (Figure 22–1).

Normal atmospheric pressure is represented in Figure 22–1 by a straight horizontal line. Sound is represented by the wavy line that crosses above and below the line. The more frequently the sound waves cross the normal atmospheric pressure line (the shorter the cycle), the higher the pitch of the sound. The greater the vertical distance above and below the atmospheric pressure line (distance X), the louder or more intense the sound.

The unit of measurement used for discussing the level of sound and, correspondingly, what noise levels are hazardous is the **decibel,** or one-tenth of a *bel.* One decibel represents the smallest difference in the level of sound that can be perceived by the human ear. Figure 22–2 shows the decibel levels for various common sounds. The weakest sound that can be heard by a healthy human ear in a quiet setting is known as the **threshold of hearing** (1 dBA). The maximum level of sound that can be perceived without experiencing pain is known as the **threshold of pain** (140 dBA).

The three broad types of industrial noise are described as follows: **Wide band noise** is noise that is distributed over a wide range of **frequencies.** Most noise from manufacturing machines is wide band noise. **Narrow band noise** is noise that is confined to a narrow range of frequencies. The noise produced by power tools is narrow band noise. Finally, **impulse noise** consists of transient pulses that can occur repetitively or nonrepetitively. The noise produced by a jackhammer is repetitive impulse noise.[1]

Figure 22–1
Sound waves.

Figure 22–2
Selected sound levels.

Source	Decibels (dBA)
Whisper	20
Quiet library	30
Quiet office	50
Normal conversation	60
Vacuum cleaner	70
Noisy office	80
Power saw, lawn mower	90
Chain saw	90
Grinding operations	100
Passing truck	100
Gunshot blast	140
Jet aircraft	150
Rocket launching	180

HAZARD LEVELS AND RISKS

The fundamental hazard associated with excessive noise is hearing loss. Exposure to excessive noise levels for an extended period can damage the inner ear so that the ability to hear high-frequency sound is diminished or lost altogether. Additional exposure can increase the damage until even lower frequency sounds cannot be heard.[2]

A number of different factors affect the risk of hearing loss associated with exposure to excessive noise. The most important of these are

- Intensity of the noise (sound pressure level)
- Type of noise (wide band, narrow band, or impulse)
- Duration of daily exposure
- Total duration of exposure (number of years)
- Age of the individual
- Coexisting hearing disease
- Nature of environment in which exposure occurs
- Distance of the individual from the source of the noise
- Position of the ears relative to the sound waves[3]

Of these various factors, the most critical are the sound level, frequency, duration, and distribution of noise (Figure 22–3). The unprotected human ear is at risk when exposed to sound levels exceeding 115 dBA. Exposure to sound levels below 80 dBA is generally considered safe. Prolonged exposure to noise levels higher than 80 dBA should be minimized through the use of appropriate personal protective devices.

Figure 22–3
Critical noise risk factors.

To decrease the risk of hearing loss, exposure to noise should be limited to a maximum eight-hour time-weighted average of 85 dBA. The following general rules should be applied for dealing with noise in the workplace:

- Exposures of less than 80 dBA may be considered safe for the purpose of risk assessment.
- A time-weighted average (threshold) of 85 dBA should be considered the maximum limit of continuous exposure over eight-hour days without protection.[4]

STANDARDS AND REGULATIONS

The primary sources of standards and regulations relating to noise hazards are OSHA, the American National Standards Institute (ANSI), and the National Institute for Occupational Safety and Health (NIOSH). OSHA regulations require the implementation of hearing conservation programs under certain conditions. OSHA's regulations should be considered as minimum standards. ANSI's standard provides a way to determine the effectiveness of hearing conservation programs such as those required by OSHA. The ANSI standard and OSHA regulations are discussed in the following sections. NIOSH bases most of its materials on OSHA regulations, although NIOSH does make recommendations that exceed OSHA regulations in some cases.

ANSI Standard

The American National Standards Institute (ANSI) published ANSI Standard S12.13, Evaluation of Hearing Conservation Programs. This standard is designed to help safety and health professionals determine if hearing conservation programs work as they are intended.[5]

Discussion Case

What Is Your Opinion?

The father and son have had this argument before. The son wants to go to a rock concert in which four different bands will play. The Megablast concert will last eight hours. It is being held in the civic auditorium in a nearby city. The father is happy that his son is attending the concert but wants him to wear earplugs to guard against excess noise. According to the father, "Eight hours of continuous loud music can damage your hearing." However, the son is adamant in refusing to wear earplugs. "I'm going to hear the music." Is the father in this case overreacting? What is your opinion?

Federal regulations require that employees be protected from excessive noise in the workplace. However, they provide no methodology for determining the effectiveness of hearing conservation programs. The primary reason for the development of ANSI S12.13 was because hearing conservation programs were not actually protecting employees but were only recording their steadily declining hearing ability.

The working group that developed the standard used **audiometric database analysis (ADBA)** to identify procedures for measuring variability in hearing threshold levels. The two procedures identified were as follows:

- *Percent worse sequential.* This procedure identifies the percentage of subjects who show a deterioration of 15 dBA or more in their ability to hear at least one test frequency (500 through 6,000 Hz) in either ear between two sequential audiograms.
- *Percent better or worse sequential.* This procedure identifies the percentage of subjects who show either a deterioration or an improvement of 15 dBA or more in thresholds for at least one test frequency (500 through 6,000 Hz) in either ear between two sequential audiograms.[6]

There are important differences between the traditional approaches to measuring the effectiveness of hearing conservation programs and those required by ANSI S12.13. The ANSI approach can be summarized as follows:

- Results of tests are compared in sequence. For example, the results of year 4 are compared with those of year 3. The results of year 3 are compared with those of year 2, and so on. In this way, a current audiogram is compared against an earlier audiogram. The results of the earlier test are used as a baseline for comparison.
- Test results from several employees in a given work unit are examined individually and compared with past results sequentially. If enough employees show hearing loss, the conclusion may be that the work unit's hearing conservation program is ineffective.[7]

OSHA Regulations

OSHA's regulations relating to occupational noise exposure and hearing conservation are found in 29 CFR 1910.95. The basic requirements generated from this standard for hearing conservation programs are as follows:

- Hearing hazards monitoring
- Engineering and administrative controls
- Audiometric evaluation
- Personal hearing protection devices
- Education and motivation
- Record keeping
- Program evaluation

Hearing Hazards Monitoring

As with any health hazard, it is important to determine accurately the nature of the hearing hazard and to identify the affected employees. Those responsible for this aspect of the program must ensure that the exposures of all employees have been properly evaluated and that reevaluations are conducted when changes in equipment or operations significantly alter working conditions. Recent evidence has indicated that aromatic solvents, metals, and petrochemicals may be associated with occupational hearing loss. Although studies are exploring the relationship between hearing loss and chemical exposures, there is insufficient information about this relationship to speculate on potential risk factors. Therefore, this section focuses on monitoring noise exposure, the major factor associated with occupational hearing losses. Hearing hazard exposure monitoring is conducted for various purposes, including

- Determining whether hazards to hearing exist
- Determining whether noise presents a safety hazard by interfering with speech communication or the recognition of audible warning signals
- Identifying employees for inclusion in the hearing loss prevention program
- Classifying employees' noise exposures for prioritizing noise control efforts and defining and establishing hearing protection practices
- Evaluating specific noise sources for noise control purposes
- Evaluating the success of noise control efforts

Various kinds of incrementation and measurement methods may be used, depending on the type of measurements being conducted. The most common measurements are area surveys, dosimetry, and engineering surveys.

In an area survey, environmental noise levels are measured using a sound level meter to identify work areas where exposures are above or below hazardous levels and where more thorough exposure monitoring may be needed. The result is often plotted in the form of a "noise map," showing noise level measurements for the different areas of the workplace.

Dosimetry involves the use of body-worn instruments (dosimeters) to monitor an employee's noise exposure over the work shift. Monitoring results for one employee can also represent the exposures of other workers in the area with similar noise exposures. It may also be possible to use task-based exposure methods to represent the exposures of other workers in different areas whose exposures result from having performed the same tasks.

Engineering surveys typically employ more sophisticated acoustical equipment in addition to sound level meters. These may include octave-band analyzers and sound level recorders that furnish information on the frequency–intensity composition of the noise being emitted by machinery or other sound sources in various modes of operation. These measurements are used to assess options for applying engineering controls.

Engineering and Administrative Controls

Engineering and administrative controls are essential to achieve an effective hearing loss prevention program. Engineering and administrative controls represent the first two echelons in the hierarchy of controls: (1) remove the hazard, and (2) remove the worker. The use of these controls should reduce hazardous exposure to the point where the risk to hearing is eliminated or at least more manageable. Engineering controls are technologically feasible for most noise sources, but their economic feasibility must be determined on a case-by-case basis. In some instances, the application of a relatively simple noise control solution reduces the hazard to the extent that the other elements of the program, such as audiometric testing and the use of hearing protection devices, are no longer necessary. In other cases, the noise reduction process may be more complex and must be accomplished in stages over a period of time. Even so, with each reduction of a few decibels, the hazard to hearing is reduced, communication is improved, and noise-related annoyance is reduced as well.

It is especially important that organizations specify low noise levels when purchasing new equipment. Many types of previously noisy equipment are now available in noise-controlled versions. Consequently, a "buy quiet" purchasing policy should not require new engineering solutions in many cases.

For hearing loss prevention purposes, *engineering controls* are defined as any modification or replacement of equipment or related physical change at the noise source or along the transmission path (with the exception of hearing protectors) that reduces the noise level at the employee's ear. Typical engineering controls involve:

- Reducing noise at the source (e.g., installing a muffler)
- Interrupting the noise path (e.g., erecting acoustical enclosures and barriers)
- Reducing reverberation (e.g., installing sound-absorbing material)
- Reducing structure-borne vibration (e.g., installing vibration mounts and providing proper lubrication)

Assessing the applicability of engineering controls is a sophisticated process. First, the noise problem must be thoroughly defined. This necessitates measuring the noise levels and developing complete information on employee noise exposure and the need for noise reduction. Next, an assessment of the effect of these controls on overall noise levels should be made. Once identified and analyzed, the preceding controls can be considered. Choices are influenced, to some extent, by the cost of purchasing, operating, servicing, and maintaining the control. For this reason, engineering, safety, and industrial hygiene personnel, as well as employees who operate, service, and maintain equipment, must be involved in the noise control plan. Employees who work with the equipment on a daily basis should be asked to provide valuable guidance on such important matters as the positioning of monitoring indicators and panels, lubrication and servicing points, control switches, and the proper location of access doors for operation and maintenance. An acoustical consultant may be hired to assist in the design, implementation, installation, and evaluation of these controls.

Administrative controls, defined as changes in the work schedule or operations that reduce noise exposure, may also be used effectively. Examples include operating a noisy machine on the second or third shift when fewer people are exposed, or shifting an employee to a less noisy job once a hazardous daily noise dose has been reached. Generally, administrative controls have limited use in industry because employee contracts seldom permit shifting from one job to another. Moreover, the practice of rotating employees between quiet and noisy jobs, although it may reduce the risk of substantial hearing loss in a few workers, may actually increase the risk of small hearing losses in many workers. A more practical administrative control is to provide for quiet areas where employees can gain relief from workplace noise. Areas used for work breaks and lunch rooms should be located away from noise.

Audiometric Evaluation

Audiometric evaluation is crucial to the success of the hearing loss prevention program in that it is the only way to determine whether occupational hearing loss is being prevented. When the comparison of audiograms shows temporary threshold shift (a temporary hearing loss after noise exposure), early permanent threshold shift, or progressive occupational hearing loss, it is time to take swift action to halt the loss before additional deterioration occurs. Because occupational hearing loss occurs gradually and is not accompanied by pain, the affected employee may not notice the change until a large threshold shift has accumulated. However, the results of audiometric tests can trigger changes in the hearing loss prevention program more promptly, initiating protective measures, and motivating employees to prevent further hearing loss.

OSHA and NIOSH presently have differing definitions of the amount of change in hearing indicated by repeated audiometry that should trigger additional audiometric testing and related follow-up. OSHA uses the term *standard threshold shift* to describe an average change in hearing from the baseline levels of 10 dB or more for the frequencies of 2,000, 3,000, and 4,000 Hz. NIOSH uses the term *significant threshold shift* to describe a change of 15 dB or more at any frequency 500 through 6,000 Hz from baseline levels that is present on an immediate retest in the same ear and at the same frequency. NIOSH recommends a confirmation audiogram within 30 days with the confirmation audiogram preceded by a quiet period of at least 14 hours. The NIOSH STS, called "15 dB twice" (same ear, same frequency), can be tested only if the baseline audiogram is available at the time of the annual audiometric test.

For maximum protection of employees, audiograms should be performed on the following occasions:

- Preemployment
- Prior to initial assignment in a hearing hazardous work area
- Annually as long as the employee is assigned to a noisy job (a time-weighted average exposure level equal to or greater than 85 dBA)
- At the time of reassignment out of a hearing hazardous job
- At the termination of employment

In addition, it is suggested that employees who are not exposed be given periodic audiograms as part of the organization's health care program. The audiograms of these employees can be compared to those of the exposed employees whenever the overall effectiveness of the hearing loss prevention program is evaluated. In an optimally effective program, the two employee groups show essentially the same amount of audiometric change.

Personal Hearing Protection Devices

A personal hearing protection device (or "hearing protector") is anything that can be worn to reduce the level of sound entering the ear. Earmuffs, ear canal caps, and earplugs are the three principal types of devices. Each employee reacts individually to the use of these devices, and a successful hearing loss prevention program should be able to respond to the needs of each employee. Ensuring that these devices protect hearing effectively requires the coordinated effort of management, the hearing loss prevention program operators, and the affected employees.[8]

There are several different types of earmuffs with which the modern safety and health professional should be familiar:

- *Passive earmuffs.* This kind of device is what most people think of when they think of earmuffs. They consist of ear cups lined with foam and block noise using nothing but the foam-lined cups. The primary weakness of passive earmuffs is they tend to block out not just unwanted noise, but also certain advantageous sounds such as voices trying to warn of danger.
- *Uniform attenuation earmuffs.* These earmuffs not only block noise, but also attenuate the noise more uniformly within several key octave bands (250 Hz–4 KHz). This allows employees wearing them to hear certain important sounds such as spoken instructions or warnings, thus reducing one of the main safety risks associated with earmuffs.
- *Electronic earmuffs.* This type of earmuff uses electronic technology to both block and modulate sound. Some of the more popular brands of electronic

earmuffs can receive AM/FM radio signals or have a wireless connection to a CD or MP3 player.

Regardless of the kind of ear protection device used, it is important to remember the four Cs: *comfort*, *convenience*, *communication* (the device should not interfere with the worker's ability to communicate), and *caring* (workers must care enough about protecting their hearing to wear the devices).

Education and Motivation

Training is a critical element of a good hearing loss prevention program. In order to obtain sincere and energetic support by management and active participation by employees, it is necessary to educate and motivate both groups. A hearing loss prevention program that overlooks the importance of education and motivation is likely to fail because employees will not understand why it is in their best interest to cooperate, and management will fail to make the necessary commitment. Employees and managers who appreciate the precious sense of hearing and understand the reasons for, and the mechanics of, the hearing loss prevention program will be more likely to participate for their mutual benefit, rather than viewing the program as an imposition.

Record Keeping

Records often get the least attention of any of the program's components. However, audiometric comparisons, reports of hearing protector use, and the analysis of hazardous exposure measurements all involve the keeping of records. Unfortunately, records are often kept poorly because there is no organized system in place, and in many cases, those responsible for maintaining the records do not understand their value. People tend to assume that if they merely place records in a file or enter them into a computer, adequate record-keeping procedures are being followed. OSHA's latest version of the Form 300 Log has a column for recording hearing loss.

Many companies have found that their record-keeping system is inadequate only when they discover that they need accurate information. This sometimes occurs during the processing of compensation claims. Problems can be avoided by implementing an effective record-keeping system, in which (1) management encourages that the system be kept active and accessible; (2) hearing loss prevention program implementers make sure that all the information entered is accurate and complete; and (3) employees validate the information.

Hearing loss prevention program records should include all items for each phase of the program: (1) hearing loss prevention audit, (2) monitoring hearing hazards, (3) engineering and administrative controls, (4) audiometric evaluation, (5) personal hearing protective devices, (6) education and motivation, (7) record keeping, and (8) program evaluation. Each phase must be considered in order to evaluate the effectiveness of the hearing loss prevention program.

Program Evaluation

The primary goal of any hearing loss prevention program must be to reduce, and eventually eliminate, hearing loss due to workplace exposures. Although management may have the best intentions of implementing this goal and a company's hearing loss prevention program may have the appearance of being complete and complying with OSHA's requirements, the program still may not achieve this goal. A thorough evaluation of the effectiveness of all the program's components is necessary to determine the extent to which the hearing loss prevention program is really working.

Management and program implementers should conduct periodic program evaluations to assess compliance with federal and state regulations and to ensure that hearing is being conserved. There are two basic approaches to following program evaluation: (1) assess the completeness and quality of the program's components, and (2) evaluate the audiometric data.

WORKERS' COMPENSATION AND NOISE HAZARDS

Hearing loss claims are being covered by state workers' compensation laws. Some states have written hearing loss into their workers' compensation law. Others are covering claims whether hearing loss is in the law or not.

Medical professionals have established a procedure for determining if there is a causal relationship between workplace noise and hearing loss. In making determinations of such relationships, physicians consider the following factors:

- Onset and progress of the employee's history of hearing loss
- The employee's complete work history
- Results of the employee's otological examination
- Results of hearing studies that have been performed
- Determination of whether causes of hearing loss originated outside the workplace

Because approximately 15 percent of all working people are exposed to noise levels exceeding 90 dBA, hearing loss may be as significant in workers' compensation costs in the future as back injuries, carpal tunnel syndrome, and stress are now.

IDENTIFYING AND ASSESSING HAZARDOUS NOISE CONDITIONS

Identifying and assessing hazardous noise conditions in the workplace involve (1) conducting periodic noise surveys, (2) conducting periodic audiometric tests, (3) record keeping, and (4) follow-up action. Each of these components is covered in the following sections.

Noise Surveys

Conducting **noise surveys** involves measuring noise levels at different locations in the workplace. The devices that are most widely used to measure noise levels are sound level meters and dosimeters. A **sound level meter** produces an immediate reading that represents the noise level at a specific instant in time.[9] A **dosimeter** provides a time-weighted average over a period of time such as one complete work shift. A dosimeter also calculates the sound as a function of sound level and duration and describes the results in terms of dose, peak level, equivalent sound level, sound exposure level, and time-weighted average.[10] The dosimeter is the most widely used device because it measures total exposure, which is what OSHA and ANSI standards specify. Using a dosimeter in various work areas and attaching a personal dosimeter to one or more employees is the recommended approach to ensure dependable, accurate readings.

Audiometric Testing

Audiometric testing measures the hearing threshold of employees. Tests conducted according to ANSI S12.13 can detect changes in the hearing threshold of the employee. A negative change represents hearing loss within a given frequency range.

The initial **audiogram** establishes a baseline hearing threshold. After that, audiometric testing should occur at least annually. Testing should not be done on an employee who has a cold, an ear infection, or who has been exposed to noise levels exceeding 80 dBA within 14 to 16 hours prior to a test. Such conditions can produce invalid results.[11]

When even small changes in an employee's hearing threshold are identified, more frequent tests should be scheduled and conducted as specified in ANSI S12.13. "For those employees found to have standard threshold shift—a loss of 10 dBA or more averaged at 2,000, 3,000, and 4,000 hertz (Hz) in either ear—the employer is required to fill out the OSHA 300 Log in which the loss is recorded as a worktime illness."[12]

Record Keeping

Figure 22–4 is an example of an audiometric form that can be used to record test results for individual employees. Such forms should be completed and kept on file to allow for sequential comparisons. It is also important to retain records containing a worker's employment history, including all past positions and the working conditions in those positions.

Follow-Up

Follow-up is critical. Failure to take prompt corrective action at the first sign of hearing loss can lead to permanent debilitating damage.

Audiometric Test Results
Johnson-Jones, Inc.

Employee Name ⎯⎯⎯⎯⎯⎯⎯⎯⎯⎯ SSN ⎯⎯⎯⎯⎯⎯⎯⎯⎯⎯⎯⎯⎯⎯⎯

Test Location ⎯⎯⎯⎯⎯⎯⎯⎯⎯⎯ Date ⎯⎯⎯⎯⎯⎯⎯ Cert. No. ⎯⎯⎯⎯⎯

Tested By ⎯⎯⎯⎯⎯⎯⎯⎯⎯⎯⎯ Date ⎯⎯⎯⎯⎯⎯⎯ Cert. No. ⎯⎯⎯⎯⎯

Frequency in Hertz

Figure 22–4
Sample audiometric test form.

Hearing loss can occur without producing any evidence of physiological damage. Therefore, it is important to follow up on even the slightest evidence of a change in an employee's hearing threshold.

Follow-up can take a number of different forms. The following would all be appropriate follow-up responses:

- Administering a retest to verify the hearing loss
- Changing or improving the type of personal protection used
- Conducting a new noise survey in the employee's work area to determine if engineering controls are sufficient
- Testing other employees to determine if the hearing loss is isolated to the one employee in question or if other employees have been affected

NOISE CONTROL STRATEGIES

Figure 22–5 illustrates the three components of a noise hazard. Noise can be reduced by engineering and administrative controls applied to one or more of these components. The most desirable noise controls are those that reduce noise at the source. The second priority is to reduce noise along its path. The

Figure 22–5
Three parts of a noise hazard.

last resort is noise reduction at the receiver using personal protective devices. The latter approach should never be substituted for the two former approaches.

The following paragraphs explain widely used strategies for reducing workplace noise at the source, along its path, and at the receiver:

- Noise can be reduced at its source by enclosing the source, altering the acoustical design at the source, substituting equipment that produces less noise, making alterations to existing equipment, or changing the process so that less noisy equipment can be used.
- Noise can be reduced along its **path** by moving the source farther away from receivers and improving the acoustical design of the path so that more sound is absorbed as it travels toward receivers.
- Noise can be reduced at the **receiver** by enclosing the worker, using personal protective devices, and changing job schedules so that exposure time is reduced.

Some of the noise reduction strategies explained in the preceding paragraphs are engineering controls; others are administrative controls. For example, enclosing a noise source and substituting less noisy equipment are both examples of engineering controls. Changing job schedules is an example of an administrative control. Safety and health professionals should be familiar with both types of controls.

Engineering Controls

Engineering controls consist of facility and equipment adjustments other than administrative and personal protection strategies made to reduce the noise level either at the source or within the worker's hearing zone. Following are some commonly used engineering controls. All these controls are designed to reduce noise at the source, along its path, or at the receiver. They focus primarily on the noise rather than the employees who are exposed to it.

Maintenance

- Replacement or adjustment of worn, loose, or unbalanced parts of machines
- Lubrication of machine parts and use of cutting oils
- Use of properly shaped and sharpened cutting tools

Substitution of Machines

- Larger, slower machines for smaller, faster ones
- Step dies for single-operation dies
- Presses for hammers
- Rotating shears for square shears
- Hydraulic presses for mechanical presses
- Belt drives for gears

Substitution of Processes

- Compression riveting for impact riveting
- Welding for riveting
- Hot working for cold working
- Pressing for rolling or forging

Reduce the Driving Force of Vibrating Surfaces by

- Reducing the forces
- Minimizing rotational speed
- Isolating

Reduce the Response of Vibrating Surfaces by

- Damping
- Additional support
- Increasing the stiffness of the material
- Increasing the mass of vibrating members
- Changing the size to change resonance frequency

Reduce the Sound Radiation from the Vibrating Surfaces by

- Reducing the radiating area
- Reducing overall size
- Perforating surfaces

Reduce the Sound Transmission through Solids by Using

- Flexible mounting
- Flexible sections in pipe runs

- Flexible-shaft couplings
- Fabric sections in ducts
- Resilient flooring

Reduce the Sound Produced by Gas Flow by

- Using intake and exhaust mufflers
- Using fan blades designed to reduce turbulence
- Using large, low-speed fans instead of smaller, high-speed fans
- Reducing the velocity of fluid flow (air)
- Increasing the cross-section of streams
- Reducing the pressure
- Reducing the air turbulence

Reduce Noise by Reducing Its Transmission through Air by

- Using sound-absorptive material on walls and ceiling in work areas
- Using sound barriers and sound absorption along the transmission path
- Completely enclosing individual machines
- Using baffles
- Confining high-noise machines to insulated rooms

Administrative Controls

Administrative controls are controls that reduce the exposure of employees to noise rather than reducing the noise itself.

Administrative controls should be considered a second-level approach, with engineering controls given top priority. Smaller companies that cannot afford to reduce noise through engineering measures may use administrative controls instead. However, this approach should be avoided if at all possible.

Hearing Protection Devices

In addition to engineering and administrative controls, employees should be required to use appropriate *hearing protection devices (HPDs)*. It should be noted, however, that such devices are effective only if worn properly. Enforcement of the proper use of HPDs is difficult in some settings. The following four classifications of HPDs are widely used: enclosures, earplugs, superaural caps, and earmuffs.

Enclosures are devices that completely encompass the employee's head, much like the helmets worn by motorcycle riders. **Earplugs** (also known as *aurals*) are devices that fit into the ear canal. Custom-molded earplugs are designed and molded for the individual employee. Premolded earplugs are generic in nature, are usually made of a soft rubber or plastic substance, and can be reused. **Formable earplugs** can be used by anyone. They are designed to be formed individually to a person's ears, used once, and then discarded.

Superaural caps fit over the external edge of the ear canal and are held in place by a headband. Earmuffs, also known as *circumaurals*, cover the entire ear with a cushioned cup that is attached to a headband. Earplugs and earmuffs are able to reduce noise by 20 to 30 dB. By combining earplugs and earmuffs, an additional 3 to 5 dB of blockage can be gained.

Figures 22–6 through 22–9 are examples of various types of HPDs. Figure 22–6 illustrates ear, face, and head protection combined into one comprehensive device. Figure 22–7 contains two types of superaural caps. Figure 22–8 shows an earmuff-style HPD equipped with an FM radio capability. Figure 22–9 displays soft, moldable earplugs.

The effectiveness of HPDs can be enhanced through the use of technologies that reduce noise levels. These **active noise reduction (ANR)** technologies reduce noise by manipulating sound and signal waves. Such waves are manipulated by creating an electronic mirror image of sound waves that tends to cancel out the unwanted noise in the same way that negative numbers cancel out positive numbers in a mathematical equation. Using ANR in conjunction with enclosure devices or earmuffs can be an especially effective strategy. Safety and health professionals should know the noise reduction rating (NRR) of all devices and technologies they recommend for employee use.

Traditional, or passive, HPDs can distort or muffle sounds at certain frequencies, particularly high-pitched sounds. *Flat-attenuation HPDs* solve this

Figure 22–6
Earmuff-style HPD.
Courtesy of ELVEX Corporation.

Figure 22–7
Canal-cap-style HPD. Courtesy of ELVEX Corporation.

Figure 22–8
Earmuff-style HPDs with FM radio capability.
Courtesy of ELVEX Corporation.

Figure 22–9
Earplug-style HPDs.
Courtesy of ELVEX Corporation.

problem by using electronic devices to block all sound frequencies equally. This eliminates, or at least reduces, the distortion and muffling problems. Flat-attenuation HPDs are especially helpful for employees in settings where high-pitched sound is present that they should be able to hear, as well as for employees who have already begun to lose their ability to hear such sounds. The ability to hear high-pitched sounds is significant because warning signals and human voices can be high-pitched.

A benefit of ANR technologies is **optimization**. The amount of noise protection can be adjusted so that employees can hear as much as they should, but not too much. Too much noise can cause employees to suffer hearing loss. Too little noise can mean that they may not hear warning signals.

VIBRATION HAZARDS

Vibration hazards are closely associated with noise hazards because tools that produce vibration typically also produce excessive levels of noise. The strategies for protecting employees against the noise associated with vibrating tools are the same as those presented so far in this chapter. This section focuses on the other safety and health hazards associated with vibration.

Eastman explains the problems associated with vibration:

Vibration related problems are not only serious, they are widespread. Donald Wasserman, author of *Human Aspects of Occupational Vibration*, says that up to

8 million workers are exposed to some type of vibration hazard. Of these, it has been estimated that more than half will show some signs of injury.[13]

The types of injuries associated with vibration depend on its source. For example, workers who operate heavy equipment often experience vibration over the whole body. This can lead to problems ranging from motion sickness to spinal injury. However, the most common vibration-related problem is known as **hand-arm vibration syndrome (HAV)**. Eastman describes HAV as follows:

> The condition, a form of Reynaud's Syndrome, strikes an alarming number of workers who use vibrating power tools day in and day out as part of their jobs. For HAV sufferers, . . . the sensations in their hands are more than just minor, temporary discomforts. They are symptoms of the potentially irreversible damage their nerves and blood vessels have suffered. As the condition progresses, it takes less and less exposure to vibration or cold to trigger the symptoms, and the symptoms themselves become more severe and crippling.[14]

Environmental conditions and worker habits can exacerbate the problems associated with vibration. For example, working with vibrating tools in a cold environment is more dangerous than working with the same tools in a warm environment. Gripping a vibrating tool tightly will lead to problems sooner than using a loose grip. Smoking and excessive noise also increase the potential for HAV and other vibration-related injuries. What all these conditions and habits have in common is that they constrict blood vessels which in turn restricts blood flow to the affected part of the body.[15]

Injury Prevention Strategies

Modern safety and health professionals should know how to prevent vibration-related injuries. Prevention is especially important with HAV because the disease is thought to be irreversible. This does not mean that HAV cannot be

Safety Fact

Do You Have a Noise Problem?

Taking measurements is an effective way to determine if a noise problem exists in the workplace, but there are also other ways. Safety and health professionals can learn a great deal by simple observation. Walk through the workplace at different times during the day. Do employees have to raise their voices to hear each other? If so, noise control and hearing protection strategies are in order. At the end of an eight-hour shift, do employees have trouble distinguishing among similar-sounding words? If so, workers should be using HPDs, and both engineering and administrative controls should be adopted, as appropriate.

treated. It can, but the treatments developed to date only reduce the symptoms. They do not cure the disease.

Following are prevention strategies that can be used by safety and health professionals in any company regardless of its size.

Purchase Low-Vibration Tools

Interest in producing low-vibration tools is relatively new but growing. Since the 1960s, only a limited number of manufacturers produced low-vibration tools. However, a lawsuit filed against three prominent tool manufacturers generated a higher level of interest in producing low-vibration tools.

According to Eastman, the suit was filed on behalf of 300 employees of General Dynamics's Electric Boat Shipyard in Connecticut, most of whom now suffer from HAV. The case claimed that three predominant tool manufacturers failed to (1) warn users of their tools of the potential for vibration-related injury and (2) produce low-vibration tools even though the technology to do so has been available for many years.[16] As a result of this lawsuit and the potential for others like it, low-vibration tools are becoming more commonplace.

Limit Employee Exposure

Although a correlation between cumulative exposure to vibration and the onset of HAV has not been scientifically quantified, there is strong suspicion in the safety and health community that such a link exists. For example, NIOSH recommends that companies limit the exposure of their employees to no more than four hours per day, two days per week.[17] Until the correlation between cumulative exposure and HAV has been quantified, safety and health professionals are well advised to apply the NIOSH recommendation.

Change Employee Work Habits

Employees can play a key role in protecting themselves if they know how. Safety and health professionals should teach employees who use vibration-producing tools the work habits that will protect them from HAV and other injuries. These work habits include the following: (1) wearing properly fitting thick gloves that can partially absorb vibration; (2) taking periodic breaks (at least 10 minutes every hour); (3) using a loose grip on the tool and holding it away from the body; (4) keeping tools properly maintained (i.e., replacing vibration-absorbing pads regularly); (5) keeping warm; and (6) using vibration-absorbing floor mats and seat covers as appropriate.[18]

Modern safety and health professionals should also encourage higher management to require careful screening of applicants for jobs involving the use of vibration-producing tools and equipment. Applicants who smoke or have other conditions that constrict blood vessels should be guided away from jobs that involve excessive vibration.

OTHER EFFECTS OF NOISE HAZARDS

Hearing loss is the principal concern of safety and health professionals relating to noise hazards. However, hearing loss is not the only detrimental effect of excess noise. Noise can also cause communication problems, isolation, and productivity problems.

Noise can also be detrimental to productivity by interfering with an employee's ability to think, reason, and solve problems. Not all employees respond to noise in this way, but many do. Can you concentrate on your studies in a noisy room? Some students can, whereas others cannot. If excessive noise makes it difficult for you to study, you will probably have the same problem on the job.

Isolation is another problem for some employees in a noisy environment. Employees can begin to feel left out and uninformed, the antithesis of the goal of a modern teamwork-oriented organization.

CORPORATE POLICY

Corporate policies relating to hearing loss prevention should be carefully planned and executed to benefit the affected employee and the employer.[19] Organizations with the most successful programs address the following areas of concern:

- The organizational environment should promote a safety culture where employees are empowered to protect their own health and that of coworkers.
- Policies should be based on effective practices rather than on minimum compliance with government regulations.
- The hearing prevention program must be a functional part of the overall company safety and health program. It should not be a stand-alone, separate-budget operation.
- A key individual (or *program implementor*) should have ultimate responsibility for the program. This person may not necessarily perform all the functions of the hearing loss prevention program but is in charge of the overall program. Experience with successful hearing loss prevention programs shows that a single individual often makes the crucial difference between success and failure. This person may be a safety and health officer, a supervisor, or a designated employee. This program implementor acts as the *conscience* and *champion* of the hearing loss prevention program.
- The program implementor should work with management and employees to develop and implement hearing loss prevention plans and policies for an effective program. As a team leader, the program implementor should be given the authority to establish hearing loss prevention provisions that meet or exceed the letter and intent of OSHA's noise control and hearing conservation regulations.
- Employee and administrative compliance with the company's hearing loss prevention program.

- Hearing loss prevention program policies should clearly describe standard operating procedures for each phase of the program. Specific policy statements should be developed for the important elements of the program. For example, company policy should require the participation of all noise-exposed employees in the audiometric program and require the consistent and proper wearing of hearing protectors in posted areas, even if employees or supervisors are only passing through these areas. These requirements should be conditions of employment. Other important policy statements should be written to cover:

> Adoption of a prescribed schedule for monitoring employee noise exposure levels and other risks, including ensuring that equipment and personnel training are appropriate to the task.
>
> Counseling of employees immediately following each audiometric test, whether it is the initial, annual, retest, threshold-shift confirmation, or termination examination.
>
> Determining the adequacy and correct use of hearing protection devices by on-site equipment checks.
>
> Educating, training, and motivating employees to support the company's hearing loss prevention program provisions; assessing employee attitudes and assessing knowledge gained from periodic training.
>
> Establishing a program of quality assurance for the performance of audiometry and management of audiometric records.
>
> Reviewing audiometric data to verify the effectiveness of the hearing loss prevention program.
>
> Encouraging employees to use company-provided hearing protectors for off-the-job exposure.
>
> Purchasing hearing protectors, audiometers, noise measuring equipment, and quieter machinery. This policy should address the reasons why the program implementor responsible for the hearing loss prevention program, not the purchasing department, should have final authority about anticipated purchases.

If services such as noise surveys, employee education, audiometric testing, medical counseling, or the fitting of hearing protection devices cannot be handled by in-house staff, outside vendors or contractors may be required. They should be selected carefully to ensure that their services complement the abilities of the company staff and functional conduct of the in-house program elements. Vendors must understand and agree to abide by the company's hearing loss prevention program policies and standards of operation. On-site personnel must supervise contractors to make sure that they carry out their obligations. Regardless of whether outside vendors or contractors are used, responsibility for the program stays with the program implementor.

Companies that issue clearly defined hearing loss prevention policies and then adhere to these policies consistently will have smoothly running hearing loss prevention programs. Employees will be fully informed, comprehend their functional role, and know what is expected of them. Equipment will be appropriate, hearing protection will be used by the right people in the right places, and the program elements will be implemented in a timely fashion.

EVALUATING HEARING LOSS PREVENTION PROGRAMS

Hearing loss prevention programs should be evaluated periodically to ensure their effectiveness.[20] Such evaluations should have at least the following components: (1) training and education; (2) supervisor involvement; (3) noise measurement; (4) engineering and administrative controls; (5) monitoring and record keeping; (6) referrals; (7) hearing protection devices; and (8) administration. Following are checklists for each of these components.

Training and Education

Failures or deficiencies in hearing conservation programs (hearing loss prevention programs) can often be traced to inadequacies in the training and education of noise-exposed employees and those who conduct elements of the program.

- Has training been conducted at least once a year?
- Was the training provided by a qualified instructor?
- Was the success of each training program evaluated?
- Is the content revised periodically?
- Are managers and supervisors directly involved?
- Are posters, regulations, handouts, and employee newsletters used as supplements?
- Are personal counseling sessions conducted for employees having problems with hearing protection devices or showing hearing threshold shifts?

Supervisor Involvement

Data indicate that employees who refuse to wear hearing protectors or who fail to show up for hearing tests frequently work for supervisors who are not totally committed to the hearing conservation programs.

- Have supervisors been provided with the knowledge required to supervise the use and care of hearing protectors by subordinates?
- Do supervisors wear hearing protectors in appropriate areas?
- Have supervisors been counseled when employees resist wearing protectors or fail to show up for hearing tests?
- Are disciplinary actions enforced when employees repeatedly refuse to wear hearing protectors?

Noise Measurement

Noise measurements, to be useful, should be related to noise exposure risks or the prioritization of noise control efforts, rather than merely filed away. In addition, the results should be communicated to the appropriate personnel, especially when follow-up actions are required.

- Were the essential or critical noise studies performed?
- Was the purpose of each noise study clearly stated? Have noise-exposed employees been notified of their exposures and apprised of auditory risks?
- Are the results routinely transmitted to supervisors and other key individuals?
- Are results entered into health or medical records of noise-exposed employees?
- Are results entered into shop folders?
- If noise maps exist, are they used by the proper staff?
- Are noise measurement results considered when contemplating procurement of new equipment? Modifying the facility? Relocating employees?
- Have there been changes in areas, equipment, or processes that have altered noise exposure? Have follow-up noise measurements been conducted?
- Are appropriate steps taken to include (or exclude) employees in the hearing loss prevention programs whose exposures have changed significantly?

Engineering and Administrative Controls

Controlling noise by engineering and administrative methods is often the most effective means of reducing or eliminating the hazard. In some cases, engineering controls will remove requirements for other components of the program, such as audiometric testing and the use of hearing protectors.

- Have noise control needs been prioritized?
- Has the cost effectiveness of various options been addressed?
- Are employees and supervisors apprised of plans for noise control measures?
- Are they consulted on various approaches?
- Will in-house resources or outside consultants perform the work?
- Have employees and supervisors been counseled on the operation and maintenance of noise control devices?
- Are noise control projects monitored to ensure timely completion?
- Has the full potential for administrative controls been evaluated? Are noisy processes conducted during shifts with fewer employees? Do employees have sound-treated lunch or break areas?

Monitoring Audiometry and Record Keeping

The skills of audiometric technicians, the status of the audiometer, and the quality of audiometric test records are crucial to hearing loss prevention program success. Useful information may be ascertained from the audiometric records as well as from those who actually administer the tests.

- Has the audiometric technician been adequately trained, certified, and recertified as necessary?
- Do on-the-job observations of the technicians indicate that they perform a thorough and valid audiometric test, instruct and consult the employee effectively, and keep appropriate records?
- Are records complete?

- Are follow-up actions documented?
- Are hearing threshold levels reasonably consistent from test to test? If not, are the reasons for inconsistencies investigated promptly?
- Are the annual test results compared to baseline to identify the presence of an OSHA standard threshold shift?
- Is the annual incidence of standard threshold shift greater than a few percent? If so, are problem areas pinpointed and remedial steps taken?
- Are audiometric trends (deteriorations) being identified, both in individuals and in groups of employees? (NIOSH recommends no more than 5 percent of workers showing 15-dB significant threshold shift, same ear, same frequency.)
- Do records show that appropriate audiometer calibration procedures have been followed?
- Is there documentation showing that the background sound levels in the audiometer room were low enough to permit valid testing?
- Are the results of audiometric tests being communicated to supervisors and managers as well as to employees?
- Has corrective action been taken if the rate of no-shows for audiometric test appointments is more than about 5 percent?
- Are employees incurring STS notified in writing within at least 21 days? (NIOSH recommends immediate notification if retest shows 15-dB significant threshold shift, same ear, same frequency.)

Referrals

Referrals to outside sources for consultation or treatment are sometimes in order, but they can be an expensive element of the hearing loss prevention program and should not be undertaken unnecessarily.

- Are referral procedures clearly specified?
- Have letters of agreement between the company and consulting physicians or audiologists been executed?
- Have mechanisms been established to ensure that employees needing evaluation or treatment actually receive the service (e.g., transportation, scheduling, reminders)?
- Are records properly transmitted to the physician or audiologist, and back to the company?
- If medical treatment is recommended, does the employee understand the condition requiring treatment, the recommendation, and methods of obtaining such treatment?
- Are employees being referred unnecessarily?

Hearing Protection Devices

When noise control measures are not feasible, or until such time as they are installed, hearing protection devices are the only way to prevent hazardous levels of noise from damaging the inner ear. Making sure that these devices are worn effectively requires continuous attention on the part of supervisors and program implementers as well as noise-exposed employees.

- Have hearing protectors been made available to all employees whose daily average noise exposures are 85 dBA or above? (NIOSH recommends requiring HPD use if noises equal or exceed 85 dBA regardless of exposure time.)
- Are employees given the opportunity to select from a variety of appropriate protectors?
- Are employees fitted carefully with special attention to comfort?
- Are employees thoroughly trained, not only initially but at least once a year?
- Are the protectors checked regularly for wear or defects and replaced immediately if necessary?
- If employees use disposable hearing protectors, are replacements readily available?
- Do employees understand the appropriate hygiene requirements?
- Have any employees developed ear infections or irritations associated with the use of hearing protectors? Are there any employees who are unable to wear these devices because of medical conditions? Have these conditions been treated promptly and successfully?
- Have alternative types of hearing protectors been considered when problems with current devices are experienced?
- Do employees who incur noise-induced hearing loss receive intensive counseling?
- Are those who fit and supervise the wearing of hearing protectors competent to deal with the many problems that can occur?
- Do workers complain that protectors interfere with their ability to do their jobs? Do they interfere with spoken instructions or warning signals? Are these complaints followed promptly with counseling, noise control, or other measures?
- Are employees encouraged to take home their hearing protectors if they engage in noisy nonoccupational activities?
- Are potentially more effective protectors considered as they become available?
- Is the effectiveness of the hearing protector program evaluated regularly?
- Have at-the-ear protection levels been evaluated to ensure that either over- or underprotection has been adequately balanced according to the anticipated ambient noise levels?
- Is each hearing protector user required to demonstrate that he or she understands how to use and care for the protector? Are the results documented?

Administration

Keeping organized and current on administrative matters will help the program run smoothly.

- Have there been any changes in federal or state regulations? Have hearing loss prevention programs' policies been modified to reflect these changes?

- Are copies of company policies and guidelines regarding the hearing loss prevention program available in the offices that support the various program elements? Are those who implement the program elements aware of these policies? Do they comply?
- Are necessary materials and supplies being ordered with a minimum of delay?
- Are procurement officers overriding the hearing loss prevention program implementer's requests for specific hearing protectors or other hearing loss prevention equipment? If so, have corrective steps been taken?
- Is the performance of key personnel evaluated periodically? If such performance is found to be less than acceptable, are steps taken to correct the situation?
- Has the failure to hear warning shouts or alarms been tied to any accidents or injuries? If so, have remedial steps been taken?

SUMMARY

1. Sound is any change in pressure that can be detected by the ear. Typically, sound is a change in air pressure. However, it can also be a change in water pressure or any other pressure-sensitive medium. The unit of measurement for sound is the decibel, or one-tenth of a bel.
2. The threshold of hearing is the weakest sound that can be detected by the human ear. The maximum sound that can be perceived without experiencing pain is known as the threshold of pain.
3. Noise is excessive sound. Industrial noise is classified as wide band noise, narrow band noise, and impulse noise.
4. Several different factors affect the risk of hearing loss from excessive noise: (a) intensity of the noise, (b) type of noise, (c) duration of daily exposure, (d) total duration of exposure, (e) age of the individual, (f) coexisting disease, (g) nature of the environment, (h) distance of the receiver from the noise source, and (i) position of the ears relative to the sound waves.
5. Noise levels of less than 80 dBA are considered safe. A level of 85 dBA should be considered the maximum limit of continuous exposure over eight hours without protection.
6. Important standards and regulations related to noise hazards are (a) ANSI S12.13, Evaluation of Hearing Conservation Programs, and (b) the Hearing Conservation Amendment to OSHA 29 CFR 1910.95.
7. With workers' compensation claims for hearing loss on the rise, the medical profession has established a procedure for identifying causal relationships between industrial noise and hearing loss. The procedure requires that physicians consider the following factors: (a) onset and progress of hearing loss, (b) the employee's complete work history, (c) results of otological examination, (d) results of hearing studies, and (e) ruling out hearing loss from sources outside the workplace.

8. Methods for identifying and assessing hazardous noise conditions include the following: (a) noise surveys, (b) audiometric tests, (c) record keeping, and (d) follow-up.

9. Noise reduction strategies are of three types: those that reduce noise at the source, those that reduce noise along its path, and those that reduce noise at the receiver.

10. Noise can be reduced by applying either engineering or administrative controls. Engineering controls attempt to reduce noise. Administrative controls limit human exposure to noise.

11. Vibration can cause physical problems ranging from motion sickness to spinal injury to hand-arm vibration syndrome (HAV). HAV is the most widespread of these problems. Because the physical damage associated with HAV may be irreversible, prevention is especially important.

12. HAV prevention strategies include purchasing low-vibration tools, limiting employee exposure, and changing employee work habits.

13. Organizations with the most successful programs adopt comprehensive corporate policies relating to hearing loss prevention.

14. Hearing loss prevention programs should be evaluated periodically in the following areas: (a) training and education, (b) supervisor involvement, (c) noise measurement, (d) engineering and administrative controls, (e) monitoring and record keeping, (f) referrals, (g) hearing protection devices, and (h) administration.

KEY TERMS AND CONCEPTS

Active noise reduction (ANR)

Administrative controls

Air pressure

ANSI S12.13

Attenuation

Audiogram

Audiometric database analysis (ADBA)

Audiometric testing

Baseline audiogram

Continuous noise

Decibel (dB)

Dosimeter

Duration

Electronic earmuffs

Earplugs

Enclosures

Engineering controls

Exchange rate

Follow-up

Formable earplugs

Frequency

Hand-arm vibration syndrome (HAV)

Hazardous noise

Hearing conservation

Hearing protection devices (HPDs)

Hearing threshold level (HTL)

Hertz (Hz)

Impulse noise

Low-vibration tools

Material hearing impairment

Medical surveillance

Narrow band noise

Noise

Noise controls

Noise-induced hearing loss

Noise surveys

Optimization

Passive earmuffs

Path

Percent better or worse sequential

Percent worse sequential

Personal protection

Receiver

Sound

Sound level meter

Sound pressure level

Source

Standard threshold shift (STS)

Superaural caps

Threshold of hearing

Threshold of pain

Time-weighted average (TWA)

Uniform attenuation earmuffs

Vibration

Wide band noise

REVIEW QUESTIONS

1. Define the term *sound*.
2. What is the difference between sound and noise?
3. Differentiate between sound and vibration.
4. Describe the relationship between the pitch of sound and the cycle of sound waves.
5. List and briefly explain the three broad types of industrial noise.
6. Describe the various physiological problems associated with excessive noise.
7. List four factors that affect the risk of hearing loss from exposure to excessive noise.
8. At what sound level is it necessary to begin using some type of personal protection?
9. Give a brief description of ANSI S12.13.
10. Give a brief description of OSHA 29 CFR 1910.95, Hearing Conservation Amendment.
11. What factors do medical professionals consider in determining causal relationships of hearing loss?
12. Define the following terms: *follow-up*, *noise survey*, and *audiometric testing*.
13. List three appropriate follow-up activities when an audiometric test reveals hearing loss in an employee.
14. Differentiate between engineering and administrative controls.
15. What is HAV? How can it be prevented?
16. Explain the four classifications of HPDs that are widely used.
17. Explain the main components of an evaluation of a hearing loss prevention program.

ENDNOTES

1. National Institute for Occupational Safety and Hearing (NIOSH), *A Practical Guide to Preventing Hearing Loss*, Publication 96–110 (Washington, DC: 2006), 2. Retrieved from http://www.cdc.gov/niosh/96-110a.html.
2. Ibid.
3. NIOSH, *A Practical Guide*, 6.
4. NIOSH, *A Practical Guide*, 4.
5. Ibid., 6.
6. Ibid.
7. J. L. Ruck, "Hidden Risk of Noise Standard," *Occupational Hazards*, April 2000, 27.
8. B. Sokol, "Ear Muffs: A Field Guide," *Occupational Health & Safety* 74, no. 6: 66–68.
9. NIOSH, *A Practical Guide*, 37.
10. NIOSH, *A Practical Guide*, 8.
11. Ibid., 12.
12. Ibid.
13. M. Eastman, "Vibration Shakes Workers," *Safety & Health* 143, no. 5; 32.
14. Ibid.
15. Ibid., 32–33.
16. Ibid., 34.
17. Ibid., 33.
18. Ibid., 35.
19. NIOSH, *A Practical Guide*, 1–2.
20. Ibid., Appendix B, 1–6.

COMPUTERS, AUTOMATION, AND ROBOTS

Major Topics

- ■ Impact of Automation on the Workplace
- ■ VDTs in Offices and Factories
- ■ Human–Robot Interaction
- ■ Safety and Health Problems Associated with Robots
- ■ Safety and Health in Office Automation
- ■ Industrial Medicine and Robots
- ■ Technological Alienation in the Automated Workplace
- ■ Minimizing the Problems of Automation
- ■ Challenge for the Future

Automation of the workplace has changed, and continues to change, how work is done. The introduction of automated processes that involve computers and robots has changed the environment of the modern workplace and what is needed to succeed in it.

IMPACT OF AUTOMATION ON THE WORKPLACE

The advent of **automation** on the workplace was the next logical step on a continuum of developments intended to enhance productivity, quality, and competitiveness. This continuum began when humans first developed simple tools to assist them in doing work. This was the age of hand tools and **manual work**. It was eventually superseded by the age of **mechanization** during the Industrial Revolution. During the age of mechanization, machines were developed to do work previously done by humans using hand tools. The 1960s saw the beginnings of broad-based efforts at automating mechanical processes and systems.

These early attempts at automation resulted in **islands of automation**, or individual automated systems lacking electronic communication with other related systems. Examples of islands of automation are a stand-alone computer numerical control milling machine or a personal computer-based word processing system, neither of which is connected to other related systems. Local area networks (LANs) for integrating personal computers are an example of integration in the office. Computer-integrated manufacturing is an example in the factory.

These developments are having an impact on the workplace. Automation and integration are having the following effects on workers:

- Changing the emphasis from physical work to automated mechanical work
- Changing the emphasis from physical to mental work
- Categorizing work as either or physical (intelligence versus labor)
- Increasing the level of stress from workers, supervisors, and managers
- Eliminating some traditional blue-collar jobs that cause morale problems and decrease employee loyalty
- Making workers feel a loss of control and the onset of helplessness and powerlessness

These various effects of automation are resulting in a marked increase in the amount of stress experienced by workers. Two factors in particular lead to increased levels of stress: rapid, continual change and an accompanying feeling of **helplessness**. With automation, the rate of change has increased. As a result, workers must continually learn and relearn their jobs with little or no relief. In addition, automated machines do more of the work that used to be done by humans. This can leave workers feeling as if they may be replaced by a machine and powerless to do anything about it. According to Ostberg,

> The development of a special curriculum known as robot medicine in Japan acknowledges the connection between emerging technology and stress. This course of study is offered to industrial doctors by the University of Occupational and Environmental Health, Japan. It is defined as the branch of medicine studying stress and other mental and physical problems of workers in automated and roboticized production.[1]

Workplace stress is a complex concept involving physiological, psychological, and social factors. People become stressed when there is an imbalance between the demands placed on them and their ability to respond.[2] Automation appears to be increasing the instances in which such an imbalance occurs. This chapter focuses on the safety and health concerns associated with computers, robots, and automation and appropriate measures for dealing with these concerns.

VDTs IN OFFICES AND FACTORIES

A safety and health concern brought about by the advent of computers has to do with the impact of **video display terminals (VDTs)**. Does prolonged use of VDTs cause safety and health problems? Are pregnant women who work at

VDTs more likely to miscarry? Are such problems as eye fatigue, muscle stiffness, and mental fatigue caused by VDT use?

The National Institute for Occupational Safety and Health (NIOSH) published a study showing that "women who work with VDTs have no greater risk of miscarriage than those who do not."[3] The study was undertaken in response to reports of a perceived high incident rate of miscarriages among women whose jobs involved regular use of VDTs. According to Castelli,

> NIOSH researchers studied telephone operators in eight southeastern states. One group did not use computer terminals; the other group, directory-assistance operators, used them seven hours a day. The researchers interviewed 2,430 operators and gathered information on 882 pregnancies that occurred among the workers. . . . About half of the pregnant workers used VDTs.[4]

The miscarriage rate among VDT users was actually lower than that for nonusers (14.8 percent for VDT users and 15.9 percent among nonusers). Both figures are close to the national average of 15 percent. As a result of this study, NIOSH plans to conduct research into two additional areas of concern: (1) whether VDT use can be tied to birth defects, premature births, or low birthweight; and (2) potential risks of extremely low frequency (ELF) radiation, a type of electromagnetic energy emitted by VDTs, power lines, electrical wires, and appliances.[5]

Not everyone agreed with the findings of the NIOSH study, however. The Service Employees' International Union (SEIU) challenged the findings as inconclusive because the effects of stress and VDT use on pregnancy were not considered. SEIU claimed that the study did not give VDTs a "clean bill of health" and that "employers should likewise not conclude from this study that they have no obligation to develop safety guidelines and better designed equipment."[6]

Regardless of which side of the miscarriage issue you believe, there is ample evidence that such problems as eye fatigue, blurred vision, eye strain, and nervousness are associated with VDT use. Research by Smith supported findings that VDT users often suffer eye-related discomfort and symptoms.[7]

The **eye strain** caused by prolonged VDT use poses safety and health problems from two different perspectives. First, the visual health of the operator experiencing the strain is impaired. Second, there is an increased likelihood of accidents caused by impaired work performance and increased psychological stress.

According to Kurimoto and associates of the Department of Ophthalmology at Japan University of Occupational and Environmental Health, the eye functions most noticeably affected by VDT use are accommodation, convergence, and lacrimation.[8] **Accommodation** is the ability of the eye to become adjusted after viewing the VDT to focusing on other objects, particularly objects at a distance. **Convergence** is the coordinated turning of the eyes inward to focus on a nearby point or object. **Lacrimation** is the process of excreting tears. The Kurimoto research confirms in nonmedical terms that prolonged VDT use can render operators unable to focus on either distant or near objects. It can also impair the tearing function, leading to dry eyes.[9]

Safety Fact

CTDs and VDT Use

Cumulative trauma disorders (CTDs) include a long list of tendon, peripheral nerve, vascular, muscular, and joint-related disorders. Those most commonly associated with VDT use are tendinitis, lateral epicondylitis (tennis elbow), and carpal tunnel syndrome. Of these, carpal tunnel syndrome is the most commonly reported.

The carpal tunnel receives its name from the eight bones in the wrist, called *carpals*, that form a tunnel-like structure. The tunnel is filled with flexor tendons, which control finger movement. It also provides a pathway for the median nerve to reach sensory cells in the hand. Repetitive flexing and extension of the wrist may cause a thickening of the protective sheaths that surround each of the tendons. The swollen tendon sheaths, or *tenosynovitis*, apply increased pressure on the median nerve and produce carpal tunnel syndrome (CTS).

The symptoms of CTS often first appear as painful tingling in one or both hands during the night, frequently painful enough to disturb sleep. Accompanying this is a feeling of uselessness in the fingers, which are sometimes described as feeling swollen, even though little or no swelling is apparent. As symptoms increase, tingling may develop during the day, commonly in the thumb, index, and ring fingers. A decreased ability to squeeze things may follow. In advanced cases, the thenar muscle at the base of the thumb atrophies, and strength is lost.

Kurimoto et al. recommend the following strategies for reducing the physiological and psychological problems associated with VDT use:

- Increasing computer response time. This is a matter of upgrading the computer's processing capability or replacing it with one that has more processing power.
- Providing more frequent breaks from VDT use or a work rotation schedule that allows users to intersperse non-VDT work in their daily routine.
- Creating a **work design** that recognizes and accommodates the need to break up continual VDT use.
- Arranging the keyboard properly so it is located in front of the user, not to the side. Body posture and the angle formed by the arms are critical factors.
- Adjusting the height of the desk. Taller employees often have trouble working at average height desks.
- Adjusting the tilt of the keyboard. The rear portion of the keyboard should be lower than the front.
- Encouraging employees to use a soft touch on the keyboard and when clicking a mouse. A hard touch increases the likelihood of injury.
- Encouraging employees to avoid wrist resting. Resting the wrist on any type of edge can increase pressure on the wrist.

- Placing the mouse within easy reach. Extending the arm to its full reach increases the likelihood of injury.
- Removing dust from the mouse ball cavity. Dust can collect, making it difficult to move the mouse. Blowing out accumulated dust once a week will keep the mouse easy to manipulate.
- Locating the VDT at a proper height and distance. The height of the VDT should be such that the top line on the screen is slightly below eye level. The optimum distance between the VDT and user will vary from employee to employee, but will usually be between 16 and 32 inches.
- Minimizing glare. Glare from a VDT can cause employees to adopt harmful postures. Glare can be minimized by changing the location of the VDT, using a screen hood, and closing or adjusting blinds and shades.
- Reducing lighting levels. Vision strain can be eliminated by reducing the lighting level in the area immediately around the VDT.
- Dusting the VDT screen. VDT screens are magnets to dust. Built-up dust can make the screen difficult to read, contributing to eye strain.
- Eliminating telephone cradling. Cradling a telephone receiver between an uplifted shoulder and the neck while typing can cause a painful disorder called *cervical radiculopathy* (compression of the cervical vertebrae in the neck). Employees who need to talk on the telephone while typing should wear a headphone.[10]

HUMAN–ROBOT INTERACTION

Every new tool developed to enhance the ability of humans to work efficiently and effectively has brought with it a new safety and health hazard. This is particularly the case with industrial robots. What makes robots more potentially dangerous than other machines can be summarized as follows: (1) their ability to acquire intelligence through programming, (2) their flexibility and range of motion, (3) their speed of movement, and (4) their power.

The often-discussed **"peopleless" factory** is still far in the future. However, robots are so widely used now that they are no longer the oddity they once were. Consequently, there is plenty of **human–robot interaction** in modern industry. According to Yamashita, "At ordinary factories ... human workers and robots coexist, creating such problems as cooperation and competition between man and machines and safety."[11]

How does human–robot interaction differ from human interaction with other machines? This is an important question for safety and health professionals. According to Lena Martensson of the Royal Institute of Technology in Stockholm, Sweden, the modern factory has or is moving toward having the following characteristics:

- Workers will supervise machine systems rather than interact with individual pieces of production equipment.

- Workers will communicate with machines via video display terminals on which complex information processed by a computer will be displayed.
- Workers will be supported by expert systems for fault identification, diagnosis, and repair.[12]

Robots and other intelligent computer-controlled machines will play an increasingly important role in modern industry. As this happens, safety and health professionals must be concerned about the new workplace hazards that will be created.

SAFETY AND HEALTH PROBLEMS ASSOCIATED WITH ROBOTS

Robots are being used in industry for such applications as arc welding, spot welding, spray painting, material handling and assembly, and loading and unloading of machines. Figures 23–1 and 23–2 show examples of modern industrial robots being used in typical applications. According to the National Safety Council the principal hazards associated with robots are:

Figure 23–1
Industrial robot in use.
Courtesy of Cincinnati Milicron.

Figure 23–2
Industrial robot in use.
Courtesy of Unimation, Inc.

- Being struck by a moving robot while inside the work envelope. The **work envelope** of a robot is the total area within which the moving parts of the robot actually move. Figure 23–3 illustrates an example of a robot's work envelope.
- Being trapped between a moving part of a robot and another machine, object, or surface.
- Being struck by a workpiece, tool, or other object dropped or ejected by a robot.[13]

Until a worker enters the work envelope of a robot, there is little probability of an accident. However, anytime a worker enters a functioning robot's work envelope, the probability becomes very high. The only logical reason for a worker to enter the work envelope of an engaged robot is to teach it a new motion. According to the NSC,

Many applications of robots use a mode of operation called the **teach mode** [emphasis added] that makes them unique among industrial equipment as regards safeguarding them. In the teach mode, an operator may be required to place himself within the operating range of the robot in order to program its movements within very close tolerance parameters. The teaching of a robot is done at a greatly reduced speed of operation; however, a hazard may arise if the robot goes out of control and moves in an unpredictable fashion at a high rate of speed. It is in the teach mode where the highest degree of hazard exists. A Japanese survey . . . showed that the greatest risk of accidents involving robots occurs during programming, teaching, and maintenance; all times when a person may be within the operating envelope of the robot.[14]

Model 762

320°

(Model 761—1500 mm)

1250

11.75°

(Model 761—13.1°)

Inaccessible area
(can be reached
in left configuration)

514 mm radius
Inaccessible to JT5
(Model 761–630 mm radius)

1263 mm radius swept by JT5
measured from center
of JT1
(Model 761–1636 mm radius)

Robot shown in
right configuration

1388 mm radius swept by
mounting flange
(Model 761–1636 mm radius)

Cylindrical volume 112 mm
diameter inaccessible to JT6
tool flange

220°

Figure 23–3
A robot's work envelope.
Courtesy of Unimation, Inc.

Minimizing the Safety and Health Problems of Robots

If human workers never had to enter a work envelope, the safety and health problems associated with robots would be minimal. However, workers must occasionally do so. Therefore safety and health professionals must be concerned with ensuring safe human–robot interaction.

The NSC recommends several strategies for minimizing the hazards associated with robots.

- Ensure a glare-free, well-lighted robot site. The recommended light intensity is 50–100 foot-candles.
- Keep the floors in and around the robot site carefully maintained, clean, and free of obstructions so that workers do not trip or slip into the work envelope.
- Keep the robot site free of associated hazards such as blinding light from welding machines or vapors from a paint booth.
- Equip electrical and pneumatic components of the robot with fixed covers and guards.
- Clear the work envelope of all nonessential objects and make sure all safeguards are in place before starting the robot.
- Apply lockout and proper test procedures before entering the work envelope.
- Remove and account for all tools and equipment used to maintain the robot before starting it.[15]

SAFETY AND HEALTH IN OFFICE AUTOMATION

According to Nishiyama, the objectives of office automation are increased efficiency, personnel reductions, economy of personnel expenditures, improved service to customers, improved planning and estimating, increased processing speed, and improved working conditions.[16] These goals are being achieved with varying degrees of success. However, automation also introduced a new set of safety and health problems into the office environment.

Morooka and Yamoda of Japan's Tokai University identified the following problems associated with office automation: eye fatigue, seeing double images and complementary colors, headache, yawny feelings, unwillingness to talk, **shoulder fatigue**, **neck fatigue**, dryness in the throat, sleepy feelings, and whole-body tiredness.[17]

As such problems began to be associated with office automation, interest grew in establishing workplace and technology standards to minimize them. Benjamin C. Amick III of Congress' Office of Technology Assessment described this situation:

> The public policy issues developing as a result of office automation range from issues of labor management relations to office, work-station, and human-computer interface standards. The current scientific research suffers from a lack of integration and clear definition of what is causing what. . . . Clearly, the current upswing in the purchasing of office automation equipment provides a unique opportunity to effect changes in the quality of worklife on a national basis. The question facing policy-makers is how best to create the policy to maintain the current level of creativity and innovation in the marketplace while not sacrificing the quality of worklife.[18]

States that have introduced legislation establishing standards relating to office automation are Oregon, Washington, California, Florida, Colorado, Missouri, Iowa, Minnesota, Wisconsin, Illinois, Indiana, Ohio, Pennsylvania, West Virginia,

Maryland, New Jersey, Connecticut, Rhode Island, Massachusetts, New Hampshire, and Maine. New Mexico established standards by executive order.

Most legislation dealing with office automation concerns standards for VDT interaction. For example, legislation introduced in Maryland requires that VDT users (1) have an eye examination every year; (2) have an adjustable chair with adjustable backrest height and tension; and (3) take a 15-minute break from the VDT every hour. Legislation introduced in other states is similar to the Maryland proposal.[19]

Public policy debate is increasing in the United States over the safety and health concerns inherent in office automation. Unfortunately, the debate has outpaced the research on this important issue. Consequently, the probability is high that policies being adopted are based on insufficient information. Regarding research that needs to be conducted, Amick recommends the following:

- Testing of the biological plausibility of health hazards from VDT work
- Ergonomic intervention research to determine the contribution of workstation design and office design to the worker's health
- Examination of the interaction of physical and **psychosocial stressors** in the high-technology workplace
- Development of prospective case-control studies to determine the temporal relationships
- Establishment of a national high-technology surveillance system for worker safety and health
- Development of organizational and job-intervention programs
- Multidisciplinary studies examining the relative contribution of various working conditions to the health and well-being of the office worker
- Programs that bring system designers and building designers into the total automation process[20]

The Japanese Association of Industrial Health developed the following set of principles upon which future safety and health measures relating to office automation should be based:

- More attention should be focused on bringing employers and employees together for the purpose of improving working conditions in automated offices.
- Because VDT use is becoming so common in so many different occupations, it should no longer be considered specialized. Consequently, jobs should be designed to accommodate VDT use.
- The amount of time spent doing VDT work exclusively should be kept short, and employees should be able to perform the work in their own way.
- Working conditions should be established that prevent safety and health problems so that management is acting instead of reacting.
- Special emphasis should be placed on education and training as the best way to prevent adverse effects from office automation.
- VDT work should not be done part time at home or contracted out because working conditions cannot be properly controlled or supervised under these circumstances.[21]

Safety Fact

Ergonomics in Office Automation

Increasing ergonomic problems resulting from office automation are being associated with cumulative trauma disorders (CTDs). The types of CTDs occurring most frequently are as follows:

Tendon-Related Disorders

- Tendinitis
- Tenosynovitis
- Stenosing tenosynovitis of the fingers (trigger finger)
- Stenosing tenosynovitis of the thumb (DeQuervain's)
- Peridentitis (strain)
- Ganglion cyst
- Lateral epicondylitis (tennis elbow)
- Bicipital tendinitis
- Rotator cuff tendinitis

Peripheral Nerve Entrapment

- Carpal tunnel syndrome
- Guyon's tunnel syndrome
- Radial tunnel syndrome
- Pronator teres syndrome
- Cubital tunnel syndrome

Vascular

- Hand-arm vibration syndrome (Raynaud's phenomena)
- Ulnar artery thrombosis

Neurovascular

- Thoracic outlet syndrome

Muscular

- Focal dystonia
- Fibromyositis
- Tension neck syndrome
- Myositis

Joint/Joint Capsule

- Osteoarthritis
- Bursitis
- Synovitis

Courtesy of Unimation, Inc.

INDUSTRIAL MEDICINE AND ROBOTS

Industrial medicine is a specialized field that is concerned with work-related safety and health issues. Practitioners of **industrial medicine** are becoming increasingly concerned with the interaction of humans and automated machines, particularly robots and computers. They are concerned about maladaptation to an automated environment.

According to Masamitsu Oshima of Japan's Medical Information System Development Center, **maladaptation** can manifest itself as an urge to quit work; **fatigue**; problems with human relations; a drop in work performance; **social pathological phenomena** such as drug use or crime; mood swings; a loss of motivation; and accidents.[22] Practitioners of industrial medicine are concerned with improving the relationship between humans and automated machines by establishing methods whereby humans can work more adequately with the machines.[23]

Oshima makes the following recommendations for improving the interaction between humans and robots, computers, and other automated machines in the high-tech workplace:

- Match the human system and the computer system.
- Position machine systems as human-supportive systems.
- Adapt human–computer interaction to human use.
- Initiate job-changing opportunities.
- Allow suitable rest periods for users of automated equipment.
- Vitalize the workplace.
- Encourage recreation.
- Promote the effective use of nonworking hours.
- Increase the contact with nature.
- Free people from hazardous, dirty, and harmful jobs.
- Shorten working hours and promote work sharing.
- Expand human contact.
- Harmonize people, things, and the environment ergonomically.[24]

Discussion Case

What Is Your Opinion?

Mountain Trek, Inc., is one of the most successful catalog ordering houses in the western United States. The company employs 115 VDT operators who work in 8-hour shifts. Customers may place orders by telephone or Internet 24 hours a day, 7 days a week. Mountain Trek has taken no steps to prevent VDT-related problems (psychological or physiological). What types of problems may the company begin to experience? How can these problems be prevented? What is your opinion?

Safety Fact

CTDs Are More than Carpal Tunnel Syndrome

The personal computer has become an all-pervasive and universal tool. Jobs from the shop floor to the executive office now involve frequent, repetitive computer use. This means that people in the workplace are typing and clicking at an unprecedented pace. Frequent, and for some, constant computer use has led to an explosion of injuries previously seen mostly in the meat-packing industry. Collectively, these injuries are known as *cumulative trauma disorders (CTDs)*.

Cumulative trauma disorders are caused by forceful or awkward hand movements repeated frequently over time. Other aggravating factors include poor posture, an improperly designed workstation, and job stress. CTDs occur to the muscles, nerves, and tendons of the hands, arms, shoulders, and neck. For years, CTDs have been incorrectly referred to as carpal tunnel syndrome, which is actually just one type of CTD (like referring to all trees as oaks).

It should be obvious from reviewing this list that it was developed with the Japanese culture in mind. However, these recommendations can also be applied in the American workplace. As human workers continually increase the amount of their interaction with automated machines, the potential for maladaptation also increases. One key to preventing maladaptation is to design automated systems around the needs of humans rather than designing them in a vacuum and then expecting humans to adapt. Another is to pay special attention to establishing and maintaining human contact for workers who interact primarily with automated machines and systems. These are areas in which attentive safety and health professionals can have a positive impact.

TECHNOLOGICAL ALIENATION IN THE AUTOMATED WORKPLACE

As technology has become more widely used in the workplace, particularly automated technology, some workers have come to resent its impact on their lives. This concept is known as **technological alienation**. According to Gary Benson of the University of Wyoming, this concept has several meanings, all of which encompass one or more of the following:

- **Powerlessness** is the feeling that workers have when they are not able to control the work environment. Powerless workers may feel that they are less important than the technology with which they work and that they are expendable.

- **Meaninglessness** is the feeling that workers get when their jobs become so specialized and so technology dependent that they cannot see the meaning in their work as it relates to the finished product or service.
- **Normlessness** is the phenomenon in which people working in a highly automated environment can become estranged from society. Normless people lose sight of societies, norms, rules, and mores.[25]

Benson investigated what he considers to be the most devastating form of technological alienation—mindlessness.[26] **Mindlessness** is the result of the process of "**dumbing down**" the workplace. This is a concept that accompanied automation. In the past, machines have been used to do physical work previously done by human workers. With the advent of computers, robots, and automation, machines began doing mental work. According to Benson,

> The net result is jobs and work environments where people do not have to use their minds or think to do their work—an environment where computers, robots, and other forms of high technology do the thinking.[27]

Mindlessness on the job should be of interest to safety and health professionals because of the other problems it can create. According to Benson, these problems can include an increase in alcoholism, drug abuse, employee theft, work-related accidents, absenteeism, sick leave abuse, turnover rates, and employee personal problems. Mindlessness can also lead to a decrease in job performance, productivity, and work quality.[28]

Each of these problems can, in turn, increase the potential for safety and health problems on the job. Employees who abuse alcohol and drugs represent a serious threat in the workplace. Absent employees force their coworkers to double up. High turnover results in a steady influx of inexperienced workers. Employees with personal problems may not be properly focused on accident prevention measures. Finally, when productivity and quality fall, supervisors can feel so much pressure to improve performance that they overlook or put aside safety precautions. These are the potentially negative safety and health effects of mindlessness in the automated workplace.

According to Benson,

> What is needed now are more employees who are aware of and willing to do something about the problem and more research into the phenomenon and causes of the solutions for mindlessness in technological alienation. This is surely a phenomenon of modern-day worklife that must be dealt with effectively—and immediately.[29]

MINIMIZING THE PROBLEMS OF AUTOMATION

The infusion of technology into the workplace has presented safety and health professionals with an entirely new set of challenges. Among the most pressing is the need to identify and minimize the new safety and health problems specifically associated with automation. Behavioral scientist A. B. Cherns developed a **sociotechnical system theory** for doing so that consists of the following components: variance control, boundary location, work group organization, management support, design process, and quality of work life.[30]

Although the sociotechnical system theory was developed in 1977, it has even more relevance now than it did then. According to Yoshio Hayashi of Japan's Keio University,

> The safety and health of workers in this high technology age cannot be discussed within the conventional framework of one worker assigned to one machine. . . . The socio-technical system may be roughly understood if the *man* and *machine* in the man-machine system are replaced by *socio* and *technical* respectively. It refers to a system composed of a work group and high technology.[31]

The various components in the sociotechnical system theory explain what must happen if humans and technological systems are going to work together harmoniously and safely. Safety and health professionals can apply the theory as they work to minimize the potential problems associated with automation in the modern workplace. These components can be summarized as follows:

- **Variance control** involves controlling the unexpected events that can be introduced by new technologies. For example, a runaway, out-of-control industrial robot introduces unexpected safety hazards at variance with the expectations of workers and management. Variance control involves bringing the situation under control and establishing appropriate preventive measures for the future.
- The concept of **boundary location** involves the classification of work. What specific tasks are included in an employee's job description? Does a robot technician just operate the robot, or is she also required to teach and maintain the robot? The accident prevention measures learned by an employee should cover all tasks in his or her job description.
- The concept of **work group organization** involves identifying the tasks that a work group is to perform and how these tasks are to be performed. The key is to make sure that all work group members have the training needed to accomplish effectively and safely all tasks assigned to them.
- **Management support** is perhaps the most important of the components of the sociotechnical system theory. It states that, in the age of high technology, managers must be willing to accept occasional temporary declines in productivity without resorting to shortcuts or improvement efforts that may be unsafe or unhealthy. Management must be willing to emphasize safety in spite of temporary declines in productivity.
- The **design process** component refers to the ability of an organization to design itself in ways that promote productivity, quality, competitiveness, safety, and health. It also involves the ability to continually redesign as technological advances and other circumstances dictate.
- **Quality of work life** involves determining ways to promote the morale and best interests of workers. The key is to ensure that technology extends the

abilities of humans and that technological systems are *human centered*. In other words, it is important to ensure that people control systems rather than vice versa.

- **Ergonomics management program** consists of activities undertaken to prevent ergonomics-related injuries and disorders. Such a program should have at least the following components: work-site analysis, hazard control, health surveillance, and training.[32]

If the sociotechnical system theory is fully applied, the safety and health hazards of the automated workplace can be minimized. Safety and health professionals can play a key role in making sure the theory is applied. To play such a role, these professionals must be technicians, diplomats, trainers, and lobbyists. They must work with the technical aspects of variance control, boundary location, work group organization, and the design process. They must be diplomats in working with supervisors and employees in promoting adherence to safe work practices. They must be trainers in order to ensure that all employees know how to apply safe work practices and appropriate accident prevention techniques. Finally, they must be lobbyists as they continually interact with management to establish and maintain management support for safety, health, and quality of life issues.

Safety Measures for Automated Systems

The sociotechnical system theory discussed in the previous section is broad and conceptual in nature. Modern safety and health professionals also need to know specific measures that can be taken to minimize the hazards associated with robots and other automated systems. Minoru Goto of the Nissan Motor Company's safety department developed specific safety measures in the categories of technological systems, auxiliary equipment, and training.[33]

Examples of safety measures that can be used at the technological systems level include

- Construction of a safety fence around the system that defines the work envelope of the system
- Control of the speed of movement of system components when working inside the work fence
- Installation of an emergency stop device colored red and placed in an easily accessible location
- Location of the control panel for the system outside the safety fence
- Establishment of automatic shutdown switches that activate any time a system component goes beyond its predetermined operational range[34]

Safety measures relating to training include training system operators to work safely within the work envelope and to work together as a team when interacting with the system. Maintenance workers should be trained on the technical aspects of maintaining all machines and equipment that make up the system. This is important because the safety level of the system is the sum of the safety levels of its individual components. A system with four properly operating components and just one faulty component is an unsafe system.[35]

CHALLENGE FOR THE FUTURE

Much more effort has gone into developing automated systems to improve productivity than has gone into the appropriate matching of people and technology. Now, with the speed of technical development being what it is, the safety and health problems associated with automation, particularly stress-related problems, are likely to increase. According to Kensaburo Tsuchiya of Japan's University of Occupational Health and Safety, the challenge for the future is "to create jobs which are free from stress and musculoskeletal overloads while at the same time being challenging and interesting for the individual."[36]

The future holds a number of problems that will have to be addressed to meet Tsuchiya's challenge. The most prominent of these are as follows:

- Increasingly intense international competition may magnify the tendency for companies to neglect safety and health precautions in favor of short-term productivity gains.
- The level of mental stress is likely to increase as the automated manipulation of information forces workers to try continually to handle too much information that is poorly understood.
- Automation and competition are likely to increase the level of anxiety as workers are required to make split-second decisions while knowing that their actions or inactions may have dire consequences.
- New **occupational diseases** relating to mental, visual, and **musculoskeletal problems** may arise whose remedies must be sought through a combination of ergonomics, psychology, occupational medicine, and design.
- There is likely to be increased introduction of robots into the workplace with even less foresight that will, in turn, introduce more unexpected safety and health risks.
- Ignorance may lead to the introduction of automation in an office or factory in forms that do not require workers to think, reason, or make judgments, giving rise to alienation and frustration.
- An aging workforce will continue to raise new issues concerning the special needs of older workers and their interaction with automated technologies.[37]

Because they are likely to face these inhibitors, the safety and health professionals of the future must be prepared to deal with them. They need to know what has to happen if the inhibitors are to be overcome. Tsuchiya suggests the following strategies for enhancing the safety and health of tomorrow's automated workplace:

- Technological systems and processes must be designed to take into account the physical, mental, and emotional needs of human workers.

- Workers will need training and continual retraining so that they can effectively and efficiently operate technological systems and interact with them from the perspective of mastery rather than inadequacy.
- Safety and health professionals, management, workers, psychologists, **ergonomists**, and practitioners of occupational medicine will have to work together as a team in all aspects of the safety and health program.
- The quality of work life and safety and health considerations will have to receive as much attention in the design and implementation of automated systems and processes as do economic and technological concerns.
- Additional research will have to be conducted to determine more clearly the psychological and physiological effects of human interaction with automated technologies.
- Much more comprehensive accident reporting will be needed. Implementation of the "critical incident" reporting system used in commercial aviation may be considered by companies for collecting safety and health data.
- Ergonomists should become involved in accident prevention. They should focus their accident prevention activities on accident and error analysis and simulation of accidents for training purposes. They should also be involved in systems design, human performance, cognitive performance, workload and methods design, the study of factors contributing to accidents and injuries related to equipment, tool, and workstation design, and issues relating to an aging and more diverse workforce.[38]

Tsuchiya sums up his thoughts regarding the future of the automated workplace as it relates to safety and health:

> A motivating, satisfying and good quality job should be consistent with a safe, healthy and efficient automated workplace. Intermediate stages of new technologies sometimes lead to repetitive and monotonous tasks and these must be replanned to minimize adverse reactions to them. Where technology reduces the number of operators, isolation should be avoided for safety and to increase social contacts to reduce stress. When people can work or act together in small groups, even for a short period of the working day, human and productivity advantages can arise (e.g., group discussion of production activities). Where technologies or products are changing rapidly a workforce with good intercommunications can be even more important in keeping productivity high.[39]

SUMMARY

1. The introduction of automation in the workplace has had several different effects on workers including reducing the amount of physical labor required, increasing the amount of mental work required (in some cases), increasing stress levels, and increasing feelings of powerlessness and helplessness.
2. There is growing concern but little solid evidence over the potential negative effects of prolonged VDT use. Safety and health professionals are concerned about various musculoskeletal and visual problems that may be associated with VDT use.

3. Strategies for reducing the physiological and psychological problems associated with VDT use include the following: faster computer response time, more frequent breaks from VDT use or a work rotation schedule that allows non-VDT work to be interspersed in the daily routine, and work design that breaks up continuous VDT use.

4. What makes robots potentially dangerous to humans is their ability to acquire intelligence through programming, their flexibility and range of motion, their speed of movement, and their power.

5. Specific safety and health risks associated with robots include the following: being struck by a moving robot while inside the work envelope, entrapment between a moving robot and another machine, and being struck by a workpiece, tool, or other object dropped or ejected by a robot.

6. There are numerous strategies for minimizing the safety and health hazards of robots and other automated machines. They include ensuring a well-lighted, glare-free robot site, maintaining good housekeeping around the robot site, keeping the robot site free of associated hazards, having fixed covers over the electrical and pneumatic components of the robot, keeping the work envelope clear of all nonessential objects, using appropriate lock-out and test procedures, and removing maintenance tools and supplies from the work envelope before starting the robot.

7. Safety and health problems associated with office automation include eye fatigue, double images, complementary colors, headaches, shoulder fatigue, neck fatigue, dryness in the throat, sleepy feelings, whole-body tiredness, and an unwillingness to talk.

8. Maladaptation to automated technologies manifests itself as an urge to quit work, fatigue, problems with human relations, a drop in work performance, social pathological phenomena such as drug use and crime, mood swings, a loss of motivation, and accidents.

9. Technological alienation is the state of mind that exists when workers resent the impact of new technologies on their lives. It is characterized by feelings of powerlessness, meaninglessness, and normlessness.

10. Mindlessness is the result of the dumbing down of the workplace so that workers are not required to use their minds in their work.

11. Problems associated with mindlessness include an increase in alcoholism, drug abuse, employee theft, work-related accidents, absenteeism, sick leave abuse, turnover rates, and employee personal problems.

12. The sociotechnical system theory consists of the following components: variance control, boundary location, work group organization, management support, design process, and quality of work life.

13. Problems that are likely to be associated with automation in the future include the following: the tendency to overlook safety and health precautions for short-term productivity gains may be exacerbated by increasingly intense international competition; the level of mental stress to which workers are subjected is likely to increase; new occupational diseases are likely to be introduced; and the tendency to dumb down the workplace is likely to continue.

14. Strategies for overcoming anticipated future problems include better design of technological systems; training and continual retraining; teaming safety and health professionals with management, workers, psychologists, ergonomists, and practitioners of occupational medicine; more research; better accident reporting; and the involvement of ergonomists in accident prevention.

KEY TERMS AND CONCEPTS

Accommodation	Musculoskeletal problems
Automation	Neck fatigue
Boundary location	Normlessness
Convergence	Occupational diseases
Design process	Peopleless factory
Dumbing down	Polarizing work
Ergonomics management program	Powerlessness
Ergonomists	Psychosocial stressors
Eye strain	Quality of work life
Fatigue	Shoulder fatigue
Helplessness	Social pathological phenomena
Human–robot interaction	Sociotechnical system theory
Industrial medicine	Teach mode
Islands of automation	Technological alienation
Lacrimation	Variance control
Maladaptation	Video display terminals (VDTs)
Management support	Work design
Manual work	Work envelope
Meaninglessness	Work group organization
Mechanization	Workplace stress
Mindlessness	

REVIEW QUESTIONS

1. Briefly summarize how automation has changed the workplace.
2. List five effects that automation of the workplace has had on workers.
3. What are the safety and health problems most widely associated with VDT use?
4. List three strategies for reducing the psychological and physiological problems associated with VDT use.

5. Explain the four factors that make robots more potentially dangerous than other machines.

6. List and explain the specific hazards associated with human–robot interaction.

7. Explain four specific strategies for minimizing the hazards associated with interacting with a robot.

8. Name six safety and health problems widely associated with office automation.

9. Explain three of the principles set forth by the Japanese Association of Industrial Health for developing safety and health measures for office automation.

10. Define the term *maladaptation*. Explain how it manifests itself in workers.

11. List five strategies for minimizing the potential for occurrences of maladaptation.

12. Define the following automation-related terms: *technological alienation*, *powerlessness*, *meaninglessness*, *normlessness*, and *mindlessness*.

13. Mindlessness in the workplace can lead to a number of other problems. Name five of them.

14. What is an ergonomic management program?

15. Define the term *sociotechnical system theory*. Explain each of its six components.

16. Explain how you would ensure that a new robot system was safe for its operators and other workers.

17. What effect may increasingly intense international competition and the need to improve productivity have on workplace safety and health in the future?

18. What role may you play as a new safety and health professional in meeting the future challenges of an automated workplace?

ENDNOTES

1. O. Ostberg, "Emerging Technology and Trends in Blue-Collar Stress," in *Occupational Health & Safety* 15.

2. Ibid., 17.

3. J. Castelli, "NIOSH Releases Results of VDT Study," *Safety & Health* 143, no. 6: 69.

4. Ibid.

5. Ibid.

6. Business Publisher, Inc., *Occupational Health & Safety Letter* 21, no. 6: 49.

7. M. J. Smith, "An Investigation of Health Complaints and Job Stress in Video Display Operations," *Human Factors* 23: 387–400.

8. S. Kurimoto, et al., "Eye Strain in VDT Work from the Standpoint of Ergophthalmology," in *Occupational Health & Safety*, 112.

9. Ibid., 112–133.

10. Ibid., 133.

11. T. Yamashita, "The Interaction Between Man and Robot in High Technology Industries," in *Occupational Health & Safety*, 140.
12. L. Martensson, "Interaction Between Man and Robots with Some Emphasis on 'Intelligent' Robots," in *Occupational Health & Safety*, 144.
13. National Safety Council, *Robots*, Data Sheet 1–717–85 (Chicago: National Safety Council, 1991), 1.
14. Ibid.
15. Ibid., 2–3.
16. K. Nishiyama, "Introduction and Spread of VDT Work and Its Occupational Health Problems in Japan," in *Occupational Health & Safety*, 251.
17. K. Morooka and S. Yamoda, "Multivariate Analysis of Fatigue on VDT Work," in *Occupational Health & Safety*, 236.
18. B. Amick III, "The Impacts of Office Automation on the Quality of Worklife: Considerations for United States Policy," in *Occupational Health & Safety*, 232.
19. Ibid., 229.
20. Ibid., 223.
21. Ibid., 261–262.
22. M. Oshima, "The Role of Industrial Medicine at the Man–Robot Interface," in *Occupational Health & Safety*, 284–285.
23. Ibid., 284.
24. Ibid., 285.
25. G. Benson, "Mindlessness: A New Dimension of Technological Alienation—Implications for the Man-Machine Interface in High Technology Work Environments," in *Occupational Health & Safety*, 326–327.
26. Ibid., 328.
27. Ibid.
28. Ibid., 332.
29. Ibid., 336.
30. A. B. Cherns, "Can Behavioral Science Help Design Organizations?" *Organizational Dynamics*, 44–64.
31. Y. Hayashi, "Measures for Improving the Occupational Health and Safety of People Working with VDTs or Robots—Small-Group Activities and Safety and Health Education," in *Occupational Health & Safety*, 383.
32. Ibid., 384.
33. M. Goto, "Occupational Safety and Health Measures Taken for the Introduction of Robots in the Automobile Industry," in *Occupational Health & Safety*, 399–417.
34. Ibid., 404–408.
35. Ibid., 411–413.
36. K. Tsuchiya, "Summary Report on the Fifth University of Occupational and Environmental Health International Symposium," in *Occupational Health & Safety*, 422.
37. Ibid., 422–426.
38. Ibid.
39. Ibid., 425–426.

BLOODBORNE PATHOGENS IN THE WORKPLACE

Major Topics

- Facts about AIDS
- Symptoms of AIDS
- AIDS in the Workplace
- Legal Concerns
- AIDS Education
- Counseling Infected Employees
- Easing Employees' Fears about AIDS
- Protecting Employees from AIDS
- Hepatitis B (HBV) and Hepatitis C (HCV) in the Workplace
- OSHA's Standard on Occupational Exposure to Bloodborne Pathogens
- Preventing and Responding to Needlestick Injuries

Acquired immunodeficiency syndrome (AIDS) has become one of the most difficult issues that safety and health professionals are likely to face today. It is critical that they know how to deal properly and appropriately with this controversial disease. The major concerns of safety and health professionals with regard to AIDS are knowing the facts about AIDS; knowing the legal concerns associated with AIDS; knowing their role in AIDS education and related employee counseling; and knowing how to ease unfounded fears concerning the disease while simultaneously taking the appropriate steps to protect employees from infection. In addition to AIDS, the modern safety and health manager must be concerned with other bloodborne pathogens including **human immunodeficiency virus (HIV)** and **hepatitis B (HBV)**.

FACTS ABOUT AIDS

Consider the following facts about HIV and AIDS:

- The total number of new AIDS cases diagnosed annually among men in the United States exceeds 33,000.[1]

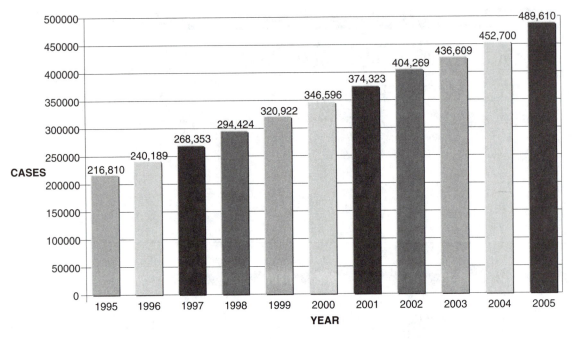

Figure 24–1
Trend for people living with AIDS in the United States.
Source: Centers for Disease Control and Prevention, National Center for HIV, STD, and TB Prevention.

- The total number of new AIDS cases diagnosed annually among women in the United States exceeds 10,000.[2]
- The estimated cumulative number of people in the United States living with AIDS has risen steadily over the years (see Figure 24–1). The increase over the years is attributed in large measure to declines in the death rate due to advances in medical technology and treatments.[3]

AIDS is feared, misunderstood, and controversial. Modern safety and health professionals need to know the facts about AIDS and be prepared to use these facts to make the workplace safer.

SYMPTOMS OF AIDS

AIDS and various related conditions are caused when humans become infected with the human immunodeficiency virus (HIV). This virus attacks the human immune system, rendering the body incapable of repelling disease-causing microorganisms. Symptoms of the onset of AIDS are

- Enlarged lymph nodes that persist
- Persistent fevers
- Involuntary weight loss

- Fatigue
- Diarrhea that does not respond to standard medications
- Purplish spots or blotches on the skin or in the mouth
- White, cheesy coating on the tongue
- Night sweats
- Forgetfulness

How AIDS Is Transmitted

The HIV virus is transmitted in any of the following three ways: (1) sexual contact, (2) blood contact, and (3) mother-to-child during pregnancy or childbirth. Any act in which body fluids are exchanged can result in infection if either partner is infected. The following groups of people are at the highest level of risk with regard to AIDS: (1) homosexual men who do not take appropriate precautions; (2) IV drug users who share needles; (3) people with a history of multiple blood transfusions or blood-product transfusions, such as hemophiliacs; and (4) sexually promiscuous people who do not take appropriate precautions.

How AIDS Is *Not* Transmitted

There is a great deal of misunderstanding about how AIDS is transmitted. This can cause inordinate fear among fellow employees of HIV-positive workers. Safety and health professionals should know enough about AIDS transmission so that they can reduce employees' fears about being infected through casual contact with an HIV-positive person.

Occupational Health & Safety magazine provides the following clarifications concerning how AIDS is *not* transmitted:

> AIDS is a blood-borne, primarily sexually transmitted disease. It is not spread by casual social contact in schools, workplaces, public washrooms, or restaurants. It is not spread via handshakes, social kissing, coughs, sneezes, drinking fountains, swimming pools, toilet facilities, eating utensils, office equipment, or by being next to an infected person.
>
> No cases of AIDS have been reported from food being either handled or served by an infected person in an eating establishment.
>
> AIDS is not spread by giving blood. New needles and transfusion equipment are used for every donor.
>
> AIDS is not spread by mosquitoes or other insects.
>
> AIDS is not spread by sexual contact between uninfected individuals—whether homosexual or heterosexual—if an exclusive sexual relation has been maintained.[4]

AIDS IN THE WORKPLACE

The first step in dealing with AIDS at the company level is to develop a comprehensive AIDS policy. Safety and health professionals should be part of the team that drafts the initial policy and updates an existing policy. If a company has no

Figure 24–2
Components of a corporate
policy for bloodborne pathogens.

- Employee rights
- Testing
- Education

AIDS policy, the safety and health professional should encourage the company to develop one. In all likelihood, most companies won't take much convincing.

According to Peter Minetos,

> Industry is doing its part to eliminate any unnecessary fear: Nearly half of companies surveyed offer their employees literature or other materials to keep them informed on the disease; more than half have an Employee Assistance Program (EAP) to deal with emotional problems concerning AIDS; two-thirds of those who have not yet addressed AIDS with employees plan to do so in the future.[5]

AIDS is having a widely felt impact in the workplace, particularly on employers. According to Minetos, employers are feeling the impact of AIDS in increased insurance premiums and health care costs, time-on-the-job losses, decreased productivity, AIDS-related lawsuits, increased stress, and related problems that result from misconceptions about AIDS.[6]

The starting point for dealing with AIDS in the workplace is the development of a company policy that covers AIDS and other bloodborne pathogens. The policy should cover the following areas at a minimum: employee rights, testing, and education (Figure 24–2).

Employee Rights

An AIDS policy should begin by spelling out the rights of employees who have tested positive for the disease. The following recommendations can be used for developing the **employee rights** aspects of an AIDS policy:

- Treat HIV-positive employees compassionately, allowing them to work as long as they are able to perform their jobs.
- Develop your company's AIDS policy and accompanying program before learning that an employee is HIV positive. This allows the company to act instead of having to react.
- Make reasonable allowances to accommodate the HIV-positive employee. The U.S. Supreme Court has recognized AIDS as a disabling condition. Consequently, reasonable allowances must include modified work schedules and special adaptations to the work environment.
- Ensure that HIV-positive employees have access to private health insurance that covers the effects of AIDS. Also, work with state and federal government insurance providers to gain their support in helping cover the costs of health care for HIV-positive employees.
- Include provisions for evaluating the work skills of employees to determine if there has been any degradation of ability caused by the disease.[7]

Testing

According to the Centers for Disease Control and Prevention (CDC), there is no single test that can reliably diagnose AIDS.[8] However, there is a test that can detect antibodies produced in the blood to fight the virus that causes AIDS. The presence of these antibodies does not necessarily mean that a person has AIDS, as the CDC states:

> Presence of HTLV-III antibodies [now called HIV antibodies] means that a person has been infected with that virus. . . . The antibody test is used to screen donated blood and plasma and assist in preventing cases of AIDS resulting from blood transfusions or use of blood products, such as Factor VIII, needed by patients with hemophilia. For people who think they may be infected and want to know their health status, the test is available through private physicians, most state or local health departments and at other sites. Anyone who tests positive should be considered potentially capable of spreading the virus to others.[9]

Whether a company can, or even should, require AIDS tests of employees or potential employees is widely debated. The issue is contentious and controversial. However, there is a growing body of support for mandatory testing. The legal and ethical concerns surrounding this issue are covered in the next section. The testing component of a company's AIDS policy should take these concerns into account.

Education

The general public is becoming more sophisticated about AIDS. People are beginning to learn how AIDS is transmitted. However, research into the causes, diagnosis, treatment, and prevention of this disease is ongoing. The body of knowledge changes continually. Consequently, it is important to have an ongoing education program to keep employees up-to-date and knowledgeable. According to Minetos,

> AIDS education campaigns can be conducted in many forms. Literature, slide shows, and video presentations are all communication vehicles. Presentations by health professionals are one of the most popular and effective methods of communicating information on AIDS. The primary purpose of each is to convey basic knowledge and, subsequently, eliminate unnecessary fear among co-workers.[10]

Once a comprehensive AIDS policy has been developed and shared with all employees, a company has taken the appropriate and rational approach for dealing with this deadly and controversial disease. If an employer has not yet taken this critical step, safety and health professionals should encourage such action immediately. It is likely that most companies either employ now, or will employ in the future, HIV-positive employees. A poll conducted by *U.S. News and World Report* found that 48 percent of the companies responding indicated that AIDS was a concern.[11]

Minetos writes, "AIDS is having an effect on the workplace. Yet . . . only 5 percent of all employers have a written corporate policy on AIDS. For a disease that according to some estimates is reaching epidemic proportions, this is an extremely low percentage."[12] Clearly, one of the major challenges facing safety and health professionals is convincing their employers to develop an AIDS policy.

LEGAL CONCERNS

There are legal considerations relating to AIDS in the workplace with which safety and health professionals should be familiar. These grow out of several pieces of federal legislation, including the Rehabilitation Act, the Occupational Safety and Health Act, and the Employee Retirement Income System Act.

The **Rehabilitation Act** was enacted in 1973 to give protection to people with disabilities (then referred to as handicaps), including workers. Section 504 of the act makes discrimination on the basis of a disability unlawful. Any agency, organization, or company that receives federal funding falls within the purview of the act. Such entities may not discriminate against individuals who have disabilities but are otherwise qualified. Through various court actions, this concept has been well defined. A person with a disability is "otherwise qualified" when he or she can perform what the courts have described as the **essential functions** of the job.

When the disability that a worker has is a contagious disease such as AIDS, it must be shown that there is no significant risk of the disease being transmitted in the workplace. If there is a significant risk, the infected worker is not considered "otherwise qualified." Employers and the courts must make these determinations on a case-by-case basis.

Another concept associated with the Rehabilitation Act is the concept of **reasonable accommodation**. In determining if a worker with a disability can perform the essential functions of a job, employers are required to make reasonable accommodations to help the worker. This concept applies to workers with any type of disabling condition, including a communicable disease such as AIDS. What constitutes "reasonable accommodation," just as what constitutes "otherwise qualified," must be determined on a case-by-case basis.

The concepts growing out of the Rehabilitation Act give the supervisor added importance when dealing with AIDS-infected employees. The supervisor's knowledge of the various jobs in his or her unit is essential in helping company officials make an otherwise qualified decision. The supervisor's knowledge that AIDS is transmitted only by exchange of body fluids coupled with his or her knowledge of the job tasks in question is helpful in determining the likelihood that AIDS may be transmitted to other employees. Finally, the supervisor's knowledge of the job tasks in question is essential in determining what constitutes reasonable accommodation and what the actual accommodations should be. Therefore, it is critical that safety and health professionals work closely with supervisors and educate them in dealing with AIDS in the workplace.

In arriving at what constitutes reasonable accommodation, employers are not required to make fundamental changes that alter the nature of the job or result in undue costs or administrative burdens. Clearly, good judgment and a thorough knowledge of the job are required when attempting to make reasonable accommodations for an AIDS-infected employee. Safety and health professionals should involve supervisors in making such judgments.

The **Occupational Safety and Health Act (OSH Act)** requires that employers provide a safe workplace free of hazards. The act also prohibits employers from retaliating against employees who refuse to work in an environment they believe may be unhealthy (Section 654). This poses a special problem for employers of AIDS-infected employees. Other employees may attempt to use Section 654 of the OSH Act as the basis for refusing to work with such employees. For this reason, it is important that companies educate their employees about AIDS and how it is transmitted. If employees know how AIDS is transmitted, they will be less likely to exhibit an irrational fear of working with an infected colleague.

Even when a comprehensive AIDS education program is provided, employers should not automatically assume that an employee's fear of working with the infected individual is irrational. Employers have an obligation to treat each case individually. Does the complaining employee have a physical condition that puts him or her at greater risk of contracting AIDS than other employees? If so, that employee's fears may not be irrational. However, a fear of working with an AIDS-infected coworker is usually irrational, making it unlikely that Section 654 of the OSH Act could be used successfully as the basis of a refusal to work.

The **Employee Retirement Income Security Act (ERISA)** protects the benefits of employees by prohibiting actions taken against them based on their eligibility for benefits. This means that employers covered by ERISA cannot terminate an employee with AIDS or who is suspected of having AIDS as a way of avoiding expensive medical costs. With ERISA, it is irrelevant whether the employee's condition is considered a disability because the act applies to all employees regardless of condition.

Discussion Case

What Is Your Opinion?

J. B. Knotts isn'st sure what to do. As safety and health director for Precision Machining and Manufacturing, he is concerned about protecting employees from workplace hazards. On the other hand, he is also concerned about protecting the privacy of his employees. Herein lies the dilemma that he now faces: He has just learned that Andrea Cobb, a lathe operator, is HIV positive. Cuts and abrasions from metal chips are common among lathe operators and other machinists. In fact, the machining supervisor had to administer minor first aid to a machine operator less than an hour ago. What if Andrea Cobb is cut or scraped? What about the blood? Knotts wonders if he should warn the other employees about Cobb's condition. What is your opinion?

Testing Issue

Perhaps the most contentious legal concern growing out of the AIDS controversy is the issue of testing. Writing in the *AAOHN Journal*, Beatrice Crofts Yorker says,

> Few topics have generated the amount of controversy that currently exists in the area of testing for Acquired Immune Deficiency Syndrome (AIDS). Proponents and opponents have strong arguments, often based on emotional reactions to this deadly epidemic. In the workplace, the issues of AIDS testing are very specific and have implications for health policies in occupational settings. Few clear laws or statutes specifically regulate AIDS testing in the workplace.[13]

Issues regarding testing for AIDS and other diseases with which safety and health professionals should be familiar are summarized as follows (see Figure 24–3):

1. *State laws* Control of communicable diseases is typically considered to be the province of the individual state. In response to the AIDS epidemic, several states have passed legislation. Some states prohibit the use of preemployment AIDS tests to deny employment to infected individuals. Because of the differences among states with regard to AIDS-related legislation, safety and health professionals should familiarize themselves with the laws of the state in which their company is located.

2. *Federal laws and regulations* The laws protecting an individual's right to privacy and due process apply to AIDS testing. These laws fall within the realm of constitutional law. They represent the primary federal contribution to the testing issue.

3. *Civil suits* **Case law** serves the purpose of establishing precedents that can guide future decisions. One precedent-setting case was taken all the way to the Supreme Court (*School Board of Nassau County v. Arline*, 1987), where it was decided that an employer cannot discriminate against an employee who has a communicable disease.

4. *Company policy* It was stated earlier that companies should have an AIDS policy that contains a testing component. This component should include at least the following: a strong rationale, procedures to be followed, employee groups

---

Health and safety professionals should be familiar with how the following factors might affect the issue of AIDS testing at their companies:

- Applicable state laws
- Applicable federal laws and regulations
- Case law from civil suits
- Company policy

Figure 24–3
Disease testing issues.

to be tested, the use and dissemination of results, and the circumstances under which testing will be done. Safety and health professionals should be knowledgeable about their company's policy and act in strict accordance with it.[14]

On one side of the testing controversy are the issues of fairness, accuracy, and confidentiality or, in short, the rights of the individual. On the other side of the controversy are the issues of workplace safety and public health. Individual rights' proponents ask such questions as: What tests will be used? How do test results relate to the maintenance of workplace safety? How will test results be used, and who will see them? Workplace safety proponents ask such questions as: What is the danger of transmitting the disease to other employees? Can the safety of other workers be guaranteed?

Bayer, Levine, and Wolf recommend the following guidelines for establishing testing programs that satisfy the concerns of both sides of the issue:

- The purpose of screening must be ethically acceptable.
- The means to be used in the screening program and the intended use of the information must be appropriate for accomplishing the purpose.
- High-quality laboratory services must be used.
- Individuals must be notified that the screening will take place.
- Individuals who are screened must have a right to be informed of the results.
- Sensitive and supportive counseling programs must be available before and after screening to interpret the results, whether they be positive or negative.
- The confidentiality of screened individuals must be protected.[15]

Facts about Testing for AIDS and Other Diseases

In addition to understanding the legal concerns associated with disease testing, safety and health professionals should also be familiar with the latest facts about AIDS tests and testing. A concerned employee may ask for recommendations concerning AIDS testing.

Ensuring the accuracy of an **HIV antibody test** (there is no such thing as an AIDS test) requires two different tests, one for initial screening and one for confirmation. The screening test currently used is the **enzyme linked immunosorbent assay (ELISA)** test. The confirmation test is the **immuno-florescent (IFA)**, or the **Western Blot test**. The ELISA test is relatively accurate, but it is susceptible to both false positive and false negative results. A **false positive** test is one that shows the presence of HIV antibodies when no such antibodies exist. A **false negative** result is one that shows no HIV antibodies in people who actually are infected. A negative result indicates that no infection exists at the time of the test. A confirmed positive result indicates that HIV antibodies are present in the blood.[16]

The American College Health Association makes the following recommendations concerning the HIV antibody test:

- The test is not a test for AIDS, but a test for antibodies to HIV, the virus that can cause AIDS.
- Talk to a trained, experienced, sensitive counselor before deciding whether to be tested.

- If you decide to be tested, do so *only* at a center that provides both pre- and posttest counseling.
- If possible, use an *anonymous* testing center.
- Be sure that the testing center uses two ELISA tests and a Western Blot or IFA test to confirm a positive result.
- A positive test result is *not* a diagnosis of AIDS. It does mean you have HIV infection and that you should seek medical evaluation and early treatment.
- A positive test result *does* mean that you can infect others and that you should avoid risky or unsafe sexual contact and IV needle sharing.
- It can take six months (and—although rarely—sometimes even longer) after infection to develop antibodies, so the test result may not indicate whether you have been infected during that period.
- A negative test result *does not* mean that you are immune to HIV or AIDS, or that you cannot be infected in the future.[17]

Safety and health professionals need to share this type of information with employees who ask about AIDS tests and testing. Employees who need more detailed information should be referred to a health care professional.

AIDS EDUCATION

The public is becoming more knowledgeable about AIDS and how the disease is spread. However, this is a slow process, and AIDS is a complex and controversial disease. Unfortunately, many people still respond to the disease out of ignorance and inaccurate information. For this reason, it is imperative that a company's safety and health program include an **AIDS education program**.

A well-planned AIDS education program can serve several purposes: (1) It can give management the facts needed to develop policy and make informed decisions with regard to AIDS; (2) it can result in changes in behavior that will make employees less likely to contract or spread the disease; (3) it can prepare management and employees to respond appropriately when a worker falls victim to the disease; and (4) it can decrease the likelihood of legal problems resulting from an inappropriate response to an AIDS-related issue. Consequently, safety and health professionals should be prepared to participate in developing AIDS education programs.

Planning an AIDS Education Program

The first step in planning an AIDS education program is to decide its purpose. A statement of purpose for an AIDS education program should be a broad conceptual declaration that captures the company's reason for providing the education program. Following is an example:

> The purpose of this AIDS education program is to deal with the disease in a positive proactive manner that is in the best interests of the company and its employees.

The next step in the planning process involves developing goals that translate the statement of purpose into more specific terms. The goals should tell specifically what the AIDS education program will do. Sample goals are as follows:

- The program will change employee behaviors that may otherwise promote the spread of AIDS.
- The program will help the company's management team develop a rational, appropriate AIDS policy.
- The program will help managers make responsible decisions concerning AIDS issues.
- The program will help employees protect themselves from the transmission of AIDS.
- The program will alleviate the fears of employees concerning working with an AIDS-infected coworker.
- The program will help managers respond appropriately and humanely to the needs of AIDS-infected workers.

Once goals have been set, a program is developed to meet the goals. The various components of the program must be determined. These components may include confidential one-on-one counseling, referral, posters, a newsletter, classroom instruction, self-paced multimedia instruction, group discussion sessions, printed materials, or a number of other approaches. Figure 24–4 shows a suggested outline for a course on AIDS.

Statement of Purpose

The purpose of this course is to give employees the knowledge they need to deal with AIDS in a positive, proactive manner.

Major Topics

- What is AIDS?
- What causes AIDS?
- How is AIDS transmitted?
- Who is most likely to get AIDS?
- What are the symptoms of AIDS?
- How is AIDS diagnosed?
- Who should be tested for AIDS?
- Where can I get an AIDS test?
- How can I reduce my chances of contracting AIDS?
- How is AIDS treated?
- Can AIDS be prevented?
- What are common myths about AIDS?

Figure 24–4
Outline for AIDS education course.

COUNSELING INFECTED EMPLOYEES

The employee who learns that he or she has AIDS will be angry, frightened, and confused. Safety and health professionals who are faced with such an employee should proceed as follows:

- Listen.
- Maintain a nonjudgmental attitude.
- Make the employee aware of the company's policy on AIDS.
- Respond in accordance with company policy.

Listen carefully as you would with an employee who comes to you with any problem. If you must ask a question for clarification, do so, but be objective, professional, and nonjudgmental. Make the employee aware of the company's policy on AIDS and respond in strict accordance with the policy.

The U.S. Public Health Service recommends the following steps for persons who have determined they are HIV positive.

- Seek regular medical evaluation and follow-up.
- Either avoid sexual activity or inform your prospective partner of your antibody test results and protect him or her from contact with your body fluids during sex. (Body fluids include blood, semen, urine, feces, saliva, and women's genital secretions.) Use a condom, and avoid practices that may injure body tissues (for example, anal intercourse). Avoid oral-genital contact and open-mouthed intimate kissing.
- Inform your present and previous sex partners, and any persons with whom needles may have been shared, of their potential exposure to HTLV-III [HIV] and encourage them to seek counseling and antibody testing from their physician or at appropriate health clinics.
- Don't share toothbrushes, razors, or other items that could become contaminated with blood.
- If you use drugs, enroll in a drug treatment program. Needles and other drug equipment must never be shared.
- Don't donate blood, plasma, body organs, other body tissue, or sperm.
- Clean blood or other body fluid spills on household or other surfaces with freshly diluted household bleach—1 part bleach to 10 parts water. (Don't use bleach on wounds.)
- Inform your doctor, dentist, and eye doctor of your positive HTLV-III status so that proper precautions can be taken to protect you and others.
- Women with a positive antibody test should avoid pregnancy until more is known about the risks of transmitting HTLV-III from mother to infant.[18]

Safety and health professionals can pass along these recommendations to employees who have contracted the virus. Any information that is requested beyond this should be provided by a qualified professional. Safety and health professionals should be prepared to make appropriate referrals. The Public Health Service's AIDS hotline number is 1-800-342-AIDS.

Employee Assistance Programs

Company-sponsored **employee assistance programs (EAPs)** should have an AIDS component so that employees can seek confidential advice and counseling about the disease. EAPs may provide on-site services or contract for them through a private organization or agency. In either case, confidential counseling, referral, the provision of AIDS-related information, seminars, and other forms of assistance should be provided. From an employee assistance perspective, AIDS should be treated like other health problems such as stress, depression, substance abuse, and so on.[19]

EASING EMPLOYEES' FEARS ABOUT AIDS

During his first inaugural address on March 4, 1933, in the depths of the Great Depression, President Franklin D. Roosevelt said, "So first let me assert my firm belief that the only thing we have to fear is fear itself."[20] This is a message that safety and health professionals should be prepared to spread among employees. Although the level of sophistication concerning AIDS is increasing, people tend to react to the disease from an emotional perspective. Typically, their fears are unfounded and are more likely to cause them problems than does AIDS.

Fear, panic, and even hysteria are all common reactions to AIDS. Fellow employees are likely to respond this way when they discover that a coworker has AIDS. Consequently, safety and health professionals need to know how to ease the fears and misconceptions associated with AIDS. According to Brown and Turner, the following strategies will help:

- Work with higher management to establish an AIDS education and awareness program that covers the following topics at a minimum: (1) how HIV is transmitted, (2) precautions that workers can take, and (3) concerns about AIDS testing.
- Conduct group round-table discussions that allow employees to express their concerns.
- Correct inaccuracies, rumors, and misinformation about AIDS as soon as they occur.[21]

PROTECTING EMPLOYEES FROM AIDS

Safety and health professionals should be familiar with the precautions that will protect employees from HIV infection on and off the job. OSHA's guidelines for preventing exposure to HIV infection identify three categories of workrelated tasks: Categories I, II, and III. Jobs that fall into Category I involve routine exposure to blood, body fluids, or tissues that may be HIV infected. Category II jobs do not involve routine exposure to blood, body fluids, or tissues, but some aspects of the job may involve occasionally performing Category I tasks. Category III jobs do not normally involve exposure to blood, body fluids, or tissues.

Most industrial occupations fall into Category III, meaning there is very little risk of contracting AIDS on the job. However, regardless of the category of their job, employees should know how to protect themselves, and safety and health professionals should be prepared to tell them how.

The U.S. Public Health Service recommends the following precautions for reducing the chances of contracting AIDS:

- Abstain from sex or have a mutually monogamous marriage or relationship with an infection-free partner.
- Refrain from having sex with multiple partners or with a person who has multiple partners. The more partners you have, the greater the risk of infection.
- Avoid sex with a person who has AIDS or whom you think may be infected. However, if you choose not to take this recommendation, the next logical course is to take precautions against contact with the infected person's body fluids (blood, semen, urine, feces, saliva, and female genital secretions).
- Do not use intravenous drugs or, if you do, do not share needles.[22]

Safety and health professionals should make sure that all employees are aware of these common precautions. They should be included in the company-sponsored AIDS education program, available through the company's EAP, and posted conspicuously for employee reading.

CPR and AIDS

It is not uncommon for an employee to be injured in a way that requires resuscitation. Consequently, many companies provide employees with CPR training. But what about AIDS? Is CPR training safe? Writing for *Safety & Health*, Martin Eastman said,

> In the early 1960s when CPR training procedures were being developed, some thought was given to the cleaning and disinfection of manikins. This was before acquired immune deficiency syndrome (AIDS) and hepatitis-B became headline diseases. In the fearful climate that has followed the spread of these diseases, practices that once appeared merely unsanitary now seem truly life-threatening.[23]

The HIV virus has been found in human saliva. Because CPR involves using your fingers to clear the airway and placing your mouth over the victim's, there is concern about contracting AIDS while trying to resuscitate someone. Although there is no hard evidence that HIV can be transmitted through saliva, there is some legitimacy to the concern.

Because of the concern, disposable face masks and various other types of personal protective devices are now being manufactured. Typical of these devices is the rescue key ring produced by Ambu, Inc. of Hanover, Maryland.[24]

The ring contains a face mask made of a transparent film material that has a oneway valve. The valve prevents the passage of potentially contaminated body fluids from the victim to the rescuer.

Safety and health professionals should ensure that such devices are used in both training and live situations involving CPR. These devices should be readily available in many easily accessible locations throughout the company.

Safety Needles

Employees who work with needles for taking blood, giving injections, or inserting intravenous systems should be considered at high risk for becoming infected with HIV. Whether inserting an IV, taking blood, or giving a vaccination, medical personnel live with the scary reality of bloodborne pathogens exposure every minute of the day. Studies consistently have shown that 50 to 85 percent of health care workers who have contracted a bloodborne disease cannot identify when or how they were exposed to bloodborne pathogens.

In addition, whereas more than one million accidental needlesticks are reported each year, it is estimated that 66 percent of accidental needlesticks go unreported. With the use of injury prevention devices, the Centers for Disease Control and Prevention estimates that the number of accidental needlesticks could be reduced by 76 percent. Not all devices that are called "safe" are safer. Injury prevention devices on syringes, IVs, catheters, blood-drawing equipment, vaccination instruments, lancets, and scalpels must be one-handed to be considered safer. Some sheathing devices require two-handed operation and often result in an increase in needlestick injuries.

Ellis writes in *Occupational Health & Safety*, "OSHA currently addresses safer needles in the Bloodborne Pathogen Standard through the Engineering and Work Practice Controls clause [1910.1030, Section d(2)(i)]. Although this clause does not specifically require the use of safer needles, inspection guidelines state that the compliance officer shall determine whether engineering controls are used. Most preferable is the use of devices which offer an alternative to needles being used to perform the procedure. Examples of such devices include . . . needle-protected systems or needleless systems."[25] Figure 24–5 contains examples of syringes with safety features.

HEPATITIS B (HBV) AND HEPATITIS C (HCV) IN THE WORKPLACE

Hepatitis B (HBV)

Although the spread of HIV receives more attention, a greater risk is from the spread of **hepatitis B (HBV)**. This bloodborne virus averages approximately 300,000 new cases per year compared with AIDS, which averages approximately 43,000 new cases. The hepatitis B virus is extremely strong compared with HIV. For example, it can live on surfaces for up to a week if it is exposed to air. Hepatitis B is also much more concentrated than HIV.

Plastic shield slides over the
needle and locks to encase
the exposed point

Syringe with protective shield

With an extra push on
the plunger...

...the needle retracts into
the syringe

Syringe with retractable needle

Figure 24–5
Examples of syringes with safety features.

Hepatitis B is caused by a double-shelled virus. It can be transmitted in the workplace in the following ways:

- Contact with blood
- Contact with body fluids including tears, saliva, and semen

The hepatitis B virus can live in body fluids for years. Carriers of the virus are at risk themselves, and they place others at risk. Persons infected with HBV may contract chronic hepatitis, cirrhosis of the liver, and primary heptocellular carcinoma. An HBV-infected individual is more than 300 times more likely to develop primary liver cancer than is a noninfected individual from the same

Safety Fact

Employees at Greatest Risk from Viral Hepatitis

Employees who are at the greatest risk of necrosis of liver cells from viral hepatitis are those who are elderly and those who have diabetes mellitus, cancer, or any other illness severe enough to require surgery and transfusions.

environment. Unfortunately, it is possible to be infected and not know it because the symptoms can vary so widely from person to person.

The symptoms of hepatitis B are varied but include

- Jaundice
- Joint pain
- Rash
- Internal bleeding

HBV Vaccination

The OSHA standard covering bloodborne pathogens requires employers to offer the three-injection hepatitis B vaccination series free to all employees who are exposed to blood or other potentially infectious materials as part of their job duties. This includes health care workers, emergency responders, morticians, first-aid personnel, law enforcement officers, correctional facilities staff, and launderers, as well as others. The vaccination must be offered within 10 days of a person's initial assignment to a job where exposure to blood or other potentially infectious materials can be "reasonably anticipated."

The hepatitis B vaccination is a noninfectious, yeast-based vaccine given in three injections in the arm. It is prepared from recombinant yeast cultures, rather than human blood or plasma. Thus, there is no risk of contamination from other bloodborne pathogens, nor is there any chance of developing HBV from the vaccine.

The second injection should be given one month after the first, and the third injection six months after the initial dose. More than 90 percent of those vaccinated develop immunity to the hepatitis B virus. To ensure immunity, it is important for individuals to receive all three injections. At this point it is unclear how long the immunity lasts, so booster shots may be required at some point in the future.

The vaccine causes no harm to those who are already immune or to those who may be HBV carriers. Although employees may opt to have their blood tested for antibodies to determine the need for the vaccine, employers may not make such screening a condition of receiving the vaccination, nor are employers required to provide prescreening.

Each employee should receive counseling from a health care professional when the vaccination is offered. This discussion will help an employee determine whether inoculation is necessary.

Hepatitis C (HCV)

Hepatitis C virus (HCV) infection is the most common chronic bloodborne infection in the United States, with approximately 36,000 new cases diagnosed every year.[26] Most of these persons are chronically infected and may be unaware of their infection because they are not clinically ill. Infected persons serve as a source of transmission to others and are at risk for chronic liver disease or other HCV-related chronic diseases during the first two or more decades following the initial infection.

Chronic liver disease is the 10th leading cause of death among adults in the United States, and accounts for approximately 25,000 deaths annually, or approximately 1 percent of all deaths. Population-based studies indicate that 40 percent of chronic liver disease is HCV related, resulting in an estimated 8,000 to 10,000 deaths each year. Because most HCV-infected persons are ages 30 to 49, the number of deaths attributable to HCV-related chronic liver disease may increase substantially during the next 10 to 20 years as this group of infected persons reaches the age at which complications from chronic liver disease typically occur.

HCV is transmitted primarily through large or repeated direct percutaneous exposures to blood. In the United States, the two most common exposures associated with transmission of HCV are blood transfusion and injecting-drug use. Figure 24–6 summarizes tests that are used to detect HCV infection.

Tests for Hepatitis C Virus (HCV) Infection

Test	Application
Hepatitis C virus antibody (anti-HCV) • EIA (enzyme immunoassay) • Supplemental assay	• Indicates past or present infection, but does not differentiate between acute, chronic, or resolved infection • All positive EIA results should be verified with a supplemental assay
HCA RNA (hepatitis C virus ribonucleic acid) **Qualitative tests** • Reverse transcriptase polymerase chain reaction (RT-PCR) amplification of HCV RNA by in-house or commercial assays	• Detect presence of circulating HCV RNA • Monitor patients on antiviral therapy
Quantitative tests • RT-PCR amplification of HCV RNA by in-house or commercial assays • Branched chain DNA (bDNA) assays	• Determine concentration of HCV RNA • Might be useful for assessing the likelihood of response to antiviral therapy
Genotype • Several methodologies available (e.g., hybridization, sequencing)	• Group isolates of HCV based on generic differences, into 6 genotypes and 90 subtypes • With new therapies, length of treatment might vary based on genotype
Serotype • EIA based on immunoreactivity to synthetic peptides	• No clinical utility

---

Figure 24–6

Various tests available for detecting the presence of HCV.

Source: Centers for Disease Control and Prevention.

Preventing and controlling HCV requires a comprehensive strategy comprised of at least the following activities:

- Primary prevention activities
    Screening and testing blood, plasma, organ, tissue, and semen donors
    Virus inactivation of plasma-derived products
    Risk-reduction counseling and services
    Implementation and maintenance of infection control practices
- Secondary prevention activities
    Identification, counseling, and testing of persons at risk
    Medical management of infected persons
- Education and training
- Monitoring the effectiveness of prevention activities to develop improved prevention methods

The next section explains OSHA's standard on occupational exposure to bloodborne pathogens. This standard applies to all bloodborne pathogens including HBV and HIV.

OSHA'S STANDARD ON OCCUPATIONAL EXPOSURE TO BLOODBORNE PATHOGENS

OSHA's standard on occupational exposure to bloodborne pathogens is contained in 29 CFR Part 1910.1030. The purpose of the standard is to limit the exposure of personnel to blood and to serve as a precaution against bloodborne pathogens that may cause diseases.

Scope of Application

This standard applies to all employees whose job duties may bring them in contact with blood or other potentially infectious material. There is no attempt on OSHA's part to list all occupations to which 1910.1030 applies. The deciding factor is the *reasonably anticipated* theory. If it can be reasonably anticipated that employees may come in contact with blood in the normal course of performing their job duties, the standard applies. It does not apply in instances of *good Samaritan* acts in which one employee attempts to assist another employee who is bleeding.

The standard covers blood and other infectious materials. These other materials include

- Semen
- Vaginal secretions
- Cerebrospinal fluid
- Synovial fluid
- Pleural fluid

- Peritoneal fluid
- Amniotic fluid
- Saliva
- Miscellaneous body fluids mixed with blood

In addition to these fluids, other potentially infectious materials include

- Unfixed human tissue, or organs other than intact skin
- Cell or tissue cultures
- Organ cultures
- Any medium contaminated by human immunodeficiency virus (HIV) or hepatitis B (HBV)

Exposure Control Plan

OSHA's 1910.1030 requires employers to have an **exposure control plan** to protect employees from exposure to bloodborne pathogens. It is recommended that such plans have at least the following major components:

Part 1 Administration

Part 2 Methodology

Part 3 Vaccinations

Part 4 Postexposure investigation and follow-up

Part 5 Labels and signs

Part 6 Information and training

Administration

This component of the plan should clearly define the responsibilities of employees, supervisors, and managers regarding exposure control. It should also designate an exposure control officer (usually the organization's safety and health manager or a person who reports to this manager). The organization's exposure control plan must be readily available to all employees, and the administration component of it must contain a list of locations where copies of the plan can be examined by employees. It should also describe the responsibilities of applicable constituent groups and individuals. These responsibilities are summarized as follows:

- *Employees.* All employees are responsible for the following: knowing which of their individual and specific job tasks may expose them to bloodborne pathogens; participating in training provided concerning bloodborne pathogens; carrying out all job duties in accordance with the organization's various control procedures; and practicing good personal hygiene.
- *Supervisors and managers.* Supervisors and managers are responsible for coordinating with the exposure control officer to implement and monitor exposure control procedures in their areas of responsibility.

- *Exposure control officer.* This individual is assigned overall responsibility for carrying out the organization's exposure control plan. In addition to the administrative duties associated with it, this position is also responsible for employee training. In small organizations, exposure control officers may double as the training coordinator. In larger organizations, a separate training coordinator may be assigned exposure control as one more training responsibility.

In either case, the exposure control officer is responsible for the following duties: overall development and implementation of the exposure control plan; working with management to develop other exposure-related policies; monitoring and updating the plan; keeping up-to-date with the latest legal requirements relating to exposure; acting as liaison with OSHA inspectors; maintaining up-to-date training files and documentation showing training; developing the needed training program; working with other managers to establish appropriate control procedures; establishing a hepatitis B vaccination program as appropriate; establishing a postexposure evaluation and follow-up system; displaying labels and signs as appropriate; and maintaining up-to-date, confidential medical records of exposed employees.

Methodology

This section of the plan describes the procedures established to protect employees from exposure. These procedures fall into one of the following five categories:

- General precautions
- Engineering controls
- Work practice controls
- Personal protection equipment
- Housekeeping controls

General precautions include such procedures as assuming that all body fluids are contaminated, and acting accordingly. Engineering controls are design and technological precautions that protect employees from exposure. Examples of engineering controls include self-sheathing needles, readily accessible handwashing stations equipped with antiseptic hand cleaners, leakproof specimen containers, and puncture-proof containers for sharp tools that are reusable.Work practice controls are precautions that individual employees take, such as washing their hands immediately after removing potentially contaminated gloves or refraining from eating or drinking in areas where bloodborne pathogens may be present. Personal protection equipment includes any device designed to protect an employee from exposure. Widely used devices include the following:

- Gloves (disposable and reusable)
- Goggles and face shields
- Respirators
- Aprons, coats, and jackets

Examples of housekeeping controls include the use, disposal, and changing of protective coverings; decontamination of equipment; and regular cleaning of potentially contaminated areas.

Vaccinations

The OSHA standard requires that employers make hepatitis B vaccinations available to all employees for whom the reasonable anticipation rule applies. The vaccination procedure must meet the following criteria: available at no cost to employees; administered at a reasonable time and place within 10 days of assignment to a job with exposure potential; and administered under the supervision of an appropriately licensed health care professional, according to the latest recommendations of the U.S. Public Health Service.

Employees may decline the vaccination, but those who do must sign a *declination form* stating that they understand the risk to which they are subjecting themselves. Employees who decline are allowed to change their minds. Employees who do must be allowed to receive the vaccination.

Part of the exposure control officer's job is to keep accurate, up-to-date records showing vaccinated employees, vaccination dates, employees who declined, and signed declination forms.

Postexposure Investigation and Follow-Up

When an employee is exposed to bloodborne pathogens, it is important to evaluate the circumstances and follow up appropriately. How did it happen? Why did it happen? What should be done to prevent future occurrences? The postexposure investigation should determine at least the following:

- When did the incident occur?
- Where did the incident occur?
- What type of contaminated substances or materials were involved?
- What is the source of the contaminated materials?
- What type of work tasks were being performed when the incident happened?
- What was the cause of the incident?
- Were the prescribed precautions being observed when the incident occurred?
- What immediate action was taken in response to the incident?

Using the information collected during the investigation, an incident report is written. This report is just like any other accident report. Keeping the exposed employee fully informed is important. The employee should be informed of the likely avenue of exposure and the source of the contaminated material, even if the source is another employee. If the source is another employee, that individual's blood should be tested for HBV or HIV, and the results should be shared with the exposed employee. Once the exposed employee is fully informed, he or she should be referred to an appropriately certified medical professional to discuss the issue. The medical professional should provide a written report to the employer containing all pertinent information and recommendations. The exposed employee should also receive a copy.

Labels and Signs

This section of the plan describes the procedures established for labeling potential biohazards. Organizations may also use warning signs as appropriate and color-coded containers. It is important to label or designate with signs the following:

- Biohazard areas
- Contaminated equipment
- Containers of potentially contaminated waste
- Containers of potentially contaminated material (i.e., a refrigerator containing blood)
- Containers used to transport potentially contaminated material

Information and Training

This section of the plan describes the procedures for keeping employees knowledgeable, fully informed, and up-to-date regarding the hazards of bloodborne pathogens. The key to satisfying this requirement is training. Training provided should cover at least the following:

- OSHA Standard 1910.1030
- The exposure control plan
- Fundamentals of bloodborne pathogens (e.g., epidemiology, symptoms, and transmission)
- Hazard identification
- Hazard prevention methods
- Proper selection and use of personal protection equipment
- Recognition of warning signs and labels
- Emergency response techniques
- Incident investigation and reporting
- Follow-up techniques
- Medical consultation

It is important to document training and keep accurate up-to-date training records on all employees. These records should be available to employees and their designated representatives (e.g., family members, attorneys, and physicians), and to OSHA personnel. They must be kept for at least three years and include the following:

- Dates of all training
- Content of all training
- Trainers' names and qualifications
- Names and job titles of all participants

Record Keeping

OSHA Standard 1910.1030 requires that medical records be kept by employers on all employees who are exposed to bloodborne pathogens. These records must be confidential and should contain at least the following information:

- Employee's name and social security number
- Hepatitis B vaccination status
- Results of medical examinations and tests
- Results of incident follow-up procedures
- A copy of the written opinions of health care professionals
- A copy of all information provided to health care professionals following an exposure incident

PREVENTING AND RESPONDING TO NEEDLESTICK INJURIES

Needlestick injuries are not a major concern in most workplaces outside the health care industry, but they are enough of a concern that safety and health professionals should know how to prevent them and how to respond when they occur. An excellent source of help for safety and health professionals concerning needlestick injuries is the National Institute for Occupational Safety and Health (NIOSH). NIOSH maintains a Web site dedicated specifically to the prevention of needlestick injuries. The Web address is as follows:

http://www.cdc.gov/niosh/topics/bbp/safer

This Web site recommends a five-step model for developing, establishing, and maintaining a needlestick-prevention program:

1. Form a sharps injury prevention team.
2. Identify priorities.
3. Identify and screen safer medical devices.
4. Evaluate safer medical devices.
5. Institute and monitor the use of the safer medical devices selected.[27]

The NIOSH site also provides detailed information that safety and health professionals can use in implementing each step in the development of a needlestick prevention program.

Responding to Needlestick Incidents

When in spite of your best efforts at prevention a needlestick injury does occur, the following steps are recommended:

1. Encourage bleeding where the skin is penetrated.
2. Wash the penetration area thoroughly with copious amounts of warm, soapy water (do not use a scrub brush).
3. If the eyes are somehow involved, wash them immediately with water.
4. If the mouth is somehow involved, rinse it immediately with water, but do not swallow.

5. Get the injured employee to the hospital as soon as possible.
6. Contact a clinical virologist.
7. Make sure that management personnel for the company in question are fully informed of the situation.[28]

SUMMARY

1. Symptoms of the onset of AIDS include enlarged lymph nodes that persist, persistent fevers, involuntary weight loss, fatigue, diarrhea that does not respond to standard medications, purplish spots or blotches on the skin or in the mouth, white cheesy coating on the tongue, night sweats, and forgetfulness.
2. AIDS is known to be transmitted in three ways: sexual contact, blood contact, and mother-to-child during pregnancy or childbirth. AIDS is not spread through casual contact such as handshakes, toilet facilities, eating utensils, or coughing.
3. High-risk groups with regard to AIDS are as follows: homosexual men who do not take appropriate precautions, IV drug users, people with a history of multiple blood transfusions (including hemophiliacs), and sexually promiscuous people who do not take appropriate precautions.
4. AIDS is having an impact on the workplace in the form of higher insurance premiums, time-on-the-job losses, decreased productivity, cost of AIDS related lawsuits, and increased stress.
5. A corporate AIDS policy should have at least three components: employee rights, testing, and education.
6. Federal legislation relating to AIDS includes the Rehabilitation Act, Occupational Safety and Health Act, and Employee Retirement Income Security Act. Other legal concerns include state laws, case law, testing, and company policy.
7. Tests for AIDS are actually tests to identify the presence of HIV antibodies. For initial screening, the ELISA test is widely used. For verification, the IFA and Western Blot tests are used.
8. A well-planned AIDS education program can serve several purposes including the following: give management the facts needed to develop policy and make informed decisions; and result in positive behavioral changes among employees.
9. An AIDS education program may consist of any or all of the following components: one-on-one counseling, referral, posters, a newsletter, classroom instruction, self-paced multimedia instruction, group discussion sessions, and printed materials.
10. When counseling AIDS-infected employees, safety and health professionals should proceed as follows: listen, maintain a nonjudgmental attitude, make the employee aware of the company's AIDS policy, and respond in accordance with company policy.

11. Safety and health professionals can use the following strategies for helping quell fears about AIDS: establish an AIDS education program; conduct group round-table discussions; and correct inaccuracies, rumors, and misinformation about AIDS as soon as they occur.

12. The following precautions will help reduce the chances of contracting AIDS: abstinence; mutually monogamous relationship or marriage with an infection-free partner; no multiple sexual partners; avoidance of sex with a high-risk person without taking proper precautions; no use of intravenous drugs or, if you do, no sharing of needles; use of a protective device while practicing live or simulated CPR; and use of safety needles.

13. Hepatitis B (HBV) poses an even greater threat than HIV. It is caused by a double-shelled virus. HBV is transmitted through blood, tears, saliva, and semen. It can stay alive for years in body fluids.

14. Hepatitis C (HCV) infection is the most common chronic bloodborne infection in the United States. Approximately 36,000 new cases are diagnosed every year. HCV is transmitted primarily through direct percutaneous exposure to blood.

15. The five-step model for preventing needlestick injuries is as follows: form a sharps injury team, identify priorities, identify and screen safer devices, evaluate the safer devices, and institute and monitor the use of safer devices.

KEY TERMS AND CONCEPTS

Acquired immunodeficiency syndrome (AIDS)

AIDS education program

AIDS policy

Blood transfusions

Body fluids

Case law

Confidentiality

Confirmation test

Counseling

Disposable face mask

Employee assistance programs (EAPs)

Employee Retirement Income Security Act (ERISA)

Employee rights

Enzyme linked immunosorbent assay (ELISA) test

Essential functions

False negative

False positive

Hepatitis B (HBV)

Hepatitis C (HCV)

HIV antibody test

HIV positive

Human immunodeficiency virus (HIV)

Immuno-florescent (IFA) test

Intravenous drugs

IV drug users

Mutually monogamous relationship

Occupational Safety and Health Act (OSH Act)

Otherwise qualified

Reasonable accommodation

Rehabilitation Act

Screening test

Sexual contact

Western Blot test

REVIEW QUESTIONS

1. What does the acronym "AIDS" mean?
2. List five symptoms of AIDS.
3. What are the three known ways that AIDS is transmitted?
4. List the groups of people who are considered high risk with regard to AIDS.
5. List five ways that AIDS is *not* transmitted.
6. Describe the ways in which AIDS is having an impact in the workplace.
7. Briefly explain the minimum components of a corporate AIDS policy.
8. Explain how the following legal concepts relate to AIDS: *otherwise qualified*, *essential functions*, and *reasonable accommodation*.
9. Briefly explain both sides of the AIDS testing controversy.
10. What are the most widely used HIV antibody tests for screening and confirming AIDS?
11. Name four purposes of a well-planned AIDS education program.
12. How should a safety and health professional proceed when faced with an employee who thinks he or she has AIDS?
13. Briefly explain the steps that a company can take to alleviate the fears of employees about AIDS.
14. Explain how OSHA categorizes work tasks relative to exposure to HIV infection.
15. List four ways to guard against contracting AIDS.
16. How can CPR be administered safely?
17. Explain why HBV poses more of a problem for safety personnel than HIV does.
18. Explain the primary prevention strategies for HCV.
19. What are the recommended steps in a needlestick prevention program?

ENDNOTES

1. Centers for Disease Control and Prevention (CDC), *HIV/AIDS Surveillance Report* 11, no. 2: 35–37.
2. Ibid.
3. Ibid.
4. "AIDS—The Basic Facts," *Occupational Health & Safety* 10, no. 3: 6.
5. P. Minetos, "Corporate America vs. AIDS," *Safety & Health* 138, no. 6: 34.
6. Ibid.
7. CDC, *HIV/AIDS Surveillance Report*, 37.
8. Centers for Disease Control and Prevention (CDC), U.S. Department of Health and Human Services, *Facts About AIDS*, Spring 2002, 6.
9. Ibid.
10. Minetos, "Corporate America vs. AIDS," 36.
11. Ibid.
12. Ibid.

13. B. C. Yorker, "AIDS Testing," *AAOHN Journal* 36, no. 5: 231.

14. Ibid.

15. R. Bayer, C. Levine, and S. M. Wolf, "HIV Antibody Screening: An Ethical Framework for Evaluating Proposed Programs," in B. C. Yorker, "AIDS Testing," *AAOHN Journal* 36, no. 5: 232.

16. American College Health Association, *The HIV Antibody Test* (Rockville, MD: American College Health Association, 1989).

17. Ibid.

18. CDC, *Facts About AIDS*, 10.

19. National Safety Council, "Is the AIDS Fear a Threat at Work?" *Safety & Health* 139, no. 6: 52.

20. G. Seldes, *The Great Quotations* (Secaucus, NJ: Castle Books, 1966), 590.

21. K. C. Brown and I. C. Turner, *AIDS: Policies and Programs for the Workplace* (New York: Van Nostrand Reinhold), 106–107, 116–117.

22. CDC, *Facts About AIDS*, 8.

23. M. Eastman, "CPR Training: Is It Still Safe?" *Safety & Health* 142, no. 5: 36.

24. Ibid., 39.

25. T. Ellis, "Toward Safer Needles," *Occupational Health & Safety* 68, no. 3: 74.

26. Centers for Disease Control and Prevention, "Recommendations for Prevention and Control of Hepatitis C Virus (HCV) Infection and HCV-Related Chronic Disease," *Morbidity and Mortality Weekly Report* 47, no. RR-19: 1–16.

27. "Web-Based Information Could Help Prevent Needlestick Injuries," *Occupational Health & Safety NEWS* 17, no. 11: 3.

28. Bhavini Lad, "Pricked by a Needle," *Occupational Health & Safety* 71, no. 4: 60.

5

MANAGEMENT OF SAFETY AND HEALTH

25 Preparing for Emergencies and Terrorism   711
26 Ethics and Safety   747
27 Hazard Analysis/Prevention and Safety Management   765
28 Promoting Safety   795
29 Environmental Safety and ISO 14000 (Environmental Management)   819
30 TSM: Safety Management in a Quality Management Setting   861
31 Establishing a Safety-Friendly Corporate Culture   878

PREPARING FOR EMERGENCIES AND TERRORISM

Major Topics

- Rationale for Emergency Preparation
- Emergency Planning and Community Right-to-Know Act
- Organization and Coordination
- OSHA Standards
- First Aid in Emergencies
- How to Plan for Emergencies
- Planning for Workers with Disabilities
- Evacuation Planning
- Customizing Plans to Meet Local Needs
- Emergency Response
- Computers and Emergency Response
- Dealing with the Psychological Trauma of Emergencies
- Recovering from Disasters
- Terrorism in the Workplace
- Resuming Business after a Disaster

Despite the best efforts of all involved, emergencies do sometimes occur. The potential for human-caused emergencies has increased significantly with the rise of worldwide terrorism. It is very important to respond to such emergencies in a way that minimizes harm to people and damage to property. To do so requires plans that can be implemented without delay. This chapter provides prospective and practicing safety and health professionals with the information they need to prepare for emergencies in the workplace. Everything in this chapter pertains to all kinds of emergencies including natural disasters and terrorism. A special section relating specifically to terrorism is included at the end of the chapter.

Figure 25–1
Elements of emergency
preparation.

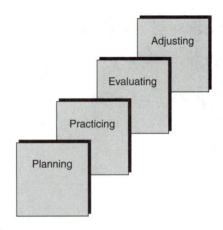

RATIONALE FOR EMERGENCY PREPARATION

An *emergency* is a potentially life-threatening situation, usually occurring suddenly and unexpectedly. Emergencies may be the result of natural or human causes. Have you ever witnessed the timely, organized, and precise response of a professional emergency medical crew at an automobile accident? While passers-by and spectators may wring their hands and wonder what to do, the emergency response professionals quickly organize, stabilize, and administer. Their ability to respond in this manner is the result of preparation. As shown in Figure 25–1, preparation involves a combination of **planning**, **practicing**, **evaluating**, and **adjusting** to specific circumstances.

When an emergency occurs, immediate reaction is essential. Speed in responding can mean the difference between life and death or between minimal damage and major damage. Ideally, all those involved should be able to respond properly with a minimum of hesitation. This can happen only if all exigencies have been planned for and planned procedures have been practiced, evaluated, and improved.

A quick and proper response—which results because of proper preparation—can prevent panic, decrease the likelihood of injury and damage, and bring the situation under control in a timely manner. Because no workplace is immune to emergencies, preparing for them is critical. An important component of preparation is planning.

EMERGENCY PLANNING AND COMMUNITY RIGHT-TO-KNOW ACT

Title III of the Superfund Amendments and Reauthorization Act of 1986 (SARA) is also known as the **Emergency Planning and Community Right-to-Know Act (EPCRA)**. This law is designed to make information about hazardous chemicals available to a community where they are being used so that residents can protect themselves in the case of an emergency. It applies to all companies that use, make, transport, or store chemicals.

Safety and health professionals involved in developing emergency response plans for their companies should be familiar with the act's requirements for emergency planning. As shown in Figure 25–2, the EPCRA includes the four major components discussed in the following paragraphs.

Emergency Planning

The emergency planning component requires that communities form **local emergency planning committees (LEPCs)** and that states form **state emergency response commissions (SERCs)**. LEPCs are required to develop emergency response plans for the local communities, host public forums, select a planning coordinator for the community, and work with the coordinator in developing local plans. SERCs are required to oversee LEPCs and review their emergency response plans. Plans for individual companies in a given community should be part of that community's larger plan. Local emergency response professionals should use their community's plan as the basis for simulating emergencies and practicing their responses.

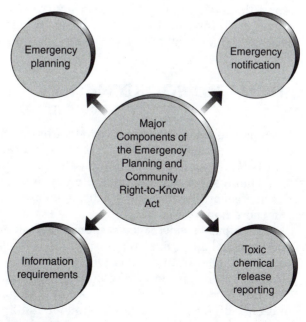

Figure 25–2
Parts of an emergency response plan.

Emergency Notification

The **emergency notification** component requires that chemical spills or releases of toxic substances that exceed established allowable limits be reported to appropriate LEPCs and SERCs. Immediate notification may be verbal as long as a written notification is filed promptly. Such reports must contain at least the following information: (1) the names of the substances released, (2) where the release occurred, (3) when the release occurred, (4) the estimated amount of the release, (5) known hazards to people and property, (6) recommended precautions, and (7) the name of a contact person in the company.

Information Requirements

Information requirements mean that local companies must keep their LEPCs and SERCs and, through them, the public informed about the hazardous substances that the companies store, handle, transport, or use. This includes keeping comprehensive records of such substances on file, up-to-date, and readily available; providing copies of material safety data sheets for all hazardous substances; giving general storage locations for all hazardous substances; providing estimates of the amount of each hazardous substance on hand on a given day; and estimating the average annual amount of hazardous substances kept on hand.

Toxic Chemical Release Reporting

The **toxic chemical release reporting** component requires that local companies report the total amount of toxic substances released into the environment as either emissions or hazardous waste. Reports go to the Environmental Protection Agency and the state-level environmental agency.

ORGANIZATION AND COORDINATION

Responses to emergencies are typically from several people or groups of people including medical, firefighting, security, and safety personnel as well as specialists from a variety of different fields. People in each of these areas have different but interrelated and often interdependent roles to play in responding to the emergency. Because of their disparate backgrounds and roles, organization and coordination are critical.

A company's **emergency response plan** should clearly identify the different personnel and groups that respond to various types of emergencies and, in each case, who is in charge. One person should be clearly identified and accepted by all emergency responders as the **emergency coordinator**. This person should be knowledgeable, at least in a general sense, of the responsibilities of each individual emergency responder and how each relates to those of all other responders. This knowledge must include the **order of response** for each type of emergency set forth in the plan.

A company's safety and health professional is the obvious person to organize and coordinate emergency responses. However, regardless of who is designated, it is important that (1) one person is in charge, (2) everyone involved knows who is in charge, and (3) everyone who has a role in responding to an emergency is given ample opportunities to practice in simulated conditions that come as close as possible to real conditions.

Union Carbide's Texas City plant employs 1,500 workers and several hundred additional contract personnel. Safety is a high priority. Consequently, emergency planning is a fundamental part of the company's safety and health program. Emergency responders include emergency medical technicians, nurses, physicians, and production workers who are assigned specific emergency response duties. They are coordinated by a designated emergency director.[1]

Union Carbide's administrator of health services stresses the need for organization and coordination of emergency response teams. According to Marley, "Often staff physicians and nurses don't know the quickest, safest route to an emergency site. That's why emergency medical teams need to be organized and trained."[2]

Workers in the plant are obviously also knowledgeable about traffic flow, access routes, exit points, emergency approaches, and shortcuts. Because of this, Union Carbide's approach to emergency response "provides a method for dispatching EMTs and ambulance drivers to the scene while nursing and medical staff remain in radio communication and available for consultation or to receive patients if necessary."[3]

Another unique aspect of emergency planning and response at Union Carbide is how responses are coordinated: "The planning and coordination of all emergency response belongs to the **emergency response management team (ERMT)**. Composed of shift emergency directors and a full-time fire chief, this group unifies all emergency groups and equipment into a single, coordinated effort" (emphasis added).[4]

OSHA STANDARDS

All OSHA standards are written for the purpose of promoting a safe, healthy, accident-free, hence emergency-free workplace. Therefore, OSHA standards play a role in emergency prevention and should be considered when developing emergency plans. For example, exits are important considerations when planning for emergencies. Getting medical personnel in and employees and injured workers out quickly is critical when responding to emergencies. The following sections of OSHA's standards deal with emergency preparedness:

Emergency Action Plan	29 CFR 1910.38
Exit arrangements	29 CFR 1910.37(e)
Exit capacity	29 CFR 1910.37(c),(d)

Exit components	29 CFR 1910.37(a)
Exit workings	29 CFR 1910.37(q)
Exit width	29 CFR 1910.37(c)
Exterior exit access	29 CFR 1910.37(g)
Occupational Health and Environmental Controls	29 CFR 1926.65, Appendix C
Hazardous Materials	29 CFR 1910.120, Appendix C

A first step for companies developing emergency plans is to review these OSHA standards. This can help safety and health personnel identify and correct conditions that may exacerbate emergency situations before they occur.

FIRST AID IN EMERGENCIES

Workplace emergencies often require a medical response. The immediate response is usually first aid. First aid consists of lifesaving measures taken to assist an injured person until medical help arrives.

Because there is no way to predict when first aid may be needed, providing first-aid training to employees should be part of preparing for emergencies. In fact, in certain cases, OSHA requires that companies have at least one employee on-site who has been trained in first aid (CFR 1910.151). Figure 25–3 contains a list of the topics that may be covered in a first-aid class for industrial workers.

Safety Tip

Failure to Warn the Community Can Be Costly

The hazards associated with accidents or incidents can extend beyond the walls of the facility in question to the surrounding community. This is why companies must have emergency plans in place for notifying the community and regulatory agencies when an accidental release poses a threat. Failure to notify its neighbors cost a food manufacturing company in Zanesville, Ohio, $43,829 in fines. When a refrigeration system's pressure relief valve malfunctioned, 820 pounds of anhydrous ammonia was released into the atmosphere. The company failed to notify local community officials and the National Response Center until three hours after the discharge.

Source: From "Didn't Warn Neighbors Promptly about Hazard," *Facility Manager's Alert* 9, no. 187: 3.

Basic First Aid

- Cardiopulmonary resuscitation
- Severe bleeding
- Broken bones and fractures
- Burns
- Choking on an obstruction
- Head injuries and concussion
- Cuts and abrasions
- Electric shock

- Heart attack
- Stroke recognition
- Moving an injured person
- Drug overdose
- Unconscious victim
- Eye injuries
- Chemical burns
- Rescue

Figure 25–3
Sample course outline for first-aid class.

First-Aid Training Program

First-aid programs are usually available in most communities. The continuing education departments of community colleges and universities typically offer first-aid training. Classes can often be provided on-site and customized to meet the specific needs of individual companies.

The American Red Cross provides training programs in first aid specifically geared toward the workplace. For more information about these programs, safety and health professionals may contact the national office of the American Red Cross at 202-639-3200 (http://www.redcross.org).

The National Safety Council also provides first-aid training materials. The First Aid and Emergency Care Teaching Package contains a slide presentation, overhead transparencies, a test bank, and an instructor's guide. The council also produces a book titled *First Aid Essentials*. For more information about these materials, safety and health professionals may contact the NSC at 800-832-0034 (http://www.nsc.org).

Beyond Training

Training employees in first-aid techniques is an important part of preparing for emergencies. However, training is only part of the preparation. In addition, it is important to do the following:

1. *Have well-stocked first-aid kits available.* First-aid kits should be placed throughout the workplace in clearly visible, easily accessible locations. They should be properly and fully stocked and periodically checked to ensure that they stay fully stocked. Figure 25–4 lists the minimum recommended contents for a workplace first-aid kit.

Safety Fact

Emergency Response Regulations

There are numerous government regulations requiring the development of emergency response plans. Following are the most important of these:

- Environmental Protection Agency's (EPA) Spill Prevention Control and Countermeasures (SPCC) and Facility Response Plans—40 CFR Part 112(d), 112.20–21.
- EPA's Resource Conservation and Recovery Act (RCRA) Contingency Plan—40 CFR Part 264, Subpart D; Part 265, Subpart D and 279.52.
- EPA's Risk Management Plan (RMP)—40 CFR Part 68.
- EPA's Emergency Planning and Community Right-to-Know (EPCRA, also known as SARA Title III).
- Comprehensive Emergency Response Compensation and Liabilities Act (CERCLA).
- Occupational Safety and Health Administration's OSHA Emergency Action Plan—29 CFR 1910.38.
- OSHA's Process Safety Standard (PSS)—29 CFR 1910.119.
- OSHA's Hazardous Waste Operations and Emergency Response (HAZWOPER) Regulation—29 CFR 1910.120.
- Department of Transportation's (DOT) Research and Special Programs Administration (RSPA) Pipeline Response Plans—49 CFR Part 194.
- U.S. Coast Guard's (USCG) Facility Response Plans—33 CFR Part 154, Subpart F.
- Minerals Management Services' (MMS) Facility Response Plans—30 CFR Part 254.

2. *Have appropriate personal protective devices available.* With the concerns about AIDS and hepatitis, administering first aid has become more complicated than in the past. The main concerns are with bleeding and other body fluids. Consequently, a properly stocked first-aid kit should contain rubber surgical gloves and facemasks or mouthpieces for CPR.

3. *Post emergency telephone numbers.* The advent of 911 service has simplified the process of calling for medical care, police, or firefighting assistance. If 911 services are not available, emergency numbers for ambulance, hospital, police, fire department, LEPC, and appropriate internal personnel should be posted at clearly visible locations near all telephones in the workplace.

4. *Keep all employees informed.* Some companies require all employees to undergo first-aid training; others choose to train one or more employees in each department. Regardless of the approach used, it is important that all employees be informed and kept up-to-date concerning basic first-aid information. Figures 25–5 and 25–6 are first-aid fact sheets of the type used to keep employees informed.

- Sterile gauze dressings (individually wrapped in sizes 4 × 4 and 8 × 10 inches)
- Triangular bandages
- Roll of gauze bandages (at least 2 inches wide)
- Assorted adhesive bandages
- Sealed moistened towelettes
- Adhesive tape
- Absorbent cotton
- Sterile saline solution
- Mild antiseptic for minor wounds
- Ipecac syrup to induce vomiting
- Powdered activated charcoal to absorb swallowed poisons
- Petroleum jelly
- Baking soda (bicarbonate of soda)
- Aromatic spirits of ammonia

- Elastic wraps
- Scissors
- Tweezers
- Needles
- Sharp knife or stiff-backed razor blades
- Medicine dropper (eye dropper)
- Measuring cup
- Oral thermometer
- Rectal thermometer
- Hot water bag
- Wooden safety matches
- Flashlight
- Rubber surgical gloves
- Face masks or mouthpieces
- Blanket
- Splint

Figure 25–4
Minimum recommended contents of workplace first-aid kits.

HOW TO PLAN FOR EMERGENCIES

Developing an **emergency action plan (EAP)** is a major step in preparing for emergencies. A preliminary step is to conduct a thorough analysis to determine the various types of emergencies that may occur. For example, depending on geography and the types of products and processes involved, a company may anticipate such emergencies as the following: fires, chemical spills, explosions, toxic emissions, train derailments, hurricanes, tornadoes, lightning, floods, earthquakes, or volcanic eruptions.

A company's EAP should be a collection of small plans for each anticipated or potential emergency. These plans should have the following components:

1. *Procedures.* Specific, step-by-step emergency response procedures should be developed for each potential emergency.

2. *Coordination.* All cooperating agencies and organizations and emergency responders should be listed along with their telephone number and primary contact person.

3. *Assignments and responsibilities.* Every person who will be involved in responding to a given emergency should know his or her assignment. Each person's responsibilities should be clearly spelled out and understood. One person may be responsible for conducting an evacuation of the affected area,

Global Technologies
500 Anchors Street • Industrial Park • Fort Walton Beach, Florida 32748

First-Aid Fact Sheet No. 16
Moving an Injured Person

If a victim has a neck or back injury, do not move him unless it must be done to prevent additional injuries. If it is absolutely essential to move the victim, remember the following rules of thumb:

1. Call for professional medical help.

2. Always pull the body lengthwise, never sideways.

3. If there is time, slip a blanket under the victim and use it to pull him to safety.

4. If the victim must be lifted, support all parts of the body so that it does not bend or jackknife. (Use a spine board and cervical collar.)

Figure 25–5
First-aid fact sheet.

Global Technologies
500 Anchors Street • Industrial Park • Fort Walton Beach, Florida 32748

First-Aid Fact Sheet No. 12
ABCs of First Aid

If a fellow employee is injured and you are the first person to respond, remember the ABCs of first aid.

A = Airway
Is the airway blocked? If so, clear it quickly.

B = Breathing
Is the victim breathing? If not, begin administering artificial respiration.

C = Circulation
Is the victim bleeding severely? If so, stop the bleeding.
Is there a pulse? If not, begin administering CPR.

Figure 25–6
First-aid fact sheet.

Discussion Case

What Is Your Opinion?

"Forget it! We are not going to train our employees in first aid. Emergencies notwithstanding, I don't want a bunch of amateur doctors running around the company doing more harm than good." Mary Vo Dinh, safety director for Gulf Coast Manufacturing, was getting nowhere trying to convince her boss that the company should have employees trained in first aid in the event of an emergency. "But John, we have had three hurricanes in just two years. Tornadoes are not uncommon here on the coast." "I will repeat myself just one more time," said her boss. "No first-aid training." Who is right in this case? What is your opinion?

Type of Emergency

____Fire   ____Explosion
____Chemical spill   ____Toxic emission
____Train derailment   ____Hurricane
____Tornado   ____Lightning
____Flood   ____Earthquake
____Volcanic eruption

Procedures for Emergency Response
1. Controlling and isolating?
2. Communication?
3. Emergency assistance?
4. First aid?
5. Shutdown/evacuation/protection of workers?
6. Protection of equipment/property?
7. Egress, ingress, exits?
8. Emergency equipment (e.g., fire extinguishers)?
9. Alarms?
10. Restoration of normal operations?

Coordination
1. Medical care providers?
2. Fire service providers?
3. LEPC personnel?
4. Environmental protection personnel?
5. Civil defense personnel (in the case of public evacuations)?
6. Police protection providers?
7. Communication personnel?

Figure 25–7
Emergency planning checklist.

another for the immediate shutdown of all equipment, and another for telephoning for medical, fire, or other types of emergency assistance. When developing this part of the EAP, it is important to assign a backup person for each area of responsibility. Doing so ensures that the plan will not break down if a person assigned a certain responsibility is one of the victims.

4. *Accident prevention strategies.* The day-to-day strategies to be used for preventing a particular type of emergency should be summarized in this section of the EAP. In this way, the strategies can be reviewed, thereby promoting prevention.

5. *Schedules.* This section should contain the dates and times of regularly scheduled practice drills. It is best to vary the times and dates so that practice drills don't become predictable and boring. Figure 25–7 is a checklist that can be used for developing an EAP.

Assignments and Responsibilities
1. Who cares for the injured?
2. Who calls for emergency assistance?
3. Who shuts down power and operations?
4. Who coordinates communication?
5. Who conducts the evacuation?
6. Who meets emergency responders and guides them to the emergency site?
7. Who contacts coordinating agencies and organizations?
8. Who is responsible for ensuring the availability and upkeep of fire extinguishers?
9. Who is responsible for ensuring that alarms are in proper working order?
10. Who is responsible for organizing cleanup activities?

Accident Prevention Strategies
1. Periodic safety inspections?
2. Industrial hygiene strategies?
3. Personal protective equipment?
4. Ergonomic strategies?
5. Machine safeguards?
6. Hand and portable power tool safeguards?
7. Material handling and storage strategies?
8. Electrical safety strategies?
9. Fire safety strategies?
10. Chemical safety strategies?

Schedules
1. Dates of practice drills:_____
2. Times of practice drills:_____
3. Duration of practice drills: _____

Figure 25–7 (*continued*)

PLANNING FOR WORKERS WITH DISABILITIES

When developing emergency action plans it is important to consider the special needs of workers with disabilities. Do not develop a separate EAP specifically for workers with disabilities—they should be included as a normal part of the organization's plan. An excellent set of guidelines is available to assist organizations in considering the needs of workers with disabilities when developing EAPs.[5]

Titled *Preparing the Workplace for Everyone*, these guidelines were developed for the use of federal agencies but are available to private organizations and other public organizations as well. To obtain a copy of the guidelines, contact the Office of Disability Employment Policy at the following address:

United States Department of Labor
Office of Disability Employment Policy
200 Constitution Ave. NW, Room S-1303
Washington, DC 20210
202-693-7880
http://www.dol.gov/odep

The guidelines are presented under numerous subheadings with questions listed relating to each subheading. What follows in the remainder of this section are categories of questions similar to and based on those presented in the actual guidelines. Safety and health professionals are encouraged to obtain a complete copy of the guidelines to ensure that their EAPs account for workers with disabilities appropriately.

Involving Key Personnel

Under this heading, the guidelines pose questions similar to the following:

- Are key personnel familiar with the EAP?
- Are personnel with disabilities involved in all aspects of the development of the EAP?
- Do senior executives support the development and updating of the EAP?
- Has the EAP been reviewed by first responders and facility personnel?
- Does any part of the plan conflict with procedures established by other applicable agencies?

Implementing Shelter-in-Place (SIP) Plans

Shelter-in-place (SIP) means that rather than evacuate in a disaster or emergency, immediate shelter is sought. SIP comes into play when attempting to evacuate might increase the risk of harm or injury. There should be a SIP provision built into the larger EAP containing provisions that speak to the following concerns:

- Are there provisions for shutting down the building's ventilation system quickly?
- Are there provisions for turning off the elevators?
- Are there provisions for closing all exits and entrances and for securing the loading dock and garage areas (as applicable)?
- Are there provisions for notifying all occupants—including visitors—of emergency procedures?
- Are there provisions for asking all occupants—including visitors—to remain in the building until it is safe to leave?

Evaluating Personnel Needs

It is important to determine the needs of workers with disabilities when developing an EAP, but to do so in strict accordance with the Rehabilitation Act. Before evaluating the needs of personnel with disabilities, consult your human resources department to develop guidelines that are consistent with the Rehabilitation Act. Then ask the following types of questions during development of the EAP:

- Has all applicable information about the needs of personnel with disabilities been collected in accordance with the Rehabilitation Act?
- Has the information collected been appropriately protected by sharing it only with those who need to know it as part of emergency planning (e.g., first-aid personnel, safety and health professionals, those responsible for carrying out the EAP during an emergency, etc.)?
- In selecting equipment, have appropriate agencies as well as individuals with disabilities been consulted as to the ability of personnel with disabilities to use the equipment?
- Have the needs of personnel with service animals been considered in the development of the plan?
- Has the establishment of personal support networks been built into the plan?
- Has the appropriate training been built into the plan for people who need assistance during an emergency as well as for people who will provide the assistance?

Distributing and Communicating the Plan

In developing an EAP, a critical element that should not be overlooked concerns how the completed plan will be distributed to personnel and how those who developed the plan will communicate with personnel about it. The following questions will help ensure proper development of this section of the plan:

- How will the EAP be communicated with the same level of detail to all personnel?
- Is the plan available on the organization's Web site as well as in hard copy? Are there text explanations of all graphic material contained in the plan?

- Are training sessions contained in the plan offered in accessible locations?
- Are the learning aids necessary to accommodate personnel with disabilities readily available in all training locations (e.g., sign-language interpreters, listening devices, etc.)?
- Are hard copies of the EAP placed in various easily accessible locations throughout the facility?
- Are appropriate sections of the plan (evacuation and SIP information) given to frequent visitors to the facility?

Balancing Employer Responsibilities and Employee Right to Self-Determination

On one hand, employers have specific responsibilities for ensuring the safety of their personnel. On the other hand, employees with disabilities have certain rights of self-determination. To ensure that these two issues do not collide when developing the EAP, ask the following questions:

- Are there employees who impede the evacuation of others during practice drills? If so, has the issue been dealt with privately and have appropriate solutions been developed?
- In determining if an employee with a disability represents a direct threat in the event of an emergency, have the following factors been considered: nature and severity of the potential harm, likelihood that the potential harm will occur during an emergency, during of the risk posed by the individual in question, the imminence of the potential harm, and the availability of reasonable accommodations that would mitigate or eliminate the risk.
- Have personnel with disabilities been included in the emergency planning process?
- Has every effort been made to ensure that personnel with disabilities have not been segregated as part of the emergency planning process?
- Have employees with disabilities made requests for reasonable accommodations in the event of an emergency? Have those accommodations been built into the EAP? If not, can the organization show clearly that the requested accommodation would pose an undue hardship?

Working with First Responders

It is important to coordinate emergency planning and the resulting EPA with first-responder agencies and organizations. These personnel and those who carry out the EAP should coordinate closely on all matters pertaining to implementation of the plan.

- Have first responders been made aware of any special issues relating to personnel with disabilities?
- Does the organization have a policy regarding evacuation built into its plan? Are all stakeholders aware of the policy and its ramifications? Has anyone

expressed opposition to the policy? Has the opposition been properly addressed?

- Have first responders been included in all steps of the emergency planning process? If not, how will they be made aware of all elements of the EAP?

Implementing an Elevator Policy

Should the organization use its elevators during an emergency or close them down? This can become a critical issue when the needs of personnel with disabilities are considered.

- Does the organization have a policy concerning elevator use during emergencies? Has the policy been built into the EAP?
- Were first responders consulted during the development of the elevator policy?
- Under what circumstances may elevators be used during an emergency? Who may operate the elevators under these conditions?
- Who gets priority in the use of elevators during an emergency?
- How are personnel, including those with disabilities, to be evacuated in the event that elevators are inoperable during an emergency?

Developing Emergency Notification Strategies

Developing the EAP is just one step in an ongoing process. Once the plan is developed, all stakeholders must be notified of its contents and what they mean. The following questions will help ensure appropriate notification:

- What steps have been built into the plan for notifying stakeholders in the event of an emergency and for ensuring that personnel with disabilities have access to the same information that any other stakeholder has?
- Have a variety of notification or communication methods been built into the EAP?
- Do notification or communication methods account for personnel who are away from their desks or the office?
- Do notification strategies take into account the presence of visitors?

Practicing and Maintaining the EAP

The emergency action plan should be viewed as a living document that is updated and maintained continually. In addition, all aspects of the plan should be practiced periodically to ensure that all parties know how to carry out their responsibilities. The following questions will help ensure that the plan is properly maintained and practiced:

- Does the organization have a policy for the regular practice and maintenance of the EAP?

- Does the organization comply with the policy and even exceed compliance requirements?
- Have first responders been involved in all practice drills and have they been consulted to ensure that all applicable equipment is appropriate and in proper working order?
- Have drills been varied in terms of time of day and type of drill?
- Have various unforeseen problems, roadblocks, and inhibitors been built into the drills so that emergency personnel have opportunities to practice improvising in a realistic way?
- Are personnel with disabilities included in all drills?
- Does the organization have a policy built into its EAP for dealing with employees or visitors who might need to leave the facility in the middle of an emergency drill?

EVACUATION PLANNING

The OSHA standard for evacuation planning is 29 CFR 1910.38. This standard requires a written plan for evacuating the facility in the event of an emergency. Critical elements of the plan are marking of exit routes, communications, outside assembly, and training.

Marking of Exit Routes

Clearly identified and marked routes of egress are critical during a time of crisis (fire, natural disaster, terrorist attack, etc.). To ensure that routes of egress and all related evacuation response items are clearly marked, safety professionals should answer the following questions about the facilities for which they are responsible:

1. Are all exit, emergency exit, and nonexit doors clearly identified and marked?
2. Are there up-to-date evacuation route maps mounted at strategic locations throughout the facility?
3. Are all egress route aisles, hallways, and stairs marked clearly and can the markings be seen in the event of darkness?
4. Are there low-level markings posted strategically throughout the facility that can be viewed in the event that smoke fills the facility?
5. Are all items of firefighting equipment clearly marked with directional signs so they can be easily located in the event of an emergency?
6. Is all emergency first-aid equipment clearly marked with directional signs so that it can be easily located in the event of an emergency?
7. Are all electrical, chemical, and physical hazards identified and clearly marked?

8. Are all physical obstructions clearly outlined?
9. Are all critical shutdown procedures and equipment identified and clearly marked?
10. Are the handrails, treads, and risers on all stairs clearly marked?
11. In the case of multicultural workforce settings, are all signs and markings provided in a bilingual or pictogram format?[6]

Ensuring that all signs, pictograms, and other markings relating to facility evacuation are visible during power outages is critical. Battery backup systems are one approach that is widely used. However, there are environments in which even the smallest spark from a battery might set off an explosion. For this reason, some facilities find the use of photoluminescent signs a better alternative.

Photoluminescent signs and markings absorb normal light energy from their surroundings and then release this energy in the form of light during periods of darkness. Some of the better photoluminescent materials will give off light for as long as 24 hours and have a maintenance-free life expectancy of up to 25 years.

Communication and Alarm Procedures

People are so accustomed to false alarms in their lives that when the *real thing* occurs, it can be difficult to convince them that it's not just another drill. In addition, people tend to trust what they can see, smell, and hear. Consequently, if they cannot physically sense an emergency, they tend to ignore the warning. The communication component of a facility's evacuation plan should include procedures for early detection of a problem, procedures for reporting an emergency, procedures for initiating an evacuation, and procedures for providing the necessary information to employees who are being evacuated.[7]

Notifying employees of the emergency is the function of the facility's alarm system. However, just pulling the alarm switch is not sufficient communication. Once the alarm has been given, verbal instruction should be broadcast so that people know that the alarm is real (not just another drill) as well as specific actions they should take immediately. There must also be procedures for informing evacuated employees that the emergency is over and they can return to their work.

External communication procedures are also important. All employees should know how to notify outside authorities of the emergency. With the advent of 911 service, this problem has been simplified. However, do not assume that all employees will remember the number when under the stress of an emergency, or that they will remember to dial "9" or some other code to gain access to an outside line. This problem can be solved by placing clearly marked signs above or near telephones containing such messages as "In an emergency dial 911" or "In an emergency dial 9911" (if it is necessary to first access an outside line).

Outside Assembly

The company's evacuation plan should include an assembly area to which employees go once evacuated.[8] This area should be well known by all employees, and employees should understand that it is critical to assemble there so that a headcount can be taken. In addition, there should be a backup assembly area known to all employees so they know where to go if the primary assembly area has been rendered inaccessible or hazardous.

Part of the evacuation plan relating to assembly areas must be devoted to transient personnel—nonemployees such as vendors, visitors, contractors, and so on. How will they know where to assemble? Who will check to see that they have been notified of the emergency? These issues should be spoken to in the evacuation plan.

Training

Training for evacuations is a critical element of the evacuation plan.[9] Developing a plan and then letting it just sit on the shelf gathering dust is a formula for disaster. Everything contained in the plan that requires action or knowledge on the part of employees should be part of the required evacuation training. Training should be provided when employees are first hired, and retraining should be provided periodically as various elements of the plan are updated.

Drills should be a major part of the training provided for employees. The old sports adage that says "What you do in practice you will do in the game" applies here. When an emergency actually occurs should not be the first time employees have gone through the action required of them in an emergency.

CUSTOMIZING PLANS TO MEET LOCAL NEEDS

Emergency plans must be **location-specific**. General plans developed centrally and used at all plant locations have limited effectiveness. The following rules of thumb can be used to ensure that EAPs are location-specific:

1. *A map of the plant.* A map of the specific plant helps localize an EAP. The map should include the locations of exits, access points, evacuation routes, alarms, emergency equipment, a central control or command center, first-aid kits, emergency shutdown buttons, and any other important elements of the EAP.

2. *Chain of command.* An organization chart illustrating the chain of command—who is responsible for what and who reports to whom—also helps localize an EAP. The chart should contain the names and telephone numbers (internal and external) of everyone involved in responding to an emergency. It is critical to keep the organization chart up-to-date as personnel changes occur. It is also important to have a designated backup person shown for every position on the chart.

3. *Coordination information.* All telephone numbers and contact names of people in agencies with which the company coordinates emergency activities should be listed. Periodic contact should be maintained with all these people so that the EAP can be updated as personnel changes occur.

4. *Local training.* All training should be geared toward the types of emergencies that may occur in the plant. In addition, practice drills should take place on-site and in the specific locations where emergencies are most likely to happen.

EMERGENCY RESPONSE

"An **emergency response team (ERT)** is a special team that responds to emergencies to ensure proper personnel evacuation and safety, shut down building services and utilities, work with responding civil authorities, protect and salvage property, and evaluate areas for safety prior to reentry" (emphasis added).[10] The ERT is typically composed of representatives from several different departments such as maintenance, security, safety and health, production and processing, and medical. The actual composition of the team depends on the size and type of company. The ERT should be contained in the assignments and responsibilities section of the EAP.

Not all ERTs are company based. Communities also have ERTs for responding to emergencies that occur outside of a company environment. Such teams should be included in a company's EAP in the coordinating organizations section. This is especially important for companies that use hazardous materials.

Most dangerous spills occur on the road and the greatest number of hazardous materials (hazmat) incidents occur on the manufacturer's loading dock. More than 80 percent of chemical releases are caused by errors in loading and unloading procedures.[11] Community-based teams typically include members of the police and fire departments who have had special training in handling hazmat emergencies.

Another approach to ERTs is the **emergency response network (ERN)**. An ERN is a network of ERTs that covers a designated geographical area and is typically responsible for a specific type of emergency. For example, the Chlorine Institute (CI), an association of more than 130 firms that produce chlorine and manufacture related products, implemented its Chlorine Emergency Plan (CHLOREP), a network for responding to chlorine emergencies. Sixty-two teams from 28 companies are assigned geographic areas where they are on call to help during chlorine emergencies.[12]

The following example shows how a CHLOREP team responds to a chlorine-related emergency. A chlorine leak was discovered in a cement pit near Arlington National Cemetery. The pit had once been used to chlorinate water for irrigating the cemetery. Initial reports were that as many as half of the ten 150-pound chlorine cylinders were leaking. Local emergency response teams asked CHLOREP for help.[13] The CHLOREP team covering Washington, DC, and vicinity responded and found that only one of the ten cylinders was leaking. They took the following action:

> The team used a recently designed recovery vessel, a coffin-like steel container for enclosing leaking cylinders. After the nine unaffected cylinders were moved aside, the 64-inch-long recovery vessel was lowered into the pit. The team slid the leaking cylinder into the coffin and secured it.[14]

Whether the ERT is a local company team or a network of teams covering a geographical region, it should be included in the EAP. In-house teams are included in the assignments and responsibilities section. Community-based teams and networks are included in the coordinating organizations section.

COMPUTERS AND EMERGENCY RESPONSE

Advances in chemical technology have made responding to certain types of emergencies particularly complicated.

Fortunately, the complications brought by technology can be simplified by technology. Expert computer systems especially programmed for use in emergency situations can help meet the challenge of responding to a mixed-chemical emergency or any other type of emergency involving multiple hazards.

An **expert system** is a computer programmed to solve problems. Such systems rely on a database of knowledge about a very particular area, an understanding of the problems addressed in that area, and expertise in solving problems in that area. Talking to an expert system is like sitting at a terminal and keying in questions to an expert who is sitting at another terminal responding to the questions. The expert in this case is a computer program that pulls information from a database and uses it to make decisions based on **heuristics** or suppositional rules stated in an if-then format.[15]

Human thought processes work in a similar manner. For example, if our senses provide the brain with input suggesting the stovetop is hot, the decision is made not to touch it. In an if-then format, this may read as follows:

> IF hot, THEN do not touch.

This similarity to human thought processes is why the science on which expert systems are based is known as *artificial intelligence.*

A modern expert system used for responding to chemical emergencies provides information such as the following:

- Personal protective equipment needed for controlling and cleaning up
- Methods to be used in cleaning up the spill or toxic release
- Decontamination procedures

Safety Fact

Responding to Chemical Spills

To respond quickly and effectively to a chemical spill, the emergency response team must have the right equipment. Standard equipment should include a portable spill cart that can be quickly rolled to the site of a chemical spill. This cart should contain at least the following items:

- Spill suppression and absorption materials such as pads, blankets, and pillows
- Mops and brooms
- Acid neutralizers
- Barricade devices or materials such as mesh or tape
- Appropriate personal protective gear (e.g., gloves, eye protection, aprons, coveralls, and so on)

- Estimation of the likelihood that employees or the community will be exposed to hazard
- Reactions that may result from interaction of chemicals
- Combustibility of chemicals and other materials on hand
- Evacuation information
- Impact of different weather conditions on the situation
- Recommended first-aid procedures[16]

Expert systems can be user-friendly so that computer novices have no difficulty interacting with them. The advantages of expert systems are they do not panic or get confused, they consider every possible solution in milliseconds, they are not biased, they do not become fatigued, and they are detailed.

DEALING WITH THE PSYCHOLOGICAL TRAUMA OF EMERGENCIES

In addition to the physical injuries and property damage that can occur in emergencies, modern safety and health professionals must also be prepared to deal with potential psychological damage. Psychological trauma among employees involved in workplace disasters is as common as it is among combat veterans. According to Johnson, "Traumatic incidents do not affect only immediate survivors and witnesses. Most incidents result in layers of victims that stretch far beyond those who were injured or killed."[17]

Trauma is psychological stress. It occurs as the result of an event, typically a disaster or some kind of emergency, so shocking that it impairs a person's sense of security or well-being. Johnson calls trauma response "the normal reactions of normal people to an abnormal event."[18] **Traumatic events** are typically unexpected and shocking, and they involve the reality or threat of death.

Dealing with Emergency-Related Trauma

The typical approach to an emergency can be described as follows: control it, take care of the injured, clean up the mess, and get back to work. Often, the psychological aspect is ignored. This leaves witnesses and other coworkers to deal on their own with the trauma they've experienced. "Left to their own inadequate resources, workers can become ill or unable to function. They may develop resentment toward the organization, which can lead to conflicts with bosses and co-workers, high employee turnover—even subconscious sabotage."[19]

It is important to respond to trauma quickly, within 24 hours if possible and within 72 hours in all cases. The purpose of the response is to help employees get back to normal by enabling them to handle what they have experienced. This is best accomplished by a team of people who have had special training. Such a team is typically called the **trauma response team (TRT)**.

Trauma Response Team

A company's trauma response team may consist of safety and health personnel who have undergone special training or fully credentialed counseling personnel, depending on the size of the company. In any case, the TRT should be included in the assignments and responsibilities section of the EAP.

The job of the TRT is to intervene as early as possible, help employees acknowledge what they have experienced, and give them opportunities to express how they feel about it to people who are qualified to help. The *qualified to help* aspect is very important. TRT members who are not counselors or mental health professionals should never attempt to provide care that they are not qualified to offer. Part of the trauma training that safety and health professionals receive involves recognizing the symptoms of employees who need professional care and referring them to qualified care providers.

In working with employees who need to deal with what they have experienced, but are not so traumatized as to require referral for outside professional care, a group approach is best. According to Johnson, the group approach offers several advantages:

- It facilitates public acknowledgment of what the employees have experienced.
- It keeps employees informed, thereby cutting down on the number of unfounded rumors and horror stories that will inevitably make the rounds.
- It encourages employees to express their feelings about the incident. This alone is often enough to get people back to normal and functioning properly.
- It allows employees to see that they are not alone in experiencing traumatic reactions (e.g., nightmares, flashbacks, shocking memories, and so on) and that these reactions are normal.[20]

Convincing Companies to Respond

Modern safety and health professionals may find themselves having to convince higher management of the need to have a TRT. Some corporate officials may not believe that trauma even exists. Others may acknowledge its presence but view trauma as a personal problem that employees should handle on their own.

In reality, psychological trauma that is left untreated can manifest itself as **posttraumatic stress disorder**, the same syndrome experienced by some veterans of Vietnam and other wars. This disorder is characterized by "intrusive thoughts and flashbacks of the stressful event, the tendency to avoid stimulation, paranoia, concentration difficulties, and physiological symptoms such as rapid heartbeat and irritability."[21]

The American Psychiatric Association included posttraumatic stress disorder in its *Diagnostic and Statistical Manual* as far back as 1980.[22] Jeffrey T. Mitchell, president of the American Critical Incident Stress Foundation, likens preventing posttraumatic stress disorder to working with cement. Wet cement can be molded, shaped, manipulated, and even washed away. However, once it hardens, there is not much one can do with it.[23] This is the rationale for early intervention.

In today's competitive marketplace, companies need all their employees operating at peak performance levels. Employees experiencing trauma-related disorders will not be at their best. Safety and health professionals should use this rationale when it is necessary to convince higher management of the need to provide a company-sponsored trauma response team.

RECOVERING FROM DISASTERS

Many organizations put a great deal of effort into planning for disaster response including emergency evacuation. But what about after the disaster? According to John Kauffman, 43 percent of U.S. companies that experience a disaster have no recovery plan.[24] Recovering quickly is the key to staying in business. Approximately 70 percent of businesses that close down for a month or more as a result of a disaster either will never reopen or will fail within three years.[25]

A comprehensive disaster recovery plan should have at least the following components: recovery coordinator, recovery team, recovery analysis and planning, damage assessment and salvage operations, recovery communications, and employee support and assistance. The overall goal of a disaster recovery plan is to get an organization fully operational again as quickly as possible.

Recovery Coordinator

There must be one person who has ultimate responsibility and authority for disaster recovery. This person must have both the ability and the authority to take command of the situation, assess the recovery needs, delegate specific responsibilities, approve the necessary resources, interact with outside agencies, and activate the organization's overall response.

Recovery Team

The recovery team consists of key personnel to whom the disaster coordinator can delegate specific responsibilities. These responsibilities include facility management, security, human resources, environmental protection (if applicable), communications, and the various personnel needed to restart operations.

Recovery Analysis and Planning

This phase involves assessing the impact of the disaster on the organization and establishing both short- and long-term recovery goals. The more recovery analysis and planning that can be done, the better. One of the ways to do this is to consider various predictable scenarios and plan for them. This is the business equivalent of the war-gaming activities that take place in the military.

Damage Assessment and Salvage

This component of the plan has two elements: preparedness and recovery. The preparedness element should include the following information: (1) a comprehensive inventory of all property at the facility in question; (2) a checklist of the items on the inventory that are essential for maintaining the facility; (3) a list of all personnel who will aid in the recovery (make sure to have fully trained and qualified backup personnel in case a primary player is not available or is injured during the emergency); (4) a list of all vendors, contractors, and so on whose assistance will be needed during the damage assessment and salvage phase of recovery; (5) a worksheet that can be used to document all actions taken during recovery operations; and (6) procedures for quickly establishing a remote operational site.

The recovery element should include procedures for securing workspace for the recovery team and coordinator; identifying areas of the facility that must be accessible; maintaining security at the facility against looting and vandalism; analyzing and inspecting damage to the facility and reporting it to the recovery coordinator; assessing the extent of damage to goods, supplies, and equipment; photographing and videotaping damage to the facility; taking appropriate action to prevent additional damage to the facility; repairing, restoring, and resetting fire detection and suppression equipment; and investigating accidents.

Recovery Communications

Communication is one of the most important considerations in disaster recovery. This component of the plan should deal with both who is to be notified and how that is to take place. The "how" aspects concern backup procedures for telephone service, e-mail, and so forth. Will cell phones be used? Will radio stations be part of the mix? Will "walkie-talkie" type radios be used for communicating on-site? The following list contains the types of entities who might have to be contacted as part of the disaster recovery operation:

- Customers
- Vendors and suppliers
- Insurance representatives
- Employees' families
- Appropriate authorities
- Media outlets (radio and television stations and newspapers)

Employee Assistance

After a disaster, employees are likely to need various types of assistance including financial, medical, and psychological. Kauffman recommends the following steps for developing this component of the disaster recovery plan:

1. Determine postdisaster work schedules and provide them to employees. Include overtime work if it will be necessary, and make sure employees know that flexibility in scheduling work hours will be important until the recovery is complete.
2. Plan for the whole range of employee-assistance services that might be needed including medical, transportation, financial, shelter, food, water, clothing, and psychological services (trauma, shock, and stress counseling).
3. Plan for the provision of grief counseling. The best way to handle this is to assign grief counseling to the company's employee assistance program provider.
4. Plan for the possible need to relocate the facility as part of disaster recovery.
5. Plan to give employees opportunities to participate in personal actions taken on behalf of fatally injured employees and their families. Work for employee consensus before deciding what to do for these families.
6. Plan to fully inform all employees about what happened, why, how the company is responding in the short term, and how it will respond in the long term. Be sure to build in ways to let employees know the company cares about them and will do everything possible to protect their safety.[26]

TERRORISM IN THE WORKPLACE

"The events of Sept. 11 and the ongoing threat of bioterrorism in American workplaces have many people living and working in fear. Unquestionably, the world has changed, and few of us are happy with the direction. It's important to put everything in perspective, however, and consider the proper role of the employer."[27] This section describes the roles that employers and their safety and health personnel can play in preparing for, preventing, and responding to terrorist attacks in the workplace.

Role of the Employer

There is no question that the threat of terrorist attacks has become an ever-present reality in today's workplace. Because this is the case, employers clearly have a role to play in preparing for terrorist attacks, taking all prudent precautions to prevent them, and in responding properly should an attack occur. Because terrorism threatens the safety and health of employees, it is more than just a security issue, it is also an occupational safety and health issue.

Chip Dawson summarizes the roles of employers and safety and health professionals relating to terrorism in the workplace:

Run a safe and caring operation. Employees watch the nightly news and read their morning newspapers. They know what is going on in their world. As a result, many are discomfited by the possibility that they or their workplace might become the target of a terrorist attack. Consequently, the first responsibility of the employer is to run a safe operation in which employees know their safety is a high priority. Many of the engineering, administrative, training, and enforcement actions taken to make the workplace safe from occupational hazards will also help mitigate the threat of terrorism.

Listen to employees. Employees are concerned about the threat of terrorism, and they have a right to be. Employers should take the concerns of employees seriously, and deal with them. Answer questions, communicate openly and frequently, and refer employees who need professional help to the employee assistance program.

Train employees. Security and safety procedures do little good unless employees know what they are and how to use them. In addition, personnel in certain positions need to have specialized knowledge relating to terrorism. For example, mailroom personnel need to be trained in how to screen incoming mail for biohazards and explosives.

Communicate. Talk with employees openly and frequently. Let them know what the company knows. It is better for employees to hear news from the company than to receive it in the form of third-party gossip and rumors. Before giving out information, however, it is a good idea to verify it. Check with local authorities or go online to one of the following Web sites: http://www.snopes.com or http://www.urbanlegends.com.

Know your personnel. Institute background checks as part of the hiring process. Make sure that supervisors get to know their direct reports well enough to sense when something is wrong with one of them. If inconsistencies in normal behavioral patterns occur, address them right away.

Empower personnel. Empower employees to back away from a situation that does not feel right or that makes them uncomfortable. Be flexible in allowing them to have time off for family activities. This can be especially important following particularly busy times for employees who have worked longer than normal hours.

Harden the site against external threats and restrict access. Insulate the workplace from negative outside influences, control who has access to the workplace, and take all necessary steps to reduce the exposure of employees to potential threats. Call in security experts if necessary to help develop and implement the necessary controls.

Remove any barriers to clear visibility around the facility. The better employees can see around them, the less likely it will be that a terrorist will be able to pull off a surprise attack. The trees, shrubs, and bushes that make the perimeter of the facility and parking lot so attractive can be used by unauthorized personnel to hide while attempting to gain access.

Have and enforce parking and delivery regulations. Arrange parking spaces so that no car is closer to a building than 100 feet. This will lessen the likelihood of damage from a car bomb. In addition, have strict delivery procedures so that terrorists posing as delivery personnel do not gain access through this means.

Make sure that visitors can be screened from a distance. Arrange the facility so that visitors are channeled into a specific area, an area in which they can be viewed by company personnel from a distance. This will lessen the likelihood of a terrorist gaining access by overpowering or killing access-control personnel.

Keep all unstaffed entrance doors locked and alarmed. Employees need to be able to get out of the building through any exit door, but access into the building should be channeled through doors staffed by access-control personnel.

Make air intakes and other utilities inaccessible to all but designated maintenance personnel. Releasing toxic material into the air intake is one way terrorists could harm the highest possible number of people. The likelihood of this happening can be decreased by locating air intakes in inaccessible locations.

Prevent access to roofs and upper stories. Terrorists who cannot gain access on the ground floor might simply gain entry through doors and other openings on the roof. Consequently, it is important to keep roof doors locked from the outside and alarmed. They should open from the inside, but doing so should trigger an alarm. It is also important to establish control procedures for emergency escape routes from the roof and upper stories so that these avenues are not used by terrorists trying to gain access into the building.

Secure trash containers. The grounds should be kept free of debris and clutter. Further, trash containers should be secured either by keeping them inside the building's wall or, if kept outside, at a distance from the building.

Ensure that employees, contractors, and visitors wear badges. Establish a system in which all employees, contractors, and visitors must wear badges in order to gain access to the facility. Require identification of all visitors and an internal sponsor before providing them with badges. Have visitors sign in and out on every visit.

Have an emergency response plan and practice it periodically. Plan for all predictable exigencies and practice the various components of the plan on a regular basis.

Be cautious of information placed on your company's Web site. Make sure that your company's Web site does not contain information that can be used by terrorists such as detailed maps, floor plans, or descriptions of hazardous materials stored on-site.

Keep up-to-date with the latest safety and security strategies. Crime Prevention Through Environmental Design (CPTED) should become part of the knowledge base of safety and health professionals. For the latest information concerning CPTED, visit the following Web site periodically: http://www.arch.vt.edu/crimeprev/pages/home.html.

Protect the integrity of your facility's key system. Terrorists can use any key from your facility to make their own master key if they know how, and many do. To make matters worse, they share information about how to make master keys via the Internet. The following procedures will help protect the integrity of your facility's key system: (a) restrict access to keys (including restroom keys); (b) consider not using master keys; and (c) switch from a key system to an electronic keycard or fingerprint recognition system.[28]

Securing Hazardous Materials

According to Dale Petroff

> You must understand that the hazmats under your control may be used by terrorists to create weapons of mass destruction. It is important to note that the first poison gas used in warfare was chlorine gas, and that fertilizer mixed with fuel oil became a devastating explosive device that destroyed the Murrah Federal Building in Oklahoma City. More recently . . . a truck used to transport natural gas was turned into a bomb in Tunisia that destroyed a synagogue and killed 17 people. All of these materials and many others that can be converted into weapons are found in large quantities throughout the United States.[29]

Clearly, one of the tactics of terrorists is to convert hazardous materials used in the workplace into weapons of mass destruction. Consequently, it is important for facilities that produce, use, or store any type of hazardous materials to develop, implement, and enforce a security program that denies terrorists access to these materials.[30] "The goal of the security plan should be to implement measures that deter, detect, delay, or defeat the threat. Deterrence can be improved by using highly visible measures and randomness. This is cost-efficient and complicates the threatening person or group's planning."[31]

A hazmat security plan should have two broad components: *personnel security* and *physical security.* The fundamental elements of the personnel component of the plan are to (1) determine who should be granted access to the materials; (2) conduct comprehensive background checks on all individuals who require access; (3) submit employees who require access to psychological

Safety Tip

Sources of Help on Terrorism for Safety and Health Professionals

- *Pentagon*
  http://www.dtic.mil/ref/biochem/Training/pent/pent_slidel.html
- *U.S. Department of Defense*
  http://www.cbiac.apgea.army.mil/
- *Centers for Disease Control and Prevention*
  http://www.bt.cdc.gov/Agent/Agentlist.asp
- *Central Intelligence Agency (CIA)*
  http://www.cia.gov/terrorism/index.html
- *U.S. State Department*
  http://www.ojp.usdoj.gov/odp/

screening to ensure stability; and (4) require identification badges with photographs or fingerprints of those authorized access.

The physical security component consists of measures taken to prevent or control access to the hazardous materials in question. Security practices in this component should be integrated and layered by using a combination of measures including fences, lights, electronic alarm systems, guards, reaction forces, and the *two-person rule* that requires two employees to be present in order to gain access to the hazardous materials. A good source of help concerning how to develop a plan for keeping hazardous materials out of the hands of terrorists is as follows:

American Chemistry Council
703-741-5000
http://www.americanchemistry.com

RESUMING BUSINESS AFTER A DISASTER

After a disaster in the workplace—regardless of whether the source of the disaster was natural, accidental, or terrorist related—a comprehensive hazard assessment should be completed before business is resumed. Before resuming business, an organization should consider the following factors:[32]

- *Structural integrity.* Has the structural integrity of the building been checked by competent engineering professionals to ensure that it is safe to enter?
- *Utility checks.* Have all utilities—gas, electricity, water, sewer—been checked by competent professionals to ensure that there are no leaks, cracks, loose wires, and so on? Have the appropriate utility companies given their approval for reopening? Remember that if reopening involves the use

of electric generators, these devices should not be used inside the building because they can create a carbon monoxide hazard.

- *Cleanup protection.* Make sure that cleanup crews are properly protected from any hazardous materials or conditions that might have been created by the disaster. Make sure they properly use the appropriate personal protective gear, and comply with all applicable safety and health regulations and procedures as they clean the facility.
- *Health and sanitation.* Kitchens, bathrooms, and any area in which food or potentially hazardous or toxic substances are stored should be checked by competent professionals and thoroughly cleaned to prevent the exposure of employees and customers to hazardous conditions.
- *Air quality.* Make sure the air quality in the facility is tested by competent professionals before allowing personnel to enter the building. Certain types of disasters such as hurricanes, for example, might have caused the proliferation of mold and mildew. The air should be checked for any potentially hazardous biological or chemical agents that could be harmful to humans.
- *Ventilation.* Make sure that all types of ductwork and ventilation have been checked for the presence of potentially harmful biological and chemical agents as well as for dust and debris that might impede airflow. Once it appears that ventilations systems are clean, have the air conditioning and heating systems started up and all ventilation systems checked again before allowing personnel back into the building. When the air-conditioning and heating systems are restarted, blow cold air through them, even in winter. This will help prevent the growth of mold in the ventilation ducts.
- *Walls, ceilings, and floors.* Check walls and ceilings to ensure that no materials are in danger of falling off—inside for occupants and outside for pedestrians. Check floors for any hazards that might contribute to slipping and falling.
- *Safety equipment.* Check all fire extinguishers, all types of alarms, and any other safety equipment to determine if it has been damaged. Make sure that all safety equipment is in proper working order before allowing personnel to reenter the facility.
- *Lighting.* Ensure that all illumination devices are in proper working order and that all personnel have the required amount of illumination to do their jobs. Employees should not return to work unless the necessary illumination is available and working.
- *Hazardous waste removal.* Any type of potentially hazardous material that is left lying around after the disaster should be collected and properly disposed of. Broken glass, debris, litter, and sharp-edged material should be removed before employees are allowed to return to work.
- *Machines and equipment.* All machines and equipment should be checked carefully before they are reenergized. All electrical, gas, hydraulic, fill, drain, and plumbing lines should be checked for leaks and proper connections before the machines are energized for use.
- *Furniture.* Check all furniture to ensure its structural integrity. Make sure that fasteners, braces, and supports have not been damaged during the emergency or that furniture has not become unstable due to water damage.

> ## Safety Fact
>
> ### Federal Guidelines for Resuming Business
>
> Businesses may obtain comprehensive federal guidelines for resuming business after a disaster from the National Institute for Occupational Safety and Health (NIOSH) or the Federal Emergency Management Agency (FEMA) at these Web sites:
>
> http://www.cdc.gov.niosh
> http://www.fema.gov

SUMMARY

1. An emergency is a potentially life-threatening situation, usually occurring suddenly and unexpectedly. Emergencies may be the result of natural or human causes.

2. Preparing for emergencies involves planning, practicing, evaluating, and adjusting. An immediate response is critical in emergencies.

3. The Emergency Planning and Community Right-to-Know Act has the following four main components: emergency planning, emergency notification, information requirements, and toxic chemical release reporting.

4. For proper coordination of the internal emergency response, it is important that one person be in charge and that everyone involved knows who that person is.

5. Because there is no way to predict when first aid may be needed, part of preparing for emergencies should include training employees to administer first aid. In certain cases, OSHA requires that companies have at least one employee on-site who has been trained in first aid.

6. In addition to providing first-aid training, it is important to have well-stocked first-aid kits readily available, have personal protective devices available, post emergency telephone numbers, and keep all employees informed.

7. A company's emergency action plan should be a collection of small plans for each anticipated emergency. These plans should have the following components: procedures, coordination, assignments and responsibilities, accident prevention strategies, and schedules.

8. The OSHA standard for evacuation planning is 29 CFR 1910.38. This standard requires a written plan for evaluating the facility in the event of an emergency. Critical elements of the plan are as follows: marking of exit routes, communications, outside assembly, and training.

9. EAPs should be customized to be location-specific by including a map, an organization chart, local coordination information, and local training schedules. They should consider the needs of all personnel, including those with disabilities.

10. An emergency response team is a special team to handle general and localized emergencies, facilitate evacuation and shutdown, protect and salvage company property, and work with civil authorities.

11. An emergency response network is a network of emergency response teams that covers a designated geographical area.

12. Computers can help simplify some of the complications brought by advances in technology. Expert systems mimic human thought processes in making decisions on an if–then basis regarding emergency responses.

13. Trauma is psychological stress. It typically results from exposure to a disaster or emergency so shocking that it impairs a person's sense of security or well-being. Trauma left untreated can manifest itself as posttraumatic stress disorder. This disorder is characterized by intrusive thoughts, flashbacks, paranoia, concentration difficulties, rapid heartbeat, and irritability.

14. A disaster recovery plan should have at least the following components: recovery coordinator, recovery team, recovery analysis and planning, damage assessment and salvage operations, recovery communications, and employee support and assistance.

15. Employers can help decrease the likelihood of a terrorist attack on their facilities by taking the following actions: run a safe and caring operation; listen to employees; train employees; communicate; know your personnel; empower personnel; harden the site against external threats and restrict access; remove any barriers to clear visibility around the facility; have and enforce parking and delivery regulations; make sure that visitors can be screened from a distance; keep all unstaffed entrance doors locked from the outside and alarmed; make air intakes and other utilities inaccessible to all but designated personnel; prevent access to roofs and upper stories; secure trash containers; ensure that employees, contractors, and visitors wear badges; have an emergency response plan and practice it on a regular basis; be cautious of what information is placed on your company's Web site; keep up-to-date with the latest safety and security strategies; and protect the integrity of your facility's key system.

16. Secure hazardous materials so that terrorists cannot gain access to them for use in making bombs and other weapons of mass destruction. A hazmat security plan should have two components: personnel security and physical security.

17. All systems, conditions, and potential hazards should be checked and corrected as appropriate before resuming business after a disaster.

KEY TERMS AND CONCEPTS

Adjusting	Emergency action plan (EAP)
Chain of command	Emergency coordinator
Elevator policy	Emergency notification

Emergency planning

Emergency Planning and Community Right-to-Know Act (EPCRA)

Emergency response management team (ERMT)

Emergency response network (ERN)

Emergency response plan

Emergency response team (ERT)

Evaluating

Expert system

First-aid training

Heuristics

Information requirements

Local emergency planning committee (LEPC)

Location-specific

Order of response

OSHA standards

Planning

Posttraumatic stress disorder

Practicing

Shelter-in-place (SIP) plans

Spill cart

State emergency response commission (SERC)

Toxic chemical release reporting

Trauma

Trauma response team (TRT)

Traumatic event

REVIEW QUESTIONS

1. Define the term *emergency*.
2. Explain the rationale for emergency preparation.
3. List and explain the four main components of the Emergency Planning and Community Right-to-Know Act.
4. Describe how a company's emergency response effort should be coordinated.
5. How do OSHA standards relate to emergency preparation?
6. Explain how you would provide first-aid training if you were responsible for setting up a program at your company.
7. Besides training, what other first-aid preparation should a company take?
8. What are the critical elements of OSHA's standard for evacuation planning?
9. Describe the essential components of an EAP and explain how to build the needs of personnel with disabilities into the plan.
10. How can a company localize its EAP?
11. Define the following emergency response concepts: ERT, ERN, and TRT.
12. What is an expert system? How can one be used in responding to an emergency?
13. What is trauma?
14. Why should a company include trauma response in its EAP?
15. Describe how a company may respond to the trauma resulting from a workplace emergency.
16. What elements should a disaster recovery plan contain?
17. How can employers prepare for the threat of terrorism?
18. Explain the precautions that should be taken before resuming business after a disaster.

ENDNOTES

1. L. Marley, "Emergency Medical Teams Need Organization and Administration," *Safety & Health* 142, no. 5: 28.
2. Ibid.
3. Ibid.
4. Ibid., 30.
5. United States Department of Labor, Office of Disability Employment Policy, "Preparing the Workplace for Everyone: Accounting for the Needs of People with Disabilities." Retrieved June 2006 from http://www.dol.gov.odep/pubs/ep/preparing/workplace_final.pdf.
6. S. Larson, "Heading for the Exits," *Occupational Health & Safety* 72, no. 2: 60.
7. C. Schroll, "Evacuation Planning: A Matter of Life and Death," *Occupational Hazards*, June 2002, 50–54.
8. Ibid., 54.
9. Ibid., 54.
10. H. Christen, *The EMS Incident Management System* (Upper Saddle River, NJ: Prentice Hall, 1998), 8.
11. Ibid., 9.
12. Ibid., 10.
13. Ibid., 16.
14. Ibid., 17.
15. Ibid., 168.
16. Ibid., 169.
17. E. Johnson, "Where Disaster Strikes," *Safety & Health* 145, no. 2: 29.
18. Ibid., 28.
19. Ibid., 29.
20. Ibid., 30.
21. National Safety Council, "Trained for Trauma," *Safety & Health* 145, no. 2: 32.
22. Ibid.
23. Ibid.
24. J. Kauffman, "Recovering from Disaster," *Occupational Hazards*, January 2003, 69–71.
25. Ibid., 71.
26. Ibid.
27. C. Dawson, "The Role of the Employer in Domestic Security," *Occupational Hazards*, January 2002, 31.
28. Ibid., 31–32.
29. D. M. Petroff, "Security of Hazardous Materials," *Occupational Health & Safety* 72, no. 4: 46.
30. Ibid., 44–48.
31. Ibid., 48.
32. American Society of Safety Engineers, "ASSE Offers Business Resumption Safety Checklist," Occupational Health & Safety Online, September 9, 2005. Retrieved from http://www.ohsonline.com/stevens/ohspub.nsf/d3d5b4f938b22b6e8625670c006bc58/6600.

ETHICS AND SAFETY

Major Topics

- ▓ An Ethical Dilemma
- ▓ Ethics Defined
- ▓ Ethical Behavior in Organizations
- ▓ Safety and Health Professionals' Role in Ethics
- ▓ Company's Role in Ethics
- ▓ Handling of Ethical Dilemmas
- ▓ Questions to Ask When Making Decisions
- ▓ Ethics and Whistle-Blowing

Practically everyone agrees that business practices of industrial firms should be above reproach with regard to ethical standards. Few people are willing to defend unethical behavior. For the most part, industry in the United States operates within the scope of accepted legal and ethical standards. According to Peter Drucker,

> Business ethics is rapidly becoming the "in" subject . . . there are countless seminars on it, speeches, articles, conferences and books, not to mention the many earnest attempts to write "business ethics" into the law.[1]

However, unethical behavior does occur frequently enough that modern safety and health professionals should be aware of the types of ethical dilemmas that they may occasionally face and should know how to deal with such issues. How to deal successfully and effectively with ethics on the job is the subject of this chapter.

AN ETHICAL DILEMMA

According to Stead, Worrell, and Stead,

> Managing ethical behavior is one of the most pervasive and complex problems facing business organizations today. Employees' decisions to behave ethically or unethically are influenced by a myriad of individual and situational factors. Background, personality, decision history, managerial philosophy, and reinforcement are but a few of the factors which have been identified by researchers as determinants of employees' behavior when faced with ethical dilemmas.[2]

Consider the following example of an ethical dilemma:

Mil-Tech Manufacturing Company is a Department of Defense contractor that produces air- and watertight aluminum containers for shipping nonnuclear munitions such as missiles, bombs, and torpedoes. Business has been good and Mil-Tech is prospering. However, the company's management team has a problem.

Mil-Tech has been awarded a contract to produce 10,000 boxes in six months. The company's maximum capacity is currently 1,000 boxes per month. Unless Mil-Tech can find a way to increase its capacity, the company will be forced to add new facilities, equipment, and personnel—an expensive undertaking that will quickly eat up the projected profits of the new contract.

The most time-consuming bottleneck in the production of the boxes is the painting process, the last step. The problem is with the paint that Mil-Tech uses. It poses no health, safety, or environmental hazards, but it is difficult to apply and requires at least two hours to dry. Clearly, the most expeditious way to increase productivity is to find a paint that is easier to apply and takes less time to dry.

The production manager has been searching frantically for a substitute paint for two weeks and has finally found one. The new paint is easy to apply, and it dries almost on contact. However, it is extremely toxic and can be dangerous to anyone exposed to it at any time before it dries. Personal protective equipment and other hazard-prevention techniques can minimize the health problems, but they must be used properly with absolutely no shortcuts. In addition, it is recommended that every employee who will work with the paint complete three full days of training.

Mil-Tech's management team is convinced that the union will not consent to the use of this paint even if the PPE is purchased and the training is provided. To complicate matters, the supplier of the paint cannot provide the training within a timeframe that meets Mil-Tech's needs. In a secret meeting, top management officials decide to purchase PPE, use the new paint, and forego the training. More importantly, the management team decides to withhold all information about the hazards associated with the new paint.

Camillo Garcia, Mil-Tech's safety and health manager, was not invited to the secret meeting. However, the decisions made during the meeting were slipped to him anonymously. Garcia now faces an ethical dilemma. What should he do? If he chooses to do nothing, Mil-Tech employees may be inappropriately exposed to an extremely hazardous substance. If he confronts the management team with what he knows, he could fall into disfavor or even lose his job. If he shares what he knows with union leaders, he may be called on to testify about what he knows. This is an example of the type of ethical dilemma that safety and health professionals face on the job.

ETHICS DEFINED

There are many definitions of the term *ethics*. However, no one definition has emerged as universally accepted. According to Paul Taylor, the concept can be defined as "inquiry into the nature and grounds of morality where morality is taken to mean moral judgments, standards, and rules of conduct."[3] According to Arlow and Ulrich, ethical dilemmas in the workplace are more complex than ethical situations in general.[4] They involve societal expectations, competition, and social responsibility as well as the potential consequences of an employee's behavior on customers, coworkers, competitors, and the public at large. The result of the often-conflicting and contradictory interests of workers, customers, competitors, and the general public is a natural tendency for ethical dilemmas to occur frequently in the workplace.

Anytime that ethics is the topic of discussion, such terms as **conscience**, **morality**, and **legality** are frequently heard. Although these terms are closely associated with ethics, they do not, by themselves, define it. For the purpose of this book, **ethics** is defined as follows:

> Ethics is the application of morality within a context established by cultural and professional values, social norms, and accepted standards of behavior.

Morality refers to the **values** that are subscribed to and fostered by society in general and individuals within society. Ethics attempts to apply reason in determining rules of human conduct that translate morality into everyday behavior. **Ethical behavior** is that which falls within the limits prescribed by morality.

How, then, does a safety and health professional know if someone's behavior is ethical? Ethical questions are rarely black and white. They typically fall into a **gray area** between the two extremes of right and wrong. Personal experience, self-interest, point of view, and external pressure often cloud this gray area further.

Guidelines for Determining Ethical Behavior

Guidelines are needed for safety and health professionals to use when trying to sort out matters that are not clearly right or wrong. First, however, it is necessary to distinguish between the concepts of *legal* and *ethical*. They are not the same thing. Just because an option is legal does not necessarily mean it is ethical.

In fact, it is not uncommon for people caught in the practice of questionable behavior to use the "I didn't do anything illegal" defense. A person's behavior can be well within the scope of the law and still be unethical. The following guidelines for determining ethical behavior assume that the behavior in question is legal (Figure 26–1):

- Apply the **morning-after test**. This test asks, "If you make this choice, how will you feel about it tomorrow morning?"
- Apply the **front-page test**. This test encourages you to make a decision that would not embarrass you if printed as a story on the front page of your hometown newspaper.
- Apply the **mirror test**. This test asks, "If you make this decision, how will you feel about yourself when you look in the mirror?"
- Apply the **role-reversal test**. This test requires you to trade places with the people affected by your decision and view the decision through their eyes.
- Apply the **commonsense test**. This test requires you to listen to what your instincts and common sense are telling you. If it feels wrong, it probably is.

Blanchard and Peale suggest their own test for deciding what the ethical choice is in a given situation.[5] Their test consists of the following three questions:

- Is it legal?
- Is it balanced?
- How will it make me feel about myself?

If a potential course of action is not legal, no further consideration is in order. If an action is not legal, it is also not ethical. If an action is balanced, it is fair to all involved. This means that safety and health professionals and their team members have responsibilities that extend well beyond the walls of their unit, organization, and company. If a course of action is in keeping with your own moral structure, it will make you feel good about yourself. Blanchard and Peale also list the following "Five P's of Ethical Power":

- *Purpose.* Individuals see themselves as ethical people who let their conscience be their guide and in all cases want to feel good about themselves.
- *Pride.* Individuals apply internal guidelines and have sufficient self-esteem to make decisions that may not be popular with others.

Figure 26–1
Guidelines to determine what is ethical.

Guidelines for Ethical Choices
1. Apply the morning-after test
2. Apply the front-page test
3. Apply the mirror test
4. Apply the role-reversal test
5. Apply the commonsense test

- *Patience.* Individuals believe right will prevail in the long run, and they are willing to wait when necessary.
- *Persistence.* Individuals are willing to stay with an ethical course of action once it has been chosen and see it through to a positive conclusion.
- *Perspective.* Individuals take the time to reflect and are guided by their own internal barometer when making ethical decisions.[6]

These tests and guidelines will help safety and health professionals make ethical choices in the workplace. In addition to internalizing the guidelines themselves, safety and health professionals may want to share these values with all employees with whom they interact.

ETHICAL BEHAVIOR IN ORGANIZATIONS

Research by Trevino suggests that ethical behavior in organizations is influenced by both individual and social factors.[7] Trevino identified three personality measures that can influence an employee's ethical behavior: (1) ego strength, (2) Machiavellianism, and (3) locus of control.

An employee's **ego strength** is his or her ability to undertake self-directed tasks and to cope with tense situations. A measure of a worker's **Machiavellianism** is the extent to which he or she will attempt to deceive and confuse others. **Locus of control** is the perspective of workers concerning who or what controls their behavior. Employees with an internal locus of control feel that they control their own behavior. Employees with an external locus of control feel that their behavior is controlled by external factors (e.g., rules, regulations, their safety and health professional, and so on).

Preble and Miesing suggest that social factors also influence ethical behavior in organizations.[8] These factors include gender, role differences, religion, age, work experience, nationality, and the influence of other people who are significant in an individual's life. Luthans and Kreitner state that people learn appropriate behavior by observing the behavior of significant role models (parents, teachers, public officials, and so on).[9] Because safety and health professionals represent a significant role model for their team members, it is critical that they exhibit ethical behavior that is beyond reproach in all situations.

SAFETY AND HEALTH PROFESSIONALS' ROLE IN ETHICS

Using the guidelines set forth in the previous section, safety and health professionals should be able to make responsible decisions concerning ethical choices. Unfortunately, deciding what is ethical is much easier than actually doing what is ethical. In this regard, trying to practice ethics is like trying to diet. It is not so much a matter of knowing you should cut down eating, it is a matter of following through and actually doing it.

Safety Fact

Models for Determining Ethical Behavior

In addition to the various tests that can be used for determining ethical behavior, there are also numerous models:

- Categorical imperative (black and white)
- Conventionalistic ethic (anything legal is ethical)
- Disclosure rule (explain actions to a wide audience)
- Doctrine of the mean (virtue through moderation)
- Golden Rule (do unto others . . .)
- Intuition rule (what is right is just known)
- Market ethic (whatever makes a profit is right)
- Means-end ethic (end justifies the means)
- Might-equals-right ethic (self-explanatory)
- Organizational ethic (loyalty to the organization)
- Practical imperative (treat people as ends, not means)
- Equal freedom (full freedom unless it deprives another)
- Proportionality ethic (good outweighs the bad)
- Professional ethic (do only what can be explained to your peers)
- Revelation ethic (answers revealed by prayer)
- Rights ethic (protect rights of others)
- Theory of justice (impartial, even-handed)

It is this fact that defines the role of safety and health professionals with regard to ethics. Their role has three parts. First, they are responsible for setting an example of ethical behavior. Second, they are responsible for helping fellow employees make the right decision when facing ethical questions. Finally, safety and health professionals are responsible for helping employees follow through and actually undertake the ethical option once the appropriate choice has been identified. In carrying out their roles, safety and health professionals can adopt one of the following approaches (Figure 26–2): the best-ratio approach, the black-and-white approach, or the full-potential approach.

Best-Ratio Approach

The **best-ratio approach** is the pragmatic approach. Its philosophy is that people are basically good and under the right circumstances behave ethically. However, under certain conditions, they can be driven to unethical behavior. Therefore, the safety and health professional should do everything possible to create conditions that promote ethical behavior and try to maintain the best possible ratio of good choices to bad. When hard decisions must be made, the appropriate choice is the one that does the most good for the most people. This is sometimes referred to as *situational ethics*.

Figure 26–2
Three basic approaches to
handling ethical problems.

Black-and-White Approach

Using the **black-and-white approach**, right is right, wrong is wrong, and circumstances are irrelevant. The safety and health professional's job is to make ethical decisions and carry them out. It is also to help employees choose the ethical route. When difficult decisions must be made, safety and health professionals should make fair and impartial choices regardless of the outcome.

Full-Potential Approach

Safety and health professionals who use the **full-potential approach** make decisions based on how the outcomes affect the ability of those involved to achieve their full potential. The underlying philosophy is that people are responsible for realizing their full potential within the confines of morality. Choices that can achieve this goal without infringing on the rights of others are considered ethical.

Decisions made may differ, depending on the approach selected. For example, consider the ethical dilemma presented at the beginning of this chapter. If the safety and health manager, Camillo Garcia, applies the best-ratio approach, he may decide to keep quiet, encourage the proper use of personal protective equipment, and hope for the best. On the other hand, if he takes the black-and-white approach, he will be compelled to confront the Mil-Tech management team with what he knows.

COMPANY'S ROLE IN ETHICS

Industrial firms have a critical role to play in promoting ethical behavior among their employees. Safety and health professionals cannot set ethical examples alone or expect employees to behave ethically in a vacuum. A company's

Discussion Case

What Is Your Opinion?

The city council members are in a real quandary about the request from International Plastics Corporation (IPC) to open a new plant on a 100-acre plot owned by the city. On one hand, the city needs the new jobs that the IPC plant would bring—badly. High unemployment is the city's most serious problem, and every member of the city council ran on a job-creation platform in the last election. On the other hand, the council members have learned that IPC is not always a good corporate citizen in spite of its claims to the contrary.

Several cities with IPC plants have had problems enforcing their safety and health regulations. The consensus among other cities is that IPC officials say all the right things until contracts are signed. Then, all of a sudden, they begin to procrastinate, stonewall, and break promises. Should the city council allow IPC to build the new plant? What is your opinion?

role in ethics can be summarized as (1) creating an internal environment that promotes, expects, and rewards ethical behavior; and (2) setting an example of ethical behavior in all external dealings (Figure 26–3).

Creating an Ethical Environment

A company creates an **ethical environment** by establishing policies and practices that ensure all employees are treated ethically and then enforcing these policies. Do employees have the right of due process? Do employees have access to an objective grievance procedure? Are there appropriate safety and health measures to protect employees? Are hiring practices fair and impartial? Are promotion practices fair and objective? Are employees protected from harassment based on race, gender, or other reasons? A company that establishes an environment that promotes, expects, and rewards ethical behavior can answer yes to all of these questions.

One effective way to create an ethical environment is to develop an **ethics philosophy** and specific, written guidelines for implementing that philosophy

Figure 26–3
Ethics cannot be practiced in a vacuum. The company has a critical role to play.

that are shared with all employees. Martin Marietta Corporation of Orlando, Florida, has a *Code of Ethics and Standards of Conduct* that is shared with all employees. The code begins with the following statement of philosophy:

> Martin Marietta Corporation will conduct its business in strict compliance with applicable laws, rules, regulations, and corporate and operating unit policies, procedures and guidelines, with honesty and integrity, and with a strong commitment to the highest standards of ethics. We have a duty to conduct our business affairs within both the letter and the spirit of the law.[10]

This statement sets the tone for all employees at Martin Marietta. It lets them know that higher management not only supports ethical behavior but also expects it. This approach makes it less difficult for safety and health professionals when they find themselves caught in the middle between the pressures of productivity and the maintenance of safe work practices.

In addition to its corporate ethics philosophy, Martin Marietta publishes a more specific *Credo and Code of Conduct*. The Martin Marietta credo is summarized as follows:

> In our daily activities we bear important obligations to our country, our customers, our owners, our communities, and to one another. We carry out these obligations guided by certain unifying principles:
>
> - Our foundation is INTEGRITY
> - Our strength is our PEOPLE
> - Our style is TEAMWORK
> - Our goal is EXCELLENCE[11]

This **ethics credo** shows employees that they have obligations extending well beyond their work units and that how they perform their work can have an impact, negative or positive, on fellow employees, the company, customers, and the country. Key concepts set forth in the credo are *integrity, people, teamwork*, and *excellence*. Safety and health professionals who stress, promote, and model these concepts will make a major contribution to ethical behavior in the workplace.

Although the emphasis on ethics in the workplace is relatively new, the concept is not. According to Shanks, Robert Wood Johnson, the leader who built Johnson & Johnson into a major international corporation, developed an ethics credo for his company as early as the mid-1940s.[12]

Johnson's credo read as follows:

- To customers and users: quality and service at reasonable prices
- To suppliers: a fair opportunity
- To employees: respect, equal opportunity, and a sense of job security
- To communities: a civic responsibility
- To the environment: protection
- To shareholders: a fair return[13]

Written philosophies and guidelines such as those developed by Martin Marietta Corporation and Johnson & Johnson are the first step in creating an ethical environment in the workplace. Safety and health professionals can play a key role in promoting ethical behavior on the job by encouraging higher management to develop written ethics philosophies, credos, and guidelines and then by modeling the behavior that they encourage.

Setting an Ethical Example

Companies that take the "Do as I say, not as I do" approach to ethics will not succeed. Employees must be able to trust their company to conduct all external and internal dealings in an ethical manner. Companies that do not pay their bills on time; companies that pollute; companies that place short-term profits ahead of employee safety and health; companies that do not live up to advertised quality standards; companies that do not stand behind their guarantees; and companies that are not good neighbors in their communities are not setting a good ethical example. Such companies can expect employees to mimic their unethical behavior.

A final word on the company's role in ethics is in order. In addition to creating an ethical internal environment and handling external dealings in an ethical manner, companies must support safety and health professionals who make ethically correct decisions. This support must be given not just when such decisions are profitable, but in all cases. For example, in the ethical dilemma presented earlier in this chapter, say Camillo Garcia decided that his ethical choice was to confront the management team with his knowledge of the hazards associated with the new paint. Management gave the order to withhold critical information. This is obviously the profitable choice in the short run. But is it the ethical choice? If Camillo Garcia does not think so, will Mil-Tech stand behind him? If not, everything else that the company does to promote ethics will fail.

HANDLING OF ETHICAL DILEMMAS

No person will serve long as a safety and health professional without confronting an ethical dilemma. How, then, should you proceed when confronting an ethical dilemma? There are three steps (Figure 26–4):

1. Apply the various guidelines for determining what is ethical presented earlier in this chapter.
2. Select one of the three basic approaches to handling ethical questions.
3. Proceed in accordance with the approach selected, and proceed with consistency.

Apply the Guidelines

In this step, as a safety and health professional, you should apply as many of the tests set forth in Figure 26–1 as necessary to determine the ethically correct decision. In applying these guidelines, attempt to block out all mitigating

Figure 26–4
Handling ethical dilemmas.

Steps for Handling Ethical Dilemmas
1. Apply the guidelines
2. Select the approach
3. Proceed accordingly and consistently

circumstances and other factors that tend to cloud the issue. At this point, the goal is only to identify the ethical choice. Deciding whether to implement the ethical choice comes in the next step.

Select the Approach

When deciding how to proceed after Step 1, you have three basic approaches. These approaches, as set forth in Figure 26–2, are the best-ratio, black-and-white, and full-potential approaches. These approaches and their ramifications can be debated ad infinitum; however, selecting an approach to ethical questions is a matter of personal choice. Factors that will affect the ultimate decision include your personal makeup, the expectations of the company, and the degree of company support.

Proceed with the Decision

The approach selected in Step 2 will dictate how you should proceed as a safety and health professional. Two things are important in this final step. The first is to proceed in strict accordance with the approach selected. The second is to proceed consistently. **Consistency** is critical when handling ethical dilemmas. Fairness is a large part of ethics, and consistency is a large part of fairness. The grapevine will ensure that all employees know how a safety and health professional handles an ethical dilemma. Some will agree, and some will disagree, regardless of the decision. Such is the nature of human interaction. However, regardless of the differing perceptions of the problem, employees respect consistency. Conversely, even if the decision is universally popular, you may lose respect if the decision is not consistent with past decisions.

QUESTIONS TO ASK WHEN MAKING DECISIONS

Safety and health professionals often must make decisions that have ethical dimensions. A constant state of tension often exists between meeting production schedules and maintaining employee safety. Safety and health professionals usually are right in the middle of these issues. Following are some questions that can and should be asked by managers when making decisions about issues that have ethical dimensions. Safety and health professionals should ask these questions themselves, and they should encourage other decision makers within their organizations to do the same.

- Has the issue or problem been thoroughly and accurately defined?
- Have all dimensions of the problem (productivity, quality, cost, safety, health, and so on) been identified?
- Would other stakeholders (employees, customers) agree with your definition of the problem?
- What is your real motivation in making this decision? Meeting a deadline? Outperforming another organizational unit, or a competitor? Self-promoting? Getting the job done right? Protecting the safety and health of employees? Some combination of these?
- What is the probable short-term result of your decision? What is the probable long-term result?
- Who will be affected by your decision and in what way? In the short term? In the long term?
- Did you discuss the decision with all stakeholders (or all possible stakeholders) before making it?
- Would your decision withstand the scrutiny of employees, customers, colleagues, and the general public?

Safety professionals should ask themselves these questions, but equally important, they should insist that other managers do so. The manager responsible for meeting this month's production quota may be so focused on the numbers that he or she overlooks safety. The manager who is feeling the pressure to cut production costs may make decisions that work in the short term but have disastrous consequences in the long term. Questions such as those posed above can help managers broaden their focus and consider the long-range impact when making decisions.

ETHICS AND WHISTLE-BLOWING

What can safety and health professionals do when their employer is violating legal or ethical standards? The first option, of course, should be to bring violations to the attention of appropriate management executives through established channels. In most cases, this will be sufficient to stop the illegal or unethical behavior. But what about those occasions when the safety and health professional is ignored or, worse yet, told to "mind your own business"? These are the types of situations that have led to the concept of **whistle-blowing**, which can be defined as follows:

> Whistle-blowing is the act of informing an outside authority or media organ of alleged illegal or unethical acts on the part of an organization or individual.

Problems with Whistle-Blowing

American society has an interesting attitude toward the concept of whistle-blowing. There seems to be an inherent uneasiness with the concept that is deeply rooted in the American psyche. Even when the illegal or unethical practice in question threatens the safety and health of employees, some people still don't like whistle-blowers. There is often a "don't tell" mentality that causes whistle-blowers to be shunned and viewed as outcasts. As children we learned not to be "tattle-tales." Many adults still seem to hold to this philosophy.

The "don't be a tattle-tale" attitude is only one of the problems that works against whistle-blowing. Others are as follows:

- *Retribution.* People who blow the whistle on their employer may be subjected to retribution. They may be fired, transferred to an undesirable location, or reassigned to an undesirable job. They may also be shunned. There are numerous ways—legal ways—for an employer to take retribution against a whistle-blower.
- *Damaged relationships and hostility.* Blowing the whistle about an illegal or unethical practice can often damage relationships. Somebody—by commission or omission—is responsible. That person or those persons may be disciplined as a result. When this happens, people tend to choose sides which, in turn, leads to damaged relationships. Damaged relationships are often manifested as hostility directed toward the whistle-blower.

Discussion Case

What Is Your Opinion?

Following are two ethical dilemmas that a safety and health professional may face. What is the right thing to do in each case? What is your opinion?

- You are the safety director for West Coast Power Company. Your son is a line worker for a branch of the company that is located in another city. He is visiting you while recuperating from a back injury for which he is collecting workers' compensation. While visiting, he jogs, lifts weights, and plays softball with friends. You finally realize that he is not really injured. What should you do?
- The manufacturing director and the union representative have just had a chin-to-chin argument about removing the new machine guards from the milling machines. They have asked you—the company's safety director—to mediate the dispute. According to the manufacturing director, "We ran these machines for five years without an accident. The only reason we put them on was because some OSHA inspector suggested it. They're fine when we are not in a hurry, but they slow us down when the rush is on. Unless we remove the guards, this job will not be shipped on time." The union representative counters by saying, "These machines are dangerous." What is the right thing to do here? What is your opinion?

- *Loss of focus.* Whistle-blowers often find that their time, energy, and attention are overtaken by the events surrounding the claim of illegal or unethical behavior. Rather than focusing on doing their jobs, they find themselves dealing with retribution, damaged relationships, and hostility.
- *Scapegoating.* Negative consequences certainly can occur as a result of whistle-blowing. Because of this, some safety and health professionals may decide to ignore the issue or to raise it to the next level of management and let it drop there. The problem with this approach is the issue of *accountability.* When an employee is injured or the environment is damaged, the actions of safety and health professionals are certain to be closely scrutinized. "Did you know about the hazardous condition?" "Did you do everything in your power to prevent the accident or incident?" These types of questions are always asked when litigation is brought, as is often the case. An irresponsible organization facing charges of negligence may begin looking for a convenient scapegoat. One obvious candidate in such situations is the organization's chief safety and health professional.

OSHA and Whistle-Blowing

OSHA dealt with the issue of whistle-blowing for certain employees when it adopted regulations governing the **employee protection** provisions of Section 211 (formerly 210) of the Energy Reorganization Act (ERA). This statute makes it illegal for an employer covered by the act to discharge or otherwise discriminate against an employee in terms of compensation, conditions, or privileges of employment because the employee or any person acting at an employee's request performs a protected activity.

Employers covered by the ERA include the following:

- Licensees of the Nuclear Regulatory Commission or an agreement state (including applicants for a license)
- A contractor or subcontractor of a licensee or applicant
- A contractor or subcontractor of the Department of Energy

Key provisions of the ERA are summarized as follows:

- *Right to raise a safety concern.* You are engaged in protected activity when you (1) notify your employer of an alleged violation of the ERA; (2) refuse to engage in any practice made unlawful by the ERA if you have identified the alleged illegality to the employer; (3) testify before Congress or at any federal or state proceeding regarding any provision or proposed provision of the ERA; (4) commence or cause to be commenced a proceeding under the ERA, or a proceeding for the administration or enforcement of any requirement imposed under the ERA; (5) testify or are about to testify in any such proceeding; or (6) assist or participate in such a proceeding or in any other action to carry out the purposes of the ERA.

- *Unlawful acts by employers.* It is unlawful for an employer to intimidate, threaten, restrain, coerce, blacklist, discharge, or in any other manner discriminate against any employee because this employee has engaged in a protected activity.
- *Complaint.* An employee or employee representative may file a complaint charging discrimination in violation of the ERA within 180 days of the discriminatory action. A complaint must be in writing and should include a full statement of facts, including the protected activity engaged in by the employee, knowledge by the employer of the protected activity and the basis for believing that the activity resulted in discrimination against the employee by the employer. A complaint may be filed in person or by mail at the nearest local office of the Occupational Safety and Health Administration (OSHA), U.S. Department of Labor; or with the Office of the Assistant Secretary, OSHA, U.S. Department of Labor, Washington, DC.
- *Enforcement.* OSHA will review the complaint to ensure that it makes an initial showing of discrimination. If not, or if the employer provides clear and convincing evidence that there was no discrimination, there will be no investigation. If the required showing is made, OSHA will notify the employer and conduct an investigation to determine whether a violation has occurred. Either the employee or the employer may request a hearing.
- *Relief.* If discrimination is found, the employer will be required to provide appropriate relief, including reinstatement (even if the period is between the decision and appeal), back wages or compensation for injury suffered from the discrimination, and attorney's fees and costs.

Discussion Case

What Is Your Opinion?

Jack Wilson, safety director for Vitacom, Inc., is facing a dilemma. His employer is seeking registration as an ISO 14000 company. Internal assessments and trial audits show that the company is ready and will probably pass the registration audit with flying colors. If so, Vitacom will be able to double its business in just two years. A European company is looking for a partner in the United States. Vitacom wants to be that partner, but ISO 14000 registration is a prerequisite.

Vitacom looks good on paper, but Wilson knows of a least two serious environmental problems that the company is covering up rather than correcting. Vitacom's Environmental Management Plan, required by ISO 14000, shows that the environmental problems have been corrected. They haven't. Worse yet, Wilson has been warned to say or do nothing that may jeopardize Vitacom's ISO 14000 registration. On the other hand, if the environmental problems in question are not corrected soon, the community's water supply could be contaminated in the future. What should Wilson do?

Noncovered Whistle-Blowing

The ERA covers a number of different types of organizations, but most employers fall outside the law's coverage. Consequently, employees in many organizations have no special protection when they blow the whistle on an unethical employer. This is one more reason why safety and health professionals should encourage their employers to develop comprehensive safety and health policies.

A key element of such a policy should be a mechanism that allows employees to raise questions about safety and health concerns. Such a mechanism should contain provisions that protect employees who raise concerns from retribution by the employer.

SUMMARY

1. Ethics is the application of morality within a context established by cultural and professional values, social norms, and accepted standards of behavior. Morality refers to the values that are subscribed to and fostered by society. Ethics attempts to apply reason in determining rules of human conduct that translate morality into everyday behavior.

2. Ethical behavior is that which falls within the limits prescribed by morality.

3. *Legal* and *ethical* are not the same. If something is illegal, it is also unethical. However, just because something is legal does not mean that it is ethical. An act can be legal but unethical.

4. To determine if a choice is ethical, you can apply the following tests: morning-after, front-page, mirror, role-reversal, and commonsense.

5. Safety and health professionals have a three-pronged role with regard to ethics. They are responsible for setting an ethical example, helping employees identify the ethical choices when facing ethical questions, and helping employees follow through and actually undertake the ethical option.

6. Safety and health professionals have three approaches available in handling ethical dilemmas: best-ratio, black-and-white, and full-potential.

7. The company's role in ethics is to create an ethical environment and to set an ethical example. An effective way is to develop a written ethics philosophy and share it with all employees.

8. Three personality characteristics that can influence an employee's ethical behavior are ego strength, Machiavellianism, and locus of control.

9. People facing ethical dilemmas should apply the tests for determining what is ethical, select one of the three basic approaches, and proceed consistently.

10. Whistle-blowing is the act of informing an outside authority or the media of alleged illegal or unethical acts on the part of an organization or individual.

KEY TERMS AND CONCEPTS

Best-ratio approach	Legality
Black-and-white approach	Locus of control
Commonsense test	Machiavellianism
Conscience	Mirror test
Consistency	Morality
Ego strength	Morning-after test
Ethical behavior	Patience
Ethical environment	Persistence
Ethics	Perspective
Ethics credo	Pride
Ethics philosophy	Purpose
Front-page test	Role-reversal test
Full-potential approach	Values
Gray area	Whistle-blowing

REVIEW QUESTIONS

1. Define the term *morality*.
2. Define the term *ethics*.
3. Briefly explain each of the following ethics tests: morning-after, front-page, mirror, role-reversal, and commonsense.
4. What is the safety and health professional's role with regard to ethics?
5. Briefly explain the following approaches to handling ethical behavior: best-ratio, black-and-white, and full-potential.
6. Briefly explain a company's role with regard to ethics.
7. Explain how one should proceed when facing an ethical dilemma.
8. Write a brief ethics philosophy for a chemical company.
9. List the individual and social factors that may influence an employee's ethical behavior.
10. List and briefly describe the Five P's of Ethical Power as set forth by Blanchard and Peale.
11. What questions should safety and health professionals ask when making decisions that have an ethical component?
12. Explain the most common problems associated with whistle-blowing.

ENDNOTES

1. P. F. Drucker, "What Is Business Ethics?" *Across the Board*, October 1981, 22–32.
2. E. W. Stead, D. L. Worrell, and J. G. Stead, "An Integrative Model for Understanding and Managing Ethical Behavior in Business Organizations," *Journal of Business Ethics* 5, no. 9: 233.
3. P. Taylor, *Principles of Ethics: An Introduction* (Encino, CA: Dickson, 1975), 96.
4. P. Arlow and T. A. Ulrich, "Business Ethics, Social Responsibility, and Business Students: An Empirical Comparison of Clark's Study," *Akron Business and Economic Review* 4, no. 3: 17–23.
5. K. Blanchard and N. V. Peale, *The Power of Ethical Management* (New York: Ballantine Books), 10–17.
6. Ibid., 79.
7. L. K. Trevino, "Ethical Decision Making in Organizations: A Person-Situation Interactionist Model," *Academy of Management Review* 11, no. 3: 601–617.
8. J. F. Preble and P. Miesing, "Do Adult MBA and Undergraduate Business Students Have Different Business Philosophies?" *Proceedings of the National Meeting of the American Institute for the Decision Sciences*, November 1984, 346–348.
9. F. Luthans and R. Kreitner, *Organizational Behavior Modification and Beyond: An Operant and Social Learning Approach* (Glenview, IL: Scott Foresman).
10. Martin Marietta Corporation, *Code of Ethics and Standards of Conduct* (Orlando, FL: Corporate Ethics Office), inside front cover.
11. Ibid., 2.
12. D. C. Shanks, "The Role of Leadership in Strategy Development," *Journal of Business Strategy*, January/February 1989, 32.
13. Ibid., 33.

HAZARD ANALYSIS/PREVENTION AND SAFETY MANAGEMENT

Major Topics

- Overview of Hazard Analysis
- Preliminary Hazard Analysis
- Detailed Hazard Analysis
- Hazard Prevention and Deterrence
- OSHA Process Safety Standard
- Risk Assessment
- Safety Management Concerns
- Best Practices in Safety Management
- Occupational Health and Safety Management Systems

There is a saying that an ounce of prevention is worth a pound of cure. This is certainly the case with workplace safety and health. Every accident that can be prevented should be. Every hazard that can be identified should be corrected or at least minimized through the introduction of appropriate safeguards. Careful analysis of potential hazards in the workplace has led to many of today's widely used safety measures and practices.

The key to preventing accidents is identifying and eliminating hazards. A **hazard** may be defined as follows:

A hazard is a condition or combination of conditions that, if left uncorrected, may lead to an accident, illness, or property damage.

This chapter provides prospective and practicing safety and health professionals with the information they need to analyze the workplace, identify hazards that exist there, and take the preventive measures necessary to neutralize the hazards.

OVERVIEW OF HAZARD ANALYSIS

If a hazard is a condition that could lead to an injury or illness, **hazard analysis** is a systematic process for identifying hazards and recommending corrective action. There are two approaches to hazard analysis: preliminary and detailed (Figure 27–1). A **preliminary hazard analysis (PHA)** is conducted to identify potential hazards and prioritize them according to the (1) likelihood of an accident or injury being caused by the hazard; and (2) severity of injury, illness, or property damage that could result if the hazard caused an accident.

The Society of Manufacturing Engineers has this to say about preliminary hazard analysis:

> Each hazard should be ranked according to its probability to cause an accident and the severity of that accident. The ranking here is relative. Some hazards identified might be placed into a category for further analysis. A useful practice for early sorting is to place catastrophic severities together, followed by critical, marginal, and nuisance hazards, respectively. Then, with each item, indicate the probability of occurrence—considerable, probable, or unlikely. Rate the correction of these hazards next by cost.[1]

Key terms in this quote are those that describe the likelihood of an accident occurring (*considerable, probable,* or *unlikely*) and those that rate the probable level of injuries that could occur (*catastrophic, critical, marginal,* and *nuisance*). A key step is rating the cost of correcting hazards. All these concepts are covered later in this chapter.

Whereas a preliminary analysis may involve just observation or pilot testing of new equipment and systems, the **detailed hazard analysis** involves the application of analytical, inductive, and deductive methods. Figure 27–2 lists some

Figure 27–1
Two approaches to hazard analysis.

Figure 27–2
Detailed hazard analysis methods.

- Failure mode and effects analysis (FMEA)
- Fault tree analysis (FTA)
- Hazard and operability review (HAZOP)
- Human error analysis (HEA)
- Risk analysis
- Technic of operation review (TOR)

of the more widely used methods for conducting a detailed hazard analysis. Each of these methods is covered at length later in this chapter.

PRELIMINARY HAZARD ANALYSIS

It is not always feasible to wait until all the data are compiled from a detailed analysis before taking steps to identify and eliminate hazards. For example, when a new system or piece of equipment is installed, management probably wants to bring it on line as soon as possible. In such cases, a preliminary hazard analysis is in order. The PHA can serve two purposes: (1) it can expedite bringing the new system on line, but at a substantially reduced risk of injuring workers; and (2) it can serve as a guide for a future detailed analysis.

Preliminary hazard analysis amounts to forming an ad hoc team of experienced personnel who are familiar with the equipment, material, substance, or process being analyzed. **Experience** and **related expertise** are important factors in conducting a preliminary review. For example, say a new piece of equipment, such as a computer numerically controlled (CNC) machining center, is installed. The safety and health professional may form a team that includes an experienced machinist, an electrician, a materials expert, and a computer control specialist.

All members of the team are asked to look over the machining center for obvious hazards relating to their respective areas of expertise. Then, they work together as a group to play devil's advocate. Each team member asks the others a series of "what if" questions: What if a cutting bit breaks? What if the wrong command is entered? What if the material stock is too long? Depending on the nature of the process being analyzed, personnel from adjacent or related processes should be added to the team.

Figure 27–3 is an example of a job hazard analysis survey adapted from one developed by the National Institute for Occupational Safety and Health (NIOSH). A preliminary analysis team would use this form to identify potential hazards associated with a spray-painting process. Key elements include the substances to which workers will be exposed, the form that those substances will take, the probable route of entry, and recommended hazard control strategies. A similar form can be developed for any process or operation that may be the focus of a preliminary hazard analysis.

Cost–Benefit Factors in Hazard Analysis

Every hazard typically has several different remedies. Every remedy has a corresponding **cost** and corresponding **benefit**. Management is not likely to want to apply $10 solutions to $1 problems. Therefore, it is important to factor in cost when recommending corrective action regarding hazards. This amounts to listing all of the potential remedies along with their respective costs and then estimating the extent to which each will reduce the hazard (its benefit).

Operation:									Date:							
Number of Employees	Job Title	Exposure Substance	Form (Type of hazard)						Route of Entry		Control Methods					
			Dust	Liquid	Vapor	Gas	Fume	Mist	Skin	Inhaled	Local Ventilation	General Ventilation	Respirator	Gloves	Face Protection	Other Protection

Figure 27–3
Sample job hazard analysis survey.

Going back to the earlier example of the CNC machining center, assume that the analysis team identified the following potential hazards:

- Lubricants sprayed on the machine operator or floor
- Flying metal chips hitting the operator or other workers
- Jammed metal stock kicking back into the operator

Cost–Benefit Analysis Matrix

Possible Remedy	Estimated Cost	Impact on Hazard		
		Spraying Lubricants	Metal Chips	Jammed Stock
Plexiglas™ door	$250	E	E	R
Flexible curtain	75	R	R	N
Acme chip/jam guard	260	N	E	E
Acme spray guard	260	N	E	E

R = Reduces the hazard

E = Eliminates the hazard

N = No effect on the hazard

I = Increases the hazard

C = Creates new hazard

Figure 27–4
Sample cost–benefit analysis matrix.

Figure 27–4 is a matrix that may be developed by the analysis team to illustrate the cost of each hazard versus the benefit of each remedy. After examining this matrix, the remedy that makes the most sense from the perspective of both cost and impact on the hazards is the Plexiglas™ door. It eliminates two of the hazards and reduces the third. The flexible curtain costs less but does not have a sufficient impact on the hazards. The third and fourth options cost more and have less impact on the hazards.

DETAILED HAZARD ANALYSIS

Typically, a preliminary hazard analysis is sufficient. However, in cases where the potential exists for serious injury, multiple injuries, or catastrophic illness, a detailed hazard analysis is conducted. A number of different methods can be used for conducting detailed analyses. The most widely used of these are as follows:

- Failure mode and effects of analysis (FMEA)
- Hazard and operability review (HAZOP)
- Technic of operations review (TOR)
- Human error analysis (HEA)
- Fault tree analysis (FTA)
- Risk analysis

Failure Mode and Effects Analysis (FMEA)

Failure mode and effects analysis (FMEA) is a formal step-by-step analytical method that is a spin-off of reliability analysis, a method used to analyze complex engineering systems. FMEA proceeds as follows:

1. Critically examine the system in question.
2. Divide the system into its various components.

Discussion Case

What Is Your Opinion?

Mike Chinchar is the new safety director at MicroTel Corporation. He completed his college degree just six weeks ago. At the moment, he is wishing he could transport himself back in time and be a college student again. Chinchar is on the hot seat. "Chinchar, you seem to think that this company should simply stop functioning every time you identify a hazard. This situation you are proposing will be expensive! Isn't there some other way to solve the problem? Did you do a cost–benefit analysis before arriving at this recommendation?" What types of factors should Chinchar have considered before recommending a solution to a hazardous situation? What is your opinion?

3. Examine each individual component and record all of the various ways in which the components may fail. Rate each potential failure according to the degree of hazard posed (0 = No hazard, 1 = Slight, 2 = Moderate, 3 = Extreme, 4 = Severe).
4. Examine all potential failures for each individual component of the system and decide what effect the failures could have.

Figure 27–5 is an example of an FMEA conducted on a direct extrusion process. The process or system is broken down into seven components: die backer, die, billet, dummy block, pressing stem, container liner, and container fillet. The types of failures that may occur are identified as corrosion, cracking, shattering, bending, and surface wear. Of the various components, only the dummy block poses an extreme hazard and a corresponding hazard to workers.

An FMEA produces an extensive analysis of a specific process or system, as illustrated in Figure 27–5. However, FMEAs have their limitations. First, the element of **human error** is missing. This is a major weakness because human error is more frequently at the heart of a workplace accident than is system or process failure. This weakness can be overcome by coupling human error analysis, which is covered later in this chapter, with an FMEA. Second, FMEAs focus on the components of a given system as if the components operate in a vacuum. They do not take into account the interface mechanisms between components or between systems. It is at these interface points that problems often occur.

Hazard and Operability Review (HAZOP)

Hazard and operability review (HAZOP) is an analysis method that was developed for use with new processes in the chemical industry. Its strength is that it allows problems to be identified even before a body of experience has been developed for a given process or system. Although originally intended for use with new processes, it need not be limited to new operations. HAZOP works equally well with old processes and systems.

HAZOP consists of forming a team of experienced, knowledgeable people from a variety of backgrounds relating to the process or system and having team members brainstorm about potential hazards. The safety and health professional should chair the team and serve as a facilitator. The chair's role is to elicit and record the ideas of team members, make sure that one member does not dominate or intimidate other members, encourage maximum participation from all members, and assist members in combining ideas where appropriate to form better ideas.

A variety of approaches can be used with HAZOP. The one recommended by the American Institute of Chemical Engineers (AICHE) is probably the most widely used. AICHE recommends the following guidewords: *no, less, more, part of, as well as, reverse,* and *other than.*[2]

These guidewords relate to the operation of a specific component in the system or a specific part of an overall operation. They describe ways in which

Plastics Extrusions, Inc.
17 Industrial Boulevard
Fort Walton Beach, Florida 32548

Department _____ Manufacturing _____   Process/System _____ Direct Extrusion _____   Date _____ November 12, 2007 _____

Component	Type of Potential Failure	Potential Effect On				0	1	2	3	4	H	M	L	U	Examination Method	Recommendation
		Component	Related Components	Process/System	Workers											
Diebacker	Corrosion	Shutdown to replace	None	Shutdown to replace	None	✓								✓	Visual	Periodic checks for corrosion
Die	Cracking	Shutdown to replace	Damage to die backer	Shutdown to replace	None			✓				✓			Visual	Periodic checks for cracks
Billet	- - -	- - -	- - -	- - -	- - -	- - -									- - -	- - - - -
Dummy block	Shattering	Shutdown to replace	Could damage others	Shutdown all	Injuries from flying metal				✓			✓			Visual	Inspect and replace periodically
Pressing stem	Bending	Shutdown to replace	None	Shutdown to replace	None	✓								✓	Visual	Inspect and replace periodically
Container liner	Surface wear	Shutdown to replace	None	None	None	✓								✓	Visual	Periodic checks for wear
Container fillet	- - -	- - -	- - -	- - -	- - -	- - -									- - -	- - - - -

Analysis Conducted by: _____

Figure 27–5
Sample FMEA.

the component may deviate from its design or its intended mode of operation. For example, if a component that should rotate 38° in a cycle fails to rotate at all, the *no* guideword applies. If it rotates fewer than 38°, the *less* guideword applies. *More* would apply if the component's rotation exceeded 38°. *Reverse* would be used if the component rotated 38° in a direction opposite of the one intended. *As well as* is similar to *more* in that it indicates an increase in an intended amount. *Other than* is used when what actually occurs is something completely different from what was intended. For example, if the component fell off rather than rotating 38°, *other than* would be used.

A HAZOP proceeds in a step-by-step manner. These steps are summarized as follows:

1. Select the process or system to be analyzed.
2. Form the team of experts.
3. Explain the HAZOP process to all team members.
4. Establish goals and timeframes.
5. Conduct brainstorming sessions.
6. Summarize all input.

Figure 27–6 is an example of a form that can be used to help organize and focus brainstorming sessions. It can also be used for summarizing the results of the brainstorming sessions. This particular example involves a plastic mixing process. Only one component in the process (flow-gate number 1) has been analyzed. If the flow-gate does not work as intended, there will be no flow, too little flow, or too much flow. Each condition results in a specific problem. Action necessary to correct each situation has been recommended. Every critical point, sometimes referred to as a *node,* in the process is analyzed in a similar manner.

HAZOPs have the same weaknesses as FMEAs—they do not factor human error into the equation. HAZOPs predict problems associated with system or process failures. However, these are technological failures. Because human error is so often a factor in accidents, this weakness must be addressed. The next section sets forth guidelines for analyzing human error.

Human Error Analysis (HEA)

Human error analysis (HEA) is used to predict human error, not to review what has occurred. Although the records of past accidents can be studied to identify trends that can, in turn, be used to predict accidents, this should be done as part of an accident investigation. HEA should be used to identify hazards before they cause accidents.

Two approaches to HEA can be effective: (1) observing employees at work and noting hazards (the task analysis approach); and (2) actually performing job tasks to get a firsthand feel for hazards. Regardless of how the HEA is conducted, it is a good idea to perform it in conjunction with FMEAs and HAZOPs. This will enhance the effectiveness of all three processes.

Anderson Chemical Company
1512 Airport Road
Crestview, Florida 32536

HAZOP SUMMARY FORM

Department ___Composites___ System/Process ___Plastic Mix___ Date ___January 7, 2007___

System/Process Component	Factor Analyzed	Guide-word	Resulting Difference	Potential Problem	Recommended Remedy
Flow-gate number one	Amount of flow	No	No flow	No mix	Make sure flow-gate is open
		Less	Insufficient flow	Weak mix	Troubleshoot and repair the flow-gate
		More	Excess flow	Too strong mix	Troubleshoot and repair the flow-gate

Figure 27–6
Sample HAZOP.

773

Technic of Operation Review (TOR)

Technic of operation review (TOR) is an analysis method that allows supervisors and employees to work together to analyze workplace accidents, failures, and incidents. It answers the question "Why did the system allow this incident to occur?" Like FMEA and HAZOP, this approach seeks to identify systemic causes, not to assign blame.

TOR is not new. It was originally developed in the early 1970s by D. A. Weaver of the American Society of Safety Engineers. However, for 20 years, user documentation on TOR was not readily available. Consequently, widescale use did not occur until the early 1990s when documentation began to be circulated. Richard G. Hallock describes TOR:

> TOR is a hands-on analytical methodology designed to determine the root system causes of an operation failure. Because it uses a work sheet written in simple-to-understand terms and follows an uncomplicated yes/no decision-making sequence, it can be used even at the lowest levels of the firm. TOR is triggered by an incident occurring at a specific time and place and involving specific people. It is not a hypothetical process. It demands careful and systematic evaluation of the real circumstances surrounding the incident, and results in isolating the specific ways in which the organization failed to prevent the occurrence.[3]

A weakness of TOR is that it is designed as an after-the-fact process. It is triggered by an accident or incident. A strength of TOR is its involvement of line personnel in the analysis. The process proceeds as follows:

1. *Establish the TOR team.* It should consist of workers who were present when the accident or incident occurred, the supervisor, and the safety and health professional. The safety and health professional should chair the team and serve as a facilitator.
2. *Conduct a roundtable discussion to establish a common knowledge base among team members.* At the beginning of the discussion, five team members may have five different versions of the accident or incident. At the end, there should be a consensus.
3. *Identify one major systematic factor that led to, or played a significant role in, causing the accident or incident.* This one TOR statement, about which there must be consensus, serves as the starting point for further analysis.
4. *Use the group consensus to respond to a sequence of yes/no options.* Through this process, the team identifies a number of factors that contributed to the accident or incident.
5. *Evaluate identified factors carefully to make sure that there is a team consensus about each.* Then, prioritize the contributing factors beginning with the most serious.
6. *Develop corrective or preventive strategies for each factor.* Include them in a final report that is forwarded through normal channels for appropriate action.

Fault Tree Analysis (FTA)

Fault tree analysis (FTA) can be used to predict and prevent accidents or as an investigative tool after the fact. FTA is an analytical methodology that uses a graphic model to display the analysis process visually. A fault tree is built using special symbols, some derived from Boolean algebra. The resultant model resembles a logic diagram or a flow chart. Figure 27–7 shows and describes the symbols used in constructing fault trees. Figure 27–8 shows how these symbols may be used to construct a fault tree. The top box in a fault tree represents the accident or incident that either could occur or has occurred.

All symbols below the top box represent events that contribute in some way to the ultimate accident or incident. The sample fault tree shown in Figure 27–8 is qualitative in nature. Fault trees can be made quantitative by assigning probability figures to the various events below the top box. However, this is rarely done because reliable probability figures are seldom available. A fault tree is developed using the following steps:

1. Decide on the accident or incident to be placed at the top of the tree.
2. Identify the broadest level of failure or fault event that could contribute to the top event. Assign the appropriate symbols.
3. Move downward through successively more specific levels until basic events are identified.

Experience, deliberate care, and systematic analysis are very important in constructing fault trees. Once a fault tree has been constructed, it is examined to determine the various combinations of failure or fault events that could lead to the top event. With simple fault trees, this can be accomplished manually; with more complex trees, this step is difficult. However, computer programs are available to assist in accomplishing this step. The final step involves making recommendations for preventive measures.

Risk Analysis

Where are we at risk? Where are we at greatest risk? These are important questions for safety and health professionals involved in analyzing the workplace for the purpose of identifying and overcoming hazards. **Risk analysis** is an analytical methodology normally associated with insurance and investments. However, risk analysis can be used to analyze the workplace, identify hazards, and develop strategies for overcoming these hazards. The risk analysis process focuses on two key questions:

- How *frequently* does a given event occur?
- How *severe* are the consequences of a given event?

The fundamental rule-of-thumb of risk analysis is that risk is decreased by decreasing the frequency and severity of hazard-related events.

Safety and health professionals should understand the relationship that exists between the **frequency** and **severity** factors relating to accidents. Historical

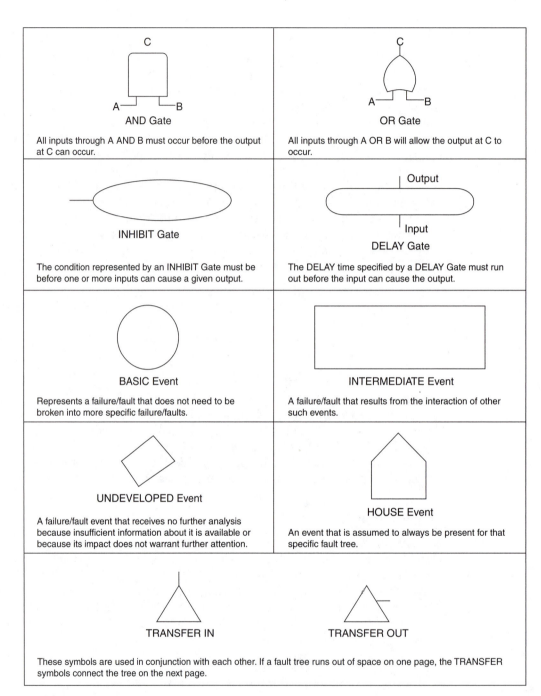

Figure 27–7
Symbols used in fault tree analysis.

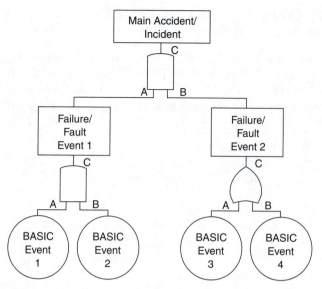

Figure 27–8
Sample fault tree.

data on accidents, injuries, and illness show that the less severe an injury or illness, the more frequently it is likely to occur. Correspondingly, the more severe an injury or illness, the less frequently it is likely to occur. For example, there are many more minor scrapes, bumps, and abrasions experienced in the workplace than major debilitating injuries such as amputations or broken bones.

A number of different approaches can be used in conducting a risk analysis. One of the most effective is that developed by Chapanis.[4] Chapanis's approach to risk analysis considers both **probability** and **impact**.

Probability levels and corresponding frequency of occurrence ratings are as follows: 1 = Impossible (frequency of occurrence: 10^{-8}/day); 2 = Extremely unlikely (frequency of occurrence: 10^{-6}/day); 3 = Remote (frequency of occurrence: 10^{-5}/day); 4 = Occasional (frequency of occurrence: 10^{-4}/day); 5 = Reasonably probable (frequency of occurrence: 10^{-3}/day); 6 = Frequent (frequency of occurrence: 10^{-2}/day).[5]

The lowest rating (1) means it is impossible that a given error will be committed or a given failure will occur. The highest rating (6) means it is very likely that a given error will be committed frequently or a given failure will occur frequently. Notice the quantification of frequency levels for each level of probability. For example, the expected frequency of occurrence for a probability level of remote is 10 to the negative fifth power per day.

Severity levels can also be rated, with the likely consequence of an accident or failure event of that severity. The least severe incidents (1) are not likely to cause an injury or damage property. The most severe incidents (4) are almost certain to cause death or serious property damage. Critical accidents (3) may cause severe injury or major loss. Marginal accidents (2) may cause minor injury, minor occupational illness, or minor damage.[6]

HAZARD PREVENTION AND DETERRENCE

All the methods and procedures discussed in this chapter have been concerned with identifying potential hazards. This section deals with using the information learned during analysis to prevent accidents and illnesses. The Society of Manufacturing Engineers recommends the following hazard control methods:

- Eliminate the source of the hazards.
- Substitute a less hazardous equivalent.
- Reduce the hazards at the source.
- Remove the employee from the hazard (e.g., substitute a robot or other automated system).
- Isolate the hazards (e.g., enclose them in barriers).
- Dilute the hazard (e.g., ventilate the hazardous substance).
- Apply appropriate management strategies.
- Use appropriate personal protective equipment.
- Provide employee training.
- Practice good housekeeping.[7]

For every hazard identified during the analysis process, one or more of these hazard control methods will apply. Figure 27–9 shows the steps involved in implementing hazard control methods. The first step involves selecting the method or methods that are most likely to produce the desired results. Once selected, the method is applied and monitored to determine if the expected results are being achieved.

Monitoring and observing are informal procedures. They should be followed by a more formal, more structured assessment of the effectiveness of the method. If the method selected is not producing the desired results, adjustments should

Figure 27–9
Steps for implementing hazard control measures.

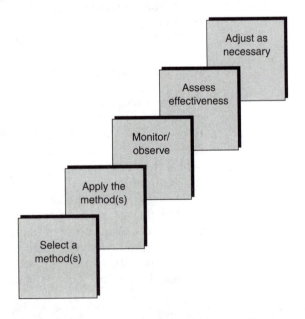

be made. This may mean changing the way in which the method is applied or dropping it and trying another.

The example of Crestview Container Corporation's (CCC) problems with toxic paint illustrates how the process works. CCC produces airtight aluminum containers for transporting electromechanical devices. The containers must be painted as the last step in the production process. Although the specified paint was supposed to be only slightly toxic—a problem that should have been resolved by using personal protective equipment—paint station operators complained frequently of various negative side effects.

The CCC safety and health professional, working with management, solved the problem by applying the following steps:

1. *Select a method.* Of the various methods available, the one selected involved eliminating the source of the hazard (the toxic paint). CCC personnel were tasked with testing various nontoxic paints until one was found that could match the problem paint in all categories (e.g., ease of application, drying time, quality of surface finish, etc.). After 40 different paints were tested, a nontoxic substitute was found.

2. *Apply the method.* The new paint was ordered and used on a partial shipment of containers.

3. *Monitor/observe.* The safety and health professional, along with CCC's painting supervisors, monitored both employee performance and employee complaints concerning the paint.

4. *Assess effectiveness.* To assess effectiveness, employee complaints were tabulated. The number of complaints was down to a negligible amount and not serious in nature. Productivity was also assessed. It was found that the new paint had no noticeable effect on productivity, negative or positive.

5. *Adjust as necessary.* CCC found that no adjustments were necessary.

Safety Fact

What to Include in a Hazards Inventory

An excellent tool for getting the work of safety and health personnel organized, prioritized, and properly focused is the *hazards inventory.* Such an inventory is a comprehensive list of all hazards associated with all processes and work tasks in a company. A hazards inventory should include at least the following information:

- Process descriptions
- Associated hazards
- Controls relating to the hazards
- Department location of each process
- Names of supervisors of all personnel who work on each process (including telephone numbers)
- Number of employees who work on each process
- Medical information relating to the hazards
- Historical information about the process and related hazards

OSHA PROCESS SAFETY STANDARD

The OSHA Process Safety Standard has relevance from the perspective of hazard prevention, relating specifically to chemical hazards. OSHA's standard for process safety is found in 29 CFR 1910.119. Its purpose is to prevent *catastrophic* accidents caused by major releases of highly hazardous chemicals. To comply with this standard, companies must have written operating procedures, mechanical integrity programs, and formal incident investigation procedures. Other key elements are as follows:

1. *Coverage.* Although the Process Safety Standard is typically associated with large chemical and petrochemical processing plants, its coverage is actually much broader than this. Any company is covered that uses the threshold amount of a chemical listed in the standard, or 10,000 pounds or more of a flammable material on-site in one location.

2. *Employee participation.* Section (c) of the standard requires that employees be involved in all aspects of the process safety management program. In addition, employees must be given access to information developed as part of the program.

3. *Process Safety Information (PSI).* Section (d) of the standard requires organizations to establish and maintain process safety information files. Information included in the files includes chemical, process, and equipment data.

4. *Process Hazard Analyses (PHAs).* Section (e) of the standard requires that companies conduct process hazard analyses for all processes covered by the standard. Like any other hazard analysis, the PHAs are supposed to identify potential problems so that prompt corrective action or preventive measures can be taken.

5. *Standard Operating Procedures (SOPs).* Section (f) of the standard requires employers to establish and maintain written standard operating procedures for using chemicals safely. The requirement applies to handling, processing, transporting, and storing chemicals.

6. *Requirements for contractors.* Section (h) of the standard describes the special requirements imposed on companies that contract portions of their work to other companies. Complying with the standard is a matter of making sure that contractors comply. The following requirements are imposed by Section (h):

- Screen contractors before issuing a contract to ensure that they have a comprehensive safety and health program in place.
- Orient contractors concerning the chemicals with which they may be required to work or be around, the emergency action plan, and other pertinent information.
- Evaluate contractors periodically to ensure that their safety performance is acceptable.
- Maintain an OSHA injury and illness log for the contractor that is separate from, and in addition to, that of the host company.

OSHA's Regulation for Chemical Spills

OSHA issues a special regulation dealing with chemical spills. The standard (29 CFR 1910.120) is called the **Hazardous Waste Operations and Emergency Response (HAZWOPER)** standard. HAZWOPER gives organizations two options for responding to a chemical spill. The first is to evacuate all employees in the event of a spill and call in professional emergency response personnel. Employers who use this option must have an emergency action plan (EAP) in place in accordance with 29 CFR 1010.38(a). The second option is to respond internally. Employers who use this option must have an emergency response plan in place that is in accordance with 29 CFR 1010.120.

1. *Emergency action plans (EAPs).* An EAP should have at least the following elements: alarm systems, evacuation plan, a mechanism or procedure for emergency shutdown of the equipment, and a procedure for notifying emergency response personnel.

2. *Emergency response plan.* Companies that opt to respond internally to chemical spills must have an emergency response plan that includes the provision of comprehensive training for employees. OSHA Standard 29 CFR 1910.120 specifies the type and amount of training required, ranging from awareness to in-depth technical training for employees who will actually deal with the spill. It is important to note that OSHA forbids the involvement of untrained employees in responding to a spill. The following topics are those covered in the HAZWOPER seminar provided by Environmental Safety Awareness, a safety and health company in Fort Walton Beach, Florida. These topics are typical of those covered in up-to-date HAZWOPER courses.

Summary of Key Federal Laws

Overview of Impacting Regulations

  Classification and Categorization of Hazardous Waste

- Definition of Hazardous Waste
- Characteristics
- TCLP
- Lists of Hazardous Wastes

  Hazardous Waste Operations

- Definitions
- Levels of Response

  Penalties for Noncompliance

- Civil Penalty Policy

  Responding to Spills

- Groundwater Contamination
- Sudden Releases
- Clean-Up Levels

- Risk Assessment
- Remedial Action

Emergency Response

- Workplan
- Site Evaluation and Control
- Site Specific Safety and Health Plan
- Information and Training Program
- Personal Protective Equipment
- Monitoring
- Medical Surveillance
- Decontamination Procedures
- Emergency Response
- Other Provisions

Contingency Plans

- Alarm Systems
- Action Plan

Personal Protective Equipment

- Developing a PPE Program
- Respiratory Equipment
- Protective Clothing
- Donning PPE
- Doffing PPE

Material Safety Data Sheets

- Introduction
- Preparing MSDSs
- MSDS Information
- Hazardous Ingredients
- Physical/Chemical Characteristics
- Fire and Explosion Hazard Data
- Reactivity Data
- Health Hazard Data
- Precautions for Safe Handling and Use
- Control Measures

Site Control

- Site Maps
- Site Preparation
- Work Zones
- Buddy System
- Site Security
- Communications
- Safe Work Practices

Hazardous Waste Containers

- Emergency Control
- Equipment
- Tools
- Safety

Decontamination

- Types
- Decontamination Plan
- Prevention of Contamination
- Planning
- Emergencies
- Physical Injury
- Heat Stress
- Chemical Exposure
- Medical Treatment Area
- Decontamination of Equipment
- Decontamination Procedures
- Sanitation of PPE
- Disposal of Contaminated Materials

RISK ASSESSMENT

Risk assessment in this context is the process of quantifying the level of risk associated with the operation of a given machine.[8] It should be a structured and systematic process that answers the following four specific questions:

- How *severe* are potential injuries?
- How *frequently* are employees exposed to the potential hazards?
- What is the *possibility* of avoiding the hazard if it does occur?
- What is the *likelihood* of an injury should a safety control system fail?

The most widely used risk-assessment technique is the decision tree, coupled with codes representing these four questions and defined levels of risk. Figure 27–10 is an example of a risk-assessment decision tree. In this example, the codes and their associated levels of risk are as follows:

S = Severity

Question 1: Severity of potential injuries
S1   Slight injury (bruise, abrasion)

S2   Severe injury (amputation or death)

F = Frequency

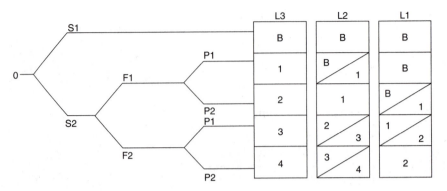

Figure 27–10
Risk-assessment decision tree.

Question 2: Frequency of exposure to potential hazards

F1   Infrequent exposure

F2   From frequent to continuous exposure

P = Possibility

Question 3: Possibility of avoiding the hazard if it does occur

P1   Possible

P2   Less possible to not impossible

L = Likelihood

Question 4: Likelihood that the hazard will occur

L1   Highly unlikely

L2   Unlikely

L3   Highly likely

Risk Levels

Associated risk factors ranging from lowest (B) to highest (4)

By applying the decision tree in Figure 27–10 or a similar device, the risk associated with the operation of a given machine can be quantified. This allows safety personnel to assign logical priorities for machine safeguarding and hazard prevention.

SAFETY MANAGEMENT CONCERNS

Students studying occupational safety might become safety engineers, technologists, technicians, specialists, or managers. Some who begin their careers serving in a technical capacity (engineer, technologist, technician, specialist, etc.) might eventually find themselves called upon to manage a team or department of safety professionals. Safety managers should be familiar with most of what is presented in this book. However, when working in a safety management position, their principal concerns will be as follows:

Strategic Planning

Safety managers work with key decision makers in organizations to ensure that safety and health are included in their strategic plans. An organization's strategic plan should have either a broad goal or a guiding principle (corporate value statement) that speaks to maintaining a safe and healthy work environment.

Safety Policy

Safety managers work with other key decision makers in organizations to develop a written policy that summarizes the organization's commitment to maintaining a safe and healthy work environment. This policy should also explain the responsibilities of managers, supervisors, and employees for maintaining a safe and healthy work environment.

Written Procedures

An organization's safety policy is translated into more specific language through the development of written policies. Safety managers should ensure that their employers put all the following in writing: job descriptions that include responsibilities for safety and health; general safety rules that apply to all employees; procedures for specialized and hazardous operations; standard operating procedures for processes, systems, and equipment (including safe operation); program for communicating with employees about safety information, issues, and concerns; emergency plan with all its subordinate plans; employee orientation program; near-miss procedures; safety suggestions procedures; procedures for job hazard analysis; and safety manual.

Employee Training

Safety managers are responsible for ensuring that both new and experienced employees receive the training they need to do their jobs safely. This involves the following types of responsibilities: arranging and scheduling training for new employees; arranging and scheduling retraining for experienced employees in need of updating; arranging training for employees whose jobs have changed in some way; identifying and assigning instructors to conduct training; monitoring and evaluating training that is provided; documenting training; and ensuring that supervisors receive the training they need in order to play their critical role in maintaining a safe and healthy work environment.

Communication

Safety managers are responsible for ensuring that employees, supervisors, and managers are fully informed about safety and health policies, practices, concerns, and other information. Some of the methods used by safety managers to communicate with stakeholders on a regular basis are safety meetings, daily personal contact (management by walking around, or MBWA), safety committees, and publications such as newsletters, e-mail, memorandums, and so on.

Human Resource Management

Safety managers work closely with human resource management personnel on personnel issues relating to safety and health. These issues include blood testing immediately following accidents, drug testing as part of the preemployment screening, developing and implementing corporate wellness programs, and monitoring injured employees who are placed in back-to-work programs.

Self-Assessments

An important responsibility of safety managers is helping supervisors and employees play their roles in maintaining a safe and healthy workplace. One of the ways safety managers do this is by developing checklists they can use in conducting safety-oriented self-assessments in their areas of responsibilities. Several such checklists are contained in various chapters of this book.

Safety and Health Promotion

Safety managers are responsible for establishing promotional programs that keep employees, supervisors, and managers focused on safety and health issues. The purpose of promotional programs is to remind people to think of safety first. The best programs encompass safety on the job as well as off. An employee injured at home might not cost the company any money in workers' compensation premiums, but does cost the company in lost time, the temporary loss of expertise the experienced employee brings to the job, and in the work added to other employees to make up for his or her absence.

Accident Investigation and Reporting

Safety managers are responsible for establishing a structured system for investigating accidents and near misses and for reporting the results of the investigations. They are also responsible for providing the training needed to allow supervisors and other personnel to participate in accident investigations in a positive, helpful manner. Included in this area of responsibility is accident cause analysis and workers' compensation filing and follow-up.

Ongoing Monitoring

Safety managers are responsible for ensuring that workplace hazards and the requirements of applicable regulatory agencies are properly monitored on a continuous basis. Hazards include noise, chemicals, smoke, fumes, dust, ergonomics, and any other potential problems that might be present in the workplace. Of course, monitoring hazards is just one aspect of the safety manager's job. The following checklist will help safety managers identify and monitor a number of issues:

1. Have you identified all the standards from OSHA and other agencies that apply to your organization? Do you have electronic access to the latest editions of these standards?
2. Are you familiar with the most recent workers' compensation legislation in your state?
3. Are you implementing workers' compensation cost-reduction strategies?
4. Are you monitoring the workplace for ergonomic hazards? Are you applying OSHA's voluntary ergonomic guidelines?
5. Are you monitoring employee stress levels and implementing appropriate stress-reduction strategies?
6. Are all machines properly guarded?
7. Have you instituted a comprehensive lockout/tagout program?
8. Have you implemented an effective slip-and-fall prevention program based on OSHA's Fall Protection Standard?
9. Have you implemented an effective ladder safety program?
10. Do all employees have and properly use the personal protective equipment called for in their jobs?
11. Have you implemented a safe-lifting program?
12. Have you implemented a forklift safety program based on OSHA's Industrial Truck standard?
13. Have you identified all hazards in the workplace relating to temperature extremes and instituted appropriate precautions?
14. Have you identified all pressure hazards in the workplace and taken appropriate precautions to protect employees?
15. Have you implemented an electrical hazards prevention program based on OSHA's Electrical Standard?
16. Are all avenues of ingress and egress properly marked and kept free of clutter and obstructions?
17. Have you implemented a fire safety program based on OSHA's Fire Safety Standard?
18. Have you identified all potentially toxic substances in the workplace and instituted appropriate procedures to protect employees from them?
19. Have you instituted a confined space safety program based on OSHA's Confined Space Standard?
20. Have you identified all radiation hazards in the workplace and implemented appropriate procedures to protect employees? Do these procedures comply with OSHA's Standards for Health and Environmental Controls?
21. Have you identified noise hazards in the workplace and instituted appropriate precautions to protect employees?
22. Have you established a comprehensive hearing loss prevention program?

23. Have you developed a comprehensive emergency plan that takes into account all possible types of emergency including terrorism? Is your plan customized to meet local needs?

24. Have you ensured that selected personnel know how to undertake safety analysis and prevention efforts? Can these personnel complete a detailed hazard analysis?

25. Does OSHA's Process Safety Standard apply to your organization? If so, have you instituted a process safety program that satisfies the standard?

26. Have you trained both safety and supervisory personnel to conduct accident investigations and complete all necessary related reports?

27. Are all required OSHA accident and injury reports and logs maintained and kept up-to-date?

28. Have you implemented a comprehensive, ongoing safety promotion campaign?

29. Are all employees provided the training they need in order to do their jobs safely?

30. Do all employees who have specialized jobs receive the training they need in order to do those jobs safely?

31. Is updated training and retraining provided as needed for all employees?

32. Have you instituted a program to protect employees from exposure to bloodborne pathogens? Does your program protect employees who have contracted HIV or other bloodborne diseases from the negative effects of hysteria?

33. Have you instituted procedures to protect employees from workplace violence?

BEST PRACTICES IN SAFETY MANAGEMENT

Safety and health texts necessarily focus on what can go wrong in the workplace and cause accidents and injuries. This is because safety and health professionals must be able to identify hazardous conditions and take the appropriate steps to keep them from harming people. However, it is also helpful for safety and health personnel to study the other side of the coin—outstanding practices that make the workplace safer. This section summarizes some of the best safety management practices of various organizations.[9]

Best Practices at the New York Power Authority

The New York Power Authority operates 17 electricity-generating plants and employs 1,500 people. When the Authority decided to make safety one of its highest priorities, lost workday injuries dropped from 438 to 31 in just one year. As a result of its efforts, the Authority has won the Best Safety Record award from the American Public Power Association several times. The safety management practices that account for this commendable performance are as follows:

1. The Authority's corporate office has an active safety committee that meets quarterly. The Authority's CEO attends these meetings without fail, to show that safety is a top priority.
2. All new employees receive one-on-one safety training from an occupational safety professional before beginning work in their new jobs.
3. Eight to 10 specialized training programs are provided for all employees each year in such key areas as ergonomics, forklift safety, and CPR. Defibrillators are made readily available at all 17 of the Authority's sites.
4. Key staff members receive periodic e-mails concerning accidents and near misses with information about how the problem has been dealt with and lessons learned. Employees are able to access this information through the Authority's intranet.
5. Staff safety and health professionals conduct periodic safety audits at all of the Authority's 17 sites. As part of the audit, they discuss their findings with site managers immediately after the audit and help them correct problems.
6. In addition to the corporate safety committee, each of the 17 sites has its own safety committee that meets monthly. The membership of these site-based committees includes union members, supervisors, and managers. These committees are responsible for identifying potential problems and taking steps to solve them immediately.
7. The Authority extends its concerns about safety beyond the doors of its various sites with various programs that encourage safety off the job, provide free flu shots and hepatitis vaccinations, a wellness program, and an annual health fair.

Best Practices at Georgia-Pacific Corporation

Georgia-Pacific Corporation is a manufacturer and distributor of tissue, paper, building products, pulp, and related chemicals. The company employs more than 70,000 people at more than 600 locations. Safety is a high priority at Georgia-Pacific. Best safety management practices of this corporation include the following:

1. Any of the corporation's facilities that work 500,000 hours without an OSHA recordable injury receive the Chairman's Safety Award. The corporation's chair visits each facility for the award presentation and personally awards the plaque to the employees.
2. The corporation participates in OSHA's Voluntary Protection Program (VPP). All the corporation's sites are encouraged to participate in the VPP. Those that already do, mentor those that are just beginning.
3. The corporation's various facilities host promotional activities such as periodic meetings that those involved refer to as "revival tent meetings" as

a way to continually reenergize employees and keep them focused on safety.

4. Daily and weekly toolbox meetings conducted at the grassroots level keep employees informed and focused on looking out for each other while doing their respective jobs safely.

5. Other elements of the corporation's safety program include training; intranet communications; employee publications; consulting with other facilities within the corporation; and regular reports to employees of safety statistics.

Best Practices at Deere & Company

Deere & Company manufactures agricultural, forest harvesting, construction, lawn care, and diesel engine products. The company operates facilities in 17 countries and employs 40,000 people. Best safety management practices at Deere & Co. include the following:

1. Each unit within the company establishes safety goals annually. Progress toward accomplishing these goals is monitored carefully, and the results are reported to employees.

2. The corporate leadership commits to providing the resources and support necessary to ensure a safe and healthy workplace in all its facilities. For example, the company employs 175 safety and health professionals.

3. The company implements safety management systems in all its facilities that ensure compliance with all applicable laws and regulations.

4. Appropriate engineering and administrative controls are established in all facilities and monitored closely.

5. Line managers and supervisors are made part of the "safety team" and are expected to make this role a high priority.

6. Safety performance is included in performance appraisals of all employees.

OCCUPATIONAL HEALTH AND SAFETY MANAGEMENT SYSTEMS

There are numerous occupational health and safety management systems (OHSMSs) on the market that are designed to provide organizations with a comprehensive model for managing all aspects of occupational safety and health. The intent and approach of these systems are similar to those of ISO 9000 for quality management and ISO 14000 for environmental management. An OHSMS typically revolves around the standard management model composed of the following essential elements: (1) assess, (2) plan, (3) implement, (4) monitor/evaluate, and (5) adjust. This model is carried out in a continuous cycle of activities that is ongoing forever.

Although the actual structure of a given OHSMS can vary slightly depending on the vendor, organization, or agency that develops the system, most contain at least the following elements:

- Assessment of risk/hazard analysis
- Planning for hazard abatement and control
- Establishment of performance outcomes (goals and objectives)
- Establishment of roles and responsibilities
- Training related to occupational safety and health
- Communication with stakeholders
- Procedures for controlling hazards and risks
- Emergency response plans and procedures
- Regular, ongoing measurement of performance (achievement of occupational safety and health goals and objectives)
- Closing the loop: acting on the results of evaluations in ways that continually improve safety and health in the workplace

When an OHSMS is effectively implemented with executive-level support, it can produce the following benefits: (1) improve the attitudes of personnel about their working environment; (2) increase the amount of participation in safety and health activities as well as the perceptions of personnel about those activities; (3) reduce lost time due to injuries and accidents and the costs associated with lost time; (4) increase productivity; and (5) enhance morale at all levels.

SUMMARY

1. A hazard is a condition or combination of conditions that, if left uncorrected, may lead to an accident, illness, or property damage.
2. Hazard analysis is a systematic process for identifying hazards and recommending corrective action. There are two approaches to hazard analysis: preliminary and detailed.
3. Hazards can be ranked as potentially catastrophic, critical, marginal, and nuisance.
4. A preliminary hazard analysis involves forming an ad hoc team of experienced personnel who are familiar with the equipment, material substance, and process being analyzed. Experience and related expertise are critical in conducting a preliminary hazard analysis.
5. Failure mode and effects analysis (FMEA) is a detailed hazard analysis methodology that involves dividing a system into its various components, examining each component to determine how it may fail, rating the probability of failure, and deciding what effect these failures would have.
6. Hazard and operability review (HAZOP) is a detailed hazard analysis methodology that was developed for use in the chemical industry. It involves forming a team of experts and brainstorming.

7. Human error analysis (HEA) is used to predict human error and its potential effects. It can be used in conjunction with FMEA and HAZOP to strengthen those approaches.

8. Technic of operation review (TOR) is a hazard analysis methodology that allows workers and supervisors to conduct the analysis. It uses a simple worksheet that allows team members to respond to a sequence of yes/no options.

9. Fault tree analysis (FTA) is a hazard analysis methodology that uses a graphic model to display the analysis process visually. The model resembles a logic diagram.

10. Risk analysis, although more commonly associated with the insurance industry, can be used for hazard and safety analysis. The process revolves around answering two questions: How frequently does a given event occur? How severe are the consequences of a given event? The fundamental rule of thumb of risk analysis is that risk is decreased by decreasing the frequency and severity of hazard-related events.

11. The fundamentals of hazard prevention and deterrence include the following strategies: eliminate the source of the hazard, substitute a less hazardous substance, reduce the hazard at the source, remove the employee from the hazard, isolate the hazard, dilute the hazard, apply appropriate management strategies, use personal protective equipment, provide employee training, and practice good housekeeping.

12. Risk assessment should answer four questions: (a) How severe are potential injuries? (b) How frequently are employees exposed to potential hazards? (c) What is the possibility of avoiding the hazard if it does occur? (d) What is the likelihood of an injury if a safety control system fails?

13. Safety management concerns include strategic planning, safety policy, written procedures, employee training, communication, human resource management, self-assessments, safety and health promotion, accident investigation and reporting, and monitoring.

14. Safety management systems provide comprehensive models for undertaking these concerns.

KEY TERMS AND CONCEPTS

Accident investigation and reporting	Fault tree analysis (FTA)
Communication	Frequency
Corrective and preventive strategies	Hazard
Cost–benefit factors	Hazard analysis
Detailed hazard analysis	Hazard and operability review (HAZOP)
Employee training	
Experience	Hazard inventory
Failure mode and effects analysis (FMEA)	Hazard prevention and deterrence

Hazardous Waste Operations and
  Emergency Response
  (HAZWOPER)
Human error
Human error analysis (HEA)
Human resource management
Impact
Monitoring
OHSMS
Preliminary hazard analysis (PHA)
Probability

Related expertise
Risk analysis
Risk assessment
Safety and health promotion
Safety policy
Self-assessments
Severity
Strategic planning
Technic of operation review (TOR)
Written procedures

REVIEW QUESTIONS

1. Define the term *hazard*.
2. What is the purpose of preliminary hazard analysis?
3. Explain why experience and related expertise are so important when conducting a preliminary hazard analysis.
4. Why is cost–benefit analysis such a critical part of hazard analysis and prevention?
5. Briefly describe the following detailed hazard analysis methodologies: FMEA, HAZOP, HEA, FTA, and TOR.
6. What is the most fundamental weakness of both FMEA and HAZOP? How can it be overcome?
7. Name and briefly explain two approaches to HEA.
8. Why did it take so long for TOR to be adopted?
9. What is the most important strength of TOR?
10. Name five widely applicable hazard prevention strategies.
11. Explain the two options given to organizations by HAZWOPER for responding to a chemical spill.
12. What is risk assessment? How is it used?
13. Explain the principal concerns of the safety manager.
14. List the major components of an OHSMS.

ENDNOTES

1. Society of Manufacturing Engineers (SME), *Tool and Manufacturing Engineers Handbook* vol. 6 (Dearborn, MI: Society of Manufacturing Engineers), 12–17.
2. American Institute of Chemical Engineers, *Guidelines for Hazard Evaluation Procedures* (Chicago: American Institute of Chemical Engineers, 1995), 13.

3. R. G. Hallock, "Technic of Operations Review Analysis: Determine Cause of Accident/Incident," *Safety & Health* 60, no. 8: 38–39, 46.

4. A. Chapanis, "To Err Is Human, to Forgive, Design," proceedings of the ASSE Annual Professional Development Conference, New Orleans, 1986, 6.

5. Ibid.

6. Ibid.

7. SME, *Tool and Manufacturing,* 12–20.

8. "Safety of Machinery—Principles of Safety Related to Control Systems," EN 954, Part I, *European Union,* 1997.

9. S. Smith, "America's Safest Companies," *Occupational Hazards* 64, no. 10: 47–62.

PROMOTING SAFETY

Major Topics

- Company Safety Policy
- Safety Rules and Regulations
- Employee Participation in Promoting Safety
- Safety Training
- Suggestion Programs
- Visual Awareness
- Safety Committees
- Personal Commitment to Workplace Safety
- Employee–Management Participation
- Incentives
- Competition
- Company-Sponsored Wellness Programs
- Teamwork Approach to Promoting Safety
- Persuasion as a Promotional Tool

One of the best ways to promote safety is to design it into the tools, machines, and technologies with which people interact in the workplace. Safety analysis can also be effective by eliminating hazards before they cause accidents and illnesses. However, even the best design or analysis cannot completely eliminate the potential for accidents. For this reason, it is important to have accident prevention procedures and make sure that employees follow them.

The purpose of safety promotion is to keep employees focused on doing their work the safe way, every day. This chapter provides prospective and practicing safety and health professionals with information that will enable them to promote safety effectively.

COMPANY SAFETY POLICY

Promoting safety begins with having a published company **safety policy**. The policy should make it clear that safe work practices are expected of all employees at all levels at all times. The safety policy serves as the foundation upon which all other promotional efforts are built.

Figure 28–1 is an example of a company safety policy. This policy briefly and succinctly expresses the company's **commitment** to safety. It also indicates clearly that employees are expected to perform their duties with safety foremost in their minds. With such a policy in place and clearly communicated to all employees, other efforts to promote safety will have solid backing.

A company's safety policy need not be long. In fact, a short and simple policy is better. Regardless of its length or format, a safety policy should convey at least the following messages:

- The company and its top managers are committed to safety and health.
- Employees are expected to perform their duties in a safe and healthy manner.
- The company's commitment extends beyond the walls of its plant to include customers and the community.

Promoting Safety by Example

After a safety policy has been implemented, its credibility with employees will be determined by the example set by management, from supervisors through executives. It is critical that managers follow the company safety policy in both letter and spirit. Managers who set a poor example undermine all the company's efforts to promote safety. The "do as I say, not as I do" approach will not work with employees today.

Okaloosa Poultry Processing, Inc.

414 Baker Highway
Crestview, FL 36710

Safety Policy

It is the policy of this company and its top management to ensure a safe and healthy workplace for employees, a safe and healthy product for customers, and a safe and healthy environment for the community. OPP, Inc. is committed to safety on the job and off. Employees are expected to perform their duties with this commitment in mind.

Figure 28–1
Sample company safety policy.

Positive examples tend to break down most frequently under the pressure of deadlines. To meet a deadline, supervisors may encourage their team members to take shortcuts or, at least, look the other way when they do. This type of behavior conveys the message that safety is not really important—something we talk about, but not something we believe in. The issue of setting a positive example is discussed further in the next section.

SAFETY RULES AND REGULATIONS

A company's safety policy is translated into everyday action and behavior by rules and regulations. Rules and regulations define behavior that is acceptable and unacceptable from a safety and health perspective. From a legal point of view, an employer's obligations regarding **safety rules** can be summarized as follows:

- Employers must have rules that ensure a safe and healthy workplace.
- Employers must ensure that all employees are knowledgeable about the rules.
- Employers must ensure that safety rules are enforced objectively and consistently.

The law tends to view employers who do not meet these three criteria as being *negligent*. Having the rules is not enough. Having rules and making employees aware of them is not enough. Employers must develop appropriate rules, familiarize all employees with them, and enforce the rules. It is this final step—enforcement—from which most negligence charges arise.

Although it is acceptable to prioritize rules and assign different levels of punishment for failing to observe them, it is unacceptable to ignore rules. If the punishment for failure to observe Rule X is a letter of reprimand, then every person who fails to observe Rule X should receive such a letter every time. Of course, consequences for repeat violations should be more than a letter of reprimand.

Objectivity and consistency are critical when enforcing rules. **Objectivity** means that rules are enforced without bias. **Consistency** means that the rules are enforced in the same manner every time with no regard to any outside factors. This means that the same punishment is assigned regardless of who commits the infraction. Objectivity and consistency are similar but different concepts in that one can be consistent without being objective. For example, one could be consistently biased. Failure to be objective and consistent can undermine the credibility and effectiveness of a company's efforts to promote safety.

Figure 28–2 contains guidelines to follow when developing safety rules. These guidelines help ensure a safe and healthy workplace without unduly inhibiting workers in the performance of their jobs. This is an important point for prospective and practicing safety and health professionals to understand. Fear of negligence charges can influence an employer in such a way that the book of safety rules becomes a multivolume nightmare that is beyond the comprehension of most employees.

- Minimize the number of rules to the extent possible. Too many rules can result in rule *overload*.
- Write rules in clear and simple language. Be brief and to the point, avoiding ambiguous or overly technical language.
- Write only the rules that are necessary to ensure a safe and healthy workplace. Do not *nitpick*.
- Involve employees in the development of rules that apply to their specific areas of operation.
- Develop only rules that can and will be enforced.
- Use common sense in developing rules.

Figure 28–2
Guidelines for developing safety rules and regulations.

Such attempts to avoid costly litigation, penalties, or fines by regulating every move that employees make and every breath that they take are likely to backfire. Remember, employers must do more than just write rules. They must also familiarize all employees with them and enforce them. This is not possible if the rulebook is as thick as an unabridged dictionary. Apply common sense when writing safety rules.

EMPLOYEE PARTICIPATION IN PROMOTING SAFETY

One of the keys to promoting safety successfully is to involve employees. They usually know better than anyone where hazards exist. In addition, they are the ones who must follow safety rules. A fundamental rule of management is *if you want employees to make a commitment, involve them from the start.* One of the most effective strategies for getting employees to commit to the safety program is to involve them in the development of it. Employees should also be involved in the implementation, monitoring, and follow-up. In all phases, employees should be empowered to take action to improve safety. The most effective safety program is one that employees view as *their* program.

SAFETY TRAINING

One of the best ways to promote safety in the workplace is to provide all employees with ongoing **safety training**. Initial safety training should be part of the orientation process for new employees. Subsequent safety training should be aimed at developing new, more specific, and more in-depth knowledge and at renewing and updating existing knowledge.

Safety training is covered at length in Chapter 12. This chapter emphasizes the importance of promoting safety by providing training on a continual basis. Training serves a dual purpose in the promotion of safety. First, it ensures that employees know how to work safely and why doing so is important. Second, it shows that management is committed to safety. Refer to Chapter 12 to see how safety training programs are organized and which topics should be covered at the different levels of training.

SUGGESTION PROGRAMS

Suggestion programs, if properly handled, promote safety and health. Well-run suggestion programs offer two advantages: (1) they solicit input from the people most likely to know where hazards exist; and (2) they involve and empower employees which, in turn, gives them **ownership** of the safety program.

Suggestion programs must meet certain criteria to be effective:

- All suggestions must receive a formal response.
- All suggestions must be answered immediately.
- Management must monitor the performance of each department in generating and responding to suggestions.
- System costs and savings must be reported.
- Recognition and awards must be handled promptly.
- Good ideas must be implemented.
- Personality conflicts must be minimized.[1]

Suggestion programs that meet these criteria are more likely to be successful than those that don't. Figure 28–3 is an example of a suggestion form that may be used as part of a company's safety program. Note that all the following must be recorded: the date that the suggestion was submitted, the date that the suggestion was logged in, and the date that the employee received a response.

This form satisfies the **formal response** and **immediate response** criteria. It also makes it easier to monitor response. Jones Petroleum Products (in Figure 28–3) publishes system costs and savings in its monthly newsletter for employees, implements good ideas, and recognizes employees with a variety of awards ranging from certificates to cash at a monthly recognition ceremony. This company's suggestion program is an example of one that promotes not just safety, but continual improvements in quality, productivity, and competitiveness.

VISUAL AWARENESS

We tend to be a visual society. This is why television and billboards are so effective in marketing promotions. Making a safety and health message *visual* can be an effective way to get the message across. Figure 28–4 shows a sign that

Jones Petroleum Products, Inc.

Highway 90 East
DeFuniak Springs, FL 32614

Suggestion Form

Name of employee: _____ Date of suggestion: _____

Department: _____

Suggested improvement: _____

_____

_____

_____

Date logged in: _____ Time: _____

Logged in by: _____

Action taken: _____

_____

Current status: _____

Date of response to employee: _____

Person responding: _____

(Signature)

Figure 28–3
Sample safety suggestion form.

Figure 28–4
Sample safety reminder sign.

Figure 28–5
Sample safety reminder sign.

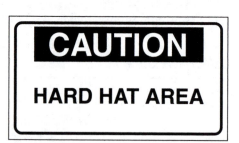

Figure 28–6
Sample safety reminder sign.

Figure 28–7
Sample safety reminder sign.

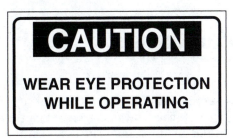

Figure 28–8
Sample safety reminder sign.

gives machine operators a visual reminder to use the appropriate machine guards. Such a sign is placed on or near the machine in question. If operators cannot activate their machines without first reading this sign, they will be reminded to use the safe way every time they operate the machine.

A sign like Figure 28–5 may be placed on the door leading into the hard hat area or on a stand placed prominently at the main point of entry if there is no door. Such a sign helps prevent inadvertent slip-ups when employees are in a hurry or are thinking about something else. Figures 28–6, 28–7, and 28–8 are additional examples of signs and posters that make a safety message visual.

Several rules of thumb can help ensure the effectiveness of efforts to make safety visual:

- Change signs, posters, and other visual aids periodically. Visual aids left up too long begin to blend into the background and are no longer noticed.

- Involve employees in developing the messages that are displayed on signs and posters. Employees are more likely to notice and heed their own messages than those of others.
- Keep visual aids simple and the message brief.
- Make visual aids large enough to be seen easily from a reasonable distance.
- Locate visual aids for maximum effect. For example, the sign in Figure 28–4 should be located on the machine in question, preferably near the on/off switch so that the operator cannot activate the machine without seeing it.
- Use color whenever possible to attract attention to the visual aid (but be sure to follow OSHA's color standards where applicable).

SAFETY COMMITTEES

Another way to promote safety through employee involvement is the **safety committee**. Safety committees provide a formal structure through which employees and management can funnel concerns and suggestions about safety and health issues. The composition of the safety committee can be a major factor in the committee's success or failure.

The most effective committees are those with a broad cross-section of workers representing all departments. This offers two advantages: (1) it gives each member of the committee a constituent group for which he or she is responsible; and (2) it gives all employees a representative voice on the committee.

There is disagreement over whether an executive-level manager should serve on the safety committee. On one hand, an executive-level participant can give the committee credibility, visibility, and access. On the other hand, the presence of an executive manager can inhibit the free flow of ideas and concerns. The key to whether an executive manager's participation will be positive or negative lies in the personality and management skills of the executive in question.

An executive who knows how to put employees at ease, interact in a non-threatening manner, and draw people out will add to the effectiveness of the committee. An executive with a threatening attitude will render the committee useless. Consequently, the author recommends the involvement of a very carefully selected executive manager on the safety committee.

The safety and health professional should be a member of the committee serving as an advisor, facilitator, and catalyst. Committee members should select a chairperson from the membership and a recording secretary for taking minutes and maintaining committee records. Neither the executive manager nor the safety and health professional should serve as chairperson, but either can serve as recording secretary. Excluding executive managers and safety and health professionals from the chair gives employees more ownership in the committee.

Safety committees work only if members are truly empowered to identify hazards and take steps to eliminate them.

Safety Committee Meetings

Every employee who serves on a safety committee has another job to do. Consequently, it is important to have meetings that are both efficient and effective. This is accomplished by having an agenda that gives both structure and direction to meetings.

A typical meeting of the safety committee should proceed as follows:

1. Call to order.
2. Record attendance.
3. Review and approve previous minutes.
4. Discuss old business.
5. Discuss new business.
6. Discuss new accidents.
7. Discuss near misses.
8. Report on inspections, subcommittee work, special assignments, safety programs, and so on.
9. Make special presentations (guests, videos/DVDs, demonstrations, and so on).
10. Make announcements.
11. Adjourn.

Do's and Don'ts of Safety Committees

For both management and employees to participate willingly and contribute effectively, safety committees must work well. In addition to having an agenda for meetings and sticking to it, the following do's and don'ts should be observed:

DO These

- Suggest strategies and options for management to improve health and safety performance.
- Train committee members so that they can successfully carry out their responsibilities.
- Give the safety committee authority commensurate with their responsibilities.
- Have goals and objectives, and measure against them to track progress.
- Encourage employee involvement by actively creating an atmosphere of trust, teamwork, respect, and partnership.
- Be patient. Be reasonable in allowing enough time for the committee to work.
- Reward progress, participation, and leadership.
- Train management on their responsibility for safety and on the support role of the safety committee.
- Stagger committee memberships to maintain a mixture of experience levels.

DON'T DO These

- Allow the safety committee to function as "safety cops." Keep management responsible for decisions and enforcements.
- Discuss topics unrelated to health and safety at safety committee meetings. Stay away from labor and personnel issues.
- Rotate members too quickly. A one-year minimum membership is the norm.
- Let any one member dominate safety committee meetings. Encourage and maintain equal participation.
- Allow safety committee members to bring just problems to the meetings. Have them bring solutions as well.
- Allow the safety committee to become scapegoats when something goes wrong. Management is responsible for safety performance.
- Punish. It creates fear, which inhibits communication and partnership. For example, blaming an injured employee can obscure other contributing causes and encourage underreporting. Do hold persons accountable for their responsibilities (which differs from punishment).[2]

PERSONAL COMMITMENT TO WORKPLACE SAFETY

If every employee is committed to working safely every day, workplace safety will take care of itself. But how does a company gain this type of personal commitment from its employees? One way is to have employees commit themselves to safety by signing on the bottom line, as Fettig states:

> Most safety superstars have formal safety programs—teams, committees, whatever the personal term. Many of these organizations also have signed commitments from their employees. They have their people's signatures on the dotted line indicating that they bought into the safety program. A signature on the dotted line is serious business. When our founding fathers signed the Declaration of Independence, they pledged their lives, their fortunes, and their sacred honor. How do you sell safety? One way is to get that signature on the dotted line.[3]

Fettig's approach to gaining a **personal commitment** from employees has merit. Ours is a society that revolves around the written signature. We sign countless documents in our lives, from credit statements to bank loans to home mortgages to college registration forms. In each of these cases, our signature is a written pledge of our commitment to meet certain responsibilities.

According to Fettig, companies gain the following three advantages from making signing on the dotted line part of their program to promote safety:

- By their signature, employees make a personal commitment.
- By their signature, employees promise to interact positively with fellow workers when they see them ignoring safety precautions.
- By their signature, employees give fellow workers permission to correct them when they ignore safety precautions.[4]

Discussion Case

What Is Your Opinion?

"I want our management team to develop a program to promote better safety practices in this company," said the CEO. "If we develop the program, it will show employees that we are committed to safety." "I like the idea of showing executive-level commitment," said the vice president for engineering. "But the program may be more readily accepted if we involve employees in developing it." Executive commitment and employee involvement. Which is the better approach? What is your opinion?

EMPLOYEE–MANAGEMENT PARTICIPATION

An excellent way to promote safety is to secure the cooperation of management and labor. For a company's safety program to succeed, **employee–management participation** and support is critical. Fortunately, employee–management agreement on workplace safety is commonplace.

When disagreement over a safety procedure does surface, the issue at the heart is usually money. Employees are likely to favor procedures that enhance workplace safety regardless of cost. Management, on the other hand, is likely to want to weigh the cost versus the benefits of safety improvement strategies. However, sometimes employees, rather than managers, question safety strategies.

An example of an employee questioning a safety enhancement strategy is the Sign Up for Safety campaign conducted by safety consultant Art Fettig at the Northwestern Region of Consumer Power Company in Muskegon, Michigan.[5] In an attempt to gain a personal commitment to safety, Fettig asks employees to sign a declaration that they will work in a safe manner. This is a technique that has met with a great deal of success.

However, in attempting to sell employees of Consumer Power Company on the strategy, Fettig ran into resistance from an employee who refused to sign his name, claiming the company's management team might use it against him. Here is Fettig's account of how he handled the situation and what eventually resulted:

> I pointed out that the document was not for the company, but for the workers. Very seldom do managers fall from their chairs, but linemen fall from ladders and poles. And I have yet to hear of a manager who was electrocuted by a computer, but I often read of such accidents that involve linemen and power lines. He . . . still wasn't sold. He said the declaration should hang at the union hall and not at the company. He's right since the declaration is for the employees. But management is also part of the team and should have a copy for the company as well. He finally signed the declaration. In fact, all of the employees signed.[6]

The employee's eventual willingness to sign the safety declaration in this case is what made the program work for Consumer Power Company. With

---

Safety Fact

Supervisor's Role in Safety

Research and experience show that the first-line supervisor plays a critical role in promoting safe and healthy work practices. The supervisor has the most frequent face-to-face contact with employees and provides the example they are most likely to follow. If supervisors are committed to safety, those who report directly to them will probably be equally committed. Therefore, time spent by safety and health professionals gaining the support, cooperation, and commitment of supervisors is time well spent.

---

management and employees on the same team, the safety program is much more likely to succeed.

INCENTIVES

If properly used, **incentives** can help promote safety. However, the proper use of incentives is a widely misunderstood concept.[7]

Safety is another issue that is being confronted using incentives. To promote safety effectively, incentives must be properly structured. The following strategies are recommended for enhancing the effectiveness of incentive programs:

1. *Define objectives.* Begin by deciding what is supposed to be accomplished by the incentive program.

2. *Develop specific criteria.* On what basis will the incentives be awarded? This question should be answered during the development of the program. Specific criteria define the type of behavior and level of performance that is to be rewarded as well as guidelines for measuring success.

3. *Make rewards meaningful.* For an incentive program to be effective, the rewards must be meaningful to the recipients. Giving an employee a reward that he or she does not value will not produce the desired results. To determine what types of rewards will be meaningful, it is necessary to involve employees.

4. *Recognize that only employees who will participate in an incentive program know what incentives will motivate them.* In addition, employees must feel it is *their* program. This means that employees should be involved in the planning, implementation, and evaluation of the incentive program.

5. *Keep communications clear.* It is important for employees to understand fully the incentive program and all of its aspects. Communicate with employees about the program, ask for continual feedback, listen to the feedback, and act on it.

6. *Use nonmonetary rewards.* Often nonmonetary incentives are more effective than money in promoting the desired results. An excellent resource

for identifying nonmonetary rewards is the book *The 1001 Rewards & Recognition Fieldbook* by Nelson and Spitzer.

   7. *Reward teams.* Rewarding teams can be more effective than rewarding individuals. This is because work in the modern industrial setting is more likely to be accomplished by a team than an individual. When this is the case, other team members may resent the recognition given to an individual member. Such a situation can cause the incentive program to backfire.[8]

Effectiveness of Incentive Programs

When handled properly, incentive programs can be an effective way to promote safety in the workplace. But how effective are incentive programs? The answer to this question is: *it depends.* According to Sandy Smith, the following statistics show the effectiveness of incentive programs on workplace safety in selected circumstances:

1. Incentive programs that target individuals improve performance by 27 percent.
2. Incentive programs that target teams improve performance by 45 percent.
3. Incentive programs have an equally positive effect on both the quality and quantity of employee performance.
4. Incentive programs work best when they are structured with employee input.
5. Long-term incentives are more effective than short-term (44 percent improvement in performance compared with 20 percent, respectively).[9]

Potential Problems with Incentive Programs

Clearly, incentive programs are an excellent way to promote safety in the workplace. However, there are potential problems that, if ignored, can undermine the effectiveness of the programs. Bill Sims states these problems as follows:

   Employee taxes. Taxes must be paid on incentives provided to employees. This means either the employer pays or the employee will be stuck with an added tax bill. In either case, taxes increase the cost of cash incentives to employers and decrease the net amount for employees.

   Injury hiding. Incentive programs that are based on statistics can undermine the real effectiveness of the safety program as opposed to the apparent effectiveness. When departments and units hide (fail to report) injuries to win incentive rewards, the integrity of the safety program is at risk. OSHA frowns on statistics-based programs because the agency thinks they cause organizations to cheat on reporting requirements.

   Unfair programs. Programs that award incentives based on yearly or quarterly improvements can undermine employee morale and support of the safety program when an accident occurs the day before the end of the period. Employees can maintain a safe working record for 364 days only to lose out on incentives when an accident occurs on day 365. It is better to base incentives on daily observation of safe behavior and work practices.

Insufficient budget. Companies typically budget too little for safety incentives. A realistic budget would be in the range of $100 per employee per year. Less will have insufficient impact to be worth the time and effort an incentive program requires.[10]

COMPETITION

Competition is another strategy that can be used to promote safety. However, if this approach is not used wisely, it can backfire and do more harm than good. To a degree, most people are competitive. A child's competitive instinct is nurtured through play and reinforced by sports and school activities. Safety and health professionals can use the adult's competitive instinct when trying to motivate employees, but competition on the job should be carefully organized, closely monitored, and strictly controlled. Competition that is allowed to get out of hand can lead to cheating and hard feelings among coworkers.

Competition can be organized between teams, shifts, divisions, or even plants. Here are some tips that will help safety and health professionals use competition in a positive way while ensuring that it does not get out of hand:

- Involve the employees who will compete in planning programs of competition.
- Where possible, encourage competition among groups rather than individuals, while simultaneously promoting individual initiative within groups.
- Make sure that the competition is fair by ensuring that the resources available to competing teams are equitably distributed and that human talent is as appropriately spread among the teams as possible.

The main problem with using competition to promote safety is that it can induce competing teams to cover up or fail to report accidents just to win. Safety and health professionals should be particularly attentive to this situation and watch carefully for evidence that accidents are going unreported. If this occurs, the best approach is to confront the situation openly and frankly. Employees should be reminded that improved safety is the first priority and winning the competition is second. Failing to report an accident should be grounds for eliminating a team from competition.

COMPANY-SPONSORED WELLNESS PROGRAMS

Faced with mounting health care and workers' compensation costs, some employers are looking for innovative ways to keep their workers safe and healthy. The annual cost of health insurance for business and industry in this country is approximately $160 billion. One innovation that is gaining in popularity as a way to promote safety and health is the company-sponsored wellness program. A **wellness program** is any program designed to help and encourage employees to adopt a healthier lifestyle.

A typical wellness program includes diet and exercise under the supervision of an appropriately qualified professional; stress management activities; and special activities designed to help high-risk employees overcome such lifestyle-related behaviors as smoking and overeating.

Industry in the United States began to experiment with company-sponsored wellness programs in the early 1980s. However, hard data on tangible financial benefits were not available in the early years, and wellness programs failed to gain broad-based acceptance, particularly among small- to mid-sized companies. However, as health care costs continued to rise and industry continued to lose valuable employees to heart attacks and strokes, corporate leaders began searching for ways to measure objectively the costs versus the benefits of company-sponsored wellness programs.

A major breakthrough in this area resulted when the Institute for Preventive Medicine (IPM), located in Houston, Texas, developed a computer-based method for measuring the success of preventive health care programs. The method was pilot-tested on a large Houston-based corporation.[11]

This company agreed to spend $46,000 to fund an in-house wellness program that included weight control, stress management, smoking cessation, exercise, and cardiac rehabilitation activities. After six months, IPM assessed the costs versus benefits of the program and arrived at the following conclusions:

> Approximately 200 employees participated in the six-month health management program, and the results were dramatic. Comparing insurance claims and absenteeism for the six months before the program and those for the six months after its inception, the IPM staff recorded an 18 percent decline in sick pay hours and a 15 percent fall in sick costs. Overall, the company saw health care costs drop 39 percent.[12]

The IPM study showed that company-sponsored wellness programs can be an effective way to promote employee safety and health. According to Myerson,

> Corporate wellness works. Participants improve their health and quality of life, and employers cut costs—while at the same time generating goodwill. It's a combination that a growing number of organizations are finding hard to resist.[13]

The good news here is that the number of companies offering wellness programs for employees is steadily increasing. The bad news is that employee participation in the programs is low.

TEAMWORK APPROACH TO PROMOTING SAFETY

Increasingly, teamwork is stressed as the best way to get work done in the contemporary workplace. Consequently, it follows that the teamwork approach is an excellent way to promote safety. Teamwork is a fundamental component of the *total safety management*, or TSM, approach explained in Chapter 31. Consequently, this section is limited to covering teamwork as it relates specifically to the promotion of safety.

Characteristics of Effective Teams

Effective teams share several common characteristics: supportive environment, team player skills, role clarity, clear direction, team-oriented rewards, and accountability.

Supportive Environment

The characteristics of a team-supportive environment are well known. These characteristics are as follows:

- Open communication
- Constructive, nonhostile interaction
- Mutually supportive approach to work
- Positive, respectful climate

Team Player Skills

Team player skills are personal characteristics of individuals that make them good team players. They include the following:

- Honesty
- Selflessness
- Initiative
- Patience
- Resourcefulness
- Punctuality
- Tolerance
- Perseverance

Role Clarity

On any team, different members play different roles. Consider the example of a football team. When the offensive team is on the field, each of the 11 team members has a specific role to play. The quarterback plays one role; the running backs, another; the receivers, another; the center, another; and the linemen, another. Each of these roles is different but important to the team. When each of these players executes his role effectively, the team performs well.

But what would happen if the center suddenly decided he wanted to pass the ball? What would happen if one of the linemen suddenly decided that he wanted to run the ball? Of course, chaos would ensue. A team cannot function if team members try to play roles that are assigned to other team members.

Role clarity means that all members understand their respective roles on the team and play those roles.

Clear Direction

What is the team's purpose? What is the team supposed to do? What are the team's responsibilities? These are the types of questions that people ask when they are assigned to teams. The team's charter should answer such questions. The various components of a team's charter are as follows:

1. *Mission.* The team's mission statement defines its purpose and how the team fits into the larger organization. In the case of a safety promotion team, it explains the team's role in the organization's overall safety program.
2. *Objectives.* The team's objectives spell out exactly what the team is supposed to accomplish in terms of the safety program.
3. *Accountability measures.* The team's accountability measures spell out how the team's performance will be evaluated.

Figure 28–9 is an example of a team charter for the safety promotion team in a manufacturing company. This charter clearly defines the committee's purpose, where it fits into the overall organization, what it is supposed to accomplish, and how the committee's success will be measured.

Team-Oriented Rewards

One of the most commonly made mistakes in organizations is attempting to establish a teamwork culture while maintaining an individual-based reward system. If teams are to function fully, the organization must adopt team-oriented rewards, incentives, and recognition strategies. For example, teams function best when the financial rewards of its members are tied at least partially to team performance. Performance appraisals that contain criteria relating to team performance, in addition to individual performance, promote teamwork. The same concept applies to recognition activities.

Accountability

There is a rule of thumb in management that says, "If you want to improve performance, measure it." **Accountability** is about being held responsible for accomplishing specific objectives or undertaking specific actions. The most effective teams know what their responsibilities are and how their success will be measured.

Potential Benefits of Teamwork in Promoting Safety

Teamwork can have both direct and indirect benefits for an organization. Through teamwork, counterproductive internal competition and internal politics are

---

Team Charter

Safety Promotion Team
MTC Corporation

Mission

The mission of the *Safety Promotion Team* at MTC Corporation is to make all employees at all levels of the company aware of the importance of safety and health, and, having made them aware, to keep them aware.

Objectives

1. Identify innovative, interesting ways to communicate the company's safety rules and regulations to employees.
2. Develop a companywide suggestion system to solicit safety-related input from employees.
3. Identify eye-catching approaches for making safety a *visible* issue.
4. Develop appropriate safety competition activities.

Accountability Measures

The quality of participation in all of the activities of this team will be assessed by the team leader and included in the annual performance appraisal of each team member. Team members are expected to be consistent in their attendance, punctual, cooperative, and mutually supportive.

Figure 28–9
Sample team charter.

replaced by collaboration. When this happens, the following types of benefits typically accrue:

- Better understanding of safety rules and regulations
- Visibility for safety
- Greater employee awareness
- Positive, productive competition
- Continual improvement
- Broader employee input and acceptance

Potential Problems with Teams

Teamwork can yield important benefits, but as with any concept, there are potential problems. The most pronounced potential problems with teams are as follows:

- It can take a concerted effort over an extended period to mold a group into an effective team, but a team can fall apart quickly.

Safety Fact

Team Building through Training

The United States is the land of the *rugged individualist*. Even people who grow up playing team sports learn early on about standing out as an individual or becoming the Most Valuable Player. Consequently, employees may not be natural team players. That's the bad news. The good news is that with the proper training, employees can learn to be good team players. An initial teamwork training program should cover the following topics:

- Rationale for team training
- Direction and understanding (team mission and objectives)
- Characteristics of team players
- Accountability measures
- Team building

- Personnel changes are common in organizations, but personnel changes can disrupt a team and break down team cohesiveness.
- Participative decision making is inherent in teamwork. However, this approach to decision making takes time, and time is often in short supply.
- Poorly motivated and lazy employees can use a team to blend into the crowd, to avoid participation. If one team member sees another slacking, he or she may respond in kind.

These potential problems can be prevented, of course. The first step in doing so is recognizing them. The next step is ensuring that all team members fulfill their responsibilities to the team and to one another.

Responsibilities of Team Members

Accountability in teamwork amounts to team members fulfilling their individual responsibilities to the team and to each other. These responsibilities are as follows:

- Active participation in all team activities
- Punctuality in attendance of meetings
- Honesty and openness toward fellow team members
- Making a concerted effort to work well with team members
- Being a good listener for other team members
- Being open to the ideas of others

If individual team members fulfill these responsibilities to each other and the team, the potential problems with teams can be overcome, and the benefits of teamwork can be fully realized. It is important for members of the safety team to understand these responsibilities, accept them, and set an example of fulfilling them. If this happens, the benefits to the organization will go well beyond just safety and health.

PERSUASION AS A PROMOTIONAL TOOL

Promotion strategies relating to safety and health have one overall goal: to gain and maintain the commitment of all personnel to working safely. In order to convince workers who are focused on concerns that can distract them from the issue to take safety seriously, professionals in the field must be persuasive. Persuasion can be viewed, in the current context, as communication that is influential, motivating, and convincing.[14]

There are two approaches to persuasive communication that might be used by safety and health professionals depending on circumstances and the immediate goal of the communication:

- *Projective communication.* This is an assertive approach in which you attempt to change people's minds, gain their commitment, move them to action, or improve their attitude. This is the approach that comes to mind for most people when the word "persuasion" is used.
- *Receptive communication.* This is a more passive approach in which you use listening more than talking to remove barriers, elicit information, show interest in the opinions of others, and identify hidden problems.

Each of these approaches has its more appropriate applications and each can be effective if properly applied. The approach to use depends on the circumstances of the situation and the immediate goals of the person who hopes to use it to persuade others. The effectiveness of the approach depends on using it in appropriate circumstances and the skills of the person using it.

Safety and health professionals should realize that using one or the other of these approaches to persuade workers to take safety seriously is not an either/or proposition. There will be times when the situation will call for first one and then the other as the circumstances change during the course of a meeting or conversation.

Projective Communication for Persuasion

This approach is most appropriate when the immediate goal is to give out information and convince the recipient to accept the information and take it seriously. To increase the chances that this goal will be achieved, try the following strategies:

- Offer the information on a personal level so that recipient can relate to it. Using personal phrases such as "What I have learned . . ." or "How this worked when I tried . . ." will help.
- Communicate—don't dictate. Use phrases such as "You might think about . . ." or "You might want to consider . . ." instead of "You must . . ." and "It is mandatory that you . . ."
- Put the other person at ease and make a personal connection. Don't just jump into the conversation about safety and health issues. Begin with small talk about something the worker is interested in such as his or her

children, grandchildren, a ballgame, or anything else that will help you connect before transitioning to the topic in question.

- Allow for give and take. Don't just give a one-way broadcast of what you have to say. Allow for a two-sided conversation. Talk a little, then stop and solicit questions or concerns. Deal with any feedback you receive before relating more of what you have to say. Repeat this process several times during the conversation.
- Reveal all you can. There may be times when your information will be incomplete. When this is the case, tell everything you know at the moment, and explain that you will get back to the recipient with more complete information as soon as you have it.

Receptive Communication for Persuasion

This approach is appropriate for use when you are trying to get workers to buy into a concept and want to identify the sources of potential resistance; gain information you don't have; let people who are concerned or opposed vent; gain feedback before solidifying an idea; or defuse anger. The following strategies will help enhance the effectiveness of this approach:

- Listen nonjudgmentally and do not interrupt.
- Resist the temptation to become defensive or angry—no matter how angry the recipient may be.
- Take notes if it is necessary, but let the recipient know in advance that you plan to take notes and that you will let him or her see everything you write down (for validation).
- Let the recipient vent his or her frustration without interruption.
- Ask open-ended questions for clarification when necessary.
- Paraphrase and repeat back in summary form what the recipient tells you.

It is not enough to simply tell people what you want them to hear or do. There are simply too many distractions in the workplace to allow this approach to be effective. In promoting occupational safety and health, it is necessary to persuade workers that safety is in their best interests so that they make a willing and voluntary commitment to the concept.

SUMMARY

1. A company's safety policy should convey the following messages: (a) a companywide commitment; (b) expectation that employees will perform their duties in a safe manner; and (c) the company's commitment includes customers and the community.
2. From a legal perspective, an employer's obligations regarding safety rules can be summarized as follows: (a) employers must have rules that ensure a safe and healthy workplace; (b) employers must ensure that all employees

are knowledgeable about the rules; and (c) employers must ensure that safety rules are enforced objectively and consistently.

3. A fundamental rule of management is: If you want employees to make a commitment, involve them from the start. This is especially important when formulating safety rules.

4. Safety training should be a fundamental part of any effort to promote safety. Safety training ensures that employees know how to work safely, and it shows that management is committed to safety.

5. Well-run suggestion programs promote safety by (1) soliciting input from the people who are most likely to know where hazards exist; and (2) involving employees in a way that lets them feel ownership in the safety program.

6. Safety committees can help promote safety if they are properly structured. The composition of the committee can be a major factor in the committee's success. The most effective committees are composed of a broad cross-section of workers representing all departments.

7. Employee–management agreement is important in promoting safety. Fortunately, safety is an issue on which employees and management can usually agree.

8. Incentives can promote safety if they are properly applied. To enhance the effectiveness of incentives, the following steps should be followed: (a) define objectives, (b) lead by example, (c) develop specific criteria, (d) make rewards meaningful, (e) keep communications clear, (f) involve employees in planning the incentives, and (g) reward teams. However, cash incentives can create tax problems, and poorly designed incentives can lead to injury hiding, unfair circumstances, and morale problems.

9. Competition can promote safety, but it can also get out of hand and do more harm than good. To keep competition positive, involve employees in planning programs of competition and encourage competition between teams rather than individuals.

10. Promotion strategies relating to safety and health have one overall goal: to gain and maintain the commitment of all personnel to working safely. The two approaches to persuasive communication are projective and receptive communication.

KEY TERMS AND CONCEPTS

Accountability	Employee participation
Commitment	Employee taxes
Competition	Formal response
Consistency	Immediate response
Employee involvement	Incentives
Employee–management participation	Injury hiding
	Insufficient budget

Mission	Safety policy
Objectives	Safety rules
Objectivity	Safety training
Ownership	Suggestion programs
Personal commitment	Unfair programs
Role clarity	Visual awareness
Safety committee	Wellness program

REVIEW QUESTIONS

1. What messages should a company's safety policy convey?
2. Explain why promoting safety by example is so important.
3. What are the employer's obligations regarding safety rules and regulations?
4. Explain the concept of negligence as it relates to a company's safety.
5. What is the significance of objectivity and consistency when enforcing safety rules?
6. Why is employee participation and involvement so critical in the promotion of safety?
7. If your task was to establish a safety committee, whom would you ask to serve on it?
8. List three benefits that companies gain from asking employees to sign a declaration of safety.
9. What are the steps for ensuring that incentives actually promote safety?
10. What problems can be caused by poorly designed incentives?
11. What are the characteristics of effective teams?
12. Explain the concept of persuasive communication.

ENDNOTES

1. B. McDermott, "Employees Are Best Source of Ideas for Constant Improvement," *Total Quality Newsletter* 1, no. 4: 5.
2. L. Cullen, "Safety Committees: A Smart Business Decision," *Occupational Hazards* 61, no. 5: 102.
3. A. Fettig, "Sign Up for Safety," *Safety & Health* 144, no. 1: 26.
4. Ibid.
5. Ibid., 27.
6. Ibid.
7. E. Robbins, "The Act of Meaningful Recognition," *Occupational Health & Safety* 74, no. 1: 48–50.
8. Ibid.

9. S. Smith, "Safety Incentives: It's Not Just a Breakfast Anymore," *Occupational Hazards* 64, no. 6: 58.

10. B. Sims, as quoted in Smith, "Safety Incentives," 58.

11. W. A. Myerson, "Wellness Is the Bottom Line," *Safety & Health* 137, no. 5: 50.

12. Ibid.

13. Ibid., 53.

14. R. Pater, "High-Level Persuasion," *Occupational Health & Safety* 74, no. 1: 24–28.

ENVIRONMENTAL SAFETY AND ISO 14000 (ENVIRONMENTAL MANAGEMENT)

Major Topics

- Safety, Health, and the Environment
- Legislation and Regulation
- Types of Environments
- Role of Safety and Health Professionals
- Hazards of the Environment
- Hazardous Waste Reduction
- Environmental Management System (EMS)
- International Organization for Standardization (ISO)
- ISO 14000
- ISO 14000 Series of Standards
- ISO 14001 Standard
- ISO 14000 Success Story

American industry and government are currently dealing with such issues as acid rain, ground-level ozone, lack of stratospheric ozone, radon gas, the need for additional cleanup of PCBs, polluted ground water, and higher-than-acceptable levels of lead in drinking water.

Industry and the environment can be compatible, mutually beneficial entities with the proper care and appropriate practices. This chapter explains the "whys" and "hows" of good environmental management and safety.

SAFETY, HEALTH, AND THE ENVIRONMENT

Much progress has been made in cleaning up and protecting the **environment** since Congress initially became interested in the 1960s. During the 1960s and 1970s, Congress passed a continual proliferation of environmental regulations aimed primarily at air, water, and ground pollution. According to the

Environmental Protection Agency (EPA), substantial progress has been made in cleaning up air and water pollution in the following areas since the 1970s:

- Carbon monoxide emissions
- Volatile organic compounds
- Sulfur oxides
- Ocean dumping of industrial wastes
- Cities without adequate sewage treatment
- Polluted rivers[1]

The progress made in cleaning up air and water pollution is encouraging. However, not all the news is good. According to the EPA, nitrogen oxide emissions and ocean dumping of sewage sludge are still problems.[2]

Much work remains to be done about environmental safety and health. The following actions are recommended by the EPA:

- *Acid rain.* Gradually reduce sulfur oxide emissions while concurrently avoiding expensive new regulation-induced clampdowns until sufficient evidence is available to justify such actions.
- *Ground-level ozone.* Develop new regulations that focus on gas stations and other small sources of ground-level ozone emissions while concurrently avoiding any new drastic restrictions on automobile emissions.
- *Global warming.* Focus more on the use of nuclear power while continuing efforts to use fossil fuels more efficiently.
- *Water pollution.* Build more sewage treatment plants throughout the nation as quickly as possible.
- *Toxic wastes.* Develop and implement incentive programs to encourage a reduction in the volume of toxic waste.
- *Garbage.* Increase the use of recycling while simultaneously reducing the overall waste stream. Burn or bury what cannot be recycled, but under strict controls.
- *Lead.* Better protective clothing and hygiene facilities should be provided for employees in lead-contaminated workplaces.
- *Mercury.* Better detection, mitigation, and elimination of mercury in work and home environments.[3]

LEGISLATION AND REGULATION

A fairly clear-cut division of authority exists for legislation and regulations concerning the environment. The Occupational Safety and Health Administration (OSHA) is responsible for regulating the work environment within an individual company or plant facility. Environmental issues that go beyond the boundaries of the individual plant facility are the responsibility of the EPA. Of course, some environmental issues and concerns do not fall clearly within the scope of either OSHA or the EPA. Therefore, these two agencies have begun to cooperate closely in dealing with environmental matters.

OSHA–EPA Partnership

The OSHA–EPA partnership has ramifications for safety and health professionals because most companies have traditionally separated environmental and workplace safety concerns into two different departments. Environmental health professionals are usually engineers or scientists who specialize in environmental matters. The workplace safety professional may be an occupational physician, industrial hygienist, or safety manager. To respond effectively and efficiently to joint efforts of OSHA and the EPA, companies may be forced to combine their environmental health and workplace safety departments.

A *memorandum of understanding (MOU)* originally signed by both agencies in 1990, and periodically updated, requires that OSHA and the EPA cooperate in developing a plan for joint enforcement and information sharing. The plan has helped eliminate the bureaucratic overlaps and inefficiencies sometimes associated with the governmental agencies. Joint training and data maintenance also have resulted.

Clean Air Act

One of the most important pieces of federal environmental legislation has been the **Clean Air Act**.

The EPA and various state regulatory agencies have a great deal of latitude in developing regulations to implement fully the intentions of Congress. The EPA had to draft more than 250 regulations to implement fully all the titles set forth in the Clean Air Act.

The Clean Air Act contains provisions that require companies to take whatever actions are necessary to prevent or minimize the potential consequences of the accidental release of pollutants into the air. It also established an independent chemical safety and hazard investigation board to investigate accidental releases of pollutants that result in death, serious injury, or substantial property damage.

The Clean Air Act, originally signed into law in 1970, has been revised several times. The Clean Air Act now contains approximately 350 pages detailing the requirements of seven titles:

Title I: Urban Air Quality
Title II: Mobile Sources
Title III: Hazardous Air Pollutants
Title IV: Acid Rain Control
Title V: Permits
Title VI: Stratospheric Ozone Provisions
Title VII: Enforcement

Figure 29–1 gives a brief summary of each title in the act.

- **Title I: Urban Air Quality**
  Establishes five classes for noncompliance with ambient air quality standards: marginal, moderate, serious, severe, and extreme.

- **Title II: Mobile Sources**
  Increases the emission standards for automobiles in two steps. Also requires the sale of reformulated gasoline in selected high-pollution cities.

- **Title III: Hazardous Air Pollutants**
  Requires EPA to use maximum control technologies to regulate the use of 189 toxic substances.

- **Title IV: Acid Rain Control**
  Allows EPA to issue acid rain allowances to existing companies. Allowances can be used, saved, or sold to other companies.

- **Title V: Permits**
  Requires major pollution emitters to obtain a special permit. Contains provisions that allow concerned citizens to petition for the revocation of a company's permit.

- **Title VI: Stratospheric Ozone Provisions**
  Bans the manufacture of methyl chloroform. The manufacture of the five most ozone-destructive chemicals were phased out completely, effective in the year 2000.

- **Title VII: Enforcement**
  Increases the ability of EPA and corresponding state agencies to impose both criminal and civil penalties against violators of the Clean Air Act.

Figure 29–1
Clean Air Act.

Economics of Environmental Regulation

A proliferation of legislation and regulations has been developed in the nation's capital in an effort to clean up and protect the environment. However, there has been very little work done to determine the costs versus benefits of this legislation and corresponding regulations.

There is ample evidence to suggest that legislation and regulations passed since 1970 have resulted in substantial improvements. Carbon monoxide emissions are down, volatile organic compounds in the environment are down, the amount of sulfur oxide is down, the extent of ocean dumping of industrial wastes is down, the number of cities without adequate sewage treatment is down, and the number of miles of rivers unfit for swimming is down. However, what no one has yet been able to produce is accurate information on the cost of benefits derived from these improvements.

There are acceptable estimates of the annual costs of implementing federal environmental regulations. It costs approximately $85 billion per year to carry out the mandates of the Clean Air Act, the Clean Water Act, and more than 20 other major federal laws regulating such things as pesticides, herbicides, drinking water contaminants, solid and hazardous wastes, and new chemicals.[4]

Each time a state or federal agency gets involved in an environmental issue, the question of the monetary benefits of proposed solutions is raised. Because it is possible to quantify the costs of federally mandated solutions, it is critical that a way be found to quantify benefits. A cleaner, safer, healthier environment has the potential to yield a variety of benefits. We need accurate data on the effect of the country's annual investment in a clean environment on the following:

- The inflation rate
- The unemployment rate
- The rate of growth of the gross national product
- The international trade balance
- Human health
- Productivity in the workplace
- Reduced damage to exposed materials
- Agricultural output
- Industrial modernization
- Research and innovation in the chemical and pharmaceutical industries

The economics of safety and health is an issue that gains importance every year. Prospective safety and health professionals need to be aware of the importance of being able to provide both costs and benefits data when safety and health measures are proposed.

However, cost–benefit analysis information should be part of the process. This is particularly necessary because opposition to environmental safety measures, just as with any other safety and health measure, will probably be based on the economic argument. By including the cost–benefit information in the decision-making process, the process is improved along with the quality of the decision.

TYPES OF ENVIRONMENTS

What do you think of when you hear the term *environment*—the air, forests, lakes, rivers, oceans, and other natural resources? These are all part of our natural environment. However, the natural environment is not the only type of environment with which we interact. We interact with the following types of environments:

1. The **natural environment** is not human-made. It is the environment we typically think of as the earth and all its natural components, including the ground, the water, flora and fauna, and the air. The natural environment is full of both beauty and hazards. Some of the hazards of nature can be controlled or at least reduced. However, there are hazards over which we have little or no control. These include hurricanes, tornadoes, lightning, floods, and other natural disasters. **Induced environments** are those that have been affected in some way by human action. For example, a big city with smog has an induced environment.

2. A **controlled environment** is a natural or induced environment that has been changed in some way to reduce or eliminate potential environmental hazards. An example of a controlled environment is a home or workplace that is

---

Safety Fact

Active Environmental Legislation

It can be difficult and confusing to keep up with the "alphabet soup" of federal environmental legislation. The major pieces of legislation explained in this chapter include the following:

- Clean Air Act
- Clean Water Act
- Resource Conservation and Recovery Act (RCRA)
- Hazardous and Solid Waste Amendments (HSWA)
- OSHA Hazardous Waste Standard (1910.120)

Note: The ISO 14000 standard is not the product of state or federal regulation. Rather, it is a volunteer standard developed by the International Organization for Standardization.

---

heated to reduce the potential hazards associated with cold. An **artificial environment** is one that is fully created to prevent definite hazardous conditions from affecting people or material. For example, the environment within a submarine is an artificial environment.

3. A **closed environment** is one that is completely or almost completely shut off from the natural environment. Both a controlled and an artificial environment must be closed. A **free environment** is one that "freely" allows the flow and free movement of air.

Safety and health professionals need to be prepared to confront potential hazards in all these types of environments. Dealing with hazards of the environment involves protecting workers within a plant from dangerous environmental conditions and protecting the general public from unsafe conditions that may be created by the operations or products of the plant. This is a critical area for modern safety and health professionals and one in which their level of sophistication will have to increase continually.

ROLE OF SAFETY AND HEALTH PROFESSIONALS

In the 1980s, workplace safety and health and environmental compliance were typically separated into two different departments. However, with the signing of the memorandum of understanding between OSHA and EPA mentioned earlier in this chapter, the lines between safety and health in the workplace and environmental compliance are no longer clear-cut.

In today's workplace, the safety and health professional will be asked to take on environmental responsibilities in addition to the traditional safety and health responsibilities. Other companies will employ one or more persons to be responsible for workplace safety and health and a separate person or persons to be responsible for environmental concerns, both positions reporting to an overall health, safety, and environmental issues manager. Such factors as the size of the company, the type of product produced, and the perceived impact of governmental regulations are also factors that determine how health, safety, and environmental departments will be structured.

A safety and health professional should be prepared to undertake the increased workload associated with environmental safety. Safety and health students should be prepared to study environmental courses or portions of courses in their college curricula. Practicing safety and health professionals should be prepared to undertake continuing education in environmental issues.

HAZARDS OF THE ENVIRONMENT

Being aware of the various hazards in the environment and the negative impact that these hazards can have on humans is another requirement for safety and health professionals. This is true whether the environment is the natural environment or that of the workplace.

Environmental hazards can have an effect on people, machines, systems, and other inhabitants of a given environment. For example, high-humidity conditions can cause mold, mildew, and damage to mechanical equipment. For this reason, manufacturers of computers, consumer electric products, and other microelectronics-based products pay particular attention to humidity control in their plant. Dust, radiation, temperature, and other environmental factors can be hazardous to people and equipment. Consequently, modern safety and health professionals must consider environmental factors in their plans for ensuring a safe and healthy workplace.

Indoor Environmental Quality (IEQ)

An important environmental factor in the modern workplace is indoor air quality. The public is probably more familiar with the terms *indoor air quality* and **sick-building syndrome**. Indoor air quality, as the name implies, simply refers to the quality of the air in a work environment. *Sick-building syndrome* is a term many people use to convey a wide range of symptoms that they believe are attributed to the building itself. Workers typically implicate the workplace environment because their symptoms are alleviated when they leave their workplace. NIOSH recommends the term **indoor environment quality (IEQ)** to describe the problems occurring in buildings and workplaces throughout the nation.[4] NIOSH investigators have found that concerns about air quality may be caused by a number of factors, encompassing much more than air contamination. Other factors such as comfort, noise, lighting, ergonomic stressors (poorly designed workstations and tasks), and job-related psychological stressors can contribute to complaints, both individually and in combination with others. Hence, IEQ more accurately describes the scope of the problem. The following paragraphs explain the most common concerns about IEQ.

- *Typical symptoms associated with IEQ.* The symptoms of IEQ-related problems are diverse and are usually not suggestive of any particular medical diagnosis. A typical spectrum of symptoms includes headaches, unusual fatigue, varying degrees of itching or burning eyes, skin irritation, nasal congestion, dry or irritated throats, and nausea.
- *Size of the IEQ problem.* Over the years, there has been a significant increase in public concern about IEQ. NIOSH scientists have completed approximately 1,300 evaluations related to the indoor work environment, and the number of these requests as a percentage of the total has risen dramatically. Requests to evaluate office environments used to make up only 8 percent of the total requests for NIOSH investigations. Today, they account for more than 50 percent of all requests.
- *Why are IEQ problems increasing?* During the 1970s, ventilation requirements were changed to conserve fossil fuels, and virtually airtight buildings emerged. At the same time, a revolution occurred in office work throughout the country. Computers and other new work technologies forced a change in office procedures and productivity, and ergonomics and organizational stress problems increased. Coupled with the conservation measures and changing technology was a dramatic increase in the number of workers in white-collar jobs. Greater awareness and increased media coverage of the potential for IEQ problems may also have contributed to the higher reporting rate of suspected problems.
- *What types of IEQ problems are typically found in the workplace?* NIOSH investigators have found IEQ problems caused by ventilation system deficiencies, overcrowding, offgassing (ventilating) from materials in the office and mechanical equipment, tobacco smoke, microbiological contamination, and outside air pollutants. NIOSH has also found comfort problems due to improper temperature and relative humidity conditions, poor lighting, and unacceptable noise levels, as well as adverse ergonomic conditions and job-related psychosocial stressors.
- *What do investigators look for during an IEQ evaluation?* NIOSH investigators typically look at four elements to determine if there is an IEQ problem.

  Pollutant sources. Is there a source of contamination or discomfort indoors, outdoors, or within the mechanical systems in the building?

  The heating, ventilating, and air conditioning (HVAC) system. Can the HVAC system control existing contaminants and ensure thermal comfort? Is it properly maintained and operated?

  Pollutant pathways and driving forces. The HVAC system is the primary pathway. Are the pressure relationships maintained between areas of the building so that the flow of the air goes from cleaner areas to dirtier areas?

  Occupants. Do the building occupants understand that their activities affect the air quality? NIOSH investigators then look at factors beyond air quality including physical factors, such as lighting, vibration, noise, ergonomic factors, and psychosocial aspects of the workplace.[5]

Safety Fact

Solving IEQ Problems In-House

Often IEQ problems can be solved with in-house expertise. To help organizations increase their expertise in this area, NIOSH and the EPA published the following guide:

- Building Air Quality: A Guide For Building Owners and Facility Managers (Stock No. 055-000-00390-4)

For information about ordering this guide, call the following number: 1-800-35-NIOSH (1-800-356-4674).

Action Plan for Ensuring IEQ

Maintaining a safe, healthy, and comfortable indoor environment that is conducive to the optimum performance of employees is a challenge. It involves the integration of many different components that make up today's complex work environment. However, it can be done. Indoor environmental problems are both preventable and solvable. This section provides an eight-step action plan that will help safety, health, and environmental professionals ensure a safe and healthy indoor work environment.

- *Step 1: Designate an IEQ manager.* The person selected to be responsible for the indoor environment can come from a number of different backgrounds. An effective IEQ manager may be a facility manager, building engineer, employee-relations manager, or the health, safety, and environmental

---

**Step Two, Part One:
Identify and Review Existing Records**

Checklist

✓ Identify, review, and familiarize yourself with construction, operating, and other documents (including blueprints, control system set-points for HVAC equipment, operations and maintenance schedules, complaint logs, and so on).

✓ Set up procedures to revise the above records as needed, but particularly with any renovation or construction.

✓ Keep on file material safety data sheets (MSDSs) for products used in the building.

✓ Determine quality of outside air entering building.

✓ Compare capacity of HVAC system with current loads to make sure there are no shortfalls in the amount of outside air provided.

Figure 29–2
Checklist for identifying and reviewing existing records.

Step Two, Part Two:
Assess the Current Situation

Checklist

✓ Conduct a whole-building walkthrough inspection and look for such problems as

- odors
- dirty or unsanitary conditions
- visible fungal growth
- mold or mildew
- moisture in inappropriate locations
- staining or discoloration of building materials
- smoke damage
- presence of toxic substances
- poorly maintained filters
- potential for soil gas entry
- unusual noises from equipment
- leaks
- uneven temperatures
- overcrowding
- personal air cleaning devices (ion and ozone generators or portable filtration units)
- personal fans
- blocked or redirected vents and diffusers

✓ Take notes during the walkthrough identifying potential or existing problems indicating a need for either close monitoring or corrective action.

✓ Inspect HVAC condition and operations: (1) list components that need to be repaired, adjusted, cleaned, or replaced; and (2) record actual control settings and operating schedules for each air handling unit.

✓ Determine if significant sources of contamination are directly exhausted to the outside or can be moved close to an exhaust fan.

Figure 29–3
Checklist for assessing the current IEQ situation.

manager. Regardless of who is selected, this person should become part of the organization's overall safety, health, and environmental management team.

- *Step 2: Develop an IEQ profile of all buildings in question.* This step is divided into two parts: (1) Existing records must be identified and reviewed; and (2) an assessment of the current situation must be conducted. The activities that are part of this step are summarized in Figure 29–2 and 29–3.
- *Step 3: Address existing and potential problems.* Any problems or potential problems are addressed in this step. Common strategies are to (1) remove or reduce any sources of problems that were identified (seal or cover the source or modify the environment); (2) improve ventilation to provide outside air to employees and to dilute or exhaust pollutants;

Step Five
Maintenance Component

Checklist

✓   Develop and follow a preventive maintenance plan that includes maintenance schedules. Activities in the plan should include:

- Inspect outside air dampers for nearby sources of contamination.
- Ensure that air dampers are clear of obstruction and operating properly.
- Regularly replace or clean air filters.
- Clean and inspect drain pans.
- Inspect and clean heating and cooling coils.
- Inspect and clean, as warranted, interior of air handling units.
- Inspect fan motors and belts.
- Regularly inspect and clean air humidification equipment and controls.
- Inspect, clean, and treat cooling towers.
- Inspect and clean, as needed, air distribution pathways and variable air volume (VAV) boxes.

✓   Update the maintenance plan when equipment is added, removed, or replaced.

Figure 29–4
Checklist for conducting preventive maintenance.

(3) improve air filtration to clean air from outside and inside the building; and (4) control occupant exposure through administrative strategies (e.g., scheduling contaminant-producing activities during unoccupied periods).

- *Step 4: Training for employees.* Employees who work in the building are often the best sources of information about problems and potential problems. Their input can be improved markedly by providing employees with training that familiarizes them with the problem and shows them how to raise an issue.

- *Step 5: Implement a plan for facility operations and maintenance.* The operations component of the plan should ensure that (1) the HVAC system is in operation during periods of significant activity; (2) the HVAC system is operated with as much outside air as possible before employees arrive for work; and (3) cleaning and housekeeping materials are properly selected and controlled. The necessary preventive maintenance activities are summarized in Figure 29–4.

- *Step 6: Manage processes that are potentially harmful.* Processes that are predictably problematic and ongoing should be managed and monitored carefully. In most organizations, these processes include remodeling, renovation, painting, pest control, smoking, and shipping and receiving.

- *Step 7: Communicate with employees about their role.* Employees need to understand their role in ensuring a quality indoor work environment. The employees' role includes the following: They should (1) raise issues and help identify problems as appropriate; (2) understand that their actions and activities can contribute to the creation of problems; and (3) know the organization's policies concerning IEQ and follow them.

Step Eight
Responding to Complaints

Checklist

 Prepare and follow clear procedures for recording and responding to IAQ complaints, including:

- Logging entries into your existing work-order system.
- Collecting information from the complainant.
- Ensuring the confidentiality of information and records obtained from complainants.
- Determining response capability of in-house staff.
- Identifying appropriate outside sources of assistance.
- Applying remedial action.
- Providing feedback to the complainant.
- Following up to ensure that remedial action has been effective.

✓ Inform building staff of these procedures.

✓ Inform building occupants and/or tenants of these procedures, and periodically remind them how to locate responsible staff and where to obtain complaint forms.

Figure 29–5
Checklist for receiving IEQ complaints and responding.

- *Step 8: Establish procedures for responding to IEQ complaints.* If employees are going to be trained to recognize potential IEQ problems and to register their related complaints, there must be a systematic procedure for receiving and responding to complaints. A recommended procedure is summarized in Figure 29–5.[6]

HAZARDOUS WASTE REDUCTION

One of the most effective ways to ensure a safe and healthy environment is to reduce the amount of hazardous waste being introduced into it. In 1976, the EPA developed a policy statement designed to encourage industrial firms to reduce their hazardous waste output. The policy contains three broad provisions. First, companies must certify on the transportation manifest that they have a program in operation to reduce the volume and toxicity of hazardous waste each time hazardous waste is transported off-site. Second, in order to qualify for permits to treat, store, or dispose of hazardous waste, companies must implement and operate a hazardous waste reduction program. Finally, companies must submit biennial plans to the EPA describing actions taken to reduce the volume and toxicity of their hazardous waste.

In 1976, the **Resource Conservation and Recovery Act (RCRA)** became the first piece of federal legislation to encourage the reduction of hazardous waste.

- Develop a safety and health program to control hazards and provide for emergency response.

- Conduct a preliminary site evaluation to identify potential hazards and select appropriate employee protection strategies.

- Implement a site control program to prevent contamination of employees.

- Train employees before allowing them to undertake hazardous waste operations or emergency response activities that could expose them to hazards.

- Provide medical surveillance at least annually and/or at the end of employment for employees exposed to greater than acceptable levels of specific substances.

- Implement measures to reduce exposure to hazardous substances to below established acceptable levels.

- Monitor air quality to identify and record the levels of hazardous substances in the air.

- Implement a program to inform employees of the names of people responsible for health and safety and of the requirements of the OSHA standard.

- Implement a decontamination process that is used each time an employee or piece of equipment leaves a hazardous area.

- Develop an emergency response plan to handle on-site emergencies.

- Develop an emergency response plan to coordinate off-site services.

Figure 29–6
General requirements of the OSHA Hazardous Waste Standard.

It was followed in 1984 by passage of the **Hazardous and Solid Waste Amendments (HSWA)**.However, the most significant federal regulation pertaining to hazardous waste reduction is the OSHA Hazardous Waste Standard, which took effect in 1990.

OSHA Hazardous Waste Standard

OSHA Standard 1910.120 sets the standard for dealing with hazardous materials. Figure 29–6 summarizes the requirements of the **OSHA Hazardous Waste Standard**. These standards require that companies identify, evaluate, and control hazardous substances; train employees in the proper accident prevention procedures and emergency response procedures; reduce exposure to hazardous substances; provide medical monitoring; keep employees informed concerning hazardous substances on the job; implement decontamination procedures; and develop both on- and off-site emergency response plans.

The goal of these and other hazardous materials regulations is to (1) encourage companies to minimize the amount and toxicity of hazardous substances that they use; (2) ensure that all remaining hazardous materials are used safely; and (3) ensure that companies are prepared to respond promptly and appropriately when accidents occur.

Source Reduction

- Management improvements in operating efficiency
- Better use of modern technology and processes
- Better selection of the materials used in production processes
- Product revisions

Recycling

- Reclamation of usable materials from hazardous waste materials
- Reconstituting of waste for reuse as an original product

Figure 29–7
Key elements in hazardous waste reduction.

Consequently, safety and health professionals need to be familiar with the concept of hazardous waste reduction, how to organize a reduction program, and how to conduct a reduction audit. **Hazardous waste reduction** can be defined as follows:

> Hazardous waste reduction is reducing the amount of hazardous waste generated and, in turn, the amount introduced into the waste stream through the processes of source reduction and recycling.

The two key elements of this definition are **source reduction** and **recycling**. Figure 29–7 breaks down these two elements into more specific subelements.

Organizing a Waste Reduction Program

There are four steps that must be accomplished in establishing a waste reduction program. The first step is to convince top-level managers that the program is not just environmentally and ethically necessary, but also cost effective. The cost–benefit data associated with waste reduction should be enough to convince the most cost-conscious manager of the feasibility of such a program. The potential for reduced costs in the areas of regulatory compliance, legal liability, and workers' compensation should be included in all briefings given to managers. The other steps involved in organizing a waste reduction program are discussed in the following paragraphs (see Figure 29–8).

Form the Waste Reduction Team

The **waste reduction team** should have representatives from every department that purchases, stores, or uses hazardous waste. In every case possible, these representatives should be supervisors to ensure maximum involvement of, and

Figure 29–8
Steps in establishing a
hazardous waste reduction
program.

1. Gain a full commitment from the executive
   level of management.
2. Form the waste reduction team.
3. Develop a comprehensive waste reduction
   plan.
4. Implement, monitor, and adjust as necessary.

commitment by, all employees. The team should be chaired by an executive-level manager to show that management supports the effort and to ensure the team access to the highest level of decision makers when recommendations are made. The appropriate role of the safety and health professional on the team is that of *ex-officio* staff member. The person who staffs the committee works closely with the chairperson to arrange meetings, record proceedings, facilitate communication, and take care of the logistical and administrative aspects of committee operation.

Develop a Comprehensive Plan

Figure 29–9 contains a basic outline for a comprehensive **waste reduction plan**. It is important that the final plan contain all these components. The statement of purpose explains briefly and succinctly the overall purpose of the waste reduction program. The goals translate the statement of purpose into specific action items with timetables for their accomplishment. Each goal should be accompanied by proposed strategies for its accomplishment. This forces the waste reduction team to go beyond just the question of *what* to a consideration of *how*. By identifying potential inhibitors at the outset, the team can avoid wasted time, effort, and resources. Inhibitors should be identified for each individual goal. The measures of success answer the question "How do we know when we've accomplished this goal?" For example, if a goal is to reduce the average monthly inventory of a given hazardous substance, the measure of success may be the monthly inventory printout. The actual measures used are less important than ensuring from the outset that everyone agrees on the measures. This will prevent disagreements over whether a goal has been achieved. To apply the measures of success, the team must be able to track and monitor hazardous material purchases, storage, use, and waste. The process

Figure 29–9
Outline of waste reduction plan.

- Statement of purpose
- Goals with timetables
- Strategies for accomplishing each goal
- Potential inhibitors associated with each goal
- Measures of success for each goal
- Tracking system
- Audit process

that will be used for ongoing audits of the program should also be explained in the plan. The program audit process is explained later in this chapter.

Implement, Monitor, and Adjust

When the plan is complete, the program can be implemented. For best results, select an implementation date and have a high-profile ceremony involving the company's top officials. The perception among employees of high-level support is critical to the success of the program. Such a ceremony preceded or followed by an intercompany memorandum of support from a top executive to all employees can establish the desired perception at the outset. At this point, the waste reduction audit begins, results are monitored by a system of continual tracking, and adjustments are made as necessary. The waste reduction audit is the most critical component of the program. The next section is devoted to this subject.

Waste Reduction Audit

A **waste reduction audit** is a process used to identify opportunities for decreasing the waste stream of a given facility.

The most important steps in a waste reduction audit are as follows: (1) target the processes; (2) analyze the processes; (3) identify reduction alternatives; (4) consider the cost–benefit ratio for each alternative; and (5) select the best options (Figure 29–10).

Target the Processes

In the first step, the waste reduction team identifies all processes that use hazardous materials or generate hazardous waste. These processes are targeted for analysis, and a targeted process list is developed.

Analyze the Processes

Each process on the list should be observed from start to finish. It is important to analyze processes in person rather than relying on flowcharts or operating manuals. Members of the waste reduction team should actually see every step in the process. A list of points in the process where waste is generated should be

---

1. Target the processes for analysis and create the target list.
2. Analyze the processes to identify sources of waste generation.
3. Identify reduction options for each waste generation source.
4. Consider the cost–benefit ratio of each reduction goal.
5. Select the best options.

Figure 29–10
Major steps in a waste reduction audit.

compiled and the types of waste recorded. In addition to the production aspects of a process, the waste reduction team should also analyze all related processes (e.g., maintenance, storage of materials, materials handling, and waste treatment).

Identify Reduction Options

For every type of waste generated, reduction options should be identified to eliminate or reduce the waste. This is both an art and a science. Options can be identified by reviewing professional literature, talking with specialists, and drawing on the expertise of professional organizations and environmental agencies. Team members should be encouraged to think creatively. Both product and process changes may be considered as options.

Consider the Costs and Benefits of Each Option

All proposed reduction options have both costs and benefits. The first step is to identify the costs of continuing the current approach. These include the cost of materials, regulatory compliance, handling and use precautions, storage, personnel, treatment, training, and disposal. The impact of each reduction option on these and all other costs should be considered.

Select the Best Option

Based on the results of a cost–benefit analysis, the best option is selected and implemented. The procedure is then repeated for each process on the target list.

ENVIRONMENTAL MANAGEMENT SYSTEM (EMS)

A management system, regardless of its application, is the component of an organization responsible for leading, planning, organizing, and controlling. An **environmental management system (EMS)** is the component of an organization with primary responsibility for these functions as they relate specifically to the impact of an organization's processes, products, or services on the environment (Figure 29–11).

An organization's EMS may be a subset of its safety and health management system (Figure 29–12) or a separate component of the organization's overall management system. Regardless of where and how it fits into an organization, the EMS should do the following:

- Establish a comprehensive environmental protection policy (planning).
- Identify all government regulations and requirements that apply to the organization's processes, products, or services (controlling).
- Establish organizationwide commitment to environmental protection (leading).
- Establish responsibility and accountability relating to environmental protection (organizing).

**Principal Elements of an
Environmental Management System**

- Management commitment
- Conforming environmental policy
- Environmental planning
- Organizational structure and responsibility
- Awareness and competence training
- Internal and external communication
- Control of environmentally related operations and documentation
- Emergency response and preparedness capability
- Checking and auditing and corrective or preventive action
- Record keeping
- Management review
- Continual improvement mechanisms

Figure 29–11
An environmental management system's minimum components.

Figure 29–12
Sample organizational structure.

- Incorporate environmental concerns in all levels of organizational planning including strategic, operational, and procedural (planning).
- Establish management processes for achieving performance benchmarks (controlling).
- Provide sufficient resources to ensure that performance benchmarks can be achieved on a continual basis (leading).
- Establish and maintain an effective emergency preparedness program (leading, planning, organizing, and controlling).
- Assess the organization's environmental performance against all applicable benchmarks and adjust as necessary (controlling).

✓ Public pressure from environmental advocacy groups

✓ Influence of environmental advocates within elected bodies and governmental organizations

✓ Potential to gain a competitive advantage in the global marketplace

✓ Improved public relations

✓ Customer demand

✓ Fear of liability and risk

✓ Preempt new and additional government oversight

✓ Reduced duplication

✓ De facto requirement to adopt a standard

✓ Get ahead of government adoption

Figure 29–13
Forces driving the interest in EMS.

- Establish a review process for auditing the EMS and identifying opportunities for improvement.
- Establish and maintain communications linkages with all stakeholders, internal and external.
- Promote the establishment of an EMS in contractors and suppliers.[7]

Rationale for the EMS Movement

Different organizations have become interested in better environmental management for different reasons (Figure 29–13). Some organizations are responding to pressure brought by environmental advocacy groups and watchdog organizations. Such groups have become increasingly proficient in winning public support for their individual causes. This support, if effectively focused, can be translated into market pressure. When this happens, better environmental management becomes a market imperative.

Environmental groups have learned how to use the political process to gain support among elected officials and in key government agencies. Some elected officials and government employees are, themselves, environmental advocates. In most industrialized countries, a number of so-called green parties even run candidates for elective offices on environmental platforms. These factors, taken together, are giving environmental advocates steadily increasing influence. Business and industry are responding to this influence with increasing interest in environmental management.

One of the key drivers behind the EMS movement is the concept of competitive advantage. Some forward-looking organizations, particularly those that must compete in the global marketplace, have begun to view having an

Safety Fact

Small Organizations and Environmental Management

Small organizations do not have the personnel to create an organizational structure such as the one shown in Figure 29–12. Instead, one individual may have to wear all three hats: safety, health, and environmental management. In such cases, the following organizations offer information, training, and consultation on environmental issues:

- American Industrial Hygiene Association (AIHA)
  PO Box 8390, 345 White Pond Dr.
  Akron, OH 44320
  216-873-2442

  Sponsors indoor air quality courses in conjunction with meetings for AIHA members only.

- American Society of Heating, Refrigerating, and Air-Conditioning Engineers (ASHRAE)
  1791 Tullie Cir. NE
  Atlanta, GA 30329
  404-636-8400

  Sponsors professional development seminars on indoor air quality.

- NIOSH Division of Training and Manpower Development & NIOSH-Funded Educational Resources Centers
  4676 Columbia Pkwy.
  Cincinnati, OH 45226
  513-533-8302

  Provides training to occupational safety and health professionals and paraprofessionals.

- OSHA Training Institute
  155 Times Dr.
  Des Plaines, IL 60018
  708-297-4913

  Provides courses to assist safety and health professionals in evaluating indoor air quality.

effective EMS as a competitive advantage. The advantage is the result of the following factors:

- Minimization of funds siphoned off into nonproductive activities such as litigation and crisis management.
- Ease of compliance with government regulations in the foreign countries that make up the global marketplace.
- Better public image in countries (especially European countries) where interest in environmental protection is high.

Improved public relations is an incentive to improve environmental management. The mood of the general public relating to the environment has become strongly pro-environment over the past several decades. Companies that are perceived as being harmful to the environment have begun to feel the public's displeasure in the marketplace, which is one reason why so many companies have begun recycling and voluntary cleanup programs.

An environmentally sensitive public often manifests itself in the form of environmentally sensitive customers. Such customers demand effective environmental management as a condition of their long-term retention. Positive environmental impact is rapidly becoming an attribute that customers consider when deciding whether to select or stay with a given vendor or supplier. Direct pressure from the marketplace (customers) is one of the most potent motivators behind the EMS movement.

Another powerful motivator is the potential threat of legal liability and the economic threat that it represents. In today's litigious business environment, a company can see its financial solvency evaporate overnight as the result of class-action lawsuits. Increasingly, individuals and advocacy groups are using the courts as their first response in disagreements over environmental issues, and the economic stakes are high. Juries are showing an increasing tendency to award not only damages but also huge punitive fines against companies that are careless with the environment.

Another motivating factor driving the EMS movement is the desire of companies to preempt additional government oversight. The idea is to adopt voluntary standards before the government establishes its own additional standards and makes them mandatory. Supporters of this approach believe that a form of voluntary peer review is inherently more effective than government oversight which, in their opinion, is often unwieldy and bureaucratic.

Reducing duplication promotes efficiency, which, in turn, promotes competitiveness. Companies that do business on a global basis are accustomed to dealing with regulations that can vary markedly from country to country. Differences in regulations often cause differences in production processes. Whenever a process must be modified to accommodate regulatory variations— or for any other reason—the price of the product is driven up. Consequently, globally competitive companies are interested in globally uniform standards— standards that help level the playing field in such key areas as environmental management.

The potential emergence of a de facto requirement to adopt an international standard in order to do business in a given market has also created interest in EMS. The International Organization for Standardization has developed a voluntary standard (ISO 14000) to promote effective environmental management systems. Potential customers may decide unilaterally to do business only with companies that adopt the standard (discussed in the next section). In such cases, the marketplace makes adoption mandatory even though the government of the country in question has no such requirement.

The corollary to a de facto requirement is an actual government mandate to adopt an international EMS standard. The potential for a government mandate has gotten many companies interested in getting a head start in the adoption of an EMS.

Potential Benefits of an EMS

Regardless of why a company becomes interested, there are many benefits that can be realized from adopting an EMS including the following:

1. *Ease of trade.* Uniform international standards knock down the barriers created by country-to-country variations.
2. *Improved compliance with regulations.* When all players must obey the same rules, regulatory compliance is improved by the resulting application of uniform accountability criteria.
3. *Credibility.* Third-party certification takes politics out of the process, thereby enhancing its credibility.
4. *Reduction in liability or risk.* Certification forces companies to focus on the issue of environmental impact. This focus on being environmentally friendly, in turn, reduces the likelihood of environmentally hazardous behavior that may lead to expensive, nonproductive litigation.
5. *Regulatory incentives.* Companies that show initiative in establishing an effective EMS can take advantage of incentives that reward organizations for showing leadership in protecting the environment.
6. *Sentencing mitigation.* Adoption of a comprehensive EMS may serve as a mitigating factor when fines are assessed for failing to comply with regulations.
7. *Pollution prevention and waste reduction.* Better environmental management results in less waste and less pollution. These, in turn, result in attendant savings. This is the ounce-of-prevention, pound-of-cure concept.
8. *Profit.* Better management of any kind—quality, human resources, time, or environmental—translates into better profits.
9. *Improved internal management.* ISO 14000 certification requires the use of management methods that can improve all aspects of internal management. This, in turn, leads to a better overall profit ratio.
10. *Community goodwill.* Good environmental management makes companies good corporate citizens in their neighborhoods which, in turn, leads to community goodwill.
11. *Retention of a high-quality workforce.* Employees live in the communities where their companies are located. Companies that are good corporate citizens find it easier to retain their best and brightest employees than do companies whose environmental practices are embarrassing or irresponsible.
12. *Insurance.* Companies with established, effective environmental management systems may have fewer problems finding insurance policies that can be written at reasonable prices.
13. *Preference in lending.* One of the key factors that lending institutions consider before making a loan is the company's ability to pay. Because costly

environmental litigation can bankrupt even the most solvent of companies, lending institutions may begin to give preference to those with established environmental management systems.[8]

INTERNATIONAL ORGANIZATION FOR STANDARDIZATION (ISO)

The **International Organization for Standardization (ISO)** is a worldwide organization of national standards bodies (Figure 29–14). The complete membership roster for ISO contains the standards bodies of 118 countries. The overall goal of ISO is

> to promote the development of standardization and related activities in the world with a view to facilitating the international exchange of goods and services and to developing cooperation in the sphere of intellectual, scientific, technological, and economic activity.[9]

Globalization of the marketplace has created a competitive environment that requires peak performance and continual improvement. The unrelenting demands of the today's marketplace have given rise to new philosophies for

---

Australia
Standards Australia
1 The Crescent
Homebush—N.S.W. 2140
PO Box 1055
Strathfield—N.S.W. 21335
Tel: +61 2 746-4700

France
Association francaise de normalisation
Tour Europe
F-92049 Paris La DeFeuse Cedex
Tel: +33 1 42 91 55 55

United Kingdom
British Standards Institution
389 Chiswick High Rd.
GB-London W4 4AL
Tel: +44 181 996 90 00

Canada
Standards Council of Canada
45 O'Connor St.
Suite 1200
Ottawa, Ontario KIP 6N7
Tel: 613-238-3222

Germany
Deutsches Institut fur Normung
Burggrafen strasse 6
D-10772 Berlin
Tel: +49 30 26 01 23 44

United States
American National Standards
Institute (ANSI)
11 W. 42nd St.
13th Floor
New York, NY 10036
Tel: 212-642-4900

Figure 29–14
Addresses of selected ISO members.

doing business, most of which fall under the broad umbrella of quality management (QM). One of the initiatives under the QM umbrella is the ISO 9000 family of quality standards. These standards contain criteria for promoting effective quality management systems.

ISO 14000

The International Organization for Standardization—the same organization that developed the ISO 9000 quality standards—has now developed the **ISO 14000** family of standards to promote effective environmental management systems.

Just as the decision to adopt the QM philosophy or the ISO 9000 standards is voluntary, adoption of ISO 14000 is based on voluntary organizational commitment to environmental protection rather than government coercion. For safety and health managers accustomed to complying with government mandates, ISO 14000 certification is a novel approach.

Shifting attitudes and greater public awareness are making it essential that business firms be good neighbors in their communities. The marketplace demands that businesses produce high-quality products at competitive prices, without harming the environment. ISO 14000 provides the framework for making effective environmental management part of the organization's overall management system.

ISO 14000 SERIES OF STANDARDS

The term *ISO 14000 Series* refers to a family of environmental management standards that cover the five disciplines shown in Figure 29–15. All the standards are in a constant state of evolution. The ISO 14000 Series contains two types of standards: specification standards and guidance standards.

Specification versus Guidance Standards

A **specification standard** contains only the specific criteria that can be audited internally or externally by a third party. A **guidance standard** explains how to develop and implement environmental management systems and principles. Guidance standards are descriptive standards that also explain how to coordinate among various quality management systems.

Guides and Technical Reports

ISO also develops **guides** and **technical reports**. A guide (not a guidance standard) is a tool to assist organizations in the improvement of environmental management. Guides are used voluntarily and do not contain criteria for certification.

_____ ✓ _____ Environmental management system

_____ ✓ _____ Environmental auditor criteria (These criteria may be used by internal auditors and external third-party auditors.)

_____ ✓ _____ Environmental performance evaluation criteria

_____ ✓ _____ Environmental labeling criteria

_____ ✓ _____ Life-cycle assessment methods

Figure 29–15
Environmental disciplines in the ISO 14000 series of standards.

Technical reports are written only when one of the following circumstances exists:

- When a technical committee of ISO cannot reach consensus on an issue. In such cases, the technical report explains why consensus was not possible.
- When an issue is still under development or when there is reason to believe that an undecided issue that cannot be immediately resolved will be resolved in the future.
- When a technical committee that is working on a standard collects information that is different from the kind normally published as a standard. This information may be published as a technical report.

Classification of ISO 14000 Standards

Figure 29–16 illustrates the two broad classifications of standards in the ISO 14000 family and their respective subclassifications. The two broad classifications are (1) process- or organization-oriented standards, and (2) product-oriented standards.

Process-oriented standards cover the following three broad areas: environmental management system, environmental performance evaluation, and environmental auditing. Product-oriented standards cover three different broad areas of concern as follows: life-cycle assessment, environmental labeling, and environmental aspects in product standards.

Subclassifications of ISO 14000 Standards

The six subclassifications shown in Figure 29–16 represent six distinct, but interrelated, areas of concern. There are actually 20 separate documents in the ISO 14000 series, all related to one of these six subclassifications. Figure 29–17 is a checklist of the process-oriented documents in the ISO 14000 series. Figure 29–18 is a checklist of the product-oriented documents. Of these 20 documents, only ISO 14000 is a standard against which a company is audited. The rest are guidance documents. This is an important point to remember.

Figure 29–16
Environmental management standards.

ISO 14001 STANDARD

The ISO 14001 standard represents an approach to protecting the environment and, in turn, a company whose processes and products may affect the environment. The standard is unique because it relies on voluntary motivation instead of mandatory compliance.[10] To understand the ISO 14001 standard requires an understanding of the following concepts: environmental management system (EMS), EMS audit, environmental aspect, continual improvement, and plan-do-check-adjust model.

1. *Environmental management system (EMS).* An organization's EMS provides the structure for implementing its environmental policy. The management system consists of the organization's structure, personnel, processes, and procedures relating to environmental management.

2. *EMS audit.* The **EMS audit** is the process used to verify that an EMS actually does what an organization says it will do. As part of the audit, results are reported to the organization's top management team.

- ISO 14001
  Environmental Management Systems—Specification with Guidance for Use
- ISO 14004
  Environmental Management Systems—General Guidelines on Principles, Systems, and Supporting Techniques
- ISO 14010
  Guidelines for Environmental Auditing—General Principles on Environmental Auditing
- ISO 14011/1
  Guidelines for Environmental Auditing—Audit Procedures—Auditing of Environmental Management Systems
- ISO 14012
  Guidelines for Environmental Auditing—Qualification Criteria for Environmental Auditors
- ISO 14014
  Initial Reviews
- ISO 14015
  Environmental Site Assessments
- ISO 14031
  Evaluation of Environmental Performance
- ISO 14020
  Goals and Principles of All Environmental Labeling
- ISO 14021
  Environmental Labels and Declarations—Self-Declaration Environmental Claims—Terms and Definitions

Figure 29–17
ISO 14000 process-oriented documents.

3. *Environmental aspect.* Any aspect of an organization's processes, products, or services that can potentially affect the environment.

4. *Continual improvement.* In the global marketplace, good enough is never good enough. Performance that is competitive today may not be tomorrow. Consequently, an EMS must be improved continually, forever.

5. *Plan-do-check-adjust model.* The plan-do-check-adjust (PDCA) model comes from the work of W. Edwards Deming in the area of quality management. The model works well in any management system that must be continually improved. Figure 29–19 lists some of the types of activities— relating to each of the model's four components—that are associated with continually improving an EMS.

- ISO 14022
  Environmental Labels and Declarations—Symbols
- ISO 14023
  Environmental Labels and Declarations—Testing and
  Verifications
- ISO 14024
  Environmental Labels and Declarations—Environmental
  Labeling
  Type 1—Guiding Principles and Procedures
- ISO 1402X
  Type III Labeling
- ISO 14040
  Life Cycle Assessment—Principles and Framework
- ISO 14041
  Life Cycle Assessment—Life Cycle Inventory Analysis
- ISO 14042
  Life Cycle Assessment—Impact Assessment
- ISO 14043
  Life Cycle Assessment—Interpretation
- ISO 14050
  Terms and Definitions—Guide on the Principles for ISO/TC
  207/SC6 Terminology Work
- ISO Guide 64
  Guide for the Inclusion of Environmental Aspects of
  Product Standards

Figure 29–18
ISO 14000 product-oriented documents.

6. *Structure and application of the ISO 14001 standards.* The ISO 14001 standard is subdivided into six broad components as follows:

   4.1   General requirements
   4.2   Environmental policy
   4.3   Planning
   4.4   Implementation and operation
   4.5   Checking and corrective action
   4.6   Management review

It is important to understand what the ISO 14001 standard does and does not do. The standard *does* allow a company voluntarily and proactively to establish a framework for involving all employees in improving environmental management. The standard does *not* establish product or performance standards, set pollutant levels, specify test methods, mandate zero emissions, nor expand on governmental regulations.

Plan

- Identify environmental aspects of processes, products, and services
- Evaluate all environmental aspects identified
- Identify legal requirements associated with the EMS
- Develop an environmental policy for the organization
- Develop performance criteria for internal audits of the EMS
- Establish EMS objectives, benchmarks, and performance targets

Do

- Provide the resources needed to support the EMS
- Assign responsibility for all aspects of the EMS
- Clarify accountability as it relates to the performance of EMS
- Provide training to ensure that all parties associated with the EMS have the knowledge and skills needed
- Communicate continually
- Develop and disseminate reports
- Document all aspects of the EMS and its performance
- Establish operational control

Check

- Monitor the performance of the EMS by making appropriate measurements
- Apply preventive action wherever possible
- Monitor EMS records continually
- Conduct audits of the EMS

Adjust

- Take corrective action when necessary
- Implement adjustments as needed continually

Figure 29–19
Plan-do-check-adjust (PDCA) model applied to an EMS.

Requirements of the ISO 14001 Standard

The ISO 14001 standard contains criteria subdivided into the following broad categories: general environmental policy; planning; implementation and operations; checking and corrective action; and management review. These broad categories are, in most cases, subdivided further.

Following is a checklist of criteria that translates ISO 14001 into a format that can be used for conducting internal audits.[11] The checklist ties into each category and subcategory of the standard.

4.1   General Requirements

1. Has the organization established an environmental management system to meet all requirements of Clause 4?
2. Is the environmental management system maintained; that is, is it in daily use, is it routinely followed, and are its requirements enforced by management?

4.2   Environmental Policy

3. Has top management defined the organization's environmental policy?
4. Is the environmental policy appropriate to the nature of the organization's activities, products, or services?
5. Is the policy appropriate to the scale of the organization's activities, products, or services?
6. Is the policy appropriate to the actual or possible environmental impacts of the organization's activities, products, or services?
7. Does the environmental policy include a management commitment to the prevention of pollution?
8. Does the policy include a management commitment to continual improvement?
9. Does the policy include a management commitment to comply with all relevant environmental laws and regulations?
10. Does the policy include a management commitment to comply with other environmental requirements to which the organization subscribes?
11. Does the environmental policy provide a workable framework for setting environmental objectives and targets?
12. Does the policy provide a workable framework for reviewing environmental objectives and targets?
13. Is the environmental policy documented?
14. Has the policy been implemented and maintained (i.e., put into daily use)?
15. Has the policy been communicated to all employees?
16. Is the environmental policy available to the public?

4.3   Planning

4.3.1   Environmental Aspects

17. Has the organization established procedures to identify the environmental aspects of its activities, products, or services that it can control and over which it can be expected to have an influence?
18. Does the organization maintain (i.e., routinely use) these procedures?
19. Has the organization determined through the use of these procedures which of its environmental aspects have, or can have, significant impacts on the environment?

20. Do the organization's procedures ensure that the aspects related to significant environmental impacts are always considered in setting environmental objectives?
21. Does the organization routinely keep information relative to its environmental aspects up-to-date?

4.3.2 Legal and Other Requirements

22. Does the organization have a procedure to identify legal and regulatory requirements applicable to the environmental aspects of its activities, products, or services?
23. Does the organization have a procedure to identify other environmental requirements to which it subscribes?
24. Does the organization have a procedure that assures access to all appropriate legal, regulatory, and other applicable environmental requirements?
25. Does the organization routinely use (maintain) these procedures?

4.3.3 Objectives and Targets

26. Does each relevant function and level of the organization have current, documented environmental objectives and targets?
27. Are the relevant functions and levels of the organization routinely working to achieve the current, documented environmental objectives and targets?
28. In establishing and reviewing its environmental objectives and targets, does the organization consider
    a. legal, regulatory, and other environmental requirements?
    b. its significant environmental aspects?
    c. its technological options?
    d. its financial, operational, and business requirements?
    e. the views of interested parties, both in and out of the organization?
29. Are the organization's objectives and targets consistent with its environmental policy?
30. Are the organization's objectives and targets consistent with its commitment to prevent environmental pollution?

4.3.4 Environmental Management Programs

31. Has the organization established programs for achieving its environmental objectives and targets?
32. Within these programs, is there a designation of responsibility for achieving objectives and targets at each relevant function and level of the organization?
33. Within these programs, is the means by which the objectives and targets are to be achieved stated and explained?
34. Within these programs, is a schedule provided for achieving the objectives and targets?
35. Are these programs routinely used and followed for achieving the organization's objectives and targets?

36. To ensure that environmental management applies to new developments, new or modified activities, new or modified products, and new or modified services, does the organization do one or both of the following:
    a. Does it develop new environmental management programs?
    b. Does it amend relevant existing programs?

4.4   Implementation and Operation

4.4.1   Structure and Responsibility

37. Has the organization defined the roles, responsibilities, and authorities of its employees in connection with environmental management?
38. Have these roles, responsibilities, and authorities been documented through organizational charts or other means?
39. Have these roles, responsibilities, and authorities been communicated to the organization's employees?
40. Has management provided the resources essential to the implementation and control of the EMS, including
    a. human resources (employees)?
    b. required specialized skills (from within or from outside)?
    c. essential technology (equipment, processes, and methods)?
    d. financial resources?
41. Has top management appointed one or more specific management representatives?
42. In addition to any other responsibilities already held, has management defined the environmental roles of the management representatives to include the responsibility and authority for
    a. ensuring that ISO 14000 EMS requirements are established?
    b. ensuring that ISO 14000 EMS requirements are implemented?
    c. ensuring that the conforming EMS is properly maintained and used?
    d. reporting on the performance of the EMS to top management for the purposes of review and as a basis for EMS improvement?

4.4.2   Training, Awareness, and Competence

43. Does the organization require that all employees whose work may create a significant impact on the environment receive appropriate training?
44. Has the organization identified training needs?
45. Does the organization have procedures to make employees at each relevant function and level aware of
    a. the importance of conformance with the organization's environmental policy and procedures and with the requirements of the EMS?
    b. the actual or potential significant environmental impacts of their work activities?
    c. the environmental benefits of improved personal performance?
    d. their roles and responsibilities in achieving conformance with the organization's environmental policy and procedures and with the requirements of the EMS?

e. their roles and responsibilities concerning the requirements for emergency preparedness and response?

f. the potential consequences of departure from specified operating procedures?

46. Are these awareness procedures routinely used?

47. Does the organization ensure that employees performing tasks that can cause significant environmental impacts be competent to perform them on the basis of education, training, and experience?

4.4.3   Communication

48. Does the organization have communication procedures for dealing internally with concerns and questions about the organization's environmental aspects and the EMS?

49. Do these procedures support communication among the various functions and levels of the organization?

50. Does the organization have communication procedures for receiving, documenting, and responding to relevant environmental communications from interested parties outside the organization?

51. Has the organization considered processes for external communication on its significant environmental aspects and whether it should implement such processes?

52. Has the organization documented these considerations and decisions?

4.4.4   Environmental Management System Documentation

53. Has the organization documented information (in paper or electronic form) that describes the core elements of the EMS and how they interact with each other?

54. Has the organization documented information (in paper or electronic form) that provides direction to EMS-related documentation? Has the organization ensured that when employees need documentation they can locate it?

55. Is this information maintained up-to-date? Is it routinely used by employees?

4.4.5   Document Control

56. Has the organization established procedures for controlling all documents required by ISO 14001?

57. Do these procedures ensure that all documents required by ISO 14001 can be located?

58. Do these procedures ensure that all documents required by ISO 14001 are periodically reviewed, revised as necessary, and approved for adequacy by authorized personnel?

59. Do these procedures ensure that the current versions of all documents required by ISO 14001 are available at all critical locations? ("Critical locations" means where work operations essential to the effective functioning of the EMS take place.)

60. Do these procedures ensure that obsolete documents are promptly removed from all points of issue and points of use, or do the procedures ensure in some other manner that obsolete documents cannot be used unintentionally?

61. Do these procedures ensure that any obsolete documents retained for any reason (e.g., legal or knowledge preservation) are suitably and clearly identified as obsolete?

62. Has the organization established procedures and responsibilities for creating and modifying the various types of documentation?

63. Do the procedures for creating and modifying documentation require and ensure that the documents
    a. are legible?
    b. carry the origination date?
    c. carry the date of any revision?
    d. are readily identifiable?
    e. are approved by signature of authorized individuals?

64. Are all these document control procedures routinely used and rigorously enforced by the organization?

65. Does the organization maintain (file, store, protect, disperse for use, retrieve, and so on) its controlled documentation in an orderly manner?

66. Does the organization retain its controlled documentation for a specified period (determined by legal or regulatory requirements or by the organization itself)?

4.4.6 Operational Control

67. To ensure that its environmental policy and its targets and objectives can be met, has the organization identified its operations and activities that are associated with its significant environmental aspects?

68. Has the organization planned these operations and activities, including maintenance, to ensure that they are carried out under specified (controlled) conditions by instituting the following processes:
    a. Establishing documented procedures to cover situations where the absence of documented procedures could lead to deviations from the environmental policy and the targets and objectives?
    b. Stipulating operating criteria (step-by-step work instructions, parametric readings, and so on) in the procedures?
    c. Establishing procedures related to identifiable significant environmental aspects of goods and services used by the organization?
    d. Communicating relevant procedures and requirements to the suppliers and contractors who provide these goods and services?

69. Are all these operational control procedures routinely used and rigorously enforced by the organization?

4.4.7 Emergency Preparedness and Response

70. Has the organization established emergency preparedness and response procedures to identify potential for environmentally related accidents and emergency situations? Has it developed procedures to respond to such accidents and emergency situations?

71. Do these procedures address the prevention or mitigation of environmental impacts that may result from such accidents?

72. Does the organization routinely use and rigorously enforce its emergency preparedness and response procedures?
73. Does the organization review (and revise when necessary) its emergency preparedness and response procedures (especially after the occurrence of accidents or emergency situations)?
74. Does the organization periodically test these procedures where practicable?

4.5   Checking and Corrective Action

4.5.1   Monitoring and Measurement

75. Does the organization have documented procedures to monitor and measure the key characteristics of its operations and activities that can have a significant impact on the environment?
76. Do these documented procedures require that monitoring and measuring of its key characteristics are done on a regular basis?
77. Do these procedures require the recording of information to track
    a. environmental performance?
    b. relevant operational controls?
    c. conformance with the organization's environmental objectives and targets?
78. Are these documented procedures routinely followed?
79. Does the organization calibrate and maintain its monitoring equipment according to an established process and schedule?
80. Does the organization have a procedure requiring the retention of records of monitoring equipment maintenance and calibration activity?
81. Are these records being retained as required by the procedure?
82. Does the organization have a documented procedure for periodically evaluating compliance with relevant environmental legislation and regulations?
83. Is this procedure routinely followed?

4.5.2   Nonconformance and Corrective and Preventive Action

84. Has the organization established procedures for defining responsibility and authority for
    a. handling and investigating nonconformance?
    b. taking action to mitigate any impacts caused by nonconformances?
    c. initiating and completing corrective action?
    d. initiating and completing preventive action?
85. Does the organization ensure that any corrective or preventive action taken to eliminate the causes of actual or potential nonconformances are
    a. appropriate to the magnitude of the problems?
    b. commensurate with the environmental impact encountered (or which is possible)?
86. When corrective or preventive actions involve documented procedures, does the organization implement and record the necessary changes to those procedures (i.e., are the procedures formally modified)?

4.5.3   Records

87. Has the organization established procedures for environmental records, including training records and the results of audits and reviews, that cover
    a. identification?
    b. maintenance (i.e., collection, indexing, filing, storage, retrieval, retention, control, and safekeeping)?
    c. disposition?
88. Do these procedures require that
    a. environmental records are legible?
    b. environmental records are readily identifiable?
    c. environmental records are traceable to the activity, product, or service involved?
    d. environmental records are protected against damage, deterioration, or loss?
    e. environmental records are readily retrievable?
    f. retention times (periods) of environmental records are established and recorded?
89. Are these procedures routinely followed and enforced?
90. Does the organization maintain records under these procedures to demonstrate conformance to the requirements of ISO 14000, including, as appropriate,
    a. legislative and regulatory requirements?
    b. relevant permits?
    c. environmental aspects and their associated impacts?
    d. product information?
    e. process information?
    f. environmental training activity?
    g. inspection, calibration, and maintenance activity?
    h. monitoring data?
    i. details of nonconformance (incidents, complaints, and follow-up action)?
    j. environmental audits (internal and external)?
    k. management reviews?
    l. supplier and contractor information?
    m. emergency preparedness and response information?

4.5.4   Environmental Management System Audit

91. Has the organization established programs and procedures for periodic internal EMS audits?
92. Are these programs and procedures designed to
    a. determine whether the EMS conforms to the organization's planned arrangements for the EMS and ISO 14000 requirements?
    b. determine whether the EMS has been properly implemented and maintained?
    c. provide audit result information to management?
93. Do the organization's internal audit programs and schedules of audits take into account the environmental importance (significance) of the activity concerned?

94. Do the organization's internal audit programs and schedules of audits take into account the results of previous audits (internal or external)?

95. Do the organization's internal EMS audit procedures cover
    a. designating the organizational activities and areas to be considered in audits?
    b. establishing the frequency of audits?
    c. establishing audit methodology?
    d. assigning responsibilities associated with managing and conducting audits (recognizing the need that persons selected as auditors must be in a position to conduct audits impartially and objectively)?
    e. ensuring auditor competence?
    f. communicating audit results?

4.6 Management Review

96. Does the organization's top management periodically review the EMS to ensure its continuing suitability, adequacy, and effectiveness?

97. Are such reviews carried out according to a schedule that top management has determined (as opposed to ad hoc reviews)?

98. Does the organization's management review process ensure that the information necessary for management's evaluation (for suitability, adequacy, and effectiveness) is collected and presented?

99. Are the management reviews documented?

100. In light of EMS audit results, the extent to which objectives and targets have been met, concerns among relevant interested parties, changing circumstances, and the commitment to continual improvement, do the organization's management reviews address the possible need for changes to the organization's
    a. environmental (or other) policy?
    b. environmental objectives or targets?
    c. other elements of the EMS (procedures, processes, and so on)?

Discussion Case

What Is Your Opinion?

The executive managers of Petroleum Processing, Inc., are debating the relative merits of seeking ISO 14000 certification. The managers have polarized around two opinions. The first opinion is that the company should proceed with preparations for certification immediately. In the words of one ISO proponent, "The sooner we get started, the sooner we will realize the benefits of ISO 14000." The second opinion is that any certification that is not required should be ignored. In the words of one ISO opponent, "Certification is time-consuming and expensive. Why bother if we don't have to? We already have to comply with OSHA. Isn't that enough?" Where do you stand on this issue? What is your opinion?

ISO 14000 SUCCESS STORY

The ISO 14000 standard has not yet achieved the widespread use in the United States that its quality counterpart—ISO 9000—enjoys. However, more and more companies are beginning to look to the ISO 14000 standard as an excellent vehicle for establishing an effective environmental management system. One such company is forestry and wood products giant Weyerhauser.[12]

As Sandy Smith wrote in *Occupational Hazards*,

> Weyerhauser, headquartered in Federal Way, Wash., sprang to life just over 100 years ago. . . . Today the company employs approximately 47,000 people in 17 countries, primarily the United States and Canada. It is one of North America's largest producers of forest products, the largest private owner and manager of merchantable softwood timber in the world, and the world's largest producer of softwood lumber, engineered wood products and hardwood lumber.[13]

Weyerhauser adopted its first environmental policy well ahead of the current worldwide focus on environmental safety. The company adopted a policy that made the following environmental commitment:

1. All employees at all levels will comply with applicable laws and regulations.
2. All employees at all levels will continuously improve environmental performance wherever the company operates.

Weyerhauser took another major step when it developed and adopted an environmental management system (EMS). Then, the company took the major step of pursuing ISO 14001 certification at its facility in Grayling, Michigan. The Grayling facility took on the challenge of piloting ISO 14001 certification and was successful. The company then committed to spread the certification throughout all its timberland operations.

As a result of developing and implementing an EMS and of pursuing ISO 14001 certification, Weyerhauser has enjoyed major improvements in the area of environmental safety. The following statistics show just how successful the company has been in this critical area:

1. The company improved its recordable incident rate for all North American operations by 25 percent.
2. The company has operated more than 140 facilities that have logged more than 20 million hours without a lost-time accident for several years in a row.
3. The company maintains a recordable incident rate of just 3.1 per 100 in the United States, which is well below the industry average (typically 9.8 per 100).
4. Several of the company's facilities now participate in OSHA's Voluntary Protection Program.[14]

SUMMARY

1. Environmental problems currently facing the United States include acid rain, ground-level ozone, global warming, water pollution, toxic wastes, and the proliferation of garbage.

2. The federal agencies most involved in environmental safety and health are OSHA and the EPA. In 1990, these two organizations signed a historic memorandum of understanding in which each agreed to work with the other to collect data, conduct inspections, and provide training.

3. One of the most important pieces of federal environmental legislation is the Clean Air Act. It contains provisions to reduce hazardous air pollutants, acid rain, and smog by significant percentages. The act deals with urban air quality, mobile sources, hazardous air pollutants, acid rain control, permits, stratospheric ozone provisions, and enforcement.

4. Measures taken to clean up and protect the environment are ethically right. They may also be cost-effective. However, research into the economics of a clean environment is limited at best. Research is needed to determine the effect of the country's investment in a clean environment not only on people's health, but also on inflation, unemployment, growth of the GNP, the international trade balance, productivity, research, innovation, and other economic factors.

5. There are different kinds of environments. These include natural, induced, controlled, artificial, closed, and free environments.

6. Safety and health professionals should be prepared to take responsibility for both workplace safety and environmental safety.

7. Some of the most common environmental hazards include high and low humidity, sunlight, high and low temperatures, airborne substances, meteorological and micrometeorological conditions, lightning, high and low air pressure, radiation, vibration, and sound.

8. Indoor environmental quality is as important as outdoor environmental quality. The following factors affect indoor environmental quality: heating, ventilation, and air-conditioning; air filtration; temperature; humidity; airborne compounds and gases; and airborne molds, bacteria, and other biological matter.

9. The goal of the OSHA Hazardous Waste Standard and other regulations is to (a) encourage companies to reduce the amount and toxicity of the hazardous substances that they use, (b) ensure that remaining hazardous materials are used safely, and (c) ensure that companies are prepared to respond promptly and appropriately when accidents occur.

10. Hazardous waste reduction involves reducing the amount of hazardous waste generated and, in turn, the amount introduced into the waste stream through the processes of source reduction and recycling.

11. Source reduction and recycling strategies include (a) management improvements in operating efficiencies, (b) better use of technology, (c) better material selection, and (d) reclamation.

12. Organizing a waste reduction program involves accomplishing the following steps: (a) gaining a commitment from executive management; (b) forming a waste reduction team; (c) developing a comprehensive plan; and (d) implementing, monitoring, and adjusting the plan.

13. The waste reduction audit is the most important part of the waste reduction program. It involves targeting processes, analyzing processes, identifying reduction options, considering cost–benefit data, and selecting the best option.

14. Trends for the future include more emphasis on prevention than cleanup, more interagency cooperation, use of economic incentives to supplement regulations, and targeting/phasing out of toxic chemicals.
15. ISO is the acronym for the International Organization for Standardization, a worldwide consortium of national standards bodies. ISO is the organization that developed the ISO 14000 family of standards.
16. A management system is that component of an organization that is responsible for leading, planning, organizing, and controlling. An environmental management system (EMS) is that component of an organization with primary responsibility for these functions as they relate specifically to an organization's impact on the environment.
17. The potential benefits of an EMS include the following: ease of international trade, improved compliance, credibility, reduction in liability/risk, regulatory incentives, sentencing mitigation, pollution prevention, waste reduction, profits, improved internal management, community goodwill, employee retention, and insurance and lending preferences.
18. The ISO 14001 standard is structured as follows: general requirements, environmental policy, planning, implementation and operation, checking and corrective action, and management review.

KEY TERMS AND CONCEPTS

Accidental release

Acid rain

Artificial environment

Clean Air Act

Closed environment

Continual improvement

Controlled environment

Cost–benefit analysis

Economic incentives

EMS audit

Environment

Environmental aspect

Environmental hazards

Environmental management system (EMS)

Environmental Protection Agency (EPA)

Free environment

Garbage

Global warming

Ground-level ozone

Guidance standard

Guide

Hazardous and Solid Waste Amendments (HSWA)

Hazardous waste reduction

Indoor environmental quality (IEQ)

Induced environment

International Organization for Standardization (ISO)

ISO 14000 Series

Natural environment

OSHA Hazardous Waste Standard

Plan-do-check-adjust (PDCA) model

Reclamation

Recycling

Reduction options

Resource Conservation and Recovery
    Act (RCRA)

Sick-building syndrome

Source reduction

Specification standard

Stratospheric ozone

Technical report

Toxic wastes

Waste reduction audit

Waste reduction plan

Waste reduction team

REVIEW QUESTIONS

1. Briefly describe the progress made in cleaning up the environment between 1970 and the present.
2. Describe the OSHA–EPA memorandum of understanding signed in 1990.
3. What are the major goals of the Clean Air Act?
4. Briefly explain the following titles of the Clean Air Act: Title I, Title II, and Title III.
5. Why is more research into the economics of a clean environment important?
6. Distinguish between the following types of environments: natural and induced; controlled and artificial.
7. How are environmental concerns changing the role of safety and health professionals?
8. List five of the most common environmental hazards.
9. What are the factors that most affect indoor environmental quality?
10. Explain the term *sick-building syndrome*.
11. Summarize briefly the requirements of the OSHA Hazardous Waste Standard.
12. What are the goals of hazardous materials regulations in general?
13. Define the term *hazardous waste reduction*.
14. List and explain three hazardous waste reduction strategies.
15. What are the steps involved in organizing a waste reduction program?
16. List the minimum contents of a hazardous waste reduction plan, and explain how you would go about developing a comprehensive plan.
17. Define the term *waste reduction audit*.
18. What are the most important steps in a waste reduction audit?
19. What are EPA's top environmental priorities for the future?
20. List four trends for the future regarding environmental protection.
21. What is an environmental management system?
22. What factors are behind the competitive advantage that can result from implementation of an EMS?
23. List five potential benefits of an EMS.
24. Distinguish between process- and product-oriented ISO 14000 standards.

ENDNOTES

1. Environmental Protection Agency (EPA), National Environmental Publications Internet Site (NEPIS), http://www.epa.gov/.
2. Ibid.
3. Ibid.
4. Ibid.
5. National Institute for Occupational Safety and Health (NIOSH), "Indoor Environmental Quality (IEQ)," *NIOSH Facts.* Retrieved from http://www.cdc.gov/niosh/ieqfs.html.
6. NIOSH/EPA, "Building Air Quality Action Plan," 1–3. Retrieved from http://www.cdc.gov.niosh/98–123a.html.
7. Environmental Protection Agency (EPA), National Environmental Publications Site (NEPIS), http://www.epa.gov/.
8. Ibid.
9. D. Goetsch and S. Davis, *ISO 14000 Environmental Management* (Upper Saddle River, NJ: Prentice Hall, 2001), 7.
10. ANSI/ISO/ASQE 14001–2004: *Environmental Management Systems Requirements with Guidance for Use* (Chicago: American Society for Quality, 2004).
11. Ibid.
12. S. Smith, "The Cutting Edge of Environmental Management," *Occupational Hazards* 64, no. 2: 33–37.
13. Ibid., 34.
14. Ibid., 36.

TSM: Safety Management in a Quality Management Setting

Major Topics

- What Is QM?
- How Does QM Relate to Safety?
- Safety Management in a QM Setting
- What Is TSM?
- Translating TSM into Action
- Fundamental Elements of TSM
- Rationale for TSM
- Implementing TSM: The Model

Quality management (QM) is an approach to doing business that began to gain wide acceptance in the United States in the late 1980s and early 1990s, and was originally called total quality management, or TQM. Various individual components of this concept had been used by forward-looking organizations for years. However, not until the 1990s were these components pulled together into a cohesive philosophy of how to do business in a competitive global environment.

The safety and health professional of today is likely to work in an organization that practices the QM philosophy. Consequently, students of safety and health should understand QM and how to apply its principles to the management of an organization's safety and health programs. After providing an overview of QM, this chapter presents a comprehensive explanation of **total safety management (TSM)**, the management of workplace safety and health according to the principles of QM. The TSM material contained in this chapter is adapted from *Implementing Total Safety Management* (Prentice Hall, 1997) by the author.

WHAT IS QM?

The QM concept goes by several different names. It is has been called TQL for *total quality leadership*, TQC for *total quality control*, TQ for *total quality* or TQM for *total quality management*. Regardless of the name used, the concept can be defined as follows:

> Quality management (QM) is an approach to doing business that maximizes the competitiveness of an organization through continuous improvement of its products, services, people, processes, and environments.[1]

Other approaches to doing business could lay claim to this or a similar definition. What distinguishes QM from the other philosophies is the following list of characteristics. These characteristics describe *how* TQM achieves its purpose.[2]

1. *Customer focus.* QM organizations continually solicit input (up front) and feedback (after the fact) from both external and internal customers.

2. *Obsession with quality.* In a QM organization, quality is every employee's job, from top management down through line employees. In such organizations, all employees learn to focus on what they can do to improve quality and competition every day.

3. *Scientific approach.* QM organizations apply scientific decision making. This means that they operate on facts rather than assumptions and apply logic rather than so-called *gut feelings*. Employees in such organizations learn to use decision-making tools such as Pareto charts, fishbone diagrams, histograms, and control charts to identify root causes clearly before solving problems. This ensures that they solve problems, rather than just attacking symptoms.

4. *Long-term commitment.* The executive management team of a QM organization is committed to this approach for the long term. QM is not a quick-fix approach that will produce overnight miracles. It requires a long-term investment of both time and effort.

5. *Teamwork.* QM organizations are teamwork oriented. Work is done by teams, and performance is measured against team, rather than individual, benchmarks.

6. *Continual process improvements.* QM organizations continually improve the processes that make their products or deliver their services. *Good enough* is never good enough. Today's improvements are just the starting point for tomorrow's improvement.

7. *Education and training.* In a QM organization, education and training are a normal part of doing business. They are used to equip all employees to be knowledgeable, skilled participants in the continual improvement of quality.

8. *Freedom through control.* A QM organization frees its employees to improve their work processes continually without relinquishing appropriate management controls. This is done by establishing the parameters within which employees must work and then empowering them to make improvements within those parameters.

9. *Unity of purpose.* Along with a long-term commitment from top management, QM organizations have unity of purpose from top to bottom. The organization's purpose is represented by its *vision statement.* All employees understand the vision, believe in it, and unite around it.

10. *Empowerment.* QM organizations tap the enormous wealth of knowledge, experience, and potential that their employees represent. They do this by involving employees in decisions that affect their work and by empowering them to do what is necessary to ensure and improve quality continually.

HOW DOES QM RELATE TO SAFETY?

QM has proven itself to be an effective way to maximize an organization's long-term competitiveness. It is also an excellent approach for maximizing the effectiveness of an organization's safety and health programs. QM can solve the same problem for safety managers that it solves for quality managers—the problem of *isolation.*

Often in a traditionally managed organization, quality is viewed as the sole responsibility of the quality department or the quality manager. The weaknesses of this approach became evident when organizations wedded to it began losing market share to foreign competition in the 1970s and 1980s. QM solved this problem by making quality everybody's job and casting the quality manager in the role of facilitator and catalyst.[3]

The same type of isolation often occurs with safety and health managers. Management and employees sometimes view safety as the responsibility of the safety department or the safety manager. In the case of quality, isolation is a prescription for failure. In the case of safety, it is a prescription for disaster. QM principles can solve the problem of isolation by making safety everybody's job and casting the safety manager in the role of facilitator and catalyst.

The isolation issue is not the only tie between QM and safety management. To fully understand how QM relates to safety, we must examine each of the distinguishing characteristics of QM from a safety and health perspective. These characteristics, as viewed from this perspective, are as follows:

1. *Customer focus.* In addition to soliciting input and feedback about the quality of its products, organizations should also collect customer data about the safety and health features of the products. After all, these features have a significant bearing on the perceived quality of a product. An unsafe product is not a quality product regardless of the attractiveness of its other characteristics. In the same way, it is important to collect input and feedback from internal customers

(i.e., the employees) concerning the safety of work processes and environments. Internal customer data are the most effective data available for continually improving safety and health procedures.

2. *Obsession with quality.* In addition to being obsessed with quality, organizations should become obsessed with safety. The two concepts are actually inseparable. An organization with unsafe processes or products will eventually pay too high a price to be able to keep up in a competitive marketplace.

3. *Scientific approach.* When dealing with accidents and incidents, it is critical to identify root causes rather than symptoms. Future accidents cannot be prevented when the action taken in response to a current accident is based on symptoms rather than the root cause. The same scientific tools used to identify root causes of quality problems can be used to identify root causes of safety and health problems.

4. *Long-term commitment.* The long-term commitment of top management is as essential to safety as it is to quality. Executive-level managers show their commitment to safety and health in the same way that they show it to quality—by making safety an organizational priority.

5. *Teamwork.* Part of making safety everybody's responsibility is taking a teamwork approach to the continuous improvement of work processes and the work environment from a safety perspective. Just as cross-functional teams are assigned specific quality improvement projects, similar teams should be formed to tackle specific safety and health issues.

6. *Continual process improvements.* QM organizations structure themselves to improve work processes continually from the perspective of quality. This can and should be done simultaneously from a safety perspective. The continual improvement of safety can be achieved in the same way as the continual improvement of quality.

7. *Education and training.* Quality and safety don't just happen. In both cases, employees at all levels develop the knowledge, skills, and attitudes needed to make a difference. This is accomplished in both cases through ongoing education and training for employees at all levels.

8. *Freedom through control.* Too often, safety precautions are viewed as unwanted inhibitors of productivity. Actually, the safe way is not always the fastest way in the short run. However, over the long run, when factoring in the downtime and expenses associated with injured employees and damaged machines, the safe way is typically the most productive way. Maximum productivity can be achieved by giving employees the freedom to apply their own initiatives in improving both quality and safety within the guidelines and parameters prescribed by management.

9. *Unity of purpose.* Unity of purpose is just as important to safety as it is to quality. The guiding principles that accompany an organization's vision statement should clearly establish safety and health as organizational priorities. Organizationwide unity concerning safety and health will go a long way toward ensuring the effectiveness of a safety program.

10. *Empowerment.* In a QM organization, some of the best ideas for quality improvements come from employees who are empowered to think, suggest, and act. The same employees, equally empowered, could put their creative energy into making safety and health improvements.[4]

QM has excellent potential as an approach to safety and health management. In fact, other ties between quality and safety are so close that it can be argued that safety and quality must be improved simultaneously.

SAFETY MANAGEMENT IN A QM SETTING

The concept of total safety management (TSM) was introduced by the author in his book *Implementing Total Safety Management.* It grew out of a need to transform safety and health management from a strict compliance orientation to a performance orientation in which compliance is an important issue, but not the only issue. A safe and healthy work environment should do more than keep employers out of trouble with regulatory agencies. Pressure from state and federal agencies in the area of workplace safety fluctuates in accordance with the prevailing political climate. However, the need to maximize the performance of employees and organizations is constant. Safety and health should be a key element in an organization's strategic plan for gaining a competitive advantage in the global marketplace. TSM is an approach to safety and health management that is rooted in organizational performance and global competitiveness. Its purpose is to give organizations the sustainable competitive advantage of a safe and healthy work environment that allows employees to achieve consistent peak performance.

WHAT IS TSM?

TSM is to safety and health management what QM is to quality management. It is safety management according to the principles of QM. QM has revolutionized the way in which organizations in the United States do business. As a result, the United States is slowly but steadily reclaiming market share lost to Japan and other countries in such critical market sectors as automobiles, consumer electronics, and computers. TSM can produce similar results for occupational safety and health, thereby making organizations that adopt it better able to gain the competitive advantage needed to compete in the global marketplace.

Just as QM involves the total organization in continually improving quality, TSM involves the total organization in establishing and maintaining a work environment that is safe and conducive to quality and productivity. Both concepts are rooted firmly in the pressures of the global marketplace.

TSM Defined

The origin of TSM can be traced back to the globalization of the marketplace that began after World War II, but really took hold in the 1970s. The need for TSM was created by the need for organizations to be competitive globally. Consequently, *TSM* is defined as follows:

> TSM is a performance and process-oriented approach to safety and health management that gives organizations a sustainable competitive advantage in the global marketplace by establishing a safe and healthy work environment that is conducive to consistent peak performance and that is improved continually forever. It involves applying the principles of QM to the management of safety and health.

This definition contains several key elements that must be understood to comprehend TSM fully. These elements are as follows:

1. *Sustainable competitive advantage.* Every organization that competes at any level, but especially those that compete at the global level, must have competitive advantages. These are capabilities or characteristics that allow them to outperform the competition. For example, if the organization in question is a baseball team, it may have such competitive advantages as an excellent pitching staff, several speedy base runners, two or three power hitters, and outstanding fielders in key positions. These advantages, if exploited wisely, help make the baseball team a winner. If these advantages can be sustained over time, they help make the team a consistent winner. This same concept applies to organizations that compete in the global marketplace. To survive and prosper, companies need as many competitive advantages as possible. Traditionally, competitive advantages have been sought in the key areas of quality, productivity, service, and distribution. However, peak-performing organizations have learned that a safe and healthy work environment is essential to gaining

competitive advantages in all these critical areas. In fact, a safe and healthy work environment is itself a competitive advantage. In today's competitive marketplace, high-performance employers are adding one more critical area to the list of those in which competitive advantages are sought. This new addition is the work environment.

Peak performance organizations are learning that a safe and healthy work environment gives them a doubly-effective competitive advantage. First, it ensures that employees work in an environment that allows them to focus all their attention, energy, and creativity on continually improving performance. Second, it prevents an organization's limited resources from being drained off by the non-value-added costs associated with accidents and injuries.

2. *Peak performance.* The primary driver behind TSM performance is organizational, team, and individual. An organization's ability to survive and prosper in the global marketplace is determined largely by the collective performance of individuals and teams. Consistent peak performance by all individuals and teams in an organization is essential to long-term success in the global marketplace. The quality of the work environment is a major determinant of the performance levels that individuals, teams, and organizations are able to achieve. A better work environment promotes better performance.

3. *Continual improvement forever.* People work in an environment, and the quality of that environment affects the quality of their work. The work environment is a major determinant of the quality of an organization's processes, products, and services. In the age of global competition, quality is an ever-changing phenomenon. Quality that is competitive today may not be tomorrow. Consequently, continual improvement is essential. If quality must be improved continually, it follows that the work environment must also be improved continually. Quality and safety are more than complementary; they are inseparable.

TRANSLATING TSM INTO ACTION

There are three fundamental components through which the TSM philosophy is translated into action on a daily basis. These three components are the **TSM steering committee**, **improvement project teams (IPTs)**, and the **TSM facilitator**.

TSM Steering Committee

The TSM steering committee oversees the organization's safety and health program. It is responsible for the formulation of safety and health policies, the approval of internal regulations and work procedures relating to safety and health, the allocation of resources, and the approval of recommendations made by IPTs.

Safety Fact

Cost of Accidents

The cost of accidents that occur on or off the job each year in the United States is approximately $150 billion. To grasp the enormity of this figure, consider that it equates to approximately $1,000 per year for every working person in the United States.

Improvement Project Teams

Improvement project teams (IPTs) are ad hoc, or temporary, teams formed by the TSM steering committee for the purpose of pursuing specific improvement projects relating to the work environment. Through the ongoing process of hazard identification, in which all employees in a TSM organization participate, potential safety and health problems are identified. When a potential problem is identified, an IPT is formed to analyze the situation and make recommendations for improvements. These recommendations are presented to the steering committee by the TSM facilitator. For example, if an organization's workers' compensation costs have increased sharply, or if the number of accidents in a given department is on the rise, IPTs may be formed to analyze these situations and make recommendations. IPTs do not supplant the regular tasks and duties of safety and health professionals. Rather, they supplement these activities.

Typically, IPTs are cross-functional. This means that all stakeholders concerned about the problem in question are represented on the IPT that will look into it. A stakeholder is any individual, team, or department that is affected by a given problem, directly or indirectly. Consequently, IPTs usually have members from several different departments. Hence, the term *cross-functional*.

However, IPTs can also be natural work teams. These are teams made up of people who work together on a daily basis in performing their jobs. How an IPT is constituted is determined by the nature and scope of the problem. If its effects are confined to a single department, the IPT should be a natural work team. If its effects cross over into several different departments, the IPT should be cross-functional.

Once an IPT fulfills the charter it is given by the steering committee, it is disbanded. While one IPT is working on an assigned project, others are formed to pursue other projects. This is an ongoing process that never stops because a fundamental aspect of TSM is continual improvement of the work environment forever.

TSM Facilitator

The TSM facilitator must be a safety and health professional from within the organization. Ideally, he or she should be the organization's chief safety and health officer. The TSM facilitator serves as the steering committee's resident expert on the technical and compliance aspects of safety and health. He or she is responsible for the overall implementation and operation of the organization's TSM program.

FUNDAMENTAL ELEMENTS OF TSM

TSM differs in several ways from traditional safety and health management. To appreciate these differences, one must understand the following fundamental elements of TSM:

- Strategically based
- Performance and process oriented
- Dependent on executive-level commitment
- Teamwork oriented
- Committed to employee empowerment and enlistment
- Based on scientific decision making
- Committed to continual improvement
- Comprehensive, ongoing training
- Unity of purpose

Strategic Basis

The fact that TSM is **strategically based** means that an organization realizes that having a safe and healthy workplace gives it a competitive advantage in the marketplace. Consequently, the organization makes maintaining a safe and healthy workplace a part of its strategic plan. Traditional organizations, on the other hand, tend to view safety and health more from a compliance perspective. TSM is strategically based because it is considered part of an organization's overall strategy for competing globally.

To compete in the global marketplace, organizations must plan for and establish sustainable competitive advantages. This is known as *strategic planning*. Once the planned advantages have been established, they must be maintained over time and exploited to the fullest possible extent. Strategic planning is a matter of answering the following questions:

- As an organization, what would we like to be? What is our dream? This is the vision component of the strategic plan.
- As an organization, what is our purpose? This is the mission component of the strategic plan.
- As an organization, what values are most important to us? This is the guiding-principles component of the strategic plan. A guiding principle is a high-priority organizational value.
- As an organization, what do we want to accomplish? This is the broad-objectives component of the strategic plan. Figure 30–1 contains an excerpt from the strategic plan for a hypothetical company, Allied Manufacturing.

Allied Manufacturing Company

Vision

Allied Manufacturing Company will be the preferred supplier of ceiling fans in the United States.

Mission

The mission of Allied Manufacturing Company is to design and manufacture high-quality ceiling fans for residential and commercial use.

Guiding Principles

The highest-priority corporate values of Allied Manufacturing Company are as follows:

- ▸ Customer satisfaction
- ▸ Product and service quality
- ▸ Cost leadership
- ▸ Employee safety and health
- ▸ Continual improvement
- ▸ Ethical business practices

Figure 30–1
Core of the strategic plan for Allied Manufacturing Company.

With TSM, a safe and healthy work environment shows up in the strategic plan as a competitive advantage that will be actively pursued and, having been established, will be maintained and continually improved over time. Safety and health concerns appear either in the guiding principles section of the strategic plan, the broad-objectives section, or both. For example, in Figure 30–1, employee safety and health is a guiding principle value of Allied Manufacturing Company.

Employee safety and health may also appear in the strategic plan as a broad objective. For example, the following broad objective may appear in an organization's strategic plan:

> To maintain a safe and healthy work environment that is conducive to consistent peak performance on the part of all employees.

Performance Oriented

In any organization that is subject to federal and state regulations, compliance is an important issue. Failure to comply with applicable regulations can lead to fines and other non-value-added expenses. Companies in such sectors as manufacturing, processing, construction, transportation, and maritime services must comply with numerous government safety regulations.

In a *compliance-driven* setting, the safety department has a regulation orientation. Such an orientation can breed a big-brother-is-watching-you mentality among employees and managers that can cause them to resent the safety department's efforts. The resentment is magnified when employees and managers are pressed to meet deadlines and they see safety regulations as slowing them down.

With TSM, the safety department is **performance oriented**, and safety and health personnel are viewed as part of the equation for continually improving performance. Compliance is still important, of course, but compliance and productivity are no longer viewed as being mutually exclusive entities. Compliance is viewed as a natural extension of an organization's performance-improvement strategies and continues to be an important responsibility of the TSM facilitator.

Executive Commitment

The TSM approach can be implemented successfully and fully only if an organization's executive management team is committed to providing a safe and healthy work environment as a performance-improvement strategy. For this reason, the safety and health professional must be able to articulate the TSM philosophy convincingly. If **executive commitment** does not exist, it will have to be generated. Naturally, responsibility for generating the necessary commitment falls to safety and health professionals.

Teamwork Oriented

In a traditional setting, workplace safety and health tend to be viewed as the responsibility of the safety department. In a **teamwork-oriented** TSM setting, although the safety department plays the key facilitating role, ensuring a safe and healthy workplace is everybody's responsibility. IPTs are formed by the TSM steering committee and given charters to improve specific aspects of the work environment. When a team satisfies its charter, it is disbanded. As the need to improve the work environment continues, new teams are chartered, and the process continues forever.

Employee Empowerment and Enlistment

Empowerment means involving employees in ways that give them a real voice. With empowerment, employees are allowed to give input concerning workplace issues, problems, and challenges. If the issue is how to make the work environment safer, empowered employees are allowed and encouraged to make suggestions, and their suggestions are given serious consideration. **Enlistment** takes the concept of empowerment one step further. With enlistment, rather than allowing employees to give input, the organization expects them to do so.

In a TSM setting, employees are viewed as invaluable sources of information, knowledge, insight, and experience that should be tapped continually. Consequently, employee input is not just wanted, it is needed; it is not just sought, it is expected. A criterion that should be included on the performance-appraisal form of modern organizations reads as follows:

To what extent does this employee provide input that is useful in making continual improvements in the workplace?

Frequently   _____
Sometimes   _____
Seldom   _____
Never   _____

Scientific Decision Making

It is no exaggeration to say that decisions about workplace safety and health can, on occasion, be life-or-death decisions. Whether they are or not, all such decisions are too important to be based on guesswork. With TSM, **scientific decision making** is the norm. This means that employees in organizations that adopt TSM must learn to use such decision-making tools as Pareto charts, statistical process control (SPC) charts, scatter diagrams, histograms, and fishbone diagrams. Such tools can take the guesswork out of decision making and problem solving and, as a result, lead to better decisions and better solutions.

Continual Improvement

With TSM, the status quo is never good enough, no matter how good it is. In a competitive marketplace, it must be assumed that the competition is improving continually. Consequently, today's competitive performance may not be tomorrow's. The state of the work environment can always be improved, and it should be. In an organization that adopts TSM, IPTs are always working to recommend accident prevention strategies that will make the workplace safer and more productive.

Comprehensive, Ongoing Training

In a TSM setting, training is an ongoing part of the job. If all employees are going to be involved in hazard identification and accident prevention, they must be trained in the basic techniques. If employees are going to be members of IPTs, they must be trained in the fundamentals of teamwork. If employees are going to use scientific tools in making recommendations for solving prob-lems, they must be given instructions in the proper use of these tools.

Training should be comprehensive and ongoing. Figure 30–2 contains a partial list of courses, seminars, and workshops to which employees in a TSM setting should have access. The specific training opportunities that should be provided are determined by the nature of the organization, its products, processes, and employees.

Unity of Purpose

In a TSM setting, all employees at all levels understand that safety and health are the responsibilities of everyone. The TSM facilitator and other safety and health professionals are responsible for planning, facilitating, coordinating, advising, and monitoring. In addition, employees are responsible for doing

Figure 30–2
TSM-related training opportunities.

- TSM: What It Is and How to Do It
- Teamwork Fundamentals
- Scientific Decision Making/Problem Solving
- Stress and Safety
- Mechanical Hazards and Body Guarding
- Lifting Hazards
- Falling, Impact, and Acceleration Hazards
- Heat and Temperature Hazards
- Pressure Hazards
- Electrical Hazards
- Fire Hazards
- Toxic Substances
- Explosive Hazards
- Radiation Hazards
- Noise and Vibration Hazards
- Hazard Analysis
- Accident Prevention
- Emergency Preparation/Response
- Facility Assessment/Hazard Identification
- Accident Investigation and Reporting
- Industrial Hygiene
- Automation-Related Hazards
- Bloodborne Pathogens and Related Hazards
- Ergonomics in the Workplace

their part to make the workplace safe and healthy. Employees satisfy their responsibilities in this regard by

- Setting a positive example of working safely
- Encouraging fellow employees to work safely
- Practicing hazard identification techniques constantly
- Recommending accident prevention strategies
- Serving effectively on IPTs

Gaining unity of purpose begins with the commitment of executive management. An organization's executive management team must commit to safety and health, make it a priority issue in the organization's strategic plan, stress its importance through both words and example, and expect it to be a priority of all employees.

RATIONALE FOR TSM

The correlation between work environment and job performance is strong. It is this correlation that forms the basis of the rationale for TSM. Globalization has increased the level of competition in the marketplace exponentially. For many organizations, adjusting to globalization has been like jumping from high school athletic competition to the Olympics.

Organizations that find themselves in these circumstances need every competitive advantage that they can muster. They need to do everything possible to improve performance continually. What every organization can, and must, do is provide employees with a safe and healthy work environment that is conducive to consistent peak performance.

IMPLEMENTING TSM: THE MODEL

Figure 30–3 contains a three-phase, 15-step model that can be used for successfully implementing TSM in any organization. The three broad phases of activity are **planning and preparation**, **identification and assessment**, and **execution**. Phase One—planning and preparation—encompasses Steps 1 to 7. Each step is completed in order. Pursuing steps out of order can throw the implementation process into disarray. For example, Step 1 is a critical preparation activity. Failure to complete Step 1 successfully can jeopardize all subsequent steps. This is because the organization's attitude toward TSM is determined primarily by the attitudes of executive managers.

Step 2 is the activity in which the TSM steering committee is established. The ideal TSM steering committee is the organization's executive management team, possibly augmented as necessary to ensure that all functional departments

---

Planning and Preparation

1. Gain executive-level commitment
2. Establish the TSM steering committee
3. Mold the steering committee into a team
4. Give the steering committee safety and health awareness training
5. Develop the organization's safety and health vision and guiding principles
6. Develop the organization's safety and health mission and objectives
7. Communicate and inform

Identification and Assessment

8. Identify the organization's safety and health strengths and weaknesses
9. Identify safety and health advocates and resistors
10. Benchmark initial employee perceptions concerning the work environment
11. Tailor implementation to the organization
12. Identify specific improvement projects

Execution

13. Establish, train, and activate improvement project teams
14. Activate the feedback loop
15. Establish a TSM culture

Figure 30–3
Model for implementation of TSM.

are well represented. Steps 3 and 4 are preparation activities that involve training. In Step 3, the TSM steering committee undergoes teamwork training and team-building activities. In Step 4, the steering committee undergoes safety and health awareness training.

Steps 5 and 6 are planning activities in which the steering committee develops a mini-strategic plan for safety and health. This plan consists of a vision, guiding principles, mission, and broad objectives, all of which focus solely on safety and health in the workplace. This mini-strategic plan becomes a subset of the organization's overall strategic plan. In Step 7, the plan is communicated to all employees.

Phase Two of the implementation model—identification and assessment— encompasses Steps 8–12. The steps in this phase allow the steering committee to identify strengths that may work in favor of the implementation and inhibitors that may work against it. Having identified strengths and inhibitors, the steering committee can tailor the implementation to exploit strengths while minimizing inhibitors.

Phase Three of the implementation model—execution—encompasses Steps 13 to 15. The steps in this phase involve actually assigning teams to specific improvement projects. As soon as a given improvement has been made, that IPT is disbanded, and a new team is formed to pursue other improvements. This process repeats itself continually forever. An outgrowth of the repetitive nature of the model is that TSM becomes ingrained in the organization's culture.

SUMMARY

1. Quality management (QM) is an approach to doing business that maximizes the competitiveness of an organization through continuous improvement of its products, services, people, processes, and environments.

2. QM achieves its purpose by developing the following characteristics in an organization: customer focus, obsession with quality, scientific approach, long-term commitment, teamwork, continual process improvement, education and training, freedom through control, unity of purpose, and empowerment.

3. QM can solve the same problems for safety managers that it solves for managers concerned about quality. QM makes quality everybody's responsibility, rather than limiting responsibility to quality personnel. It can do the same for safety managers. In addition, a quality product is a safe product, and the best environment in which to produce quality products is a safe and healthy environment.

4. TSM is an approach to safety and health management that gives organizations a sustainable competitive advantage in the global marketplace. This is accomplished by involving all employees in establishing, maintaining, and continually improving the work environment so that it is conducive to consistent peak performance.

5. The fundamental elements of TSM are its strategic basis, performance orientation, executive commitment, teamwork orientation, employee empowerment and enlistment, scientific decision making, continual improvement, comprehensive and ongoing training, and unity of purpose.

6. The rationale of TSM can be found in the connection between job performance and the work environment. To compete in the global marketplace, organizations need all employees performing at peak levels on a consistent basis. A safe and healthy workplace promotes peak performance.

7. The model for implementing TSM consists of 15 steps arranged in three phases. These phases are planning and preparation, identification and assessment, and execution.

KEY TERMS AND CONCEPTS

Assessment

Comprehensive, ongoing training

Continual improvement

Continual process improvement

Customer focus

Education and training

Employee empowerment and enlistment

Empowerment

Enlistment

Execution

Executive commitment

Freedom through control

Identification

Improvement project teams (IPTs)

Long-term commitment

Obsession with quality

Peak performance

Performance oriented

Planning

Preparation

Quality management (QM)

Scientific approach

Scientific decision making

Steering committee

Strategically based

Sustainable competitive advantage

Teamwork

Teamwork oriented

Total safety management (TSM)

TSM facilitator

TSM steering committee

Unity of purpose

REVIEW QUESTIONS

1. Define *QM*.
2. What are the critical characteristics of QM?
3. How does QM relate to safety?
4. Define the term *total safety management*.
5. List and briefly describe the fundamental elements of TSM.
6. What is the rationale for TSM?
7. What are the three broad phases that make up the model for implementing TSM?
8. What is the most important step in the TSM implementation model? Why?

ENDNOTES

1. D. L. Goetsch and S. Davis, *Quality Management* (Upper Saddle River, NJ: Prentice Hall, 2006), 6.
2. Ibid., 585.
3. E. J. White, "Saturn's Implementation of Deming's Fourteen Points," an undated training document, 1.
4. Ibid., 3.

ESTABLISHING A SAFETY-FRIENDLY CORPORATE CULTURE

Major Topics

▪ Safety-Friendly Corporate Culture Defined
▪ Importance of Having a Safety-Friendly Corporate Culture
▪ Globalization of Competition and Safety
▪ How Corporate Cultures Are Created
▪ What a Safety-Friendly Corporate Culture Looks Like
▪ Ten Steps for Establishing a Safety-Friendly Corporate Culture

Safety-related disasters that have occurred over time, such as the Chernobyl nuclear power plant meltdown, the explosion of the space shuttle *Challenger* on takeoff from Cape Canaveral, and the disintegration of the space shuttle *Columbia* as it attempted reentry from space, have given rise to the term *safety culture*. This term is often heard now when something major goes wrong relating to occupational safety and health. Organizations that experience major safety or health failures that lead to accidents or disasters are sometimes said to be lacking a safety culture.

In reality the term *safety culture* is a misnomer in that it implies that safety is a stand-alone, nonintegrated concept that can occur in a vacuum—that it is not part of a larger corporate culture. This is not the case. An organization's safety culture or lack of it is an important part of its larger corporate culture. For this reason, this chapter introduces the term **safety-friendly corporate culture.**

SAFETY-FRIENDLY CORPORATE CULTURE DEFINED

We could initiate a lively discussion among business experts, safety and health professionals, and academicians by raising the question: What is meant by the term *safety-friendly corporate culture*? It sometimes appears that there are as many definitions of the concept as there are people who use the term. Safety and health professionals should understand what the concept means. The concept can be defined as follows:

> A safety-friendly corporate culture exists when the tacit assumptions, beliefs, values, attitudes, expectations, and behaviors that are widely shared and accepted in an organization support the establishment and maintenance of a safe and healthy work environment for all personnel at all levels.

Evidence of an organization's corporate culture includes:

1. *Its priorities.* Are safety and health top priorities in the organization?
2. *How people in the organization succeed.* Are personnel recognized and rewarded for working safely?
3. *How decisions are made in the organization.* Is safety a major consideration when decisions are made?
4. *Expectations management has of employees.* Do executives and management personnel make it clear that safe behavior is the expected behavior in all cases?
5. *Expectations employees have of management.* Are employees encouraged to make their views known about the quality of the work environment?
6. *Effects of internal peer pressure on safety.* Does peer pressure among workers support or undermine safety?
7. *Unwritten rules that are widely accepted.* Do the organization's unwritten rules support or undermine safety?
8. *How conflict about safety is handled.* When conflicts arise between productivity and safety, are they settled in favor of safety?

These questions make the critical point that the cultural elements of safety and health are part of an organization's larger corporate culture, not some separate and distinct component that stands alone. Safety and health should be so fully integrated into an organization's culture that they are seen to be critical elements in the organization's ability to compete in the global marketplace.

IMPORTANCE OF HAVING A SAFETY-FRIENDLY CORPORATE CULTURE

If asked to summarize in just one word why it is so important for organizations to establish and maintain a safety-friendly corporate culture, the appropriate word would be *competition.* The market force behind the need for a safety-friendly corporate culture is competition. There are also the moral obligation of employers to provide a safe and healthy work environment for workers and the practical obligations that grow out of regulatory compliance, but competition is the foundational driver behind the need. From a business perspective, competition is the key driver because in order to survive in the global marketplace, businesses must be competitive. A business that fails to provide a work environment that is conducive to peak performance and continual improvement will not—in the long run—be able to compete, and if it cannot compete it will fail.

When the work environment is not safe, organizations begin to experience problems that sap their financial and intellectual strength, making them less able to compete in the global marketplace. These problems include accidents and injuries that erode the organization's talent pool, poor morale among employees, regulatory penalties, negative publicity, increased medical costs, increased workers' compensation costs, and lawsuits.

The competitive value of a safe and healthy workplace is an important concept for modern safety and health professionals to understand and be able to articulate because it is often the pressures of competition that cause organizations to cut corners when it comes to safety and health concerns. Such an approach is shortsighted, ill-advised, and costly in the long run.

Competition has always been an important issue for organizations, but the concept of globalization has magnified the nature and intensity of competition. It has also magnified the need for a safe and healthy work environment. In order to survive and thrive in a globally competitive environment, organizations must be innovative, adaptable, flexible, lean, productive, and able to continually improve their processes, products, people, and services. The undeniable truth is that over the long run, these necessary traits cannot be sustained in an unsafe, unhealthy work environment.

In order to win in the global marketplace, organizations must hire and retain the best workers available, and they must support them in ways that ensure consistent peak performance and continual improvement of performance. A critical aspect of this support is the provision of a safe and healthy work environment—the best type of environment for promoting, encouraging, and allowing peak performance and continual improvement.

GLOBALIZATION OF COMPETITION AND SAFETY

The term *globalization* is relatively new, but the concept is not. In its most primitive form, globalization began when ships from one country approached the shores of other country for the purpose of trade. At that time in history, ships were the *enablers* of globalization. The next phase in the ongoing development of the concept was the establishment of multinational companies—companies that opened facilities in other countries to gain the benefits of market proximity, less expensive labor, and—in some cases—less government regulation.

Enablers during the early years of the second phase of globalization were the mechanization of work that occurred as a result of the industrial revolution, improved transportation capabilities brought about by the invention of the steam engine, and improved long-distance communication based on telegraphy technology. Over the years, enhancements in these key areas—the nature of work, transportation, and communication—continued to enable ever-increasing levels of global competition. They also increased the number and types of hazards present in the workplace, making safety and health an increasingly important element in an organization's ability to compete.

With the advent of the World Wide Web, globalization entered its third and current phase. The Internet has become the ultimate enabler of globalization and, as a result, e-commerce has changed both the nature and intensity of global competition. Global fiber-optic networks and innovative software applications have transformed how work is done, where it is done, and who does it. Local companies that once competed with only local and regional rivals now find themselves in the same boat with multinational corporations competing against rivals located in such countries as India, Pakistan, China, Taiwan, South Korea, Indonesia, Bangladesh, Brunei, Singapore, Thailand, Japan, and Malaysia. These countries have substantial labor forces, comparatively low wages, fewer government regulations, and competitive education systems. In this third phase of globalization, terms such as *open-sourcing, outsourcing, off-shoring*, and *supply-chaining* are now widely used, and the concepts they represent are widely practiced. Global pressure makes it more difficult to compete, leading some organizations to respond by cutting corners, taking shortcuts, and making short-sighted decisions. This approach then undermines the quality of the work environment relating to safety and health and ultimately renders organizations unable to compete.

Many business executives in the United States who are facing dwindling markets due to global competition attribute the difficulties they face exclusively to lower wages and fewer government regulations in offshore locations. This is a serious mistake because these factors—although important—do not tell the whole story of what is behind the rise in global competition. Companies in the United States that focus solely on these specific competitive disadvantages run the risk of missing the boat when it comes to identifying strategies that will give them a competitive advantage.

Companies in emerging countries around the world have observed the so-called Japanese miracle and learned from their observations that long-term competitiveness in the global marketplace is ultimately about consistently providing superior value to customers, as illustrated in Figure 31–1. Consequently, they have adopted a whole array of quality management practices including Six Sigma, ISO 9000 registration, and the awarding of national quality prizes like Japan's Deming Prize and America's Malcolm Baldrige Award. More and more, companies in emerging industrialized nations see their future as being tied directly to the concepts of quality and value, not low wages and government regulation. Consequently, the key to outperforming competitors from these countries is to provide superior value to customers. As can be seen from Figure 31–1, superior value is a combination of superior cost, quality, and service—all of which are achieved more easily in a safe and healthy work environment and all of which are undermined by anything less.

Safety as a Cultural Imperative

Jack Welch, the longtime CEO of General Electric, was well known for making the point to GE personnel throughout the world that value is a combination of cost, quality, and service. Companies that can consistently outperform the competition in these three areas—in other words, that can consistently provide

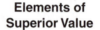

Elements of Superior Value
■ Superior quality
■ Superior cost
■ Superior service
Note: *All three of these essential elements in the formula for superior value require a safe and healthy work environment. They cannot be sustained in the long run without such an environment.*

Figure 31–1
Superior value and the quality of the work environment are inseparably intertwined.

superior value for customers—will win the daily battle of the global marketplace. This is the good news. The bad news is that few companies will be able to meet this challenge without first undergoing major cultural change. An important aspect of that change will have to be the incorporation of safety as a fully integrated, fully instilled component of a competitive corporate culture.

Consistently doing the things necessary to provide superior value—things such as providing a safe and healthy work environment—requires a certain type of corporate culture. Just as a coach cannot turn a losing team into a winner without first instilling a winning attitude, executives, managers, supervisors, and safety and health professionals cannot turn an organization into a global competitor without first instilling a winning corporate culture—one that fully integrates safety and health. Global competitiveness in business is built on a cultural foundation. Unless an organization first lays a foundation for success by establishing a winning corporate culture, its efforts to compete in the global arena are doomed from the outset. Companies that attempt to adopt the world-class practices needed to compete globally without first establishing a world-class culture are like hastily built houses constructed without a solid foundation. Furthermore, companies that fail to integrate safety and health as key elements in their larger corporate culture—a world-class practice itself—will eventually fail in their efforts to compete globally.

HOW CORPORATE CULTURES ARE CREATED

Many factors contribute to the creation of an organization's corporate culture. The value systems of executive-level decision makers are often reflected in their organization's culture. How managers treat employees and how employees at all levels interact are also factors that contribute to the organizational culture. What management expects of employees and what employees in turn expect of management are factors that contribute to an organization's culture. The stories passed along from employee to employee typically play a major role in the establishment and perpetuation of an organization's culture. All these factors can either help or hurt an organization.

If supervisors push workers to take shortcuts on safety procedures when management personnel are not looking, it is not likely that there will be a safety-friendly corporate culture. On the other hand, if supervisors insist on the safe and healthy approach in all cases regardless of who is watching, it is more likely that there will be a safety-friendly corporate culture. If none of the organization's corporate heroes are people who built a reputation for safety, it is not likely that there will be a safety-friendly corporate culture. On the other hand, if the stories that are passed down through generations of workers about the organization's corporate heroes include stories about managers, supervisors, or employees who earned a reputation for safety, it is more likely that there will be a safety-friendly corporate culture.

Corporate cultures in organizations are established based on what is expected, modeled, passed on during orientation, taught by mentors, included in training, monitored and evaluated, and reinforced through recognition and rewards. If safety is expected by management personnel and individual workers, if it is modeled by people in positions of authority, if it is stressed during the worker's initial orientation to the organization, if it is taught as the right way to do things by mentors, if it is stressed through training, if it is monitored and evaluated by supervisors, and if it is reinforced by management personnel through recognition and rewards, safety will become a fully integrated part of the organization's corporate culture.

WHAT A SAFETY-FRIENDLY CORPORATE CULTURE LOOKS LIKE

Part of the process of establishing a safety-friendly corporate culture is developing an understanding of what one looks like. This is a lot like a person who wants to lose weight taping a picture of a role model to the bathroom mirror. The picture serves not only as a constant reminder of the desired goal, but also as a measurement device that indicates when a goal has been met. If a picture of an organization with a safety-friendly corporate culture could be taped to an organization's wall for all employees to see, it would have the following characteristics:

- Widely shared agreement among key decision makers that providing a safe and healthy work environment is an essential competitive strategy.
- Emphasis on the importance of human resources to the organization and the corresponding need to protect them from hazards.
- Ceremonies to celebrate safety and health–related successes.
- Widely shared agreement that the work environment that is most conducive to peak performance and continual improvement is a safe and healthy work environment.

- Recognition and rewards given to high-performing workers and teams include safety and health–related performance on the job.
- Strong customer focus that includes product safety as a critical concern.
- Insistence on safety and health as part of supplier relations.
- Effective internal network for communicating safety and health information and expectations.
- Informal rules of behavior that promote safe and healthy work practices.
- Strong pro-safety corporate value system as set forth in the strategic plan.
- High expectations and standards for performance relating to safety and health.
- Employee behavior that promotes safe and healthy work practices.

TEN STEPS FOR ESTABLISHING A SAFETY-FRIENDLY CORPORATE CULTURE

The process for establishing a safety-friendly corporate culture in an organization consists of 10 broad steps. Those steps are as follows:

1. *Understand* the need for a safety-friendly corporate culture.
2. *Assess* the current corporate culture as it relates to safety.
3. *Plan* for a safety-friendly corporate culture.
4. *Expect* appropriate safety-related behaviors and attitudes.
5. *Model* the desired safety-related behaviors and attitudes.
6. *Orient* personnel to the desired safety-friendly corporate culture.
7. *Mentor* personnel in the desired safety-related behaviors and attitudes.
8. *Train* personnel in the desired safety-related behaviors and attitudes.
9. *Monitor* safety-related behavior and attitudes at all levels.
10. *Reinforce and maintain* the desired safety-friendly corporate culture.

Understand the Need for a Safety-Friendly Corporate Culture

Personnel at all levels need to be shown that providing a safe and healthy work environment is an important responsibility of management. Everyone from the CEO of the organization to the newest employee should understand and be able to articulate the following factors that support the need for a safety-friendly corporate culture in organizations.

- An organization's corporate culture determines the normal and accepted way things are done in the organization. Consequently, if the normal and accepted way is to be the safe and healthy way, the organization must have a safety-friendly corporate culture.
- In the same way that the work practices of individuals become habitual, the work practices of organizations become cultural; that is, they become ingrained and codified in the organization's unwritten rules. They become

the way things are done when "the boss isn't looking." In order for the way people work when not closely supervised to be the safe and healthy way, an organization must establish and maintain a safety-friendly corporate culture.

Assess the Current Corporate Culture as It Relates to Safety and Health

Does an organization have a safety-friendly corporate culture? This is a good question to ask and the answer should never be assumed—either yes or no. The answer to this question, no matter how good or bad the organization's safety and health record might be, should be the result of a thorough assessment. Figure 31–2 is an example of an assessment instrument that can be used as is or modified to fit the specific needs of an individual organization in conducting an assessment of the safety component of its corporate culture.

An instrument such as the one in Figure 31–2 should be widely distributed among personnel at all levels of the organization. Although management personnel should complete the instrument, do not make the mistake of having only management personnel complete it. All personnel should be given the opportunity to complete the instrument anonymously to ensure truthful answers.

The organization's chief safety and health professional should be the facilitator of the assessment process. It is a good idea to ask respondents to indicate whether they are a manager, supervisor, or employee so that the perceptions of the various employment categories can be compared. It is not uncommon for management to view things one way and employees to view them quite another. Larger organizations might want to take the time to create an electronic distribution, collection, and tabulation system for conducting the assessment.

Plan for a Safety-Friendly Corporate Culture

Once the organization's chief safety and health professional has collected and tabulated the results of the assessment shown in Figure 31–2, the results should be used as the basis for planning for the establishment of a safety-friendly corporate culture or for enhancing one that already exists but has weaknesses. For example, assume that the results of the assessment show that employees are not recognized and rewarded for working safely. A planning goal to correct this deficiency might read as follows:

Revise the organization's employee recognition and reward system to include safety and health criteria as part of the selection process.

The plan for establishing a safety-friendly corporate culture or for enhancing one that has weaknesses should be based on the results of a comprehensive and thorough assessment.

Assessment Instrument
Safety-Related Aspects of the Corporate Culture

The executive management team is attempting to determine if our organization has a *safety-friendly* corporate culture. Please assist us in the endeavor by indicating your responses to the following questions using the code provided below:

> CF = Completely false
> SF = Somewhat false
> CT = Completely true
> ST = Somewhat true
> XX = Do not know or not applicable

_____ Safety and health are high priorities in this organization.

_____ Key decision makers view providing a safe and healthy work environment as a positive competitive strategy.

_____ Our organization considers employees to be valuable assets who should be protected from workplace hazards.

_____ Employees are recognized and rewarded for working safely.

_____ Ceremonies are held to celebrate excellence in maintaining a positive safety and health record.

_____ Safety and health are critical factors in all decisions made in our organization.

_____ Personnel at all levels agree that the most conducive environment for peak performance and continual improvement is a safe and healthy work environment.

_____ Managers and supervisors make it clear by their words and actions that the safe way is the right way.

_____ Employees are encouraged to speak out when they have concerns about safety and health issues.

_____ Internal peer pressure and unwritten rules support working safely.

_____ Managers and supervisors are good role models for safe behavior on the job.

_____ Our organization focuses on product safety for customers.

Figure 31–2
Always assess the existing culture—never presume to know what it is.

Expect Appropriate Safety-Related Behaviors and Attitudes

If you want people to perform well, you must first have high expectations. It is the same with safety-related behaviors and attitudes. If you want people to work safely, you have to let them know that safety is expected. There are several ways that organizations can let employees know what is expected of them, including safety-related expectations. These include job descriptions, team charters, and the examples set by supervisors and managers.

Job descriptions should contain at least one statement that clearly shows that the employees are expected to work safely and to help the team as well as the larger organization maintain a safe and healthy work environment. Such a statement might read as follows:

> Employees in this position are expected to (1) comply with all applicable safety and health–related rules, regulations, and procedures; (2) work safely themselves and help their teammates work safely; (3) assist supervisors and managers in identifying and mitigating potentially hazardous conditions; and (4) help the organization maintain a safe and healthy work environment.

Team charters are also an excellent way to show employees that safe and healthy work practices are expected. Not all organizations use team charters, but all should. A team charter explains the mission of the team in question, enumerates its overall goals, and lists the expectations of team members. For example, Figure 31–3 illustrates a team charter for the Safety and Health Department of a private sector company that competes in the global marketplace.

You can see that any new employee assigned to this company's Safety and Health Department (team) would understand from the outset the team's mission, overall goals, and expectations. Under "Expectations of Team Members" you can see that team members are not only expected to work safely themselves, but also to be role models of safe work behavior.

Model the Desired Safety-Related Behaviors and Attitudes

One of the worst mistakes supervisors and managers can make is to say to employees with either verbally or through actions, "Do as I say, not as I do." Nothing speaks louder to employees than the examples—good or bad—set by supervisors and managers. Consequently, if people in positions of authority want employees to work safely, they must set a consistently positive model for doing so themselves. In addition, they must set an example of making decisions, expending resources, and managing the organization in ways that clearly say "Safety is a priority here."

Orient Personnel to the Desired Safety-Related Behaviors and Attitudes

Organizations often miss out on an excellent opportunity to get employees started off on the right foot after they are hired. This opportunity is the organization's orientation for new employees. Too often orientations for new employees are little more than filling out forms, choosing insurance program options, and learning how to navigate the facility successfully. This is unfortunate because

---

Team Charter

Safety and Health Department

Mission

The mission of the Safety and Health Department at ABC, Inc., is to assist our company in achieving its vision of global market dominance by assisting management in providing a safe and healthy workplace that promotes peak performance and continual improvement.

Overall Team Goals

1. Prevent work-related accidents, injuries, and illnesses.
2. Minimize the costs of workers' compensation.
3. Provide workstations and conditions that are ergonomically sound.
4. Provide a workplace that is as stress free as possible.
5. Ensure that management personnel consider the safety and health ramifications of policies, procedures, processes and decisions before implementing them.
6. Ensure that all employees understand how to perform their jobs in a safe and healthy manner.
7. Ensure that supervisors know how to monitor daily work practices from the perspective of safety and health and that they follow through and consistently monitor these practices.
8. Ensure that all personnel know how to do their part in accident prevention and hazard management.

Expectations of Team Members

1. All team members are expected to interact with each other in mutually supportive ways that promote peak performance and continual improvement.
2. All team members are expected to be role models whose behavior on the job is an exemplary role model of safe and healthy work practices.
3. All team members are expected to be critical thinkers.
4. All team members are expected to ensure that when they disagree with others on the job, they do not become disagreeable.
5. All team members are expected to be able to articulate in all cases why the safe way is the productive, competitive way to perform job tasks.

Figure 31–3
Sample team charter.

the only time the organization can make a first impression on new employees is during its orientation.

Anything and everything that is relevant to the organization's corporate culture should be introduced and explained during the new employee orientation sessions—including expectations relating to safety and health. Human resources personnel who conduct the orientation sessions should be encouraged to emphasize that in this organization the right way is the safe way. In addition, a representative of the organization's safety and health department should participate in the sessions.

Mentor Personnel in the Desired Safety-Related Behaviors and Attitudes

Once personnel have completed a comprehensive orientation, the next step is to assign them an experienced mentor who exemplifies the desired safety-related behaviors and attitudes. Mentors help guide inexperienced personnel until they gain the experience necessary to work safely without assistance, but even more important, they also help them develop a safety-first attitude. Mentors answer questions, make suggestions, and provide guidance, but the most important responsibility of mentors is to set a positive example, an example that includes a safety-first approach.

Train Personnel in the Desired Safety-Related Behaviors and Attitudes

There are two fundamental principles of good management that apply when trying to establish a safety-friendly corporate culture. The first is that you should never expect employees to do anything they have not been trained to do. The second is that you should never assume that employees know how to do anything without having been trained. If you want employees to work safely, teach them how. If you want employees to have the right attitudes toward safety and health, teach them what such an attitude looks like and the practical applications of such an attitude. Do not assume that employees know how to work safely—teach them how.

Monitor and Evaluate Safety-Related Behaviors and Attitudes at All Levels

A principle of effective supervision is "You get the behavior you accept." Supervisors and managers who allow their direct reports to get away with unsafe work practices are saying, "Your unsafe behavior is acceptable to me." Letting unsafe work practices go unchallenged and uncorrected is the same thing as approving of them. Consequently, it is critical that supervisors and managers monitor their direct reports and correct all unsafe work practices immediately. Another reason for monitoring employees and correcting them immediately is that work practices become habitual. Once people become accustomed to doing a task a given way, that way becomes a habit and habits are hard to break. If the habitual way is the unsafe way, the employee in question is heading down a one-way street to disaster.

In addition to monitoring on a daily basis, it is important to make safety and health–related behaviors part of formal performance evaluations. The performance appraisal instrument should have at least one criterion about safety and health such as the one shown in Figure 31–4.

There is a management principle that says, "If you want performance to improve, measure it." This is why it is important to evaluate the safety and health–related performance of workers. By measuring how well they work in

Figure 31–4
Sample safety-related criterion for a performance appraisal instrument.

> This employee exemplifies our organization's motto that "the safe way is the right way."
>
> ___ Always
>
> ___ Usually
>
> ___ Seldom
>
> ___ Never

terms of safety and health, supervisors and managers have the hard data needed in order to make improvements.

Reinforce and Maintain the Safety-Friendly Corporate Culture

Just as employees should never stop working safely no matter how good their team's safety record might be, organizations should never stop doing what is necessary to maintain a safety-friendly corporate culture. Such a culture is not a goal an organization achieves and then moves on to other matters. It is a state of being that must be reinforced constantly or it will be lost. What follows are some strategies organizations can use to reinforce and maintain their safety-friendly corporate culture once it has been established:

1. Reward safe work behavior by making it an important factor when promoting workers to higher positions.
2. Reward safe work behavior by making it an important factor when giving workers wage increases.
3. Reward safe behavior by making it an important factor when giving performance incentive awards to workers.
4. Recognize safe work behavior by making it an important criterion when singling out workers or teams for recognition awards.
5. Encourage supervisors to verbally and publicly recognize workers who are doing their jobs safely every day as they monitor work performance.

By applying the strategies explained in this section, organizations can establish and maintain a safety-friendly corporate culture that will contribute greatly to making them more competitive in the global marketplace.

SUMMARY

1. A safety-friendly corporate culture is one in which the tacit assumptions, beliefs, values, attitudes, expectations, and behaviors that are widely shared and accepted in an organization support the establishment and maintenance of a safe and healthy work environment for all personnel at all levels.

2. Evidence of a safety-friendly corporate culture exists in an organization's priorities, how people in the organization succeed, how decisions are made, expectations management has of employees, expectations employees have of management, effects of internal peer pressure, unwritten rules that are widely accepted, and how conflict about safety is handled.

3. Having a safety-friendly corporate culture is important to organizations because it is the right thing to do and because it contributes to more effective regulatory compliance. However, the most fundamental reason for having a safety-friendly corporate culture is competition. The most effective way for an organization to succeed in the global marketplace is to consistently provide superior value to its customers. This is achieved by consistently providing superior quality, cost, and service. All three of these critical elements of superior value require a safe and healthy work environment and cannot be achieved in the long run without such an environment.

4. The globalization phenomenon has only served to increase the intensity of the competition organizations are subjected to on a daily basis. As competition continues to become more and more intense, some organizations take the shortsighted approach and begin to cut corners and take shortcuts when it comes to safety and health. In the long run such an approach only makes it more difficult to compete as regulatory penalties, accidents and injuries, insurance costs, workers' compensation costs, lawsuits, poor morale among workers, negative publicity, and all the other costs of an unsafe work environment begin to sap the financial and intellectual strength of an organization. Consequently, companies that attempt to adopt the world-class practices needed to compete in the global marketplace without first establishing a world-class corporate culture—a major element of which is making safety and health high priorities—are like houses built without a foundation.

5. Corporate cultures are established in organizations based on what is expected, modeled, passed on during orientation, taught by mentors, included in training, monitored and evaluated, and reinforced through recognition and rewards.

6. When a safety-friendly corporate culture exists in an organization there is widely shared agreement among decision makers that a safe and healthy work environment is essential to success; an emphasis on protecting valuable human resources from on-the-job hazards; ceremonies to celebrate safety and health-related successes; widely shared agreement that a safe and healthy work environment is conducive to peak performance and continual improvement; recognition of safe work behavior; rewards for safe work behavior; a strong customer focus that includes product safety as a critical concern; insistence on safety as part of supplier relations; an effective internal network for communicating safety and health information and expectations; informal rules of behavior that promote safe and healthy work practices; a strong prosafety value system set forth in the strategic plan; high expectations and high standards for performance relating to safety and health; and employee behavior that promotes safe and healthy work practices.

7. The 10 steps for establishing a safety-friendly corporate culture are understand, assess, plan, expect, model, orient, mentor, train, monitor, and reinforce and maintain.

KEY TERMS AND CONCEPTS

Assess	Plan
Attitudes	Reinforce
Behaviors	Safety culture
Beliefs	Safety-friendly corporate culture
Competition	Superior cost
Expect	Superior quality
Expectations	Superior service
Globalization	Superior value
Mentor	Tacit assumptions
Model	Train
Monitor	Understand
Orient	Values

REVIEW QUESTIONS

1. Define the concept of the safety-friendly corporate culture.
2. Why is the term *safety culture* a misnomer?
3. List at least five ways (evidence) to tell if an organization has a safety-friendly corporate culture.
4. Why is it important for organizations to have a safety-friendly corporate culture?
5. How does the concept of globalization factor into the need for a safety-friendly corporate culture?
6. How does the concept of superior value factor into the need for a safety-friendly corporate culture?
7. Describe how corporate cultures are established.
8. What does a safety-friendly corporate culture look like?
9. List and explain each of the 10 steps for establishing a safety-friendly corporate culture.

GLOSSARY

abatement period The amount of time given an employer to correct a hazardous condition that has been cited.

ability to pay Applies when there are a number of defendants in a case but not all have the ability or means to pay financial damages.

absorption Passage through the skin and into the bloodstream.

acceleration Increase in the speed of a falling object before impact.

accident/incident theory Theory of accident causation in which overload, ergonomic traps, or a decision to err lead to human error.

accident prevention The act of preventing a happening that may cause loss or injury to a person.

accident rate A fixed ratio between the number of employees in the workforce and the amount who are injured or killed every year.

accident report Records the findings of an accident investigation, the cause or causes of an accident, and recommendations for corrective action.

accident scene The area where an accident occurred.

accidents Unexpected happenings that may cause loss or injuries to people who are not at fault for causing the injuries.

acclimatization Process by which the body becomes gradually accustomed to heat or cold in a work setting.

accommodation The ability of the eye to become adjusted after viewing the VDT to be able to focus on other objects, particularly objects at a distance.

adjustable guard A device that provides a barrier against a variety of different hazards associated with different production operations.

administrative controls Procedures that are adopted to limit employee exposure to hazardous conditions.

aerosols Liquid or solid particles so small that they can remain suspended in air long enough to be transported over a distance.

aerospace engineering Program of study incorporating a solid foundation of physical and mathematical fundamentals that provides the basis for the development of the engineering principles essential to the understanding of both atmospheric and extra-atmospheric flight.

agreement settlement   The injured employee and the employer or its insurance company work out an agreement on how much compensation will be paid and for how long.

air dose   Dose measured by an instrument in the air at or near the surface of the body in the area that has received the highest dosage of a toxic substance.

altitude sickness   A form of hypoxia associated with high altitudes.

ampere   The unit of measurement for current.

analysis and evaluation   All potential solutions to a problem are subjected to scientific analysis and careful evaluation.

ancestry   A person's line of descent.

anesthetics   In carefully controlled dosages, anesthetics can inhibit the normal operation of the central nervous system without causing serious or irreversible effects.

appeals process   The process of challenging an OSHA standard as it goes through the adoption process and after formal adoption.

artificial environment   One that is fully created to prevent a definite hazardous condition from affecting people or material.

aseptic necrosis   A delayed effect of decompression sickness.

asphyxiants   Substances that can disrupt breathing so severely as to cause suffocation.

associate degree   A two-year college degree.

assumption of risk   Based on the theory that people who accept a job assume the risks that go with it. It says employees who work voluntarily should accept the consequences of their actions on the job rather than blaming the employer.

audiogram   The results of an audiometric test to determine the noise threshold at which a subject responds to different test frequencies.

audiometric testing   Measures the hearing threshold of employees.

auto-ignition temperature   The lowest temperature at which a vapor-producing substance or a flammable gas will ignite even without the presence of a spark or a flame.

automatic ejection   A system that ejects work pneumatically and mechanically.

automatic feed   A system that feeds stock to the machine from rolls.

bachelor's degree   A four-year college degree.

barometer   Scientific device for measuring atmospheric pressure.

bends   Common name for decompression sickness.

best-ratio approach   People are basically good and under the right circumstances behave ethically.

biological hazards   Hazards from molds, fungi, bacteria, and insects.

black-and-white approach   Right is right, wrong is wrong, and circumstances are irrelevant.

blackball   To ostracize an employee.

body surface area (BSA)   The amount of surface area that is covered with burns.

bonding   Used to connect two pieces of equipment by a conductor. Also, involves eliminating the difference in static charge potential between materials.

Boyle's law   States that the product of a given pressure and volume is constant with a constant temperature.

cancer   A malignant tumor.

carcinogen   Any substance that can cause a malignant tumor or a neoplastic growth.

carpal tunnel syndrome (CTS)   An injury to the median nerve inside the wrist.

case law   Serves the purpose of establishing precedents that can guide future decisions.

causal relationship   A situation in which an action leads to a certain result.

ceiling   The level of exposure that should not be exceeded at any time for any reason.

central factor   The main issue or factor in a problem or act.

chemical burn injuries   Burn damage to the skin caused by chemicals such as acids and alkalies.

chemical engineer   An engineer concerned with all the physical and chemical changes of matter to produce economically a product or result that is useful to humans.

chemical hazards   Include mists, vapors, gases, dusts, and fumes.

chokes   Coughing and choking, resulting from bubbles in the respiratory system.

circadian rhythm   Biological clock.

circuit tester   An inexpensive piece of test equipment with two wire leads capped by probes and connected to a small bulb.

claim notice   Notice filed to indicate an expectation of workers' compensation benefits owed.

closed environment   One that is completely or almost completely shut off from the natural environment.

closing conference   Involves open discussion between the compliance officer and company and employee representatives.

code   A set of standards, rules, or regulations relating to a specific area.

Code of Hammurabi   Developed around 2000 B.C. during the time of the Babylonians, the ruler Hammurabi developed this code which encompassed all the laws of the land at that time. The significant aspect is that it contained clauses dealing with injuries, allowable fees for physicians, and monetary damages assessed against those who injured others.

coefficient of friction   A numerical correlation of the resistance of one surface against another surface.

cold stress   Physical or mental stress that results from working in cold conditions.

combination theory   The actual cause of an accident may be explained by combining many models.

combustible substance   Any substance with a flash point of 37.8°C (100°F) or higher.

combustion   A chemical reaction between oxygen and a combustible fuel.

combustion point   The temperature at which a given fuel can burst into flame.

commonsense test   Requires person to listen to what instincts and common sense are telling him or her.

competitiveness   The ability to succeed and prosper consistently in the marketplace whether it is local, regional, national, or global.

conduction   The transfer of heat between two bodies that are touching or from one location to another within a body.

conductors   Substances that have many free electrons at room temperature and can pass electricity.

conference method   A problem-solving teaching method in which the trainer serves as a facilitator rather than a teacher.

confined space   An area with limited means of egress that is large enough for a person to fit into, but is not designed for occupancy.

consistency   The rules are enforced in the same manner every time with no regard to any outside factors.

continuity tester   May be used to determine whether a conductor is properly grounded or has a break in the circuit.

contributory negligence   An injured worker's own negligence contributed to the accident. If the actions of employees contributed to their own injuries, the employer is absolved of any liability.

controlled environment   A natural or induced environment that has been changed in some way to reduce or eliminate potential environmental hazards.

convection   The transfer of heat from one location to another by way of a moving medium.

convergence   The coordinated turning of the eyes inward to focus on a nearby point or object.

Cooperative Safety Congress (CSC)   The CSC was a result of the Association of Iron and Steel Electrical Engineers' (AISEE) desire to have a national conference on safety. The first meeting took place in Milwaukee in 1912. The meeting planted the seeds for the eventual establishment of the National Safety Council (NSC).

corium   The inner layer of human skin.

cost   The amount of money needed to produce a product, not to be confused with "price," which is the amount of money needed to purchase a product after the cost has been marked up.

cost allocation   Spread the cost of workers' compensation appropriately and proportionately among industries ranging from the most to the least hazardous.

creeps   Caused by bubble formation in the skin, which causes an itchy, crawling, rashy feeling in the skin.

critical burns   Second-degree burns covering more than 30 percent of the body and third-degree burns covering over 10 percent are considered critical.

crushing   Occurs when a part of the body is caught between two hard surfaces that progressively move together, thereby crushing anything between them.

cutis   The inner layer of human skin.

cutting   Occurs when a body part comes in contact with a sharp edge.

Dalton's law of partial pressures   States that in a mixture of theoretically ideal gases, the pressure exerted by the mixture is the sum of the pressures exerted by each component gas of the mixture.

damages   Financial awards assigned to injured parties in a lawsuit.

DBBS   See *Division of Biology and Biomedical Science.*

death rates   A fixed ratio between the number of employees in the workforce and the number that are killed each year.

decibel   The unit applied when measuring sound. One-tenth of a bel. One decibel is the smallest difference in the level of sound that can be perceived by the human ear.

decompression sickness   Can result from the decompression that accompanies a rapid rise from sea level to at least 18,000 feet or a rapid ascent from around 132 to 66 feet underwater.

demonstration method   The instructor shows students how to perform certain skills or tasks.

dermis   The inner layer of human skin.

design flaw   A defect in a product.

design process   A plan of action for reaching a goal.

direct settlement   The employer or its insurance company begins making what it thinks are the prescribed payments.

discovery period   Period in which evidence is collected, depositions are taken, and products are examined.

Division of Biology and Biomedical Science (DBBS)   Conducts research in the areas of toxicology, behavioral science, ergonomics, and the health consequences of various physical agents.

Division of Training and Manpower Development (DTMD)   Implements Section 21 of the OSH Act, which sets forth training and education requirements.

document and communicate   Engineering drawings, detailed calculations, and written specifications document the design of a product and communicate its various components to interested parties.

domino theory   Injuries are caused by the action of preceding factors. Removal of the central factor negates the action of the preceding factors and, in so doing, prevents accidents and injuries.

dose   The amount of ionizing radiation absorbed per unit of mass by part of the body or the whole body.

dose threshold   The minimum dose required to produce a measurable effect.

dosimeter   Provides a time-weighted average over a period such as one complete work shift.

double insulation   A means of increasing electrical equipment safety.

drowning   The act of suffocating due to submersion in water.

DTMD   See *Division of Training and Manpower Development.*

dusts   Various types of solid particles that are produced when a given type of organic or inorganic material is scraped, sawed, ground, drilled, heated, crushed, or otherwise deformed.

dysbarism   The formation of gas bubbles due to rapid ambient pressure reduction.

ego strength   An employee's ability to undertake self-directed tasks and to cope with tense situations.

electrical engineering A science-oriented branch of engineering primarily concerned with all phases and development of the transmission and utilization of electric power and intelligence.

electrical hazards Potentially dangerous situations related to electricity (e.g., a bare wire).

electrical system grounding Achieved when one conductor of the circuit is connected to the earth.

electricity The flow of negatively charged particles called electrons through an electrically conductive material.

electrolytes Minerals that are needed for the body to maintain the proper metabolism and for cells to produce energy.

electromechanical devices Contact bars that allow only a specified amount of movement between the worker and the hazard.

electrons Negatively charged particles.

emergency action plan (EAP) A collection of small plans for every anticipated emergency (e.g., fire, hurricane, chemical spill, and so on).

emergency coordinator A person who is clearly identified in a company emergency response plan as the responsible party for a specific type of emergency situation.

emergency notification Requires that chemical spills or releases of toxic substances that exceed established allowable limits be reported to appropriate LEPCs and SERCs.

emergency planning Requires that communities form local emergency planning committees and that states form state emergency response commissions.

emergency response management team Composed of shift emergency directors and a full-time fire chief. Unifies all emergency groups and equipment into a single, coordinated effort.

emergency response network A network of emergency response teams that covers a designated geographical area and is typically responsible for a specific type of emergency.

emergency response plan A written document that identifies the different personnel or groups that respond to various types of emergencies and, in each case, who is in charge.

emergency response team A special team that responds to general and localized emergencies to facilitate personnel evacuation and safety, shut down building services and utilities as needed, work with responding civil authorities, protect and salvage company property, and evaluate areas for safe reentry.

employee A person who is on the company's payroll, receives benefits, and has a supervisor.

employee responsibilities Specific obligations of employees relating to safety and health as set forth in OSHA 2056 (revised).

Employee Retirement Income Security Act (ERISA) Protects the benefits of employees by prohibiting actions taken against them based on their eligibility for benefits.

employee rights   Protections that an employee has against punishment for complaining to an employer union, OSHA, or any governmental agency about hazards on the job.

employer-biased laws   A collection of laws that favored employers over employees in establishing a responsibility for workplace safety.

employer liability   In 1877, the Employer's Liability Law was passed and established the potential for employers to be liable for accidents that occurred in the workplace.

employer responsibilities   Specific responsibilities of employers as specified in OSHA 2056 (revised).

employer rights   The rights of an employer as specified in OSHA 2056 (revised).

environment   The aggregate of social and cultural conditions that influence the life of an individual.

environmental engineer   An individual whose job is to protect and preserve human health and the well-being of the environment.

environmental factors   Characteristics of the environment in which an employee works that can affect his or her state of mind or physical conditions such as noise or distractions.

environmental heat   Heat that is produced by external sources.

environmental management system (EMS)   That component of an organization with primary responsibility for leading, planning, organizing, and controlling as these functions relate specifically to an organization's processes, products, or services and the impact that they have on the environment.

enzyme linked immunosorbent assay (ELISA)   Screening test currently used to ensure the accuracy of an HIV antibody test.

epidemiological theory   Holds that the models used for studying and determining epidemiological relationships can also be used to study causal relationships between environmental factors and accidents or diseases.

epidermis   The outer layer of human skin.

ergonomic hazards   Workplace hazards related to the design and condition of the workplace. For example, a workstation that requires constant overhead work is an ergonomic hazard.

ergonomic trap   An unsafe condition unintentionally designed into a workstation.

ergonomics   The science of conforming the workplace and all its elements to the worker.

ERISA   See *Employee Retirement Income Security Act.*

ethical behavior   That which falls within the limits prescribed by morality.

ethics   The application of morality within a context established by cultural and professional values, social norms, and accepted standards of behavior.

expert system   A computer programmed to solve problems.

expiration   Occurs when air leaves the lungs and the lung volume is less than the relaxed volume, increasing pressure within the lungs.

explosion   A very rapid, contained fire.

explosive range   Defines the concentrations of a vapor or gas in air that can ignite from a source.

exposure ceiling   Refers to the concentration level of a given substance that should not be exceeded at any point during an exposure period.

exposure threshold   A specified limit on the concentration of selected chemicals. Exposure to these chemicals that exceeds the threshold is considered hazardous.

extraterritorial employees   Those who work in one state but live in another.

failure mode and effects analysis   A formal, step-by-step analytical method that is a spin-off of reliability analysis, a method used to analyze complex engineering systems.

falls   Category of accidents in which an employee unintentionally drops under the force of gravity from a surface that is elevated.

false negative   A result that shows no HIV antibodies in people who, in reality, are infected.

false positive   A result that shows the presence of HIV antibodies when, in reality, no such antibodies exist.

fault tree analysis   An analytical methodology that uses a graphic model to display visually the analysis process.

feedback   Employee's opinions of a project or an employer's opinion about an employee's job performance.

fellow servant rule   Employers are not liable for workplace injuries that result from negligence of other employees.

fire   A chemical reaction between oxygen and a combustible fuel.

fire hazards   Conditions that favor the ignition and spread of fire.

fire point   The minimum temperature at which the vapors or gas in air can ignite from a source of ignition.

fire-related losses   Loss of life caused by fires and instances related to fires such as burns, asphyxiation, falls, and so on.

first-degree burn   Results in a mild inflammation of the skin known as erythema.

fixed guards   Provide a permanent barrier between workers and the point of operation.

flame-resistant clothing   Special clothing made of materials or coated with materials that are able to resist heat and flames.

flammable substance   Any substance with a flash point below 37.8°C (100°F) and a vapor pressure of less than 40 pounds per square inch at this temperature.

flash point   The lowest temperature for a given fuel at which vapors are produced in sufficient concentrations to flash in the presence of a source of ignition.

flashing back   When a flame consumes a flammable material such as a gas traveling quickly back to its source.

flexible hours   A work / time scheduling system in which employees are allowed to adopt a work schedule that more closely matches their personal needs.

fluid loss   Depletion of necessary body fluids, primarily through perspiration.

foreign object   Any object that is out of place or in a position to trip someone or to cause a slip.

foreseeability   Concept that a person can be held liable for actions that result in damages or injury only when risks could have been reasonably foreseen.

four-step teaching method Preparation, presentation, application, and evaluation.

four-to-one ratio Base one foot away from the wall for every four feet between the base and the support point of a ladder.

free environment One that does not interfere with the free movement of air.

freeze The inability to release one's grip voluntarily from a conductor of electricity.

frequency The number of cycles per second.

friable asbestos Asbestos that is in a state of crumbling deterioration. When asbestos is in this state, it is most dangerous.

front-page test Encourages persons to make a decision that would not embarrass them if it were printed as a story on the front page of their hometown newspaper.

full-potential approach People are responsible for realizing their full potential within the confines of morality.

fuses Consist of a metal strip or wire that will melt if a current above a specific value is conducted through the metal.

gases Formless fluids that are airborne and can be toxic.

gates Provide a barrier between the danger zone and workers.

global marketplace The worldwide economic market in which many companies must compete for business.

good housekeeping Proper cleaning and maintenance of a work area.

ground fault When the current flow in the hot wire is greater than the current in the neutral wire.

ground fault circuit interrupter Can detect the flow of current to the ground and open the circuit, thereby interrupting the flow of current.

grounded conductor Neutral wire.

grounding conductor The groundwire.

hand-arm vibration (HAV) syndrome A form of Raynaud's syndrome that afflicts workers who use vibrating power tools frequently over time.

harmful equipment A work environment in which physical or psychological factors exist that are potentially hazardous.

hazard A condition with the potential of causing injury to personnel, damage to equipment or structures, loss of material, or lessening of the ability to perform a prescribed function.

hazard analysis A systematic process for identifying hazards and recommending corrective action.

hazard and operability review An analysis method that was developed for use with new processes in the chemical industry.

hazardous condition A condition that exposes a person to risks.

hazardous waste reduction Reducing the amount of hazardous waste generated and, in turn, the amount introduced into the waste stream through the process of source reduction and recycling.

health physicist Concerned primarily with radiation in the workplace.

hearing conservation Systematic procedures designed to reduce the potential for hearing loss in the workspace. Employers are required by OSHA to implement hearing conservation procedures in settings where the noise level exceeds a time-weighted average of 85 dBA.

heart disease   An abnormal condition of the heart that impairs its ability to function.

heat burn injuries   Burn damage to the skin caused by high temperatures from some of the following activities: welding, cutting with a torch, and handling tar or asphalt.

heat cramps   A type of heat stress that occurs as a result of salt and potassium depletion.

heat exhaution   A type of heat stress that occurs as a result of water or salt depletion.

heat rash   A type of heat stress that manifests itself as small raised bumps or blisters that cover a portion of the body and give off a prickly sensation that can cause discomfort.

heat stroke   A type of heat stress that occurs as a result of a rapid rise in the body's core temperature.

heat transfer   The spread of heat from a source to surrounding materials or objects by conduction, radiation, or convection.

high-radiation area   Any accessible area in which radiation hazards exist that could deliver a dose in excess of 100 millirem within one hour.

horizontal work area   One that is designed and positioned so that it does not require the worker to bend forward or to twist the body from side to side.

human error   A mistake that is made by a human, not a machine.

human error analysis   Used to predict human error and not as an after-the-fact process.

human factors theory   Attributes accidents to a chain of events ultimately caused by human error.

humidification   Adding moisture to the air to reduce electrical static.

hyperoxia   Too much oxygen or oxygen breathed under too high a pressure.

hypothermia   The condition that results when the body's core temperature drops to dangerously low levels.

ignition temperature   The temperature at which a given fuel can burst into flame.

impact accidents   Involve a worker being struck by or against an object.

impulse noise   Consists of transient pulses that can occur repetitively or non-repetitively.

inanimate power   Power that is lacking life or spirit. During the Industrial Revolution, humans and animals were replaced with inanimate power (e.g., steam power).

inappropriate activities   Activities undertaken with disregard for established safety procedures.

inappropriate response   A response in which a person disregards an established safety procedure.

income replacement   Replacement of current and future income (minus taxes) at a ratio of two-thirds (in most states).

independent contractor   A person who accepts a service contract to perform a specific task or set of tasks and is not directly supervised by the company.

indirect costs   Costs that are not directly identifiable with workplace accidents.

induced environments   Environments that have been affected in some way by human action.

industrial hygiene   An area of specialization in the field of industrial safety and health that is concerned with predicting, recognizing, assessing, controlling, and preventing environmental stressors in the workplace that can cause sickness or serious discomfort. Concerns environmental factors that can lead to sickness, disease, or other forms of impaired health.

industrial hygiene chemist   Often hired by companies that use toxic substances to test the work environment and the people who work in it. Industrial hygiene chemists have the ability to detect hazardous levels of exposure or unsafe conditions so that corrective and preventive measures can be taken at an early stage.

industrial hygienist   A person having a college or university degree in engineering, chemistry, physics, medicine, or related physical or biological sciences who, by virtue of special studies and training, has acquired competence in industrial hygiene.

industrial medicine   A specialized field that is concerned with work-related safety and health issues.

industrial place accidents   Accidents that occur at one's place of work.

industrial safety engineer   An individual who is a safety and health professional with specialized education and training. Responsible for developing and carrying out those aspects of a company's overall safety and health program relating to his or her area of expertise.

industrial safety manager   An individual who is a safety and health generalist with specialized education and training. Responsible for developing and carrying out a company's overall safety and health program including accident prevention, accident investigation, and education and training.

industrial stress   Involves the emotional state resulting from a perceived difference between the level of occupational demand and a person's ability to cope with this demand.

infection   The body's response to contamination by a disease-producing microorganism.

ingestion   Entry through the mouth.

inhalation   Taking gases, vapors, dust, smoke, fumes, aerosols, or mists into the body by breathing in.

inspection tour   A facility tour by an OSHA compliance officer to observe, interview, and examine as appropriate.

inspiration   When atmospheric pressure is greater than pressure within the lungs, air flows down this pressure gradient from the outside into the lungs.

instructional approach   A brief action plan for carrying out the instruction.

interlocked guards   Shut down the machine when the guard is not securely in place or is disengaged.

interlocks   Automatically break the circuit when an unsafe situation is detected.

internal factors   Factors that can add a burden on a person and interfere with his or her work, such as personal problems.

ionizer   Ionizes the air surrounding a charged surface to provide a conductive path for the flow of charges.

ionizing radiation   Radiation that becomes electrically charged or changed into ions.

irritants   Substances that cause irritation to the skin, eyes, and the inner lining of the nose, mouth, throat, and upper respiratory tract.

islands of automation   Individual automated systems lacking electronic communication with other related systems.

job autonomy   Control over one's job.

job descriptions   Written specifications that describe the tasks, duties, reporting requirements, and qualifications for a given job.

job safety analysis   A process through which all the various steps in a job are identified and listed in order.

job security   A sense—real or imagined—of having the potential for longevity in a job and having a measure of control concerning that longevity.

kinetic energy   The energy resulting from a moving object.

lacrimination   The process of excreting tears.

learning objectives   Specific statements of what the learner should know or be able to do as a result of completing the lesson.

let-go current   The highest current level at which a person in contact with a conductor of electricity can release the grasp of the conductor.

liability   A duty to compensate as a result of being held responsible for an act or omission.

lifting hazard   Any factor that if not properly dealt with may lead to an injury from lifting.

lightning   Static charges from clouds following the path of least resistance to the earth, involving very high voltage and current.

line authority   The safety and health manager has authority over and supervises certain employees.

litigation   The carrying on of legal matters by judicial process.

load   A device that uses currents.

lockout/tagout system   A system for incapacitating a machine until it can be made safe to operate. "Lockout" means physically locking up the machine so that it cannot be used without removing the lock. "Tagout" means applying a tag that warns employees not to operate the machine in question.

locus of control   The perspective of workers concerning who or what controls their behavior.

lost time   The amount of time that an employee was unable to work due to an injury.

lost wages   The amount that an employee could have earned had he or she not been injured.

Machiavellianism   The extent to which an employee will attempt to deceive and confuse others.

malpractice   Negligent or improper practice.

material safety data sheets (MSDSs)   These sheets contain all the relevant information needed by safety personnel concerning specific hazardous materials.

meaninglessness   The feeling that workers get when their jobs become so specialized and so technology-dependent that they cannot see the meaning in their work as it relates to the finished product or service.

means of egress   A route for exiting a building or other structure.

mechanical engineering   The professional field that is concerned with motion and processes whereby other energy forms are converted into motion.

mechanical hazards   Those associated with power-driven machines, whether automated or manually operated.

mechanical injuries   Injuries that have occurred due to misuse of a power-driven machine.

medical expenses   Money paid to cover the costs of emergency medical response and follow-up treatment when an employee is injured.

Merit Program   Seen as a stepping-stone to recognize companies that have made a good start toward Star Program recognition.

metabolic heat   Produced within a body as a result of activity that burns energy.

mindlessness   The result of the process of dumbing down the workplace.

minor burns   All first-degree burns are considered minor as well as second-degree burns covering less than 15 percent of the body.

mirror test   Encourages people to make choices based on how they will feel about their decision when they look in the mirror.

mists   Tiny liquid droplets suspended in air.

moderate burns   Second-degree burns covering less than 30 percent of the body and third-degree burns covering less than 10 percent are considered moderate.

monetary benefits   Actual money owed to injured employees or their relatives under workers' compensation laws.

morning-after test   Encourages people to make choices based on how they will feel about their decision the next day.

MSDSs   See *Material safety data sheets.*

NACOSH   See *National Advisory Committee on Occupational Safety and Health.*

narrow band noise   Noise that is confined to a narrow range of frequencies.

National Advisory Committee on Occupational Safety and Health (NACOSH)   Makes recommendations for standards to the secretary of health and human services and to the secretary of labor.

National Council of Industrial Safety (NCIS)   Established in 1913, one year after the first meeting of the CSC. In 1915, the NCIS changed its name to the National Safety Council. It is now the premier safety organization in the United States.

National Electrical Code (NEC)   Specifies industrial and domestic electrical safety precautions.

National Institute for Occupational Safety and Health (NIOSH)   This organization is part of the Centers for Disease Control and Prevention (CDC) of the

Department of Health and Human Services (DHHS). It is required to publish annually a comprehensive list of all known toxic substances. It will also provide on-site tests of potentially toxic substances so that companies know what they are handling and what precautions to take.

natural disasters   Incidents prompted by nature such as earthquakes, hurricanes, floods, and tornadoes.

natural environment   Not human-made. It is the environment that we typically think of as Earth and all its natural components, including the ground, the water, flora and fauna, and the air.

NCIS   See *National Council of Industrial Safety.*

NEC   See *National Electrical Code.*

negative pressures   Caused by pressures below atmospheric level.

negligence   Failure to take reasonable care or failure to perform duties in ways that prevent harm to humans or damage to property.

negligent manufacture   The maker of a product can be held liable for its performance from a safety and health perspective.

neoplastic growth   Cancerous tissue or tissue that might become cancerous.

neutrons   Particles that neutralize the negative charge of electrons. They act as temporary energy repositories between positively charged particles called protons and electrons.

NIOSH   See *National Institute for Occupational Safety and Health.*

noise   Unwanted sound.

nonionizing radiation   Radiation on the electromagnetic spectrum that has a frequency of 10^{15} or less and a wavelength in meters of 3×10^{-7} or less.

nonskid footwear   Shoes that have special nonskid soles.

normlessness   The phenomenon in which people working in a highly automated environment can become estranged from society.

notice of contest   A written note stating that an employer does not wish to comply with a citation, an abatement period, or a penalty.

notice of proposed rule making   Explains the terms of the new rule, delineates the proposed changes to existing rules, or lists rules that are to be revoked.

nuclear engineering   Concerned with the release, control, and safe utilization of nuclear energy.

objectivity   Rules are enforced equally regardless of who commits an infraction from the newest employee to the chief executive officer.

occupational diseases   Pathological conditions brought about by workplace conditions or factors.

occupational health nurse   One whose job it is to conserve the health of workers in all occupations.

Occupational Safety and Health Act (OSH Act)   Passed by the United States Congress in 1970 and updated periodically since that time.

Occupational Safety and Health Administration (OSHA)   The government's administrative arm for the Occupational Safety and Health Act. It sets and revokes health and safety standards, conducts inspections, investigates problems, issues citations, assesses penalties, petitions the courts to take appropriate action against unsafe employers, provides safety training, provides injury prevention consultation, and maintains a database of safety and health statistics.

Occupational Safety and Health Review Commission (OSHRC) An independent board whose members are appointed by the president and given quasi-judicial authority to handle contested OSHA citations.

ohms Measure resistance.

opening conference An initial meeting in which an OSHA compliance officer informs company officials and employee representatives of the reason that an on-site inspection is going to occur and what to expect.

organized labor A group of employees who joined together to fight for the rights of all employees (i.e., unions).

OSHA See *Occupational Safety and Health Administration.*

OSH Act See *Occupational Safety and Health Act.*

OSHA Form 101 *Supplementary Record of Occupational Injuries and Illnesses.*

OSHA Form 300 *Log of Work-Related Injuries and Illnesses.*

OSHA Form 300A *Summary of Workplace Injuring and Illnesses.*

OSHA Form 301 *Injury and Illness Incident Report.*

OSHA Poster 2203 Explains employee rights and responsibilities as prescribed in the OSH Act.

OSHRC See *Occupational Safety and Health Review Commission.*

overexertion The result of employees working beyond their physical limits.

overload Amounts to an imbalance between a person's capacity at any given time and the load that person is carrying in a given state.

oxygen limit The amount of oxygen that must be present in a given substance in order for an explosion to occur.

participatory ergonomics An approach to intervention that combines outside experience and inside experience to design ergonomic interventions that are tailored to a specific workplace.

patent defect One that occurs in all items in a manufactured batch.

permanent partial disability The condition that exists when an injured employee is not expected to recover.

permanent total disability The condition that exists when an injured employee's disability is such that he or she cannot compete in the job market.

permanent variance An exemption from an OSHA standard awarded to an organization that can show it already exceeds the standards.

permissible exposure limit (PEL) OSHA-established exposure threshold.

personal monitoring devices Devices worn or carried by an individual to measure radiation doses received.

personal protective equipment (PPE) Any type of clothing or device that puts a barrier between the worker and the hazard (e.g., safety goggles, gloves, hard hats, and so on).

Petition for Modification of Abatement (PMA) Available to employers who intend to correct the situation for which a citation was issued, but who need more time.

photoelectric devices Optional devices that shut down the machine anytime the light field is broken.

photoelectric fire sensors   Detect changes in infrared energy that is radiated by smoke, often by the smoke particles obscuring the photoelectric beam.

physical hazards   Include noise, vibration, extremes of temperature, and excessive radiation.

PMA   See *Petition for Modification of Abatement.*

point-of-operation guards   Machine guards that provide protection right at the point where the user operates the machine.

poisoning   To be injured or killed by a harmful substance.

powerlessness   The feeling that workers have when they are not able to control the work environment.

preceding factors   Factors that led up to an accident.

predispositional characteristics   Human personality characteristics that can have a catalytic effect in causing an accident.

preliminary hazard analysis   Conducted to identify potential hazards and prioritize them according to (1) the likelihood of an accident or injury being caused by the hazard; and (2) the severity of injury, illness, or property damage that could result if the hazard caused an accident.

pressure   The force exerted against an opposing fluid or thrust distributed over a surface.

pressure hazard   A hazard caused by a dangerous condition involving pressure.

private insurance   Workers' compensation coverage purchased from a private insurance company.

problem identification   Engineers must draft a description of a problem before anything else can be done.

produce and deliver   Shop or detail drawings are developed, and the design is produced, usually as a prototype. The prototype is analyzed and tested. Design changes are made. The product is then produced and delivered.

product literature   Tells users about hazards that cannot be removed by design or controlled by guards and safety devices.

product safety auditor   Responsible for evaluating the overall organization and individual departments within it.

product safety committee   Consists of representatives within a company having a wide range of expertise. Offers support to the product safety coordinator on safety issues.

product safety coordinator   Responsible for coordinating and facilitating a product safety management program in all departments of a company.

product safety program   Purpose is to limit as much as possible a company's exposure to product liability litigation and related problems.

productivity   The concept of comparing output of goods or services to the input of resources needed to produce or deliver them.

professional societies   Typically formed for the purpose of promoting professionalism, adding to the body of knowledge, and forming networks among colleagues in a given field.

proof pressure tests   Tests in which containers are "proofed" by subjecting them to specified pressures for specified periods.

property damage   Facilities, equipment, or other nonpersonnel items damaged as a result of an accident.

proposed penalty   The initial penalty proposed by the OSHA compliance officer after an inspection tour.

protons   Positively charged particles.

proximate cause   The cause of an injury or damage to property.

psychophysiological techniques   Require simultaneous measurement of heart rate and brain waves, which are then interpreted as indexes of mental workload and industrial stress.

psychosocial questionnaires   Evaluate workers' emotions about their jobs.

public hearing   If an injured worker feels that he or she has been inadequately compensated or unfairly treated, a public hearing can be requested.

pullback devices   Pull the operator's hands out of the danger zone when the machine starts to cycle.

puncturing   Results when an object penetrates straight into the body and pulls straight out, creating a wound in the shape of the penetrating object.

quality   A measure of the extent to which a product or service meets or exceeds customer expectations.

quality management (QM)   A way of managing a company that revolves around a total and willing commitment of all personnel at all levels to quality.

quality product   One that meets or exceeds customer standards and expectations.

rad   A measure of the dose of ionizing radiation absorbed by body tissue stated in terms of the amount of energy absorbed per unit of mass of tissue.

radiant heat   The result of electromagnetic nonionizing energy that is transmitted through space without the movement of matter within that space.

radiation   Consists of energetic nuclear particles and includes alpha rays, beta rays, gamma rays, X-rays, neutrons, high-speed electrons, and high-speed protons.

radiation area   Any accessible area in which radiation hazards exist that could deliver doses (1) within one hour, a major portion of the body could receive more than 5 millirem; or (2) within five consecutive days, a major portion of the body could receive more than 100 millirem.

radiation control specialist   Monitors the radiation levels to which workers may be exposed, tests workers for levels of exposure, responds to radiation accidents, develops companywide plans for handling radiation accidents, and implements decontamination procedures when necessary.

radioactive material   Material that emits corpuscular or electromagnetic emanations as the result of spontaneous nuclear disintegration.

radio-frequency devices   Capacitance devices that brake the machine if the capacitance field is interrupted by a worker's body or another object.

reasonable risk   Exists when consumers (1) understand risk, (2) evaluate the level of risk, (3) know how to deal with the risk, and (4) accept the risk based on reasonable risk–benefit considerations.

receptacle wiring tester   A device with two standard plug probes for insertion into an ordinary 110-volt outlet and a probe for the ground.

reclamation   A process whereby potentially hazardous materials are extracted from the by-products of a process.

rehabilitation   Designed to provide the needed medical care at no cost to the injured employee until he or she is pronounced fit to return to work.

Rehabilitation Act   Enacted to give protection to people with disabilities.

rem   A measure of the dose of ionizing radiation to body tissue stated in terms of its estimated biological effect relative to a dose of one roentgen of X-rays.

repeat violation   A violation of any standard, regulation, rule, or order where, upon reinspection, a substantially similar violation is found.

repetitive motion   Short-cycle motion that is repeated continually.

repetitive strain injury (RSI)   A broad and generic term that encompasses a variety of injuries resulting from cumulative trauma to the soft tissues of the body.

resistance   A tendency to block flow of electric current.

response time   The amount of time between when an order is placed and the product is delivered.

restraint devices   Hold back the operator from the danger zone.

restricted area   Any area to which access is restricted in an attempt to protect employees from exposure to radiation or radioactive materials.

retrofit   Renovating rather than replacing.

risk analysis   An analytical methodology normally associated with insurance and investments.

role ambiguity   The condition that occurs when an employee is not clear concerning the parameters, reporting requirements, authority, or responsibilities of his or her job.

role-reversal test   Requires a person to trade places with the people affected by the decision that he or she made and to view the decision through their eyes.

safeguarding   Machine safeguarding was designed to minimize the risk of accidents of machine–operator contact.

safety and health movement   Began during World War II when all the various practitioners of occupational safety and health began to see the need for cooperative efforts. This movement is very strong today.

safety and health professional   An individual whose profession (job) is to be concerned with safety and health measures in the workforce.

safety-friendly corporate culture   Exists when the tacit assumptions, beliefs, values, attitudes, expectations, and behaviors that are widely shared and accepted in an organization support the establishment and maintenance of a safe and healthy work environment for all personnel at all levels.

safety policy   A written description of an organization's commitment to maintaining a safe and healthy workplace.

safety trip devices   Include trip wires, trip rods, and body bars that stop the machine when tripped.

SARA   See *Superfund Amendments and Reauthorization Act.*

second-degree burn   Results in blisters forming on the skin.

self-insurance   Workers' compensation coverage in which a company insures itself by building its own fund.

semiautomatic ejection   A system that ejects the work using mechanisms that are activated by the operator.

semiautomatic feed   A system that uses a variety of approaches for feeding stock to the machine.

semiconductors   Substances that are neither conductors nor insulators.

shift work   Employees work at different times of the day instead of during the same hours.

shock   A depression of the nervous system.

short circuit   A circuit in which the load has been removed or bypassed.

short-term exposure limit   The maximum concentration of a given substance to which employees may be safely exposed for up to 15 minutes without suffering irritation, chronic or irreversible tissue change, or narcosis to a degree sufficient to increase the potential for accidental injury, impair the likelihood of self-rescue, or reduce work efficiency.

sick-building syndrome   An internal environment that contains unhealthy levels of biological organisms in the air. A common cause is the introduction of unhealthy outdoor air that is brought in and circulated through the cooling system.

simulation   Involves structuring a training activity that simulates a line situation.

situational characteristics   Factors that can change from setting to setting and can have a catalytic effect in causing an accident.

situational factors   Environmental factors that can affect an employee's safety and that can differ from situation to situation.

smoke   The result of the incomplete combustion of carbonaceous materials.

social environment   The general value system of the society in which an individual lives, works, grows up, and so on.

sound   Any change in pressure that can be detected by the ear.

sound level meter   Produces an immediate reading that represents the noise level at a specific instant in time.

spraining   The result of torn ligaments.

staff authority   The safety and health manager is the staff person responsible for a certain function, but he or she has no line authority over others involved with that function.

standards and testing organizations   Conduct research, run tests, and establish standards that identify the acceptable levels for materials, substances, conditions, and mechanisms to which people may be exposed in the modern workplace.

standpipe and hose systems   Provide the hose and pressurized water for firefighting.

Star Program   Recognizes companies that have incorporated safety and health into their regular management system so successfully that their injury rates are below the national average for their industry.

state funds   Workers' compensation coverage provided by the state.

static electricity   A surplus or deficiency of electrons on the surface of a material.

step and fall   An accident that occurs when a person's foot encounters an unexpected step down.

straining   The result of overstretched or torn muscles.

stress   A pathological, and therefore generally undesirable, human reaction to psychological, social, occupational, or environmental stimuli.

stress claims   Workers' compensation claims that are based on stress-induced disabilities.

stressors   Stimuli that cause stress.

stump and fall   An accident that occurs when a worker's foot suddenly meets a sticky surface or a defect in the walking surface.

subjective ratings   Ratings that are less than objective and can be affected by emotions, human biases, presumptions, and perceptions.

Superfund Amendments and Reauthorization Act (SARA)   Act designed to allow individuals to obtain information about hazardous chemicals in their community so that they are able to protect themselves in case of an emergency.

synthesis   Second step in the design process wherein engineers combine systematic, scientific procedures with creative techniques to develop initial solutions.

systems theory   Views a situation in which an accident may occur as a system comprised of the following components person (host), machine (agency), and environment.

technic of operation review   An analysis method that allows supervisors and employees to work together to analyze workplace accidents, failures, and incidents.

technological alienation   The frame of mind that results when employees come to resent technology and the impact that it has on their lives.

technology access   Access to time- and work-saving devices, processes, or equipment that are up to date.

temporary emergency standards   OSHA standards that can be adopted on a temporary basis without undergoing the normal adoption procedures.

temporary partial disability   The injured worker is incapable of certain work for a period of time but is expected to recover fully.

temporary total disability   The injured worker is incapable of any work for a period of time but is expected to recover fully.

temporary variance   Employers may ask for this when they are unable to comply with a new standard but may be able to if given time.

thermal expansion detectors   Use a heat-sensitive metal link that melts at a predetermined temperature to make contact and ultimately sound an alarm.

third-degree burn   Penetrates through both the epidermis and the dermis. It may be fatal.

three E's of safety   Engineering, education, and enforcement.

threshold limit values (TLVs)   The levels of exposure to which all employees may be repeatedly exposed to specified concentrations of airborne substances without fear of adverse effects. Exposure beyond TLVs is considered hazardous.

threshold of hearing   The weakest sound that can be heard by a healthy human ear in a quiet setting.

threshold of pain   The maximum level of sound that can be perceived without experiencing pain.

time-weighted average   The level of exposure to a toxic substance to which a worker can be repeatedly exposed on a daily basis without suffering harmful effects.

tort   An action involving a failure to exercise a reasonable care that may, as a result, lead to civil litigation.

total safety management (TSM)   The principles of total quality management (TQM) applied to safety management.

toxic substance   One that has a negative effect on the health of a person or animal.

trade associations   Promote the trade that they represent.

trip and fall   An accident that occurs when a worker encounters an unseen foreign object in his or her path.

two-hand controls   Require the operator to use both hands concurrently to activate the machine.

ultraviolet detectors   Sound an alarm when the radiation from fire flames are detected.

Underwriters Laboratories (UL)   Determines whether equipment and materials for electrical systems are safe in the various NEC location categories.

unreasonable risk   Exists when (1) consumers are not aware that a risk exists; (2) consumers are not able to judge adequately the degree of risk even when they are aware of it; (3) consumers are not able to deal with the risk; and (4) risk could be eliminated at a cost that would not price the product out of the market.

unrestricted area   Any area to which access is not controlled because no radioactivity hazard is present.

unsafe act   An act that is not safe for an employee.

unsafe behavior   The manner in which people conduct themselves that is unsafe to them or to another.

useful consciousness   A state of consciousness in which a person is clear-headed and alert enough to make responsible decisions.

vacuum mentality   Workers think that they work in a vacuum and don't realize that their work affects that of other employees and vice versa.

vacuums   Caused by pressures below atmospheric level.

value added   The difference between what it costs to produce a product and the value the marketplace puts on it.

vapor   A mist state into which certain liquids and solids can be converted (e.g., gasoline fumes are vaporized petroleum).

vertical work area   One that is designed and positioned so that workers are not required to lift their hands above their shoulders or bend down in order to perform any task.

vocational rehabilitation   Involves providing the education and training needed to prepare the worker for a new occupation.

volatility   The evaporation capability of a given substance.

voltage   Measures the potential difference between two points in a circuit.

wage-loss theory   Requires a determination of how much the employee could have earned had an injury not occurred.

water hammer   A series of loud noises caused by liquid flow suddenly stopping.

wellness program   Any program designed to help and encourage employees to adopt a healthier lifestyle.

whistle-blowing   Act of informing an outside authority or media organ of alleged illegal or unethical acts on the part of an organization or individual.

whole-person theory   What the worker can do after recuperating from the injury is determined and subtracted from what he or she could do before the accident.

wide band noise   Noise that is distributed over a wide range of frequencies.

willful or reckless conduct   Involves intentionally neglecting one's responsibility to exercise reasonable care.

work envelope   The total area within which the moving parts of a robot actually move.

work injuries   Injuries that occur while an employee is at work.

work stress   A complex concept involving physiological, psychological, and social factors.

worker negligence   Condition that exists when an employee fails to take necessary and prudent precautions.

workers' compensation   Developed to allow injured employees to be compensated appropriately without having to take their employer to court.

workplace accidents   Accidents that occur at an employee's place of work.

workplace inspection   An on-site inspection conducted by OSHA personnel.

workplace stress   Human reaction to threatening situations at work or related to the workplace.

INDEX

AAOH Journal, 688

AAOHN. *See* American Association of Occupational Health Nurses (AAOHN)

Abatement, asbestos, 563

Abatement period, 133

ABC model, 56–57

ABCO model, 56

ABET. *See* Accreditation Board for Engineering and technology (ABET)

ABIH. *See* American Board of Industrial Hygiene (ABIH)

Ability to pay, 167

ABOHN. *See* American Board for Occupational Health Nurses (ABOHN)

Absenteeism, 276

Absorption, 554

Abuse
  incidents of, 372
  workers' compensation and, 199–200

Acceleration, hazards in, 424–433

Acceleration objects, protection from, 424

Access control, in risk reduction, 360–361

Accidental deaths in the United States, 24–25

Accident-analysis report, 212

Accident causation
  accident/incident theory of, 46–48
  behavioral theory of, 55–56
  combination theory of, 54–55
  depression and, 58–59
  domino theory of, 41–43
  drugs and, 57–58
  epidemiological theory of, 48–50
  human factor theory of, 45–46
  management failures and, 59–61
  obesity and, 61
  systems theory of, 50–54

Accident/incident theory, of accident causation, 46–48

Accident investigation and reporting
  conducting, 216–219
  importance of, 212–213
  interviewing witnesses, 219–222
  mistakes to avoid in, 227–229
  personnel in, 216
  purpose of, 213
  reporting, 222–227
  safety managers and, 786
  types of, 212

Accident prevention, 22
  developing procedures for, 338
  development of programs on, 1314
  strategies of, 723
  under workers' compensation, 176

Accident rates, 26

Accident reports, 212
  forms, 223, 225–226

Accident(s), 22. *See also* Accident causation; Accident prevention; Motor vehicle accidents; Workplace accidents
  burn-related, 465
  catastrophic, 780
  causes of, 214–215
  costs of, 23–24, 32–35, 868
  defined, 211
  ergonomics, apparent trends in, 278
  impact, 28, 35–36
  industrial, 3, 11, 40
  investigating, importance of, 211
  mining, 3
  motor vehicle, 22, 24, 25
  reporting, 222–227
  slip and fall, 412
  step and fall, 412
  stump and fall, 412
  unreported, reasons for, 224–227
  vs. other causes of death, 25

Accident scene
  isolating, 217
  photographing or
    videotaping, 218
Acclimatization, 454, 456
Accommodation, 661
  reasonable, 686
Accountability, 760, 811, 813
Accreditation Board for
    Engineering and
    technology (ABET), 88
Acetone, 561
Acetylene hydrocarbons, 561
Acid rain, 820
Acquired immunodeficiency
    syndrome (AIDS), 681
  cardiopulmonary
    resuscitation and, 694–695
  counseling infected
    employees, 692–693
  course outline on, 691
  easing employees' fears
    about, 693
  education, 685–686, 690–691
  facts about, 681–682
  legal concerns on, 686–690
  protecting employees from,
    693–695
  symptoms of, 682–683
  testing for, 685
  transmission of, 683
  in workplace, 683–686
Active environmental
    legislation, 824
Active noise reduction (ANR),
    644
Activities, inappropriate,
    45, 46
Activity support, 361–362
Acts, unsafe, as cause of
    accidents, 215
Actual penalty, 131
ADBA. See Audiometric
    database analysis (ADBA)
Adjustable guards, 394
Adjusting
  in emergency preparation,
    712
  in waste reduction program,
    834

Adjustments and
    modifications, in
    administrative controls,
    370
Administration
  of hearing loss prevention,
    654–655
  workers' compensation and,
    197–198
Administrative controls,
    362, 576
  in confined place
    management policy,
    592–593
  in hearing loss prevention,
    634–635, 652
  noise control strategies
    and, 643
  workplace violence and,
    369–370
Adult illiteracy, 342–343
Advanced Safety Certificate,
    341
AEP. See Associate Ergonomics
    Professional (AEP)
Aerosol cans, 482
Aerosols, 559
Aerospace engineering, 77–78
Affected employee, 401
Age, decompression sickness
    and, 472
Agencies
  dealing with safety and
    health, 145
  government, 76
  for safety and health
    managers, 73–75
Aging
  ergonomics of, 278–279
  of workforce, 104
Agreement settlement, 197–198
Agricultural employees,
    workers' compensation
    coverage of, 176, 177
AICHE. See American Institute
    of Chemical Engineers
    (AICHE)
AIDS. See Acquired
    immunodeficiency
    syndrome (AIDS)

AIHA. See American Industrial
    Hygiene Association
    (AIHA)
Airborne contaminants,
    558–559
Airborne emissions, hazardous,
    560
Airborne toxics, 559–561
  asphyxiants, 560
  irritants, 559–560
  narcotics and anesthetics,
    561
Air dose, 606
Air Hygiene Foundation, 9
Air safety program elements,
    580–581
AISEE. See Association of Iron
    and Steel Electrical
    Engineers (AISEE)
Alarm procedures, in
    evacuation planning,
    729
Alarm stage of, 300
Allied Manufacturing
    Company, 870–871
Alpha Meat Packing Company,
    205
Alpha particles, 605
Alternating current (AC
    current), 489
Alternating current (AC) fields,
    617
Altitude sickness, 478
Alzamora, Lori, 56
Alzheimer's disease, 619
Ambu, Inc., 694
American Association of
    Occupational Health
    Nurses (AAOHN), 86
American Board for
    Occupational Health
    Nurses (ABOHN), 88,
    92, 93
American Board of Industrial
    Hygiene (ABIH),
    88, 91
American Chemical Council,
    741
American College Health
    Association, 689

American Conference of Governmental Industrial Hygienists (ACGIH) classifications of TLVs and BEIs, 570–571
description of TLVs, 569
developing TLVs, 570
American Conference of Governmental industrial Hygienists (ACGIH), 453, 457, 458, 569, 570, 617
American Council of Government Industrial Hygienists (ACGIH), 90, 570
American Critical Stress Foundation, 735
American Industrial Hygiene Association (AIHA), 90
American Institute of Chemical Engineers (AICHE), 770
American National Standards Institute (ANSI), 426, 427, 432, 440
confined space standard, 594
Fall Protection Standard, 418
hard hat standard, 425
indoor air quality/HVAC standard, 566–567
noise standard, 631–632
S12.13, 631–632
American Psychiatric Association, 735
American public Power, 788
American Red Cross, 717
American Social Science (Gagilardo), 179
American Society of Heating, refrigerating and Air-Conditioning Engineers, 567
American Society of Mechanical Engineers, 474
American Subcontractors Association, 343
American workers, homicide as cause of death in, 352
Amick, Benjamin C. III, 667
Amine process, 476
Amine vessels, 476

Ammonia vessels, 476
Amperes, 489
Analysis and evaluation, in design process, 79
Ancestry, 42
Anchorage, 417, 418
Andrews, Jane, 49–50
Anesthetics, 561
Anger management, 373, 376–377
Anoxia, 478
ANR. *See* Active noise reduction (ANR)
ANSI. *See* American National Standards Institute (ANSI)
Anthracene oil, 561
Antibody test, 685
Antifatigue mats, 437
Appeals
employee, 133
employer, 133
formal, 133
process, 116, 133–134
Application, as step in teaching method, 321
Application assignments list, 323
Arc flash injuries, 505, 506
Arco Chemical Company, 26
Arcs
electric, 532–533
and spark hazards, 492–493
Area, unrestricted, 606
Arlington National Cemetery, 731
Arsenic, 561
Artificial environment, 824
Artificial intelligence, 732
Asbestos
abatement of, 563
encapsulation of, 564
enclosure of, 564
friable, 562
hazards with, 561–565
medical records and examination of, 564–565
personal protective equipment for removal of, 564

removal and containment of, 562–563
safety movement and, 9–10, 13
Asbestos-containing material (ACM), 562–563
Aseptic necrosis, 479
ASP. *See* Associate Safety professional (ASP)
Asphalt, 5611
Asphyxiants, 560
Assessment
of current corporate culture, 885
of hand protection, 439–440
of hazardous noise conditions, 638–640
in implementing total safety management model, 874–875
toxic mold, 568–569
of workplace for fire hazards, 534
Associate degrees, 72
Associate Ergonomics Professional (AEP), 92
Associate Safety professional (ASP), 89
Association of Iron and Steel Electrical Engineers (AISEE), 6, 16
"Assumption of risk", 11, 178
Atmospheric pressure, 471, 629
Atom, basic structure of, 488
Atomic Energy Act, 610
Atomic Energy Commission, 609, 611
Attenuation, 627
Audiogram, 639
Audiometric database analysis (ADBA), 632
Audiometric evaluation, 635–636
Audiometric test form, 640
Audiometric testing, 639
Audiometry monitoring, in hearing loss prevention, 652–653
Aurals. *See* Earplugs
Authorized employees, 401

Auto-ignition temperature, 517, 536
Automatic ejection, 400
Automatic feed, 399
Automatic sprinkler systems, 522
Automation
    challenge for future, 675–676
    impact of, on workplace, 659–660
    islands of, 659
    minimizing problems of, 672–675
    technological alienation and, 671–672
    video display terminals in, 660–663
Awareness, 850
    in fire safety programs, 534–535
    public, 130
    safety, 6–7
    visual, in safety promotion, 799–802
Axioms of industrial safety, 41
Axioms of Industrial Safety (Heinrich), 41

Bachelor's degree, 72
Back, lower, injuries of, 433–434
Back disorder risk factors, 268
Back safety/lifting program, 434–435
Barometer, 471
Baseline audiogram, 627
BASF Corporation, 26
Baxter, Garner, 107
BBS. *See* Behavior-based safety (BBS)
BCSP. *See* Board of Certified Safety professionals (BCSP)
Bednarowski, Jack, 311
Behavior
    psychotic, 362
    unsafe, 46
Behavioral theory, of accident causation, 55–56

Behavioral time sharing, in measuring workplace stress, 301
Behavior-based safety (BBS), 55–57
BEIs. *See* Biological exposure indices (BEIs)
Bends, 479
Benefits
    in hazard analysis, 767
    of waste reduction program, 835
Benson, Gary, 671–672
Benzene, 561
Beryllium, 561
Best practices
    at Deere & Company, 790
    at Georgia-Pacific Corporation, 789–790
    at the New York Power Authority, 788–789
Best-ration approach, 752
Best Safety Record Award, 788
Beta particles, 605
Bhopal tragedy, 10
Biases, in accident investigation, 228
Bicipital tendinitis, 288–289
Biological exposure indices (BEIs), 570–571
Biological hazards, 551, 553
Bitumen waxes, 561
Black-and-white approach, 753
Blackball, 178
Black spit, 11
Blastomogen, 561
Bloodborne pathogens. *See also* Acquired immunodeficiency syndrome (AIDS); Hepatitis B virus (HBV); Hepatitis C virus (HCV)
    OSHA's standard on occupational exposure to, 699–704
    preventing and responding to needlestick injuries, 704–705
Bloodborne Pathogen Standard, 695

Blood contact, 683
Blood transfusions, 683, 698
BLS. *See* Bureau of Labor Statistics (BLS)
BLS: *Record-Keeping Guidelines for Occupational Injuries and Illnesses.* September 1986, 275
BNF. *See* British Nuclear Fuels (BNF)
Board of Certification in Professional Ergonomics, 88
Board of Certified Safety professionals (BCSP), 88
Body, in structured lecture, 326
Body belt, 417
Body fluids, 683, 692, 694, 696
Body harness, 417
Body surface area (BSA), 464
Boilers, pressure hazards and, 473
Bonding, 498, 585
Boundary location, 673
Boyle's law, 471
Breach of warranty, 234
Breaking, as type of mechanical injury, 387, 389
Breast cancer, 619
British Nuclear Fuels (BNF), 604
BSA. *See* Body surface area (BSA)
Budget, insufficient, 808
Bureau of Labor Statistics (BLS), 5, 275, 352, 426
Bureau of Mines, 6
Bureau of Workers' Compensation, 196
Burn-related accidents, 465
Burns
    chemical, 465–467
    severity of, 463–464
Bush, George W., 260, 261
By-products of combustion, 512

California Occupational Safety and Health Administration (CAL/OSHA), 437
California Workers' Compensation Institute (CWCI), 183

CAL/OSHA. *See* California Occupational Safety and Health Administration (CAL/OSHA)

Cancellation approach, electromagnetic fields in workplace and, 619–620

Cancer, 25, 615, 618

Carbon, 513

Carbon dioxide, 513, 520

Carbon monoxide, 513, 520, 560, 822

Carcinogens, effects of, 561

Cardiopulmonary resuscitation (CPR), 501

AIDS and, 694–695

Care, 165–166

Carpal tunnel syndrome (CTS), 31, 32, 283, 289, 662, 671

Case law, 688

Case management, workers' compensation and, 197–198

Casual employees, workers' compensation coverage of, 176, 177

Catastrophic accidents, 780

Causal relationship, 46

Causes, of accident investigation, 214–215

Caution signs and labels, 608

CD-ROM technology, as presentation method, 330–331

Centers for Disease Control and Prevention (CDC), 16, 146, 685, 695

Central factor, 42

Centrality, in SPICE model, 204

Certification board, 73, 74

Certified Industrial Hygienist (CIH), 88, 90

Certified Occupational Health Nurse/Case Manager (COHN/CM), 93

Certified Occupational Health Nurse (COHN), 88, 92–93

Certified Occupational Health Nurse-Safety Manager (COHN-SM), 93

Certified Occupational health Nurse-Specialist/Case Manager (COHN-S/CM), 93

Certified Occupational health Nurse-Specialist (COHN-S), 92, 93

Certified Professional Ergonomist (CPE), 88, 91–92

Certified safety professional (CSP), 88–90

Cervical radiculopathy, 32, 289

CFR. *See* Code of Federal Regulations (CFR)

Chain of command, 730

Chalk, William, 79

Charitable employees, workers' compensation coverage of, 176, 177

Checklists

for emergency planning, 722–723

for employee orientation verification, 337

for ergonomics, 267

for evaluating training programs, 333

for incorporating workplace-violence prevention, 367

of OSHA's advisory guidelines on workplace violence, 365

for safe selection and use of hand tools, 291

for training employees working in cold environment, 462

for workplace-violence risk reduction, 361

Chemical burns, 465–467

effects of, 466–467

first aid for, 467

injuries of, 29–30

Chemical engineers, safety and, 82–83

Chemical environment, contact lenses in, 431

Chemical hazards, 551

Chemical process guidelines, OSHA's, 581

Chemical protective clothing, selecting, 577

Chemicals, harmful effects of, 466

Chemical spills

OSHA's regulation for, 548–550, 781–783

responding to, 733

Cherns, A. B., 672

Chinchar, Mike, 769

Chlorine Emergency plan (CHLOREP), 731

Chlorine Institute (CI), 731

Chloroform, 561

Chokes, 479

Chromium, 561

CIH. *See* Certified Industrial Hygienist (CIH)

Circadian rhythm, 302

Circuit

short, 491

three-wire, 490

Circuit tester, 496

Circumaurals. *See* Earmuffs

Citations and penalties, 131–133

Claim notices, 197

Claims, stress, 199

Claims and character, of workers' compensation, 189

Clark Processing Company, 223

Clean Air Act, 581–582, 821, 822, 824

Cleaning solvents, 536

Clean Water Act, 822, 824

Clear direction, in teamwork approach to promotion safety, 811

Clinical depression, 58

Clinton Administration, 258

Closed environment, 824

Closing, in structured lecture, 326–327

Closing conference, 129

Clothing

of brigade members, 525–526

caustic chemical burns and, 467

Clothing (*Contd.*)
  chemical protective,
    selecting, 577
  flame-resistant, 533
  heat stress and, 455–456
Cloth streamers, in detecting
  pressure hazards, 481
CNC. *See* Computer
  numerically controlled
  (CNC)
Coal Mine Health and Safety
  Act, 1969, 12
Coal-miner's pneumoconiosis
  (CWP), 12
Coal tar, 561
Cobalt, 561
Cobb, Andrea, 687
Code, 163
*Code of Ethics and Standards
  of Conduct* (Martin
  Marietta Corporation), 755
Code of Ethics of the American
  Academy of Industrial
  Hygiene, 545
Code of Federal Regulations
  (CFR), 112, 116, 274, 315,
  605, 613
Coefficient of friction, 413
COE injury. *See* Course of
  employment (COE) injury
Coffee, William R. Jr., 227
COHN. *See* Certified
  Occupational Health Nurse
  (COHN)
COHN/CM. *See* Certified
  Occupational Health
  Nurse/Case Manager
  (COHN/CM)
COHN-S. *See* Certified
  Occupational Health
  Nurse-Specialist (COHN-S)
COHN-SM. *See* Certified
  Occupational health Nurse-
  Safety Manager (COHN-
  SM)
Cold stress, 458–462
Combination theory, of
  accident causation, 54–55
Combustible and explosive
  materials, 493

Combustible substance, 535
Combustion, 512
  major chemical products
    of, 520
  spontaneous, 515
Combustion engines, internal,
  536–537
Combustion point, 512
Commercial diving operations,
  157–158
Comminuted fracture, 389
Commitment, 796
  long term, in quality
    management, 862, 864
  personal, 804–805
  in product safety program,
    240
  safety and health managers
    and, 69, 70
  in total safety management,
    871
Committee on Industrial
  Ophthalmology, 467
Commonsense test, 750
Communication(s), 851
  alarm procedures in
    evacuation planning, 729
  in design process, 79
  in evacuation planning, 729
  outside, confined space and,
    590
  projective, 814
  receptive, 814
  recovery, 736–737
  in safety management,
    785–786
  in team-building strategies,
    243, 244
Community, failure to warn,
  716
Community Right-to-Know Act,
  237–238, 312
  emergency notification, 237
  emergency planning, 237
  reporting requirements, 238
  toxic chemical release
    reporting, 238
Company-sponsored wellness
  programs, in safety
  promotion, 808–809

Companywide commitment, to
  safety and health, 69
Comparative negligence, 164
Compensatory damages, 167
Competence, 850
Competition
  safety-friendly corporate
    culture and, 879–880
  globalization of, safety and,
    880–883
  as strategy to promote safety,
    808
Competitive advantage
  EMS movement and,
    837–838
  sustainable, in total safety
    management, 866–867
Competitiveness, 70, 71, 98
  defined, 99–100
  global, 882
  global, safety and, 106
  productivity and, 100–103
  quality and, 103
  role of safety and health in
    improving, 104–107
Complete fracture, 389
Compliance, 244
Compliance -driven setting, 871
Compound fracture, 389
Comprehensive, ongoing
  training, in total safety
  management, 872–873
Compressed gas/air, 155
Computer numerically
  controlled (CNC), 767
Computers, emergency
  response and, 732–733
Conditioning period follow-up,
  273
Conditions, unsafe, as cause of
  accidents, 215
Conduction, 452–453, 514
Conductive loss, 627
Conductors, 488
Conference
  closing, 129
  opening, 129
Conference method,
  328–329
Confidentiality, 689

Confined spaces
    hazards of, 585–587
    management policy, 592–595
    purging, 590
    standard of, 118–119,
        585–587
    ventilation of, 590
Confirmation test, 689
Conflict guidelines,
    establishing, 373
Conflicting goals, allowing, in
    investigations, 228
Conflicting rights, 358
Conflict management
    counterproductive, 373
    productive, 373
Conflict prevention and
    resolution skills,
    developing, 374
Conflict resolution, workplace
    violence and, 373–377
Congressional Review Act
    (CRA), 260
Congress Office of Technology
    Assessment, 103, 667
Connector, 417
Conners, Mark, 242
Conscience, 749
Consistency, 757, 797
Construction Products
    Company (CPC), 42–43
Construction standards, of
    OSHA, 162
Consultation services, of
    OSHA, 137, 138
Consultation Services for the
    Employer, publication
    3407, 137
Consumer Power Company, 805
Consumer Product Safety Act,
    235, 236
Consumer Product Safety
    Commission (CPSC), 236
Contact lenses
    in chemical environment,
        431
    respirators and, 580–581
Contamination by dirt, 481
Contested cases, 198
Continual feedback, 320

Continual improvement, 99,
    845, 872–873
Continual improvement
    forever, in total safety
    management, 867
Continual process
    improvement, in quality
    management, 862, 864
Continuous noise, 627
Contractors and subcontractors,
    workers' compensation
    coverage of, 176, 177
Contract Workers and Safety
    Standards Act, 7
Contributory negligence, 11,
    164, 178
Controlled environment, 823
Controlled pace, 304
Control(s). See also
    Administrative controls
    of access in risk reduction,
        360–361
    of hazardous energy, 400–406
    as source of workplace stress,
        297–298
    two-hand, 398
Convection, 453, 515
Convergence, 661
Conveyor belts, 492
Cooperative Safety Congress
    (CSC), 6, 16
Coordination
    in emergency preparation,
        714–715
    in planning for emergencies,
        719
Coordination information, 731
Corium, 462
Corporate culture. See Safety-
    friendly corporate culture
Corporate image, 106
Corporate policy
    for bloodborne pathogens,
        684
    relating to hearing loss
        prevention, 649–650
Corrective action, taking
    ineffective, 227
Corrosion, in detecting pressure
    hazards, 481

Cost allocation, under workers'
    compensation, 176, 198
Cost-benefit analysis, in
    economics of
    environmental regulation,
    823
Cost benefit factors, in hazard
    analysis, 767–769
Cost-benefit studies, inhibitors
    of, 286–287
Cost estimation method, 32–34
Cost-reduction strategies, under
    workers' compensation,
    202–205
Cost(s), 70
    of accidents, 23–24, 32–35,
        868
    of back injuries with workers'
        compensation, 183
    in hazard analysis, 767
    insured, 32–33
    productivity and, 105
    uninsured, 32–33
    of waste reduction program,
        835
    of work accident, 26
    workers' compensation,
        controlling rising, 202
Cote, Arthur E., 527
Council of Industrial Health of
    the American Medical
    Association, 467
Counseling, 692, 697
Counterproductive conflict, 373
Course of employment (COE)
    injury, 186–187
Course outline, in planning
    instruction, 321–322
CPE. See Certified Professional
    Ergonomist (CPE)
CPR. See Cardiopulmonary
    resuscitation (CPR)
CPTED. See Crime Prevention
    Through Environmental
    Design (CPTED)
Credo and Code of Conduct
    (Martin Marietta
    Corporation), 755
Creeps, 479
Creosote oil, 561

Crestview Container
  Corporation (CCC), 779
Crestview Grain Corporation
  (CGC), 54–55
Crime Prevention Through
  Environmental Design
  (CPTED), 360, 740
Critical burns, 464
Cross functional, 868
Cross-functional safety
  committee, 534–535
Crushing, as type of mechanical
  injury, 387, 388
CSP. See Certified safety
  professional (CSP)
CTDs. See Cumulative trauma
  disorders (CTDs)
CTS. See Carpal tunnel
  syndrome (CTS)
Cubital tunnel syndrome,
  32, 289
Culture
  language and national origin
    and, 346–347
  safety and, 205
Cumulative radiation records,
  612
Cumulative trauma disorders
  (CTDs), 262
  classification of, 287–290
  defined, 287
  ergonomic problems and,
    276, 669
  evaluation, treatment, and
    follow-up of, 274
  health care providers
    protocols in, 274
  meat packing and poultry
    processing and, 261
  medical management of,
    271–272
  monitoring trends of, 274
  preventing, 290–292
  reporting symptoms of, 274
  risk factors of, 268
  video display terminals use
    and, 662
  vs. Carpal tunnel Syndrome,
    671
Current, let-go, 495

Customer feedback, 246
Customer focus, in quality
  management, 862, 863–864
Customer(s), 354
  superior values and, 881–882
Custom-molded earplugs, 643
Cutis, 462
Cutting, as type of mechanical
  injury, 387
CWCI. See California Workers'
  Compensation Institute
  (CWCI)

Dalton's law of partial
  pressures, 471
Damage assessment and
  salvage, 736
Damaged relationships,
  whistle-blowing and, 759
Damages, 167
Davis-Bacon Act, 162
Dawson, Chip, 738
Deaeration process, 476
Deaerator vessels, 476
Death and burial benefits,
  under workers'
  compensation, 193,
  194, 195
Deaths
  accidental, 22, 24–25, 26
  accidents vs. other causes
    of, 25
  from occupational diseases,
    35
  rates by industry, 28–29
  in work accidents, 27
  from work-related
    injuries, 35
Decibel, 629
Decision to work unsafely, as
  cause of accidents, 214
Decompression procedures,
  479–480
Decompression sickness, 472
"Deep pockets," 167
Deere & Company, best
  practices at, 790
Degenerative joint disorder
  (osteoarthritis), 32
Delegating responsibility, 304

Demand for strength/power,
  large vs. small, 259
Deming, W. Edwards, 103
De minimis violation, 131
Demonstration method,
  327–328
Demonstration Program,
  139–140
Department of Health and
  Human Services (DHHS),
  16, 146, 577
Depression, accident causation
  and, 58–59
DeQuervain's disease, 31, 288
De Re Metallica (Agricola), 4
Dermis, 462
Design flaws, 412
Design process, 79–80, 673
Design tools, in workstation
  analysis, 271
Destructive testing, 481
Detailed hazard analysis
  failure mode and effects
    analysis, 769–770, 771
  fault tree analysis, 775,
    776, 777
  hazard and operability
    review, 770, 772, 773
  human error analysis, 772
  risk analysis, 775, 777
  technic of operation
    review, 774
Detectors
  thermal expansion, 521
  ultraviolet (infrared), 521
Diagnostic and Statistical
  Manual (APA), 735
Dictatorial management,
  363–364
Dip tanks, 547
Direct current (DA current), 489
Direct settlement, 197
Disabilities
  permanent partial, 190–192
  temporary, 188–190
  temporary partial, 188
  temporary total, 188–190
  types of, 188
  workers' compensation and,
    188–192

workers with, planning for, 724–728

Disaster recovery operation, entities to be contacted in, 737

Disasters
preparations for, 526
recovering from, 735–737
resuming business after, 741–742

Discipline and accident reporting, 227

Discourse on the Diseases of Workers (Ramazzini), 5

Discovery period, 233

The Diseases of Workers (Ramazzini), 85

Disposable face masks, 694

Division of Respiratory Disease Studies (DRDS), 148–149

Division of Surveillance, Hazard Evaluations, and Field Studies (DSHEFS), 149

Division of Training and Manpower Development (DTMD), 149–150

Documentation
in design process, 79
of environmental management system, 851

Document control, 851–851

Domestic employees, workers' compensation coverage of, 176, 177

Domino theory, of accident causation, 41–43

Dose, 557
lethal, 558

Dose-response relationship, 557–558

Dose threshold, 557

Dosimeters, 627, 633–634, 639

Dosimetry, 633–634

Double insulation, 500

Doubling rate. *See* Exchange rate

Draper, Alfred III, 568–569

Drawing, in identifying ergonomic problems, 278

Drawing power, 248

DRDS. *See* Division of Respiratory Disease Studies (DRDS)

Drills, 729, 730

Drowning, accidental deaths from, 24

Drucker, Peter, 747

Drug dependence, 363

Drug-free workplace programs, establishing, 58

Drugs, accident causation and, 57–58

DSHEFS. *See* Division of Surveillance, Hazard Evaluations, and Field Studies (DSHEFS)

DTMD. *See* Division of Training and Manpower Development (DTMD)

Due care, 167

"Dumbing down", 672

Dupruyten's contracture, 32

Duration, as noise risk factor, 631

Dusts, 558

Duty to warn, 235

DVD/videotape method, in workstation analysis, 269

Dyes, 561

Dyktor, H. G., 17

Dysbarism, 479

EAPs. *See* Emergency action plans (EAPs); Employee assistance programs (EAPs)

Earmuffs, 636–637

Earplugs, 643

EASE. *See* Ergonomic Assist and Systems Equipment (EASE)

Eastman, Martin, 694

Economics
of environmental regulation, 822–823
of ergonomics, 285–287

Economic threat, EMS movement and, 839

Education. *See also* Training
on AIDS, 685–686, 690–691
on back safety/lifting program, 434
effective ergonomics program and, 275–276
of employees, 274
of fire brigades, 524
in hearing loss prevention, 637, 651
job safety analysis as technique in, 338, 340
in quality management, 862, 864
requirements in safety and health, 315–319
of safety and health managers, 72–73
workplace violence and, 371

Educational aspect, of safety program, 13

Educational Resource Center, 340

Education and Research Centers (ERCs), 340

Education services, occupational Safety and Health Administration, 140

Effective current, 489

Ego strength, 751

Egress, means of, 151–152, 531–532

Electrical current, 489

Electrical engineering, 78

Electrical equipment, 617
failure of, 494
hazardous locations for, 495

Electrical hazard guard, 433

Electrical hazards
defined, 487–491
detection of, 496–497
to humans, 495–496
OSHA's standards on, 501–502
reducing, 497–501
safety program for, 502–503
self-assessment in, 503–504
sources of, 491–495

Electrical shock, causes of, 491

Electrical standards, OSHA's, 501–502

Electrical system grounding, 498

Electrical systems design, 157

Electric and magnetic fields (EMFs), 617

Electric arcs, 532–533

Electricity, 488

Electrocution, 493, 502
occurrence of, 490
workplace deaths from, 498

Electromagnetic fields (EMFs), 618

Electromagnetic fields (EMFs) in workplace
cancellation approach and, 619–620
health problems and, 618–619
reducing, 621
safety professionals and, 619
shielding approach and, 620–621

Electromagnetic radiation (EMR), 615

Electromechanical devices, 397

Electronic earmuffs, 636–637

Electrons, 487, 488

Electrostatic discharge, sources of, 491–492

Electrostatic hazards, 491–492

Elevator policy, implementing, 727

ELF radiation. *See* Extremely low frequency (ELF) radiation

ELISA test. *See* Enzyme linked immunosorbent assay (ELISA) test

Embarrassment, as reason for not reporting accident, 224

Emergency action plans (EAPs), 548, 719
in chemical spills, 781
developing emergency notification strategies, 727
distributing and communicating, 725–726

employers, employees, and, 726

first responders, 626–627

implementing elevator policy, 727

key personnel, 724

personal needs, 725

practicing and maintaining, 727–728

shelter-in-place plans, 724–725

Emergency coordinator, 714

Emergency(ies). *See also* Emergency preparation
dealing with psychological trauma of, 733–735
external communication procedures in, 729
first aid in, 716–717
planning for, 719–723

Emergency notification, 237–238, 582–583, 714

Emergency notification strategies, developing, 727

Emergency planning, 237, 582, 597, 722–723

Emergency planning and Community Right-to-Know Act (EPCRA), 10, 712–714
emergency notification, 714
emergency planning, 713
information requirements, 714
toxic chemical release reporting, 714

Emergency preparation
organization and coordination, 714–715
OSHA standard in, 715–716
rationale for, 712

Emergency preparedness and response, 852–853

Emergency preparedness plan, 378

Emergency procedures, 212

Emergency-related trauma, dealing with, 734

Emergency response, 731
computers and, 732–733

Emergency response management team (ERMT), 715

Emergency response plans, 713, 714
in chemical spills, 548, 781

Emergency response regulations, 718

Emergency response team (ERT), 731–732

EMFs. *See* Electric and magnetic fields (EMFs); Electromagnetic fields (EMFs)

Employee appeals, 133

Employee assistance, in disaster recovery, 737

Employee Assistance Professionals Association, 59

Employee assistance programs (EAPs), 59, 684, 693

Employee empowerment enlistment, in total safety management, 871–872

Employee footwear program, 416

Employee-generated changes, 277

Employee involvement, workplace violence and, 365–366

Employee-management participation, in safety promotion, 805–806

Employee protection, 760

Employee-related outsider, 354

Employee Retirement Income System Act (ERISA), 686–687

Employee rights, 684
under Occupational Safety and Health Act, 142–143
violent, 358–359

Employee right-to-know, 596–597

Employee(s). *See also* Employee training
acclimatized, 457
affected, 401

assistance after disaster, 737
authorized, 401
changing work habits of, 648
communication of
     information to, 128
complaints of, 276–277
counseling AIDS infected,
     692
defined, 353
easing fears on AIDS, 693
empowered, 243
exposure to radiation,
     606–607
exposure to vibration, 648
improving job satisfaction of,
     204–205
injured, risk and, 178
morale of, 26
not-acclimatized, 457
notices to, radiation hazards
     and, 612–615
productivity and, 375
protecting from, 693–694
reporting symptoms
     of CTD, 274
responsibilities for exposure
     control, 700
responsibilities under OSHA
     Act, 143
safety promotion
     participation in, 798–799
viral hepatitis and, 696
workers' compensation and,
     176–177, 187
Employee surveys, workplace
     violence and, 368
Employee taxes, 807
Employee training
and education, 274
in first-aid techniques, 717
of new and transferred,
     335–337
program, in OSHA's hazard
     communication standard,
     597
in safety management, 785
Employer-biased laws, 11
Employer liability, 6
Employer(s)
appeals, 133

hazardous waste standards
     and, 119–120
liability for workplace
     violence, 359
premium rates of, 185
responsibilities under OSHA
     Act, 141–142
rights, under OSHA Act,
     140–141
role in workplace terrorism,
     738–740
Employer's Liability Law, 6
Empowerment
in quality management, 863,
     865
in team-building strategies,
     243, 244
in total safety management,
     871–872
EMR. See Electromagnetic
     radiation (EMR)
EMS. See Environmental
     management system (EMS)
Enamels, 561
Encapsulation of asbestos, 564
Enclosure
of asbestos, 564
in hearing protection, 643
Endothermic, 512
Energy control procedure, 401
Energy control program, 401
Energy-isolating devices, 401,
     403
Energy Reorganization Act
     (ERA), 760
Enforcement, in administrative
     controls, 370
Enforcement aspect, of safety
     program, 14
Engineering
aerospace, 77–78
electrical, 78
industrial, 78–79, 81
mechanical, 78
nuclear, 79
Engineering and Work Practice
     Controls, 695
Engineering aspect, of safety
     program, 13
Engineering controls

in hearing loss prevention,
     634–635, 652
industrial hygiene and,
     573–574
noise control strategies and,
     641–643
workplace violence and, 369
Engineering metals, 18
Engineer(s)
chemical, 82–83
environmental, 82
industrial safety, 18
process, 83
safety, 75–79
English as a second language,
     training issues of, 345–347
Enlistment, 872
Environment, 51
artificial, 824
closed, 824
controlled, 823
creating ethical, 754–756
free, 824
harmful, 186
hazards of, 552
induced, 823
legislation and regulation
     concerning, 820–823
natural, 823
person-machine, 54
social, 42
vibration and, 647
work, 278
Environmental aspect, 845,
     848–849
Environmental engineers,
     safety and, 82
Environmental factors, 44, 260
violence and, 363–364
Environmental hazards, 552
Environmental heat, 453
Environmental management,
     small organizations and,
     838
Environmental management
     programs, 849–850
Environmental management
     system (EMS)
in an organization, 844
audit in, 844, 854–855

Environmental management
    system (EMS) (*Contd.*)
  defined, 835
  documentation of, 851
  function of, 835–837
  plan-do-check-adjust model
    applied to, 847
  potential benefits of, 840–841
  rationale for movement,
    837–840
Environmental policy, 848
Environmental Protection
    Agency (EPA), 238, 550, 714
  Clean Air Act, 581–582
  indoor environmental quality
    information, 567
  OSHA partnership with, 821
  progress in cleaning up air
    and water pollution, 820
Environmental regulation,
    economics of, 822–823
Environmental safety, 299
  environmental management
    safety, 835–837
  hazardous waste reduction,
    830–835
  hazards of environment,
    825–830
  health and, 819–820
  ISO and, 841–842
  ISO 14000 and, 842–856
  legislation and regulation,
    820–823
  role of safety and health
    professionals, 824–825
  types of environments,
    823–824
Environmental Safety
    Awareness, 781
Environmental stressors,
    552–553
Enzyme linked immunosorbent
    assay (ELISA) test, 689
EPCRA. *See* Emergency
    Planning and Community
    Right-to-Know Act
    (EPCRA)
Epicondylitis, 32, 289
Epidemiological theory, of
    accident causation, 48–50

Epidermis, 462
Equipment
  broken-down, repairing, 511
  failure of, 494–495
  firefighting, 525
ERA. *See* Energy
    Reorganization Act (ERA)
Ergonomic Assist and Systems
    Equipment (EASE), 437
Ergonomic guidelines for
    Manual Material Handling
    (EGMMH), 436–437
Ergonomic hazards, 268,
    548, 553
  human factors and, 257–258
Ergonomic job hazard analyses,
    268–269
Ergonomics
  of aging, 278–279
  benefits of, 256
  defined, 256
  economics of, 285–287
  hazard prevention and
    control, 270–2271
  hazards, human factors and,
    257–258
  identifying specific
    problems, 277–279
  indicators of problems,
    276–277
  keys to successful program,
    266
  medical management
    program, 271–275
  in office automation, 669
  OSHA's voluntary guidelines
    on, 261–265
  participatory, 292
  physical stress and,
    258–260
  politics and, 260–261
  problem-solving strategies,
    279–285
  training and education,
    275–276
  worksite analysis program
    for, 266–270
  workstation checklist for
    video display terminals,
    264–265

Ergonomics management
    program, 674
Ergonomics risk factors, in
    worksite analysis, 268
Ergonomic traps, 46
Ergonomists, 676
ERISA. *See* Employee
    Retirement Income System
    Act (ERISA)
ERMT. *See* Emergency response
    management team (ERMT)
ERT. *See* Emergency response
    team (ERT)
Erythema, 463
Essential functions, 686
Ether, 561
Ethical behavior
  defined, 749–751
  guidelines for determining,
    749–751
  managing, 748–749
  models for determining, 752
  in organizations, 751
Ethical dilemmas
  faced by safe and health
    professional, 759
  handling, 748–749, 756–757
Ethical environment, creating,
    754–756
Ethical example, setting, 756
Ethical power, 750–751
Ethical reasons, for training,
    311–315
Ethics
  company's role in, 753–756
  questions for making
    decisions, 757–758
  safety and health professionals'
    role in, 751–753
  situational, 752
  whistle-blowing and,
    758–762
Ethics credo, 755
Ethics philosophy, 754–755
Evacuation planning
  communication and alarm
    procedures, 729
  customizing, to meet local
    needs, 730–731
  marking exit routes, 728–729

outside assembly, 730
training for, 730
Evacuation warning signal, 609
Evaluation
audiometric, 635–636
of CTD, 274
in emergency preparation, 712
hazard, 572–573
of hearing loss prevention
programs, 638, 651–655
of lockout/tagout programs,
405–406
of personnel needs, 725
of product safety program,
240–241
of safety and health
instruction, 331–333
of safety-related behaviors
and attitudes, 889–890
as step in teaching
method, 321
workplace violence and,
371–373
Evaluation methodology, 323
Evans, Annette, 328–329
Evidence, recording concerning
accident, 217
Evolved gas effects, 479
Excello Corporation, 56–57
Exchange rate, 627
Exclusivity provision, 359
Execution, in implementing
total safety management
model, 874–875
Executive commitment, in total
safety management, 871
Exhaust, local, ventilation and,
590
Exhaustion, stage of, 301
Exhaust ventilation, 574–574
Exothermic, 512
Expectancy, in SPICE model, 204
Expectations of safety-related
behaviors and attitudes,
886–887
Experience, 767
Experienced miner, 318
Experience rating, 185
Expert system, 732
Expiration, 471

Explosion, 514
Explosive hazards, 535–537
explosives-related concepts,
535–536
health hazards of explosive
materials, 537
uses of flammable and
combustible substances,
536–537
Explosive materials, 493, 537
Explosive range, 517
Exposure control officer, 701
Exposure control plan, 700
Exposure to pressure,
decompression sickness
and, 472
Extension cords, do's and
don'ts of, 504
External communication
procedures, in
emergencies, 729
Extrahazardous occupations,
184
Extraterritorial employees,
workers' compensation
coverage of, 176, 177
Extremely low frequency (ELF)
radiation, 617, 661
Eye appeal of product
literature, 248
Eye hazards, assessing
workplace for, 429
Eye injuries, first aid for, 431
Eye protection, 426–428
Eye strain, 661
Eyewash stations, general
requirements for, 576

Face protection, 426–428
Facilitators, TSM, 868–869
Facility alarm system, 729
Factories, inspection of, 5
Failure mode and effects
analysis (FEMA), 769–770,
771
Failure to abate prior to
violation, 131
Fainting, 454, 455
Falling objects, protection
from, 424

Fall Protection Standard
(ANSO Z359.1: Safety
Requirements for Personal
Fall Arrest Systems,
Subsystems, and
Components), 418
Falls
accidental deaths from, 24
aftermath, 423
as cause of injury, 186
causes of, 411–412
kinds of, 412
monitoring protection
equipment, 424
prevention programs, 416
Falls protection
equipment, monitoring, 424
language of, 417
OSHA's recommendations
for, 420
False negative result, 689
False positive test, 689
Family problems, as source of
workplace stress, 298
Fatigue, 670
Fault finding, 213
Fault tree analysis (FTA), 775,
776, 777
Fear factor, 178
Federal Coal Mine and Safety
Act, 7
Federal Employer's Liability
Act, 177
Federal laws, on chemical
spills, 781–783
Federal Metal and Non-metallic
Mine Safety Act, 7
Federal Mine Safety and
Health Amendments Act
of 1977, 148
Federal Railroad
Administration (FRA), 150
Feedback
in administrative controls,
370
immediate and continual,
320
lack of, 225–226, 297
in workstation analysis,
269–270

Feeding and ejection systems, 399–400
Feero, William E., 619–620
Fellow servant rule, 11, 178
FEMA. *See* Failure mode and effects analysis (FEMA)
Fern, Bruce, 56
Fettig, Art, 804, 805
Fibromyalgia, 32
Financial awards, for safety and health training, 314
Finger pointers, 363
Firearms, accidental deaths from, 25
Fire brigades, 538
 OSHA's regulations for, 524–526
Fire dangers to humans, 520
Fire extinguishers
 characteristics of, 523
 portable, 522
Fire extinguishing systems, 522–523
Firefighting equipment, 525
Fire hazards
 defined, 512–516
 detection of, 521
 identification of, 518
 reducing, 521–527
 sources of, 516–519
Fire point, 517
Fire protection, 154–155
 self-assessment in, 539
 Web sites for, 532
Fire ratings
 for doors, 531
 for floors, 531
 for walls, 531
Fire-related injuries, accidental deaths from, 24
Fire-related losses, 23–24
Fire safety programs, 533–535
 assessment in, 534
 awareness and prevention in, 534–535
 planning in, 534
 response in, 535
Fire safety standards, development of, 527–528

Fire standards, OSHA's, 528–529
First aid
 for chemical burns, 467
 for corrosive materials injuries, 154
 in emergencies, 716–717
 for eye injuries, 431
 fact sheet, 720–721
 recommended contents of, 719
 training program, 717
First Aid Essentials (NSC), 717
First-degree burns, 463
Fixed guards, 394
Flame-resistant clothing, 532–533
Flammable and Combustible Liquids Code (NFPA), 595
Flammable Fabrics Act, 235
Flammable range, 517
Flammable substance, 535
Flashing back, 518
Flash point, 517, 535
Flat-attenuation HPDs, 644–645
Flexible hours, in reducing workplace stress, 305
Flexor tenosynovitis (trigger finger), 31, 288
Fluid loss, 466
Focal dystonia (writer's cramp), 32
Focus groups, workplace violence and, 368
Follow-up
 after hearing loss, 639–640
 on bloodborne pathogens postexposure, 702
 of CTD, 274
 inspections, 129–130
 in training employees, 336–337
 in workstation analysis, 269–270
Food, Drug, and Cosmetics Act, 235
Foot protection, 432
Foot rails, 439
Footwear
 nonskid, 414

OSHA's regulations relating to, 432–433
 proper, 439
 terms related to, 433
Forearm tendinitis, 32, 289
Foreign object, 412
Foreseeability, 167–168
Forklift hazards, ambient noise in increasing, 447
Forklift safety (Powered industrial trucks), 444–448
 initial training program content, 446–447
 OSHA's training requirements, 446
 refresher training program content, 447–448
Forklift Standard, OSHA's, 347
Formable earplugs, 643
Formal education, for safety and health managers, 73
Formal instruction, 347
Formal response, 799
Form 300, OSHA's, 123, 141, 275, 276
Form 300A, OSHA's, 122, 128
Form 301, OSHA's, 122, 124, 212
Four-step teaching method, 321
Four-to-one ratio, 421
FRA. *See* Federal Railroad Administration (FRA)
Fractures, 387
 comminuted, 389
 complete, 389
 compound, 389
 incomplete, 389
 oblique, 389
 simple, 389
 transverse, 389
Fraud, workers' compensation and, 199–200
Freedom through control, in quality management, 863, 864
Free environment, 824
Freeze, 495
Frequency, 489, 629, 775
Friable asbestos, 562
Front-page test, 750

FTA. *See* Fault tree analysis (FTA)
Fuels, solid, 516–517
Full ergonomics program, 263
Full-potential approach, 753
Fumes, 559
Fuses, 500

Gamma radiation, 605
Ganglion cysts, 32, 289
Garbage, 820
Garcia, Camillo, 748, 753, 756
Garner, Amanda, 242
Gas compression, 482
Gases, 559
Gas exchange, 471
Gas leaks, causes of, 481
Gates, 398
Gauley Bridge disaster, 12
GE Aircraft Engines, 205
Gear certification, 161
Geller, E. Scott, 55–56
General Dynamics, 648
General Electric, 881
General environmental controls, 153–154
General observation, 278
General strategies, workers' compensation and, 202–204
Georgia-Pacific Corporation, best practices at, 789–790
GFCI. *See* Ground fault circuit interrupter (GFCI)
GFI. *See* Global fault interrupter (GFI)
Gice, Jon, 204
Global competitiveness, safety and, 106
Global fault interrupter (GFI), 498
Globalization, 880–883
Global marketplace, 100, 103
Global Technologies, 720–721
Global warming, 820
Gloves materials, common, 440–441
Good housekeeping, 576
reducing fall hazards with, 413, 414

Goto, Minoru, 674
Government agencies, 73, 76
workplace safety and, 15
Gray area, 749
Great care, 165, 167
Great Depression, 7
Greenman v. Yuba products, Inc., 234
Gross negligence, 164
Grounded conductor, 490
Ground fault, 498
Ground fault circuit interrupter (GFCI), 498
Ground fault interrupter (GFI), 498
Grounding conductor, 490
Ground-level ozone, 820
Ground wire, 490
Guards point-of-operation, 393–396
Guidance standard, 842
Guides, 842
Gulf Coast Manufacturing, 722
Guyon's canal syndrome, 32, 289

H. R. Johnson Company, 53–54
Halcarz, Joseph, 345, 346
Hallock, Richard G., 774
Halogenated refrigerants, 566
Hamilton, Alice, 85
Hammurabi, 4
Hand-arm vibration syndrome (HAV), 647
Handbook for Electrical Safety in the Workplace (National Fire Protection Association), 505
Handles, in workstation analysis, 271
Hand protection, 439–441
common glove materials, 440–441
Hand tools, 156
checklist for safe selection and use of, 291
work with, 283
Hard hats, 425–426, 427
Harmful environment, 186
Hartshorn, Dan, 211

Hate group, membership in, 362
HAV. *See* Hand-arm vibration syndrome (HAV)
Hawk's Nest Tragedy, 7–9
Hayashi, Yoshio, 673
Hazard analysis
cost-benefit factors in, 767–769
detailed, 769–777
overview of, 766–767
preliminary, 767–769
Hazard and operability review (HAZOP), 770, 772, 773
Hazard communication programs, 312
Hazardous airborne emissions, 560
Hazardous and Solid Waste Amendments (HSWA), 824
Hazardous condition, 42
Hazardous energy (lockout/tagout systems), control of, 400–406
Hazardous locations for electrical equipment, 495
Hazardous materials, 152–153
OSHA's standards for, 595–596
securing, 740–741
transportation of, Uniform Safety Act and, 583–584
Hazardous noise, 627
Hazardous noise conditions, identifying and assessing, 638–640
Hazardous product inventory, 596
Hazardous substances, highly, 554
Hazardous Substances Act, 235
Hazardous waste operations and emergency response (HAZWOPER) standard, 548, 781
Hazardous waste reduction, 830–835
Hazardous waste standard, 119–120, 831–832

Hazardous work employees, workers' compensation coverage of, 176, 177

Hazard prevention and control design of work methods, 270–271
tool design and handles, 271
workplace violence and, 368–369
workstation design, 270

Hazard protection devices, 430

Hazard(s)
biological, 551, 553
chemical, 551
defined, 470, 765
of the environment, 552, 825–830
ergonomic, 548, 553
inventory, 779
levels and risk, 630–631
lifting, 269, 433–437
lighting, 493
mechanical, 386
personal protection from, 574
physical, 551
potential, identifying, 338
prevention and deterrence, 778–779
recognition and evaluation, 572–573
reducing, VDTs and, 284–285
standing, 437–439
vibration, 646–648
in workplace, 551–553

HAZOP. See Hazard and operability review (HAZOP)

HAZWOPER standard. See Hazardous waste operations and emergency response (HAZWOPER) standard

HBV. See Hepatitis B virus (HBV)

HCV. See Hepatitis C virus (HCV)

HEA. See Human error analysis (HEA)

Head protection, 425–426

Health, workplace stress and, 302

Health and environmental controls, 152, 621

Health and Morals of Apprentices Act, 1802, 5

Health and safety managers, 435

Health care providers, 271
monitoring trending for CTD, 275
protocols for, 274
role in education and training of employees, 274
trained and available, 272

Health Partnership Programs (HPPs), 196

Health physicist, 83–84

Health problems
electromagnetic fields and, 618–619
role of specific, 11–13

Health programs, safety and, 78

Health surveillance, 273

Hearing, threshold of, 629

Hearing hazards monitoring, 633–634

Hearing loss, factors affecting risk of, 630

Hearing loss prevention
corporate policy and, 649–650
engineering and administrative controls and, 634–635
terms related to, 626–628

Hearing loss prevention programs, evaluation of, 638, 651–655

Hearing protection devices (HPDs), 643–646, 653–654

Hearing protector, 636

Hearing threshold level (HTL), 627

Heart disease, 25

Heat
environmental, 453
metabolic, 453
radiant, 453

Heat burn injuries, 30–31

Heat cramps, 454

Heat exhaustion, 453
symptoms of, 455

Heating, ventilation, and air conditioning (HVAC), 499

Heat rash, 454

Heat strain
defined, 454
recognizing, 455

Heat stress
clothing and, 455–456
defined, 453–454
guidelines and materials, 458
management of, 456–458

Heat syncope or fainting, 4554

Heat transfer, 514

Heidel, Stephen, 58

Heinrich, Herbert W., 41
domino theory of accident causation, 41–43

Helplessness, 660

Henningson v. Bloomfield Motors, Inc., 234

HEPA. See High-efficiency particle absolute (HEPA)

Hepatitis B virus (HBV), 681, 695–697, 697

Hepatitis C virus (HCV), 697–699

Hertz (Hz), 627–628

Heuristics, 732

Hidden costs, estimating, 34–35

High-efficiency particle absolute (HEPA), 563

High-mass impact test, 428

High-pressure systems, hazards of, 475

High-radiation area, 606

High-risk occupations, workplace violence and, 353

Highspeed electrons, 605

High-speed protons, 605

High surface contact, low vs. high, 260

High-temperature water hazards, 473–474

High-temperature water (HTW), 473–474

High-velocity impact test, 428

Hispanics, on-the-job fatality rate of, 345

HIV. *See* Human immunodeficiency virus (HIV)

HIV antibody test, 689

Hole, 417

Home problems, as source of workplace stress, 298

Homicide, as cause of death, 352

Honeywell (GE), 573

Horizontal work area, good *vs.* bad, 259

Hose systems, 522–523

Hostility, unattended, 364

Hot wire, 490

How, in interviewing witnesses, 219–222

HPDs. *See* Hearing protection devices (HPDs)

HPPs. *See* Health Partnership Programs (HPPs)

HSWA. *See* Hazardous and Solid Waste Amendments (HSWA)

HTL. *See* Hearing threshold level (HTL)

HTW. *See* High-temperature water (HTW)

Human Aspects of Occupational Vibration (Wasserman), 646

Human error, 43, 44, 46, 52, 770

Human error analysis (HEA), 772

Human factors
in action, 257
defined, 257
ergonomic hazards and, 257–258
safety and, 258

Human factor theory of accident causation, 43–46

Human immunodeficiency virus (HIV), 681

Human nature, accounting for, in conducting interviews, 228

Human reactions, workplace stress and, 300–301

Human resource management (HRM)
in safety management, 786
as source of workplace stress, 299

Human-robot interaction, 663–664

Humans
electrical hazard to, 495–496
fire dangers to, 520
pressure dangers to, 478–479

Human skin, 462–463

Human stress response, stages of, 300–301

Humidification, 498–499

Huttle, Joe Don, 48

HVAC. *See* Heating, ventilation, and air conditioning (HVAC)

Hydrogen, 513

Hydrogen cyanide, 560

Hydrogen sulfide, 560

Hydrostatic test, 521

Hyperbaric chambers, 478

Hypergolic, 515

Hyperoxia, 472, 478

Hypoxia, 478

Hypothermia, 459

IAQ. *See* Indoor air quality (IAQ)

Identification, in implementing total safety management model, 874–875

IEQ. *See* Indoor environmental quality (IEQ)

IFA test. *See* Immuno-florescent (IFA) test

IGC Contracting, Inc., 135

Ignition temperature, 512

Ignorance, as reason for not reporting accident, 224

Illiteracy
fighting, industry's role in, 344–345
impact on industry, 343
OSHA and problems of, 343–344

problem of, 104
safety and, 342–347

ILO. *See* International Labour Organization (ILO)

Image
company's, 72
competitiveness and, 106–107
corporate, 106

Immediacy
in interviewing witnesses, 219
in SPICE model, 204

Immediate feedback, 320

Immediate response, 799

Immuno-florescent (IFA) test, 689

Impact, 777

Impact accidents, 28

Impact hazards, 424–433

Implementing, in waste reduction program, 834

Implementing Total safety Management (Prentice Hall), 861

Improvement project teams (IPTs), 868
steering committee, 867–868

Impulse noise, 629

Impulsive noise, 628

Inanimate power, 5

Inappropriate activities, 45, 46

Inappropriate responses, 44, 46

Incentives, in safety promotion, 806–808

Incidence rates, 126, 267

Incident(s)
analysis of, 368
notification of, in radiation hazards, 611
OSHA's definition of, 611

Incipient fire brigades, 538

Income replacement, in workers' compensation, 175

Incomplete fracture, 389

Independent contractors, 187

In-depth investigation, failing to conduct, 228

Indirect costs, 23–24

Individual factors, violence and, 362–263
Indoor air quality/HVAC standard, ANSI, 566–567
Indoor air quality (IAQ), 565, 566
  toxic mold and, 567–569
Indoor environmental quality (IEQ), 825–830
  action plan for ensuring, 827–830
  solving problems in-house, 827
Indoor environmental quality information, 567
Induced environment, 823
Industrial accidents, 3, 11, 40
Industrial employees, working conditions for, 3
Industrial engineering, 78–79, 81
Industrial engineers, safety and, 81–82
Industrial hygiene, 48–49
  airborne contaminants in, 558–559
  airborne toxics in, 559–561
  asbestos hazards in, 561–565
  carcinogens in, 561
  confined space hazards in, 585–592
  defined, 544
  entry points for toxic agents in, 553–556
  general safety precautions, 584–585
  hazard recognition and evaluation in, 572–573
  indoor air quality in, 565–567
  NIOSH and, 577–578
  NIOSH guidelines for respirators, 578–581
  OSHA (OSH Act) and, 546–550
  OSHA's hazard communication standard, 596–597
  OSHA's standards for toxic and hazardous materials, 595–596

  overview of, 545
  prevention and control in, 573–577
  relationship of doses and responses in, 557–558
  standards and regulations in, 545–546, 581–584
  threshold limit values in, 569–571
  toxic mold and indoor air quality in, 567–569
  toxic substances in, 553, 556–557
  workplace hazards in, 551–553
Industrial hygiene chemist and/or engineer, 17
Industrial hygienist, 83
Industrial medicine, robots and, 670–671
Industrial Revolution, 4–5, 659
Industrial safety engineer, 18
Industrial Safety Equipment Association (ISEA), 440
Industrial safety managers, 18
Industrial sector, productivity in, 102
Industry
  death rates by, 28–29
  impact of illiteracy on, 343
  role in fighting illiteracy, 344–345
Infection, 466
Information
  about indoor environmental quality, 567
  coordination of, 731
  on hazards of bloodborne pathogens, 703
Information requirements, 583, 714
Information sources, in records analysis and tracking, 266
Infrared radiation, 616
Infrared thermal imaging, 497
Ingestion, 555
Inhalation, 554
Initial training program content, 447–448
Injection, 556

Injury hiding, 807
Injury(ies). See also Mechanical injuries; Work injuries
  arc flash, prevention of, 505
  associated with vibration, 647
  back, 433–434
  from chemical burns, 29–30
  eye, first aid for, 431
  fire-related, 24
  to foot and toes, 432
  heat burn, 30–31
  impact, 35–36
  needlestick, 704–705
  nonscheduled, 190–191
  repetitive-motion/ soft tissue, 28
  repetitive strain/soft-tissue, 31–32
  workers' compensation and, 175–176, 183, 186
Innovations in American Government Award, 139
Input, 100
In-service training, for safety and health managers, 73
Inspection(s), 614
  follow-up, 129–130
  of ladders, 421
  programmed, 130
  safety, 434
  tour, 129
  of workplace, 128–129
Inspiration, 471
Institute for Corporate Competitiveness, 99
Institute for Preventive Medicine (IPM), 809
Institute for Work & Health, 292
Instruction
  formal, 347
  planning, 321–323
  programmed, 330
Instructional approach, 323
Insulation, double, 500
Insulation failure, 494
Insulators, 488
Insurance
  administration of, 23–24
  workers' compensation, 183–185

Insurance organizations, 165
Insured costs, 32–33
Interactive video training, 330–331
Interior structural fire brigades, 538
Interlocked guards, 394
Interlocks, 500
Internal combustion engines, 536–537
Internal factors, 44
International Conference of Building Officials and Code Administration (BOCA), 527
International Labour Organization (ILO), 35, 36
International Medical Commission, 10
International Organization for Standardization (ISO). *See* ISO
International Plastics Corporation (IPC), 754
International Safety Equipment Association (ISEA), 445
International Union v. Micro Manufacturing, Inc., 359
Interviews
   identifying ergonomic problems, 278
   witnesses, 219
Inventory, of hazardous product, 596
Investigation techniques, failing to learn, 228
Involvement, in team-building strategies, 243, 244
Ionization sensors (radiation sensors), 521
Ionizers, 499
Ionizing radiation, 552
   terms and concepts of, 605–606
IPM. *See* Institute for Preventive Medicine (IPM)
IPTs. *See* Improvement project teams (IPTs)
Irritants, 559–560

ISEA. *See* Industrial Safety Equipment Association (ISEA)
Island of automation, 659
ISO, 841–842
ISO 14000, 842
   process-oriented documents, 845, 846
   success story, 856
ISO 14001 standard
   concepts of, 844–847
   requirements of, 847–855
ISO 14000 series of standards
   classification of, 843
   guides and technical reports, 842–843
   specification *vs.* guidance standards, 842
   subclassifications of, 843
Isolation issue, in quality management, 863–865
IV drug users, 683
Ives v. South Buffalo Railway Company (201 N. Y. 271, 1911), 179–180

Japanese Association of Industrial Health, 668
Japanese miracle, 881
Japanning and drying ovens, 536
Job autonomy, 303
Job descriptions, 304, 887
Job hazard analysis, 269, 767, 768
Job safety analysis (JSA)
   sample form, 339
   training opportunities available, 340–342
   as training technique, 338–340
Job satisfaction, 204–205
Job security, as source of workplace stress, 298
Job-specific procedures, 336
Johnson, Linda F., 405
Johnson & Johnson, 755

Kaliokin, Alex, 434
Kauffman, John, 735

Keltronics, inc., organizational structure of, 836
Keypone tragedy, 573
Kinetic energy, 514
Kitchen Manufacturing Incorporated (KMI), 45–46
Kitchens, preventing slips and, 415
Knotts, I. B., 687

LaBar, Gregg, 311
Labeling
   of hazardous substances, 597
   of potential biohazards, 703
Lacrimation, 661
Ladders
   do's and don'ts of use, 421
   inspecting, 421
   safety of, 421, 423
Lamp black, 561
LANs. *See* Local area networks (LANs)
Lanyard, 417
Lasers, 617
Latent defects, 237
Lateral protection, hard hats and, 426
Laws and liability, 163–168
   ability to pay, 167
   care, 165, 167
   damages, 167
   foreseeability, 167–168
   legal principles of, 163–164
   liability, 165
   negligence, 164–165
   proximate cause, 167
   tort, 167
   willful/reckless conduct, 167
Lawsuit process, product liability and, 233
Lockout device, 401, 402
Lead, 820
Leak detectors, in detecting pressure hazards, 481
Learning, principles of, 320
Learning objectives, 323
Lecture-discussion method, 325–327
Legal concerns on AIDS in the workplace, 686–690

Legality, 749

Legality, ethics and, 749

Legal liability, EMS movement and, 839

Legal principles, of law and liability, 163–164

Legal reasons, for training, 311–315

Legionnella, 553

Legislation, of workers' compensation, 179–180

LEPCs. *See* Local emergency planning committees (LEPCs)

Lesson plans, 322–323

Let-go current, 495

Lethal concentration, 558

Lethal dose, 558

LFL. *See* Lower flammable limit (LFL)

Liability, 165
  employer, for workplace violence, 359
  manufacturer's, 234
  of products, 232

Lifeline, 417, 590

Life safety, 529–532
  basic requirements of, 529–531
  means of egress, 531–532

Life Safety Code (National Fire Protection Association), 529, 532

Liftline, 417

Lifting hazards
  back safety/lifting program, 434–435
  NIOSH guidelines, 436–437
  proper techniques, 435–436
  in workstation analysis, 269

Lifting techniques, proper, 435–436

Light-duty jobs, keeping list of, 273

Lightning hazards, 493
  control of, 495

Lignite, 561

Lin authority, 68

Liquid penetration test, 477

Literacy training, role of safety and health professionals in, 345

Litigation, 182

Liver disease, chronic, 698

Load, 490

Lobbies, building, preventing slips and, 415

Local area networks (LANs), 660

Local emergency planning committees (LEPCs), 237, 238, 582, 713, 714

Local exhaust, ventilation and, 590

Localizing emergency action plan, 730–731

Location-specific, 730

Lockout/tagout devices, 402

Lockout/tagout language, 401–402

Lockout/tagout programs, evaluating, 405–406

Lockout/tagout standard, 154

Locus of control, 751

Longshoring, 160–161

Lorenzo Textile Mill, Inc., 135

Loss of focus, whistle-blowing and, 760

Loss of wage-earning capacity theory, 191–192

Lost time, work injuries and, 27

Lost wages, 23–24

Lower flammable limit (LFL), 586

Low-slope roof, 417

Low-vibration tools, 648

Lung disease
  in 1800, 11
  respirable silica exposure and, 564

Machiavellianism, 751

Machine guarding
  OSHA's requirements for, 390
  self-assessment in, 391, 399

Machine operation, in risk assessment, 390–392

Machinery and machine guarding, 155–156

Machine safeguards
  basic program content, 407
  feeding and ejection systems, 399–400
  general precautions in, 406–407
  point-of-operation devices, 397–398
  requirements for, 390
  robot safeguards, 400
  taking corrective action, 407–408

MacPherson v. Buick Motor Company, 234

MACT. *See* Maximum achievable control technology (MACT)

Magnetic circuit breakers, 500

Magnetic particle test, 477

Maine OSHA's top 200 Program, 139

Maladaptation, 670

Malpractice, 182

Managed care, integrated, 205

Managed care organization (MCO), 196

Management, accidents and, 221

Management by walking around (MBWA), 786

Management commitment, workplace violence and, 365–366

Management failures, accident causation and, 59–61

Management policy of confined space, 592–595

Management support, 673

Managers
  employee stress and, 306
  responsibility for exposure control, 700

Manual material handling, 277

Manual rating, 185

Manual work, 659

Manufacturer liability, 234

Marine terminals, 160

Maritime employees, workers' compensation coverage of, 176, 177

Maritime standards, of OSHA, 159–161

Marshall v. Barlow's Inc., 129

Martensson, Lena, 663

Martin Marietta Corporation, 755

Material hearing impairment, 628

Material safety data sheets (MSDs), 238, 312, 551–552

Material safety data sheets (MSDSs) file, 597

Materials handling and storage, 155

Maximum achievable control technology (MACT), 581–582

Mazda Corporation, 246

MBWA. *See* Management by walking around (MBWA)

McGuire, Maxine, 203

Meaninglessness, 672

Mechanical engineering, 78

Mechanical hazards, 42, 386

Mechanical injuries
breaking, 387, 389
common, 386–389
crushing, 387, 388
cutting and tearing, 387
puncturing, 389
shearing, 387
straining and spraining, 389

Mechanical suffocation, accidental deaths from, 25

Mechanization, 659

Medical complications, 25

Medical costs, workers' compensation and, 182–183

Medical examinations, of ACM employees, 564–565

Medical expenses, 23–24

Medical management
of corrosive materials injuries, 154
of workplace injuries, 196

Medical management program, 271–272

Medical records, of ACM employees, 564–565

Medical rehabilitation, 195

Medical reports, 372

Medical treatment, 193–195

Memorandum of understanding (MOU), 821

Mentoring personnel, in desired safety-related behaviors and attitudes, 889

Mercury, 12–13, 820

Merit Program, 139

Metabolic heat, 453

Metal and Non-Metallic Mine Safety Act, 150

Metatarsal guard, 433

Methyl-ethyl-ketone, 561

Methyl isocyanate (MIC), 10

MIC. *See* Methyl isocyanate (MIC)

Micro Electronics Manufacturing (MEM), 374

MicroTel Corporation, 769

Microwave (MW) radiation, 616

Microwave oven, 232–233

Mil-Tech Manufacturing Company, 748, 753, 756

Mindlessness, 672

Mine Safety and Health Act, 235

Mine Safety and Health Administration (MSHA), 150

Minetos, Peter, 684, 685, 686

Mining accidents, 3

Minor burns, 464

Minors, workers' compensation coverage of, 176, 177

Minter, Stephen, 57

Minutes of safety meetings, 372

Mirror test, 750

Mismatch, as cause of accidents, 215

Mission of teams, 811, 812

Mistrust, as inhibitor in safety and health training, 314

Mists, 559

Mobile *vs.* stationary jobs, 259

Modeling safety-related behaviors and attitudes, 887

Moderate burns, 464

Moisture vapor transfer rate (MVTR), 454

Monetary benefits, of workers' compensation, 193

Monitoring
in administrative controls, 370
in safety management, 786–788
safety-related behaviors and attitudes, 889–890
in waste reduction program, 834

Morale, employee, 26

Morning-after test, 750

Mortality, 749

Motivation, in hearing loss prevention, 637

Motor vehicle accidents
accidental deaths from, 24, 25
cost of, 22

Mountain Trek, Inc., 670

MSDs. *See* Material safety data sheets (MSDs); Musculoskeletal disorders (MSDs)

MSHA. *See* National Mine Health and Safety Academy (MSHA)

MTC Corporation, 812

Multiple risk factors, 268

Multitask analysis strategy, 436

Muscle damage, 31

Musculoskeletal disorders (MSDs), 262, 263, 675

Mutually monogamous relationship, 694

MVTR. *See* Moisture vapor transfer rate (MVTR)

Myofacial muscle damage, 287–288

NACOSH. *See* National Advisory Committee on Occupational Safety and Health (NACOSH)

Nader, Ralph, 236

Narcotics, 561

Narrow band noise, 629
National Advisory Committee on Occupational Safety and Health (NACOSH), 114–115
National Center for education Statistics, 344
National Coal Workers' Autopsy Program, 148
National Council of Industrial Safety (NCIS), 6, 16
National Electrical Code (NEC), 157, 490, 501, 503
National Electrical Manufacturers Association, 499
National Electrical Safety Code, 503
National Electronic Information Surveillance System (NEISS), 236
National Fire Protection Association (NEPA), 490, 504, 505, 518, 532
National Institute for Occupational Safety and Health (NIOSH), 276, 312
  on chemical protective clothing, 577
  description of, 16, 115, 146
  division of Biomedical and Behavioral Science, 146–148
  Division of Surveillance, Hazard Evaluations, and Field Studies, 149
  Division of Training and Manpower Development, 149–150
  Division of Respiratory Disease Studies, 148
  ergonomic guidelines for manual material handling and, 436–437
  on establishing electrical safety program, 502–503
  Federal Railroad Administration, 150
  on fire protection issues, 532

  guidelines for categorizing injury, 359–360
  guidelines for respirators, 578–579
  on heat stress guidelines and materials, 458
  industrial hygiene and, 577–578
  on job hazard analysis, 767, 768
  Mine Safety and Health Administration, 150, 315, 318–319
  Occupational Safety and Health Review Commission, 150
  on prevention of needlestick injuries, 704
  training opportunities at, 340–341
  on vibration exposure of employees, 648
  on video display terminals, 661
National Institute of Mental Health, 59
National Mental Health Association, 59
National Mine Health and Safety Academy (MSHA), 341
National Response Center, 716
National Safety Council (NSC), 3, 23
  about injured eyes by chemicals, 467
  on accident rates, 26
  accident report forms, 223–224
  authority of product safety coordinator, 239
  on characteristics of trainers, 319
  on conducting accident investigation, 213
  on consumer feedback collection and analysis, 246
  definition of safeguarding, 389

  on failing to follow safety procedures, 311
  first-aid training of, 717
  formation of, 6, 16
  on photographing and videotaping accident scene, 218
  on physiological process in chemical burns, 465
  on product liability, 232
  on product safety program record keeping, 244
  recommendations in inspecting ladders, 421
  recommendations on minimizing robots hazards, 666–667
  on responsibility of product safety auditor, 239–240
  on robots hazards, 664–665
  on safety and health professional role in product safety and liability, 241–242
  safety course for supervisors, 334–335
  statistical records of injuries, 186
  on supervisor safety training, 333–334
  training opportunities available, 341–342
  on transportation of toxic substances, 583
  on uses of flammable and combustible substances, 536
National Silicosis Conference, 12
National Traffic and Motor Vehicle Safety Act, 235
Natural disasters, deaths from, 22–23
Natural environment, 823
Natural surveillance, in risk reduction, 360
NCIS. See National Council of Industrial Safety (NCIS)
Near misses, failing to investigate, 227

NEC. *See* National Electrical Code (NEC)

Neck fatigue, 667

Need for safety-friendly corporate culture, 884–885

Needlestick injuries, preventing and responding to, 704–705

Negative charge, 487, 488

Negative pressures or vacuums, 482

Negligence, 164–165
  comparative, 164
  contributory, 11, 164, 178
  gross, 164
  worker, 178

Negligent design, 234

Negligent manufacture, 233, 234

NEISS. *See* National Electronic Information Surveillance System (NEISS)

Neoplastic growth, 561

NEPA. *See* National Fire Protection Association (NEPA)

Nerve and circulation disorders, 32

Neurotransmitters, 58

Neutral wire, 490

Neutrons, 488, 605

New employees, safety and health training for, 335–337

New York Court of Appeals, 179–180

New York Power Authority, best practices at, 788–789

Nickel compounds, 561

Nissan Motor Company, 674

Nitrogen narcosis, 472

Noise, 552. *See also* Noise control strategies; Noise hazards
  continuous, 627
  hazardous, 627
  identifying and assessing hazardous conditions, 638–640
  impulse, 629

impulsive, 628
  measurement of, in hearing loss prevention, 651–652
  narrow band, 629
  problems, taking measurement in, 647
  standards and regulations, 631–638
  wide band, 629

Noise control strategies, 640–646
  administrative controls, 643
  engineering controls, 641–643
  hearing protection devices, 643–646

Noise dose, 627

Noise hazards
  effects of, 649
  workers' compensation and, 638

Noise-induced hearing loss, 628

Noise reduction rating (NRR), 644

Noise surveys, 639, 650

Noncovered whistle-blowing, 762

Nondestructive testing, 481

Nonionizing radiation, 615–617

Nonmetals, 18

Nonrepetitive motion, repetitive motion *vs.*, 260

Nonscheduled disabilities, 190–191

Nonskid footwear, 414

Normlessness, 672

Notetaking, in interviews, 220

Notice of contest, 134

Notice of proposed rule making, 115

NRC. *See* Nuclear Regulatory Commission (NRC)

NRR. *See* Noise reduction rating (NRR)

NSC. *See* National Safety Council (NSC)

Nuclear engineering, 79

Nuclear Regulatory Commission (NRC), 607, 612–615, 760

NUS Training Corporation, 330

Obesity, accident causation and, 61

Objectives of teams, 811, 812

Objectivity, safety rules and, 797

Oblique fracture, 389

Obsessions
  with quality, 862, 864
  romantic, 362

Occupational diseases, 112, 178, 675

Occupational Hazards, 134, 135, 345

Occupational Health and Safety Management Systems (OHSMSs), 790–791

Occupational health nurse (OHN), 86–87, 271

Occupational Health & Safety, 695

Occupational noise exposure, 152

Occupational physician, 84–86

Occupational Safety and Health Act (OSH Act), 16, 151. *See also* Occupational Safety and Health Administration (OSHA)
  coverage of, 113–114
  employee rights under, 142–143
  employer responsibilities under, 141–143
  employer rights under, 140–141
  industrial hygiene and, 546–550
  legislation of, 111
  product liability laws, 235
  rationale for, 112
  on record keeping and reporting, 120–127
  requirements for free hazards workplace, 686, 687
  safety and health manager and, 66
  on safety and health training, 311–312

Occupational Safety and Health
  Administration (OSHA),
  16, 17. *See also*
  Occupational Safety and
  Health Administration
  (OSHA) regulations;
  Occupational Safety and
  Health Administration
  (OSHA) standards
assessment for eye hazards,
  429
citations and penalties,
  131–133
on confined space, 587–590,
  592
on construction, 418, 420
on emergency fire safety plan
  requirements, 534
enhanced enforcement policy
  of, 129
on fall protection, 420
firefighting options, 537–538
form 300, 123, 141, 275, 276
form 300A, 122, 128
form 301, 122, 124, 212
on health and environmental
  controls, 152
on heat stress guidelines and
  materials, 458
on investigating toxic mold
  in workplace, 567–568
keeping employees informed
  on, 128
keeping up-to-date on,
  143–144
lockout/tagout systems,
  400–406
on log of injury and illness,
  371–372
on machine guarding
  requirements, 390
on material safety data
  sheets, 312
mission and purpose,
  112–113
on occupational exposure to
  bloodborne pathogens,
  699–704
partnership with EPA, 821
poster 2203, 122

on prevention measures of
  boiler and pressure
  hazards, 473
problems with, 144
Process Safety Standard, 547
on record-keeping and
  reporting exceptions,
  127–128
on record keeping forms,
  274–275
on record keeping
  requirements, 121–126,
  222–223
regulations *vs.*, 115
services available from,
  137–140
site-specific targeting
  process, 130
standards *vs.*, 115
state-level programs, 134–137
on vision protection devices
  criteria, 430
voluntary ergonomics
  guidelines, 261–265
voluntary guidelines,
  364–373
Web site of, 532
whistle blowing and,
  760–761
Occupational Safety and Health
  Administration (OSHA)
  general industry standards
Subpart C: General Safety
  and Health Provisions, 151
Subpart D: Walking-Working
  Surfaces, 151, 418
Subpart E: means of Egress,
  151–152
Subpart F: powered
  Platforms, 152, 418
Subpart G: health and
  Environmental Controls,
  152
Subpart H: Hazardous
  Materials, 152–153
Subpart I: Personal Protective
  Equipment, 153
Subpart J: General
  Environment Controls,
  153–154

Subpart K: Medical and First
  Aid, 154
Subpart L: Fire Protection,
  154–155
Subpart M: Compressed
  Gas/air, 155
Subpart N: Materials
  Handling and Storage, 155
Subpart O: Machinery and
  Machine Guarding,
  155–156
Subpart P: Hand
  Tools/Portable Power
  Tools, 156
Subpart Q: Welding, Cutting,
  and Brazing, 156
Subpart R: Special Industries,
  156–157, 418
Subpart S: Electrical, 157
Subpart T: Commercial
  Diving Operations,
  157–158
Subpart Z: Toxic and
  Hazardous Substances,
  158–159
Occupational Safety and Health
  Administration (OSHA)
  regulations
for chemical spills, 548–550,
  781–783
for fire brigades, 524–526
on hazard communication,
  312–313
relating to footwear, 432–433
relating to noise exposure,
  633
Occupational Safety and Health
  Administration (OSHA)
  standards, 114–129
adoption, amendment or
  revocation of, 115
appeal, 116–117
confined space, 118–119
construction, 162
for control of hazardous
  energy, 403–405
electrical, 501–502
emergency preparation,
  715–716
ergonomics, 260–262

fire, 528–529
Forklift, 347
hazard communication,
   596–597
hazardous waste, 119–120,
   831–832
maritime, 159–161
mission and purpose,
   112–113
reading, 115–116
requesting variance, 117–118
safety, 780
temporary emergency, 116
toxic and hazardous
   materials, 595–596
and training, 347
Occupational Safety and Health
   Review Commission
   (OSHRC), 134, 150
Occupational Safety and
   Workplace violence, 353
Occupational violent crime
   (OVC), 353
Octave-band analyzers, 634
Office automation
   ergonomics in, 669
   safety and health in, 667–668
Office fires, preventing,
   526–527
Office of Disability
   Employment Policy, 724
Offices, video display terminals
   in, 660–663
Off-shoring, 881
Ohio Division of Safety and
   Hygiene, 205
Ohms, 489
Ohm's law, 489
OHN. *See* Occupational health
   nurse (OHN)
OHSMSs. *See* Occupational
   Health and Safety
   Management Systems
   (OHSMSs)
Oil burners, 536
Oncogen, 561
Ongoing monitoring, in safety
   management, 786–788
Online training (E- learning),
   331

On-the-job fatality rate, for
   Hispanics, 345
*On the Miners' Sickness and
   Other Miners' Diseases*
   (Aureolus), 4
Open-ended interview, 219
Open ground, 493
Opening, 326, 417
Opening conference, 129
Open-sourcing, 881
Operational control, 852
Operator errors, gas leaks
   and, 481
Optimization, 646
Order of response, in emergency
   preparation, 714
Organizational structure, 836
Organizations
   corporate culture of, 879
   dealing with safety and
      health, 145
   developing standards and
      codes, 164–165, 166
   in emergency preparation,
      714–715
   ethical behavior in, 751
   ethics role in, 753–756
   for safety and health
      managers, 73–75
   small, environmental
      management and, 838
Organized culture, 298
Organized labor, role of,
   10–11
Orientation, in training
   employees, 336, 337
OSHA. *See* Occupational Safety
   and Health Administration
   (OSHA)
OSH Act. *See* Occupational
   Safety and Health Act
   (OSH Act)
OSHA Hazardous Waste
   Standard (1910.120), 824
OSHA regulations. *See*
   Occupational Safety and
   Health Administration
   (OSHA) regulations
OSHA standards. *See*
   Occupational Safety and

Health Administration
   (OSHA) standards
OSHRC. *See* Occupational
   Safety and Health Review
   Commission (OSHRC)
Other-than-serious violation,
   131
Out of employment (AOE)
   injury, 186–187
Output, 100
Outside assembly, in
   evacuation planning, 730
Outsider, 353
Outsourcing, 881
OVC. *See* Occupational violent
   crime (OVC)
Overexertion, 28, 186
Overload, 43–44, 45–46
   as cause of accidents, 215
Overpressurization, gas leaks
   and, 481
Ownership, 799
Oxygen limit, 536

Pain
   cost of, 259
   threshold of, 629
Paints, 561
Paraffin oils, 561
Parcel Delivery Service (PDS),
   49–50
Partial pressures, 471
Participation safety information
   (PSI), 547
Participatory ergonomics (PE),
   282
Passive earmuffs, 636
Patent defect, 237
Path, 641
Patience, 751
PDCA model. *See* Plan-do-
   check-adjust (PDCA) model
PDS. *See* Parcel Delivery
   Service (PDS)
PE. *See* Participatory
   ergonomics (PE)
Peak performance, 99, 867
PELs. *See* Permissible exposure
   limits (PELs)
Penalty, 131

"Peopleless" factory, 663
Percent better or worse sequential, 632
Percent worse sequential, 632
Perez, Juan, 54–55
Perfluorocarbons, 566
Performance, peak, 99
Performance oriented, in total safety management, 871
Periodic ergonomic surveys, in workstation analysis, 269
Periodic health surveillance, 273–274
Periodic workplace walk-through, 272
Peripheral nerve entrapment, 669
Permanent partial disability, 190–192
Permanent total disability, 192
Permanent variance, 117–118
Permissible exposure limits (PELs), 158, 562
Permit requirement, in confined space standard, 118
Permitting procedures, for organizations exposing workers to confined spaces, 593–594
Persistence, 751
Personal beliefs and feelings, as causes of accidents, 214–215
Personal commitment, 804–805
Personal fall arrest system, 417
Personal hearing protection devices, 636–637
Personal medical devices, threshold limit values and, 620
Personal monitoring devices, 606
Personal monitoring precautions, radiation hazards and, 607
Personal protective equipment (PPE), 505
  applicable standards, 153
  description of, 701–702

hazards and, 564
staying up-to-date on, 454
strategies for using, 442–444
use of, 574
Person-machine environment, 54
Personnel
  instructing and informing about radiation, 609–610
  rescue, 588
  training for, confined space and, 593
Personnel orientation to safety-related behaviors and attitudes, 887–888
Perspective, 751
Persuasion, in safety promotion, 814–815
Pesticides, 561
Peterson, Dan, 46
Petition for modification of abatement (PMA), 133–134
Petroff, Dale, 740
Petroleum Processing, Inc., 855
PHA. See Preliminary hazard analysis (PHA)
PHAs. See Process hazard analyses (PHAs)
Photoelectric devices, 397
Photoelectric fire sensors, 521
Photographing accident scene, 218
Photography, in identifying ergonomic problems, 278
Photoluminescent signs, 729
Physical agents, 571
Physical fitness, decompression sickness and, 472
Physical hazards, 42, 551
Physical stress
  factors associated with, 258–260
  reducing, 303–304
  with standing job, 281
  work with hands above chest and, 283
  work with hand tools and, 283
Physician(s)
  employers and selection of, 185
  occupational, 84–86

Pitch, 561
Plan-do-check-adjust (PDCA) model, 845, 847
Planning
  AIDS education program, 690–691
  for emergencies, 237, 582, 597
  for emergency fire safety, 534
  in emergency preparation, 712, 719–723
  for evacuation, 728–730
  in implementing total safety management model, 874–875
  recovery analysis and, 736
  safety and health instruction, 321–324
  for safety-friendly corporate culture, 885–886
  for workers with disabilities, 724–728
Planning instruction, 321–323
Plant map, 730
Platform, 417
Platforms, powered, 152
PMA. See Petition for modification of abatement (PMA)
Point-of-operation devices, 397–398
Point-of-operation guards, 393–396
Poisoning, accidental deaths from, 24
Poison Prevention Packaging Act, 235
Polaroid, 344
Politics
  ergonomics and, 260–261
  investigation and, 228
Portable fire extinguishers, 522
Portable power tools, 156
Positioning device system, 417
Positive charge, 488
Poster 2203, OSHA's, 122, 128
Postexposure investigation, bloodborne pathogens and, 702–703

Postincident response, workplace violence and, 370–371

Posttraumatic stress disorder, 735

Potter, Mark, 56–57

Poultry Processing, Inc. (PPI), 67, 68

Poultry Processing Corporation (PPC), 46–48

Powered platforms, 152

Powerlessness, 671

The Power of Ethical Persuasion (Rusk), 374

PPE. *See* Personal protective equipment (PPE)

Practicing, in emergency preparation, 712

Preceding factors, 42

Precision Machining and Manufacturing, 687

Precision Tooling Company (PTC), 52–54

Predispositional characteristics, 49

Preemployment screening, 434

Pregnancy, work and, 44

Preliminary hazard analysis (PHA), 766

Premium discounting, 185

Premium rates, 198
of employers, 185

Premolded earplugs, 643

Preparation
for emergencies, 237, 582, 597
in implementing total safety management model, 874–875
rescue, 590–592
as step in teaching method, 321

Preparing the Workplace for Everyone, 724

Presentation, as step in teaching method, 321

Presentation methods, in safety and health instruction, 324–331

Pressure
atmospheric, normal, 629
extremes of, as hazard in workplace, 552–553

Pressure hazards, 470–483
boilers and, 473
cracking, in vessels, 475–476
defined, 470–471
of high-pressure systems, 475
high-temperature water hazards, 473–474
measurement of, 480–481
reduction of, 482–483
sources of, 472–473
of unfired vessels, 474

Pressure vessels
cracking hazards in, 475–476
hazards of unfired, 474
nondestructive testing of, 477–478
storage of, 482

Prevention
in fire safety programs, 534–535
of hazard, 778–779

Prevention program, for back safety, 435

Pride, 750

Primary witnesses, 218

Principal's office syndrome, 219

Principles of learning, 320

Privacy, no respect for, 364

Private insurance, 184

Probability levels, 777

Problem identification, in design process, 79

Process engineers, 83

Processes
analyzing, in waste reduction program, 834
targeting, in waste reduction program, 834

Process hazard analyses (PHAs), 547

Processing areas, preventing slips and, 415

Process-oriented standards, 843

Process safety information (PSI), 780

Process Safety Standard, OSHA's, 547

Produce and deliver, in design process, 79

Production *vs.* safety, 69

Productive conflict, 373

Productivity
competitiveness and, 100–103
cost and, 105
defined, 100
employees and, 375
global view of, 101–103

Product liability, 232

Product liability laws, 233
applying, 236–237
burden-of-proof for, 236
community Right-to-Know Act, 237
history of, 234–235
statutory, 235–236
strict liability in tort, 235

Product literature and safety, 246–248

Product-oriented standards, 843

Product safety and liability
developing, 238–240
evaluating product safety program, 240–241
product liability and the law, 233–238
product literature and safety, 246–248
product safety program, 238–240
product safety program record keeping, 244–246
quality management and, 242–244
role of safety and health professional, 241–242
user feedback collection and analysis, 246

Product safety auditor, 239–240

Product safety committee, 239

Product Safety Coordinator, 238–239

Product safety program
developing, 238–240

Product safety program (*Contd.*)
  evaluating, 240–241
  record keeping, 244–246
Product safety team, 245
Professional associations, 164
Professional societies, 73, 74
Program implementor,
  corporate policy and, 649
Programmed inspections, 130
Programmed instruction, 330
Projective communication for
  persuasion, 814–815
Proof pressure tests, 481
Property damage, 23
Proper work practices, in
  administrative controls, 370
Proposed penalty, 131
Protocols, for health care
  providers, 274
Protons, 488
Proximate cause, 167
Proximity, in SPICE model,
  203–204
Prudent man concept, 237
PSI. *See* Process safety
  information (PSI)
Psychological support, as source
  of workplace stress, 299
Psychological trauma, dealing
  with in emergencies,
  733–735
Psychophysiological
  techniques, in measuring
  workplace stress, 301
Psychosocial questionnaires, in
  measuring workplace
  stress, 302
Psychosocial stressors, 668
Psychotic behavior, 362
Public awareness, 130
Public hearing, 198
Pullback devices, 397
Pulp digester vessels, 47
Puncture-resistant guard, 433
Puncturing, as type of
  mechanical injury, 389
Punitive damages, 167
Purchases training programs,
  checklist for evaluating,
  333

Purging confined spaces, 590
Purpose, 750
Pyrophor hypergolic
  fuels, 516

QM. *See* Quality management
  (QM)
Quality, 70. *See also* Quality
  management (QM)
  competitiveness and, 105
  defined, 103
  global view of, 103, 105
  poor, 277
Quality management (QM),
  103, 842
  concept of, 862–863
  defined, 861
  elements of, 243
  product safety and, 242–244
  relationship to, 863–865
  setting, safety management
  in, 865
Quality of work life, 673–674
Quality product, 242
Questionnaires, for identifying
  ergonomic problems, 278

Rad, 606
Radial tunnel syndrome, 32
Radiant heat, 453
Radiation, 514
  extremely low frequency,
  617
  infrared, 616
  ionizing, 552, 605–606
  microwave, 616
  radio frequency, 616
  ultraviolet, 616
  visible, 616
Radiation area, 606
Radiation control specialists,
  17–18
Radiation exposure, record of,
  614
Radiation hazards
  caution signs and labels, 608
  electromagnetic fields in
  workplace, 618–621
  evacuation warning
  signal, 609

exposure of employees to,
  606–607
  instructing and informing
  personnel, 609–610
  ionizing, 605–606
  nonionizing, 615–617
  notices to employees,
  612–615
  notification of incidents, 611
  OSHA standards for health
  and environmental
  controls, 621
  precautions and personal
  monitoring, 607
  reports and records of
  overexposure, 611–612
  storage and disposal of
  radioactive material, 610
Radiation sensors, 521
Radioactive material, 605, 610,
  613
Radioactive source, 613
Radio-frequency devices, 397
Radio frequency radiation, 616
Radio frequency (RF), 616
Radiography test, 477–478
Railroad workers, workers'
  compensation coverage of,
  176, 177
Ramazzini, Bernardino, 5, 85
Rameses II, 4
Rates
  incidence, 126, 267
  of insurance, of employers,
  185
  of work accidents, 26
Rating, premium, of employers,
  185
Raynaud's disease, 32, 289
RCRA. *See* Resource
  Conservation and Recovery
  Act (RCRA)
Real-world baseline audiogram,
  627
Reasonable accommodation,
  686
Reasonable care, 165
Reasonable risk, 237
Receiver, 641
Receptacle wiring tester, 497

Receptive communication for persuasion, 814
Recognition, hazard, 572–573
Recommended weight limit (RWL), 436
Record analysis and tracking, 266
Record keeping
  of audiometric tests, 639
  on employees exposed to bloodborne pathogens, 704
  exceptions to, 126
  in hearing loss prevention program, 637
  monitoring, in hearing loss prevention, 652–653
  OSHA's forms, 274–275
  OSHA's requirements for, 121–126, 222–223
  product safety, federal regulations requiring, 245
  product safety program, 244–246
  workplace violence and, 371–373
Record of training programs, 372
Records
  access to, 125
  cumulative radiation, 612
  keeping of, 120
  monitoring and tracking of, 366–367
Record-spoiling, as reason for not reporting accident, 224
Recovery analysis and planning, 736
Recovery coordinator, 735–736
Recovery team, 736
Recycling, 832
Red tape, as reason for not reporting accident, 224
Reduction, 88
Reduction options, identifying, in waste reduction program, 835
Reenactment technique, 222
Referrals, in hearing loss prevention, 653

Reflex sympathetic dysfunction, 32
Reform, in workers' compensation, 200–201
Refresher training program content, 446–447
Refrigerator Safety Act, 235
Registry of Toxic Effects of Chemical Substances, 312
Rehabilitation, workers' compensation and, 195
Rehabilitation Act, 686, 725
Rehabilitation program, in workers' compensation, 175–176
Reinforcement, in team-building strategies, 244
Reinforcing and maintaining safety-related corporate culture, 890
Related expertise, 767
Religious employees, workers' compensation coverage of, 176, 177
Rem, 606
Remediation of toxic mold, 568–569
Repeat violations, 131
Repercussions, fear of, as reason for not reporting accident, 225
Repetitive motion, nonrepetitive *vs.*, 260
Repetitive strain injury (RSI), 31
Repetitive-strain/soft tissue injuries, 28
Reporting requirements
  in Community Right-to-Know Act, 238
  OSHA's, 121, 122–123
Reports/reporting
  of accidents in investigation, 222–227
  exceptions to, 126
  of radiation overexposure, 611–612
  toxic chemical release, 238, 583, 714
Rescue personnel, 588

Rescue preparation, 590–592
Resistance, 489
  stage of, 300
Resource Conservation and Recovery Act (RCRA), 824
Resources, lack of, 70
Respect, in team-building strategies, 244
Respirable silica exposure, lung disease and, 564
Respirators, 592
  contact lenses and, 580–581
Respiratory protective devices, 526
Response, in fire safety programs, 535
Response time, 70, 105–106
Responsibility(ies)
  delegating, 304
  of employees, 143
  of employers, 141–142
  for safety and health training, 313–315
  as source of workplace stress, 298
Restraint devices, 397–398
Restrooms, preventing slips and, 415
Resuming business
  after disaster, 741–742
  federal guidelines for, 743
Retribution, whistle blowing and, 759
Retrofit, 414
Retrospective rating, 185
Return-to-work program, in workers' compensation, 203
Revere Copper Products, Inc., 447
Reversed polarity, 493
Rewarding teams, 807
RF. *See* Radio frequency (RF)
Richfield Toys, 242
Rights
  conflicting, 358
  of employees, 142–143, 685
  of employers, 140–141
  of violent employees, 358–359

Risk. *See also* Risk assessment; Risk factors
  assumption of, 11, 178
  defined, 87
  hazard, 630–631
  injured employee and, 178
  management of, 87–88
  reasonable, 237
  unreasonable, 237
Risk analysis, 775, 777
Risk assessment
  defined, 390
  in machine operation, 390–392
  of potential hazards, 783–784
Risk factors
  back disorder, 268
  cumulative trauma disorders (CTDs), 268
  ergonomics, 268
  in machine operations, 392
  multiple, 268
Risk management, 87
Risk manager, 87–88
Risk-reduction strategies, 360–362
Robbins Engineering Corporation, 406
Robot medicine, 660
Robots
  industrial medicine and, 670–671
  safety and health problems associated with, 664–667
Robot safeguards, 400
Rockwell International, 344
Role ambiguity, 297, 364
Role clarity, in teamwork approach to promotion safety, 810–811
Role-reversal test, 750
Romantic obsessions, 362
Roosevelt, Franklin D., 693
Roosevelt, Theodore, 179
Rope grab, 417
Rotator cuff tendonitis, 289
Royal Institute of Technology in Stockholm, 663
RSI. *See* Repetitive strain injury (RSI)

Run-in point hazards, 387
Rusk, Tom, 374

Safeguarding, defined, 389
Safeguards
  requirements for, 392–393
  robot, 400
Safety. *See also* Safety and health; Safety management; Safety movement; Safety promotion
  chemical engineers and, 82–83
  as cultural imperative, 881–882
  culture and, 205
  engineers and, 75–79
  environmental engineers and, 82
  globalization of competition and, 880–883
  health programs and, 78
  human factors and, 258
  illiteracy and, 342–347
  improvements in, 3
  industrial engineers and, 81–82
  inspecting, 434
  life, 529–532
  production *vs.*, 69
  supervisor's role in, 60
  three E's of, 14
Safety and health
  integrated approach to, 17–18
  in office automation, 667–668
  problems associated with robots, 664–667
  role of competitiveness in improving, 104–107
  states with approved plans, 136
Safety and health managers, 66–75
  education and training for, 72–73
  job of, 67–68
  problems faced by, 69–72
  role in company hierarchy, 68–69

Safety and health movement, 17–18
Safety and health problems, robots and, 664–667
Safety and health profession, 19
Safety and health professionals, 18
  certification of, 88–93
  on dealing with confined spaces, 587
  ethical dilemmas faced by, 759
  language of lockout/tagout terms, 401–402
  role in environmental safety, 824–825
  role in ethics, 751–753
  role in industrial hygiene, 545
  role in literacy training, 345
  role in product safety and liability, 241–242
  sources on safety for, 741
Safety and health promotion, 786
Safety and health teams, 65–67
Safety and health training
  applying, 331
  English as a second language training issues, 345–347
  evaluating, 331–333
  legal and ethical reasons for, 311–315
  for new and transferred employees, 335–337
  preparing, 321–324
  presenting, 324–331
  rationale for, 311–315
  requirements of, 315–319
  safety and health professionals as trainers, 319–321
  for training supervisors, 333–335
Safety awareness, 6–7
Safety committees, 802–804
Safety course for supervisors, 334–335
Safety engineer, 80–81

Safety-friendly corporate
    culture
    creation of corporate culture,
        882–883
    defined, 878–879
    globalization of competition
        and safety, 880–882
    importance of having,
        879–880
    steps for establishing,
        884–890
Safety glasses, 427, 428
Safety management
    accident investigation and
        reporting in, 786
    best practices in, 788–790
    communication in,
        785–786
    employee training in, 785
    human resource management
        in, 786
    ongoing monitoring in,
        786–788
    policy in, 785
    in QM setting, 865
    safety and health promotion
        in, 786
    self-assessments in, 786
    strategic planning in, 785
    written procedures in, 785
Safety movement
    development of accident
        prevention programs,
        13–14
    developments before
        Industrial Revolution,
        4–5
    milestones in, 5–7, 8
    new materials, new
        processes, and new
        problems, 18
    organized labor in, 10–11
    rapid growth in profession, 19
    safety organizations in, 14–16
    specific health problems in,
        11–13
    tragedies that have changed,
        7–10
    war and, 15
Safety needles, 695

Safety organizations,
    development of, 14–16
Safety Performance Solutions,
    Inc., 55
Safety policy, 785, 796
Safety professionals,
    electromagnetic fields and,
    619
Safety promotion
    committees in, 802–804
    company policy in, 796–797
    company-sponsored wellness
        programs in, 808–809
    competition in, 808
    employee-management
        participation in, 805–806
    employee participation in,
        798–799
    incentives in, 806–808
    personal commitment in,
        804–805
    persuasion as a tool in,
        814–815
    role of supervisors in, 806
    rules and regulations in,
        797–798
    suggestion programs in, 799
    teamwork approach to,
        809–813
    training in, 798–799
    visual awareness in, 799–802
Safety-related behaviors and
    attitudes
    employees expectations to,
        886–887
    mentoring personnel in, 889
    modeling for, 887
    monitoring and evaluating,
        889–890
    orientation of personnel to,
        887–888
    orientations for employees,
        887–88
    training personnel in, 889
Safety-related work practices,
    157
Safety rules and regulations,
    797–798
Safety training, 798–799
Safety trip devices, 398

SARA. See Superfund
    Amendments and
    Reauthorization Act (SARA)
SBCC. See Southern Building
    Code Congress
    International (SBCC)
Scapegoating, whistle-blowing
    and, 760
Scents, in detecting pressure
    hazards, 481
Schedule disabilities, 190
Schedule rating, 185
Schedules of drills, 723
Scherer, Robert, 305–306
School Board of Nassau
    County v. Arline, 1987, 688
Scientific approach, in quality
    management, 862, 864
Scientific decision making, in
    total safety management,
    872
Scientific standards and testing
    organizations, 73, 74
Screening surveys, in worksite
    analysis, 267
Screening test, 689
Seated control work, 280–281
Seated repetitive work with
    light parts, 279–280
Seated work with larger parts,
    280
Secondary witnesses, 218
Second-degree burns, 463
Second Restatement of Torts
    (American law Institute),
    235
Section 11(b) summary
    enforcement orders, 130
Security analysis, 368
SEIU. See Service Employees'
    International union (SEIU)
Self-assessment
    for electrical hazards,
        503–504
    in fire protection, 539
    machine guarding, 399
    in safety management, 786
Self-insurance, 184
Self-protection strategies,
    576–577

Self-retracting lifeline/lanyard, 417

Selikoff, Irving J., 9

Semiautomatic ejection, 400

Semiautomatic feed, 399–400

Semiconductors, 488, 489

Sensorineural loss, 627

Sensors
 ionization (radiation), 521
 photoelectric fire, 521

Septicemia, 466

SERCs. *See* State emergency response commissions (SERCs)

Serious violation, 131

Service, 72, 106

The Service Contract Act of 1965

Service Employees' International union (SEIU), 661

Settlements, 130

Severity levels, 775

Sexual contact, 683

Shalimar Machine Tool Company, 328–329

Shearing, as type of mechanical injury, 387

Sheathes, 287

Shelter-in-place (SIP), 724–725

Shielding approach, electromagnetic fields in workplace and, 620–621

Shift work, 302
 reducing stress associated with, 303

Shipbuilding, 159–161

Shipyard employment, 160

Shock, 466–467

Shoe inserts, 438–439

Short circuit, 491

Short-cycle operations, 304

Shoulder fatigue, 667

Shoulder tendinitis, 31, 288–289

SHRM. *See* Society for Human Resource Management (SHRM)

SIC codes. *See* Standard Industrial Classifications (SIC) codes

Sick-building syndrome, 565–567, 825

Significant threshold shift, 627, 635–636

Signs, for potential biohazards, 703

Silicosis, 7–9

Simmonds, Rollin H., 32

Simple fracture, 389

Simplicity, in SPICE model, 203

SIP. *See* Shelter-in-place (SIP)

Sit/stand chairs, 439

Sitting, standing *vs.*, 259

Situational characteristics, 49

Situational ethics, 752

Situational factors, 44

Sketching, in identifying ergonomic problems, 278

Skidmore, Hubert, 9

Slight care, 167

Slip and fall accidents, 412

Slip resistant guard, 433

Slips prevention
 general strategies for, 413–414
 programs for, 416
 specific strategies for, 415–416

SME. *See* Society of Manufacturing Engineers (SME)

Smith, Sandy, 856

Smoke, 559

Smoking
 hand-arm vibration syndrome and, 647
 prohibiting, 584

Social environment, 42

Social pathological phenomena, 670

Society for Human Resource Management (SHRM), 59, 356

Society of Manufacturing Engineers (SME), 7, 13, 16, 178, 211, 213, 766, 778

Sociotechnical system theory, 673–674

Solid fuels, 516–517

Soot, 561

SOPs. *See* Standard operating procedures (SOPs)

Sound
 characteristics of, 628–629
 in detecting pressure hazards, 481

Sound level meter, 639

Sound level recorders, 634

Sound levels, selected, 630

Sound waves, 629

Source reduction, 832

Sources, on terrorism, 741

Southern Building Code Congress International (SBCC), 527

Spark, 493

Spark-resistant tools, using, 585

SPC. *See* Statistical process control (SPC)

Specification *vs.* guidance standards, 842

"SPICE" model, 203, 204

Spill cart, 733

Spillover violence, 356

Spontaneous combustion, 515

Spraining, as type of mechanical injury, 389

Spray-painting booths, 537

Squeeze-point hazards, 387

Staff authority, 68

Stage of alarm, 300

Stage of exhaustion, 301

Stage of resistance, 300

Standard Industrial Classifications (SIC) codes, 126

Standard operating procedures (SOPs), 547, 780

Standards, 163
 for confined spaces, ANSI, 594
 industrial hygiene, 545–546
 Occupational Safety and Health Administration, 347

Standard threshold shift (STS), 635

Standing
 for heavy lifting and carrying, 282
 sitting *vs.*, 259

Standing hazards
  antifatigue mats, 437
  foot rails, 439
  proper footwear, 439
  shoe inserts, 438–439
  sit/stand chairs, 439
  workplace design, 439
Standing work, 281
Standpipe and hose systems, 522
Star Program, 139
State-dependent, 301
State emergency response commissions (SERCs), 237, 238, 714
State funds, 184
State-level OSHA programs, 134–137
Statement of philosophy of Martin Marietta code of ethics, 755
Statement of purpose, 323
  in AIDS education course outline, 691
Static electricity, 488–489
  eliminating, 585
Stationary *vs.* mobile jobs, 259
Statistical process control (SPC), 344
Steel toe, 414, 433
Stenosing tenosynovitis, 31, 288
Step and fall accidents, 412
Stimulation, as presentation method, 329
Storage and disposal, of radioactive material, 620
Straining, as type of mechanical injury, 389
Strategic basis, in total safety management, 869–870
Strategic planning
  in safety management, 785
  total safety management and, 869–870
Strategies
  accident prevention, 723
  cost-reduction, 202–205
  developing emergency notification, 272

ergonomic problem-solving, 279–285
injury prevention, 647–648
risk-reduction, 360–362
self-protection, 576–577
of teamwork in quality management, 243–244
Stress
  employees, managers and, 306
  in safety managers, 305–206
  workers' compensation and, 186, 306–307
  workplace, 296–297, 660
Stress claims, 199, 306
Stressors, 52
Strict liability, 165
  in tort, 234, 235
Strokes, 25
Structural integrity, after disaster, 741
STS. *See* Standard threshold shift (STS)
Stump and fall accidents, 412
Subjective ratings, in measuring workplace stress, 301
Substances
  highly hazardous, 554
  toxic, 556–557
Suffocation, accidental deaths from, 25
Suggestion programs, 799
Suicides, 619
Sulcus ulnaris syndrome, 32
Sulfur hexafluoride, 566
Summer-injury epidemic, 43
Sunburn, 463
Superaural caps, 644
Superfund Amendments and Reauthorization Act (SARA) of 1986, 237, 582, 712
Superior value, customers and, 881–882
Supervision, partial, inconsistent, 364
Supervisors
  do's and don'ts for, 377–378

in hearing loss prevention, 651
responsibility regarding exposure control, 700
role in safety promotion, 806–808
role in workplace safety and health, 60
safety course for, 334–335
training, 333–335
Supply-chaining, 881
Surface tests, 477
Surface traction, measuring, 413
Surveys
  annual, 273
  instituting, 272–273
  periodic ergonomic, 269
  screening, 267
  symptom, 272
Symptoms reporting, health care providers and, 274
Symptoms survey, 272
Synthesis, in design process, 79
Systematic cost-reduction program, 205
Systems failure, as cause of accidents, 215
Systems theory, of accident causation, 50–54

Tagout, 402
Tagout device, 402
Taping interviews, in accident investigation, 220, 222
Task analysis, in ergonomics, 277
Task complexity, as source of workplace stress, 297
Taylor, Frederick, 6
Taylor, Paul, 749
TCE. *See* Trichloroethylene (TCE)
TCLP. *See* Toxic characteristics leaching procedure (TCLP)
Teaching method, four-step, 321
Teach mode, 665
Team charters, 245, 812, 887, 888

Team-oriented rewards, 811
Teams. *See also* Teamwork
  building through training, 813
  characteristics of effective, 810
  potential problems with, 812–813
  recovery, 736
  responsibilities of members, 813–814
  rewarding, 807
  safety and health, 66
  trauma response, 734
Teams players skills, in promotion safety, 810
Teamwork
  approach to safety promotion, 809–813
  orientation, in total safety management, 871
  potential benefits in promoting safety, 811–812
  in quality management, 243, 862, 864
Tearing, as type of mechanical injury, 387
Technical and trade associations, 164–165
Technical reports, 842
Technic of operation review (TOR), 774
Technological alienation, 671
Technology access, in reducing workplace stress, 305
Temperatures
  auto-ignition, 517
  control of, in environmental hazards, 552
  decompression sickness and, 472
  excessive, gas leaks and, 481
Temporary disability, 188–190
Temporary emergency standards, 116
Temporary partial disability, 188
Temporary total disability, 188–190
Temporary variance, 117

Tendinitis, 3, 287
  forearm, 289
  shoulder, 31, 288–289
Tendon-related disorders, 667
Tenosynovitis, 31, 288, 662
Territoriality, establishment of, 361
Terrorism in workplace
  role of employer, 738–740
  securing hazardous materials, 740–741
Tertiary witnesses, 218
Testing for AIDS, 685, 688–690
Testing organizations, 165
Tests for hepatitis C virus (HCV) infection, 698
Textile industry, 573
Thermal comfort, 452–453
Thermal expansion detectors, 521
Thermal imaging, infrared, 497
Third-degree burns, 463–464
Thoracic outlet syndrome, 32, 289
Threshold limit value-ceiling (TLV-C), 571
Threshold limit value-short-term exposure limit (TLV-STEL), 571
Threshold limit values (TLVs), 459, 587, 617
  ACGIH's classification of, 570–571
  description of, 569–570
  personal medical devices and, 620
Threshold limit value-time-weighted average (TLV-TWA), 570–571
Threshold of hearing, 629
Threshold of pain, 629
Time loss, work injuries and, 27
Time-weighted averages (TWAs)
  calculating, 571
  threshold limit value, 570–571
Time-weighted average (TWA), 628

Tints, 561
TLVs and BEI: Threshold Limit Values for Chemical Substances and Physical Agents and Biological Exposure Indices, 453
Tool design and handles, in workstation analysis, 271
Tools, low-vibration, 648
TOR. *See* Technic of operation review (TOR)
Tort, 167
  strict liability in, 235
Total quality control (TQC). *See* Quality management (QM)
Total quality leadership (TQL). *See* Quality management (QM)
Total quality management (TQM). *See* Quality management (QM)
Total quality (TQ). *See* Quality management (QM)
Total Safety Management (Goetsch), 865
Total safety management (TSM)
  concept of, 865–867
  defined, 861, 866–867
  facilitator in, 868–869
  fundamental elements of, 869–874
  implementing the model, 874–876
  introduction of, 865
  rationale for, 874
  translating into action, 867–869
Toxic agents, entry points for, 553–556
Toxical chemical release reporting, 238, 583, 714
Toxic and hazardous substances, 158–159
Toxic characteristics leaching procedure (TCLP), 548
Toxic chemical release reporting, 238, 714
Toxic materials, OSHA's standards for, 595–596

Toxic mold
- assessment and remediation, 568–569
- indoor air quality and, 567–569

Toxic substances
- defined, 553
- effects of, 556–557

Toxic Substances Control Act, 236

Tracer gas techniques, 565

Trade association, 73, 77

Traditional training methods, 346

Trainers, safety and health professionals as, 319–321

Training. *See also* Education; Safety and health training
- on back safety/lifting program, 434
- for effective ergonomics program, 275–276
- of employees, 274, 403–404
- for evacuations, 730
- of fire brigades, 524
- on hazards of bloodborne pathogens, 703
- in hearing loss prevention, 651
- insufficient, 364
- job safety analysis as technique in, 338, 340
- local, in emergencies, 731
- OSHA's services for, 140, 316–318, 347, 446–448
- for personnel of confined space, 593
- in quality management, 244, 862, 864
- for safety and health managers, 72–73
- in safety promotion, 798–799
- team building through, 813
- in use of vision protection devices, 430
- workplace stress and, 303–303
- workplace violence and, 371

Training aids, 323

Training methods, modifying, 346

Training programs, first aid, 717

Training Project Grants, 340

Transference, 88

Transferred employees, safety and health training for, 335–337

Transportation of toxic substances, 583584

Transverse fracture, 389

Trapped gas effects, 478

Traps, as cause of accidents, 215

Trauma, 733–734

Trauma response team (TRT), 734

Travelers insurance company, 41

Treatment of CTD, 274

Trend analysis, in workstation analysis, 270

Trend monitoring, workplace violence and, 368

Triangle Shirtwaist Factory, 190

Trichloroethylene (TCE), 560

Trip and fall accidents, 412

TRT. *See* Trauma response team (TRT)

TSM. *See* Total safety management (TSM)

Tsuchiya, Kensaburo, 675, 676

Tumorigen, 561

Tunnel syndromes, 289

Two-hand controls, 399

U. S. Chamber of Commerce, 181

U. S. Court of Appeals, 116–117, 134

U. S. Department of Justice, 353, 356

U. S. Department of labor, 9, 12, 112, 115, 117, 134, 146, 193, 275, 344, 364

U. S. Public Health Service, 692, 694, 702

U. S. Supreme Court, 129, 684

UFL. *See* Upper flammable limit (UFL)

Ulnar nerve disorders, 32

Ultrasonic test, 478

Ultraviolet (infrared) detectors, 521

Ultraviolet radiation, 616

Unattended hostility, 364

Uncontested cases, 197–198

Uncooperative people, failing to account for effects of, 229

Underwriters Laboratories (UL), 490

Unfair programs, 807–808

Unfired pressure vessels, hazards of, 474

Uniform attenuation earmuffs, 636

Uniform Safety Act, hazardous materials transportation and, 583–584

Uninsured costs, 32–33

Union Carbide India Ltd., 10

Union Carbide Texas City plant, 715

United States
- AIDS cases in, 681–682
- cost of accidents, 868
- death from chronic liver disease, 698
- productivity in, 102

Unity of purpose, 863, 864, 873–874

Unreasonable risk, 237

Unrestricted area, 606

Unsafe acts, as cause of accidents, 42, 215

Unsafe behavior, 46

Unsafe conditions, as cause of accidents, 215

Upper flammable limit (UFL), 586

Useful consciousness, 478

User feedback collection and analysis, 246

USX, 26

Utility checks, after disaster, 740–741

Vaccination, Hepatitis B virus (HBV), 702
Vacuum mentality, 243
Vacuums, 482
Value added, 100
Values, 749
Van der Mark v. For Motor Company, 234–235
Vapors, 513, 559
Variance control, 673
Variances
 other, 118
 permanent, 117–118
 temporary, 117
VDTs. *See* Video display terminals (VDTs)
Ventilation
 of confined spaces, 590
 industrial hygiene and, 573–574
 local exhaust and, 590
 of space with airborne contaminants, 588
Vertical incidence tribometers (VITs), 414
Vertical work areas, good *vs.* bad, 260
Vibration hazards, 646–648
Vibrationnoise-induced hearing loss, 627–629
Vibration syndrome, 289
Video display terminals, workstation checklist for, 264–265
Video display terminals (VDTs), 617
 cumulative trauma disorders and, 662
 in offices and factories, 660–663
 working with, 283–285
Videotapes/DVDs, as presentation method, 329–330
Videotaping, in identifying ergonomic problems, 278
Videotaping accident scene, 218
Violations, 613, 614
 de minimis, 131

failure to abate prior to, 131
 other-than-serious, 131
 repeat, 131
 serious, 131
 willful, 131
Violence
 record of, 362
 spillover, 356
Visible radiation, 616
Vision protection devices, requirements when choosing, 430
 training in, 430
Visual awareness, in safety promotion, 799–802
Visual examination, 477
Vitacom, Inc., 761
VITs. *See* Vertical incidence tribometers (VITs)
Vocational rehabilitation, 195
Volatility, 536
Voltage, 489
Volumetric tests, 477
Voluntary ergonomics guidelines, OSHA's, 261–265
Voluntary guidelines, OSHA's, 364–373
Voluntary Protection Programs (VPPs), 137–138, 789
VPPs. *See* Voluntary Protection Programs (VPPs)

Wage-loss theory, 191
Wagner v. White Star Candy Company, 217 N. Y. Supp. 173, 179
Walking and slipping, 412–416
 factors decreasing traction, 413
 measuring surface traction, 413
 strategies for preventing slips, 413–416
Walking-working surfaces, 151
War, safety movement and, 15
Wasserman, Donald, 646
Waste reduction audit, 834
Waste reduction plan, 833

Waste reduction program, organizing, 832
Waste reduction team, forming, 832–833
Water, handling equipment exposed to, 499
Water hammer, 482
Water pollution, 820
Weaver, D. A., 774
Web site(s)
 employee Assistance Professionals Association, 59
 for fire protection issues, 532
 National Institute of Mental Health, 59
 National Mental Health Association, 59
 of NIOSH, 532
 of OSHA, 532
 in workplace terrorism, 740
Welch, Jack, 881
Welding, cutting, and brazing, 156
Wellness program, 809
West Coast Power Company, 759
Western Blot test, 689
Wet bulb globe temperature (WBCT), 456
Wet hydrogen sulfide vessels, 476
What, in accident investigation, 213–215
When, in accident investigation, 212–213
When, in interviewing witnesses, 219
Where, in interviewing witnesses, 219
Whistle-blowing
 defined, 758
 ethics and, 758–762
 noncovered, 762
 OSHA and, 760–761
 problems with, 759
Who, in accident investigation, 216
Whole-person theory, 191

Why, in accident investigation, 227–229
Wide band noise, 629
Will-Burt Company, 311
Willful/reckless conduct, 167
Willful violation, 131
Wilson, Jack, 761
Wind, cooling effect of, 460
Windchill, 459
Wiring, improper, 493–494
Witnesses
    identifying, 218–219
    interviewing, 219–222
    primary, 218
    secondary, 218
    tertiary, 218
Work
    with hands above the chest, 283
    with hand tools, 283
    pregnancy and, 44
    with VDTs, 283–285
Work areas
    horizontal, 259
    vertical, 260
Work design, 662
Work envelope, 400, 665
Work environment, measuring, 278
Worker habits, vibration and, 647
Worker negligence, 178
Workers' compensation, 6
    accident prevention under, 176
    administration and case management, 197–198
    claims and character, 189
    complaints about, 306
    concept of, 174–175
    cost allocation under, 176, 198
    cost-reduction strategies under, 202–205
    coverage methods for selected states, 184
    disabilities and, 188–192
    future of, 200–202
    historical perspective of, 177–179

income replacement under, 175
    injuries and, 183, 186
    insurance, 183–185
    legislation, 179–180
    medical management of workplace injuries, 196
    medical treatment and rehabilitation, 193–195
    modern, 180–183
    monetary benefits of, 193
    noise hazards and, 638
    objectives of, 175
    opponents of, 182
    problems with, 198–199
    rehabilitation and, 195
    rehabilitation of injured employee under, 175–176
    resolution of disputes, 185–186
    scams in Florida, 201
    spotting fraud and abuse, 199–200
    stress and, 186, 306–307
    workers covered by, 176–177
Workers' Compensation Act, 236
Work group organization, 673
Work habits, changing, of employees, 648
Work hardening, 435
Work injuries
    death rate from, 3
    injuries to specific body parts, 29
    medical management of, 196
    time loss and, 27
    by type of accident, 27–28
Workload demands, as source of workplace stress, 299
Work methods, design of, 270–271
Work overload, workplace stress and, 297
Workplace. See also Workplace stress; Workplace violence; Workplace safety and health

Acquired immunodeficiency syndrome in, 683–686
assessment for eye hazards, 429
automated, technological alienation in, 671–672
design of, 439
development in Western world, 255
emergencies in, 716
hazards in, 551–553
impact of automation on, 659–660
inspections and enforcement in, 128–129
OSHA's process for investigating toxic mold in, 567–568
recommended contents of first-aid kits, 719
reducing EMF exposure in, 621
terrorism in, 737–741
Workplace accidents, 23, 112, 178
    costs and rates of, 23, 26
    deaths by industry, 27, 28–29
    deaths in, 27
    estimating costs of, 34
    hidden costs in, 34–35
    parts of the body injured on the job, 29
Workplace analysis, 366, 367
Workplace homicide, 352
Workplace safety and health
    supervisor's role in, 60
    value of, 880
Workplace stress, 660
    common causes of, 299–300
    defined, 296–297
    health and, 302
    improving safety by reducing, 303–305
    measurement of, 301–302
    sources of, 297–300
Workplace violence
    cases in, 354–356
    conflict resolution and, 373–377

Workplace violence *(Contd.)*
contributing social and
cultural factors, 362–363
defining, 353
do's and don'ts for
supervisors, 377–379
emergency preparedness
plan, 378
high-risk occupations and, 353
legal considerations in,
358–359
record keeping in, 371–373
size of problem, 356–357
steps for preventing, 363
work-related determinations
in, 359–360

Work-related determinations,
making, 359–360
Work relationship, as source
of workplace stress,
298–299
Work schedules, as source of
workplace stress, 298
Worksite analysis program, for
ergonomics, 266–270
Workstation analysis,
269–270
Workstation design, 270
Work-team requirements,
confined spaces workers
and, 595
Work tolerance time (WTT), 454

Written hazard communication
program, 596
Written procedures, in safety
management, 785

X-radiation, 605

Yohay, Stephen C., 358, 359
Yorker, Beatrice Crofts, 688

Zeglin, John, 330
Zero potential, 490